The
Rhabdoviruses

THE VIRUSES

Series Editors
HEINZ FRAENKEL-CONRAT, *University of California*
Berkeley, California

ROBERT R. WAGNER, *University of Virginia School of Medicine*
Charlottesville, Virginia

THE VIRUSES: Catalogue, Characterization, and Classification
Heinz Fraenkel-Conrat

THE ADENOVIRUSES
Edited by Harold S. Ginsberg

THE HERPESVIRUSES
Volumes 1–3 • Edited by Bernard Roizman
Volume 4 • Edited by Bernard Roizman and Carlos Lopez

THE PAPOVAVIRIDAE
Volume 1 • Edited by Norman P. Salzman
Volume 2 • Edited by Norman P. Salzman and Peter M. Howley

THE PARVOVIRUSES
Edited by Kenneth I. Berns

THE PLANT VIRUSES
Volume 1 • Edited by R. I. B. Francki
Volume 2 • Edited by M. H. V. Van Regenmortel and Heinz Fraenkel-Conrat
Volume 3 • Edited by Renate Koenig

THE REOVIRIDAE
Edited by Wolfgang K. Joklik

THE RHABDOVIRUSES
Edited by Robert R. Wagner

THE TOGAVIRIDAE AND FLAVIVIRIDAE
Edited by Sondra Schlesinger and Milton J. Schlesinger

THE VIROIDS
Edited by T. O. Diener

The Rhabdoviruses

Edited by
ROBERT R. WAGNER
University of Virginia School of Medicine
Charlottesville, Virginia

SPRINGER-SCIENCE+BUSINESS MEDIA, LLC

Library of Congress Cataloging in Publication Data

The Rhabdoviruses.

(The Viruses)
Includes bibliographies and index.
1. Rhabdoviruses. I. Wagner, Robert R., 1923–   . II. Series. [DNLM: 1. Rhab-
doviridae. QW 168.5.R2 R4681]
QR415.R47 1987   576′.6484   87-12279
ISBN 978-1-4684-7034-5   ISBN 978-1-4684-7032-1 (eBook)
DOI 10.1007/978-1-4684-7032-1

© 1987 Springer Science+Business Media New York
Originally published by Plenum Press, New York in 1987
Softcover reprint of the hardcover 1st edition 1987

Contributors

Fred Brown, Wellcome Biotechnology Ltd., Surrey GU24 0NQ, England

Nancy L. Davis, Department of Microbiology and Immunology, School of Medicine, The University of North Carolina, Chapel Hill, North Carolina 27514; *present address:* Department of Microbiology, North Carolina State University, Raleigh, North Carolina 27650

Suzanne Urjil Emerson, Department of Microbiology, University of Virginia School of Medicine, Charlottesville, Virginia 22908

R. I. B. Francki, Department of Plant Pathology, Waite Agricultural Research Institute, The University of Adelaide, Glen Osmond, South Australia 5064

John J. Holland, Department of Biology, University of California at San Diego, La Jolla, California 92093

A. O. Jackson, Department of Plant Pathology, University of California at Berkeley, Berkeley, California 94720

Ranajit Pal, Department of Microbiology and Cancer Center, University of Virginia School of Medicine, Charlottesville, Virginia 22908

John Patton, Department of Microbiology and Immunology, School of Medicine, The University of North Carolina, Chapel Hill, North Carolina 27514; *present address:* Department of Biology, University of Southern Florida, Tampa, Florida 33620

Craig R. Pringle, Department of Biological Sciences, University of Warwick, Coventry CV4 7AL, England

John Rose, Departments of Pathology and Cell Biology, Yale University School of Medicine, New Haven, Connecticut 06510

Manfred Schubert, Laboratory of Molecular Genetics, National Institute of Neurological and Communicative Disorders and Stroke, Bethesda, Maryland 20892

Robert E. Shope, Yale Arbovirus Research Unit, Department of Epidemiology and Public Health, Yale University School of Medicine, New Haven, Connecticut 06510

Robert B. Tesh, Yale Arbovirus Research Unit, Department of Epidemiology and Public Health, Yale University School of Medicine, New Haven, Connecticut 06510

Robert R. Wagner, Department of Microbiology and Cancer Center, University of Virginia School of Medicine, Charlottesville, Virginia 22908

Gail W. Wertz, Department of Microbiology and Immunology, School of Medicine, The University of North Carolina, Chapel Hill, North Carolina 27514; *present address*: Department of Microbiology, University of Alabama, Birmingham, Alabama 35294

William H. Wunner, The Wistar Institute of Anatomy and Biology, Philadelphia, Pennsylvania 19104

Douwe Zuidema, Department of Plant Pathology, University of California at Berkeley, Berkeley, California 94720

Preface

The viruses of the family Rhabdoviridae have an exceedingly broad host range and are widely distributed throughout the animal and plant kingdoms. Animal rhabdoviruses infect and often cause disease in insects, fish, and mammals, including man. The prototype rhabdovirus, vesicular stomatitis virus (VSV), has been extensively studied and provides perhaps the best model system for studying negative-strand viruses.

The popularity of VSV as a model system is to a considerable extent due to its relative simplicity and to its rapid growth, generally to high titer, in many cell types ranging from yeast to human. The nucleocapsids of these viruses also carry transcriptional and replicative functions that are expressed in cell-free systems. The first RNA-dependent RNA polymerase was described in VSV and its G protein provided an early model system for studying the synthesis, processing, and membrane insertion of mammalian glycoproteins. VSV is also highly cytopathogenic and has been studied quite extensively for its capacity to kill cells and to shut off cellular macromolecular synthesis. Even earlier, VSV was discovered to be highly susceptible to the action of interferons and has served ever since as a means for quantitating the activity of interferons.

To my way of thinking, the spark that ignited the explosion of research in this field was struck at the First International Colloquium on Rhabdoviruses, attended by 30 or so participants in Roscoff, France, in June 1972. Recent years have witnessed the rapid development of insights into the structure and function of the subunit components of rhabdoviruses, based largely on concomitant advances in molecular biology, genetics, and protein chemistry. In 1975, a single review chapter in Volume 4 of *Comprehensive Virology*, the series antedating *The Viruses*, was deemed sufficient to encompass much of the knowledge about rhabdoviruses. The time now seems ripe to produce a revised and greatly expanded treatise on *The Rhabdoviruses*. The expansion in our knowledge over the past decade or so mandated a complete volume rather than a single chapter. Moreover, it was inconceivable that the breadth and depth of relevant information could be reviewed and analyzed adequately by a single author. Hence, the current volume contains reviews, by authors versed in the intricacies of specific areas of rhabdovirus research, of those subjects to which they have made major contributions. For all of us and for many other colleagues throughout the world, it has been a labor of love.

Robert R. Wagner

Charlottesville, Virginia

Contents

Chapter 3

Rhabdovirus Membrane and Maturation

Ranajit Pal and Robert R. Wagner

Chapter 4

Rhabdovirus Genomes and Their Products

John Rose and Manfred Schubert

Chapter 5

Rhabdovirus Genetics

Craig R. Pringle

Chapter 6

Transcription of Vesicular Stomatitis Virus

Suzanne Urjil Emerson

Chapter 7

The Role of Proteins in Vesicular Stomatitis Virus RNA Replication

Gail W. Wertz, Nancy L. Davis, and John Patton

Chapter 8

Defective Interfering Rhabdoviruses

John J. Holland

Chapter 9

Rabies Viruses—Pathogenesis and Immunity

William H. Wunner

Chapter 10

Biology, Structure, and Replication of Plant Rhabdoviruses

A. O. Jackson, R. I. B. Francki, and Douwe Zuidema

Chapter 11

The Ecology of Rhabdoviruses That Infect Vertebrates

Robert E. Shope and Robert B. Tesh

CHAPTER 1

The Family Rhabdoviridae
General Description and Taxonomy

FRED BROWN

I. INTRODUCTION

The occurrence of rhabdoviruses in vertebrate, invertebrate, and plant hosts makes them a subject of interest not only to animal and plant virologists, but also to those interested in comparative virology and evolution. The structure of the viruses infecting the different kinds of host is remarkably similar, and it has been suggested that some rhabdoviruses that infect plants may also infect vertebrates (Johnson et al., 1969). Moreover, one member, vesicular stomatitis virus, infects several vertebrate hosts, multiplies in Aedes mosquitoes, and grows in leafhoppers, which are the natural vector of maize mosaic virus, a plant rhabdovirus. Consequently, the rhabdoviruses offer great opportunity to study the evolution of viruses within a family. Unfortunately, this opportunity does not seem to have been grasped so far, but the availability of nucleic acid-sequencing techniques should now make this study attractive.

The term "rhabdovirus" was first suggested by Melnick and Mc-Coombs in 1966 and the International Committee on Nomenclature of Viruses recommended its adoption in 1970 (Wildy, 1971). In the universal taxonomic scheme adopted by the International Committee on Taxonomy of Viruses, the term Rhabdoviridae is used. The name rhabdovirus is derived from the Greek "rhabdos," meaning rod, but in fact the viruses have a bullet shape or bacilliform morphology.

FRED BROWN ● Wellcome Biotechnology Ltd., Surrey GU24 0NQ, England.

The first morphological studies on a member belonging to the family were made by Chow *et al.* (1954) and Bradish *et al.* (1956), working with vesicular stomatitis virus. The distinctive bullet-shaped morphology was also found for rabies virus (Almeida *et al.*, 1962), and since that time large numbers of morphologically similar viruses have been found in vertebrates, invertebrates, and plants. More than 100 viruses are presently classified as members or possible members of the family on the basis of their morphology (Brown *et al.*, 1979). Without the distinctive shape, it is highly likely that most of the viruses would have remained as unclassified, lipid-solvent-sensitive viruses.

While the reliance on electron microscopy is probably acceptable with a group of viruses possessing such a distinctive morphology, the result has been that many possible members have not been examined and characterized in a satisfactory way. Consequently, with the exception of a small number of viruses that infect animals, fish, and plants, very few structural details are known for most of the so-called rhabdoviruses. Only a few serological relationships have been worked out and even the most rudimentary examination of the virus nucleic acid and proteins is lacking for most putative members.

Most information is available for vesicular stomatitis virus. There are historic reasons for this situation. First, the virus causes a disease that is clinically similar to, if not indistinguishable from, one caused by foot-and-mouth disease virus. Consequently, there were important economic reasons for studying vesicular stomatitis virus. Second, there was the discovery by Cooper and Bellett in 1959 that the virus exhibited the autointerference phenomenon which even today has not been explained in molecular terms. Moreover, the importance of the phenomenon in the natural regulation and outcome of the disease has led to several studies with the interfering particle of vesicular stomatitis virus (see Chapter 8). Third, the ease with which the virus can be grown in a variety of tissue culture systems and subsequently purified has made it an attractive model for biochemists interested in replication and in the dissection of viruses into biologically active subunits. Fourth, and possibly of greatest significance, the observation that its RNA was not infectious and the subsequent demonstration that an RNA-dependent RNA polymerase (Baltimore *et al.*, 1970) within the virus particle was required for infectivity led to the major division of RNA-containing viruses into positive- and negative-strand viruses (Baltimore, 1971).

The recognition that rabies virus had many structural features in common with vesicular stomatitis virus, stemming from the electron microscopic examination of the particles, provided an added stimulus to the study of the rhabdoviruses because of the importance of rabies. This has led in turn to the more detailed study of a few bullet-shaped viruses which cause other important diseases of animals, fish, and plants. Consequently, some pattern has emerged regarding the family of rhabdoviruses.

It is probably fair to say that the relationships between those viruses infecting vertebrates are more clearly defined, probably because virologists working with them have used serological methods to a much greater extent than have plant virologists. Thus the situation exists within the plant rhabdoviruses that isolates of the same virus have been given different names, and few comparisons have been reported between different isolates (Milne *et al.*, 1986). The failure to use such powerful tools on a wide scale means that relationships between plant rhabdoviruses are less well-defined. The value of the serological approach has been emphasized by the recent demonstration that rhabdoviruses bearing the names Moroccan wheat rhabdovirus and wheat rosette stunt virus (from China) are in fact strains of barley yellow striate mosaic virus from Italy (Milne *et al.*, 1986).

II. VIRUS STRUCTURE

A. Electron Microscopy

All the rhabdoviruses occurring in vertebrates are bullet-shaped and consist of a helically wound ribonucleocapsid surrounded by a unit membrane envelope layer which in turn is surrounded by an outer layer through which spike projections protrude. The surface projections are clearly discernible, spaced at 4- to 5-nm intervals, on several members of the family. Although in most viruses there is no apparent arrangement of the projections, with some, particularly rabies virus, there appears to be some symmetry in their positioning. The honeycomb appearance of rabies virus and Klamath virus and the number of copies of each of the major structural proteins of vesicular stomatitis virus suggested to Cartwright *et al.* (1972) that there was a hexagonal arrangement of the matrix protein. This arrangement in turn suggested that the surface projections passed through the lipid membrane and came into close juxtaposition with the matrix protein. Although there is no direct electron microscopic evidence for this arrangement with intact particles, micrographs of vesicular stomatitis virus from which most of the lipid layer has been removed with phospholipase C provide good evidence for the close contact of the surface projections with the matrix protein because they appear much longer than in the untreated particles (Cartwright *et al.*, 1969). Moreover, virus particles which had been treated with low concentrations of formaldehyde were no longer disrupted with sodium dodecyl sulfate (Brown *et al.*, 1974). Although the lipid was completely removed, the skeleton of the virus, to which the surface projections were still attached, remained. This evidence suggested that the surface projections had cross-linked to the matrix protein or ribonucleoprotein.

In confirmation of this structure Mudd (1974) showed that trypsin, which shaves the surface projections from the virus particles, leaves a protein long enough to traverse the lipid bilayer. The balance of evidence thus points to a structure in which the surface projections penetrate the lipid bilayer and come into close juxtaposition with the membrane protein in an ordered array.

B. Chemical Composition

Detailed chemical analyses are available for only a few of the members of the family. Where these are available, however, there is a remarkable consistency. The RNA, which makes up about 2% of the virus particle, is single-stranded with a sedimentation coefficient of 38 to 45S, corresponding to a molecular weight of approximately 4×10^6, and is noninfectious, being of negative polarity. There is one copy of the RNA in each virus particle. An RNA-dependent RNA polymerase which converts the virion into five messenger RNAs is associated with the particle. Each of the messenger RNAs codes for a protein found in the virus particle.

Most of the rhabdoviruses contain five proteins: large protein (L; mol. wt. 190×10^3), glycoprotein (G; 70×10^3), nucleoprotein (N; 50×10^3), nonstructural protein (NS; $40–45 \times 10^3$), and matrix protein (M; $20–30 \times 10^3$). The G protein is the surface projection which is so prominent in electron micrographs of the virus. All the other proteins are located inside the viral envelope. The N protein is closely associated with the RNA and although the two molecules are not covalently linked (being separable by 0.1% sodium dodecyl sulfate), the RNA is protected from the action of ribonuclease by the protein. The ribonucleoprotein can be extracted from virus particles with sodium deoxycholate as a stringlike molecule, but within the virus particle the ribonucleoprotein is in the form of a helix, presumably maintained in this configuration by the matrix protein M. From calculations of the number of molecules of the G, N, and M proteins in a single virus particle, Cartwright *et al.* (1972) proposed a model that accounts for most of the structural features of the virus and viral subunits.

The other two proteins, L and NS, are present in only small amounts. They form the complex that, with the ribonucleoprotein as template, provides the transcriptase activity of the virus. This enzyme transcribes the virus RNA into the five messenger RNAs already referred to (see review by Hunt *et al.*, 1979). These observations account for the fact that the nucleocapsid produced from the virus particle by removal of the glycoprotein and lipid envelope with mild detergents is still infectious, albeit at a much lower level than the virus particle. Similarly, virus particles from which the cell attachment protein G has been removed by treatment with trypsin still possess a low level of infectivity. The rate of attachment of these particles and the nucleocapsid to cells is different from that of the virus particles, indicating that the low infectivity that can be detected is not due to residual virus particles.

The glycoprotein G is the only one of the five virus proteins which contains carbohydrate. It can be removed from the surface of the virus particle by trypsin, but it leaves a stub that is long enough to pass through the membrane (Mudd, 1974). This evidence confirmed the view, based on the radiochemical analysis of the fragments obtained by detergent treatment of virus particles which had been labeled with precursors of protein, sugar, and lipid and fixed with formaldehyde, that the surface projections form a bridge across the membrane to interact directly with the core (Brown et al., 1974). This evidence in turn supported the earlier observation that phospholipase C, although removing a considerable proportion of the lipid of the virus, did not remove any of the surface projections, as demonstrated by electron microscopy, and left the immunizing activity of the particles unimpaired (Cartwright et al., 1969).

The matrix protein is located inside the viral membrane. It is still associated with the ribonucleoprotein in particles which have had their membrane removed by treatment with Tween–ether or Nonidet. However, treatment of the virus or nucleocapsid with sodium deoxycholate removes the M protein and releases the ribonucleoprotein as a stringlike structure (Cartwright et al., 1970).

C. Relationships between Different Members

1. Ribonucleic Acid

The relationships between the RNAs of different rhabdoviruses have not been the subject of extensive study. Base composition analyses, summarized by Clewley and Bishop (1979), showed a remarkably consistent pattern for rabies, vesicular stomatitis, and pike fry disease viruses, with a distinctly high percentage of uracil. The 11,000 or so bases could code for about 3700 amino acid residues or 370,000 daltons of protein, corresponding to only slightly more than the sum of the molecular weights of the five viral proteins. It seems, therefore, that the five viral proteins L, G, N, NS, and M are the only proteins coded for by the virion RNA.

Sequence homology between several rhabdoviruses has been analyzed by hybridizing labeled virus RNA of one virus to an excess of unlabeled complementary RNA of another virus and determining the resistance to various ribonucleases under stringent and nonstringent conditions. Under stringent conditions very little homology has been found between the RNAs of the vesicular stomatitis virus strains Indiana, New Jersey, and Cocal, and Piry, Chandipura, rabies, and spring viremia of carp viruses (Repik et al., 1974). Using nonstringent conditions there was some inexact homology between the three vesicular stomatitis viruses. Hybridization studies between different isolates belonging to the New Jersey serotype of vesicular stomatitis virus revealed the existence of two subgroups containing the Concan, Ogden, and Guatemala isolates and the Hazelhurst and Missouri isolates, respectively (Reichmann et al., 1978).

Fingerprinting of the RNAs of the five isolates following ribonuclease T1 digestion supported the conclusion that there are two subgroups of this virus. The RNAs of different isolates of the Indiana serotype of vesicular stomatitis virus are readily distinguishable from each other.

2. Proteins

The relationships between the proteins of the viruses have been established almost entirely by the use of serological methods. A variety of methods has been used, but unfortunately these have been applied to only a small number of viruses. The most detailed work has been done with the vertebrate viruses and two major serogroups have been defined, the Vesiculovirus genus and the Lyssavirus genus. The Vesiculovirus genus contains three groups: (1) the vesicular stomatitis viruses, related to the Indiana serotype; (2) those belonging to the New Jersey serotype; and (3) Piry, Chandipura, and Isfahan. The viruses in group 1 are clearly related, even when the neutralization test is used and these relationships are emphasized when the neutralization reaction between the infective nucleocapsid and serum from infected guinea pigs is used for the comparisons (Cartwright and Brown, 1972). These results indicate that the major cross-reactive antigen is the ribonucleoprotein N, whereas the type specificity is associated with the surface glycoprotein G. The cross-challenge experiments conducted by Federer et al. (1967) showed that, in general, neutralizing antibody levels gave a good indication of the outcome of challenge with the homologous and heterologous strains. The neutralization results obtained with the New Jersey isolates indicate that, in contrast to the conclusions made from the RNA homology data, the isolates represent geographical variants rather than subtypes.

There is some debate regarding the relationship of Piry and Chandipura viruses to other members of the Vesiculovirus genus. Whereas Murphy and Shope (1971) found some degree of cross-reaction between Piry and Chandipura viruses and between these two viruses and the Indiana serotype viruses, Cartwright and Brown (1972) were unable to demonstrate any cross-reaction in neutralization tests, even when the infective nucleocapsids were used. Recent evidence provided by Wilks and House (1984) has shown that Piry virus is quite unrelated pathogenically to the New Jersey and Indiana strains of vesicular stomatitis virus.

The Lyssavirus genus contains three serogroups of which the most important is that containing rabies, Mokola, Lagos bat, and Duvenhage viruses (Frazier and Shope, 1979). The importance of rabies as a disease agent is generally recognized, but Mokola and Duvenhage have also been associated with clinical disease and each has caused fatal infection. Although all the members of the group are clearly closely related, they are sufficiently different in cross-neutralization and cross-protection tests to warrant serious consideration of their inclusion in a cocktail vaccine containing all members of the group.

The other members of the Lyssavirus genus, Obodhiang and koton-kan, are much more distantly related to the rabieslike viruses, and there are doubts whether they should be included in this genus. Clearly, the situation is sufficiently obscure to warrant further investigation of their relatedness to the other members of the genus.

As mentioned above, the relatedness between the many plant rhabdoviruses has not been investigated by serological methods to the same extent as those viruses infecting vertebrate and invertebrate hosts. It is clear from the example cited in Section I that much interesting information on the relationships between the plant rhabdoviruses themselves and between the viruses infecting plant, vertebrate, and invertebrate hosts could be gained from serological studies, particularly of the ribonucloprotein.

III. CONCLUSIONS

With the development of methods for the sequencing of DNA during the last few years, it is now possible to determine the relationship between viruses at the most fundamental level. Although it is highly unlikely that there will be any comprehensive study of the rhabdoviruses using these methods, it is clear that only by that means will meaningful taxonomic relationships between the viruses be established. At the more practical level it is also clear that the relationships between the viruses causing rabieslike illness will only be established by reference to the nucleic acid sequences of the different agents. Such information would be invaluable to epidemiologists and to control authorities alike. In the context of this chapter, however, if the nucleic acid sequences were available, it would help to end the often tiresome speculations on the evolution of viruses.

REFERENCES

Almeida, J. D., Howatson, A. F., Pinteric, L., and Fenje, P., 1962, Electron microscope observations on rabies virus by negative staining, *Virology* **18**:147.

Baltimore, D., Huang, A. S., and Stampfer, M., 1970, Ribonucleic acid synthesis of vesicular stomatitis virus. II. An RNA polymerase in the virion, *Proc. Natl. Acad. Sci. U.S.A.* **66**:572.

Baltimore, D., 1971, Expression of animal virus genomes, *Bacteriol. Rev.* **35**:235.

Bradish, C. J., Brooksby, J. B., and Dillon, J. F., 1956, Biophysical studies of the virus system of vesicular stomatitis, *J. Gen. Microbiol.* **14**:2980.

Brown, F., Smale, C. J., and Horzinek, M. C., 1974, Lipid and protein organization in vesicular stomatitis and Sindbis viruses, *J. Gen. Virol.* **22**:455.

Brown, F., Bishop, D. H. L., Crick, J., Francki, R. I. B., Holland, J. J., Hull, R., Johnson, K., Martelli, G., Murphy, F. A., Obijeski, J. F., Peters, D., Pringle, C. R., Reichmann, M. E., Schneider, L. G., Shope, R. E., Simpson, D. I. H., Summers, D. F., and Wagner, R. R., 1979, Rhabdoviridae, *Intervirology* **12**:1.

Cartwright, B., and Brown, F., 1972, Serological relationships between different strains of vesicular stomatitis virus, *J Gen. Virol.* **16**:391.

Cartwright, B., Smale, C. J., and Brown, F., 1969, Surface structure of vesicular stomatitis virus, *J. Gen. Virol.* **5**:1.

Cartwright, B., Smale, C. J., and Brown, F., 1970, Dissection of vesicular stomatitis virus into the infective ribonucleoprotein and immunizing components, *J. Gen. Virol.* **7**:19.

Cartwright, B., Smale, C. J., Hull, R., and Brown, F., 1972, Model for vesicular stomatitis virus, *J. Virol.* **10**:256.

Chow, T. L., Chow, F. H., and Hanson, R. P., 1954, Morphology of vesicular stomatitis virus, *J. Bacteriol.* **68**:724.

Cooper, P. D., and Bellett, A. J. D., 1957, A transmissible interfering component of vesicular stomatitis virus preparations, *J. Gen. Microbiol.* **21**: 485.

Federer, K., Burrows, R., and Brooksby, J. B., 1967, Vesicular stomatitis virus—the relationship between some strains of the Indiana serotype, *Vet. Res. Sci.* **8**:103.

Frazier, C. L., and Shope, R. E., 179, Serologic relationships of animal rhabdoviruses, in: *Rhabdoviruses*, Vol. I (D. H. L. Bishop, ed.), p. 43, CRC Press, Boca Raton, Florida.

Hunt, D. M., Mellon, M. G., and Emerson, S. U., 1979, Viral transcriptase, in: *Rhabdoviruses*, Vol. I (D. H. L. Bishop, ed.), p. 169, CRC Press, Boca Raton, Florida.

Johnson, K. M., Tesh, R. B., and Peralta, P. H., 1969, Epidemiology of vesicular stomatitis virus: Some new data and a hypothesis for transmission of the Indiana serotype, *J. Am. Vet. Med. Assoc.* **155**:2133.

Melnick, J. L., and McCoombs, R. M., 1966, Classification and nomenclature of animal viruses, *Prog. Med. Virol.* **8**:400.

Milne, R. G., Masenga, V., and Conti, M., 1986, Serological relationships between the nucleocapsids of some planthopper-borne rhabdoviruses of cereals, *Intervirology* **25**:83.

Mudd, J. A., 1974, Glycoprotein fragment associated with vesicular stomatitis virus after proteolytic digestion, *Virology* **62**:573.

Murphy, F. A., and Shope, R. E., 1971, Bridging groups of viruses, in: *Proceedings of the Second International Congress for Virology* (J. L. Melnick, ed.), p. 261, S. Karger, Basel.

Reichmann, M. E., Schnitzlein, W. M., Bishop, D. H. L., Lazzarini, R. A., Beatrice, S. T., and Wagner, R. R., 1978, Classification of the New Jersey serotype of vesicular stomatitis virus into two sub-types, *J. Virol.* **25**:446.

Repik, P., Flamand, A., Clark, H. F., Obijeski, J. F., Roy, P., and Bishop, D. H. L., 1974, Detection of homologous RNA sequences among six rhabdovirus genomes, *J. Virol.* **13**:150.

Wildy, P., 1971, Classification and nomenclature of viruses. First report of the International Committee on Nomenclature of Viruses, *Mono. Virol.* **5**:1.

Wilks, C. R., and House, J. A., 1984, Susceptibility of various animals to the vesiculovirus Piry, *J. Hyg.* **93**:147.

CHAPTER 2

Rhabdovirus Biology and Infection
An Overview

ROBERT R. WAGNER

I. INTRODUCTION

Chapter 1 describes the general features of the family Rhabdoviridae, and Chapter 3 provides a moderately detailed analysis of the rhabdovirus membrane. This chapter is designed to set the stage for the in-depth chapters that follow by providing a general description of the structural components, infectivity, host reactivity, and immunology of rhabdoviruses. As is true of all but Chapters 1 and 9–11, this chapter is concerned primarily with the best-studied and prototypic rhabdovirus, vesicular stomatitis virus (VSV). It seems reasonable to assume that the structural components that comprise the VSV virion are similar to those of other rhabdoviruses, but comparisons will be made, where appropriate, with the rabies subgroup and rhabdoviruses of other animals. It should be stated from the outset that productive infection by VSV of mammalian and avian cells far exceeds that of rabies and other animal rhabdoviruses, such as those of fish. Nothing will be said in this chapter about the structure and infectivity of the plant rhabdoviruses, which are discussed in Chapter 10.

In some respects, this chapter is an updated sequel to the one on the reproduction of rhabdoviruses I wrote for Vol. 4 of our earlier series entitled *Comprehensive Virology* (Wagner, 1975). A similar review was written by Emerson (1976), and three volumes on the subject of rhabdoviruses

ROBERT R. WAGNER • Department of Microbiology and Cancer Center, University of Virginia School of Medicine, Charlottesville, Virginia 22908.

were edited by Bishop (1979). Rather striking advances in our knowledge of rhabdoviruses have taken place since the publication of these earlier reviews. This explosion in our knowledge merits this complete volume, rather than a single chapter. Also recommended is a more recent and detailed analysis of rhabdovirus cytopathology and effects on cellular macromolecular synthesis (Wagner *et al.*, 1984), which also includes a brief summary of the then current knowledge of VSV reproductive strategies.

As indicated in Chapter 1, the mammalian Rhabdoviridae are classified in two genera, *Vesiculovirus* and *Lyssavirus* (rabies). There are at least two serotypes of the *Vesiculovirus* genus, designated Indiana and New Jersey, and there are several strains of each serotype (Clewley *et al.*, 1977; Reichmann *et al.*, 1978). Unless otherwise indicated, the basic descriptions of VSV referred to in this chapter will be of the San Juan strain of the VSV-Indiana virus, by far the best studied of all VSV strains.

II. THE VIRION AND ITS COMPONENTS

A. Morphology

Figure 1 shows a schematic representation of the morphology and structural components of the virion of vesicular stomatitis virus (VSV). The infectious units of all rhabdoviruses, except possibly the plant rhabdoviruses, are structurally similar by electron microscopy to that of the prototype VSV (Howatson, 1970; Wagner, 1975). The infectious (standard) virion is bullet-shaped, round at one end and flat at the other, hence the term B particles; they measure about 180 nm in length and 65 nm in width. These dimensions vary somewhat for different strains of VSV and for different animal rhabdoviruses. In fact, Orenstein *et al.* (1976) contend that the bullet shape of VSV is an artifact of fixation and staining and that the infectious particles are bacilliform with two rounded ends, although they agree that the internal nucleocapsid is itself bullet-shaped. Exceedingly common, particularly in uncloned preparations, are truncated (T) or defective interfering (DI) particles that are about the same width as standard B particles but vary in length from 80 to 50 nm, depending on the amount of viral RNA that has been deleted (Huang and Baltimore, 1977). The nature and properties of DI particles are described in Chapter 8.

By negative-staining electron microscopy, all rhabdoviruses exhibit protruding spikes that measure about 10 nm in length and are frequently penetrable at the blunt end by phosphotungstic acid (Fig. 2). These electron-microscopic analyses have led to the conclusion that VSV and all other rhabdoviruses are composed of a tightly coiled nucleocapsid surrounded by a membrane (envelope) with protruding spikelike structures that are readily removed by exposure to proteases. Disruption of the

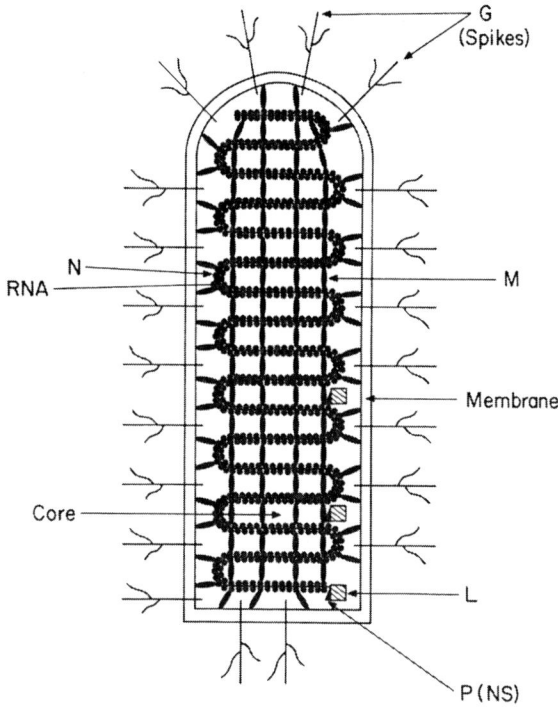

FIGURE 1. Schematic representation of the morphology and structural components of VSV. The two major structural components are (1) the nucleocapsid core, which contains single-stranded RNA tightly encased by the major N protein and two minor proteins, L and P (formerly and commonly designated NS), which collectively comprise the RNA polymerase, and (2) the envelope or limiting membrane (outer layer of predominantly choline phospholipids and inner layer of predominantly amino phospholipids) associated with two proteins, the integral, externally oriented glycoprotein [G (spikes)] and the peripheral matrix (M) protein, which lines the inner surface of the membrane in close association with the nucleocapsid core. (See Chapter 3 for greater detail.)

envelope by detergents results in release of the nucleocapsid, which can retain its tightly coiled structure in the absence of salt but is uncoiled in solutions of high ionic strength (Newcomb and Brown, 1981). These and other studies have led to the hypothesis that the secondary structure of the VSV nucleocapsid is due to electrostatically bound matrix protein, which dissociates from the nucleocapsid in hypertonic solutions (Emerson and Wagner, 1972; Newcomb and Brown, 1981; Newcomb et al., 1982). The extended nucleocapsid of infectious VSV is approximately 3.5 μm long (Howatson, 1970), whereas the nucleocapsids of DI particles vary from one-third to one-half the length of standard B virion nucleocapsids.

The approximate composition of VSV, probably like that of all other rhabdoviruses, is 74% protein, 20% lipid, 3% carbohydrate, and 3% RNA (McSharry and Wagner, 1971; Wagner, 1975). The lipid is present entirely in a limiting external membrane in the classic form of a lipoprotein

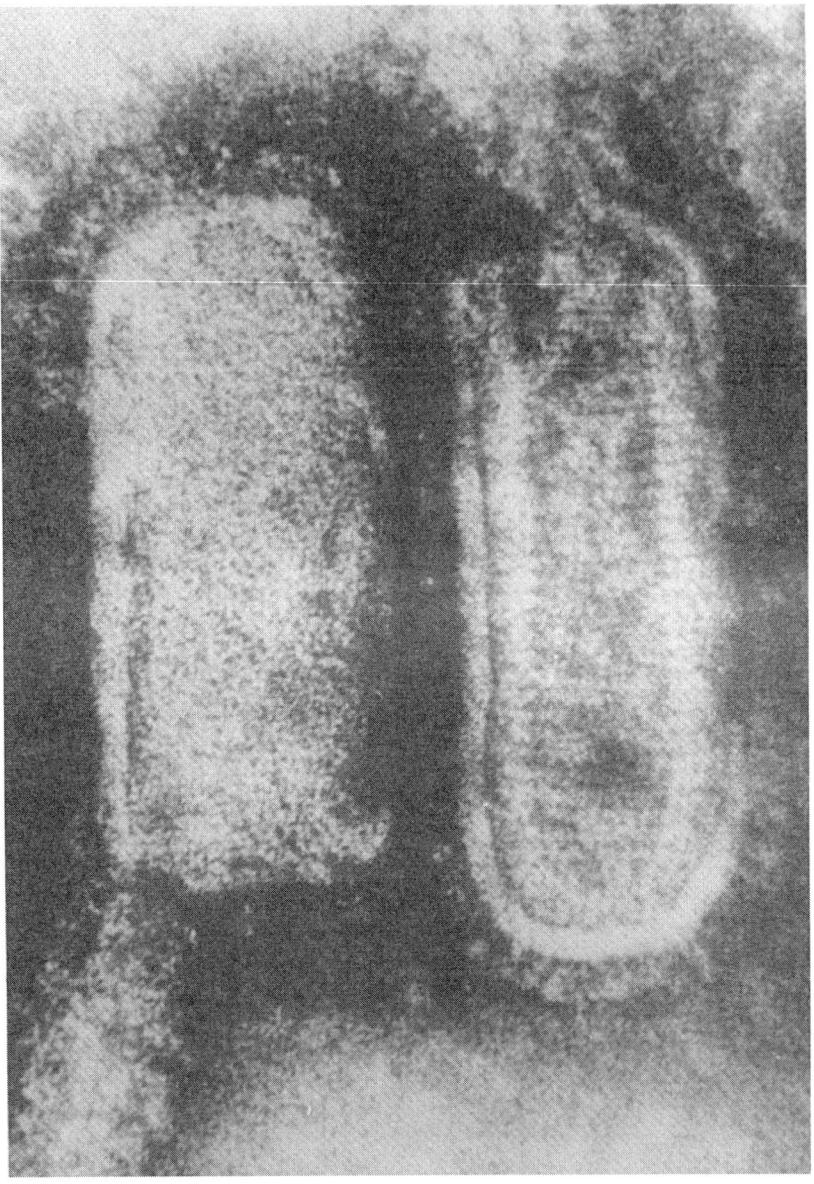

FIGURE 2. Electron micrograph of VSV, negatively stained with phosphotungstic acid and partially disrupted in distilled water to show internal nucleocapsid, external membrane, and protruding spikes.

bilayer, about half of which is composed of proteins. The other structural component of the virion is the nucleocapsid, which contains genomic RNA tightly encapsidated with protein. Figure 1 depicts diagrammatically the two components of VSV, indicating current concepts of the location of the proteins, RNA, and lipids.

B. Virion Nucleocapsid

The genetic information of all rhabdoviruses is contained in an unsegmented single strand of RNA that cannot serve as messenger (negative-strand viruses). The single strand of RNA in infectious VSV-Indiana virions has a molecular weight of approximately 3.68×10^6 and appears to contain 11,162 nucleotides (Schubert et al., 1984). The entire nucleotide sequence and coding potentials are described in Chapter 4. The RNA is not infectious (Huang and Wagner, 1966b).

Associated with the virion RNA are three proteins to form the viral nucleocapsid, which is infectious (Szilagyi and Uryvayev, 1973; Emerson and Yu, 1975) at an efficiency of 10^{-5} to 10^{-6} that of the intact virion; higher efficiency of nucleocapsid infectivity requires adjuvants such as DEAE–dextran or calcium phosphate (Szilagyi and Uryvayev, 1973; Bishop et al., 1974). The major nucleocapsid protein is designated N; it is present in approximately 1600 copies and contains 422 amino acids (Gallione et al., 1981). The N protein is exceedingly insoluble and is so tightly complexed with viral RNA that it is readily dissociable only with sodium dodecyl sulfate (SDS) (Emerson, personal communication). The RNA present in the VSV nucleocapsid is completely resistant to digestion with various ribonucleases even after exposure to high salt (Emerson and Wagner, 1973).

The infectious nucleocapsid also contains two other minor proteins that are present in approximately 50 and 150 copies each (Bishop and Roy, 1972) and together with the nucleocapsid template form the endogenous RNA-dependent RNA polymerase (Emerson and Yu, 1975). The larger of the two polymerase proteins is designated L, $M_r = 241,012$, and the gene contains 6380 nucleotides, representing 60% of the coding potential of the entire genome (Schubert et al., 1984); the L protein is very labile to heat and sheer forces and must be handled very gently. The other protein required for polymerase activity has been designated NS (P) and in the Indiana serotype contains 222 amino acids (Gallione et al., 1981); VSV-New Jersey NS protein has 274 amino acids and exhibits only 41% base sequence homology to that of VSV-Indiana NS protein (Gill and Banerjee, 1985). The NS protein is highly phosphorylated and appears to exist in two forms depending on degree of phosphorylation (Clinton et al., 1978b; Kingsford and Emerson, 1980). Phosphorylation of the VSV NS protein is carried out by host-cell protein kinases (Imblum and Wagner, 1974; Clinton et al., 1982), which also determine distribution of phosphorylation among serine, threonine, and tyrosine residues of the protein (Clinton and Huang, 1984; Clinton and Finley-Whelan, 1984). The original name, NS, mistakenly indicating nonstructural, will soon be changed to P protein because of its highly phosphorylated state. The L and NS proteins can be dissociated from the nucleocapsid under varying conditions of high ionic environment, which results in loss of transcriptase activity and infcctivity. Transcription and infectivity can be restored by reconstituting the L and NS proteins with denuded nucleocapsids in a low-ionic-strength environment (Emerson and Wagner, 1972, 1973). Both the L and NS proteins, as well as the N protein–RNA complex, are essential for transcriptase activity (Emerson and Yu, 1975). A detailed description of the enzymology and regulation of transcriptase is presented in Chapter 6.

C. Virion Membrane

Chapter 3 presents a detailed analysis of the structure, dynamics, function, and origin of the membranes of VSV and other rhabdoviruses. Suffice it to say here that the VSV membrane is composed of approximately 50% lipid and 50% protein (McSharry and Wagner, 1971). The lipids are derived entirely from the host cell, but are selected in somewhat different proportions than those in the host-cell plasma membrane; the principal lipid differences of the VSV membranes are the larger proportion of cholesterol and, among the phospholipids, much less phosphatidylcholine compared to sphingomyelin and larger amounts of amino phospholipids (McSharry and Wagner, 1971; Patzer *et al.*, 1979). This altered lipid composition contributes significantly to the greater viscosity of the VSV membrane compared to that of the host-cell membrane from which it is derived (Barenholz *et al.*, 1976).

The VSV membrane, and probably that of all rhabdoviruses, contains two proteins: an externally oriented, integral glycoprotein (G) and a peripheral matrix (M) protein that lines the inner surface of the virion membrane (Patzer *et al.*, 1979; Zakowski and Wagner, 1980). On the basis of cloned complementary DNA (cDNA) sequences of their messenger RNAs (mRNAs), the Indiana serotype VSV G protein is composed of 511 amino acids and is N-glycosylated to two separate asparagine residues, whereas the M protein is composed of 229 amino acids and is not glycosylated (Rose and Gallione, 1981). The G protein is the major antigenic determinant responsible for type specificity and gives rise to neutralizing antibody (Kelley *et al.*, 1972; Volk *et al.*, 1982). The M protein appears to serve as the "glue" that attaches the nucleocapsid to the cell plasma membrane where the G protein is inserted; the M protein is quite basic (pI ~ 9.1) and inhibits transcription by binding to the nucleocapsid (Carroll and Wagner, 1979; Wilson and Lenard, 1981) (see Chapter 6). The basic M protein also binds to acidic phospholipid headgroups, by which means it appears to attach the nucleocapsid–M protein complex to phosphatidylserine residues that line the inner surface of the virion membrane (Zakowski *et al.*, 1981; Wiener *et al.*, 1983).

The mass and molecular composition of the VS virion were recently reported in an elegant study by D. Thomas *et al.* (1985). Using dark-field scanning transmission electron microscopy, these authors reported masses of 265.6 ± 13.3 megadaltons (Md) for the native VS virion and 69.4 ± 4.9 Md for the nucleocapsid. The lipid content of the virion was estimated as 56.1 Md, and the G protein spikes averaged 1205 molecules per virion compared to 1826 molecules of the M protein. The nucleocapsid was calculated to contain 3.7 Md of RNA, 1258 molecules of N protein, 466 molecules of NS (P) protein, and 50 molecules of L protein. Using four different electron-microscopic procedures, the nucleocapsid was calculated to be 3.5–3.7 μm in length and to comprise a strand of repeating units with a center-to-center spacing of 3.3 nm. These precise measurements are not very different from other estimates made by less exact techniques.

III. SUSCEPTIBILITY TO INFECTION

A. Distribution in Nature

Rhabdoviruses are widely distributed among genera of the animal and plant kingdoms. Plant rhabdoviruses cause widespread diseases among many plant species and are apparently always transmitted by arthropods that feed on the plants (see Chapter 10). Among the animal rhabdoviruses, many of those that belong to the genus *Vesiculovirus* infect insects, and perhaps other arthropods, but it is uncertain whether they transmit infection to vertebrates; Schnitzlein and Reichmann (1985) recovered identical vesicular stomatitis virus (VSV)-New Jersey from black flies and diseased horses during the 1982 epizootic in Colorado. The VSVs, which can be divided into two species (serotypes) called VSV-Indiana and VSV-New Jersey (Cartwright and Brown, 1982b) (see also Chapter 1), appear to infect insects and mammals. It is of interest that the various subtypes of VSV-Indiana [e.g., Indiana, Brazil, Argentina, Cocal (see Chapter 1)] are more commonly isolated from insects than are the VSV-New Jersey strains, which are enzootic in Central America and the Caribbean basin and cause widespread epizootics in cattle and swine in Central America, Mexico, and the western United States. Various substrains of the Hazelhurst strain of VSV-New Jersey are apparently responsible for the widespread infections of cattle and swine in North America and northern South America. The epidemiology of this disease is not very well understood. Vesicular stomatitis in man is a severe influenzalike disease that is rarely fatal and is almost invariably caused by laboratory infection or by transmission from infected animal carcasses.

There may be a number of viruses of the vesiculovirus group that are confined to insects, the best known of which is the sigma virus of *Drosophila* (Printz, 1973). VSV usually does not cause a fatal infection in insects and usually does not grow to high titer. However, Gillies and Stollar (1980a) reported the production of high yields of infectious VSV in *Aedes albopictus* cells compared with its replication in baby hamster kidney (BHK)-21 cells. Defective interfering (DI) particles of VSV were readily generated in *A. albopictus* cells (Gillies and Stollar, 1980b), and significant biochemical changes were noted in these mosquito cells, particularly if they were deprived of methionine (Gillies and Stollar, 1981).

The other major genus of Rhabdoviridae is *Lyssavirus*, which comprises the rabies and rabies like viruses (see Chapter 1). These viruses have also been isolated from insects, but they are usually transmitted by bites of rabid animals, almost invariably causing a fatal disease of the central nervous system (Murphy, 1977). Our understanding of the pathogenicity and neural dissemination of rabies virus is expanding by means of incisive techniques (Kucera *et al.*, 1985). The nature, pathogenicity, and immunology of rabies viruses are described in detail in Chapter 9, and the general epidemiology of the rhabdoviruses is presented in Chapter 11.

An interesting and economically important group of rhabdoviruses belong to those that infect fish, particularly in pisciculture hatcheries, where the mortality is very high (McAllister, 1979). The salmonid rhabdoviruses appear to be more closely related to *Lyssavirus* than to *Vesiculovirus*, at least on the basis of their protein composition (McAllister and Wagner, 1975). Of great interest is the fact that salmonid rhabdoviruses grow best in cells at the usual ambient temperature of fish (15–18°C) and contain an RNA-dependent polymerase that functions optimally at 15–18°C and is virtually inactive at 30°C or higher (McAllister and Wagner, 1977). The spring viremia of carp rhabdorivus more closely resembles VSV, and its molecular biology has been studied quite extensively (Kiuchi and Roy, 1984). A detailed description of the fish rhabdoviruses and other fish viruses can be found in the predecessor to this series, *Comprehensive Virology*, Vol. 14 (McAllister, 1979).

B. Host Range

In general terms, variation in cellular susceptibility to viral infection depends either on availability of cell-surface receptors for attachment of the invading virus or on the intracellular environment required to support viral replication. In the case of VSV, there is little evidence for variation in cell-surface factors as determinants of susceptibility. In fact, VSV has a very wide host range, varying from insect to mammalian cells (Wagner, 1975). Infected insect cells usually yield much lower titers of VSV than do mammalian cells, but Wyers *et al.* (1980) reported high yields of VSV from certain selected *Drosophila* cell lines, and Gillies and Stollar (1980b) were able to clone *Aedes albopictus* cells that support growth of VSV to high yield. However, the quality of the virus produced in insect cells may differ; *Drosophila* cells produce much less G_0 and G_1 protein (Wyers *et al.*, 1980) and *A. albopictus* cells give rise to vSV deficient in L as well as G protein, although excess G protein mRNA is produced in these cells (Gillies and Stollar, 1980b). Incubation temperature probably does not contribute significantly to virus yield or quality of progeny viruses because our laboratory routinely produces our highest-titered VSV stocks by growth in mammalian cells at a temperature of 31°C, rather than 37°C.

Mammalian cells can also vary considerably in their degree of susceptibility to infection with VSV. We have routinely used BHK-21 cells to produce yields of progeny frequently 10 times greater than that produced by L cells (Wagner, unpublished data), even though L cells are considerably more susceptible to inhibition of cellular RNA synthesis by VSV than are BHK-21 cells (Weck and Wagner, 1978).

As described in a previous review (Wagner et al., 1984), cell differentiation may play a role in cellular susceptibility to viral infection, as illustrated by the finding that VSV replication is restricted in one human lymphoblastoid cell line but not in another or in HeLa cells (Nowakowski et al., 1973). It has been known for some years that VSV will not replicate in resting lymphocytes, which will become susceptible to infection, however, after induction by mitogens or by specific antigens (Bloom et al., 1974). Mouse spleen cells, which cannot normally support VSV growth, become susceptible to infection after intraperitoneal propagation of syngeneic or allogeneic tumors (Hecht and Paul, 1981). Robertson and Wagner (1981) also found that HeLa cells and L cells are more permissive for VSV growth than an end-stage myeloma cell line, MPC-11, which, in turn, is more permissive than the Abelson-virus-transformed 18-81 pre-B cell that does not secrete immunoglobulin. The converse relationship was found when we tested these same cells for the capacity of VSV to shut off cellular RNA synthesis; the susceptibility of these cells to VSV inhibition of cellular RNA synthesis could be ranked in the order 18-81 > MPC-11 > L cells > HeLa cells (Robertson and Wagner, 1981). This difference in susceptibility to inhibition of cellular macromolecular synthesis does not appear to be related to the specific proteins produced by differentiated cells. Myeloma cells infected with VSV exhibit far less inhibition of the synthesis of immunoglobulin than of other cellular proteins (Nuss et al., 1975). Similarly, globin synthesis in differentiated Friend erythroleukemia cells is more resistant to shut off by VSV infection than are other proteins of the same cell (Nishioka and Silverstein, 1978). Cellular factors of unknown nature were also found to determine the ability of mengovirus to inhibit the synthesis of VSV protein or of VSV to inhibit the synthesis of mengovirus proteins in doubly infected HeLa, Chinese hamster ovary (CHO), or L-929 cells (Otto and Lucas-Lenard, 1980). VSV also appears to inhibit uptake of uridine in infected chick-embryo cells (Genty, 1975), but not in other cells (Genty and Berreur, 1975; Weck and Wagner, 1978). Quite obviously, the factors that control cellular susceptibility to infection with VSV, or with other viruses, for that matter, remain unclear.

The role of various subcellular organelles in regulating viral replication or susceptibility to VSV infection is not readily apparent, but is not likely to be of major significance. The nucleus of mammalian cells is not a factor in VSV replication, which takes place quite normally in enucleated cells (Follett et al., 1974). On the other hand, rabies virus undergoes an incomplete cycle of replication in enucleated TC-7 cells compared to normal yields of VSV from similarly infected enucleated cells (Wiktor and Koprowski, 1974). One possible factor controlling viral replication and yield is the site of budding of VSV progeny. Wide variation has been noted among mammalian cells, in which VSV can bud exclusively from the cytoplasmic membrane, from internal endoplasmic reticulum, or from both (Zee *et al.*, 1970), which could obviously determine the yield of VSV emerging from infected cells. No other cytoplasmic components have been implicated in determining cell susceptibility or virus yield.

Tissue tropism may well play some role in rhabdovirus infection of the intact host as well as causation of disease. It has been observed that mice are unaffected by enormous doses of virulent VSV administered intraperitoneally, but readily succumb and die after intracerebral inoculation of only a few plaque-forming units (Wagner, 1974). VSV replicates quite readily in the bronchial epithelium of mice infected intranasally, and death occurs after neurological manifestations due to dissemination of progeny virus from the lung to the brain (Wagner, 1974). It has recently been suggested that the body temperature of the host may play a role in susceptibility to disease and virus recovery on the basis of the finding that a hypothermia-inducing neuropeptide (bombesin) resulted in enhanced recovery of VSV from brains of mice infected up to 90 days previously (Hughes *et al.*, 1985). Certain temperature-sensitive mutants of VSV can persist for long periods in mice and cause fatal brain disease (Rabinowitz *et al.*, 1976), presumably owing to cell-immunity hypersensitivity reactions, rather than to the presumed direct, acute lethal effect of rapidly replicating wild-type virus. In all likelihood, the neurological disease caused by rabies virus resembles more closely the late pathogenic effects induced by the less acutely virulent forms of VSV (see Chapter 9).

C. Virus Variants

A critical point that must always be kept in mind in evaluating all research done with rhabdoviruses is the extreme genetic instability of these viruses. Most in-depth studies with VSV are done with the Indiana serotype, but different investigators often use different strains that are usually antigenically indistinguishable but can vary greatly in other biological properties, presumably because of nucleotide substitutions or deletions in the genome. As an example, the sequence of the M-protein gene of the Glasgow strain of VSV-Indiana differed by 11 nucleotides from that of the San Juan strain determined by Rose and Gallione (1981); other significant substitutions in nucleotides and amino acids were evident in the base-sequence homology of the M genes of the Orsay and Glasgow wild-type strains (Gopalakrishna and Lenard, 1985). In their herculean sequencing of the L gene of VSV, Schubert *et al.* (1984) detected 16 point mutations among cDNA clones prepared from viral mRNAs spanning the 6380 nucleotides of the L gene; this extremely high frequency of base substitutions can be attributed either to extreme mutability of the genome, to frequent mistakes by the VSV transcriptase, to infidelity of reverse transcription, or to any combination of these factors. Indirect evidence for high mutability is provided by the extremely large number of mutations that arose spontaneously in the genetic studies of Flamand (1970), particularly in the L gene, which represents 57% of the VSV-genome coding potential (Schubert *et al.*, 1984), but also in cistrons for the other four complementation groups. Additional evidence for high degrees of mutability or other variations due to infidelities in transcription and replication is the high frequencies of temperature-sensitive mutants and DI virus particles (Holland *et al.*, 1980; Youngner and Preble, 1980) that arise on passage in tissue culture or in persistent VSV infections in cell culture or animals.

Later chapters in this volume discuss rhabdovirus genomes and their products (Chapter 4), rhabdovirus genetics (Chapter 5), and DI rhabdoviruses (Chapter 8). Our interest here is in the genotypic and phenotypic variations in rhabdoviruses that lead to alterations in infectivity. Even highly cloned populations of VSV contain particles with differing levels of infectivity. We observed some years ago that the San Juan strain of VSV-Indiana contained two populations of virus, one that gave rise to large plaques and another that give rise to small plaques with significantly reduced capacity to multiply in mouse L cells (Wagner *et al.*, 1963). More stable small-plaque variants with enhanced virulence have also been described (Wertz and Levine, 1973). It has been known for many years that most stocks of vSV, perhaps even all, contain a 5- to 10-fold preponderance of non-plaque-forming and hence noninfectious particles that are morphologically and biochemically indistinguishable from the plaque-forming infectious virions. Recent evidence by Schubert *et al.* (1984) suggests that these noninfectious progeny particles are DI virions that arise from frequent mutations in the L gene that give rise to a defective polymerase that exhibits infidelity in its capacity to transcribe or replicate the genome of the majority of progeny virions.

Also exceedingly common in VSV stocks are DI particles with large deletions of the genome (Huang *et al.*, 1966; Huang and Wagner, 1966a). These DI particles are of two major types: In one type, half to three quarters of the 3' end of the VSV genome is deleted; such 5' DI particles are transcriptionally inert except for synthesis of a small 46-nucleotide leader RNA (Emerson *et al.*, 1977) and can replicate only in the presence of a helper standard (B) virion (Huang and Wagner, 1966a). The other type of DI particle, which is far less common, contains a genome in which approximately 50% of the 5' end is deleted; these 3' DI particles contain a template that can transcribe messengers of all four genes that have not been deleted (Johnson *et al.*, 1979). Both types of DI particles interfere with replication of standard infectious virus, but the 3' DI effectively interferes only with homotypic VSV. Meier *et al.* (1984) have provided convincing evidence for a copy-choice mechanism of replicating for generation of DI-particle RNA of the four types described by Lazzarini *et al.* (1981). Details concerning the nature and biological activities of DI particles are presented in Chapter 8.

Conditional lethal mutants of VSV, usually identified as temperature-sensitive (*ts*) mutants, can be induced by various mutagenic agents, but they also arise spontaneously at relatively high frequency (10^{-6} to 10^{-8}). As described in detail in Chapter 5, *ts*-mutant growth in various cells is restricted at approximately 40°C, but all functions are relatively normal, as is the yield of viral progeny, at the permissive temperature of approximately 30°C. Rhabdovirus genetics, particularly for the prototype VSV-Indiana, is quite clear-cut. All *ts* mutants fall into five complementation groups, each of which has been mapped to one of five specific cistrons on the genome, and each mutation results in a phenotypically defective function in one of the five structural proteins, usually but not always characterized by thermolability. It seems likely that each mutant is the result of a single base substitution resulting in replacement of a single amino acid, as has been demonstrated for *ts* mutants of the M protein of VSV-Indiana coomplementation group III (Gopalakrishna and Lenard, 1985). It seems likely that the changed phenotype of each *ts* mutant is the result of a conformational change in the affected protein resulting from a single amino acid substitution, possibly over a wide region of the polypeptide chain (Gopalakrishna and Lenard, 1985). The function of each protein is altered at restrictive temperature, with lesions ranging from defective transcription in the case of lesions in the L-protein polymerase gene (Hunt *et al.*, 1976) to block in assembly and budding of virions in the case of lesions in the G-protein and M-protein genes (Martinet *et al.*, 1979). In any case, the hallmark of the conditional temperature-sensitive phenotype in abortive infection is markedly reduced yield of progeny at restrictive temperature, a phenomenon that can be readily reversed by temperature down-shift. These *ts* mutants have provided enormously useful tools for probing and mapping the diverse functions of VSV and, more recently, other rhabdoviruses (see Chapter 5).

IV. CYCLE OF INFECTION

A. Sequence of Events

When an infectious virion of the family Rhabdoviridae encounters a susceptible host cell, the result is often a series of events that terminates in release of progeny virions and, frequently, death of the cell. Although each event is not necessarily a precursor to succeeding events but can proceed simultaneously, it is convenient to consider the process of infection depicted in Fig. 3 as a linear series in which each event depends

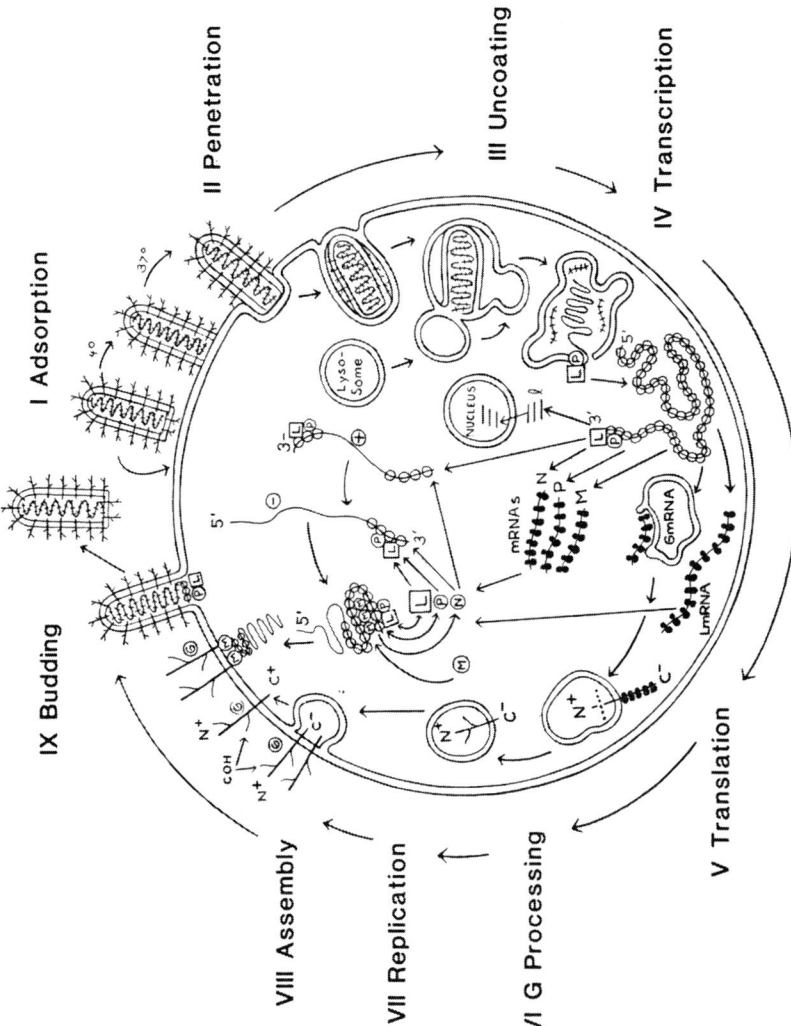

FIGURE 3. Schema of the sequence of events in the cycle of infection of a susceptible cell with VSV from the onset of attachment to the cell surface to release of progeny virions. It should be kept in mind that events such as transcription and translation as well as replication and translation proceed simultaneously.

on occurrence of the preceding event in the following order: adsorption, penetration, uncoating, transcription, translation, replication, assembly, and budding. Figure 4 illustrates the end result of mature virions budding from the surface of an infected cell. Subsequent chapters are devoted to detailed descriptions of transcription (Chapter 6) and replication (Chapter

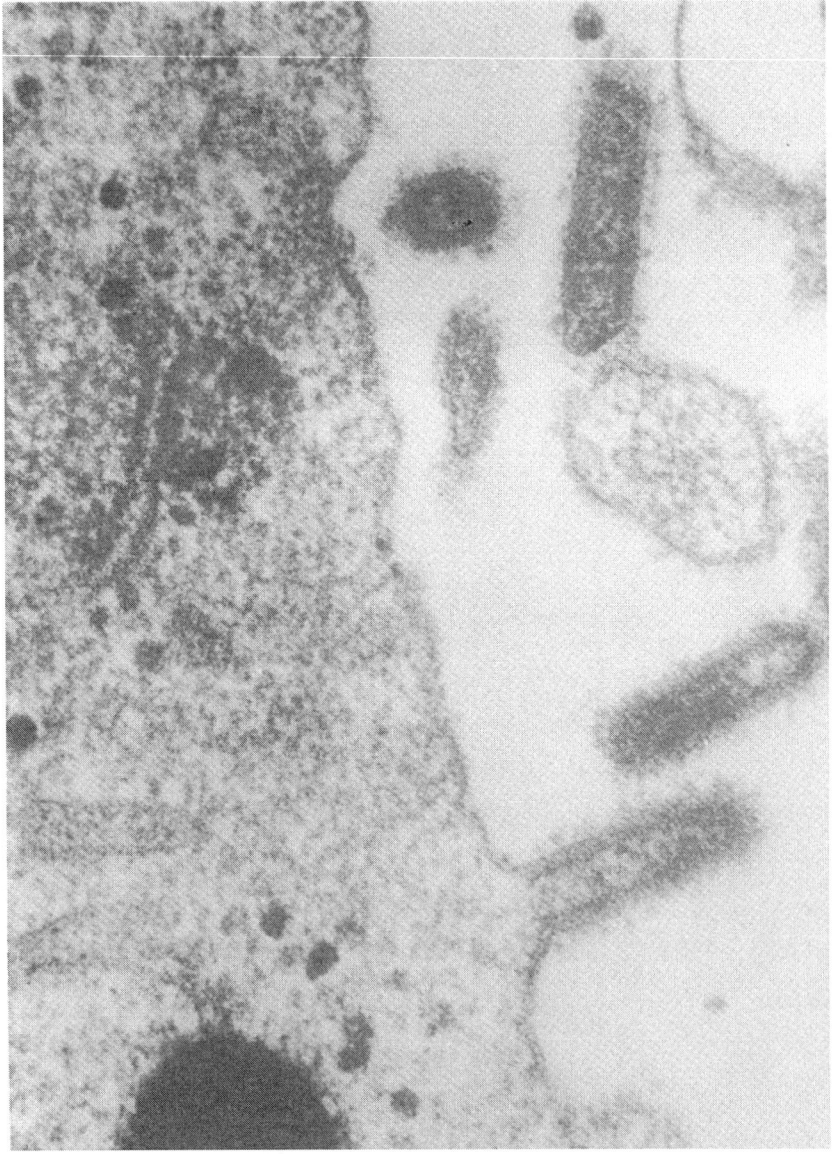

FIGURE 4. Thin section of electron micrograph of L cells infected with VSV. Fixed with glutaraldehyde–OsO$_4$ and stained with uranyl acetate–lead. Note the budding and released VS virions. Reproduced from Wagner (1975) with permission.

7), which will therefore be discussed only briefly here. By the same token, subsequent chapters on rhabdovirus genomes and their products (Chapter 4), genetics (Chapter 5), and budding (Chapter 3) will cover in detail these critical principles for understanding rhabdovirus functions that lead to successful infection. Also in this chapter and elsewhere, by far the greatest emphasis will be placed on vesicular stomatitis virus (VSV), which serves as the prototype for most in-depth studies on the molecular events in infection by rhabdoviruses. Certain properties and behavioral characteristics of other rhabdoviruses deserve special attention and are discussed in other chapters, particularly those on rabies (Chapter 9) and plant rhabdoviruses (Chapter 10).

As is true for all productive viral infections, VSV infection of susceptible cells results in a characteristic growth curve such as that shown on page 45 in Vol. 4 of the previous series, *Comprehensive Virology* (Wagner, 1975). Under standard conditions, VSV infection of a monolayer or suspended-cell culture begins with a latent period of about 2 hr, during which time no progeny virus is detected, followed by exponential multiplication and release of progeny to a peak of approximately 1000 infectious virions per cell at 6–8 hr postinfection. Maximal viral growth depends, of course, on optimal environmental conditions, such as proper media and temperature (replication is faster at 37°C, but yields are often greater at 31°C), as well as on input multiplicity of 5–10 plaque-forming units (PFU)/cell to ensure infectivity of more than 99% of the cell population. Also critical to the success of such experiments and the optimal yield of progeny is the use of freshly cloned seed virus free of defective interfering (DI) particles. It is also wise to keep in mind that cells vary in their capacity to support VSV replication; in our experience, for example, BHK-21 cells routinely yield 5–10 times more progeny than do mouse L cells, which, however, are more susceptible to cytopathic effects. Rates of replication and yields also depend on serotypes and strains within serotype; e.g., VSV-Indiana routinely replicates faster and to higher yield than does VSV-New Jersey (Grinnell and Wagner, 1983).

With these basic principles in mind, the following sections will highlight each of the events in standard virus–cell interaction that occur during the cycle of VSV infection, which is illustrated by Fig. 3.

B. Adsorption

As is the case with all viral infections, adsorption of VSV, or of any other rhabdovirus, depends on the presence of the proper cell-attachment organ on the surface of the virion and the existence of the proper receptor on the surface of the host cell. Clearly, both the virus and the host cell must be considered in measuring adsorption. It should be stated at the outset that VSV adsorption is quite inefficient and, in comparison with other virus–cell systems, is rather difficult to quantitate with the accu-

racy one would like to see. The ratio of physical particles to infectious
viral particles is rarely less than 5:1. However, more is known about the
cell-attachment organ of VSV than is known about the cell receptor. Early
experiments revealed that adsorption of VSV is not an energy-dependent
event, since the reaction occurs readily at 4°C and equally for chick-
embryo cells and mouse L cells (Wagner et al., 1963). These studies also
revealed that the kinetics of adsorption are exponential, at least for plaque-
forming virus adsorption, which was the technique used at that time.
Somewhat more reliable results can probably be obtained by using radio-
actively labeled virus, a technique that will of course measure both plaque-
forming and the majority of noninfectious virus.

Detailed kinetic analyses of the adsorption of [^{35}S]methionine-labeled
VSV to Madin–Darby canine kidney (MDCK) cells by Matlin et al. (1982a)
revealed very efficient binding of virus at 0°C compared with other enveloped viruses. VSV binding was found to be exceedingly pH-dependent;
about 10 times more VSV attached to cells at pH 6.5 than at higher pH
levels. Even at optimal pH, binding failed to reach equilibrium, and the
results were not very reproducible. At least 90% of the adsorbed virus
was removable by protease treatment, and only a limited amount of the
bound virus could be internalized by warming the virus–cell complex to
37°C. These studies highlight the difficulties in studying cellular adsorption and penetration of VSV, a problem that probably extends to all rhabdoviruses.

1. Infectious Virion

The cell-attachment organ of VSV is the glycoprotein (G) spike, removal of which by proteases reduces infectivity more than 10^5-fold (Cartwright et al., 1969; Schloemer and Wagner, 1975b); partitioning intact G
protein (removed from virions by detergent) into the membrane of spikeless virions can restore infectivity approximately 100-fold (Bishop et al.,
1975). The concept that the terminal sialic acid in the carbohydrate chains
of VSV G protein is responsible for efficient infectivity and adsorption to
cells (Schloemer and Wagner, 1975a,b) has been refuted (Cartwright and
Brown, 1977); the erroneous interpretation was probably due to aggression
of virus particles, the charge repulsion of which was altered by exposure
to neuraminidase, resulting in plaque reduction due to infection of a single
cell by multiple particles in the virus aggregate. Thimmig et al. (1980)
found that isolated G protein, free of virions, attached readily to BHK-21
cells; they estimated the attachment as 3×10^5 G-protein molecules per
cell. This, of course, would not correspond to the number of cell receptors,
because VSV G protein free of detergent forms micellar aggregates of
numerous G protein molecules (Petri and Wagner, 1980). These studies
of Thimmig et al. (1980) are intrinsically interesting, but they shed little
light on the cell-receptor site or the cell-attachment organ of intact VS
virions, since the isolated G protein did not preempt BHK-21 cell sites

for subsequent attachment of VSV; neither did intact VSV saturate the cell receptors for adsorption of the isolated G protein. These studies do demonstrate a great difference between the G protein in intact virions and the isolated G protein, presumably due to conformational changes under these different conditions.

It is likely that the lipid bilayer that comprises the membrane of VSV also plays some role in its infectivity, since exposure to phospholipase reduces infectivity, but not nearly as much as exposure to trypsin (Cartwright et al., 1969). The isolated nucleocapsid of VSV, free of membrane and its G protein, can also infect cells and produce plaques in monolayers, but at very low efficiency (10^{-5} to 10^{-6}); nucleocapsid infectivity depends on the presence of active polymerase and is considerably augmented by adjuvants in the adsorption medium such as DEAE–dextran (Szilagyi and Uryvayev, 1973; Bishop et al., 1974). It is also of interest that the presence of DEAE–dextran in the adsorption medium increases the infectivity of intact VS virions by 4-fold (Bailey et al., 1984).

Viral membrane components that contribute to adsorption, other than the G protein, have not been studied in detail. However, Flamand and Bishop (1973) found no difference in adsorption to BHK-21 cells of VSV grown in BHK-21 or chick-embryo cells, a finding that suggests that cell-specific membrane components, such as glycolipids (Cartwright and Brown, 1972a), do not influence VSV adsorption to host cells. The cell-attachment function of the VSV G protein is also not affected by altering its secondary structure with impermeable sulfhydryl reducing agents (with or without alkylation) that disrupt the G-protein disulfide bonds. These impermeable reducing agents effectively disrupted disulfide bonds without affecting agglutination of goose erythrocytes or viral infectivity (Beatrice and Wagner, 1980a).

The rather finicky hemmagglutinating activity of rhabdorivuses provides another method for studying the virion component responsible for adsorption to cells. Goose erythrocytes under very restricted conditions of pH, ionic strength, and temperature can be agglutinated by rather massive concentrations of rabies virus, VSV, and other rhabdoviruses (Kuwert et al., 1968; Arstila et al., 1969). The hemagglutinating activity of VSV was shown by Arstila (1972, 1973) to be due to the spike G protein. These results were confirmed and extended by McSharry et al. (1978), who found that G protein removed from VSV virions by Triton X-100 was capable of hemagglutinating goose erythrocytes, whereas the G protein-denuded virions were not. VSV antibody blocks hemagglutination by isolated G protein or whole virions. It is also of interest that the hemagglutinating activity of whole VSV virions or solubilized G protein is greater per infective virion for virus grown in hamster cells (HAK and BHK) than for that grown in MDBK cells, suggesting that a cell component, presumably carbohydrate, contributes to hemagglutinating activity. Varying patterns of glycosylation of the G protein are likely to be the basis for cell modification of the hemagglutination capacity of VSV virions

(McSharry *et al.*, 1978). In this regard, it is of interest that neuraminidase can inactivate the hemagglutinating activity of VSV virions (Schloemer and Wagner, 1975b). However, since the terminal sialic acid on the carbohydrate chains of VSV G protein is now known not to affect VSV infectivity (Cartwright and Brown, 1977), it seems likely that two separate functions of the VSV G protein are responsible for adsorption to goose erythrocytes and to host cells.

2. Host-Cell Receptors

Since adsorption of VSV virions to the surface of all cells tested thus far is relatively inefficient, it has not been easy to design experiments to identify unequivocally the cell receptors responsible for binding the virion. Therefore, unlike myxoviruses and paromyxoviruses, which recognize cell-surface sialic acids as the adsorption receptor, identification of the cell-surface receptor(s) for rhabdovirus attachment has been far more elusive. Exposure of L cells to neuraminidase does not affect adsorption of VSV, a finding that tends to eliminate sialoglycoproteins or sialoglycolipids as major receptors for VSV (Schloemer and Wagner, 1975b). However, some evidence has been presented that neuraminidase removes receptors for rabies virus from the surface of chick cells (Superti *et al.*, 1984b). The same authors also reported phospholipase-sensitive receptors for rabies virus (Superti *et al.*, 1984a). Moreover, prolonged treatment of L cells with large amounts of trypsin did not significantly impair attachment of VSV, thus suggesting that readily accessible surface proteins are not prime candidates as cell receptors for VSV adsorption (Schloemer and Wagner, 1975b).

Similar studies with Vero monkey cells also revealed that exposure of cells to trypsin did not affect binding of VSV at 4°C (Schlegel *et al.*,1982b). These authors also studied the kinetics of VSV binding to untrypsinized cells and came to the conclusion that adsorption is a two-phase phenomenon. Monolayers of Vero cells exposed to 1 ml of [^{35}S]methionine-labeled VSV attached virus slowly and in a nonlinear fashion, reaching a saturation plateau at 12 hr; by this technique, only 12% of the input virus was adsorbed to cells, and this binding was reversible by elution of 47% of the attached virus when cells were warmed to 37°C for 20 min (Schlegel *et al.*, 1982b). It was concluded that almost none of the VSV was internalized after adsorption to cells at 4°C, on the basis of the finding that 94% of the surface virus could be removed by trypsinization. Dose–response curves and Scatchard plots of binding affinities, studied under conditions of excess and nonexcess input, revealed two distinct interactions between Vero cells and VSV, indicating separate saturable and nonsaturable sites for attachment. Prior exposure to trypsin did not diminish nonsaturable binding. There apparently is no competition for the two separate binding sites. Further studies showed that addition of antibody sufficient to neutralize the infectivity of VSV did not impede its binding to Vero cells at

4°C, nor did the antibody reduce internalization of VSV in cells warmed to 37°C (Schlegel and Wade, 1983). In fact, antibody-inactivated VSV was taken up at 37°C faster than was VSV in the absence of antibody. The authors hypothesize that antibody-neutralized VSV binds to a separate or additional cell-binding site, which strengthens the argument that there are two separate cell receptors (saturable and nonsaturable).

Schlegel et al. (1983), in an extension of their studies, set out to identify the nonsaturable cell-surface receptor responsible for binding VSV. They first extracted Vero-cell membranes with the dialyzable, nonionic detergent octyl-β-D-glucopyranoside and found that the dialyzed extract specifically inhibited the saturable, high-affinity binding of [^{35}S]methionine-labeled VSV to Vero cells. The binding inhibitor was resistant to protease and neuraminidase, but was inactivated by phospholipase C, suggesting that it is a phospholipid. Of all phospholipids tested, only phosphatidylserine (PS) totally inhibited the high-affinity binding to VSV to Vero cells and also inhibited VSV plaque formation by 80–90%, but did not block herpesvirus plaque formation (Schlegel et al., 1983). The authors suggested that PS may therefore be the VSV receptor on Vero cells. They also reported in an admittedly not quantitative electron-microscopic study that mixed vesicles of PS and phosphatidylcholine (PC) bind to VSV better than do vesicles made of PC alone. However, other investigators have reported exceedingly high binding of PC vesicles to VSV with intact G protein spikes, but not to spikeless VSV (Moore et al., 1978). Nevertheless, these experiments by Schlegel et al. (1982b, 1983) are intriguing and deserve further investigation. Some preliminary studies by Superti et al. (1984a) indicate decreased infectivity by VSV and rabies virus of chick-embryo-related (CER) cells previously tested with phospholipases A_2, C, and D as well as sphingo-myelinase, as determined by immunofluorescence. They also reported restoration of viral receptors 5 hr after treatment of CER cells with phospholipase A_2. Large amounts of phosphatidylethanolamine, phosphatidylinositol, and PS, but not PC, inhibited VSV infection when added to virus or cells. These experiments are difficult to interpret, particularly since phospholipids can cause profound nonspecific, detergent like effects on viruses and cells.

Evidence for a common rhabdovirus receptor on host-cell surfaces was provided by Wunner et al. (1984), who characterized the saturable binding sites for rabies virus. Attachment of the ERA strain of rabies virus to BHK cells followed the laws of mass action and was much more efficient at pH 6.0 than at higher pH levels. Binding was enhanced by DEAE–dextran, but was not affected by prior exposure of cells to protease or neuraminidase. There was clear evidence for saturable receptors on BHK cells and mouse neuroblastoma 1300 cells. Competition for ERA binding occurred with a nonpathogenic rabies strain and with VSV, but not with reovirus type 3 and only slightly with West Nile virus, suggesting a common virus receptor for rhabdoviruses different from that for other viruses. Octylglucoside extracts of BHK-cell membrane yielded a chloroform–methanol-soluble component or components that blocked rabies virus attachment and that appear to be either phospholipid or glycolipid. More detailed information on the pathogenicity of rabies virus is provided in Chapter 9.

A variety of other experiments, largely by indirect phenomenological techniques, have been performed in an attempt to identify the cell receptors for rhabdoviruses. For example, human serum lipoproteins, mostly the very-low-density variety, have been found to inhibit the infectivity of VSV-Indiana and a fixed rabies virus, ostensibly by decreasing viral attachment to host cells (Sesanti *et al.*, 1983). Concanavalin A also affects the infectivity, maturation, and cytopathogenicity of VSV, presumably because of its lectin-binding activity to viral carbohydrates (Takehara, 1979). Acetylcholine receptors on myotubules have been postulated to be receptor sites for attachment of rabies virus on the basis of reduced infectivity after exposure to α-bungarotoxin and α-tubocurarine (Lentz *et al.*, 1982); neurotransmitters as possible binding sites may explain the neurotropism of rabies virus (Lentz *et al.*, 1983). Rabies virus also decreases agonist binding to opiate receptors of mouse neuroblastoma–rat glioma hybrid cells (Munzel and Koschel, 1981).

C. Penetration and Uncoating

Penetration of host cells by VSV closely follows adsorption (see Fig. 3) by a series of mechanisms that have long been in dispute. An event that occurs simultaneously or soon after penetration is called uncoating, or more properly in the case of VSV, envelope removal. These two events are not readily dissociable and probably should be considered together. Whereas adsorption can occur efficiently at 4°C, entry of the virus into the cell (penetration) is an energy-dependent event generated by the cell and requires a physiological temperature. Penetration of VSV differs only in minor ways from penetration of other viruses, particularly other enveloped viruses. In fact, it has been postulated that the VSV and α_2-microglobulin share the same receptor for endocytosis by coated pits (Dickson *et al.*, 1983).

Detailed recent reviews of the phenomena of virus entry into cells have been provided by Howe *et al.* (1980), Bukrinskaya (1982), and White *et al.* (1983), among others. The closely related phenomenon of virus–erythrocyte interaction has also been reviewed by Howe and Lee (1972). Strong evidence for interaction of VSV G protein with the surface membrane of cells comes from experiments on fusion from within, by which one VSV-infected cell can fuse with an uninfected cell by virtue of the G protein inserted in the plasma membrane, particularly when virion maturation is blocked in cells infected with temperature-sensitive mutants restricted in the M protein (Storey and Kang, 1985).

Electron-microscopic studies of VSV penetration yielded conflicting results. Heine and Schnaitman (1969; 1971) presented evidence that the membrane of VSV can fuse with the surface cytoplasmic membrane, thereafter discharging the nucleocapsid into the cytoplasm. Simpson *et al.* (1969), on the other hand, could find evidence for entry of only intact VS virions in phagocytic vesicles. In an attempt to resolve this dispute, Dahlberg (1974) attempted quantitative electron-microscopic studies and found that VSV particles entered cells almost always by viropexis and only rarely by fusion to plasma membrane. When he performed experiments in a manner similar to that of Heine and Schnaitman (1971) by cocentrifugation of virus and cells, somewhat greater numbers of VSV particles fused with plasma membrane, but the major proportion appeared to enter cells in an intact form. In an exhaustive series of studies, Helenius and his colleagues have come to the conclusion that enveloped viruses adsorb to cell surfaces at the site of histocompatibility antigens and coated pits; ingestion of the adsorbed virus then appears to occur by endocytosis of coated vesicles, a process that applies to VSV (Matlin *et al.*, 1982b) and other enveloped viruses, particularly the well-studied Semliki Forest virus (Helenius *et al.*, 1980). After endocytosis, the coated vesicle fuses in a succession of events at pH < 6 with lysosomes, thus resulting in release of the nucleocapsid (Marsh *et al.*, 1983). Certain aspects of this series of events have not been confirmed by other investigators. Oldstone *et al.* (1980) reported that cells lacking H-2 or HLA histocompatibility antigens were readily penetratable by Semliki Forest virus. Nevertheless, it seems likely that the Helenius hypothesis is largely true and that VSV and other enveloped viruses can enter cells by endocytosis and are uncoated (de-membraned) by reaction of endocytic vesicles with lysosomes at low pH. Another possibility is that VSV and other enveloped viruses can use the alternative pathways of endocytosis or cytoplasmic membrane fusion to deposit the membrane-stripped nucleocapsid into the host-cell cytoplasm. In either case, the virus membrane is removed by fusion with surface or internal cell membranes.

1. Endocytosis

The discrepancy in data on the mechanisms of viral penetration and their interpretation is undoubtedly due to variations in experimental conditions and the nature of the virus used. In a very nice study by Fan and Sefton (1978), specific viral antibodies and complement were used to detect the presence of viral antigens on the surface of host cells that had engulfed either Sendai virus, Sindbis virus, or VSV. Cells infected with moderate amounts of Sendai virus were lysed by antibody and complement, whereas cells infected with Sindbis virus or VSV were not unless enormous input amounts of these latter viruses were used. Fan and Sefton (1978) concluded that the membrane of Sendai virus, which contains a protein fusion factor (Scheid and Choppin, 1977), customarily fuses with the plasma membrane of the host cell, leaving its membrane antigens on the surface, whereas VSV and Sindbis virus, membrane and nucleocapsid, are engulfed by the host cell and do not leave their antigens on the cell surface except at enormous input multiplicity. These data go a long way to explain the electron-microscopic observations of Heine and Schnaitman (1969, 1971).

The critical conditions for VSV penetration into cells that has been amply documented by Helenius and his colleagues is the requirement for low pH of the surrounding medium (White *et al.*, 1981; Matlin *et al.*, 1982b). When VSV, previously adsorbed to MDCK cells at 0°C, is shifted to 37°C, 40–50% of the prebound virus becomes resistant to protease if the external pH is below 6 for at least 30 sec. On the basis of these and other experiments, these investigators conclude that "VSV fuses to the MDCK cell plasma membrane at low pH" and further suggest that "the virus enters cells by endocytosis in coated pits and coated vesicles where the low pH triggers a fusion reaction ultimately leading to the transfer of the genome into the cytoplasm" (Matlin *et al.*, 1982a,b). Technical details are presented in *Methods in Enzymology* (Marsh *et al.*, 1983). Similar fusogenic activity at low pH occurs when cells transfected with cDNA vectors expressing VSV G protein, which is subsequently inserted into the plasma membrane, fuse with adjacent cells (Florkiewicz and Rose, 1984; Riedel *et al.*, 1984).

In an effort to define the reactive components in VSV and the target-cell membrane that lead to fusion, Eidelman *et al.* (1984) designed a system in which unilammelar PC vesicles into which VSV G protein had been inserted by the method of Petri and Wagner (1979) were tested for their capacity to fuse with other vesicles containing different phospholipids. Fusion was monitored by electron microscopy and fluorescence energy transfer. Eidelman *et al.* (1984) found that G-protein vesicles fused only with vesicles that contained acidic phospholipids (PS or phosphatidic acid); the fusion reaction was pH-dependent with a pK of about 5 and an apparent energy of activation for the fusion reaction of 16 ± 1 kcal/mole.

2. Hemolysis

Another method for studying cell penetration by VSV is to measure hemolysis of erythrocytes to which the virus has adsorbed, a procedure that lends itself to quantitation as well as to measuring the kinetics of the fusion reaction. Human erythrocytes exposed to certain proteases will adsorb VSV or rabies virus and will rapidly undergo hemolysis. The extent of the hemolytic reaction was found to depend almost entirely on the pH of the reaction mixture and was optimal at pH 5.0 (Manen *et al.*, 1982). The hemolytic reaction could be completely blocked by exposure of VSV to anti-VSV serum (Mifune *et al.*, 1982). Hemolysis of human erythrocytes by VSV is maximal at pH $\leq$ 5.0 and is negligible at pH $\geq$ 6.0 (Bailey *et al.*, 1981). The polycation DEAE–dextran greatly enhanced the hemolytic activity of VSV for human erythrocytes and for those of other species, in a manner quite similar to its capacity to increase the binding capacity and infectivity of VSV for BHK cells (Bailey *et al.*, 1984). These effects were maximal at pH 5.0 and at 37°C. Vesicles containing G protein and VSV membrane lipids displayed about 40% of the hemolytic activity of intact VSV virions and were completely inactive at pH 6.0, as were spike-less VSV virions.

Schlegel and Wade (1984) surmised that the VSV G protein amino-terminal segment, which has clusters of positively charged and hydrophobic amino acids in many VSV strains (Rose and Gallione, 1981; Kotwal *et al.*, 1983), may be the VSV factor responsible for hemolysis. To test this hypothesis, they prepared a synthetic oligopeptide corresponding to the 25 amino-terminal amino acids of the VSV G protein and tested its hemolytic activity. They found that this oligopeptide had pH-dependent hemolytic activity quite comparable to that of the intact VSV G protein. Antibody prepared against the synthetic oligopeptide did not neutralize VSV infectivity and could bind only to SDS-denatured G protein, suggesting that the G protein is not hemolytic in the native conformation (Schlegel and Wade, 1984). In an extension of this study, Schlegel and Wade (1985) found that only the 6 terminal amino acids of the 25-amino-acid oligopeptide were essential for causing hemolysis. The hemolytic effect was retained if the amino-terminal lysine of this peptide was replaced by another positively charged amino acid, arginine, but not when it was replaced by glutamic acid. The 25-amino-acid oligopeptide was also found to be cytotoxic and to affect gross changes in permeability in cultured Vero cells, confirming the membrane-destabilizing properties of this VSV G protein domain. However, when the N-terminal lysine was replaced by glutamic acid by oligodeoxynucleotide mutagenesis, this mutated G protein retained its pH-dependent cell-fusion activity (Woodgett and Rose, 1986). It would seem unwise, therefore, to equate hemolytic activity of G protein with its cell-fusion activity.

3. Membrane Perturbation

There have been limited biophysical studies on perturbation of cell membranes undergoing penetration by VSV. By means of fluorescence depolarization using the hydrophobic compound 1,6-diphenyl-1,3,5-hexatriene inserted into the membrane of foreskin cells, Levanon *et al.* (1979) observed reduced membrane microviscosity (increased fluidity) at 37°C, a temperature at which penetration was presumably progressing. These authors noted the same results with other enveloped and nonenveloped viruses and suggested this as a diagnostic procedure for identifying viruses. Altstiel and Landsberger (1981) used spin-label electron spin resonance (ESR) of a nitroxide derivative of stearic acid to study structural changes at 37°C in the plasma membrane of BHK-21 cells exposed to VSV at a multiplicity of 100 PFU/cell. These experiments showed an increased rigidity (decreased fluidity) of the cell membrane that was attributed to cross-linking of receptors in the plane of the plasma membrane that did contain surface VSV antigen detectable by immunofluorescence. The monovalent water-soluble G_s protein of VSV did not cause this effect, but G_s plus anti-G serum did, an effect that was reversed by colchicine. Young et al. (1983) measured the conductance of planar lipid bilayers after interaction with VSV, Sendai, influenza, and Semliki Forest virus (SFV). VSV, influenza virus, and SFV all increased conductance in the bilayer at pH 5.2 but not at pH 7.0, which was optimal for Sendai virus. The presence of cholesterol in the planar bilayer enhanced the conductance changes for all the viruses except VSV. Gutierrez et al. (1984) studied synchronized entry of VSV into spin-labeled cultured human cells and found an increase in the rigidity of the cell membrane as measured by ESR spectroscopy. This effect was not altered by treatment of the cells with α-interferon, despite its antiviral activity. It is difficult to explain or reconcile the discrepancy in results obtained by Altstiel and Landsberger (1981) and Gutierrez et al. (1984) with those of Levanon et al. (1979). Quite obviously, more detailed biophysical experiments are required to provide conclusive data on the dynamics of cell-membrane interaction with rhabdoviruses, or with other viruses, for that matter.

4. Uncoating

For infection to proceed following penetration of a rhabdovirus, the membrane must be removed (uncoating) to permit viral transcription. Since endocytosis appears to be the predominant mode of VSV entry into the cytoplasm, the cytoplasm appears to be the logical site for uncoating. This event still requires fusion of viral membrane with a cellular membrane, which in the case of VSV and other rhabdoviruses appears to be the membranes of endocytic vesicles. The sequence of events appears to be similar for rhabdoviruses and other enveloped viruses, with the exception of those paramyxoviruses that contain an active fusion-factor protein (Lenard and Miller, 1982; White et al., 1983). The intact virus apparently enters the cell by means of a coated pit, progressing to a coated vesicle and thence to larger uncoated collecting vacuoles and finally to secondary lysosomes. Fusion of the VSV virion membrane with the vacuole membrane requires quite a low pH that ostensibly can be provided only by lysosomes. Much of the evidence for the function of the lysosome comes from the use of antiviral lysosomotropic amines that are highly lipophilic and can diffuse in their uncharged form across lysosomal membranes, where they are protonated. These basic lysosomotropic amines can cause a rapid rise in the pH of the lysosomal contents; it is by this series of events that they are thought to abort viral infections (Lenard and Miller, 1982). In theory, the higher pH prevents uncoating of VSV or other enveloped viruses, thus terminating the cycle of infection.

The capacity of lysosomotropic agents to interrupt an intracellular step required for completing the cycle of VSV infection has been studied by various investigators (Shimizu *et al.*, 1972; Miller and Lenard, 1980, 1981; Matlin *et al.*, 1982a,b). The lysosomal inhibitor chloroquine, when added with the infecting virus, completely inhibited VSV infection at all multiplicities of infection and caused 50% inhibition when added up to 1.5 hr postinfection (Miller and Lenard, 1980). Quite a large number of lipophilic amines of diverse chemical structure, such as local anesthetics (dibucaine, tetracaine, lidocaine, procaine), antihistaminics (pyrilamine maleate, chlorpheniramine, promethazine-HCl), an antipyretic (amino-pyrine), and miscellaneous amines (dansylcadaverine, ethylenediamine, 1-propylamine, imidazole, methylamine), as well as chloroquine, were all found to exhibit similar kinetics, but at somewhat different doses, in inhibiting VSV infection of BHK cells (Miller and Lenard, 1981). The inhibitory effects of these amines were found to occur at a stage following transfer of virus from the cell surface to an intracellular site but prior to primary transcription of the viral genome. It was logical for Miller and Lenard (1981) to assume that the lysosome was the intracellular site of the inhibitory action of these lysosomotropic amines and that the effect occurs at the stage of viral uncoating in a manner similar to that for infection with Sendai virus, influenza virus and SFV (White *et al.*, 1981, 1983).

Similar but not identical results were obtained by Schlegel *et al.* (1982a), who found that dansylcadaverine and amantadine inhibited the uptake of VSV by mouse Swiss 3T3 cells as determined by immunoflu-orescence, electron microscopy, [^{3}H]-VSV uptake, and subsequent viral RNA synthesis. These compounds were also found to inhibit uptake of α_2-macroglobulin, which ostensibly binds to the same receptor as that for VSV, but these compounds do not block uptake of concanavalin A or insulin, which may have different receptors. Schlegel *et al.* (1982b) pos-tulate one cellular target for amantadine and dansylcadavarine that ap-pears to result from the clustering of membrane-bound ligands or particles in clathrin-coated pits. In somewhat disparate results, McCooms *et al.* (1981) found that inhibition of endocytosis by cytochalasin B did not influence infection with Sindbis virus or VSV and that the lipophilic amine chloroquine did not block ingestion of these viruses, but still greatly reduced their yields. These results led McCooms *et al.* (1981) to conclude that "endocytosis is not essential for the infection of cultured cells by Sindbis virus or vesicular stomatitis virus." Clearly, more data are re-quired to resolve the exact nature of VSV penetration and uncoating events more definitively.

Two lysosomotropic agents, ammonium chloride and chloroquine, were found by Tsiang and Superti (1984) to inhibit infection of murine neuroblastoma cells by rabies virus, suggesting to these authors that ra-bies virus infects neuronal cells by means of endosomes and a pH-de-pendent pathway. In a companion study, this laboratory noted by electron-microscopic and fluorescence techniques that rabies virus enters CER cells by means of adsorptive endocytosis, presumed to be independent of cellular metabolic processes. Certain evidence was also presented to suggest that ammonium chloride and chloroquine, the latter more slowly, inhibit virus-membrane fusion at a prelysosomal step (Superti *et al.*, 1984b). Specific sites of entry of rabies viruses into peripheral nerves of infected mice have been described to be at motor endplates as determined by cholinesterase-positive and immunofluorescent staining (Watson *et al.*, 1981).

D. Transcription

Transcription is the first viral metabolic event after cells are penetrated by rhabdoviruses and their uncoated nucleocapsids are released into the cytoplasm. Details of the transcriptional events and their regulation are presented in Chapter 6. Suffice it to say here that VSV transcription is independent of host-cell functions and takes place intracellularly in the presence of actinomycin D and inhibitors of protein synthesis, such as cycloheximide; this has been called "primary transcription" (Huang and Manders, 1972), which takes place on input (parental) nucleocapsids (see Fig. 3). The early literature on *in vivo* transcription was reviewed by Wagner (1975) and also more recently in the preceding series, *Comprehensive Virology*, Vol. 19, by Wagner *et al.* (1984), some of which is reproduced below.

The deproteinized RNA of VSV, unlike that of poliovirus RNA, is not infectious (Huang and Wagner, 1966b). Baltimore *et al.* (1970) first described the virion-associated RNA-dependent RNA polymerase and worked out the basic conditions for VSV transcription *in vitro*. The enzymology of the VSV transcriptase was described by Emerson and Wagner (1972, 1973) with the absolute requirements for N-protein-encapsidated genome RNA serving as a template for reconstitution with both the L and NS (P) proteins; L and NS (P) proteins, from homologous or heterologous virions, are both required to initiate the transcriptase reaction (Emerson and Yu, 1975; Mellon and Emerson, 1978). Under these conditions, the nucleocapsid is infectious (Bishop *et al.*, 1974) and codes for all five VSV monocistronic mRNAs *in vitro* (Bishop *et al.*, 1974) and *in vivo* (Banerjee *et al.*, 1977). The gene order of the five VSV cistrons was worked out by Ball and White (1976) and by Abraham and Banerjee (1976a) as 3'-N-NS-M-G-L-5', which also represents the decreasing molar ratios in which the five mRNAs are synthesized, based on their proximity to the 3'-genome terminus. Colono and Banerjee (1977) described a 47-nucleotide leader RNA sequence that, unlike the five mRNAs, is neither capped, polyadenylated, nor translated, but initiates the transcription process and is made in larger molar amounts than any mRNA. All five VSV mRNAs are capped and polyadenylated, but no one, to date, has identified specific capping or adenylating enzymes (Abraham and Banerjee, 1976b).

Another VSV protein that regulates transcription is the M protein, which binds to nucleocapsids (Wilson and Lenard, 1981) and inhibits transcription by approximately 80% (Clinton *et al.*, 1978a; Carroll and Wagner, 1978a, 1979).

E. Translation

Translation of each of the mRNAs of VSV, but not the leader RNA sequence, proceeds immediately after, and in fact is coupled with, transcription. All five VSV proteins are synthesized throughout the cycle of infection, but the amount of each protein synthesized is roughly in decreasing order of the linear sequence of genome cistrons: N, NS, M, G, and L proteins (Wagner *et al.*, 1970; Mudd and Summers, 1970; Hsu *et al.*, 1979). Figure 3 offers a brief schematic representation of the process. The G protein is synthesized from mRNA on endoplasmic reticulum membrane-associated polyribosomes by means of a signal sequence, step wise glycosylation, and migration of vesicles to the cytoplasmic membrane for fusion and insertion of the fully glycosylated G protein (Knipe *et al.*, 1977; Rothman *et al.*, 1980). Chapters 3 and 4 describe in detail the synthesis, processing, and membrane insertion of the VSV G protein. The other four VSV proteins are synthesized from monocistronic messengers on cytoplasmic polyribosomes (Morrison and Lodish, 1975). Outlined below are some VSV translational strategies abstracted from a previous short review of this subject (Wagner *et al.*, 1984).

The mRNAs of VSV appear to be translated in a manner quite similar to that of other eukaryotic mRNAs (Breindl and Holland, 1975). The VSV mRNAs are polyadenylated and are capped *in vitro* as well as *in vivo* by incorporating the methyl group of *S*-adenosyl-L-methionine during synthesis by the virion-associated polymerase (Rhodes *et al.*, 1974). If methylation and capping are inhibited by the analogue *S*-adenosylhomocysteine, the uncapped VSV mRNAs are poorly translated in a wheat-germ system (Both *et al.*, 1975). However, Rose and Lodish (1976) found that removal of the 5'-terminal 7-methylguanosine did not greatly reduce the translational activity of VSV mRNAs in a reticulocytelysate cell-free system. In addition, certain VSV proteins are phosphorylated to varying degrees, particularly the P protein and to a lesser extent the M protein, by a kinase, probably of host-cell origin (Imblum and Wagner, 1974). The L protein is thought to have kinase activity and possibly regulates the degree of phosphorylation of the P protein (Sanchez *et al.*, 1985), a finding at variance with the studies of Clinton *et al.* (1982), who consider the protein kinase to be of cellular origin. The degree of phosphorylation of two species of VSV NS (P) protein appears to determine its role in VSV transcriptase activity (Kingsford and Emerson, 1980). These and other properties of VSV proteins and their specific mRNAs are potentially involved in competition for translation of cellular proteins. Although different VSV mRNAs appear to be equally effective in binding to ribosomes and in initiating and elongating polypeptides, the N-protein mRNA appears to be more efficient in translation than the G-protein mRNA under conditions of inhibition by high salt concentrations or by aurintricarboxylate, suggesting differences in requirements for factors that initiate translation (Lodish and Froshauer, 1977). Lodish and Porter (1980) have reported that VSV mRNAs in general out-compete cellular mRNAs for binding to ribosomes, although they do not differ significantly in initiation of translation in reticulocyte lysates. These authors concluded that this competitive advantage of VSV mRNAs for ribosomes explains the capacity of the virus to reduce cellular protein synthesis, but, as will be indicated later, evidence by other techniques controverts such a conclusion (Dunigan and Lucas-Lenard, 1983; Schnitzlein et al., 1983).

F. Replication

A detailed description of replicative strategies for VSV is presented in Chapter 7. The summary presented here is largely abstracted from our recent short review in *Comprehensive Virology*, Vol. 19 (Wagner et al., 1984).

Unlike transcription, replication of VSV requires coupled translation (Huang and Manders, 1972; Wertz and Levine, 1973; Davis and Wertz, 1982) and probably one or more host factors (Hill et al., 1981; Patton et al., 1983). In all likelihood, the same endogenous polymerase functions for replication as well as transcription. Although all full-length plus-strand and negative-strand RNA molecules and some messengers and leader RNAs are encapsidated with N protein (Blumberg et al., 1981), partially double-stranded replicative intermediates can be found in infected cells (Wertz, 1978); as described later, these replicative intermediates may be involved in inhibiting cellular protein synthesis (J.R. Thomas and Wagner, 1982). The viral N protein is thought to modulate transcription and replication by its ability to bind to nascent leader RNA, thus promoting read-through of the termination signals as the full-length RNA is assembled into nucleocapsids (Schubert et al., 1982). Experiments by Blumberg et al. (1981) have led to the hypothesis that interaction of VSV leader RNA and nucleocapsid protein may control VSV genome replication. Proteins N, L, and P (NS) rapidly associate with nascent progeny RNA to form a ribonuclease-resistance ribonucleocapsid; even the presence of cycloheximide does not block replication for at least 20 min, since the existing pool of N, L, and P proteins continues to assemble with nascent RNA chains (Rubio et al., 1980). The same events apparently occur *in vitro* in a coupled replication–translation system (Wertz, 1983).

It is becoming increasingly evident that the VSV polymerase serves the dual purpose of messenger transcription and replication of the entire VSV genome. The apparent site of entry of the polymerase is the 3′ terminus of the VSV genome (Emerson, 1982), which codes for the plus-strand leader RNA. It has been proposed that the 5' terminus of the leader RNA serves as the nucleation site for binding of newly synthesized N protein to form the ribonucleocapsid, thus regulating the switch from transcription to replication (Blumberg et al., 1981, 1983). The La protein of systemic lupus erythematosus has been found complexed with both the plus-strand leader RNA (Kurilla and Keene, 1984) and the minus-strand leader (Wilusz et al., 1984) and has been proposed as a host-cell factor that may control the level of N protein, which ostensibly effects the switch from VSV transcription to replication.

Assuming that the leader gene is the site of entry for the VSV polymerase, it has been proposed that the VSV leader RNA sequence regulates the switch from transcription to replication (Blumberg et al., 1981, 1983). This led Kolakofsky and co-workers to speculate that the switch involves the leader sequence and the specific concentration of N protein. Wilusz and Keene (1984) have proposed that formation of an La–protein–leader RNA complex may serve as an in vivo attenuator of transcription to ensure adequate levels of viral protein accumulation before the switch to replication takes place. Evidence for displacement of La protein by viral N protein on the plus-strand leader RNA is consistent with this possibility (Kurilla and Keene, 1984).

The leader sequence is thought to contain the nucleation site for nucleocapsid assembly within the first 18–20 nucleotides of both the wild-type and DI leader RNAs. Of the five strains of VSV examined, there is a remarkable homology in the first 18 nucleotides with the consensus sequence 3'-UGCUUN-UNNUNNUUUGU-5' (Giorgi et al., 1983). The nucleocapsid assembly signal is thought to be a 5-times-repeated A residue in every 3rd position at the 5' end of the leader RNA (Giorgi et al., 1983). This repeated pattern in the proposed encapsidation signal was pointed out by Blumberg and co-workers (personal communication) to be analogous to the encapsidation signal for tobacco mosaic virus. Perhaps, as suggested by Rose (1980), these genes may have a common ancestry. The critical role of the N protein in VSV replication and nucleocapsid assembly is emphasized by studies of Arnheiter et al. (1985), who showed that a monoclonal antibody that reacts only with cytoplasmic unbound N protein, and not with nucleocapsid N protein, selectively inhibits genome replication and not messenger transcription. A separate monoclonal antibody to N protein inhibits VSV transcription and not replication. These studies strongly point to two conformational forms of N protein that selectively regulate replication and transcription.

G. Assembly and Budding

Certain of the events in assembly and budding of VSV and other rhabdoviruses occur independently and others are tightly coupled. As indicated in the preceding section, replication of viral progeny RNA and translation proceed simultaneously, and the newly synthesized N, L, and P proteins bind to the newly synthesized progeny RNA to form newly assembled ribonucleoprotein cores (Rubio *et al.*, 1980). As illustrated in Fig. 3, independently synthesized M protein apparently has marked affinity for associating with the cell membranes (Morrison and McQuain, 1978) as well as progeny nucleocapsids (Wilson and Lenard, 1981); as the M protein binds to these progeny nucleocapsids, it appears to result in the formation of a tightly coiled "skeleton" that is the final nucleocapsid structure of the rhabdovirus virion (Newcomb and Brown, 1981; Newcomb *et al.*, 1982). At the same time as the nucleocapsid RNA and proteins are being synthesized and assembled, the VSV G protein is being synthesized independently on membrane-associated poly-ribosomes as well as being processed and glycosylated in endoplasmic reticulum–Golgi structures. Coated vesicles then migrate to the plasma membrane for fusion and insertion of the G protein (Rothman *et al.*, 1980). It is believed that the nucleocapsid–M protein complexes then migrate to regions of the cytoplasmic membrane that contain the newly inserted but randomly distributed VSV G protein (Wagner *et al.*, 1971). The nucleocapsid–M protein complex appears to bind to the cytoplasmic surface of the plasma membrane at a region rich in inserted G protein and PS phospholipids; there is some evidence that the positively charged M protein (Carroll and Wagner, 1979) binds to negatively charged headgroups of PS (Zakowski *et al.*, 1981). The tightly coiled nucleocapsid is apparently next enveloped in the G-protein-converted cytoplasmic membrane, leading to budding and release of fully formed and infectious VS virions. The evidence for these assembly and budding steps is largely circumstantial and far from complete, but certain data by Jacobs and Penhoet (1982) indicate that viral nucleocapsid–M protein complexes promote lateral diffusion of G protein in the plane of the membrane bilayer to those regions of the cytoplasmic membrane that are devoid of cellular proteins and destined for nucleocapsid envelopment and budding of completed membrane-enclosed VS virions. A more detailed analysis of the budding of VSV is presented in Chapter 3. Synthesis and processing of the VSV G protein are discussed in Chapter 4.

V. CELLULAR RESPONSES TO INFECTION

Rhabdoviridae, particularly those of the genus *Vesiculovirus*, are often highly virulent and can cause rapid changes in metabolism of the host cell leading to death. Other members of the family Rhabdoviridae can also cause acute and severe diseases, probably by somewhat different pathogenetic mechanisms; subsequent chapters review the pathogenesis of plant rhabdoviruses (Chapter 10) and of rabies viruses (Chapter 9). Detailed reviews of the pathogenesis and cytopathology of vesicular stomatitis virus (VSV) were published in vol. 19 of *Comprehensive Virology*, the predecessor to this series (see Wagner, 1984; Wagner *et al.*, 1984). Some of this material, updated, is reproduced here to provide a complete analysis in one volume.

In addition to its cytopathic effect, VSV also inhibits cellular nucleic-acid and protein synthesis. These effects on cellular macromolecular synthesis occur very rapidly and long before cell death (see Wagner *et al.*, 1984). The pervading question is whether these effects of VSV on cellular macromolecular synthesis are the cause(s) of cell death. There is reasonable evidence that a similar property of VSV inhibits both RNA and DNA synthesis, but suppression of protein synthesis is probably caused by a different VSV function (McGowan and Wagner, 1981). Though they are related, these various cellular responses to VSV infection will be discussed separately.

A. Cytopathology

VSVs generally cause rapid cytopathic effects in vertebrate cells and to a lesser extent in invertebrate cells, but it is wise to keep in mind that there can be considerable variation in pathogenesis caused by different VSV types and strains. The review by Bablanian (1975) provides an excellent description of the cytopathology caused by VSV. Clearly, there are two distinct types of cytopathic effects resulting from VSV infection: (1) a rapid cellular response at high multiplicity, characterized by cell rounding by 1 hr postinfection (Baxt and Bablanian, 1976a), which was once thought not to require active viral synthetic functions and was dubbed "cytotoxic"; (2) a slower response that appears to require active VSV replication, usually accompanied by release of progeny virions from the infected cell. Quite characteristic of the rapid cellular response is early inhibition of cellular RNA, DNA, and protein synthesis (McGowan and Wagner, 1981). These early events were originally considered to be due to structural components of input parental virions because they were noted at high multiplicity and with nonreplicating defective interfering (DI) particles and UV-inactivated wild-type virus, as well as in the presence of cycloheximide (Baxt and Bablanian, 1976b). It is now quite clear that very high multiplicities of VSV DI 5' particles, when completely free of standard wild-type VSV, do not kill cells (Marcus and Sekellick, 1974) and do not inhibit cellular RNA synthesis (Weck and Wagner, 1979a). By the same token, UV irradiation at moderate doses does not eliminate all biological activity of VSV, which can still retain the capacity to kill cells (Marvaldi *et al.*, 1977) and to inhibit cellular nucleic-acid synthesis (McGowan and Wagner, 1981) and protein synthesis (Marvaldi *et al.*, 1978). Therefore, the most compelling data suggest that certain viral gene functions must retain their activity in order to kill cells. In fact, it seems quite clear from genetic studies by Marcus and Sekellick (1975) and Marcus *et al.* (1977) that VSV can kill cells only if the infecting virion retains a certain degree of transcriptase activity. Moreover, only a single standard

(B) virus particle is sufficient to kill a cell, though VSV DI 5' particles cannot (Marcus and Sekellick, 1976).

It has also long been known that cells vary greatly in their susceptibility to viral infection, even when no differences can be demonstrated in their surface receptors for virus adsorption. Moreover, the same virus clone can inhibit macromolecular synthesis in one cell type to a greater extent than in another, even though virus yields may not differ significantly. For example, Baxt and Bablanian (1976b) showed that VSV inhibits nucleic-acid synthesis in BHK-21 cells more readily than it does in LLC-MK2 cells. Weck and Wagner (1978) also reported that MPC-11 mouse myeloma cells were more susceptible to VSV shut off of cellular RNA synthesis than were BHK-21 or mouse L cells.

Cell differentiation may play a role in cellular susceptibility to viral infection, as illustrated by the finding that VSV replication is restricted in one human lymphoblastoid-cell line but not in another or in HeLa cells (Nowakowski et al., 1973). Robertson and Wagner (1981) also found that HeLa cells and L cells are more permissive for VSV growth than an end-stage myeloma-cell line, MPC-11, which, in turn, is more permissive than the Abelson-virus-transformed 18-81 pre-B cell that does not secrete immunoglobulin. The converse relationship was found when we tested these same cells for capacity of VSV to shut off cellular RNA synthesis; the susceptibility of these cells to VSV inhibition of cellular RNA synthesis could be ranked in the order 18-81 > MPC-11 > L cells >HeLa cells (Robertson and Wagner, 1981). This difference in susceptibility to inhibition of cellular macromolecular synthesis does not appear to be related to the specific proteins produced by differentiated cells. Myeloma cells infected with VSV exhibit far less inhibition of the synthesis of immunoglobulin than of other cellular proteins (Nuss and Koch, 1976). Similarly, globin synthesis in differentiated Friend erythroleukemia cells is more resistant to shutoff by VSV infection than are other proteins of the same cell (Nishioka and Silverstein, 1978). Cellular factors of unknown nature were also found to determine the ability of mengovirus to inhibit the synthesis of VSV protein or of VSV to inhibit the synthesis of mengovirus proteins in doubly infected HeLa, CHO, or L-929 ells (Otto and Lucas-Lenard, 1980). VSV also appears to inhibit uptake of uridine in infected chick-embryo cells (Genty, 1975), but not in other cells (Weck and Wagner, 1978). Quite obviously, the factors that control cellular susceptibility to infection with VSV remain unclear.

B. Vesicular Stomatitis Virus Inhibition of Cellular RNA Synthesis

The mechanisms by which VSV shuts off RNA synthesis in the infected host cell has been reviewed in some detail (Wagner et al., 1984). It seems clear that the major aspects of this problem are: (1) the nature of the viral product that is responsible for inhibiting cellular RNA synthesis and (2) the cell target. Earlier studies suggested that structural components of the virion, presumably proteins, were responsible for inhibiting cellular RNA synthesis; this conclusion was based on evidence that cellular RNA synthesis could be inhibited by heavily UV-irradiated preparations of standard VSV (Huang and Wagner, 1965; Yaoi et al., 1970) or by high multiplicities of noninfectious DI particles (Huang and Wagner, 1965; Baxt and Bablanian, 1976b). In retrospect, these conclusions were erroneous because subsequent experiments showed that only enormous doses of UV irradiation could destroy all VSV functions (Weck et al., 1979) and DI particles truly free of standard VSV do not shut off cellular RNA synthesis even at enormous input multiplicities (Weck and Wagner, 1979a). Another question that arises is whether VSV inhibition of cellular RNA synthesis is a primary effect or is secondary to VSV inhibition of DNA synthesis. Essentially circumstantial evidence that inhibition of cellular RNA synthesis is not secondary to its inhibition of cellular protein synthesis was provided by McGowan and Wagner (1981), who showed different kinetics of VSV inhibition of cellular nucleic-acid synthesis and protein synthesis. However, these authors also described identical kinetics for inhibition of cellular RNA and DNA synthesis, a finding that suggests that a single viral function is responsible for compromising cellular RNA and DNA synthesis. McGowan and Wagner (1981) found that equivalent doses of UV irradiation were required to ablate the RNA- and DNA-synthesis-inhibitory activities of VSV, compared to much lower UV doses required to inhibit the effect of VSV on cellular protein synthesis (Marvaldi et al., 1977).

This statement still applies despite a study by Dunigan and Lucas-Lenard (1983) that provides evidence for two sites of UV inactivation of protein-synthesis inhibition, one of which may be identical to the leader sequence presumably responsible for inhibiting cellular RNA synthesis (Weck et al., 1979; McGowan et al., 1982). In any case, the kinetic data and UV-inactivation data seem to implicate the same or closely related viral functions in inhibiting both cellular RNA and DNA synthesis, but not inhibition of cellular protein synthesis, at least not entirely.

1. Identifying the Viral Product Responsible for Inhibiting Cellular
   RNA Synthesis

The evidence cited above suggests that structural components of the invading virion are not responsible for shutting off cellular RNA synthesis. The most telling evidence for this suggestion is that enormous multiplicities of VSV DI particles, the proteins and other structural components of which are identical to those of fully infections VSV (Wagner et al., 1969), do not significantly affect cellular RNA synthesis (Weck and Wagner, 1979a). Moreover, 5' DI particles synthesize no RNA, even though they contain a functional polymerase (Emerson and Wagner, 1972) and therefore synthesize no viral proteins. These and other experiments strongly suggest that products newly synthesized by fully infectious invading virions are responsible for inhibiting cellular RNA synthesis. The only products encoded by the VSV genome are a 47-nucleotide leader RNA (Colonno and Banerjee, 1977) and five monocistronic RNAs that are translated into five proteins *in vitro* (Bishop *et al.*, 1974) and *in vivo* (Banerjee *et al.*, 1977). It seemed logical, therefore, to devise experiments designed to identify a specific viral RNA or protein newly synthesized by the infectious, cytopathic VSV as the putative inhibitor(s) of cellular RNA synthesis. In the absence of a cell-free system in which each of the components could be tested independently, it was essential to use indirect techniques *in vivo*. The use of specific inhibitors of protein synthesis had drawbacks because all compounds tested by us, including cycloheximide, puromycin, and several amino acid analogues, also inhibited cellular RNA synthesis as well as cellular and viral protein synthesis, at least in the MPC-11 myeloma cells used in our experiments (Weck and Wagner, unpublished data). Very recent studies by Poirot *et al.* (1985) using mouse L cells led to the conclusion that VSV protein synthesis was at least partially required to inhibit cellular RNA synthesis because removal of cycloheximide resulted in inhibition of host-cell RNA synthesis or, at least, the approximately 50% residual RNA synthetic activity not affected by cycloheximide. These results may be valid, but the experiment suffers from the drawback that cycloheximide, pactamycin, or emetine had already reduced cellular RNA synthesis considerably, possibly wiping out that part of the RNA-synthetic machinery susceptible to another viral function.

Other investigators resorted to different indirect techniques to study *in vivo* inhibition of RNA synthesis in cells infected with VSV. The two principal methods used have been to test temperature-sensitive conditional lethal mutants restricted in transcription and to inactivate by UV irradiation the specific regions of the wild-type viral genome required for transcription of each viral gene. Some of these experiments are recounted in Chapter 5 and will be briefly summarized here.

Weck and Wagner (1979a) analyzed RNA metabolism in MPC-11 cells infected with various temperature-sensitive (*ts*) mutants and 5' DI particles, which cannot synthesize mRNA. A complementation group I mutant, *ts*G114, restricted in transcriptional activities (Hunt *et al.*, 1976), failed to shut off host-cell RNA metabolism in MPC-11 cells incubated at the restrictive temperature of 39°C for 4 hr. At the permissive temperature (31°C), all mutants (including *ts*G114) were as effective as the wild-type virus in the shutoff of RNA synthesis. This and other experiments suggested that primary transcription of the VS viral genome is essential to compromise host-cell RNA synthesis. Temperature shift-up experiments with *ts*G114(I) infection of MPC-11 cells revealed that the cellular inhibitory factor of VSV is synthesized maximally during the first hour of infection at permissive temperature (Weck and Wagner, 1979a). Similar results in L cells infected with *ts* mutants from all five complementation groups were obtained by Wu and Lucas-Lenard (1980), who also reported an effect on cellular RNA synthesis by the putative NS(P) protein mutant *ts*G22(II), but not those of other complementation groups.

Since 5' DI particles fail to replicate without a helper and transcribe only a 46-nucleotide leader RNA (Emerson *et al.*, 1977), they provide an ideal means of examining the effects of nonreplicating truncated particles on cellular RNA metabolism. Weck and Wagner (1979a) infected MPC-11 cells with purified DI particles derived from the 5' end of the VSV genome and found no significant reduction in host-cell RNA synthesis even at a multiplicity of infection (MOI) equivalent to 10,000 particles per MPC-11 cell. Thus, even if a very short leader sequence (Emerson *et al.*, 1977; Schubert *et al.*, 1978) is transcribed by the 5' DI particles, it apparently plays no role in the inhibition of cellular RNA synthesis (Weck and Wagner, 1979a,b). In contrast DI particles derived from the 3' end of the genome are capable of transcribing plus-strand leader and functional mRNAs for N, NS, M, and G proteins and can inhibit cellular protein synthesis (Marcus *et al.*, 1977) and cellular DNA synthesis (McGowan and Wagner, 1981).

These experiments strongly indicate that VSV transcript(s) can inhibit cellular nucleic-acid synthesis directly or by means of their translated products. The genetic evidence that transcription of VSV is required to shut off cellular RNA synthesis led two laboratories to reinvestigate the effects of UV-irradiated VSV on cellular RNA synthesis (Weck *et al.*, 1979; Wu and Lucas-Lenard, 1980). Such UV-inactivation studies are based on evidence for linear order of cistrons [3'-leader-N-NS(P)-M-G-L-5'] that have been shown to be differentially susceptible to increasing doses of UV light (Ball and White, 1976; Abraham and Banerjee, 1976b; Testa *et al.*, 1980). Weck *et al.* (1979) compared the UV doses for 37% (1/e) survival levels for VSV infectivity, *in vitro* transcriptase activity, viral RNA synthesis in infected MPC-11 cells, and shutoff of host RNA synthesis. Heavily UV-irradiated virus (72,000 ergs/mm^2) retained 37% of its capacity to shut off cellular RNA synthesis and could transcribe a very limited portion of the viral genome; the transcription products were low-molecular-weight, nonadenylated RNA molecules. On the basis of the data obtained, Weck *et al.* (1979) concluded that perhaps the inhibitor of cell RNA synthesis could be the plus-strand leader RNA transcribed from the 3' end of the wild-type genome.

Wu and Lucas-Lenard (1980) did similar studies on UV irradiation of VSV to determine the target size of the VSV genome segment required for inhibition of RNA synthesis in L cells; the approximate target size in their studies appeared to be the VSV N gene. The difference obtained in target size of the VSV genome for shutoff activity may be dependent on the type of cell or other experimental conditions used by different investigators (Robertson and Wagner, 1981; Wu and Lucas-Lenard, 1980).

Subsequent studies on UV doses required to inactivate the capacity of VSV to inhibit RNA synthesis in MPC-11 cells revealed 37% (1/e) survival rates of approximately 52,000 ergs/mm² for VSV-Indiana (Mc-Gowan and Wagner, 1981) and approximately 12,000 ergs/mm² for both VSV-Indiana and VSV-New Jersey (Grinnell and Wagner, 1983). The lower inactivating dose of UV irradiation reported by Grinnell and Wagner (1983) was obtained using exactly the same UV-light source and the same techniques as reported by Weck *et al.* (1979); these dose differences can only be attributed to the fact that Grinnell and Wagner (1983) used highly purified, serum-free preparations of VSV, essentially free of UV-absorbing nonviral proteins and other impurities. The figure of approximately 12,000 ergs/mm² of UV irradiation required to reduce the MPC-11 cell RNA-synthesis activity to 37% appears to be the most reliable one. Moreover, Grinnell and Wagner (1983) showed a direct correlation between the UV dose required for reduction of VSV-Indiana and VSV-New Jersey leader RNA synthesis *in vivo* and *in vitro* and the capacity of these two viruses to shut off RNA synthesis in MPC-11 cells.

These data clearly indicate the potential relationship between synthesis of VSV leader RNA and inhibition of cellular RNA synthesis, at least in MPC-11 myeloma cells. Using a value of 104 ergs/mm² for the 37% (1/e) survival dose for VSV infectivity and a VSV genome size of 12,000 nucleotides, Grinnell and Wagner (1983) calculated the VSV genome UV-target size for 37% survival of cellular RNA-synthesis inhibition as approximately 85 nucleotides, compared with the UV-target size for synthesis of the VSV leader as 150 nucleotides, rather than the actual size of 47 nucleotides for the leader RNA. Such values are probably well within the range of error for such methods, but are close enough to relate the two phenomena of synthesis of leader RNA and shutoff of cellular RNA synthesis by VSV. In this regard, it is interesting to comare the recent data of Dunigan and Lucas-Lenard (1983), who calculated UV-target sizes of 42 and 373 nucleotides for the two VSV-genome segments required to transcribe products that inhibit cellular protein synthesis; the smaller genome segment (42 nucleotides) is presumed to code for the leader RNA sequence and the larger (373 nucleotides) for the N-protein mRNA (see Section V.D.1). A follow-up study (Dunigan *et al.*, 1986) claimed a lack of correlation between the accumulation of plus-strand RNA and the inhibition of protein and RNA synthesis in VSV-infected mouse L cells.

2. Cellular Targets for VSV Inhibition of RNA Synthesis

Inhibition by VSV of cellular RNA synthesis does not appear to be related to membrane exclusion of nucleotide precursors. Although Genty (1975) described decreased uptake of uridine by VSV-infected chick-embryo cells, presumably reducing the intracellular nucleotide pool available for cell RNA synthesis, a similar effect was not noted in L cells or HeLa cells infected with VSV. Moreover, Weck and Wagner (1978) examined the acid-soluble and acid-insoluble pools of [³H]uridine in VSV-infected and uninfected MPC-11 cells and found that the transport of uridine was unaffected in VSV-infected cells. Very large input multiplicities of VSV mutant $tsG114(I)$ at restrictive temperature or of DI particles at doses that would possibly alter membrane permeability of uridine had no effect on RNA synthesis by MPC-11 cells. Besides not having any significant effect on cellular uptake of RNA precursors, VSV does not appear to affect processing, stability, or transport of newly synthesized cellular RNA from nucleus to cytoplasm. The rate of prelabeled [³H]-RNA degradation was not significantly different in uninfected and VSV-infected cells (Nishioka and Silverstein, 1978; Weck and Wagner, 1978). The VSV-infected cells also showed no impairment of polyadenylation or transport of cellular mRNA to the cytoplasm (Weck and Wagner, 1978).

VSV infection appears to inhibit synthesis of all three species of cellular RNA. Nuclei isolated from MPC-11 cells 2 hr after VSV infection were found to have approximately 50% reduction in their capacity to synthesize ribosomal, messenger, and transfer RNAs compared to nuclei of uninfected cells (Weck and Wagner, 1978). Further studies by Weck and Wagner (1979b) using solubilized polymerase and chromatin templates from VSV-infected and uninfected MPC-11 cells revealed that the polymerases *per se* were not the targets of the virus. Instead, the number of active polymerase molecules was reduced by perhaps 50% in infected cells, but those polymerases that survived the infection retained their ability to transcribe the requisite number of rRNA, mRNA, and tRNA molecules. These experiments suggested that RNA-chain elongation is not affected by VSV infection, but the number of polymerases able to initiate transcription is reduced in the infected cell.

To obtain more definitive data on the hypothesis that VSV inhibits initiation of transcription, McGowan et al. (1982) exploited the cell-free transcription system described by Manley et al. (1980). These studies are described in some detail in a previous review (Wagner et al., 1984) and will be briefly summarized here. In these experiments, we used HeLa-cell extracts to supply polymerases and cofactors as well as excised clones of adenovirus major late promoter (MLP) or virus-associated (VA) genes, which require polII and polIII, respectively. To these in vitro transcription systems, we added at various times wt leader RNA, DI leader RNA, and several synthetic oligodeoxynucleotides. The wt leader RNA quite efficiently, and at relatively low concentrations, shut off transcription of the MLP and VA genes, whereas DI leader RNA and VSV mRNAs did not (McGowan et al., 1982). Comparison of the nucleotide sequences of wt and DI leaders revealed considerable homology for the first 17 nucleotides at the 5' ends, but marked differences in nucleotides 18–30 from the 5' ends. The hypothesis was advanced that the wt leader sequence 5'-AU-UAUUAUCAUUA-3' constitutes the transcription inhibitor (McGowan *et al.*, 1982). Grinnell and Wagner (1983) found that the *wt* leader RNA transcribed by VSV-New Jersey has a similar but extended AUUAUU region, also beginning at nucleotide 18, that is even more effective than the VSV-Indiana *wt* leader in inhibiting transcription *in vitro* of the adenovirus MLP and VA genes. Grinnell and Wagner (1983) also noted, however, that VSV-Indiana was somewhat more effective in inhibiting cellular RNA synthesis *in vivo*, presumably because it synthesized approximately 4 times more leader RNA than did VSV-New Jersey in infected MPC-11 cells. Kurilla *et al.* (1982) found that VSV leader RNA made in the cytoplasm migrates to the nucleus, undoubtedly the site at which cellular RNA synthesis is inhibited. These studies were confirmed by Grinnell and Wagner (1983), who also found a temporal correlation between *in vivo* synthesis of VSV leader RNA and the degree of cellular RNA-synthesis inhibition. Of considerable interest is the fact that the VSV-Indiana leader RNA forms a ribonucleoprotein with the cellular La protein in a manner quite similar to association of the La protein with endogenous small cellular RNAs (Kurilla and Keene, 1984).

Also of interest was the finding that a *Pst*I-excised cDNA copy of the VSV leader RNA cloned in pBR322 was as effective in inhibiting MLP- and VA-gene transcription as was the leader RNA itself (McGowan and Wagner, unpublished data). In fact, both the complementary single strands of DNA were about equally effective, whereas, the double-stranded leader DNA had little or no effect. These studies prompted Grinnell and Wagner (1984, 1985) to embark on a series of studies designed to test the oligonucleotide sequences responsible for inhibiting *in vitro* transcription of MLP and VA genes. A synthetic oligodeoxynucleotide homologous to residues 18–30 of the VSV *wt* leader, ATTATTATCATTA, was found to inhibit transcription *in vitro* of both MLP and VA genes, but not as well as equimolar amounts of native leader RNA (Grinnell and Wagner, 1984). Synthetic oligodeoxynucleotides with base substitutions (e.g., T → G) resulted in complete or almost complete loss of capacity to inhibit transcription. Longer oligodeoxynucleotides, homologous to the same leader region but including flanking sequences, were more efficient in transcription inhibition (Grinnell and Wagner, 1985). Additional evidence from studies with fragments of native leader RNA and with endonuclease-cleaved fragments of leader cDNA indicates that stem-loop structures of the AU(AT)-rich region plus flanking sequences are the most effective transcription inhibitors. In sharp contrast, RNA or DNA strands homologous to the 3' and 5' thirds of the VSV *wt* leader RNA were as devoid of transcription-inhibitory activity as is the DI leader RNA (Grinnell and Wagner, 1983, 1984, 1985). It seems clear from these studies that the AU-rich region of the VSV *wt* leader of two separate serotypes is responsible, at least in large measure, for inhibiting DNA-dependent transcription.

The major unanswered question in VSV inhibition of cellular RNA synthesis is the target(s) at which the inhibitor is directed. A closely related question is the mechanism by which an inhibitor could affect transcription primed by both polII and polIII, for which there is no precedent. Negative data thus far indicate that the VSV *wt* leader and homologous DNA do not bind to MLP or VA gene templates and do not react directly with polII (McGowan, Grinnell, and Wagner, unpublished data). The only remaining potential targets for inhibition of DNA-dependent transcription by leader RNA are cellular protein cofactors that could act at the level of promoters or enhancers. Grinnell and Wagner (1985) explored this possibility by screening HeLa-cell proteins for their reactivity with *wt* leader RNA and synthetic oligodeoxynucleotides. HeLa-cell proteins were subjected to polyacrylamide gel electrophoresis, transferred onto nitrocellulose filters, and tested for their capacity to bind leader RNA. These experiments revealed specific binding by *wt* leader RNA and not DI leader RNA to a 65K protein present in the HeLa-cell extract; additional HeLa-cell proteins bound nonspecifically to both *wt* and DI leader RNAs. Synthetic oligodeoxynucleotides homologous to the AU-rich region of the leader also bound the 65K protein. Very preliminary studies also reveal that only those HeLa-cell fractions containing the 65K protein that binds to *wt* leader RNA are capable of reversing the transcription-inhibitory activities of the VSV *wt* leader RNA or of homologous DNAs.

Far more studies are required to uncover the factors and mechanisms that result in cellular-transcription inhibition by VSV, including the possibility that active viral protein synthesis is also required for VSV inhibition *in vivo* of host-cell RNA synthesis (Poirot *et al.*, 1985).

C. Vesicular Stomatitis Virus Inhibition of Cellular DNA Synthesis

In their studies on the effect of VSV-Indiana on interferon synthesis, Wagner and Huang (1966) noted that the incorporation of [³H]thymidine into acid-precipitable material was drastically inhibited in VSV-infected compared to uninfected Krebs-2 ascites tumor cells. Yaoi and Amano (1970) infected chick-embryo cells with VSV-New Jersey at an MOI of 500 PFU/cell and observed an inhibition of DNA synthesis within 2 hr after infection. The same inhibitory effect was observed with UV-inactivated VSV, but not with VSV heated at 56°C for 20 min. In much later studies, McGowan and Wagner (1981) examined the kinetics of VSV inhibition of cellular DNA synthesis at MOIs of 1, 10, and 100 in exponentially growing MPC-11 and L cells. Inhibition of cellular DNA synthesis in unsynchronized MPC-11 cells was more rapid at each MOI tested than in VSV-infected L cells. Studies by Yaoi *et al.* (1970) suggesting that VSV inhibited DNA synthesis in chick-embryo cells just before or just after they entered the S phase of the growth cycle were not confirmed by

McGowan and Wagner (1981), who found that VSV inhibited DNA synthesis in highly synchronized MPC-11 or mouse L cells at any stage of the growth cycle.

In attempts to identify the viral component responsible for inhibiting DNA synthesis in VSV-infected cells, McGowan and Wagner (1981) repeated many of the experiments previously reported by Weck and Wagner (1979a,b) for identifying the viral product responsible for inhibiting cellular RNA synthesis. It was found that input virion components were not responsible for the effect on cellular DNA synthesis because extremely high multiplicities of 5′ DI particles did not significantly alter cell DNA synthesis. However, the transcribable 3′ DI particle HR-LT did inhibit cellular DNA synthesis, whereas the *ts* mutant *ts*G114(I), restricted in transcription, did not significantly block cellular DNA synthesis at nonpermissive temperature (McGowan and Wagner, 1981). These experiments strongly suggest that inhibition of cellular DNA synthesis by VSV requires viral transcription, in a manner quite similar to that for VSV inhibition of cellular RNA synthesis. That a similar viral product is required for inhibition of both DNA and RNA synthesis is also supported by evidence for identical kinetics of inhibition for cellular DNA and RNA synthesis (McGowan and Wagner, 1981) as well as by the heat lability of the endogenous VSV transcriptase (Yaoi and Amano, 1970). Moreover, similar doses of UV irradiation are required to suppress the capacity of VS virions to shut off cellular DNA and RNA synthesis (45,000–52,000 ergs/mm^2), which tends to implicate the VSV *wt* leader RNA transcript as the inhibitor of cellular DNA synthesis (McGowan and Wagner, 1981). Recent studies by Remenick and McGowan (1986) have shown that the isolated VSV *wt* leader RNA, but not the DI leader RNA, is capable of blocking adenovirus DNA replication in a cell-free system similar to that designed by Challberg and Kelly (1979a,b), thus strongly suggesting that the same viral leader RNA product is responsible for inhibiting both cellular transcription and replication.

The cellular target for VSV inhibition of cellular DNA synthesis has not yet been identified, but studies summarized by McGowan and Wagner (1981) and reviewed by Wagner *et al.* (1984) appear to rule out a viral effect on nucleoside transport, DNA degradation, innactivation of DNA polymerase and thymidine kinase, or premature termination of already initiated DNA chains. By exclusion, we hypothesize that VSV inhibits DNA synthesis by blocking initiation of DNA replication perhaps in a manner similar to its inhibition of RNA transcription (McGowan *et al.*, 1982). Preliminary studies by Remenick and McGowan (1986) indicate that nuclear extracts of adenovirus-infected or uninfected HeLa cells can reverse the inhibition of DNA synthesis by *wt* leader RNA, thus implicating a cell protein as the target for replication inhibition. Further studies are necessary to test the hypothesis that a VSV product, such as the plus-strand leader RNA, inhibits initiation of both RNA transcription and DNA replication by reacting with critical cellular cofactors.

D. Vesicular Stomatitis Virus Inhibition of Cellular Protein Synthesis

The recent review on rhabdovirus cytopathology by Wagner *et al.* (1984) provides a moderately detailed analysis of the effects of VSV on host-cell protein synthesis. Two long-standing questions, still not completely resolved, are: (1) whether structural components of the invading virus or newly synthesized viral products are responsible for inhibiting cellular protein synthesis and (2) whether inhibition of cellular protein synthesis is a primary effect or secondary to other cell functions, such as host-cell RNA synthesis. In an attempt to resolve the question of input virion components as factors that contribute to inhibition of host-cell macromolecular synthesis, Wertz and Youngner (1972) hypothesized that "Two mechanisms may be involved in the inhibition of host protein synthesis by VSV: (1) an initial multiplicity-dependent and ultraviolet-sensitive inhibition, and (2) a progressive ultraviolet-sensitive inhibition." Despite considerable evidence now available and presented below favoring newly synthesized viral products as the major or sole mechanism for inhibition of cellular protein synthesis, the role of input virion toxic components has not been ruled out completely, even though enormous amounts of isolated VSV G protein were found to have no effect on cellular protein synthesis (McSharry and Choppin, 1978). The likelihood that the effect of VSV on cellular protein synthesis is a primary effect of the virus was proposed by Bablanian (1975), and confirmed by McGowan and Wagner (1981), largely on the basis of kinetic data showing parallel reduction in both RNA and protein synthesis early after VSV infection.

Although most, if not all, cells are susceptible to protein-synthesis inhibition by VSV, there can be considerable variation in the degree of susceptibility among different cells (Wertz and Youngner, 1970; Otto and Lucas-Lenard, 1980; McGowan and Wagner, 1981) as well as variations among virus strains (Lodish and Porter, 1981). Several studies have also shown differential inhibition of various proteins in host cells infected with VSV; in particular, differentiated cell proteins, such as immunoglobulins in myeloma cells (Nuss and Koch, 1976) and hemoglobin in Friend erythroleukemia cells (Nishioka and Silverstein, 1978), are far less susceptible to inhibition by VSV infection than are other cell proteins. In contrast, VSV infection of undifferentiated mouse L cells resulted in rapid and continuous reduction in synthesis of all classes of cellular proteins throughout the 5-hr period of infection (McAllister and Wagner, 1976).

1. Mechanism(s) by Which VSV Inhibits Cellular Protein Synthesis

Experimental procedures similar to those used for studying VSV inhibition of RNA synthesis have been used to investigate the mechanisms by which VSV shuts off host-cell protein synthesis. In an attempt to resolve the question whether structural proteins of invading virions or newly synthesized viral products are required to shut off cellular protein synthesis, investigators have tested intrinsically inert viral particles, such as DI particles and conditional lethal mutants, or infectious virions inactivated by physical or chemical means. The salient experiments can be briefly summarized:

Schnitzlein et al. (1983) detected little or no inhibition of BHK-cell protein synthesis by various types of highly purified 5' DI particles, even at high MOI. These DI particles fully retained the capacity to inhibit replication of infectious B particles, and, as will be discussed later, the 5' DI particles did not affect the capacity of infectious B virions to inhibit cellular protein synthesis (Schnitzlein et al., 1983). It seems safe to conclude that 5' DI particles of VSV contain no structural components or biological activity that will inhibit cellular protein synthesis.

McAllister and Wagner (1976) compared the synthesis of cellular proteins, differentiated from VSV proteins by polyacrylamide gel electrophoresis, in L cells infected with wt VSV and various ts mutants. Two group I mutants restricted in transcription (Hunt et al., 1976) and, particularly, tsG114(I) completely failed to inhibit cellular protein synthesis at nonpermissive temperature. In sharp contrast, a group IV RNA⁻ mutant and a group II RNA$^\pm$ mutant, neither of which is restricted in primary transcription, shut off cellular protein synthesis at nonpermissive temperatures, as did the wild-type VSV (McAllister and Wagner, 1976). Marvaldi et al. (1977), using the same and some additional ts mutants, came to similar conclusions; they found that several group I ts mutants failed to inhibit cellular protein synthesis, but the leaky mutant ts05(I) did, even at restrictive temperature. Somewhat in contrast to the results of McAllister and Wagner (1976), Marvaldi et al. (1977) reported that one ts mutant in complementation group IV did inhibit protein synthesis at restrictive temperature, but another group IV mutant, tsG41(IV), and a group II mutant, tsG22(II), failed to inhibit cellular protein synthesis under restrictive conditions. This latter result was not confirmed by Schnitzlein et al. (1983), who found that replication mutant tsG22(II) did inhibit protein synthesis in BHK-21 cells infected at nonpermissive temperature. There is general agreement among all these investigators that VSV transcription is the minimal event required for inhibition of host-cell protein synthesis. However, none of these experiments rules out synthesis of VSV proteins as potential inhibitors of cell protein synthesis; in fact, Marvaldi et al. (1977) were inclined to favor implication of VSV mRNAs and perhaps proteins N and NS on the basis of their genetic analysis.

Stanners et al. (1977) have described a ts mutant derived from the VSV HR strain that contains a second non-ts mutation in the viral polymerase gene (which they call "P") purported to be responsible for inhibition of cellular protein synthesis. This mutant appeared to inhibit initiation of translation in some cells, but not in others. This mutant (T1026) and its revertant R1 have interesting properties for probing VS viral effects on cellular protein synthesis, but Lodish and Porter (1981) have strongly questioned the validity of a specific genetic "P" function that phenotypically expresses inhibition of cellular protein synthesis.

VSV virions inactivated by UV irradiation or heat have been tested for their capacity to inhibit cellular protein synthesis. Baxt and Bablanian (1976b) have shown that heat-inactivated VSV-Indiana does not shut off cellular protein synthesis, but UV-irradiated VSV does. By comparing the relative doses of UV irradiation required to produce a given loss of infectivity with that required to give the same loss in protein-synthesis inhibition, one can determine a target size for protein-synthesis inhibition that represents that portion of the genome required to effect protein-synthesis inhibition. This has been done by Marvaldi et al. (1978) for VSV infection of mouse L cells at MOIs of 10 and 100. Their results indicate the transcription of one fifth of the VSV genome, the N gene, and possibly the NS gene, is required for inhibition of host-cell protein synthesis. These results have been confirmed in our laboratory by dose–responses to UV irradiation of VSV used to infect mouse L cells (Thomas, Carroll, and Wagner, unpublished results). Similarly, others have determined the target sizes required for nucleic-acid-synthesis inhibition by VSV; these sizes appear to be approximately 1/40th the size of that required for protein-synthesis inhibition (Weck et al., 1979; McGowan and Wagner, 1981). This finding would argue that VSV inhibition of host-cell protein synthesis is a separate phenomenon from that of nucleic-acid-synthesis inhibition by VSV.

Dunigan and Lucas-Lenard (1983) have published the most definitive study to date on the effect of UV irradiation on the capacity of VSV to shut off protein synthesis in L cells. They present evidence for a biphasic UV inactivation curve, "suggesting that transcription of two regions of the viral genome is necessary for the virus to become inactivated in this capacity" to shut off cellular protein synthesis. They calculated UV-target sizes of 373 nucleotides for one transcription product that inhibits protein synthesis and 42 nucleotides for the other. UV inactivation of the larger transcription product left the virus with 60–65% of its total capacity to shut off protein synthesis; it required more than 20,000 ergs/mm² to eliminate this second smaller protein-synthesis-inhibitory product, a UV dose equivalent to that required for inactivating the leader RNA and VSV inhibition of cellular RNA synthesis (McGowan and Wagner, 1981; Grinnell and Wagner, 1983). Dunigan and Lucas-Lenard (1983) also tested the R1 revertant of the T1026 mutant of Stanners et al. (1977) and found that its lesser capacity to inhibit cellular protein synthesis was inactivated by low-dose of UV irradiation, suggesting that the 42-nucleotide small transcript inhibitor of cellular protein synthesis is not functional in the T1026 revertant. Cell-free translation by extracts of cells infected with UV-irradiated wt VSV also revealed two VSV-genome targets for protein-synthesis inhibition similar to that in the in vivo system. These authors postulate that the larger (373-nucleotide) VS transcription product that inhibits protein synthesis could be the N protein mRNA and the smaller protein-synthesis inhibitor could be the plus-strand leader RNA of wt VSV, possibly affecting protein synthesis secondary to inhibition of cellular RNA synthesis (McGowan *et al.*, 1982). These intriguing data of Dunigan and Lucas-Lenard (1983) point the way to testing these hypotheses, principally by assaying the protein-synthesis-inhibitory capacity of VSV leader RNA and N-gene mRNA in tightly controlled cell-free translation systems.

Lodish and Porter (1980) have suggested an alternative hypothesis: that inhibition of host-cell protein synthesis by VSV is due to the successful competition by large amounts of VSV transcripts for a limited number of ribosomes. They concluded from *in vivo* studies that viral and cellular mRNA are about equivalent in their efficiencies of translational initiation, but viral transcripts are simply in large excess and, for this reason, can successfully out-compete the cellular transcripts for available ribosomes. In a follow-up publication, Lodish and Porter (1981) described a correlation between the concentration of intracellular VSV mRNA and the extent to which cellular protein synthesis is inhibited by VSV wild-type and various mutants. Although this mechanism can successfully explain the switch from host to viral protein synthesis, it does not explain the overall reduction in total protein synthesis.

There are a number of other experiments that fail to support the hypothesis that excess VSV mRNAs out-compete cellular messages for available ribosomes, thus inhibiting cellular protein synthesis: (1) Jaye *et al.* (1982) found that viral and cellular RNAs from infected cells are translated equally well by an *in vitro* reticulocyte extract; (2) despite a marked decline in synthesis of viral mRNA, VSV subjected to moderate doses of UV irradiation is as efficient as fully transcribable unirradiated VSV in shutting off cellular protein synthesis (Dunigan and Lucas-Lenard, 1983); and (3) the most telling argument against the Lodish–Porter hypothesis is presented by Schnitzlein *et al.* (1983), who reported quite conclusive studies that marked suppression of wt VSV mRNA synthesis by coinfecting DI particles has a negligible effect on the capacity of the wt VSV to shut off cellular protein synthesis. It seems likely, therefore, that competition between VSV and cell mRNAs does not completely, or even partially, explain VSV inhibition of cellular protein synthesis.

From the foregoing analyses of data, it seems safe to assume that a newly synthesized VSV product is the principal inhibitor of protein synthesis in infected cells. The fact that VSV transcription is required to inhibit cellular protein synthesis (McAllister and Wagner, 1976; Marvaldi *et al.*, 1977) limited the search to newly synthesized viral RNAs and proteins. Circumstantial evidence obtained from very early studies with interferon and other inhibitors appeared to rule out viral proteins as primary inhibitors of cellular protein synthesis (Yamazaki and Wagner, 1970; Wertz and Youngner, 1970; reviewed by Bablanian, 1975). Careful studies by UV inactivation of selected VSV transcripts appear to rule out the mRNAs for the M, G, and L proteins (Marvaldi et al., 1977; Dunigan and Lucas-Lenard, 1983). This would seem to leave the VSV leader RNA, the N protein mRNA, or the NS protein mRNA, or some combination thereof, as the principal candidates for inhibitors of cellular protein synthesis or possibly one or all of these transcripts associated with template nucleocapsid. Since the likely target for protein-synthesis inhibition is initiation of translation (Nuss et al., 1975), it seems clear that the viral inhibitory component(s) can be studied adequately only in cell-free translation systems that permit reconstitution of mRNAs, ribosomes, and all the individual components required for initiating translation, as well as the putative inhibitors.

As far as we are aware, ours is the only published study (J.R. Thomas and Wagner, 1982) that was focused entirely on an attempt to identify a VSV gene product that interrupts initiation of in vitro translation, and this study is far from being completely satisfactory. In this research, we examined cell-free protein synthesis in a reticulocyte lysate to which was added polyadenylated RNA fractions from VSV-infected HeLa cells or VSV transcripts made in vitro. Free VSV mRNA by itself had no effect on reticulocyte lysate translation, but both polyadenylated and nonpolyadenylated VSV mRNA associated with nucleocapsid templates did inhibit translation of endogenous mRNAs and exogenously added globin mRNA. This double-stranded (ds) VSV RNA inhibitor of translation exhibited the characteristics of other dsRNA inhibitors in reticulocyte cell-free systems (Hunter et al., 1975) and could be inactivated by melting and by micrococcal nuclease (J.R. Thomas and Wagner, 1982). It is well known that synthetic or natural dsRNAs can serve as potent inhibitors of protein synthesis in various in vitro translation systems derived from mammalian cells (reviewed by Kaempfer, 1984).

2. Cellular Target for Inhibition of Protein Synthesis

Most of the circumstantial evidence indicates that inhibition of protein synthesis in VSV-infected cells takes place at the level of initiation of translation (Nuss and Koch, 1976; Stanners et al., 1977; Centrella and Lucas-Lenard, 1982). Other investigators studying inhibition of protein synthesis by quite different positive-strand RNA viruses have used cell fractionation and in vitro reconstitution techniques with some success to pinpoint the target for inhibition of protein synthesis; in general, these studies reveal alterations in initiation factor(s) associated with ribosomal proteins (Ehrenfeld, 1984). J.R. Thomas and Wagner (1983) applied similar techniques of fractionation and reconstitution of the cellular protein-synthesizing machinery to compare mock-infected and VSV-infected mouse L cells. Their results are similar to those found for other virus–host cell systems and are as follows: (1) Postmitochondrial lysates of VSV-infected L cells show a reduction in their ability to synthesize proteins compared to mock-infected control cells; (2) this inhibition occurs at the level of initiation of translation; (3) reconstitution studies reveal that the factor or factors associated with impaired protein synthesis appears in the fraction released from infected-cell ribosomes washed with 0.5 M KCl; (4) the responsible initiation factor or factors are present in the 0–40% ammonium sulfate precipitate that contains predominantly eIF-3 and eIF-4B. Further studies indicated that the effect on protein synthesis due to the altered factor(s) can be reversed by purified initiation factors eIF-3 and eIF-4B from mock-infected cells. Clearly, however, more refined experiments are required to make a more definitive identification of the target for VSV inhibition of translation.

VI. RHABDOVIRAL IMMUNOLOGY

Vertebrates acquire immunity to specific rhabdoviruses in much the same manner as they do to other acute viral infections. Primary infection results in the development of a humoral immune response within a week after first exposure. Cellular immune responses also occur following infection, but these are probably of greater importance in subacute infections, such as those with rabies virus. Antibodies generated during the humoral immuno response are directed to two major antigens, designated group-specific and type-specific (Brown and Cartwright, 1966). Kang and Prevec (1970) found that the group-specific antibody is directed to the nucleoprotein, and the type-specific antibody to membrane protein. Cartwright and Brown (1972b) were able to determine that nucleocapsid (N) protein is the group-specific antigen of vesicular stomatitis virus (VSV) and glycoprotein (G protein) the type-specific antigen. By this means, Cartwright and Brown (1972b) were able to divide the vesiculoviruses into two major serotypes, designated Indiana and New Jersey. Kelley *et al.* (1972) identified the type-specific VSV G protein as the viral antigen that gives rise to and reacts with neutralizing antibody, and Wiktor *et al.* (1973) identified the same functions for the rabies virus G protein. About 70% of the complement-fixing antibody produced by VSV infection reacts with the G-protein antigen, and virtually all the remaining 30% reacts with the N protein (Cartwright and Brown, 1972b).

A. Antigens and Antibodies

All five of the structural proteins of VSV, and probably all those of other rhabdoviruses, are highly antigenic; their injection into rabbits, mice, and other animals results in circulating antibody of relatively high titer. The one possible exception had been the rather labile larger (L) protein of the two minor nucleocapsid protein, which, until recently, had defied attempts to raise antibody. This problem has been solved by Harmon and Summers (1982) who have managed to produce relatively high titered anti-L serum in rabbits. Many laboratories have been successful in producing polyclonal antibodies reactive with each of the five VSV proteins; these polyclonal antibodies have provided interesting means for assigning specific biological or biochemical functions to the various rhabdoviral proteins. More recently, several investigators have used the hybridoma technique to produce monoclonal antibodies to serve as selective probes for antigenic domains of several rhabdoviral proteins.

1. Specific Polyclonal Antibodies

Injection of rabbits with a purified VSV protein, such as the G protein, results in production of antiserum that reacts by immunoprecipitation or immunodiffusion with the G-protein antigen (Kelley *et al.*, 1972). Similar results were obtained by Dietzschold *et al.* (1974), who raised specific, polyclonal non-cross-reacting antibodies to all three of the VSV major proteins: N, G, and matrix (M). Harmon and Summers (1982) extended these studies to all five of the VSV proteins, which produced in rabbits five distinct antisera that immunoprecipitated each homologous protein in the cytoplasm of VSV-infected cells and reacted specifically by immunodiffusion. It should also be noted, however, that minute amounts of cellular antigens, not visible by gel electrophoresis, are present in virions released from infected cells. Cartwright and Pearce (1968) found evidence for host-cell components in VSV that were later found to be glycolipids from host cells as determined by complement fixation (Cartwright and Brown, 1972a). At least one of these host-cell components can be histocompatability (H-2) antigens derived from mouse L cells (Hecht and Summers, 1976). In a detailed study, Little *et al.* (1983) found that antisera made against uninfected HeLa cells could immunoprecipitate, but not neutralize, the infectivity of VSV grown in HeLa cells, but not in murine cells; moreover, anti-Vero cell serum did not immunoprecipitate VSV grown in HeLa cells. These studies emphasize the unique character and monospecificity of the five protein antigens of VSV; the cellular components are minor constituents trapped in virions during the budding process.

Monospecific antibodies can frequently affect the biological or biochemical functions, or both, of the specific antigenic proteins. Antibodies to the G proteins of VSV and rabies virus neutralize the infectivity of these viruses (Kelley *et al.*, 1972; Wiktor *et al.*, 1973). A more detailed analysis of neutralizing antibodies and specific antigenic sites of the G protein is presented below. Also, as recounted in Section IV.C.2, a specific antibody to the G protein prevents hemolysis of erythrocytes (Mifune *et al.*, 1982). Antiserum directed against the VSV G protein also prevents maturation and budding of virus from cells exposed to anti-VSV serum and, in fact, results in degradation of the G protein inserted in the surface membrane of the infected cell (O'Rourke *et al.*, 1983).

Specific, polyclonal antibodies against each of the three VSV nucleocapsid proteins all affect viral transcription. Imblum and Wagner (1975) raised antibodies in rabbits inoculated with purified nonstructural (NS) protein [soon to be renamed P (for highly phosphorylated)] protein that reacted specifically with the cytoplasmic and the virion-associated NS protein of the homotypic virus. Immunoglobulin prepared from this anti-NS serum rapidly shut off *in vitro* transcription of the homotypic VSV-Indiana nucleocapsids, but not appreciably that of VSV-New Jersey; the kinetics of the reaction suggested that anti-NS γ-globulin inhibited chain elongation in the transcription reaction. Specific antibody directed against the N protein of either VSV-Indiana or VSV-New Jersey also exhibited marked inhibition of *in vitro* transcription of each homotypic virion as well as partial inhibition of heterotypic transcription (Carroll and Wagner, 1978b). Anti-RNP-Indiana immunoglobulin G (IgG) did not inhibit transcription of Chandipura virus, whereas anti-RNP-New Jersey did to a limited extent, indicating some degree of antigenic relatedness of the group-specific N antigens of these three vesiculoviruses.

Somewhat more detailed experiments were performed by Harmon and Summers (1982), who tested immunoglobulins directed against each of the five VSV proteins for their capacity to inhibit *in vitro* transcription of nucleocapsids from VSV virions or VSV-infected cells. They showed for the first time marked transcription inhibition by anti-L IgG and reported that large amounts of anti-G, anti-M, and anti-N immunoglobulins surprisingly had some effect on *in vitro* transcription of detergent-disrupted virions, but only anti-L and anti-NS IgG efficiently inhibited transcription by cytoplasmic ribonucleocapsids; they interpreted these results as evidence for nonspecific inhibition of transcription except for anti-L and anti-NS sera, which specifically and efficiently inhibited transcription of cytoplasmic ribonucleoproteins, whereas the other antisera had little or no effect. Electropherograms of the virion-associated ribonucleoproteins revealed considerable contamination with membrane proteins. The capacity of anti-N IgG to inhibit transcription by detergent-disrupted virions described by Carroll and Wagner (1978b) was interpreted by Harmon and Summers (1982), who reported the same effect, as due to conformational differences in virion nucleocapsids compared to cytoplasmic nucleocapsids, the transcription of which is not inhibited by anti-N serum. Harmon and Summers (1982) also reported greater inhibition of RNP transcription by anti-L IgG compared with that by anti-NS IgG. Recent experiments by Arnheiter *et al.* (1985) reveal that a monoclonal antibody that binds to VSV nucleocapsids also markedly inhibits transcription of detergent-disrupted virions, whereas another monoclonal antibody that binds poorly to nucleocapsids but reacts with free cytoplasmic N protein does not inhibit virion transcription. These experiments suggest not only different conformational changes of N protein but also exposure of different epitopes reactive with immunoglobulins, which presumably determine the extent of transcription inhibition.

2. Monoclonal Antibodies

The recently developed technology of producing cloned hybridoma cells that produce monoclonal antibodies (MAb's) has provided the most incisive means for studying antigenic relatedness among viruses and for probing the biologically functional sites of specific viral proteins. In the case of rhabdoviruses, Flamand *et al.* (1980a,b) have produced hybridomas that secrete MAb's to eight different antigenic determinants (epitopes) of the N protein and to four different epitopes of the G protein of rabies viruses. These techniques have provided important tools for classifying rabies and rabieslike rhabdoviruses, which are discussed in detail in Chapters 9 and 11. Portner *et al.* (1980) used MAb's reactive with the VSV G protein to analyze the frequency of antigenic variation, which was found to be quite similar to frequencies of antigenic variation on the HN protein of Sendai virus and the HA protein of influenza virus. Volk *et al.* (1982) analyzed 19 hybridomas producing MAb's to the VSV-Indiana G protein and were able to identify 11 different non-cross-reacting antigenic determinants, 4 of which reacted with specific neutralizing MAb's; of the 19 MAb's, 2 in the same epitope exhibited strong binding to VSV M protein, but the other 17 did not.

Detailed analyses were made by LeFrancois and Lyles (1982a,b), who also identified four nonoverlapping epitopes on the VSV-New Jersey G protein that reacted with neutralizing MAb's. They also found four neutralizing-antibody-reactive epitopes on the VSV-Indiana G protein, but in the latter case, there were varying degrees of overlap attributed either to close proximity or to allosteric modification of the VSV-Indiana epitopes. A fifth neutralizing MAb that could bind to a putative fifth epitope of the VSV-Indiana G protein was found to cross-react with the VSV-New Jersey G protein, but there was no competitive binding to heterotypic G proteins among the other neutralizing MAb's of VSV-Indiana or VSV-New Jersey. On the other hand, 6 of 11 nonneutralizing MAb's to the G proteins of VSV-Indiana and VSV-New Jersey cross-reacted with G proteins of both serotypes, defining three nonoverlapping cross-reactive epitopes on each (LeFrancois and Lyles, 1982b). Three nonneutralizing MAb's reacted only with VSV-New Jersey G protein and recognized unique epitopes, whereas one nonneutralizing MAb specific for VSV-Indiana G protein did cross-react with another MAb. The existence of four epitopes involved in MAb neutralization of either VSV-New Jersey or VSV-Indiana was confirmed by selecting G-protein variants of VSV grown in the presence of neutralizing MAb's to each of the four epitopes (LeFrancois and Lyles, 1983a). Virus variants with reduced capacity to bind a specific MAb occurred at a frequency of 10^{-5} to 10^{-6} for most epitopes, but a few did not convert to nonreactivity. In general, variation in one epitope resulted in reduced binding by all MAb's to that determinant, but a single epitope (designated A) for both serotypes exhibited subregions for binding different epitope

A MAb's. Antigenic variants did not appear in a few epitopes. In an elegant study, Seif *et al.* (1985) isolated mutants of rabies virus resistant to neutralization by MAb's to a specific region (site III) of the rabies G protein. These MAb-resistant mutants could be divided into five groups, one of which lost its pathogenicity for mice and contained a G protein with a single amino acid substitution at residue 333. Vandepol *et al.* (1986) sequenced by primer extension the G genes of VSV-Indiana variants resistant to neutralization by monoclonal and polyclonal antibodies; they found mutations over extended regions of the G gene.

In later studies, Bricker *et al.* (1987) examined MAb's to VSV-New Jersey G protein made by 25 hybridoma clones and were able to identify nine epitopes and several overlapping epitopes; once again, MAb's specific for four epitopes neutralized the infectivity of VSV-New Jersey to varying degrees and MAb's to other epitopes were nonneutralizing. VSV-Indiana G protein was found to possess three epitopes in common with VSV-New Jersey G protein. Once again, all these serotypically cross-reacting epitopes could bind only nonneutralizing MAb; the neutralizing MAb's for VSV-New Jersey and VSV-Indiana were specific for the G protein of the respective serotype.The degree of MAb cross-reactivity for G proteins of the two serotypes is consistent with the finding of approximately 50% base-sequence homology between the two G-protein genes (Rose and Gallione, 1981). Bricker *et al.* (1987) also found that reduction of disulfide bonds in the VSV-New Jersey G protein resulted in complete loss of its capacity to bind all neutralizing MAbs and most nonneutralizing MAb's, emphasizing the critical role of secondary and probably tertiary structure in availability of antigenic determinants for antibody binding. It was also found by the technique of using MAb for protection against protease digestion that all nine of the epitopes of the VSV-New Jersey G protein are clustered in a stretch of approximately 100 amino acids near the middle of the G protein (Bricker *et al.*, 1987).

MAb's to rhabdoviral proteins other than G protein also provide useful probes for studying their functions. Of only two MAb's to the VSV N protein studied by Arnheiter *et al.* (1985), one binds to completely assembled nucleocapsids and probably to free cytoplasmic N protein as well as inhibits *in vitro* VSV transcription, whereas the other MAb binds only to free cytoplasmic N protein and does not inhibit transcription, but delays the appearance of newly formed nucleocapsids when microinjected into VSV-infected cells. Very preliminary studies on monoclonal antibodies to NS (P) protein of VSV reveal differences in binding to different regions of the NS protein and differential capacity to inhibit VSV transcription; preliminary mapping of the NS protein epitopes has been done with expressed cDNA clones in which various sequences have been deleted (Williams *et al.*, 1987).

Detailed experiments have been done with the MAb's to the M protein of both VSV-Indiana (Pal *et al.*, 1985a) and VSV-New Jersey (Ye *et al.*, 1985). There is surprisingly little cross-reaction of MAb's to heterotypic M proteins, but each M serotype possesses at least four antigenic determinants, some of which show varying degrees of overlap. Distinct MAb's to two epitopes on the VSV-New Jersey M protein are capable of reversing inhibition of transcription caused by M protein of the homotypic VSV-New Jersey, but MAb's to two other epitopes not only failed to reverse transcription inhibition by M protein but also enhanced the transcriptional inhibitory effect (Ye *et al.*, 1985). Pal *et al.* (1985b) have made a detailed analysis of the capacity of MAb's to affect the ability of VSV-Indiana M protein to regulate transcription. MAb to M epitope-1 reversed transcription inhibition, whereas MAb's to two other unrelated epitopes (2 and 3) either failed to reverse or actually increased transcription inhibition by homotypic M protein. Among four group III (M protein) temperature-sensitive (*ts*) mutants of VSV-Indiana, the M proteins of which have lost the capacity to inhibit transcription, one (*ts*023III) contained an M protein that completely lacks the capacity to bind MAb to epitope 1 (which reverses transcription inhibition by wild-type VSV M protein). The M-protein genes of wild-type and *ts*023(III) have been sequenced by primer extension by Gopalakrishna and Lenard (1985), who found only a single base change in the mutant that led to a single amino acid substitution (Gly → Glu), at residue 21. Pal *et al.* (1985b) also studied four revertants of *ts*023(III) and found that only one (R11) completely reverted to the wild-type M phenotype by reacquiring its capacity to bind MAb to epitope 1 as well as its capacity to inhibit transcription; the M proteins of the other three *ts*023III revetants (R12, R13, R14) completely failed to bind MAb to epitope 1 could inhibit transcription only partially. These studies, although quite incomplete, can only lead to the conclusion that specific amino acids at crucial sites lead to conformational changes in M protein, or perhaps in any other protein, that determine the biological and antigenic properties of the protein, and in the case of VSV, these mutations are frequent.

B. Immunity to Vesicular Stomatitis Virus

Comments here will be brief and limited to VSV because Chapter 9 discusses immunity to rabies viruses in detail. As is the case with other viruses, VSV induces both humoral and cellular immunity. The appearance of circulating antibody requires 3–5 days, as noted by Wagner (1974) and Beatrice and Wagner (1980b), who inoculated mice intranasally with nonreplicating *ts* mutants; these mice actually became resistant to challenge with virulent virus by the same route within a day after intranasal vaccination, the resistance presumably being related to appearance in bronchial secretions of an antiviral substance with some of the properties of interferon (Beatrice and Wagner, 1980b). Quite clearly, however, the major antiviral substance that protects animals against VSV infection is neutralizing antibody. Without question, the VSV G protein contains the antigenic sites that give rise to and react with neutralizing antibody (Kelley et al., 1972).

The mechanisms by which specific immunoglobulins neutralize the infectivity of a virus have long been a matter of dispute (Dimmock, 1984), but the concept that a single antibody molecule blocks a single viral cell-attachment site is no longer tenable. As mentioned above, the G proteins of VSV-Indiana or VSV-New Jersey possess at least four distinct antigenic determinants that react with different MAb's (Volk et al., 1982; LeFrancois and Lyles, 1982a,b; Bricker et al., 1987). Volk et al. (1982) have found that a mixture of two separate MAb's will neutralize VSV infectivity more effectively than either one alone. LeFrancois (1984) studied the capacity of MAb's to the G protein of either VSV-New Jersey or VSV-Indiana to immunize mice against the neuropathic and lethal effects of the respective viruses. It was noted that neutralizing MAb's inoculated 24 hr prior to or simultaneous with homotypic VSV protected the mice, but MAb administered after challenge did not. Of some interest was the finding that nonneutralizing MAb did provide some protection against subsequent challenge, and MAb's also conferred a certain amount of passive immunity to challenge with heterotypic VSV. However, it should be noted that passive immunity was conferred only by (Fab)$_2$ fragments of neutralizing antibody, not by (Fab)$_2$ fragments of nonneutralizing MAb. These data, which are not very quantitative, are of interest but a bit difficult to interpret mechanistically. An obvious possible explanation for passive immunity conferred by nonneutralizing homotypic or heterotypic MAb's is lysis by the complement system of VSV bound by MAb.

Cellular immune responses to viruses have been shown to be mediated by polymorphic major histocompatability antigens (Zinkernagel, 1979). Viral infection generates cytotoxic T lymphocytes (CTL) specific for both a viral antigen and a self-component coded by the K or D locus, or both, of the H-2 histocompatability complex (Doherty et al., 1976). VSV has provided a nice model for such studies because the well-characterized G protein appears to be the antigen that mediates the CTL response and is its major target (Hale et al., 1978; Zinkernagel et al., 1978). Of considerable interest is the finding that cells infected with either VSV-Indiana or VSV-New Jersey are both lysed by effector T cells generated by the heterotypic virus (Rosenthal and Zinkernagel, 1980). This cross-reactivity of CTL is quite different from non-cross-reactivity of neutralizing antibody to the two serotypes, suggesting that the epitopes common to the VSV-Indiana and VSV-New Jersey G proteins that bind heterotypic MAb's (LeFrancois and Lyles, 1982b) are the targets for CTL. Attempts to block CTL lysis of VSV-infected cells with anti-G sera yielded conflicting results. VSV polyclonal antisera raised in mice did not block CTL activity, but hyperimmune VSV rabbit antisera did inhibit the cytotoxic effect by CTL (Zinkernagel et al., 1978; Rosenthal et al., 1981). MAb's to VSV G proteins did not block lysis of VSV-infected cells by primary, secondary, or long-term cultured CTL cells specific for VSV-

Indiana or VSV-New Jersey (Rosenthal *et al.*, 1981; Zinkernagel and Rosenthal, 1981). However, Sethi and Brandis (1980) were able to demonstrate MAb blocking of homotypic but not heterotypic lysis by CTL, whereas LeFrancois and Lyles (1983b) were able to block the lysis of both VSV-Indiana and VSV-New Jersey cells by homotypic and heterotypic CTL by cross-reactive MAb's to the two heterotypic G proteins. They obtained better blocking results with a combination of MAb's, but the results clearly show that MAb to VSV-Indiana G protein inhibited lysis by CTL of cells infected with either VSV serotype (LeFrancois and Lyles, 1983b). Although their data were difficult to analyze, it seems likely that the cross-reactive nonneutralizing MAb's were the primary blockers of CTL, suggesting that these shared G-protein epitopes are the primary targets of CTL.

One of the likely effects of antibody on the G protein in VSV-infected cells is the redistribution of the G protein and H-2 antigens, particularly $H-2K_b$ (Geiser *et al.*, 1979), which could markedly affect the action of CTL. The nature of the H-2 haplotype in association with the VSV G protein, as shown with reconstituted vesicles, also determines the specificity of the CTL activation that is generated (Loh *et al.*, 1979). Another possible factor that contributes to the degree of T-effector-cell cytolysis is the carbohydrate moiety of the G protein in the infected target cell; Black *et al.* (1981) showed a reduced effect by CTL on cells infected with VSV in the presence of tunicamycin. Clearly, many diverse factors influence the action of CTL in cellular immunity to VSV infection and the relationship with histocompatability antigens. The enormous literature on this subject is only superficially recounted here.

ACKNOWLEDGMENTS. The research from my laboratory reported herein was supported by grants from the National Institute of Allergy and Infectious Diseases (AI-11112 and AI-21652) and the American Cancer Society (MV-9). I am deeply indebted to Sandy Shiflett for excellent preparation of the manuscript.

REFERENCES

Abraham, G., and Banerjee, A. K., 1976a, Sequential translation of the genes of vesicular stomatitis virus, *Proc. Natl. Acad. Sci. U.S.A.* **73**:1504.

Abraham, G., and Banerjee, A. K., 1976b, The nature of the RNA products synthesized *in vitro* by subviral components of vesicular stomatitis virus, *Virology* **71**:230.

Altstiel, L. D., and Landsberger, F. R., 1981, Structural changes in BHK cell plasma membrane caused by the binding of vesicular stomatitis virus, *J Virol.* **39**:82.

Arnheiter, H., Davis, N. L., Wertz, G., Schubert, M., and Lazzarini, R. L., 1985, Role of the nucleocapsid protein in regulating vesicular stomatitis virus RNA synthesis, *Cell* **41**:259.

Arstila, P., 1972, Two hemagglutinating components of vesicular stomatitis virus, *Acta Pathol. Microbiol. Scand. Sect. B* **80**:33.

Arstila, P., 1973, Small-sized hemagglutinin of vesicular stomatitis virus released spontaneously by Nonidet P40, *Acta Pathol. Microbiol. Scand. Sect. B* **81**:27.

Arstila, P., Halonen, P. E., and Salmi, A., 1969, Hemagglutinin of vesicular stomatitis virus, *Arch Gesamte Virusforsch.* **27**:198.

Bablanian, R., 1975, Structural and functional alterations in cultured cells infected with cytocidal viruses, *Prog. Med. Virol.* **19**:40.

Bailey, C., Miller, D., and Lenard, J., 1981, Hemolysis of human erythrocytes by vesicular stomatitis virus, *J. Cell Biol.* **91**:111a.

Bailey, C. P., Miller, D., and Lenard, J., 1984, Effect of DEAE–dextran on infection and hemolysis by VSV: Evidence that non-specific electrostatic interactions mediate binding of VSV to cells, *Virology* **133**:111.

Ball, L. A., and White, C. N., 1976, Order of transcription of genes of vesicular stomatitis virus, *Proc. Natl. Acad. Sci. U.S.A.* **73**:442.

Baltimore, D., Huang, A. S., and Stampfer, M., 1970, Ribonucleic acid synthesis of vesicular stomatitis virus. II. An RNA polymerase in the virion, *Proc. Natl. Acad. Sci. U.S.A.* **66**:572.

Banerjee, A. K., Abraham, G., and Colonno, R. J., 1977, Vesicular stomatitis virus: Model of transcription, *J. Gen. Virol.* **34**:1.

Barenholz, Y., Moore, N. F., and Wagner, R. R., 1976, Enveloped viruses as model membrane systems: Microvioscosity of vesicular stomatitis virus and host cell membranes, *Biochemistry* **15**:3563.

Baxt, B., and Bablanian, R., 1976a, Mechanisms of vesicular stomatitis virus-induced cytopathic effect. I. Early morphologic changes induced by infectious and defective-interfering particles, *Virology* **72**:370.

Baxt, B., and Bablanian, R., 1976b, Mechanisms of vesicular stomatitis virus-induced cytopathic effect. II. Inhibition of macromolecular synthesis induced by infectious and defective-interfering particles, *Virology* **72**:383.

Beatrice, S. T., and Wagner, R. R., 1980a, Effect of sulfhydryl reagents on the infectivity of vesicular stomatitis virus, *Virology* **100**:246.

Beatrice, S. T., and Wagner, R. R., 1980b, Immunogenecity in mice of temperature-sensitive mutants of vesicular stomatitis virus: Early appearance in bronchial secretions of an interferon-like inhibitor, *J. Gen. Virol.* **47**:529.

Bishop, D. H. L., 1979, *Rhabdoviruses*, 3 vols., CRC Press, West Palm Beach.

Bishop, D. H. L., and Roy, P., 1972, Dissociation of vesicular stomatitis virus and relation of the virion proteins to the viral transcriptase, *J. Virol.* **10**:234.

Bishop, D. H. L., Emerson, S. U., and Flamand, A., 1974, Reconstitution of infectivity and transcription activity of homologous and heterologous viruses: Vesicular stomatitis (Indiana serotype), Chandipura, vesicular stomatitis (New Jersey serotype) and Cocal virus, *J. Virol.* **14**:139.

Bishop, D. H. L., Repik, P., Obijeski, J. F., Moore, N. F., and Wagner, R. R., 1975, Restitution of infectivity to spikeless vesicular stomatitis virus by solubilized virus components, *J. Virol.* **16**:75.

Black, P. L., Vitetta, E. S., Forman, J., Kang, C. Y., May, R. D., and Uhr, J. W., 1981, Role of glycosylation in the H-2-restricted cytolysis of virus-infected cells, *Eur. J. Immunol.* **11**:48.

Bloom, B. R., Stoner, G., Fiscetti, V., Nowakowski, M., Muschel, R., and Rubinstein, A., 1974, Products of activated lymphocytes (PALs) and the virus plaque assay, in *Progress in Immunology II*, Vol. 3 (L. Brent and J. Holborow, eds.), pp. 133–146, Elsevier North-Holland, New York.

Blumberg, B. M., Leppert, M., and Kolakofsky, D., 1981, Interaction of VSV leader RNA and nucleocapsid protein may control VSV genome replication, *Cell* **23**:837.

Blumberg, B. M., Giorgi, M. C., and Kolakofsky, D., 1983, VSV N protein selectively encapsidates VSV leader RNA *in vitro*, *Cell* **32**:559.

Both, G. W., Banerjee, A. K., and Shatkin, A. J., 1975, Methylation-dependent translation of viral messenger RNA *in vitro*, *Proc. Natl. Acad. Sci. U.S.A.* **72**:1189.

Breindl, M., and Holland, J. J., 1975, Coupled *in vitro* transcription and translation of vesicular stomatitis virus messenger RNA, *Proc. Natl. Acad. Sci. U.S.A.* **72**:2545.

Bricker, B. J., Snyder, R. M., Fox, J. W., Volk, W. A., and Wagner, R. R., 1986, Monoclonal antibody footprinting of the glycoprotein of vesicular stomatitis virus (New Jersey serotype), *J. Virol.* (in press).

Brown, F., and Cartwright, B., 1966, The antigens of vesicular stomatitis virus. II. The presence of two low molecular weight immunogens in virus suspension, *J. Immunol.* **97**:612.

Bukrinskaya, A. G., 1982, Penetration of viral genetic material into host cells, *Adv. Virus Res.* **27**:141.

Carroll, A. R., and Wagner, R. R., 1978a, Reversal by certain polyanions of an endogenous inhibitor of the vesicular stomatitis virus-associated transcriptase, *J. Biol. Chem.* **253**:3361.

Carroll, A. R., and Wagner, R. R., 1978b, Inhibition of transcription by immunoglobulins directed against the ribonucleoprotein of homotypic and heterotypic vesicular stomatitis virus, *J. Virol.* **25**:675.

Carroll, A. R., and Wagner, R. R., 1979, Role of the membrane (M) protein in endogenous inhibition of *in vitro* transcription of vesicular stomatitis virus, *J. Virol.* **29**:134.

Cartwright B., and Brown, F., 1972a, Glycolipid nature of the complement-fixing host cell antigen of vesicular stomatitis virus, *J. Gen. Virol.* **15**:243.

Cartwright B., and Brown, F., 1972b, Serological relationships between different strains of vesicular stomatitis virus, *J. Gen. Virol.* **16**:91.

Cartwright B., and Brown, F., 1977, Role of sialic acid in infection with vesicular stomatitis virus, *J. Gen. Virol.* **35**:197.

Cartwright B., and Pearce, C. A., 1968, Evidence for a host cell component in vesicular stomatitis virus, *J. Gen. Virol.* **2**:207.

Cartwright B., Smale, C. J., and Brown, F., 1969, Surface structure of vesicular stomatitis virus, *J. Gen. Virol.* **5**:1.

Centrella, M., and Lucas-Lenard, J., 1982, Regulation of protein synthesis in vesicular stomatitis virus-infected mouse L-929 cells by decreased protein synthesis initiation factor 2 activity, *J. Virol.* **41**:781.

Challberg, M. D., and Kelly, T. J., Jr., 1979a, Adenovirus DNA replication *in vitro*, *Proc. Natl. Acad. Sci. U.S.A.* **76**:655.

Challberg, M. D., and Kelly, T. J., Jr., 1979b, Adenovirus DNA replication *in vitro*: Origin and direction of daughter strand synthesis, *J. Mol. Biol.* **135**:999.

Clewley, J. P., Bishop, D. H. L., Kang, C. Y., Coffin, J., Schnitzlein, W. M., Reichmann, M. E., and Shope, R. E., 1977, Oligonucleotide fingerprints of RNA species obtained from rhabdoviruses belonging to the vesicular stomatitis virus subgroup, *J. Virol.* **23**:152.

Clinton, G. M., and Finley-Whelan, J., 1984, Tyrosyl kinase acquired from anchorage-independent cells by a membrane-enveloped virus, *J. Cell Biol.* **99**:788.

Clinton, G. M., and Huang, A. S., 1984, Distribution of phosphoserine, phosphothreonine, and phosphotyrosine in proteins of vesicular stimatitis virus, *Virology* **108**:510.

Clinton, G. M., Little, S. P., Hagen, F. S., and Huang, A. S., 1978a, The matrix (M) protein of vesicular stomatitis virus regulates transcription, *Cell* **15**:1455.

Clinton, G. M., Burge, B. W., and Huang, A. S., 1978b, Effects of phosphorylation and pH on the association of NS protein with vesicular stomatitis virus cores, *J. Virol.* **27**:240.

Clinton, G. M., Guerina, N. G., Guo, H.-Y., and Huang, A. S., 1982, Host-dependent phosphorylation and kinase activity associated with vesicular stomatitis virus, *J. Biol. Chem.* **257**:3313.

Colonno, R. J., and Banerjee, A. K., 1977, Mapping and initiation studies on the leader RNA of vesicular stomatitis virus, *Virology* **77**:260.

Dahlberg, J. E., 1974, Quantitative electron microscopic analysis of the penetration of VSV into L cells, *Virology* **58**:250.

Davis, N. L., and Wertz, G. W., 1982, Synthesis of vesicular stomatitis virus negative-strand RNA *in vitro*: Dependence on viral protein synthesis, *J. Virol.* **41**:821.

Dickson, R., Willingham, M., and Pastan, I., 1983, α_2-Microglobulin adsorbs to colloidal gold: A new process in the study of receptor-mediated endocytosis, *J. Cell Biol.* **89**:29.

Dietzschold, B., Schneider, L. G., and Cox, S. H., 1974, Serological characterization of the three major proteins of vasicular stomatitis virus, *J. Virol.* **14**:1.

Dimmock, N. J., 1984, Mechanisms of neutralization of animal viruses, *J. Gen. Virol.* **65**:1015.

Doherty, P. C., Blanden, R. V., and Zinkernagel, R. M., 1976, Specificity of virus-immune effector T cells for H-2K and H-2D compatible interactions: Implications for H-antigen diversity, *Transplant. Rev.* **29**:89.

Dunigan, D. D., and Lucas-Lenard, J. M., 1983, Two transcription products of the vesicular stomatitis virus genome may control L-cell protein synthesis, *J. Virol.* **45**:618.

Dunigan, D. D., Baird, S., and Lucas-Lenard, J. J., 1986, Lack of correlation between the accumulation of plus-strand leader RNA and the inhibition of protein and RNA synthesis in vesicular stomatitis virus infected mouse L cells, *Virology* **150**:231.

Ehrenfeld, E., 1984, Picornavirus inhibition of host cell protein synthesis, in: *Comprehensive Virology*, Vol. 19 (H. Fraenkel-Conrat and R. R. Wagner, eds.), pp. 177–221, Plenum Press, New York.

Eidelman, O., Schlegel, R., Tralka, T., and Blumenthal, R., 1984, pH-Dependent fusion induced by vesicular stomatitis virus glycoprotein reconstituted into phospholipid vesicles, *J. Biol. Chem.* **259**:4622.

Emerson, S. U., 1976, Vesicular stomatitis virus: Structure and function of virion components, *Curr. Top. Microbiol. Immunol.* **73**:1.

Emerson, S. U., 1982, Reconstitution studies detect a single polymerase entry site on the vesicular stomatitis virus genome, *Cell* **31**:635.

Emerson, S. U., and Wagner, R. R., 1972, Dissociation and reconstitution of the transcriptase and template activities of vesicular stomatitis B and T virions, *J. Virol.* **10**:297.

Emerson, S. U., and Wagner, R. R., 1973, L protein requirement for *in vitro* RNA synthesis by vesicular stomatitis virus, *J. Virol.* **12**:1325.

Emerson, S. U., and Yu, Y. H., 1975, Both NS and L proteins are required for *in vitro* RNA synthesis by vesicular stomatitis virus, *J. Virol.* **15**:1348.

Emerson, S. U., Dierks, P. M., and Parsons, J. T., 1977. *In vitro* synthesis of a unique RNA species by a T particle of vesicular stomatitis virus, *J. Virol.* **23**:708.

Fan, D. P., and Sefton, B. M., 1978, The entry into host cells of Sindbis virus, vesicular stomatitis virus and Sendai, *Cell* **15**:985.

Flamand, A., 1970, Etude génétique du virus de la stomatite vesiculaire: Classement de mutants thermosensible spontanées en groupe de complementation, *J. Gen. Virol.* **8**:187.

Flamand, A., and Bishop, D. H. L., 1973, Primary *in vivo* transcription of vesicular stomatitis virus and temperature-sensitive mutants of five vesicular stomatitis virus complementation groups, *J. Virol.* **12**:1238.

Flamand, A., Wiktor, T. J., and Koprowski, H., 1980a, Use of hybridoma monoclonal antibodies in the detection of antigenic differences between rabies and rabies-related virus proteins: The nucleocapsid protein, *J. Gen. Virol.* **48**:97.

Flamand, A., Wiktor, T. J., and Koprowski, H., 1980b, Use of hybridoma monoclonal antibodies in the detection of antigenic differences between rabies and rabies-related virus proteins: II. The glycoprotein, *J. Gen. Virol.* **48**:105.

Florkiewicz, R., and Rose, J. K., 1984, A cell line expressing vesicular stomatitis virus glycoprotein fuses at low pH, *Science* **225**:771.

Follett, E. A. C., Pringle, C. R., Wunner, W. H., and Skehel, J. J., 1974, Virus replication in enucleated cells: Vesicular stomatitis virus and influenza virus, *J. Virol.* **13**:394.

Gallione, C. J., Greene, J. R., Iberson, L., and Rose, J. K., 1981, Nucleotide sequences of the mRNA's encoding the vesicular stomatitis N and NS proteins, *J. Virol.* **39**:529.

Geiser, B., Rosenthal, K. L., Klein, J., Zinkernagel, R. M., and Singer, S. J., 1979, Selective and unidirectional membrane redistribution of an H-2 antigen with an antibody-clustered viral antigen: Relationship to mechanisms of cytotoxic T-cell interactions, *Proc. Natl. Acad. Sci. U.S.A.* **76**:4603.

Genty, N., 1975, Analysis of uridine incorporation in chick embryo cells infected with vesicular stomatitis virus and its temperature-sensitive mutants: Uridine transport, *J. Virol.* **15**:8.

Genty, N., and Berreur, P., 1975, Metabolisme des acides ribonucleique et des proteines de cellule de l'embryon de poulet infectées per le virus de la stomatite vesiculaire: Etudes des effects de mutants thermosensible, *Ann. Microbiol. (Inst. Pasteur)* **124A:**135.

Gill, P. S., and Banerjee, A. K., 1985, Vesicular stomatitis virus NS proteins: Structural similarity without extensive sequence homology, *J. Virol.* **55:**60.

Gillies, S., and Stollar, V., 1980a, Generation of defective interfering particles of vesicular stomatitis virus in *Aedes albopictus* cells, *Virology* **107:**497.

Gillies, S., and Stollar, V., 1980b, The production of high yields of infectious vesicular stomatitis virus in *Aedes albopictus* cells and comparisons with replication in BHK-21 cells, *Virology* **107:**509.

Gillies, S., and Stollar, V., 1981, Biochemical characterization of vesicular stomatitis virus-infected *Aedes albopictus* cells deprived of methionine, *Virology* **112:**318.

Giorgi, C., Blumberg, B., and Kolakofsky, D., 1983, Sequence determinations of the (+) leader regions of the Chandipura, Coccal and Piry serotype genomes, *J. Virol.* **46:**125.

Gopalakrishna, Y., and Lenard, J., 1985, Sequence alterations in temperature-sensitive M protein mutants (complementation group III) of vesicular stomatitis virus, *J. Virol.* **56:**655.

Grinnell, B. W., and Wagner, R. R., 1983, Comparative inhibition of cellular transcription by vesicular stomatitis virus serotypes New Jersey and Indiana: Role of each viral leader RNA, *J. Virol.* **48:**88.

Grinnell, B. W., and Wagner, R. R., 1984, Nucleotide sequence and secondary structure of VSV leader RNA and homologous DNA involved in inhibition of DNA-dependent transcription, *Cell* **36:**533.

Grinnell, B. W., and Wagner, R. R., 1985, Inhibition of DNA-dependent transcription by the leader RNA of vesicular stomatitis virus: Role of specific nucleotide sequences and cell protein binding, *Mol. Cell. Biol.* **5:**2502.

Gutierriez, P. L., Davis, L. H., and Pottathil, R., 1984, Vesicular stomatitis virus-induced membrane changes: A spin label study, *Life Sci.* **35:**747.

Hale, A. H., Witte, O. N., Baltimore, D., and Eisen, H. N., 1978, Vesicular stomatitis virus glycoprotein is necessary for H-2 restricted lysis of infected cells by cytotoxic T lymphocytes, *Proc. Natl. Acad. Sci. U.S.A.* **75:**970.

Harmon, S. A., and Summers, D. F., 1982, Characterization of monospecific antisera against all five vesicular stomatitis virus-specific proteins: Anti-L and anti-NS inhibit transcription *in vitro*, *Virology* **120:**194.

Hecht, T. T., and Paul, W. E., 1981, Replication of vesicular stomatitis virus in mouse spleen cells, *Infect. Immun.* **32:**1014.

Hecht, T. T., and Summers, D. F., 1976, Interactions of vesicular stomatitis virus with murine cell surface antigens, *J. Virol.* **19:**833.

Heine, J. W., and Schnaitman, C. A., 1969, Fusion of vesicular stomatitis virus with cytoplasmic membrane of L cells, *J. Virol.* **3:**619.

Heine, J. W., and Schnaitman, C. A., 1971, Entry of vesicular stomatitis virus into L cells, *J. Virol.* **8:**786.

Helenius, A., Kartenbeck, J., Simons, K., and Fries, E., 1980, On the entry of Semliki Forest virus into BHK-21 cells, *J. Cell Biol.* **59:**530.

Hill, V. M., Marnell, L., and Summers, D. F., 1981, *In vitro* replication and assembly of vesicular stomatitis virus nucleocapsids, *Virology* **113:**109.

Holland, J. J., Kennedy, S. I. T., Semler, B. L., Jones, C. J., Roux, L., and Grabau, E. A., 1980, Defective-interfering RNA viruses and the host-cell response, in: *Comprehensive Virology*, Vol. 16 (H. Fraenkel-Conrat and R. R. Wagner, eds.), pp. 137–192, Plenum Press, New York.

Howatson, A. F., 1970, Vesicular stomatitis and related viruses, *Adv. Virus Res.* **16:**195.

Howe, C., and Lee, L. T., 1972, Virus–erythrocyte interaction, *Adv. Virus Res.* **17:**1.

Howe, C., Coward, J. E., and Fenger, R. W., 1980, Viral invasion: Morphological, biochemical, and biophysical aspects in: *Comprehensive Virology*, Vol. 16 (H. Fraenkel-Conrat and R. R. Wagner, eds.), pp. 1–71, Plenum Press, New York.

Hsu, C.-H., Kingsbury, D. W., and Murti, K. G., 1979, Assembly of vesicular stomatitis virus nucleocapsids *in vivo:* A kinetic analysis, *J. Virol.* **32**:304.

Huang, A. S., and Baltimore, D., 1977, Defective-interfering animal viruses, in: *Comprehensive Virology,* Vol. 10 (H. Fraenkel-Conrat and R. R. Wagner, eds.), pp. 73–116, Plenum Press, New York.

Huang, A. S., and Manders, E., 1972, Ribonucleic acid synthesis of vesicular stomatitis virus. IV. Transcription of standard virus in the presence of defective-interfering particles, *J. Virol.* **9**:909.

Huang, A. S., and Wagner, R. R. 1965, Inhibition of cellular RNA synthesis by non-replicating vesicular stomatitis virus, *Proc. Natl. Acad. Sci. U.S.A.* **54**:1579.

Huang, A. S., and Wagner, R. R., 1966a, Defective T particles of vesicular stomatitis virus. II. Biological role in homologous interference, *Virology* **30**:173.

Huang, A. S., and Wagner, R. R., 1966b, Comparative sedimentation coefficients of RNA extracted from plaque-forming and defective particles of vesicular stomatitis virus, *J. Mol. Biol.* **22**:381.

Huang, A. S., Greenawalt, J. W., and Wagner, R. R., 1966, Defective T particles of vesicular stomatitis virus. I. Preparation, morphology, and some biological properties, *Virology* **30**:161.

Hughes, J. V., Doll, S. C., and Johnson, T. C., 1985, Hypothermia-inducing peptide promotes recovery of vesicular stomatitis virus from persistent animal infections, *J. Virol.* **53**:781.

Hunt, D. M., Emerson, S. U., and Wagner, R. R., 1976, RNA⁻ temperature-sensitive mutants of vesicular stomatitis virus: L protein thermosensitivity accounts for transcriptase restriction of group I mutants, *J. Virol.* **18**:596.

Hunter, T., Hunt, T., and Jackson, R. J., 1975, The characteristics of inhibition of protein synthesis by double-stranded ribonucleic acid in reticulocyte lysates, *J. Biol. Chem.* **250**:409.

Imblum, R. L., and Wagner, R. R., 1974, Protein kinase and phosphoproteins of vesicular stomatitis virus, *J. Virol.* **13**:113.

Imblum, R. L., and Wagner, R. R., 1975, Inhibition of viral transcription by immunoglobulin directed against the nucleocapsid NS protein of vesicular stomatitis virus, *J. Virol.* **15**:1357.

Jacobs, B. L., and Penhoet, E. E., 1982, Assembly of vesicular stomatitis virus: Distribution of the glycoprotein on the surface of infected cells, *J. Virol.* **44**:1047.

Jaye, M. C., Godehaux, W., and Lucas-Lenard, J., 1982, Further studies on the inhibition of cellular protein synthesis by vesicular stomatitis virus, *Virology* **116**:148.

Johnson, L. D., Binder, M., and Lazzarini, R. A., 1979, A defective interfering vesicular stomatitis virus particle that directs synthesis of functional proteins in the absence of helper virus, *Virology* **99**:203.

Kaempfer, R., 1984, Regulation of eukaryotic translation, in: *Comprehensive Virology,* Vol. 19 (H. Fraenkel-Conrat and R. R. Wagner, eds.), pp. 99–175, Plenum Press, New York.

Kang, C. Y., and Prevec, L., 1970, Proteins of vesicular stomatitis virus. II. Immunological comparisons of viral antigens. *J. Virol.* **6**:20.

Kelley, J. M., Emerson, S. U., and Wagner, R. R., 1972, The glycoprotein of vesicular stomatitis virus is the antigen that gives rise to and reacts with neutralizing antibody, *J. Virol.* **10**:1231.

Kingsford, L., and Emerson, S. U., 1980, Transcriptional activity of different phosphorylated species of NS protein purified from vesicular stomatitis virus and cytoplasm of infected cells, *J. Virol.* **33**:1097.

Kiuchi, A., and Roy, P., 1984, Comparison of the primary structure of spring viremia of carp virus M protein with that of vesicular stomatitis virus, *Virology* **134**:238.

Knipe, D. M., Baltimore, D., and Lodish, H. F., 1977, Separate pathways of maturation of the major structural proteins of vesicular stomatitis virus, *J. Virol.* **21**:1128.

Kotwal, G., Capone, J., Irving, R., Rhee, S., Bilan, P., Toneguzzo, F., Hofmann, T., and Ghosh, H., 1983, Viral membrane glycoproteins: Comparison of the amino-terminal amino acid sequences of the precursor and mature glycoproteins of three serotypes of vesicular stomatitis virus, *Virology* **129**:1.

Kucera, P., Dolivo, M., Coulon, P., and Flamand, A., 1985, Pathways of the early propagation of virulent and avirulent rabies strains from the eye to the brain, *J. Virol.* **55**:158.

Kurilla, M. G., and Keene, J. D., 1984, The leader RNA of vesicular stomatitis virus is bound by a cellular protein reactive with anti-La lupus antibody, *Cell* **34**:837.

Kurilla, M. G., Piwnica-Worms, H., and Keene, J. D., 1982, Rapid and transient localization of the leader RNA of vesicular stomatitis virus in the nuclei of infected cells, *Proc. Natl. Acad. Sci. U.S.A.* **79**:5240.

Kuwert, E., Wiktor, T. J., Sokol, F., and Koprowski, H., 1968, Hemagglutination by rabies virus, *J. Virol.* **2**:1381.

Lazzarini, R. A., Keene, J. D., and Schubert, M., 1981, The origins of defective-interfering particles of the negative-strand RNA viruses, *Cell* **26**:145.

LeFrancois, L., 1984, Protection against lethal viral infection by neutralizing and nonneutralizing monoclonal antibodies: Distinct mechanisms of action *in vivo, J. Virol.* **51**:208.

LeFrancois, L., and Lyles, D. S., 1982a, The interaction of antibody with the major surface glycoprotein of vesicular stomatitis virus. I. Analysis of neutralizing epitopes with monoclonal antibodies, *Virology* **121**:157.

LeFrancois, L., and Lyles, D. S., 1982b, The interaction of antibody with the major surface glycoprotein of vesicular stomatitis virus. II. Monoclonal antibody to nonneutralizing and cross-reactive epitopes of Indiana and New Jersey serotypes, *Virology* **121**:168.

LeFrancois, L., and Lyles, D.S., 1983a, Antigenic determinants of vesicular stomatitis virus: Analysis with antigenic variants, *J. Immunol.* **130**:394.

LeFrancois, L., and Lyles, D.S., 1983b, Cytotoxic T lymphocytes reactive with vesicular stomatitis virus: Analysis of specificity with monoclonal antibodies directed to the viral glycoprotein, *J. Immunol.* **130**:1408.

Lenard, J., and Miller, D. K., 1982, Uncoating of enveloped viruses, *Cell* **28**:5.

Lentz, T. L., Burrage, T. G., Smith, A. L., Crick, J., and Tignor, G. H., 1982, Is the acetylcholine receptor a rabies virus receptor?, *Science* **215**:182.

Lentz, T. L., Burrage, T. G., Smith, A. L., and Tignor, G. H., 1983, The acetylcholine receptor as a cellular receptor for rabies virus, *Yale J. Biol. Med.* **56**:315.

Levanon, A., Inbar, M., and Kohn, A., 1979, Fluorescence polarization of DPH-labeled cells adsorbing viruses and its diagnostic potential, *Arch. Virol.* **59**:223.

Little, L. M., Lanman, G., and Huang, A. S., 1983, Immunoprecipitating human antigens associated with vesicular stomatitis virus grown in HeLa cells, *Virology* **129**:127.

Lodish, H. F., and Froshauer, S., 1977, Relative rates of initiation of translation of different vesicular stomatitis messenger RNAs, *J. Biol. Chem.* **252**:8804.

Lodish, H. F., and Porter, M., 1980, Translational control of protein synthesis after infection by vesicular stomatitis virus, *J. Virol.* **36**:719.

Lodish, H. F., and Porter, M., 1981, Vesicular stomatitis virus mRNA and inhibition of translation of cellular mRNA—Is there a P function in vesicular stomatitis virus?, *J. Virol.* **16**:1351.

Loh, D., Ross, A. H., Hale, A. H., Baltimore, D., and Eisen, H. N., 1979, Synthetic phospholipid vesicles containing a purified viral antigen and cell membrane proteins stimulate the development of cytotoxic T lymphocytes, *J. Exp. Med.* **150**:1067.

Manen, K., Ohuchi, M., and Mifune, F., 1982, pH-Dependent hemolysis and cell fusion of rhabdoviruses, *Microbiol. Immunol.* **26**:1035.

Manley, J. L., Fire, A., Cano, A., Sharp, P. A., and Gefter, M. L., 1980, DNA-dependent transcription of adenovirus genes in a soluble whole cell extract, *Proc. Natl. Acad. Sci. U.S.A.* **77**:3855.

Marcus, P. I., and Sekellick, M. J., 1974, Cell-killing by viruses. I. Comparison of cell-killing, plaque-forming and defective-interfering particles of vesicular stomatitis virus, *Virology* **57**:321.

Marcus, P. I., and Sekellick, M. J., 1975, Cell killing by viruses. II. Cell killing by vesicular stomatitis virus: A requirement for virion-derived transcription, *Virology* **63**:176.

Marcus, P. I., and Sekellick, M. J., 1976, Cell killing by vesicular stomatitis viruses. III. The interferon system and inhibition of cell killing by vesicular stomatitis virus, *Virology* **63**:378.

Marcus, P. I., Sekellick, M. J., Johnson, L. D., and Lazzarini, R. A., 1977, Cell killing by
    viruses. V. Transcribing defective-interfering particles of vesicular stomatitis virus func-
    tion as cell-killing particles. *Virology* **82**:242.

Marsh, M., Helenius, A., Matlin, K., and Simons, K., 1983, Binding, endocytosis, and deg-
    radation of enveloped animal viruses, *Methods Enzymol.* **98**:260.

Martinet, C., Combard, A., Printz-Ané, C., and Printz, P., 1979, Envelope proteins and
    replication of vesicular stomatitis virus: *In vivo* effects of RNA$^+$ temperature-sensitive
    mutations on viral RNA synthesis, *J. Virol.* **29**:123.

Marvaldi, J., Lucas-Lenard, J., Sekellick, M. J., and Marcus, P. I., 1977, Cell killing by viruses.
    IV. Cell killing and the inhibition of cell protein synthesis requires the same gene
    functions of vesicular stomatitis virus, *Virology* **79**:267.

Marvaldi, J., Sekellick, M. J., Marcus, P. I., and Lucas-Lenard, J., 1978, Inhibition of mouse
    cell protein synthesis by ultraviolet-irradiated vesicular stomatitis virus requires viral
    transcription, *Virology* **84**:127.

Matlin, K. S., Reggio, H., Helenius, A., and Simons, K., 1982a, Pathway of vesicular sto-
    matitis virus entry leading to infection, *J. Mol. Biol.* **156**:609.

Matlin, K. S., Reggio, H., Helenius, A., and Simons, K., 1982b, The entry of enveloped
    viruses into an epithelial cell line, *Prog. Clin. Biol. Res.* **91**:599.

McAllister, P. E., 1979, Fish viruses and viral infections, in: *Comprehensive Virology*, Vol.
    14 (H. Fraenkel-Conrat and R. R. Wagner, eds.), pp. 401–470, Plenum Press, New York.

McAllister, P. E., and Wagner, R. R., 1975, Structural proteins of two salmonid rhabdovi-
    ruses, *J. Virol.* **15**:733.

McAllister, P. E., and Wagner, R. R., 1976, Differential inhibition of host protein synthesis
    in cells infected with RNA$^-$ temperature-sensitive mutants of vesicular stomatitis virus,
    J. Virol. **18**:550.

McAllister, P. E., and Wagner, R. R., 1977, Virion RNA polymerases of two salmonid
    rhabdoviruses, *J. Virol.* **22**:839.

McCoombs, K., Mann, E., Edwards, J., and Brown, D. T., 1981, Effects of chloroquine and
    cytochalasin B on the infection of cells by Sindbis virus and vesicular stomatitis virus,
    J. Virol. **37**:1060.

McGowan, J. J., and Wagner, R. R., 1981, Inhibition of cellular DNA synthesis by vesicular
    stomatitis virus, *J. Virol.* **38**:356.

McGowan, J. J., Emerson, S. U., and Wagner, R. R., 1982, The plus-strand leader RNA of
    VSV inhibits DNA-dependent transcription of adenovirus and SV40 genes in a soluble
    whole cell extract, *Cell* **28**:325.

McSharry, J. J., and Choppin, P. W., 1978, Biological properties of the VSV glycoprotein. I.
    Effects of the isolated glycoprotein on host macromolecular synthesis, *Virology* **84**:172.

McSharry, J. J., and Wagner, R. R., 1971, Lipid composition of purified vesicular stomatitis
    virus, *J. Virol.* **7**:59.

McSharry, J. J., Ledda, C. A., Freiman, H., and Choppin, P. W., 1978, Biological properties
    of VSV glycoprotein. II. Effects of the host cell and of the glycoprotein carbohydrate
    composition on hemagglutination, *Virology* **84**:183.

Meier, E., Harmison, G., Keene, J. D., and Schubert, M., 1984, Sites of copy choice replication
    involved in generation of vesicular-stomatitis virus defective-interfering particle RNAs,
    J. Virol. **51**:515.

Mellon, M. G., and Emerson, S. U., 1978, Rebinding of transcriptase components (L and NS
    proteins) to the nucleocapsid template of vesicular stomatitis virus, *J. Virol.* **27**:560.

Mifune, K., Ohuchi, M., and Manen, K., 1982, Hemolysis and cell fusion by rhabdoviruses,
    FEBS Lett. **137**:293.

Miller, D. K., and Lenard, J., 1980, Inhibition of vesicular stomatitis virus infection by spike
    glycoprotein: Evidence for an intracellular G protein-requiring step, *J. Cell Biol.* **84**:430.

Miller, D. K., and Lenard, J., 1981, Antihistaminics, local anesthetics, and other amines as
    antiviral agents, *Proc. Natl. Acad. Sci. U.S.A.* **78**:3605.

Moore, N. F., Patzer, E. J., Shaw, J. M., Thompson, T. E., and Wagner, R. R., 1978, Interaction
    of vesicular stomatitis virus with lipid vesicles: Depletion of cholesterol and effect on
    membrane fluidity and infectivity, *J. Virol.* **27**:320.

Morrison, T. G., and Lodish, H. F., 1975, Sites of synthesis of membrane and non-membrane proteins of vesicular stomatitis virus, *J. Biol. Chem.* **250**:6955.

Morrison, T. G., and McQuain, C. D., 1978, Assembly of viral membranes: Nature of the association of vesicular stomatitis virus proteins to membranes, *J. Virol.* **26**:115.

Mudd, J. A., and Summers, D. F., 1970, Protein synthesis in vesicular stomatitis virus-infected HeLa cells, *Virology* **42**:928.

Munzel, P., and Koschel, K., 1981, Rabies virus decreases agonist binding to opiate receptors of mouse neuroblastoma–rat glioma hybrid cells 108-CC-15, *Biochem. Biophys. Res. Commun.* **101**:1241.

Murphy, F. A., 1977, Rabies pathogenesis, *Arch. Virol.* **54**:279.

Newcomb, W. W., and Brown, J. C., 1981, Role of vesicular stomatitis virus matrix protein in maintaining the viral nucleocapsid in the condensed form found in virions, *J. Virol.* **39**:295.

Newcomb, W. W., Tubin, G. J., McGowan, J. J., and Brown, J. C., 1982, *In vitro* assembly of vesicular stomatitis virus skeletons, *J. Virol.* **41**:1055.

Nishioka, Y., and Silverstein, S., 1978, Alteration in the protein synthesis apparatus of Friend erythroleukemia cells infected with vesicular stomatitis virus or herpes simplex virus, *J. Virol.* **25**:422.

Nowakowski, M., Bloom, B. R., Ehrenfeld, E., and Summers, D. F., 1973, Restricted replication of vesicular stomatitis virus in human lymphoblastoid cells, *J. Virol.* **12**:1272.

Nuss, D. L., and Koch, G., 1976, Translation of individual host cell mRNAs in MPC-11 cells is differentially suppressed by vesicular stomatitis virus, *J. Virol.* **19**:572.

Nuss, D. L., Opperman, H., and Koch, G., 1975, Selective blockage of initiation of host protein synthesis by RNA virus-infected cells, *Proc. Natl. Acad. Sci. U.S.A.* **72**:1258.

Oldstone, M. B. A., Tishon, A., Dutko, F. J., Kennedy, S. I. T., Holland, J. J., and Lampert, P. W., 1980, Does the major histocompatibility complex serve as a specific receptor for Semliki Forest virus?, *J. Virol.* **34**:256.

Orenstein, J., Johnson, L., Shelton, E., and Lazzarini, R. A., 1976, The shape of vesicular stomatitis virus, *Virology* **71**:291.

O'Rourke, E. J., Guo, W. H. Y., and Huang, A. S., 1983, Antibody-induced modulation of protein in vesicular stomatitis virus-infected fibroblasts, *Mol. Cell Biol.* **3**:1580.

Otto, M. J., and Lucas-Lenard, J., 1980, The influence of the host cell on the inhibition of virus protein synthesis in cells infected with vesicular stomatitis virus and mengovirus, *J. Gen. Virol.* **50**:29.

Pal, R., Grinnell, B. W., Snyder, R. M., Wiener, J. R., Volk, W. A., and Wagner, R. R., 1985a, Monoclonal antibodies to the M protein of vesicular stomatitis virus (Indiana serotype) and to a cDNA M-gene expression product, *J. Virol.* **55**:298.

Pal, R., Grinnell, B. W., Synder, R. M., and Wagner, R. R., 1985b, Regulation of viral transcription by the matrix protein of vesicular stomatitis virus probed by monoclonal antibodies and temperature-sensitive mutants, *J. Virol.* **56**:386.

Patton, J. T., Davis, N. L., and Wertz, G. W., 1983, Cell-free synthesis and assembly of vesicular stomatitis virus nucleocapsids, *J. Virol.* **45**:155.

Patzer, E. J., Wagner, R. R., and Dubovi, E. J., 1979, Viral membranes: Model systems for studying biological membranes, *CRC Crit. Rev. Biochem.* **6**:165.

Petri, W. A., Jr., and Wagner, R. R., 1979, Reconstitution into liposomes of the glycoprotein of vesicular stomatitis virus by detergent dialysis, *J. Biol. Chem.* **254**:4313.

Petri, W. A., Jr., and Wagner, R. R., 1980, Glycoprotein micelles isolated from vesicular stomatitis virus spontaneously partition into sonicated phosphatidylcholine vesicles, *Virology* **107**:543.

Poirot, M. K., Schnitzlein, W. M., and Reichmann, M. E., 1985, The requirement of protein synthesis for VSV inhibition of host cell RNA synthesis, *Virology* **140**:91.

Portner, A., Webster, R. G., and Bean, W. J., 1980, Similar frequency of antigenic variants in Sendai, vesicular stomatitis and influenza viruses, *Virology* **104**:235.

Printz, P., 1973, Relationship of Sigma virus to vesicular stomatitis virus, *Adv. Virus Res.* **18**:143.

Rabinowitz, S. G., Dal Canto, M. C., and Johnson, T. C., 1976, Comparison of central nervous system disease produced by wild-type and temperature-sensitive mutants of vesicular stomatitis virus, *Infect. Immun.* **13**:1242.

Reichmann, M. E., Schnitzlein, W. M., Bishop, D. H. L., Lazzarini, R. A., Beatrice, S. T., and Wagner, R. R., 1978, Classification of the New Jersey serotype of vesicular stomatitis virus subtypes, *J. Virol.* **25**:446.

Remenick, J., and McGowan, J. J., 1986, A small RNA transcript of vesicular stomatitis virus inhibits the initiation of adenovirus replication *in vitro, J. Virol.* **59**:660.

Rhodes, D. P., Moyer, S. A., and Banerjee, A. K., 1974, *In vitro* synthesis of methylated messenger RNA by the virion-associated RNA polymerase of vesicular stomatitis virus, *Cell* **3**:169.

Riedel, H., Kondor-Koch, C., and Garoff, H., 1984, Cell surface expression of fusogenic vesicular stomatitis virus G protein from cloned DNA, *Eur. Mol. Biol. Org. J.* **3**:1477.

Robertson, B. H., and Wagner, R. R., 1981, Host range variation in response to vesicular stomatitis virus inhibition of RNA synthesis, in: *The Replication of Negative Strand Viruses* (D. H. L. Bishop and R. W. Compans, eds.), pp. 955–963, Elsevier/North-Holland, Amsterdam.

Rose, J. K., 1980, Complete intergenic and flanking sequences from the genome of vesicular stomatitis virus, *Cell* **19**:415.

Rose, J. K., and Gallione, C. J., 1981, Nucleotide sequences of the mRNAs encoding the vesicular stomatitis virus G and M proteins determined from cDNA clones containing the complete coding regions, *J. Virol.* **39**:519.

Rose, J. K., and Lodish, H. F., 1976, Translation *in vitro* of vesicular stomatitis virus mRNA lacking 5′-terminal 7-methylguanosine, *Nature (London)* **262**:32.

Rosenthal, K. L., and Zinkernagel, R. M., 1980, Cross-reactive cytotoxic T cells to serologically distinct vesicular stomatitis virus, *J. Immunol.* **124**:2301.

Rosenthal, K. A., Oldstone, M. B. A., and Zinkernagel, R. M., 1981, Long term cytotoxic T cell cultures are cross-reactive for distinct serotypes of vesicular stomatitis virus, *Fed. Proc. Fed. Am. Soc. Exp. Biol.* **40**:956.

Rothman, J. E., Bursztyn-Pettegrew, H., and Fine, R. E., 1980, Transport of the membrane glycoprotein of VSV to the cell surface in two stages, *J. Cell Biol.* **86**:162.

Rubio, C., Kolakofsky, D., Hill, V. M., and Summers, D. F., 1980, Replication and assembly of VSV nucleocapsids: Proteins associated with RNPs and effects of cycloheximide on replication, *Virology* **105**:123.

Sanchez, A., De, B. P., and Banerjee, A. K., 1985, *In vitro* phosphorylation of NS protein by the L protein of vesicular stomatitis virus, *J. Gen. Virol.* **66**:1025.

Scheid, A., and Choppin, P. W., 1977, Two disulfide-linked polypeptide chains constitute the active F protein of paramyxoviruses, *Virology* **80**:54.

Schlegel, R., and Wade, M., 1983, Neutralized vesicular stomatitis virus binds to host cells by a different "receptor," *Biochem. Biophys. Res. Commun.* **114**:774.

Schlegel, R., and Wade, M., 1984, A synthetic peptide corresponding to the NH_2 terminus of vesicular stomatitis virus glycoprotein is a pH-dependent hemolysin, *J. Biol. Chem.* **259**:4691.

Schlegel, R., and Wade, M., 1985, Biologically active peptides of the vesicular stomatitis virus glycoprotein, *J. Virol.* **53**:319.

Schlegel, R., Dickson, R., Willingham, M., and Pastan, I., 1982a, Amantadine and dansyl-cadaverine inhibit vesicular stomatitis virus uptake and receptor-mediated endocytosis of α_2-macroglobulin, *Proc. Natl. Acad. Sci. U.S.A.* **79**:2291.

Schlegel, R., Willingham, C., and Pastan, I. H., 1982b, Saturable binding sites for vesicular stomatitis virus on the surface of Vero cells, *J. Virol.* **43**:871.

Schlegel, R., Tralka, T. S., Willingham, M. C., and Pastan, I., 1983, Inhibition of VSV binding and infectivity by phosphatidylserine: Is phosphatidylserine a VSV binding site?, *Cell* **32**:639.

Schloemer, R. H., and Wagner, R. R., 1975a, Sialoglycoprotein of vesicular stomatitis virus: Role of the neuraminic acid in infection, *J. Virol.* **14**:270.

Schloemer, R. H., and Wagner, R. R., 1975b, Cellular adsorption function of the sialogly-coprotein of vesicular stomatitis virus and its neuraminic acid, *J. Virol.* **15**:882.

Schnitzlein, W. M., and Reichmann, M. E., 1985, Characterization of New Jersey vesicular stomatitis virus isolates from horses and black flies during the 1982 outbreak in Colorado, *Virology* **142**:426.

Schnitzlein, W. M., O'Banion, M. K., Poirot, M. K., and Reichmann, M. E., 1983, Effect of intracellular vesicular stomatitis virus mRNA concentration on the inhibition of host cell protein synthesis, *J. Virol.* **45**:206.

Schubert, M., Keene, J. D., Lazzarini, R. A., and Emerson, S. U., 1978, The complete sequence of a unique RNA species synthesized by a DI particle of VSV, *Cell* **15**:103.

Schubert, M., Harmison, G. G., Sprague, J., Condra, C., and Lazzarini, R. A., 1982, In vitro transcription of vesicular stomatitis virus: Interaction with GTP at a specific site within the N cistron, *J. Virol.* **43**:166.

Schubert, M., Harmison, G., and Meier, E., 1984, Primary structure of the vesicular stomatitis virus polymerase (L) gene: Evidence for a high frequency of mutations, *J. Virol.* **51**:505.

Seif, G., Coulon, P., Rollin, P. E., and Flamand, A., 1985, Rabies virulence: Effect on pathogenicity and sequence characterization of rabies virus mutations affecting antigenic site III of the glycoproteins, *J. Virol.* **53**:926.

Sesanti, L., Grassi, M., Mastromarino, P., Pana, A., Superti, F., and Orsi, N., 1983, Activity of human serum lipoproteins on the infectivity of rhabdoviruses, *Microbiologica* **6**:91.

Sethi, K. K., and Brandis, H., 1980, The role of vesicular stomatitis virus major glycoprotein in determining the specificity of virus-specific and H-2 restricted cytolytic T cells, *Eur. J. Immunol.* **10**:268.

Shimizu, Y., Yamamoto, S., Hana, M., and Ishida, Z. N., 1972, Effect of chloroquine on the growth of animal viruses, *Arch. Gesamte Virusforsch.* **36**:93.

Simpson, R. W., Hauser, R. E., and Dales, S., 1969, Viropexis of vesicular stomatitis virus by L cells, *Virology* **37**:285.

Stanners, C. P., Francoeur, A. M., and Lam, T., 1977, Analysis of VSV mutant with attenuated cytopathogenicity in viral function P for inhibition of protein synthesis, *Cell* **11**:273.

Storey, D. G., and Kang, C. Y., 1985, Vesicular stomatitis virus-infected cells fuse when the intracellular pool of functional M protein is reduced in the presence of G protein, *J. Virol.* **53**:374.

Superti, J., Seganti, L., Tsiang, H., and Orsi, N., 1984a, Role of phospholipids in rhabdovirus attachment to CER cells: Brief report, *Arch. Virol.* **81**:321.

Superti, F., Derer, M., and Tsiang, H., 1984b, Mechanism of rabies virus entry into CER cells, *J. Gen. Virol.* **65**:781.

Szilagyi, J. F., and Uryvayev, L., 1973, Isolation of an infectious ribonucleoprotein from vesicular stomatitis virus containing an RNA transcriptase, *J. Virol.* **11**:279.

Takehara, M., 1979, Effect of concanavalin A on viral infectivity, maturation, and cytopathogenicity in vesicular stomatitis virus-infected cells, *Kobe Med. Sci.* **25**:205.

Testa, D., Chanda, P. K., and Banerjee, A. K., 1980, Unique model of transcription in vitro by vesicular stomatitis virus, *Cell* **21**:267.

Thimmig, R. L., Hughes, J. V., Kinders, R. J., Milenkovic, A. G., and Johnson, T. C., 1980, Isolation of the glycoprotein of vesicular stomatitis virus and its binding to cell surfaces, *J. Gen. Virol.* **50**:279.

Thomas, D., Newcomb, W. W., Brown, J. C., Wall, J. S., Hainfeld, J. F., Trus, B. L., and Steven, A. C., 1985, Mass and molecular composition of vesicular stomatitis virus: A scanning transmission electron microscopy analysis, *J. Virol.* **54**:598.

Thomas, J. R., and Wagner, R. R., 1982, Evidence that vesicular stomatitis virus produces double-stranded RNA that inhibits protein synthesis in a reticulocyte lysate, *J. Virol.* **44**:189.

Thomas, J. R., and Wagner, R. R., 1983, Inhibition of translation in lysates of mouse L cells infected with vesicular stomatitis virus: Presence of a defective ribosome-associated factor, *Biochemistry* **22**:1540.

Tsiang, H., and Superti, F., 1984, Ammonium chloride and chloroquine inhibit rabies virus infection in neuroblastoma cells, *Arch. Virol.* **81**:377.

Vandepol, S. B., LeFrancois, L., and Holland, J. J., 1986, Sequences of the major antibody binding epitopes of the Indiana serotype of vesicular stomatitis virus, *Virology* **148**:312.

Volk, W. A., Snyder, R. M., Benjamin, D. C., and Wagner, R. R., 1982, Monoclonal antibodies to the glycoprotein of vesicular stomatitis virus: Comparative neutralizing activity, *J. Virol.* **42**:220.

Wagner, R. R., 1974, Pathogenicity and immunogenicity for mice of temperature-sensitive mutants of vesicular stomatitis virus, *Infect. Immunol.* **10**:309.

Wagner, R. R., 1975, Reproduction of rhabdoviruses, in: *Comprehensive Virology*, Vol. 4 (H. Fraenkel-Conrat and R. R. Wagner, eds.), pp. 1–94, Plenum Press, New York.

Wagner, R. R., 1984, Cytopathic effects of viruses: A general review, in: *Comprehensive Virology*, Vol. 19 (H. Fraenkel-Conrat and R. R. Wagner, eds.), pp. 1–63, Plenum Press, New York.

Wagner, R. R., and Huang, A. S., 1966, Inhibition of RNA and interferon synthesis in Krebs-2 cells infected with vesicular stomatitis virus, *Virology* **28**:1.

Wagner, R. R., Levy, A. H., Synder, R. M., Ratcliff, G. A., and Hyatt, D. F., 1963, Biologic properties of two plaque variants of vesicular stomatitis virus (Indiana serotype), *J. Immunol.* **91**:112.

Wagner, R. R., Schnaitman, T. C., and Synder, R. M., 1969, Structural proteins of vesicular stomatitis virus, *J. Virol.* **3**:395.

Wagner, R. R., Snyder, R. M., and Yamazaki, S., 1970, Proteins of vesicular stomatitis virus: Kinetics and cellular sites of synthesis, *J. Virol.* **5**:548.

Wagner, R. R., Heine, J. W., Goldstein, G., and Schraitman, C. A., 1971, Use of anti-viral–antiferritin hybrid antibody for localization of viral antigen in plasma membrane, *J. Virol.* **7**:274.

Wagner, R. R., Thomas, J. R., and McGowan, J. J., 1984, Rhabdovirus cytopathology: Effects on cellular macromolecular synthesis, in: *Comprehensive Virology*, Vol. 19 (H. Fraenkel-Conrat and R. R. Wagner, eds.), pp. 223–295, Plenum Press, New York.

Watson, H. D., Tignor, G. H., and Smith, A. L., 1981, Entry of rabies virus into the peripheral nerves of mice, *J. Gen. Virol.* **56**:372.

Weck, P. K., and Wagner, R. R., 1978, Inhibition of RNA synthesis in mouse myeloma cells infected with vesicular stomatitis virus, *J. Virol.* **25**:770.

Weck, P. K., and Wagner, R. R., 1979a, Transcription of vesicular stomatitis virus is required to shut off cellular RNA synthesis, *J. Virol.* **30**:410.

Weck, P. K., and Wagner, R. R., 1979b, Vesicular stomatitis virus infection reduces the number of active DNA-dependent RNA polymerases in myeloma cells, *J. Biol. Chem.* **254**:5430.

Weck, P. K., Carroll, A. R., Shattuck, D. M., and Wagner, R. R., 1979, Use of UV irradiation to identify the genetic information of vesicular stomatitis virus responsible for shutting off cellular RNA synthesis, *J. Virol.* **30**:746.

Wertz, G. W., 1978, Isolation of possible replicative intermediate structures from vesicular stomatitis virus infected cells, *Virology* **85**:271.

Wertz, G. W., 1983, Replication of vesicular stomatitis virus defective-interfering particle RNA *in vitro*: Transition from synthesis of defective-interfering leader RNA to synthesis of full length defective-interfering RNA, *J. Virol.* **46**:513.

Wertz, G. W., and Levine, M., 1973, RNA synthesis by vesicular stomatitis virus and a small plaque mutant: Effect of cycloheximide, *J. Virol.* **12**:253.

Wertz, G. W., and Youngner, J. S., 1970, Interferon production and inhibition of host synthesis in cells infected with vesicular stomatitis virus, *J. Virol.* **6**:476.

Wertz, G. W., and Youngner, J. S., 1972, Inhibition of protein synthesis in L cells infected with vesicular stomatitis virus, *J. Virol.* **9**:85.

White, J., Matlin, K., and Helenius, A., 1981, Cell fusion by Semliki Forest, influenza and vesicular stomatitis viruses, *J. Cell Biol.* **89**:674.

White, J., Kielian, M., and Helenius, A., 1983, Membrane fusion proteins of enveloped animal viruses, Q. Rev. Biophys. **16**:151.

Wiener, J. R., Pal, R., Barenholz, Y., and Wagner, R. R., 1983, Influence of the peripheral matrix protein of vesicular stomatitis virus on the membrane dynamics of mixed phospholipid vesicles: Fluorescence studies, Biochemistry **22**:2162.

Wiktor, T. J., and Koprowski, H., 1974, Rhabdovirus replication in enucleated host cells, J. Virol. **14**:300.

Wiktor, T. J., Gyorgy, E., Schlumberger, H. D., Sokol, F., and Koprowski, H., 1973, Antigenic properties of rabies virus components, J. Immunol. **110**:269.

Williams, P. M., Schubert, M., Herman, R., and Emerson, S. U., 1987, Production and analysis of monoclonal antibodies directed to the NS protein of vesicular stomatitis virus, J. Virol. (in press).

Wilson, T., and Lenard, J., 1981, Interaction of wild-type and mutant M protein of VSV with nucleocapsids in vitro. Biochemistry **20**:1349.

Wilusz, J., and Keene, J. G., 1984, Interactions of plus and minus strand leader RNAs of the New Jersey serotype of vesicular stomatitis virus with the cellular La protein, Virology **135**:65.

Wilusz, J., Kurilla, M. G., and Keene, J. D., 1984, La protein binds to a unique species of minus sense leader RNA during the replication of vesicular stomatitis virus, Proc. Natl. Acad. Sci. U.S.A. **80**:5827.

Woodgett, C., and Rose, J. K., 1986, Amino terminal mutation of the vesicular stomatitis glycoprotein does not affect its cell fusion activity, J. Virol. **59**:486.

Wu, F. S., and Lucas-Lenard, J. M., 1980, Inhibition of ribonucleic acid accumulation in mouse L cells infected with vesicular stomatitis virus requires viral ribonucleic acid transcription, Biochemistry **19**:804.

Wunner, W., Reagan, K., and Koprowski, H., 1984, Characterization of saturable binding sites for rabies virus, J. Virol. **50**:691.

Wyers, F., Richard-Molard, C., Blondel, D., and Dezeles, S., 1980, Vesicular stomatitis virus growth in Drosophila melanogaster cells: G protein deficiency, J. Virol. **33**:411.

Yamazaki, S., and Wagner, R. R., 1970, Action of interferon: Kinetics and differential effects on viral functions, J. Virol. **6**:421.

Yaoi, Y., and Amano, M., 1970, Inhibitory effect of ultraviolet-irradiated vesicular stomatitis virus on inhibition of DNA synthesis in cultured chick embryo cells, J. Gen. Virol. **9**:69.

Yaoi, Y., Mitsui, H., and Amano, M., 1970, Effect of UV-irradiated vesicular stomatitis virus on nucleic acid synthesis in chick embryo cells, J. Gen. Virol. **8**:165.

Ye, Z., Pal, R., Ogden, J. R., Synder, R. M., and Wagner, R. R., 1985, Monoclonal antibodies to the matrix protein of vesicular stomatitis virus (New Jersey serotype) and their effects on transcription, Virology **142**:657.

Young, J. D., Young, G. P., Cohn, Z. A., and Lenard, J., 1983, Interaction of enveloped viruses with planar bilayer membranes: Observations on Sendai, influenza, vesicular stomatitis and Semliki Forest viruses, Virology **128**:186.

Youngner, J. S., and Preble, O. T., 1980, Viral persistence: Evolution of viral populations, in: Comprehensive Virology, Vol. 16 (H. Fraenkel-Conrat and R. R. Wagner, eds.), pp. 73–135, Plenum Press, New York.

Zakowski, J. J., and Wagner, R. R., 1980, Localization of the membrane-associated proteins in vesicular stomatitis virus by use of hydrophobic membrane probes and cross-linking reagents, J. Virol. **36**:93.

Zakowski, J. J., Petri, W. A., Jr., and Wagner, R. R., 1981, Role of matrix protein in assembling the membrane of vesicular stomatitis virus: Reconstitution of matrix protein with negatively charged phospholipid vesicles, Biochemistry **20**:3902.

Zee, Y., Hackett, A. J., and Talens, L., 1970, Vesicular stomatitis virus maturation sites in six different host cells, J. Gen. Virol. **7**:95.

Zinkernagel, R. M., 1979, Cellular immune responses to viruses and the biological role of polymorphic major transplantation antigens, in: *Comprehensive Virology*, Vol. 15 (H. Fraenkel-Conrat and R. R. Wagner, eds.), pp. 171–204, Plenum Press, New York.

Zinkernagel, R. M., and Rosenthal, K. L., 1981, Experiments and speculation on antiviral specificity of T and B cells, in: *Immunological Reviews*, Vol. 8 (G. Moller, ed.), pp. 131–155, Munksgaard, Copenhagen.

Zinkernagel, R. M., Adler, B., and Holland, J. J., 1978, Cell-mediated immunity to vesicular stomatitis virus infection in mice, *Exp. Cell Biol.* **46:**53.

CHAPTER 3

Rhabdovirus Membrane and Maturation

RANAJIT PAL AND ROBERT R. WAGNER

I. INTRODUCTION

Rhabdoviruses, particularly vesicular stomatitis (VSV), have provided an incisive and widely used system for studying the far more complicated biological membranes of eukaryotic cells. In addition to its simplicity, the VSV membrane can be produced in large amounts and is readily purified to homogeneity free of contaminating cell membranes. Lipids of the VSV membrane are derived from the plasma membrane of the host cell and form a lipid bilayer that assumes many of the characteristics of a biological unit membrane. The rhabdovirus membrane contains only two proteins, one integral and one peripheral; these viral proteins have interesting properties and have been widely used by biochemists to study the synthesis and mode of action of integral and peripheral membrane proteins of cells. Furthermore, the synthesis and assembly of the VSV membrane proteins in infected host cells have shed considerable light on the intricacies of cellular membrane-protein biogenesis.

The membrane structure of enveloped viruses has attracted considerable attention over the years, and several excellent reviews have been written on this subject (Lenard and Compans, 1974; Rott and Klenk, 1977; Lenard, 1978; Compans and Klenk, 1979; Patzer *et al.*, 1979). These reviews have provided a unified concept of the virion membrane as a struc-

RANAJIT PAL AND ROBERT R. WAGNER • Department of Microbiology and Cancer Center, University of Virginia School of Medicine, Charlottesville, Virginia 22908.

tural and to some extent as a functional entity. In this review, we offer a comprehensive analysis of the molecular organization of the membrane of rhabdovirions. Attempts will be made to discuss the structural as well as the functional aspects of lipids and proteins in the viral membrane and the role played by these two components in determining the dynamics of the membrane matrix.

Before the individual components of the VSV membrane are discussed, it is important that its general structure and organization be considered. The morphology of this virion is discussed in Chapter 2. Figure 1 presents a schematic representation of the VSV membrane illustrating the location of the two membrane proteins in the envelope. Lipids in the membrane are arranged as a bilayer structure and consist primarily of phospholipids and cholesterol. The envelope of VSV contains two proteins. The glycoprotein (G) is an externally oriented transmembrane protein that completely spans the envelope by a sequence of 20 consecutive hydrophobic amino acids (Rose *et al.*, 1980; Rose and Gallione, 1981). In contrast, the viral matrix (M) protein is not exposed on the exterior surface of the intact virion, as determined by its being refractory to protease digestion and to lactoperoxidase iodination (Schloemer and Wagner, 1975; McSharry, 1977). Use of various monofunctional and bifunctional cross-linking reagents has demonstrated that the M protein lines the inner leaflet of the lipid bilayer in close proximity to the nucleocapsid (N) protein core (Zakowski and Wagner, 1980). The envelope of VSV was found to dissociate from the N protein core under conditions that disassemble the virus structure. Thus, disruption of VSV by a combination of freezing and thawing, osmotic shock, and sonic treatment led to the formation of lipid vesicles containing G proteins and the same

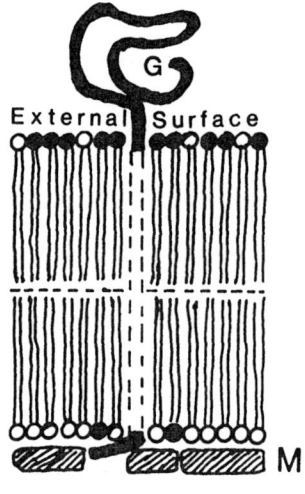

FIGURE 1. Schematic representation of VSV membrane and the position of G and M proteins in the bilayer. (●) Choline headgroups; (○) amino headgroups of the phospholipids; (---) 20 hydrophobic amino acids of the G protein that spans the lipid bilayer.

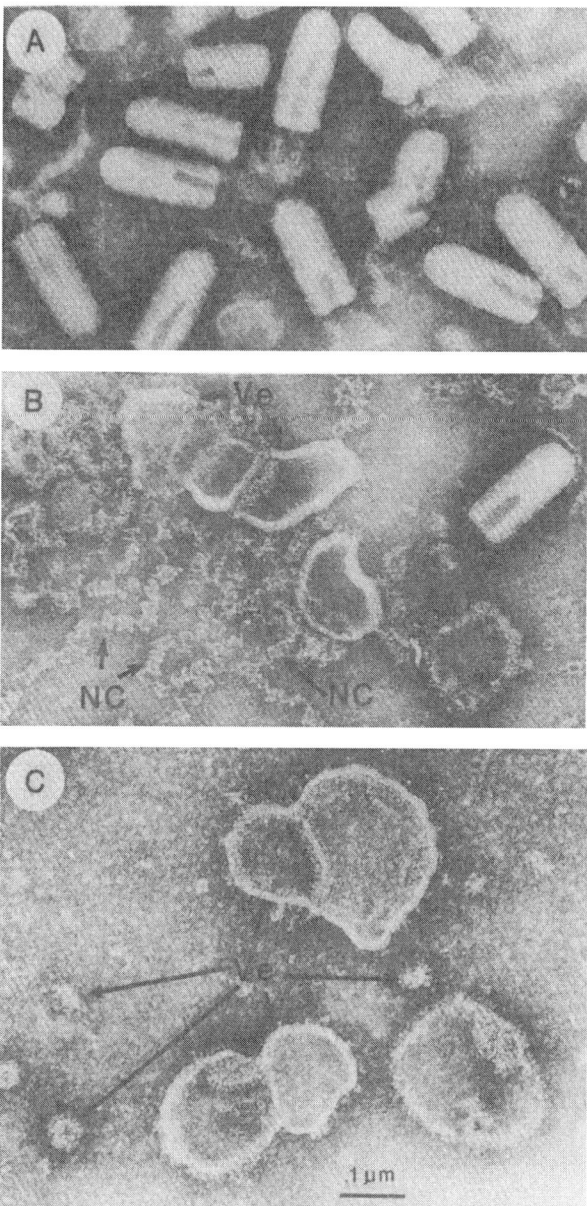

FIGURE 2. Electron microscopy by negative staining with phosphotungstic acid of intact VSV virions (A), VSV virions (50 μg) treated for 10 min with 0.2 ml pardaxin (20 μg) (B), and lipid vesicles generated form VSV virions (300 μg) treated with pardaxin (120 μg) (C) and floated to the top of a 10–50% sucrose gradient by density centrifugation. Arrows point to vesicles (Ve) with spikes and to nucleocapsids (NC) × 140,000. Reprinted from Pal et al. (1981c) by permission of the Journal of Biological Chemistry.

phospholipid composition as that in the intact virions (Taube and Roth-field, 1978). More recently, Pal *et al.* (1981b,c) have demonstrated that the interaction of VSV with Pardaxin, a hydrophobic toxin of the Red Sea flat fish, resulted in the dissociation of the virus membrane from the ribonucleocapsid protein (RNP) cores, which could be readily separated by density-gradient centrifugation. Electron micrographs of these lipo-protein vesicles show G-protein spikes in the same orientation as those in intact virions (Fig. 2).

The discussion of rhabdovirus membrane in this chapter is divided into five sections. Section II deals with the lipids present in viral membrane; Sections III and IV deal with the two membrane proteins (G and M) that are intimately associated with the lipid bilayer. Section V is concerned mainly with the dynamic properties of viral membranes and Section VI with how lipids and proteins are assembled to form the mature virion. Unless indicated otherwise, all properties described here refer to the membrane of VSV (Indiana serotype), since this virus serves as the prototypic model of all rhabdoviruses.

II. VIRION MEMBRANE LIPIDS

The approximate composition of the vesicular stomatitis virion is 74% protein, 20% lipid, 3% carbohydrate and 3% RNA. The VSV membrane contains approximately 50% lipid and 50% protein. The lipid composition of the VSV membrane has been analyzed in detail (McSharry and Wagner, 1971a; Patzer *et al.*, 1978a). Phosphatidylcholine (PC), phosphatidylethanolamine (PE), phosphatidylserine (PS), and sphingomyelin (SPM) are the major phospholipids present in the virion envelope. Other phospholipids, such as phosphatidylinositol (PI), phosphatidic acid (PA), and lysophosphatidylcholine (LPC), are present as minor components comprising only 3% of the total phospholipids. Among neutral lipids of the viral membrane, cholesterol represents the major component and can account for as much as 35–40% of total lipids in the envelope. Free fatty acids and glycerides are also detected in the VSV membrane, although they represent very minor constituents. In another member of the family Rhabdoviridae, the lipid composition of a highly purified Flury strain of rabies virus was found to contain PE, PC, and SPM as the major phospholipid components in the envelope, but the content of PS was found to be less than that observed in VSV (Blough *et al.*, 1977). Cholesterol was also present as the major neutral lipid component in rabies virus.

The origin of the lipids present in the VSV membrane has been determined by metabolic studies and electron microscopic observations. It has been observed that rhabdoviruses replicating in animal cells obtain their membranes by budding through the plasma membrane of the infected cells (Lenard and Compans, 1974). Metabolic labeling experiments have revealed that predominantly preformed cellular lipids are incorporated into viral membranes. In general, the lipid composition of viral membranes resembles quite closely that of the plasma membranes of infected host cells. However, the cholesterol and aminophospholipid concentrations were found to be higher in viral membrane than in cellular plasma membrane (Patzer *et al.*, 1978a). Thus, it has been demonstrated that the cholesterol/phospholipid ratio of the VSV membrane is in the range of 0.72, whereas that of the plasma membrane of the host cell is 0.54. Similarly, the ratio of aminophospholipid to total phospholipid content is 0.49 for VSV compared to 0.31 for the plasma membrane. Electron spin resonance and fluorescence spectroscopic studies have demonstrated that the membrane of VSV is more rigid than that of the host-cell plasma membrane from which it buds (Landsberger and Compans, 1976; Barenholz*et al.*, 1976; Patzer *et al.*, 1978a). The higher content of cholesterol in the membrane of VSV may well be responsible for such increased membrane rigidity compared to that of the host-cell plasma membrane.

A. Phospholipids

Phospholipids are the major lipid constituent of the VSV membrane. About 48% of total phospholipids in the envelope of VSV grown in baby hamster kidney (BHK)-21 cells are composed of PC and SPM, the two being present in almost equal proportions (Patzer *et al.*, 1978a). The envelope is rich in PE, which constitutes nearly 31% of the total lipid, while the PS content is almost 18%. A similar PE concentration had been noted in VSV grown in L-929 cells (McSharry and Wagner, 1971a), suggesting the possibility that VSV may preferentially select PE during budding from the plasma membrane of infected cells that contain much less PE.

1. Bilayer Distribution of Membrane Phospholipids

Phospholipids present in the membrane of VSV exhibit asymmetrical orientation in the bilayer. Figure 3 illustrates the distribution of phospholipid headgroups and fatty acyl chains in the two layers of the VSV membrane. The technique used to determine their location included reaction of the free amino group of PE with the membrane-impermeable reagent 2,4,6-trinitrobenzene sulfonic acid (TNBS), exposure to phospholipases (which hydrolyze various portions of the phospholipid molecule in the external half of the bilayer), and also treatment with phospholipid-exchange proteins (which catalyze the exchange of radiolabeled phospholipids from viral membrane into interacting lipid vesicles). Using phospholipase C exposure and TNBS labeling of VSV virions, Patzer *et al.* (1978a) demonstrated that approximately 94% PC and 80% SPM but only 38–47% PE reside in the external half of the VSV membrane, with the remainder of each phospholipid in the inner leaflet. Similar distribution of PE in the VSV bilayer had also been observed by Fong *et al.*

(1976), who employed TNBS to label the aminophospholipids in intact
or spikeless virions. The location of PC in the VSV membrane was also
studied by using a phospholipid-exchange protein; this experiment revealed that only 70% of PC was present in a rapidly exchangeable pool
in the VSV membrane, presumably due to its external orientation in the
bilayer (Shaw et al., 1979). This discrepancy between the accessibility of
PC to phospholipase C and to exchange protein might be due to a larger
pool available to phospholipase C but partially shielded from the exchange
protein by other membrane components (Patzer et al., 1979). In this context, it should be noted that membranes of other viruses also exhibit
phospholipid asymmetry in the bilayer. For example, the inner membrane
leaflet of influenza and Semliki Forest viruses, much like that of VSV, is
enriched in amino phospholipids, whereas the choline phospholipids exhibit no preferential location in the viral membranes (Tsai and Lenard,
1975; Rothman et al., 1976; Gahmberg et al., 1972; van Meer et al., 1981).

The fatty acyl chains of VSV phospholipids are also distributed asymmetrically in the bilayer (Fig. 3). A large proportion of saturated fatty acids
were found to be present in the external monolayer of the VSV membrane,
while virtually all the polyunsaturated fatty acids were located in the
inner monolayer (Patzer et al., 1978a). The enrichment of palmitic acid
in the external half of the bilayer is presumably due to the high content
of SPM in the external leaflet of the VSV membrane. Asymmetrical distribution of fatty acyl chains in PE was also determined by Fong and

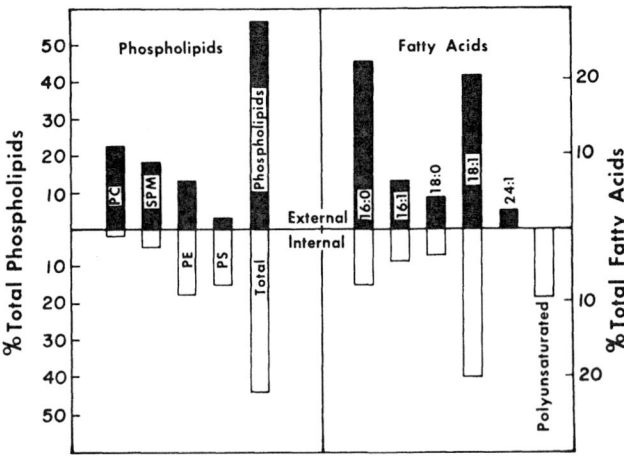

FIGURE 3. Putative location of phospholipids and fatty acids in the two halves of the
membrane bilayer of VSV. The phospholipids were assigned to external or internal orientation on the basis of their availability in intact virions to reaction with phospholipase C
and to labeling with TNBS in the case of PE and PS. The fatty acids of the hydrolyzed and
unhydrolyzed phospholipids were analyzed by gas–liquid chromatography and expressed as
percentages of the control fatty acids. Reprinted from Patzer et al. (1978a) by permission
of the Journal of Biological Chemistry.

Brown (1978). PE molecules present in the inner leaflet of the VSV envelope were also found to contain a significantly higher proportion of unsaturated fatty acyl chains. The functional significance, if any of phospholipid asymmetry of either headgroups or fatty acyl chains is unknown. As will be discussed later, however, the presence of amino phospholipids, particularly PS, in the inner layer of the viral membrane might be involved in the interaction of matrix (M) protein with the lipid bilayer. This interaction may play an important role in virus assembly at the plasma membrane of the host cell.

The asymmetrical distribution of phospholipids in the viral membrane also raises the question of transmembrane movement of lipids across the bilayer and the maintenance of lipid asymmetry under such conditions. By measuring the kinetics of phospholipid exchange from trypsinized VSV into lipid vesicles catalyzed by a phospholipid-exchange protein, Shaw *et al.* (1979) have demonstrated that the PC in VSV membranes has a half-time of 7–11 hr for transmembrane migration. Interestingly, the PC and SPM in influenza virions showed a half-time for transmembrane movement of more than 10 days and more than 30 days, respectively (Rothman *et al.*, 1976). This difference in the rate of transbilayer movement of phospholipids in these two enveloped virions is difficult to understand. It is quite possible that the nature of the viral proteins interacting with the lipid bilayer may determine the phospholipid asymmetry and transbilayer movement in the viral membranes.

2. Phospholipid Functions

The functions of phospholipids in viral membrane are not clearly understood. It has long been known that the infectivity of VSV is decreased markedly by solubilization of the lipid membrane by ethyl ether, detergents, or other lipid solvents. Two general functions of the lipid membrane seem clear: First, the lipid bilayer serves as a permeability barrier to protect the enclosed nucleocapsid; second, it serves as a base to anchor the viral glycoprotein (G), which plays a critical role in attachment of the virus to the host-cell plasma membrane. Lipid-dependent functions of virions have been determined by *in vivo* and *in vitro* modification of the phospholipids in the viral envelope; virions with different lipid compositions are then examined for a measureable change in physical and biological activities, such as membrane rigidity and infectivity.

Phospholipase C has been used to alter the headgroup composition of phospholipids in the VS viral membrane. Studies on incubation of enveloped virions with phospholipase C yielded conflicting data on loss of infectivity. Cartwright *et al.* (1969) reported a minimal loss of VS viral infectivity on removal of phospholipid headgroups by phospholipase C, whereas Simpson and Hauser (1966) claimed a 4–5 log drop in infectivity of myxoviruses due to phospholipase C treatment. More recently, Moore *et al.* (1977a) demonstrated that only a minor reduction in viral infectivity occurred when nearly 55% of phospholipids in the membrane were hydrolyzed to diglycerides by phospholipase C. The integrity of the viral membrane was not affected under such conditions, although the fluidity of the lipid envelope (as measured by fluorescence depolarization) was increased after treatment with phospholipase C. In sharp contrast, when the fatty acyl chains of VSV phospholipids were hydrolyzed by phospholipase A_2 to lyso derivatives, the infectivity of the virions decreased 100- to 1000-fold (Moore et al., 1977a). This decrease in infectivity might be due to the lytic action of lysophospholipids, formation of which could affect the integrity of the viral membrane. The effect of phospholipases on the stability of the envelope in other viral systems was also examined. Thus, Friedman and Pastan (1969) found that phospholipase C apparently affected the integrity of Semliki Forest virus membrane only when it was incubated with the enzyme for a long period of time. However, hydrolysis of nearly 60% of the virion phospholipids had no significant effect on the infectivity of Semliki Forest virus.

Phospholipid headgroups as well as fatty acid compositions of VS viral membrane have also been altered in vitro by growing the virus in host cells loaded with foreign lipid precursors. Fluorescent and photoreactive fatty acids were incorporated in vivo into the VSV envelope by infecting cells previously grown in the presence of these fatty acid precursors (Capone et al., 1983; Stoffel et al., 1978). These lipids with photosensitive groups as chemical probes have also been utilized for localization of the G and M protein in the membrane of VSV (Capone et al., 1983; Stoffel et al., 1978). Pal et al. (1980b) also altered the headgroups and fatty acid composition of VSV phospholipids by growing the virus in BHK-21 cells supplemented with certain fatty acids and dimethylethanolamine. When VS virions were harvested from infected BHK-21 cells fed the choline analogue dimethylethanolamine, 29% of the membrane phospholipids were present as phosphatidyldimethylethanolamine. More than 80% of this new phospholipid species was located in the external monolayer of the viral membrane, as determined by phospholipase C hydrolysis. Exogenous fatty acids added to the medium of BHK-21 cells infected with VS virions comprised up to 30% of the fatty acyl chains of the viral glycerophospholipids. The presence of phosphatidyldimethylethanolamine or unusual fatty acyl chains in the VSV membrane had no significant effect on viral infectivity. Controlled alteration of the headgroups of VS viral phospholipids was also achieved by Pessin and Glaser (1980), who harvested VSV from chick-embryo fibroblasts (CEF) fed with the choline analogue N,N-dimethylethanolamine or 1,2-amino-1-butanol. The phospholipid composition of the VSV membrane after such manipulation was found to be different from that of the plasma membrane of infected CEF, which led the authors to conclude that VSV buds from localized lipid regions of the plasma membrane that do not reflect the average properties of the plasma membrane. All these studies on the manipulation of the phospholipid composition of the VSV membrane, both in vivo and in *vitro*, demonstrate that the phospholipids as such do not play a significant role in viral infectivity or other properties.

B. Cholesterol

1. Bilayer Distribution of Cholesterol

The transbilayer distribution of cholesterol in the VSV envelope has been determined by measuring the spontaneous transfer of cholesterol between virions and interacting sonicated lipid vesicles without the aid of catalytic protein. This research provided a means for determining the availability of cholesterol for transfer from the external surface of the membrane. Patzer *et al.* (1978b) studied the distribution and transbilayer movement of cholesterol in the VSV membrane by following the kinetics of depletion and exchange of cholesterol from the intact virion envelope into lipid vesicles. These studies revealed the presence of two pools of cholesterol in the viral membrane, presumably representing the two halves of the bilayer. Cholesterol was found to be distributed asymmetrically in the membrane leaflets, with 70% of it present in the outer monolayer (erroneously assigned to the inner monolayer in the original report). Spike-less virions were used in these studies because lipid vesicles were found to attach irreversibly to G protein spikes. Similar spontaneous exchange of VSV-membrane cholesterol was also noted by Sefton and Gaffney (1979), who observed that symmetrically loaded viral cholesterol undergoes complete transfer both to lipid vesicles and to serum lipoproteins. More recently, Pal *et al.* (1981a) used serum lipoproteins enriched with phospholipids to deplete cholesterol from the VSV envelope. Unlike the reaction of virions with phospholipid vesicles, nonspecific adherence of lipoprotein and exogenous lipids to the envelope of the virions was found to be minimal. By also studying the exchange of cholesterol from the VSV membrane with polyvinylpyrrolidone (PVP), complexed with bovine serum albumin (BSA), PC, and cholesterol, it was concluded that cholesterol is distributed asymmetrically in the viral membrane, with nearly 75% of it present in the outer monolayer and 25% located in the inner monolayer. In this context, it should be noted that cholesterol in the membrane of influenza virus appears to be more symmetrically distributed between the two halves of the bilayer (Lenard and Rothman, 1976).

Transbilayer movement of cholesterol across the VSV membrane was also measured by following the kinetics of its transfer between intact VSV and lipid vesicles or serum lipoproteins. The transmembrane movement of cholesterol was found to be rapid, with a half-time for equilibration between the inner and outer monolayers of 4–6 hr (Patzer *et al.*, 1978b: Pal *et al.*, 1981a). Similar fast equilibration of cholesterol was also noted in the membrane of Sindbis virus (Sefton and Gaffney, 1979). In sharp contrast, the transmembrane movement of cholesterol in influenza virus was assigned a half-time of more than 13 days (Lenard and Rothman, 1976). Since similar techniques and incubation conditions were used to determine the asymmetry and transbilayer movement of cholesterol in VSV and influenza virions, the conflicting results on distribution of cholesterol and its membrane movement in the respective bilayers might be due to inherently different properties of each viral membrane system.

2. Contribution of Cholesterol to Viral Infectivity

Two experimental approaches have been used to alter cholesterol content in the VSV envelope and to determine what effect, if any, this alteration has on viral infectivity. In the first method, VSV virions were exposed to reagents that react specifically with sterols in the biological membrane; in the second method, the cholesterol content of the VSV membrane was either enriched or depleted both *in vivo* and *in vitro*. To investigate the potential biological function of cholesterol in the VSV membrane, infectivity of VSV virions after manipulation of cholesterol was measured by plaque assay.

The polyene antibiotic filipin, which has been shown to interact with sterol in phospholipid–sterol vesicles, was incorporated into the VSV membrane in the same molar ratio as viral cholesterol (Bittman *et al.*, 1976; Majuk *et al.*, 1977). The infectivity of VSV virions was reduced 500-fold following incorporation of filipin into the lipid bilayer. The sterol glycoside digitonin, which complexes with cholesterol, resulted in the disruption of viral membranes and released G and M protein into the medium (Wagner and Schnaitman, 1970). In another study, cholesterol in the membrane of phospholipase-C-treated VSV was oxidized by exposure of virions to cholesterol oxidase, resulting in a 4–5 log loss of viral infectivity (Moore *et al.*, 1977a). This loss of infectivity is complicated, however, by the formation of H_2O_2 as a by-product of the oxidation reaction; H_2O_2 added externally was found to cause a similar loss in viral infectivity, but only at a concentration 10-fold higher than that of the H_2O_2 generated internally during the enzymatic reaction.

The cholesterol concentration in the VSV membrane can be increased or decreased by incubating intact virions with lipid vesicles or with serum lipoproteins complexed with exogenous lipids. Pal *et al.* (1980a) observed that the membrane of VSV could be readily enriched with exogenous cholesterol by exposing virion suspensions to serum lipoproteins or to PVP–BSA complexed with cholesterol. Although these procedures increased the level of cholesterol in the VSV membrane from approximately 35 to approximately 60 mole%, no significant effect on the infectivity of these VS virions was observed after such enrichment. Nearly 80% of cholesterol (both endogenous and exogenous) in such cholesterol-enriched virions was susceptible to oxidation by the enzyme cholesterol oxidase, whereas in the normal virions, cholesterol in the viral envelope was not exposed to the enzyme. In fact, similar results were obtained with a model membrane system of mixed unilamellar vesicles, suggesting that at higher concentrations, cholesterol may form discrete aggregates or domains in the lipid bilayer that are readily accessible to oxidation by the enzyme cholesterol oxidase.

Depletion of cholesterol from the VSV membrane was achieved by interacting virions with cholesterol-free phospholipid vesicles (Moore *et al.*, 1978; Patzer *et al.*, 1978b). The depletion of cholesterol from the envelope of VSV was found to cause an approximate 1000-fold decrease in the infectivity of the virions. This loss in infectivity was complicated by finding lipid vesicles adhering to the G protein spikes of the virions, but adhering lipid vesicles, which did not deplete virion cholesterol, caused only a 10-fold loss in viral infectivity. More recently, Pal *et al.* (1981a) used serum lipoproteins or PVP beads complexed with BSA to deplete cholesterol from the membrane of VSV. These procedures altered the morphology of the virions, but not its membrane integrity; adherence of exogenous lipids to the G-protein spikes was found to be minimal under such conditions. Depletion of cholesterol from the virion membrane by serum lipoproteins or PVP resulted in a significant drop in viral infectivity; no effect on viral infectivity was observed when virion cholesterol was merely exchanged without any reduction in the concentration of viral membrane cholesterol. Part of the loss in infectivity following depletion of cholesterol could be restored by reincorporation of cholesterol in the depleted viral membrane. These experiments collectively indicate that cholesterol in the VS viral membrane may partly contribute to the infectivity of the virus.

The cholesterol content of the VSV envelope can also be altered *in vivo* by growing the virus in different cell lines fed various sterol precursors. A unique feature of sterol maturation in L cells was exploited by Bates and Rothblat (1972) to study the sterol requirements of VSV. In the absence of exogenously added cholesterol, L cells synthesized desmosterol, which is the precursor of cholesterol. When VSV was grown in L cells containing both cholesterol and desmosterol, the virus incorporated a higher percentage of cholesterol than was present in the whole cells. The presence of desmosterol or a desmosterol–cholesterol mixture in the VSV membrane did not affect the stability or infectivity of the virions. More recently, Pal *et al.* (1980b) manipulated the cholesterol content in the membrane of VSV by growing the virions in a sterol mutant of Chinese hamster ovary (CHO) cells. VSV released from the infected CHO MI sterol auxotroph cells grown in delipidated serum had a 50% lower cholesterol/phospholipid ratio and an 80% drop in infectivity (plaque formation) compared with VSV release from infected CHO MI cells grown in fetal calf serum. Although this drop in infectivity is not very large, it still tends to suggest that among all the viral membrane lipids, cholesterol may play some role in the infectivity of the virus.

C. Glycolipids

Glycolipids represent only a very minor constituent of the rhabdo-virus membrane, and little is known about their role in organization or function of the viral membrane. The glycolipid composition of enveloped viruses in general reflects the glycolipid composition of the cells in which the virions are grown. Hematoside (neuraminosyllactosylceramide) has been shown to be the only glycolipid present in the envelope of VSV grown in BHK-21 cells (Klenk and Choppin, 1971). The glycolipid content of rabies virus released from hamster fibroblasts (Nil-2 cells) grown either in monolayer or in suspension culture reflected the overall glycolipid content of the host cells. Neutral glycolipids represented close to 80% of all the glycolipids in rabies virus grown in monolayer cells and about 50% in virus grown in suspension cells; the rest of the glycolipids were shown to be gangliosides (Portoukalian et al., 1977).

The orientation of glycolipid in the viral membrane has been deter-mined by neuraminidase treatment of intact virions. In VSV, hematoside was found to be oriented exclusively in the outer leaflet of the lipid bilayer, as evidenced by its quantitative conversion to lactosylceramide after neuraminidase treatment (Stoffel et al., 1975; Stoffel and Sorgo, 1976). Cartwright and Brown (1972) demonstrated that VSV virions con-tain the glycolipids unique for the host cells in which they were grown, on the basis of the capacity of the virions to react by complement fixation with antibody to the host glycolipids. All these studies suggest that the glycolipids in the membrane of rhabdoviruses are localized predominantly on the external surface of the lipid bilayer and are a very minor component of the virion scavenged from the host cell.

III. RHABDOVIRUS GLYCOPROTEINS

Vesicular stomatitis virus (VSV) and probably all rhabdoviruses con-tain five structural proteins, of which the glycoprotein (G) and the matrix (M) protein are associated with the lipid bilayer, while the major nucleo-capsid (N) protein and the two minor nucleocapsid proteins, the non-structural (NS)—which is actually a misnomer [see Chapter 2 (Section II.B)]—and the large (L), are complexed with the negative-strand RNA to form the enzymatically active ribonucleoprotein core of the virion (Wag-ner, 1975). As described in Section II, the viral lipids are derived from the cellular plasma membrane, whereas the membrane proteins are coded by the viral genome. As noted in detail elsewhere, VSV virions have an endogenous RNA-dependent RNA polymerase that transcribes the viral genome from the 3' end, resulting in the sequential formation of a 47-nucleotide leader RNA and five messenger RNAs (mRNAs) for the N, NS, M, G, and L proteins. Translation of each mRNA but not of the leader RNA results in synthesis of all viral proteins during the infectious cycle.

The G protein is synthesized on membrane-bound polyribosomes, whereas the other viral proteins are made on cytoplasmic polyribosomes. This section is limited to a discussion of the structure and function of the G protein of VSV.

The VSV G protein has served as a model for studying the mode of synthesis, transport, glycosylation, and processing of cellular membrane glycoproteins. Although the G protein is translated from virus-specific mRNAs, all subsequent events in maturation of the protein are similar to those of the host-cell membrane proteins. Of great advantage, the viral G protein and the corresponding mRNA can be readily purified in large quantities. This has provided the means for making monoclonal antibodies and cloned complementary DNA (cDNA) of the G protein gene. Moreover, reconstitution of the G protein with defined lipid vesicles has provided simple models to probe the interaction of an integral membrane protein with the lipid bilayer.

A. Structure of Vesicular Stomatitis Virus Glycoprotein

The G protein constitutes the spikelike projection that protrudes from the virion membrane (Cartwright *et al.*, 1969) and reacts with neutralizing antibody (Kelley*et al.*, 1972). The complete nucleotide sequence of the VSV mRNA that encodes the G protein has been determined from cDNA clones that contain the entire coding sequence of the mRNA (Rose and Gallione, 1981). The G protein mRNA is composed of 1665 nucleotides, excluding the polyadenylic acid tail, and encodes a protein of 511 amino acids including a signal peptide of 16 amino acids. The VSV-Indiana G protein is glycosylated and acylated, but not phosphorylated. Two sites of glycosylation are predicted in asparagine residues at positions 178 and 335 (Rose and Gallione, 1981). The G protein is anchored in the viral envelope by a nonglycosylated protease-resistant fragment of $M_r \approx 5200$ (Schloemer and Wagner, 1975). Rose *et al.* (1980) have demonstrated that this fragment is composed of the carboxy-terminal 45 amino acid residues of the polypeptide chain. Residues 1–29 from the carboxy terminus constitute a hydrophilic peptide domain that protrudes beyond the interior surface of the viral membrane, and residues 30–49 form a sequence of 20 consecutive hydrophobic amino acids that spans the lipid bilayer. The remaining 462 amino acids form the spikelike projection that protrudes from the viral membrane; the carbohydrate residues are attached in this region.

The nucleotide sequence of the mRNA that encodes the G protein of the VSV New Jersey serotype has also been determined from a cDNA clone containing the entire coding region (Gallione and Rose, 1983). The VSV-New Jersey G protein was found to contain 517 amino acids, with 50.9% sequence homology between the G proteins of the two serotypes. However, the position and size of transmembrane domains, signal sequence, and glycosylation sites are identical for both proteins, except that the G protein of the New Jersey serotype is not acylated. The glycoprotein of rabies virus has also been sequenced; it has 524 amino acids, with approximately 20% base-sequence homology with the VSV glycoprotein (Rose *et al.*, 1982; Anilionis *et al.*, 1981).

Stable expression of the VSV-Indiana G protein has been achieved in mammalian cells transfected with a hybrid expression vector (Rose and Bergman, 1982), and the G protein expressed in these cells was found to be processed, glycosylated, and transported to the cell surface in a normal fashion. Conditional expression of the VSV G protein gene in *Escherichia coli* was also achieved by Rose and Shafferman (1981). Detailed study of the expression of the VSV G protein gene is described in Chapter 4.

The external location of the G protein in VSV virions has been established by a number of techniques. It was observed that lactoperoxidase-catalyzed iodination of intact virions labeled predominantly G protein (Moore *et al.*, 1974; McSharry, 1977). Similar results were obtained employing a labeling procedure with pyridoxal phosphate in which only the G protein was found to be radiolabeled to a significant extent (Eger *et al.*, 1975). Treatment of VSV virions with a number of membrane-impermeable proteases, such as trypsin (Mudd, 1974), pronase (McSharry *et al.*, 1971), and bromelain, chymotrypsin, or thermolysin (Bishop *et al.*, 1975; Schloemer and Wagner, 1975), removed only the G -protein spikes present at the surface of the virus.

By using bivalent cross-linking reagents, Dubovi and Wagner (1977) demonstrated the presence of G–G homodimers as well as trimers and G–M heterodimers in intact virions, suggesting that the G-protein spikes are composed of more than one protein molecule and that the protein traverses the membrane bilayer to interact with the M protein at the inner leaflet. The hydrophobic tail fragment of the G protein has also been specifically labeled by the photoactivatable hydrophobic aryl azide probes [^{125}I]-5-iodonaphthyl-1-azide and [^{3}H]pyrenesulfanylazide in intact or protease-treated virions (Zakowski and Wagner, 1980). These studies again support the contention that the G protein completely spans the membrane. Furthermore, photoactivated cross-linking to the VSV G protein of photosensitive phospholipids incorporated *in vivo* into the VSV membrane also supports the location of G protein in the membrane (Capone *et al.*, 1983; Stoffel *et al.*, 1978).

B. Glycoprotein Acylation

A new type of posttranslational modification of the VSV G protein has been reported by Schmidt and Schlesinger (1979, 1980). The VSV-Indiana G protein was found to contain 1–2 moles of fatty acid (usually palmitic acid) per mole of the protein; this linkage was shown to be resistant to extraction with organic solvents and to boiling in sodium dodecyl sulfate (SDS), but the fatty acid was released by mild alkali treatment of the G protein. The fatty acid attachment to the protein has been described as posttranslational event because it was shown to take place nearly 15 min after complete synthesis of the complete polypeptide, presumably during maturation of the protein as it moves to the cell plasma membrane. Much smaller amounts of [^{3}H]fatty acid were found to acylate the G protein of a temperature-sensitive (ts) group V mutant (ts045) at nonpermissive temperatures; no [^{3}H]fatty acid was bound to the protein synthesized at 37°C in cells pretreated with tunicamycin, an inhibitor of glycosylation (Schmidt and Schlesinger, 1979, 1980).

The covalently attached fatty acid of the membrane G protein of VSV has also been fluorescently labeled *in vivo* by isolating VSV from infected BHK-21 cells that had been grown in the presence of 16-(9-anthroyloxy)-palmitate (Petri *et al.*, 1981a). Steady-state fluorescence anisotropy of G protein reconstituted into dipalmitoylphosphatidylcholine vesicles indicated that the fatty acid attached to G protein was located in a lipid domain that did not undergo the fluid–gel phase transition (Petri *et al.*, 1981a). Similar labeling of VSV G protein by a photoreactive fatty acid, ω-[9-^{3}H]diazirinophenoxy nonanoate, was also achieved by Capone *et al.* (1983).

The determination of the site of binding of fatty acid to the G protein has been the subject of many studies. Fatty acid attached to the G protein was shown to be located solely in the tail fragment of the G protein left in the viral membrane after thermolysin digestion (Petri and Wagner, 1980; Capone *et al.*, 1982). Recently, it has been demonstrated that the fatty acid could be released from the G protein with 1 M hydroxylamine at pH 8.0 when a significant fraction of the G protein was converted to disulfide-linked dimers (Magee *et al.*, 1984). These data implicate a cysteinyl group in the protein as the site(s) involved in the fatty acid acylation. Since the fatty acid was found to reside in the membrane-anchored domain, it was suggested that the cysteine residue on the cytoplasmic side of the membrane might be the site of acylation by the fatty acid (Magee *et al.*, 1984). Rose *et al.* (1984) have recently obtained expression in eukaryotic cells of mutagenized cDNA clones that encode VSV G proteins lacking portions of the cytoplasmic domain; labeling of these truncated proteins with [^{3}H]palmitate indicated that the palmitate might be linked to an amino acid within the first 14 residues on the carboxy-terminal side of the transmembrane domain. Using oligonucleotide site-directed mutagenesis, it was found that the linkage of palmitate to G protein was through the cysteine in the cytoplasmic domain. Interestingly, the G protein lacking the palmitate was glycosylated and transported normally to the cell surface (Rose *et al.*, 1984).

The function of this fatty acid in the G protein is not understood. It has been postulated that the fatty acid could act as a lipophilic anchor, stabilizing the G protein in the membrane during its migration (Schmidt and Schlesinger, 1979, 1980). Zilberstein *et al.* (1980) described a *ts* mutant of VSV defective in the transit of its G protein from the Golgi complex to the plasma membrane at nonpermissive temperature. Such defects in the movement of the G protein from the sites of synthesis to the plasma membrane have been reported to be due to the lack of fatty acid acylation, indicating a possible role for the covalently bound fatty acid in the transport of the protein. However, later studies by Lodish and Kong (1983) demonstrated that some nonacylated G protein was also transported to the cell surface but not incorporated into the virions, suggesting that the fatty acid from the G protein was essential for the assembly and budding of the virus from the host-cell plasma membrane. In a separate study, it was shown that cerulenin, an antibiotic that inhibits *de novo* fatty acid and cholesterol biogenesis, effectively inhibited the formation and release of virus particles from CEF infected with VSV (M. J. Schlesinger and Malfer, 1982). Nonacylated G protein was found to accumulate inside and on the surface of cerulenin-treated cells, suggesting that the fatty acid acylation is not essential for intracellular transport of this membrane protein, but that it may play an important role in the interaction of G protein with membrane during virus assembly and budding (M. J. Schlesinger and Malfer, 1982). However, recent data showing that the G proteins of the New Jersey serotype and Cocal strain of VSV do not contain the covalently linked fatty acid suggest that the fatty acid acylation of the viral membrane G protein may not be a general requirement for maturation and budding of the virus (Gallione and Rose, 1983; Kotwal and Ghosh, 1984). Furthermore, expression in eukaryotic cells of a mutagenized cDNA clone that encodes VSV G protein lacking the fatty acid was found to be glycosylated and to be transported normally to the cell surface (Rose *et al.*, 1984). All these results tend to suggest that the fatty acid acylation of G protein may not play a very critical role in the transport of the G protein to the cell surface and in the subsequent budding process; however, it may have an effect on the long-term stability of G protein in the viral membrane.

C. Carbohydrate Chains of Vesicular Stomatitis Virus Glycoprotein

1. Structure

The carbohydrate portion of the G protein comprises approximately 10% of the G protein and is present in an oligosaccharide structure with an apparent molecular weight of 4000. The oligosaccharide chains have a complex structure and are composed of mannose, galactose, N-acetylglucosamine, sialic acid, and fucose (McSharry and Wagner, 1971b). The G protein appears to have two glycosylation sites, which were shown to involve N-acetylglucosamine and asparagine, presumably at amino acid residues 178 and 335 (Rose and Gallione, 1981). Besides these two major carbohydrate chains, there has been some evidence for additional gly-cosylation sites in the protein. For example, Moyer *et al.* (1976) reported four major and several minor tryptic peptides labeled with [³H]glucosamine. Kingsford *et al.* (1980) have demonstrated that when cyanogen bromide cleaved G protein into 11 peptide fragments, the major oligosaccharide chains were attached to 2 different CNBr-cleaved peptides. In addition, 6 other peptides contained small amounts of sialic acid, fucose, and man-nose, indicating that the G protein may contain more carbohydrate than the two major chains linked to asparagine residues. In this context, it is to be noted that rabies virus appears to contain at least three different oligosaccharide chains (Dietzschold, 1977).

Both the major oligosaccharide chains of G protein are of the complex type and are identical in VSV grown in BHK cells. The decisive analysis of the carbohydrate structure of VSV glycoprotein was carried out by Reading *et al.* (1978). The three sialoyl-*N*-acetyllactosamine branches are attached to a mannose-rich core pentasaccharide. Fucose is attached to position C_6 of the first *N*-acetylglucosamine residue. The following struc-ture was deduced:

$$\alpha\text{NeuNAc-}^3\beta\text{Gal-}^4\beta\text{GlcNAc} \qquad\qquad \alpha\text{Fuc}$$
$$\qquad\qquad\qquad \downarrow_4 \qquad\qquad\qquad\qquad \downarrow_6 \qquad |$$
$$\alpha\text{NeuNAc-}^3\beta\text{Gal-}^4\beta\text{GlcNAc-}^2\alpha\text{Man-}^6\ \beta\text{Man-}^4\beta\text{GlcNAc-}^4\ \text{GlcNAc-Asn}$$
$$\qquad\qquad\qquad\qquad\qquad \uparrow_3 \qquad\qquad\qquad\qquad |$$
$$\alpha\text{NeuNAc-}^3\beta\text{Gal-}^4\beta\text{GlcNAc-}^2\alpha\text{Man}$$

The degree of sialylation produces heterogeneity in the VSV G protein oligosaccharide chain (Etchison and Holland, 1974; Moyer and Summers, 1974; J. S. Robertson *et al.*, 1976; Sefton, 1976). Another source of het-erogeneity is the content of fucose in the oligosaccharide chain. Although the *N*-acetylglucosamine residue linked to the polypeptide is often sub-stituted with fucose, there is evidence for only one fucose residue per two oligosaccharide side chains (Etchison and Holland, 1974; J. S. Rob-ertson *et al.*, 1976). Further evidence for heterogeneity within the oli-gosaccharide structure was detected by endoglycosidase enzyme; the G protein obtained from VSV grown in BHK-21 cells transformed by po-lyoma virus contained glycopeptides that showed increased resistance to endo-β-*N*-acetylglucosaminidase, indicating structural changes in the core region of the oligosaccharides (Moyer and Summers, 1974). Similar gly-cosidase-resistant glycopeptides were also detected in VSV grown in un-transformed HeLa cells (Moyer *et al.*, 1976). In a more recent study, Hunt *et al.* (1983) observed unusual heterogeneity in the glycosylation of the G protein of VSV-New Jersey (Hazelhurst strain), mostly in the content of an acidic-type structure in the oligosaccharide chains.

The limited coding capacity of the VSV genome makes it obvious that the virion must depend on the host cell for the synthesis of the

carbohydrate chains. As expected, host-dependent differences in the glycosylation patterns were noted in the oligosaccharide structure of VSV
(Etchison and Holland, 1974; Moyer and Summers, 1974; Moyer et al.,
1976). Furthermore, the G protein of VSV was shown to reflect the glycosylation defect of various lectin-resistant CHO cells (M. A. Robertson
et al., 1978; Stanley et al., 1984). M. A. Robertson et al. (1978) also
identified G protein specifically lacking sialic acid and terminal N-acetylglucosamine–galactose–sialic acid sequences, but possessing an increased number of mannose residues in the core as well as differential
types of linkages in the oligosaccharide chains.

Although the foregoing studies point out the host-specific determination for glycosylation of the viral G protein, the virus itself also appears
to play a role in determining the structure of its G protein oligosaccharides. Thus, Sefton (1976) demonstrated that Sindbis virus, VSV, and Rous
sarcoma virus grown in primary chick embryo cells acquire a different
set of oligosaccharide side chains. To examine the extent to which the
polypeptide structure of a virus G protein contributes to the overall structure and composition of its carbohydrate moieties, J. S. Robertson et al.
(1982) made detailed analyses of the structures of the oligosaccharide
chains of wild-type VSV G protein compared with that of a G-protein-
defective mutant of VSV, tl-17. Characterization of the oligosaccharides
by ion-exchange and gel-filtration chromatography, after sequential enzymatic degradation, revealed similar structures for the wild-type and
mutant G proteins. However, the altered polypeptide structure of the tl-
17 G protein affected the extent of addition of sialic acid and fucose, both
of which are added late in the maturation of the G protein. It has been
suggested that the increased sialic and fucose content and the appearance
of a small quantity of an additional branched structure in the tl-17 oligosaccharide side chains might be due to an altered configuration in the
polypeptide backbone of the tl-17 G protein, which in turn might result
in partial unfolding of the protein molecule. This perturbation of the
tertiary structure of the tl-17 G protein in the neighborhood of the carbohydrate moieties might be responsible in part for the thermal instability
of the tl-17 viral G protein (J. S. Robertson et al., 1982).

2. Functions

No biological role has been definitively assigned to the oligosaccharide chains of the VSV G protein. Two experimental approaches have
been used in attempts to assign biological function to the G-protein carbohydrates. One method, selective hydrolysis in vitro of individual sugars,
such as the terminal neuraminic acid, has been used to test residual VSV
infectivity, adsorption, and hemagglutination; this literature is discussed
in Chapter 2. A second approach has been to study the maturation of the
G protein in vivo after inhibiting glycosylation of nascent polypeptide

chains; some results and interpretations of such experiments are discussed here.

The glycosylation of VSV G protein has been inhibited *in vivo* by the antibiotic tunicamycin, which inhibits the formation of the *N*-acetylglucosamine–lipid intermediate (Leavitt *et al.*, 1977). The unglycosylated form of G protein was readily detected in VSV-infected host cells, but no evidence was found for the transfer of the protein to the outer surface of the plasma membrane. In a subsequent study, Gibson *et al.* (1978) detected a low level of infectious VSV virions released from tunicamycin-treated cells; these unglycosylated virions were found to have a specific infectivity comparable to that of VSV containing glycosylated G protein. The yield of these unglycosylated virions increased when the temperature of infection was lowered from 37 to 30°C. These results tend to suggest that glycosylation of G protein is probably not an absolute requirement for transport and insertion of the viral G protein into the plasma membrane, although the process may be less efficient in the absence of glycosylation. In another study, Morrison *et al.* (1978) observed that the maturation of the G protein was affected in the presence of tunicamycin. Tunicamycin had no effect on the attachment of the G protein to intracellular membranes or on the transport of the protein to the lumen of the endoplasmic reticulum. However, it prevented the migration of the G protein from the rough endoplasmic reticulum to smooth intracellular membranes. Of course, tunicamycin may also be affecting cell functions other than glycosylation, thus reducing the yield of G protein in a compromised cell.

The requirement for carbohydrate in morphogenesis of two different VSV strains of the Indiana sereotype was studied by Gibson *et al.* (1979). At 30°C, in the presence of the glycosylation inhibitor tunicamycin, cells infected with the San Juan strain released only 5–10% of the expected yield of virus particles, whereas the Orsay strain released between 45 and 70% of the expected number of infectious virions. These data suggested that unlike the case of the San Juan strain, carbohydrate is not a critical factor for efficient morphogenesis of the Orsay strain at 30°C. Interestingly, the carbohydrate requirement for release of virions of both serotypes was found to be temperature-sensitive; at 37°C, tunicamycin inhibited the assembly of VSV virions by 85–95% for both strains. Alternation in the physical properties of the nonglycosylated G protein was detected in an *in vitro* assay that measured the extent of protein aggregation that occurred when guanidine hydrochloride was removed from the denatured protein molecule (Gibson *et al.*, 1979). It was observed that the levels of virus assembly and the absence of glycosylation were inversely correlated with the extent of aggregation of the nonglycosylated G protein in the infected cells. Thus, at elevated temperature (39°C), the G proteins of both strains were found to aggregate, whereas at 30°C, the G protein of the San Juan strain aggregated more than did that of the Orsay-strain virions. These results led the authors to conclude that the failure to glycosylate the nascent G protein could affect the folding of the molecule, and this, in turn, results in increased sensitivity to temperature during polypeptide folding.

The effect of oligosaccharide chains of different sizes on the maturation and physical properties of VSV G protein was also studied by Gibson *et al.* (1981), who observed that the size of the oligosaccharides present on the folding G protein was a determining factor in attaining a proper conformation; the extent of this effect ostensibly depends on the primary structure of the polypeptide. The denaturation and renaturation of pure G protein containing no sialic acid, or no complex or no altered oligosaccharide chains, were also studied by Crimmins and Schlesinger (1982). Fluorescence quenching using acrylamide showed no marked difference between the native and denatured states of G protein due to its sialic acid content. However, attempts to renature G protein with shorter oligosaccharide chains led to extensive aggregation.

A more detailed study on the requirement of carbohydrates for the morphogenesis of VSV virions was reported by Chatis and Morrison (1981). They observed that G proteins from complementation group V mutants *ts*O44 and *ts*O45, like their Orsay wild-type parent, did not require carbohydrate for efficient morphogenesis. However, the G protein of another group V mutant, *ts*O110, was totally dependent on carbohydrate addition for migration to the cell surface. Furthermore, they isolated a pseudo-revertant of *ts*O44 (*ts*O44R) that, unlike the Orsay parent, no longer exhibited a requirement for carbohydrates at 39.5°C; the unglycosylated G protein migrated to the cell surface very efficiently. All these studies led the authors to conclude that simple mutational changes, as opposed to many alterations in the protein molecule, might be sufficient to alter the carbohydrate requirement for morphogenesis of the virions.

D. Biogenesis of Vesicular Stomatitis Virus Glycoprotein

Considerable research has been done on the biogenesis and maturation of the VSV G protein; this literature has been reviewed quite extensively (Klenk and Rott, 1980; Ghosh, 1980; Lodish *et al.*, 1981; Zilberstein *et al.*, 1982; Kääriäinen and Pesonen, 1982). The biogenesis of the G protein is described in some detail in Chapter 4. In the following section, we summarize only the information most pertinent to the VSV membrane and its assembly.

A number of studies have demonstrated that the VS viral G protein is synthesized on polyribosomes initially bound to endoplasmic reticulum and probably sequestered in inside-out vesicles. This process eventuates in migration of the complex to the plasma membrane, where the mature glycoprotein is inserted (Morrison and Lodish, 1975; Atkinson *et al.*, 1976;

Knipe *et al.*, 1977a,b; Morrison and McQuain, 1978). Partial confirmation of this mechanism of the maturation of the VSV G protein came from studies of *ts* mutants in complementation group V that were shown to be restricted in insertion and transmembrane migration of the G protein at nonpermissive temperature (Lafay, 1974; Knipe *et al.*, 1977b).

The sequence of events in the biosynthesis, processing, and membrane insertion of the VSV G protein has been studied extensively in a cell-free system devised by Rothman and Lodish (1977) and Toneguzzo and Ghosh (1977, 1978). A model for the synthesis, glycosylation, and transmembrane insertion of the VSV G protein has been presented by Lodish *et al.* (1981). When mRNA coded by the G-protein gene is translated in a cell-free system derived either from wheat germ or from HeLa cells, a nonglycosylated protein (G_0) with an apparent molecular weight of 63,000 is synthesized. However, in the presence of microsomal membrane vesicles, a glycosylated protein (G_1) with a molecular weight of 67,000 was obtained, suggesting that the presence of membrane is important for the glycosylation of the protein. About 3000 daltons of this protein was found to be digested by proteolytic enzymes, indicating that most of the G protein had been sequestered into the lumen of the vesicles. A signal sequence of 16 amino acids has been identified at the amino terminus of the G_0 protein, which is cleaved during its synthesis in the presence of the microsomal membranes. This conclusion is based on the signal hypothesis of Blobel, which predicts that all secretory proteins have a signal sequence that leads the nascent polypeptide chain through the membrane of the endoplasmic reticulum into the cisternal space (Lingappa *et al.*, 1978). Recently, Kotwal *et al.* (1983) determined the NH_2-terminal amino acid sequences of the envelope G proteins and the *in vitro* synthesized, nonglycosylated precursors of the G proteins of three VSV strains, namely, Indiana, Cocal, and New Jersey, relegated to two serotypes. A comparison of the sequences of these three G proteins showed little homology in the signal peptides present in the nonglycosylated precursors except for their high content of hydrophobic amino acids. However, the NH_2-terminal amino acid sequences of the mature G proteins revealed extensive homology among these different VSV strains.

After synthesis of the G protein in endoplasmic reticulum, it is transported into the plasma membrane via the Golgi apparatus. The importance of the Golgi complex in this process has been demonstrated elegantly in an *in vitro* system described by Rothman and his associates (Rothman and Fries, 1981; Fries and Rothman, 1981; Dunphy *et al.*, 1981). The polypeptides move first to the Golgi apparatus, where glycosylation is completed. Another posttranslational modification of the G protein, which is probably carried out in the Golgi complex, is the covalent attachment of several molecules of fatty acid to the polypeptide chain. The mechanism of channeling of integral membrane proteins from the site of synthesis to the Golgi apparatus and then to the surface is not clearly understood. There is apparently some controversy concerning whether this intracellular transport is mediated by coated vesicles (Rothman and Fine, 1980; Wehland et al., 1982).

The glycosylation of the G protein starts in the rough endoplasmic reticulum during the synthesis of the nascent G-protein molecule. The biosynthesis of the complex-type asparagine-linked oligosaccharides is a multistep process that involves both addition and removal of specific sugar residues. A branched oligosaccharide precursor containing three glucose, nine mannose, and two N-acetylglucosamine residues is preformed on a lipid carrier molecule localized in the rough endoplasmic reticulum. This oligosaccharide chain was found to be transferred to the nascent G protein. It was also observed that one of the carbohydrate chains is added to the nascent G protein when it is about one-third complete, whereas the other was added when it was 70% complete. Immediately after the transfer to the polypeptide, one or two glucose residues are removed from the oligosaccharide chains. Further processing of the oligosaccharide was found to take place only 10–20 min after the synthesis of the G protein, presumably at a time when the protein is transferred to the Golgi apparatus. In a stepwise formation, the remaining glucose residues and six of the nine mannose residues are removed from the oligosaccharide complex, and the peripheral sugar residues N-acetylglucosamine, galactose, sialic acid, and fucose are added. The processing of oligosaccharides is apparently complete 10 min before the G protein reaches the cell surface. It has been observed recently by S. Schlesinger et al. (1984) that conditions that prevent the removal of glucose residues from the synthesized oligosaccharide chain of the G protein can arrest the growth of the virus.

The process of intracellular transport of G protein has been studied extensively by the use of VS viral ts mutants in complementation group V, assigned to the coding region of the viral glycoprotein. Several ts mutants in this structural gene (V) have been isolated, and it has been observed that cells infected with these ts mutants at nonpermissive temperature produce G protein normally, but the protein fails to mature at the cell surface (Lafay, 1974; Lodish and Weiss, 1979; Knipe et al., 1977a,b; Zilberstein et al., 1980). These group V ts mutants of VSV G protein have been subdivided into two classes based on the stage at which posttranslational processing of the glycoprotein is blocked. The mutants ts-L513(V), tsM501(V), and tsO45(V) encode G-protein molecules that are blocked at nonpermissive temperatures at an early (pre-Golgi) step in G protein maturation (Zilberstein et al., 1980). However, this defect in transport of G protein for these mutants was shown to be reversible; lowering the temperature to a permissive level leads to the movement of G protein from the rough endoplasmic reticulum to the Golgi complex and thence to the plasma membrane (Lodish and Kong, 1983; Bergmann et al., 1981; Knipe et al., 1977b).

The second class of ts mutants, namely tsL511(V), was shown to be specifically blocked in the fatty acid acylation step at nonpermissive temperatures (Zilberstein et al., 1980; Lodish and Kong, 1983). It was observed that whereas a significant fraction of this G protein reached the plasma membrane, no incorporation of the protein into the virus particle could be detected. However, an appreciable fraction of the tsL511(V) G protein, which accumulates in the cells at 40°C, was found to mature into virions shifted down to permissive temperature (Lodish and Kong, 1983).

Several different factors have been shown to affect the synthesis, transport, and insertion of the VSV G protein into the cell-surface membrane. In an attempt to study the effect of ionophores on the maturation of viral glycoproteins, Johnson and Schlesinger (1980) showed that the two cationic ionophores monensin and A23187 arrested the movement of viral G protein from the Golgi apparatus to the cell surface, where budding and release of the virus is destined to occur. Furthermore, it was observed that the VSV G protein is synthesized and glycosylated in the presence of chloroquine, but the drug prevents the expression of the glycoprotein at the cell surface during the final stages of G-protein assembly (Dille and Johnson, 1982). The insertion of VSV G protein into the plasma membrane was also shown to be affected when the composition of the host LM cell membrane phospholipids is modified by loading the cells with choline analogues (Maeda et al., 1980).

The role of various domains of the G protein concerned with its synthesis and maturation has been studied by a number of workers. A soluble form of the G protein was found to be released from infected cells into the extracellular medium (Kang and Prevec, 1971; Little and Huang, 1977, 1978; Irving and Ghosh, 1982; Chatis and Morrison, 1983). The properties of this soluble G protein and the cellular site at which it was generated were characterized by Chatis and Morrison (1983). They observed that a segment between 5000 and 6000 daltons at the carboxy-terminal end of the G protein is cleaved to generate the Gs molecule, and this truncated molecule contained no fatty acid. Moreover, the soluble glycoprotein was found to result from proteolytic cleavage of cell-surface G protein distal to the membrane-spanning region of the protein. Indeed, a similar truncated form of G protein, lacking the 79 amino acids from the COOH terminus, was released from mammalian cells transfected with a simian virus 40 expression vector (Rose and Bergmann, 1982). Such an expressed protein was glycosylated normally and secreted slowly from the cells into the medium. In a separate study, Garreis-Wabnitz and Kruppa (1984) detected in VSV-infected BHK cells an intracellular G protein that was found to lack the membrane-anchoring oligopeptide of the viral G protein.

Cytoplasmic domains of the G protein inserted in the plasma membrane have been shown to contain 29 amino acid residues, 10 of which are basic and 2 acidic (Rose and Gallione, 1981). This hydrophilic domain was shown to play an important role in the biosynthesis and maturation of G protein. It was noted that recombinant G proteins, with various deletions and additions in this cytoplasmic domain, either accumulate in the endoplasmic reticulum in a pre-Golgi state or may be transported to cell surface at a reduced rate (Rose and Bergmann, 1982, 1983). In another study, Arnheiter *et al.* (1984) observed that when polyclonal antibodies to a synthetic 22-amino-acid oligopeptide homologous to the G-protein carboxy terminus were microinjected into monolayer BHK cells before or shortly after infection with wild-type VSV, G protein accumulated in large intracellular patches and little G was observed in the Golgi complex or at the cell surface. In sharp contrast, no effect on the maturation of G protein was observed when antibodies to an oligopeptide representing the first 11 carboxy-terminal amino acids were injected into the infected cells.

E. Self-Assembly of Glycoproteins and Liposomes

The G protein of VSV is anchored in the lipid bilayer by a stretch of 20 hydrophobic amino acids (Rose and Gallione, 1981). Reconstitution of G protein in defined lipid vesicles serves as an important model to study the interaction of an integral protein with a lipid bilayer. In addition, the function of G protein in various lipid environments can be investigated with these reconstituted vesicles.

Reconstitution of the VS viral glycoprotein into liposomes has been achieved by a number of laboratories in recent years. Petri and Wagner (1979) were able to selectively liberate G protein from the virion membrane by the dialyzable nonionic detergent, β-D-octylglucoside. The isolated viral G protein, rendered free of viral phospholipids and detergent, forms tail-to-tail G protein micelles in the form of rosettes when viewed by electron microscopy. It was also noted that when a mixture of viral G protein and phosphatidylcholine (PC) was dialyzed free of octylglucoside, G-protein vesicles were formed spontaneously with spikes protruding from the exterior of the lipid membrane (Petri and Wagner, 1979). Evidence for external orientation of the VSV G protein inserted in these lipid vesicles was substantiated by proteolytic digestion with thermolysin, which gave rise to a hydrophobic 4- to 5-kilodalton G-protein tail fragment in both the reconstituted vesicles and whole virions. Similar selective octylglucoside extraction of VSV G protein and viral lipids was also reported by Miller *et al.* (1980), who noted the spontaneous formation of G-protein vesicles 250–1000 Å in diameter when dialyzed free of the detergent. These G-protein-reconstituted vesicles were predominantly unilamellar and sealed; both phosphatidylethanolamine and gangliosides incorporated in these reconstituted vesicles were symmetrically distributed in the two leaflets of the bilayer. The G protein was found to be asymmetrically oriented in the outer layer, with about 80% accessible to digestion by exogenous protease (Miller *et al.*, 1980).

Reconstitution of VSV G protein was also achieved with preformed vesicles. Petri and Wagner (1980) demonstrated that G-protein micelles, isolated by octylglucoside extraction and purified free of detergent and phospholipid, spontaneously partition into preformed sonicated PC vesicles when the G-protein micelles and the lipid vesicles are incubated together at 37°C. The hydrophobic tail fragment of the G protein became resistant to protease digestion when the G protein was inserted into the vesicle bilayer. The ability of the micellar form of G protein to partition into preformed vesicles distinguished it from other membrane G proteins, such as Sendai HN or F glycoproteins, which do not partition into preformed vesicles (Hsu *et al.*, 1979). In a separate study, Altstiel and Landsberger (1981) extracted the G protein from intact VSV particles by solubilization of the viral envelope with lysolecithin; when the extracted protein was added to sonicated PC vesicles and the detergent was removed with BSA, the G protein was found to reconstitute with the lipid bilayer in the same molar ratio as that in the intact virus envelope. The G protein was found to be incorporated into the outer surface of the reconstituted vesicles, which were then able to hemagglutinate goose erythrocytes under the same conditions as VSV.

Eidelman *et al.* (1984) have recently used these G-protein-reconstituted vesicles to study the mechanism of low-pH-dependent fusion induced by VSV G protein. Incubation of G-protein-reconstituted vesicles with small unilamellar vesicles containing negatively charged phospholipids resulted in the fusion of the vesicles at low pH. This process of fusion did not cause any leakage of a vesicle-encapsulated aqueous marker that could be monitored by energy-transfer measurements.

IV. MATRIX PROTEINS

Many enveloped viruses contain a matrix (M) protein. The molecular weight of this protein varies among different viruses (Lenard, 1978). In rhabdoviruses and orthomyxoviruses, the molecular weight of the M protein was shown to be close to 26,000, whereas for paramyxoviruses, it was in the range of 38,000–41,000. Nearly 30% of the total protein of vesicular stomatitis virus (VSV) is represented by this highly basic, nonglycosylated polypeptide. Attempts have been made in recent years in several laboratories to investigate the structure and function of the VSV M protein. Aiding in this effort are temperature-sensitive (*ts*) VSV mutants in complementation group III that have lesions in the M protein (see Chapter 5). The M protein is considered to be indispensable in the budding process of the VSV virions; phenotypic mixing experiments between the *ts* mutants of VSV and retroviruses have demonstrated quite clearly that the synthesis of functional M protein is required for VSV maturation (Weiss and Bennett, 1980). Moreover, in studies on the production of virions from cells infected with several *ts* mutants in complementation group III, Schnitzer and Lodish (1979) have shown that the M protein is the only polypeptide involved in the budding process of the virus. It has been suggested that the M protein acts as a "glue" that helps in assembling the ribonucleocapsid with the viral lipid envelope into a compact structure. To perform such a function, the protein must undergo interactions with the lipid bilayer and the ribonucleocapsid protein (RNP) cores. In this section, we describe the functions of the M protein and examine how the protein is involved in assembling the virus into a biologically functional unit.

A. Primary Structure of Vesicular Stomatitis Virus Matrix Protein

The M protein of VSV is composed of a single polypeptide chain that has a molecular weight of approximately 26,000 as determined by SDS–polyacrylamide gel electrophoresis (Wagner, 1975). The complete sequence of the M protein of VSV of the Indiana serotype (San Juan strain) has been determined from cDNA clones made from mRNA that contained the complete coding sequence (Rose and Gallione, 1981). The mRNA that encodes the VSV M protein has been found to be 831 nucleotides long and encodes a protein of 229 amino acids. The protein has a pI of 9.1 as determined by isoelectric focusing (Carroll and Wagner, 1979). The basic nature of the protein is attributed to the fact that the protein has 21 lysine, 10 arginine, and 8 histidine residues in a total of 229 amino acids. The amino-terminal segment of the M protein is highly basic, containing 8 lysine residues within the first 19 amino acids. The protein was shown to be nonglycosylated (Wagner et al., 1969) and nonacylated (Schmidt and Schlesinger, 1979), but is phosphorylated (Imblum and Wagner, 1974). In addition to phosphorylated serines and threonines, the M protein contains phosphorylated tyrosine residues. Clinton et al. (1982) identified $pp60_{src}$ as one of the protein kinases in VS virions grown in Rous-sarcoma-virus-transformed cells, a finding confirmed by Bell et al. (1984). Endogenous phosphorylation, however, has not yet been shown to affect the various functions of M protein.

VSV ts mutants in complementation group III encode a defective M protein (Lafay, 1974). Most of these ts mutants described in the literature were derived either from the Glasgow or the Orsay wild-type VSV strain (Flamand, 1970; Pringle, 1970). Nucleotide sequences of the coding regions of the M-protein genes of the Glasgow and Orsay wild-type VSV-Indiana serotype and of two group III (M protein) mutants derived from each wild-type were determined (Gopalakrishna and Lenard, 1985). Both Glasgow and Orsay wild-type M-protein genes differed in 13 bases from the nucleotide sequence of the San Juan strain Indiana serotype determined by Rose and Gallione (1981). Of these base changes, 6 resulted in amino acid substitutions, whereas 7 were degenerate codons. The Orsay and Glasgow wild-type M-protein sequences resembled each other more closely than either resembled the sequence of the San Juan strain M protein, differing from each other in 8 nucleotides representing 4 amino acid changes. Each of the four mutants, however, differed from its parent wild-type in only 1 or 2 nucleotides, in each case resulting in one amino acid substitution (two in one case). Every mutation in the protein caused a change of charge of the M protein, reflected in the pI of the polypeptide molecule as measured by isoelectric focusing (Gopalakrishna and Lenard, 1985). In a separate study, Kiuchi and Roy (1984) sequenced the M protein of the spring viremia of carp virus, a fish rhabdovirus structurally similar to VSV. This M protein was 710 nucleotides long, with 223 amino acids, and was found to·contain no significant amino- or carboxy-terminal hydrophobic domains. The predicted amino acid sequence of the M protein of spring viremia of carp virus, when aligned with that of VSV, revealed nearly 28% homology. However, the carboxy-terminal regions of the two protein exhibited much less homology than did their amino termini.

The primary structures of M proteins from different VSV serotypes have been compared by enzymatic and chemical cleavages. By comparing tryptic peptide maps of VS viral protein Doel and Brown (1978) observed that many peptides are common for M proteins of the Indiana and New Jersey serotypes. On the other hand, Brown and Prevec (1978) noted very little homology between the tryptic maps of New Jersey and Indiana M protein. In a more detailed study, Burge and Huang (1979) observed that when these two serologically distinct VSV isolates were digested by V8 protease or by chymotrypsin, the peptide maps of the New Jersey and Indiana M proteins were found to be different in all but four peptides. By selective cleavage at tryptophan residues with N-chlorosuccinimide, Brown and Prevec (1979, 1982) observed considerable conservation of tryptophan number and location in the M proteins of four strains of two VSV serotypes. In a more recent study, the sequence of the M protein of VSV-New Jersey (Ogden strain) has been deduced from a cDNA clone that contained the complete coding sequence of the mRNA that encodes the matrix protein (A. Banerjee, personal communication). A large degree of homology (>50%) has been observed between the nucleotide and amino acid sequences of M proteins from the two serotypes.

Although the VSV and influenza virions are members of different virus families, they both contain M proteins with similar molecular weights and apparently similar functions. The sequences of M proteins of these two viruses were compared, and a considerable degree of homology was predicted (Rose et al., 1982). It was noted that the single cysteine residue of VSV M protein was aligned with a cysteine residue in influenza M protein, and both proteins were found to start with the sequence Met-Ser-X-Lys and end with an identical dipeptide. It has been suggested that this significant relatedness between these two matrix proteins may indicate common ancestry. A greater evolutionary constraint might have been applied to the structure of M proteins to enable them to perform a crucial role in virus maturation (Rose *et al.*, 1982). However, one must keep in mind that the M protein of influenza virus, unlike the VSV M protein, has long stretches of hydrophobic amino acids and is not so positively charged (Gregoriades, 1980; Gregoriades and Frangione, 1981).

B. Virion Location of the Matrix Protein

The location of the M protein in the VSV virion has been determined by exposure to proteolytic enzymes and by various cross-linking reagents. The M protein was found to be refractory to proteolytic cleavages when virions were treated with trypsin, pronase, thermolysin, chymotrypsin, or bromelain (Mudd, 1974; McSharry *et al.*, 1971; Bishop *et al.*, 1975; Schloemer and Wagner, 1975). Moreover, no labeling of M protein was observed when the intact virions were carefully iodinated by the lacto-peroxidase-catalyzed reaction in the absence of EDTA (McSharry, 1977). These findings suggest that the matrix protein is not exposed at the surface of the virus.

1. Membrane Association

Labeling studies of M protein in intact virions with various hydrophilic and lipophilic cross-linking reagents have led to the hypothesis that the M protein is located in close proximity to the interior leaflet of the viral membrane. Zakowski and Wagner (1980) demonstrated that the M protein is preferentially cross-linked to phospholipids by the surface-membrane reagent tartryldiazide, but is refractory to phospholipid cross-linking by the hydrophobic reagent 4,4'-dithiobisphenylazide inserted in the virion membrane. In addition, the photoactivated arylazide probe [³H]pyrenesulfonylazide labeled the M protein to a greater extent than did the more hydrophobic probe [¹²⁵I]-5-iodonaphthylazide, suggesting that the M protein does not penetrate the membrane very deeply. Interestingly, Mancarella and Lenard (1981) observed increased labeling with [¹²⁵I]-5-iodonaphthylazide of the M protein of a group III *ts* mutant, suggesting that in this mutant virus the M protein appears to penetrate the membrane more deeply than does the wild-type M protein. In a separate study with photoreactive lipid probes, such as 16-azido(9,10-[³H]₂)palmitate, Stoffel *et al.* (1978) were able to demonstrate extensive labeling of glycoprotein (G), but no labeling of M protein. Indeed, similar results were obtained by Capone *et al.* (1983), who also failed to observe any labeling of M protein with viral phospholipid containing ω[9,³H]diazirinophenoxy nononoate, suggesting that the M protein does not penetrate the lipid bilayer to a significant extent. By the same token, Pepinsky and Vogt (1979) showed that the VSV virion M protein could be cross-linked to membrane phosphatidylethanolamine by the superficial free amino acid cross-linking reagent dimethylsuberimidate. These results are consistent with the postulate that the M protein does not penetrate, but lies in close proximity to, the inner leaflet of the VS virion membrane.

The interaction of M protein with a lipid bilayer was demonstrated *in vitro* by its reconstitution of isolated M protein with negatively charged phospholipid vesicles (Zakowski *et al.*, 1981). Reconstitution of the basic M protein with the lipid bilayer occurred only in the presence of negatively charged phospholipids, such as phosphotidylserine (PS), phosphatidic acid, or phosphatidylinositol; no significant reconstitution was observed with phospholipid vesicles made of 100% neutral phospholipids, such as phosphatidylcholine (PC). An additional indication of the electrostatic nature of M-protein binding to the vesicles was the fact that M protein could not be reconstituted in the presence of 0.5 M NaCl. Nonelectrostatic forces also appeared to be involved in the association of the M protein with the bilayer, since previously reconstituted M protein remained associated with the vesicle membrane on subsequent exposure to high salt. In a later study, Wiener *et al.* (1983a) showed that the M protein could be reconstituted with preformed small unilamellar vesicles or with preformed fused vesicles composed of PC and phosphatidylglycerol (PC/PG) or PC/PS. Comparative buoyant-density studies of detergent-dialyzed dimyristoylphosphatidylserine/dipalmitoylphosphatidylcholine (DMPS)/(DPPC) vesicles or sonicated DPPC/dipalmitoylphosphatidylglycerol (DPPG) vesicles reconstituted in the absence or presence of VSV M protein are shown in Fig. 4. The PC/PG vesicles were shown to possess a greater capacity to bind M protein than were the PC/PS vesicles when an equivalent amount of M protein was added to the preformed vesicles. Binding of M protein to more vesicles resulted in a significant increase in vesicle density as determined by sucrose density-gradient centrifugation. Proteolysis of isolated vesicles reconstituted with M protein indicated that 95% of the protein was accessible to and degradable by the thermolysin, suggesting that the M protein in these reconstituted vesicles is interacting in a peripheral fashion with the lipid bilayer. In this context, it is to be noted that the influenza M protein, which may be related to VSV M protein in terms of their location within the virion and their presumed functions in viral assembly, could be reconstituted with uncharged phospholipid vesicles, and such reconstitution was unaffected by the presence of salt in the system (Bucher *et al.*, 1980; Gregoriades, 1980; Gregoriades and Frangione, 1981). These studies tend to suggest that the M protein of influenza virus penetrates more deeply into the lipid bilayer, presumably due to hydrophobic interaction between the protein and the lipid bilayer, in contrast to that observed with the VSV M protein.

2. Ribonucleoprotein Association

The M protein has been shown to interact strongly with RNP cores of the virus. Treatment of intact VS virions with penetrating bivalent

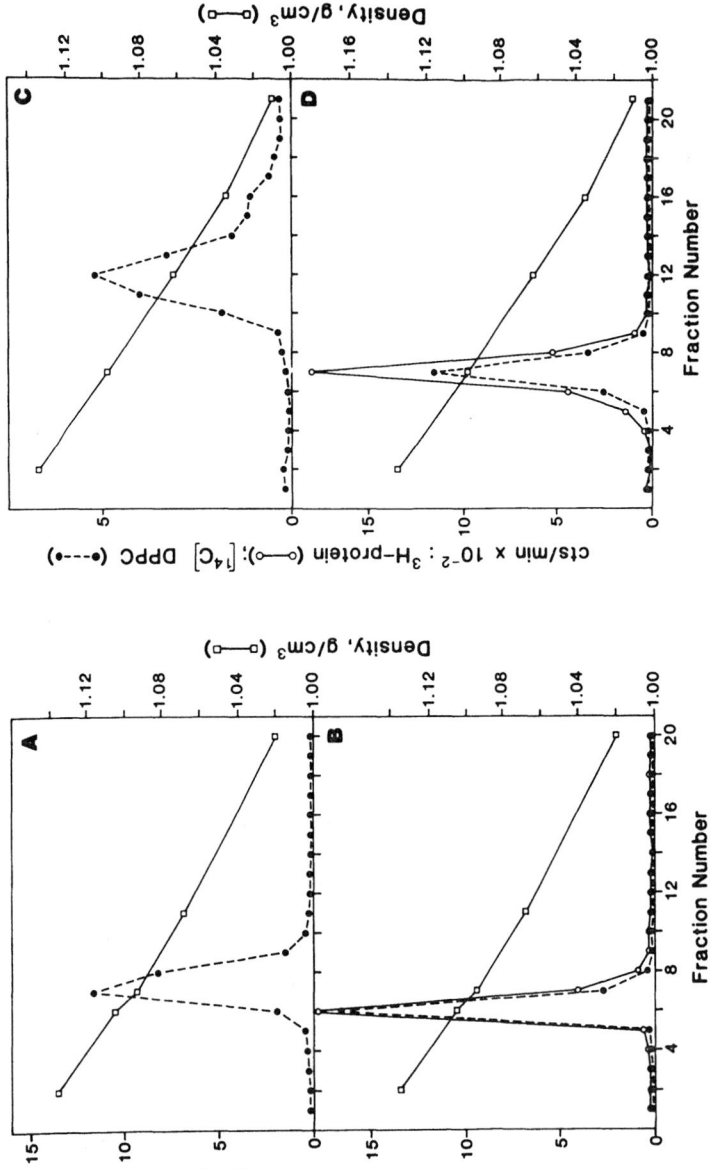

FIGURE 4. Comparative buoyant density of detergent-dialyzed DMPS/DPPC (1 : 1) vesicles or sonicated DPPG/DPPC (1 : 1) vesicles (1 : 1) reconstituted in the absence or presence of M protein of VSV. (A) DMPS/DPPC vesicles alone; (B) DMPS/DPPC vesicles containing 0.5 mole% M protein; (C) DPPG/DPPC vesicles alone; (D) DPPG/DPPC vesicles containing 0.5 mole% M protein. All vesicle preparations contained [^{14}C]-DPPC in trace quantity to label the lipid, and the M protein was metabolically labeled with [^{3}H]amino acids. Buoyant density was determined by flotation upward through a 0–30% sucrose gradient subjected to centrifugation at 200,000g at 46°C for 16 hr. Gradients were fractionated and 0.4-ml aliquots counted for ^{3}H and ^{14}C. (○) ^{3}H-labeled protein; (●)[^{14}C]-DPPC; (□), density (g/cm^3). Reprinted from Wiener et al. (1983a) by permission of Biochemistry.

protein cross-linking reagents showed the presence of both M–M homodimers and M–N heterodimers (Dubovi and Wagner, 1977), suggesting
strong interaction of matrix protein with RNP cores. Solubilization of
VSV virions with detergent in the absence of salt resulted in the formation
of RNP cores in which M protein was found to be associated with the
nucleocapsids (Carroll and Wagner, 1979; Wilson and Lenard, 1981). It
has also been demonstrated that extraction of VS virions with octylglucoside detergent in the absence of salt yields subviral structures called
"skeletons" (Newcomb and Brown, 1981). The skeletons contained nucleocapsids (N), L, NS, and M proteins, but lacked the G protein completely. Morphologically, these skeleton structures resembled VSV virions except that they were smaller in diameter than the intact virus. In
a subsequent study, Newcomb *et al.* (1982) observed that extraction of
VSV with octylglucoside in the presence of 0.5 M NaCl yielded highly
extended nucleocapsids in which N protein was predominant. Dialysis
of VSV disrupted with detergent and salt to remove NaCl led to the
reassociation of nucleocapsids with the M protein. These reassociated
structures were found to be highly condensed and resembled "native"
VSV skeletons produced by the extraction of virions with detergent in a
low-ionic-strength buffer. Similar observations were noted by De *et al.*
(1982), who observed condensation of RNP cores by M protein in the
presence of low-salt buffer. Heggeness *et al.* (1980) also found that the
conformations of the helical nucleocapsids of paramyxoviruses and the
rhabdovirus VSV varied extensively with changes in salt concentration.
These studies suggest that the interaction of M protein with the RNP
core is electrostatic in nature and that the protein plays a functional role
in vitro in condensing the viral nucleocapsid to a compact structure. This
raises the possibility of a similar role of M protein *in vivo* during assembly
of the VS virions. In a separate study, Pal *et al.* (1981b,c) also observed
that disassembly of VS virions with pardaxin, a toxin from the Red Sea
flat fish, resulted in the formation of viral envelopes and nucleocapsid
cores that were readily separable by density-gradient centrifugation. The
M protein was found to be associated more strongly with the nucleocapsid
core than with the membrane after such dissociation of the virion structure. This study again supports the strong interaction of M protein with
RNP cores of the virus, imputing an important role to the M protein in
condensation of the ribonucleocapsid into a tightly coiled structure.

C. Matrix Protein Regulation of Vesicular Stomatitis Virus Transcription

The strong interaction of M protein with RNP cores raises the possibility that the M protein is an endogenous regulator of vital transcription. The presence of an endogenous inhibitor of *in vitro* viral transcription was first observed by Perrault and Kingsbury (1974), who noticed

that strong inhibition of RNA synthesis *in vitro* occurred when VSV virions were present at high concentrations. The inhibitor was considered to be an internal protein and showed some degree of serotype specificity. In a subsequent study, Breindl and Holland (1976) reported that transcribing VSV RNP cores appeared to lack the inhibitory activity found in complete virions. Later, Carroll and Wagner (1978) observed that certain polyanions, in particular poly(L-glutamic acid), were able to reverse the transcriptase-inhibitory activity observed at high virus concentration. In a subsequent study, Carroll and Wagner (1979) demonstrated that the M protein served as the endogenous inhibitor of *in vitro* viral transcription in VSV virions. Purified M protein was found to inhibit transcription by ribonucleocapsid cores, an effect that was partially reversed by polyglutamic acid. The M protein, being a basic protein, was able to bind to polyglutamic acid, thus resulting in a reversal of inhibition of transcription. It was also observed that the group III *ts* mutants of VSV (*ts*O23 and *ts*G31) with lesions in the M protein exhibited little or no endogenous inhibitory activity compared with the two wild-type strains and a group V mutant (*ts*O45) with a lesion in the G protein. In a separate study, Clinton *et al.* (1978) observed a 2- to 5-fold increase in transcription in cells infected with VSV *ts* mutants belonging to complementation group III at the nonpermissive temperature. Comparison of the transcriptional and translational products from mutant-infected cells revealed an overall increase in each of the viral mRNA species concomitant with degradation of the M protein at the nonpermissive temperature. The increase in mRNA, however, did not lead to increased synthesis of viral proteins. It was concluded from the quantitation of individual mRNA species that M protein was acting as a direct inhibitor of transcription as well as an attenuator of sequential transcription. Similar observations of M protein as a regulator of *in vitro* viral transcriptase were reported from other laboratories (Combard and Printz-Ané, 1979, De *et al.*, 1982; Wilson and Lenard, 1981; Pinney and Emerson, 1982). Wilson and Lenard (1981) noted that the interaction of M protein with the RNP core is electrostatic in nature on the basis of evidence that increasing salt concentration reversed transcription inhibition of RNP cores complexed with M protein. This interaction of M protein with RNP cores was markedly lower in VSV virions of group III *ts* mutants with lesions in the M protein. However, some revertants of several group III *ts* mutants were shown to possess M proteins with inhibitory activity similar to that of wild-type VSV virions (Wilson and Lenard, 1981). In another study, De *et al.* (1982) observed that wild-type VSV M protein significantly inhibited the RNA synthesis *in vitro* at low ionic strength. This inhibition was thought to be at the level of RNA chain elongation and not at the initiation step. Parallel electron-microscopic studies revealed that inhibition of transcription by M protein is accompanied by a profound structural change in the transcribing nucleocapsid, progressing from an extended structure to a highly compacted form. The mode of inhibition of *in vitro* VS viral transcription by M protein was also investigated by Pinney and Emerson (1982), who observed that the M protein restricts VSV RNA synthesis to short chains, 11–14 nucleotide long, that are transcribed from the 3' end of the N gene. The synthesis of these N-gene-coded oligonucleotides was found to increase under conditions of inhibited viral transcription by the M protein and was not observed with group III *ts* mutants. The authors suggested that the inhibition of transcription by M protein is a necessary step in preparing the viral nucleocapsid for budding.

It seems likely, therefore, that M protein interacts with the RNP core to cause the release of nascent RNA chains and prevent further movement of the polymerase along the template (see Chapter 6). In a different kind of study, Rosen *et al.* (1983) observed an M-protein-cleavage product with a molecular weight of approximately 17,500 that accumulates in VSV-infected CHO cells late in infection. It has been suggested that degradation of M protein in infected cells may be an important factor in the regulation of VS viral transcription *in vivo*.

D. Monoclonal Antibodies to the Matrix Protein of Vesicular Stomatitis Virus

Monoclonal antibodies have been extensively used in recent years to probe the antigenic properties of viral proteins (Yewdell and Gerhard, 1981). To probe the functional domains of the VSV M protein, monoclonal antibodies have recently been raised against the VSV M protein of both the Indiana and New Jersey serotypes (Pal *et al.*, 1985b; Ye *et al.*, 1985). Little, if any, antibody cross-reactivity could be found between the M proteins of the two serotypes. A large number of these antibodies was surprisingly found to be of the IgM isotype. By measuring the competitive binding of ^{125}I-labeled monoclonal antibodies, four antigenic determinants were identified for the M protein of each VSV serotype. Some antigenic determinants were unique, but others exhibited varying degrees of overlaps. In this regard, it should be noted that van Wyke *et al.* (1984) also reported three antigenic determinants in the M protein of influenza virus, with some overlap in two epitopes.

Monoclonal antibodies directed to a single epitope of the M protein of VSV-Indiana and to two M-protein epitopes of VSV-New Jersey were found to reverse the inhibition of transcription by VSV ribonucleocapsid cores (Pal *et al.*, 1985c; Ye *et al.*, 1985). In-depth studies have been performed to date only with the VSV-Indiana transcription system. Pal *et al.* (1985c) observed that antibodies to one antigenic determinant (epitope 1) were able to reverse the transcription inhibition by M protein of VSV-Indiana RNP cores. This epitope 1 was shown to be absent in the M protein of a group III *ts* mutant (*ts*O23), which also exhibits no capacity to inhibit VSV transcription. However, monoclonal 1 antibodies readily detected epitope 1 in M proteins of other group III *ts* mutants that also have considerable, but lesser, inhibitory activity for viral transcription. The M protein of a revertant (R11) isolated from *tsO23* virions was found to recover all its capacity to inhibit viral transcription and to bind monoclonal antibodies to epitope 1, whereas the M protein of three other revertants remained restricted in their capacity to inhibit transcription and to bind monoclonal antibodies to epitope 1. These studies indicate that exposure of epitope 1 on the surface of VSV-Indiana M protein appears to be essential for inhibiting transcription by VS viral RNP cores. Definite identification of the M-protein site responsible for transcription inhibition of VSV RNP cores must await detailed biochemical and genetic studies, which may be difficult to perform if secondary or tertiary structure of the M protein is the dominant feature. The amino-terminal region of the M protein is rich in hydrophilic amino acids and particularly lysine residues that contribute to the basic nature of the protein. Rose and Gallione (1981) postulated that the amino terminus of the M protein might be responsible for its binding to the negatively charged RNP cores. Monoclonal antibodies that recognize these or adjacent domains could provide useful probes for studying the interaction of M protein with the nucleocapsid complex, which may be an important component of the assembly of the virus.

V. MEMBRANE DYNAMICS

The simplicity of viral membranes makes them excellent model systems for studying the dynamic properties of biological membranes. In this section, we discuss the fluidity of the intact vesicular stomatitis virus (VSV) virion membrane compared with that of the plasma membrane of host cells; in this context, we also examine the effect of an integral compared with a peripheral protein on the dynamic properties of lipid vesicles reconstituted with either the glycoprotein (G) or the matrix (M) protein of VSV. The techniques generally used to measure the dynamics of lipids in virion and model membranes are electron spin resonance (ESR), fluorescence depolarization, and nuclear magnetic resonance (NMR) spectroscopy. In each of these techniques, a probe molecule of either endogenous or exogenous origin is incorporated within the lipid bilayer. In a few experiments, the thermotropic phase behavior of lipids in reconstituted vesicles has also been measured by differential scanning calorimetry.

A. Fluidity of the Virion Membrane

Analysis of the fluidity of the lipid bilayer of the intact VSV virion has demonstrated that the viral membrane is invariably more rigid than that of the host-cell plasma membrane from which it buds. The ESR spectrum of the VSV membrane probed with spin-labeled fatty acids indicated a considerably less fluid environment than that of the host-cell plasma membrane (Landsberger and Compans, 1976). This finding was confirmed by fluorescence depolarization studies of the VSV virion envelope using the hydrophobic fluorophore 1,6-diphenyl1-1,3,5-hexatriene (DPH) (Barenholz et al., 1976; Patzer et al., 1978a). The apparent fluidity of the membrane of intact virions, vesicles composed of viral lipids or protease-treated virions, was always found to be lower than that of the cell plasma membrane, thus suggesting that the lipid content rather than the presence of membrane protein was the major factor in determining the overall membrane dynamics. Similar differences in membrane fluidity between a viral membrane and infected host cells were also observed with a fish rhabdovirus, infectious hematopoietic necrosis virus (IHNV), and its host CHSE-214 cells (Moore et al., 1976). A comparison of membrane dynamics by fluorescence depolarization studies of two rhabdoviruses, VSV and fish IHNV, each grown in different cell lines at characteristic growth temperatures, led to some interesting results (Moore et al., 1976; Barenholz et al., 1976). At a given temperature, both IHNV and its host CHSE-214 cells were more fluid than the membrane of VSV or its host L cells. On the other hand, when the fluidity of the two systems was compared at their respective growth temperature, i.e., 18°C for IHNV and CHSE-214 cells and 37°C for VSV and L cells, then the fluidity of the viruses was similar and that of the cells was similar. These results tend to suggest that the host cells may be capable of altering their lipids to maintain a constant fluidity at their characteristic growth temperature and thereby cause an alteration in the fluidity of the viral membrane derived from these cells.

The intrinsic composition of the constituent lipids contributes significantly to the dynamics of the VSV envelope. Fluorescence depolarization studies of DPH embedded in the VSV membrane indicated that the removal of phospholipid headgroups by phospholipase C increases the fluidity of the lipid bilayer (Moore et al., 1977a). Another major contributing factor is the very large proportion of cholesterol in the VSV virion membrane. Fluorescence depolarization studies using various fluorophores such as DPH, trimethylammonium (TMA)–DPH and hydroxycoumarine demonstrated that depletion of cholesterol from the VSV envelope increased the fluidity of the membrane (Moore et al., 1978; Pal et al., 1981a, 1983). By use of the fluorophore transparinaric acid, which has an inherent property of partitioning into the gel phase of the membrane, it was observed that the depletion of cholesterol resulted in the formation of a gel phase in the VSV envelope (Pal et al., 1983). Formation of such gel phases in viral membranes might be due to the presence in the VSV envelope of 25 mole% sphingomyelin, which undergoes fluid-gel transition around 37°C.

The effect of viral proteins on the dynamic properties of the VSV lipid bilayer has been studied with intact virions, spikeless virus, and lipid vesicles derived from the virion. The ^{31}P NMR spectrum of the intact VSV membrane has shown a shorter spin-lattice relaxation time (T_1) for the phospholipid headgroups than for the extracted viral lipids in sonicated vesicles or in multilamellar dispersions (Moore et al., 1977b). This finding suggests a dramatic increase in the mobility of the phospholipid headgroups following removal of the G protein from the envelope. In a separate study, the headgroups of phosphatidylcholine (PC) and sphingomyelin in the outermost regions of the VSV membrane were metabolically labeled with [^{13}C]choline during growth, and their mobility was determined by spin-lattice relaxation time (T_1) using ^{13}C NMR. On the basis of these studies, Stoffel and Bister (1975) and Stoffel et al. (1976) reported that no change in the mobility of the viral lipids was detected in protease-treated virions or in the liposomes prepared from the extracted lipids free of viral proteins. The effects of proteins exerting their effects within the interstices of viral membranes have also been examined by ESR and NMR spectroscopy. At positions slightly deeper in the membrane, mainly at position C_3 and C_5 of phospholipid fatty acyl chains, there was a detectable increase in the fluidity of lipids in protease-treated virions or in protein-free viral lipid vesicles (Landsberger and Compans, 1976; Stoffel et al., 1976). Even at position C_{11} of the fatty acyl chain of viral membrane phospholipids, there was only a slight increase in lipid mobility in the absence of the proteins (Stoffel et al., 1976). However, both ESR and fluorescence spectroscopic studies have demonstrated that the most hydrophobic region of the bilayer of the VSV membrane, near the C_{16} position of fatty acyl chains, was unaffected by proteins (Patzer et al., 1978a; Barenholz et al., 1976; Landsberger and Compans, 1976). However, Stoffel et al. (1976), using ^{13}C NMR, reported a small increase in the fluidity in the core region of the VSV membrane in the absence of protein, but only in the extracted lipids labeled at the C_{16} position of the phospholipids. All these experiments suggest that the VSV membrane G protein tends to affect the dynamics of the virion lipid bilayer to varying degrees in different locations in the membrane. Penetration at the deeper regions of the viral membrane results in progressively less influence of the viral G protein on the dynamics of lipids in the bilayer.

B. Effects of Vesicular Stomatitis Virus Proteins on the Dynamics of Reconstituted Vesicles

1. G Protein

Reconstitution of the G protein and the M protein into well-defined lipid vesicles has offered simple model systems to study the interactions of VSV membrane proteins with lipid bilayers. As mentioned in the previous section, the G protein of VSV can be readily reconstituted into well-defined lipid vesicles by the detergent-dialysis method (Petri and Wagner,

1979). In these vesicles, the G protein was found to be present in the membrane in the same exterior orientation as that found in the native viral envelope. Analysis of the thermotropic behavior of dipalmitoyl-phosphatidylcholine (DPPC) vesicles reconstituted with the G protein has shown that the insertion of G protein in the bilayer decreased both the fluid–gel transition temperature and the enthalpy of transition (Petri et al., 1980). This decrease in transition enthalpy was found to be directly proportional to the mole percentage of the G protein in the bilayer. When this change in enthalpy by the G-protein concentration was extrapolated to zero, it was concluded that each G-protein molecule removes 270 ± 150 phospholipid molecules from the phase transition. Subsequent studies using DPPC vesicles reconstituted with G protein have shown that the G protein lowered the lipid-phase transition temperature and disordered the gel state significantly when monitored by the steady-state fluorescence depolarization of four membrane probes of varying hydrophobicity (Petri et al., 1981b). Such steady-state fluorescence depolarization studies with the four hydrophobic membrane probes residing in different regions of the bilayer are shown in Fig. 5. This study was further extended by Pal et al. (1983), who observed by differential polarized-phase fluorimetric techniques that the G protein disordered the lipid bilayer in the gel state and ordered it to some extent in the fluid state. Moreover, the G protein reconstituted with total viral lipids showed no effect on the dynamics of lipids, mostly because of the presence of a high concentration of cholesterol in the viral membrane (Pal et al., 1983). In a separate study, Allstiel and Landsberger (1981) noticed that most egg PC vesicles reconstituted with the G protein were found to exhibit greater membrane rigidity when monitored by ESR spectroscopy of a lipid-associated stearic acid spin-label.

2. M Protein

The M protein of VSV reconstitutes with preformed or detergent-dialyzed vesicles containing acidic phospholipids; at least 95% of the M protein peripherally attached to these reconstituted vesicles is susceptible to complete proteolytic digestion with thermolysin (Zakowski et al., 1981). The mode of interaction of M protein with lipid bilayers has been studied extensively by differential scanning calorimetry, steady-state fluorescence anisotropy, and differential polarized-phase fluorimetry of probes such as DPH, TMA-DPH, and trans-parinaric acid inserted into the bilayer.

Steady-state fluorescence depolarization studies with fluorophores DPH and TMA-DPH in DPPC/DPPG vesicles containing various proportions of M protein are shown in Fig. 6. In mixed vesicles containing DPPC/DPPG (1 : 1 molar ratio), the M protein was found to increase the fluid–gel transition temperature by 7–8°C and to increase the order of the lipids in the gel state to a significant extent (Wiener et al., 1983a). In

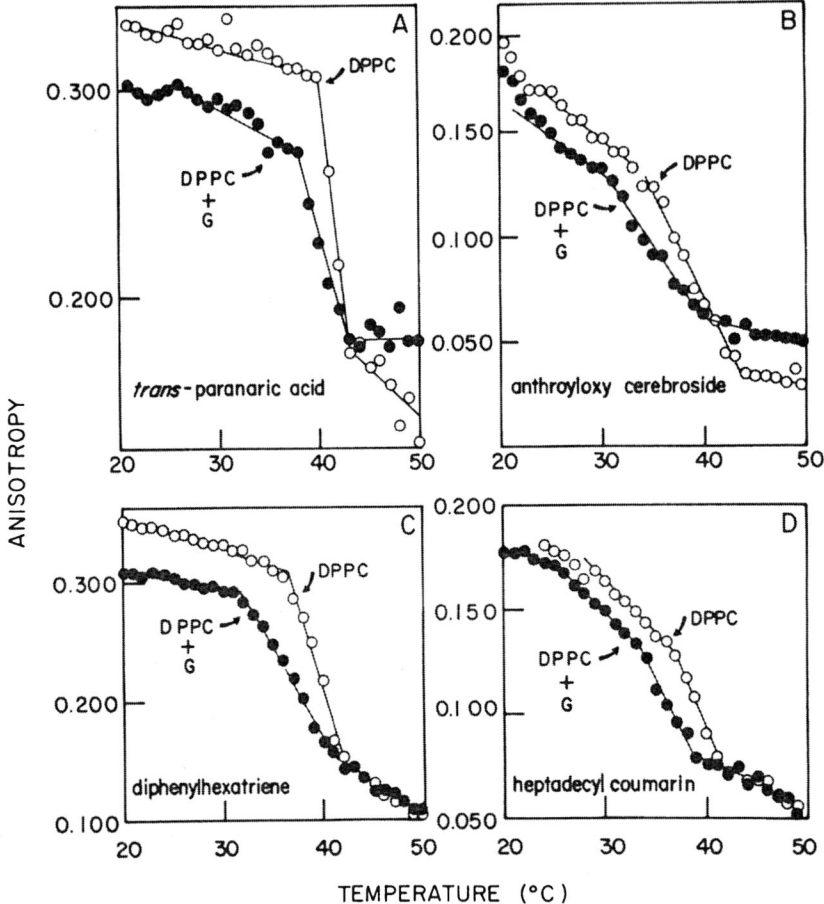

FIGURE 5. Fluorescence anisotropy of *trans*-paranaric acid (A), 16-(9-anthroy-loxy)palmitoylglucocerebroside (B), 1,6-diphenyl-1,3,5-hexatriene (C), and 4-heptadecyl-7-hydroxycoumarin (D) in G protein–DPPC vesicles as a function of temperature. The DPPC vesicles contained either no glycoprotein (○) or 0.45 mole% (A), 0.34 mole% (B), 0.56 mole% (C), or 0.40 mole% (D) G protein (●). Reprinted from Petri *et al.* (1981b) by permission of *Biochemistry.*

a subsequent report on studies of the thermotropic phase behaviors of M-protein-reconstituted DPPC/DPPG fused vesicles, it was observed that the M protein induced an upward shift in the phase-transition tempera-ture at M protein/phospholipid molar ratios as low as 1 : 12,000 (Wiener *et al.*, 1983b). At higher bilayer concentrations of M protein, phase sep-aration of lipids into acidic and neutral lipid components was observed (Wiener *et al.*, 1983b). Despite the effect of M protein on the transition temperature and the shape of the heat-capacity function, the enthalpy change associated with the transition remained constant at 7.1 kcal/mole, indicating that the lipids bound by the M protein are able to participate

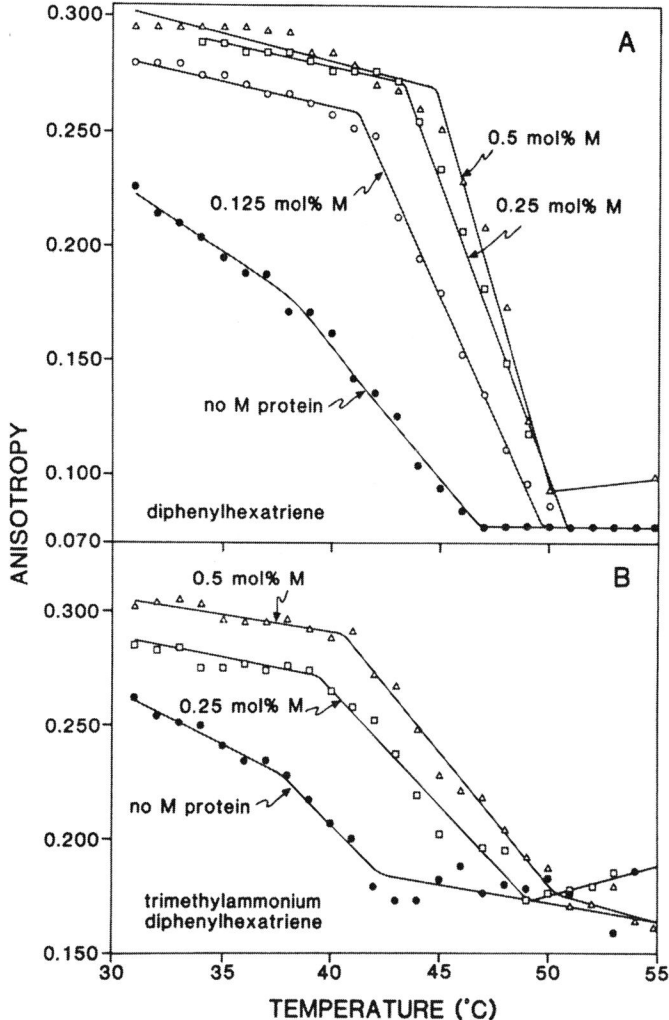

FIGURE 6. Fluorescence anisotropy as a function of temperature of 1,6-diphenyl-1,3,5-hexatriene (A) and 1-[4-(trimethylammonio)phenyl]-6-phenyl-1,3,5,-hexatriene (B) in sonicated DPPG/DPPC (1 : 1) vesicles reconstituted with varying concentrations of VSV M protein. The vesicles were reconstituted with no M protein (●), 0.125 mole% M protein (○), 0.25 mole% M protein (□), or 0.5 mole% M protein (△). Reprinted from Wiener *et al.* (1983a) by permission of *Biochemistry*.

in the phase transition. Evidence that M protein also induces phase separation of bilayer lipids was further confirmed by using pyrene phospholipid analogues as probes (Wiener *et al.*, 1985). These results, based on excimer formation of pyrene phospholipids in the bilayer reconstituted with M protein, suggested that the binding of M protein to membranes containing acidic phospholipid headgroups induces lateral reorganization of lipids in the membrane plane. This effect was far more pronounced during the transition of membrane lipids from the liquid–crystalline to the gel state.

The results of these studies on the interaction of G and M proteins with lipid bilayers clearly suggest that the two proteins affect the dynamics of membrane lipids in two different ways. The G protein interacts with the bilayer in an integral fashion and disorders the membrane in the gel state. On the other hand, the M protein interacts with the bilayer in a peripheral fashion and in the process increases the order of the lipids in the gel state and induces lateral phase separation of lipids into an M-protein-bound-lipid phase and a free-lipid phase.

VI. ASSEMBLY AND BUDDING OF VESICULAR STOMATITIS VIRUS

The final assembly of vesicular stomatitis virus (VSV) from its components takes place at the plasma membrane of infected host cells (Wagner et al., 1971). Pulse–chase experiments suggested that the three major VS viral proteins take independent paths to the site of assembly (Knipe et al., 1977a,b). Viral nucleocapsid (N) protein is first synthesized as a soluble protein in the cytosol and then preassembles with RNA to form the nucleocapside in the cytosol of the infected cell (Atkinson, 1978; David, 1977; Knipe et al., 1977a). The matrix (M) protein is synthesized on non-membrane-bound polysomes, but becomes associated with membrane structures with the same density as whole virus (Atkinson, 1978; David,.1977; Knipe et al., 1977a). The glycoprotein (G), on the other hand, is synthesized on membrane-bound ribosomes (Both et al., 1975; Morrison and Lodish, 1975) and migrates through internal membranous structures to the plasma membrane (Rothman et al., 1980). The nucleocapsids and the M protein then associate with the plasma membrane and exit the cell, picking up the G-protein-containing lipid bilayer in the process (Knipe et al., 1977a).

The molecular interactions that occur during the assembly of VSV virions are not well understood. A number of models can be proposed to describe such an event. In one model, the M protein may associate with the negatively charged phospholipids in the inner layer of the membrane and then bind to the carboxy terminus of the G protein protruding through the plasma membrane. Viral nucleocapsids could presumably recognize these G–M pairs, bind to them, and induce clustering of G protein by the multivalent nature of the nucleocapsids. This process could lead to budding of the virus. Association of M protein with the lipid bilayer at the cytoplasmic domain is supported by the fact that M protein binds to the plasma membrane of infected cells (Morrison and McQuain, 1978). M protein can be reconstituted with lipid bilayers containing acidic phospholipids, particularly phosphatidylserine (PS), which is asymmetrically oriented in the inner layer of the viral membrane (Zakowski et al., 1981;

Wiener *et al.*, 1983a). The matrix protein can be cross-linked to phospholipids in intact virions by various cross-linking reagents (Zakowski and Wagner, 1980). Moreover, the ribonucleocapsid protein (RNP) cores of VSV virions containing only N, L, and NS proteins were found to associate with phospholipid vesicles reconstituted with the M protein (Ogden *et al.*, 1986; Capone and Ghosh, 1984). In another more likely model, the M protein binds first to nucleocapsids, and this M–nucleocapsid structure may bind to acidic PS headgroups or the carboxy terminus of G protein, or both. In either model, the M protein functions as an intermediate with dual recognition sites for G protein and viral nucleocapsid structures. The strong interaction of M protein with viral nucleocapsids is electrostatic in nature and is responsible for inhibition of viral transcription both *in vivo* and *in vitro* (Clinton *et al.*, 1978; Carroll and Wagner, 1979).

The interaction of G protein with M protein and RNP during assembly of VSV is exemplified by clustering of G protein in areas of the plasma membrane destined for viral assembly. Wagner *et al.* (1971) demonstrated by electron-microscopic analysis of budding intermediates during assembly of VSV that the viral G protein is widely distributed on the surface of the plasma membrane of infected cells active in viral assembly. More recently, Jacobs and Penhoet (1982) demonstrated that the G protein of VSV is clustered in the plasma membrane of infected Chinese hamster lung cells during morphogenesis; they suggested that viral nucleocapsids are required for this clustering. Using a temperature-sensitive (*ts*) mutant VSV-New Jersey virus (*ts*E-1), which is temperature-sensitive for the synthesis of viral nucleocapsids, Jacobs and Penhoet (1982) observed that early in infection at permissive temperatures, the G protein was randomly distributed on the surface of membrane ghosts. Later, clusters of G protein molecules were seen on the surface of the cell. In contrast, ghosts prepared from virus-infected cells maintained at a restrictive temperature always had a random distribution of viral G protein. In an early study by the fluorescence photobleaching recovery technique to measure the mobility of G protein on the surface of infected cells, Reidler *et al.* (1981) demonstrated that mutations in the M gene lead to an increased fraction of G protein being mobile at a restrictive temperature as compared with the fraction that was mobile at the permissive temperature. A similar technique was used by Johnson *et al.* (1981) to study the envelopment of Sindbis virus and VSV by the plasma membrane of infected cells; they suggested that the newly synthesized G protein moved to the cell surface and mixed freely with a large pool of mobile membrane G protein. A small fraction of the mobile G protein then aggregated at nucleation sites where budding occurred. The nucleation sites might be represented by those regions in which VSV M protein interacted with the cytoplasmic tail segment of G protein and thus aggregated these G protein molecules into patches. Envelopment results as these G-protein-containing membrane regions interact with helical ribonucleocapsid structures. In contrast, the surface G proteins of Sindbis-virus-infected host cells were found to be partially mobile early after infection and immobile at later times when greater amounts of these proteins reached the cell surface. These results suggest that unlike VSV, Sindbis-virus assembly initiates at an intracellular vesicle where G proteins aggregate and bind nucleocapsids.

Lodish and Porter (1980a,b) demonstrated that an interaction between a precise number of surface G proteins with either the M or the N protein in infected cells may not be an essential prerequisite for VSV maturation and, in fact, the VSV G protein may not be essential for the formation of a budding virion. They reported that cells infected at nonpermissive temperature by VSV mutants (in complementation group V) defective in maturation of the G protein, released at low levels noninfectious particles with the shape and physical properties of normal VSV. These particles contained no appreciable amount of G protein, but had VSV RNA and all other VSV proteins in normal proportions (Lodish and Weiss, 1979). Lodish and Porter (1980b) also isolated VSV particles, formed at different times after infection, that contained varying proportions of G protein but a constant amount of M and N polypeptides. The ratio of G protein to either M or N protein was found to vary over a 6-fold range, indicating that the interaction between the precise number of G proteins with either of the underlying M or N proteins may not be a prerequisite for budding of infectious virus particles from the cell surface. In contrast, the association of M and N proteins had been shown to be an important interaction for virus assembly. By the use of various *ts* mutants in complementation group III (the M-protein gene), noninfectious virus particles were produced that contained the viral M and G proteins, but less than 10% of the normal proportion of N protein or RNA (Schnitzer and Lodish, 1979). Since these *ts* mutants had ostensibly a defective structural gene for the M protein, these findings seem to indicate that an association between the matrix protein and the nucleocapsid N protein may be of primary importance in the budding of virus particles from the plasma membrane of infected cells. The key role of M protein in virus assembly was also demonstrated by phenotypic mixing experiments between the mutants of VSV and retroviruses (Weiss and Bennett, 1980).

Although VSV maturation sometimes occurs on internal cell membranes (Zee et al., 1970), the budding of VSV virions usually takes place at the plasma membrane of infected host cells. For the most part, the lipid composition of the VSV membrane is a reflection of lipids present in the plasma membrane. However, the envelope is relatively enriched in amino phospholipids and cholesterol (Patzer et al., 1978a). This suggests the possibility that the VSV virions may bud from localized regions of the cellular plasma membrane. The origin of the envelope lipids acquired by VSV during budding from the plasma membrane of chick-embryo fibroblasts was examined by Pessin and Glasar (1980). When the phospholipid composition of the cells was modified by growing them in the presence of the choline analogue N,N-dimethylethanolamine or 1,2-amino-1-butanol, the phospholipid composition of the virus was subsequently altered, but in a very different manner than that of the plasma membrane. These results were interpreted as indicating that VS virions bud from localized lipid regions of the plasma membrane that do not always reflect the average composition of lipids in the plasma membrane. It is quite likely that viral membrane proteins may be involved in such lipid recruitment during budding of the virus.

The budding of VSV virions and other viruses from the apical or basolateral regions of polarized epithelial cells has been extensively studied, but will not be covered in detail here. The initial observation of preferential budding of VSV virions and influenza virus in polarized epithelial cells was made by Rodriguez-Boulan and Sabatini (1978), who observed that when Madin–Darby canine kidney (MDCK) cells were infected with influenza virus or with VSV, progeny influenza virus budded from the apical plasma membrane, whereas VSV progeny were formed by budding through the basolateral plasma membrane. This observation was subsequently confirmed and extended in various laboratories (Fuller *et al.*, 1984; van Meer and Simons, 1982; Roth and Compans, 1981; Rodriguez-Boulan and Pendergast, 1980; Rodriguez-Boulan, 1983). Unique and different phospholipid compositions were noted in the membranes of influenza virus and VSV budding from apical and basolateral membrane regions, respectively (van Meer and Simons, 1982). Influenza virus was enriched in phosphatidylethanolamine and phosphatidylserine, whereas VSV contained higher proportions of phosphatidylcholine, sphingomyelin, and phosphatidylinositol. When MDCK cells were trypsinized after infection and then replated, but did not form confluent monolayers, the phospholipid composition of the two viruses was very similar. This finding led the authors to conclude that the two virions obtained their lipids from the plasma membrane passively and, presumably, by the same mode; therefore, the different phospholipid compositions of the two viruses grown in polarized cells merely reflect differences in the phospholipid composition of the two plasma-membrane domains. Immuno-electron-microscopic examination of MDCK cells doubly infected with VSV and influenza virus demonstrated that the influenza hemagglutinin (HA) and the VSV G protein traversed the same Golgi apparatus and even the same Golgi cisternae, whence they migrated to the cell surface (Rindler *et al.*, 1984). These data indicate that the pathways of the two viral glycoproteins migrating to the plasma membrane did not diverge before passage through the Golgi apparatus. Therefore, the critical sorting steps in channeling the proteins to either the apical or the basolateral surface presumably takes place during or after passage of the individual glycoproteins through the Golgi organelle. In a separate study, Rodriguez-Boulan and Pendergast (1980) observed that only the asymmetrical distribution of viral envelope proteins, rather than a polarized delivery of nucleocapsids, directed the polarized budding of the two viruses in MDCK cells. Interestingly, certain ionophores were shown to inhibit completely the budding of VSV G protein from MDCK cells without affecting the expression of influenza virus HA on the surface of these cells (Alonso-Caplen and Compans, 1983).

The carbohydrate moieties of the glycoprotein of influenza virus and VSV apparently do not serve as the signals that determine the polarized distribution of viral envelope proteins on the surface of epithelial cells. Roth *et al.* (1979) examined the budding of VSV and influenza virus in MDCK cells treated with tunicamycin at a concentration that completely inhibited glycosylation of viral glycoproteins; they found that polarity in the maturation sites of both viruses was maintained in tunicamycin-treated MDCK cells, suggesting that glycosylation of viral glycoprotein may not be a requirement for the determination of the cellular maturation site of these enveloped viruses. A similar result was also obtained by Green *et al.* (1981), who studied the maturation of VSV and influenza virus from concanavalin-A- and ricin-resistant mutants of MDCK cells. Several laboratories are continuing intensive studies to uncover the biological properties of epithelial cells that control the diverse polarized functions that also regulate differential budding of VSV and influenza virus from different membrane domains separated by tight junctions.

ACKNOWLEDGMENTS. The research from our laboratory reported herein was supported by grants from the American Cancer Society (MV-9) and from the National Institute of Allergy and Infectious Diseases (AI-21652 and AI-11112). We are deeply indebted to Sandy Shiflett for patience and excellent preparation of the manuscript.

REFERENCES

Alonso-Caplen, F. V., and Compans, R. W., 1983, Modulation of glycosylation and transport of viral membrane glycoprotein by a sodium ionophore, *J. Cell Biol.* **97**:659.

Altstiel, L. D., and Landsberger, F. R., 1981, Lipid–protein interactions between the surface glycoprotein of vesicular stomatitis virus and the lipid bilayer, *Virology* **115**:1.

Anilionis, A., Wunner, W. H., and Curtis, P. J., 1981, Structure of the glycoprotein gene in rabies virus, *Nature (London)* **294**:275.

Arnheiter, H., Dubois-Dalcq, M., and Lazzarini, R. A., 1984, Direct visualization of protein transport and processing in the living cell by microinjection of specific antibodies, *Cell* **39**:99.

Atkinson, P. H., 1978, Glycoprotein and protein precursors to plasma membranes in vesicular stomatitis virus infected HeLa cells, *J. Supramol. Struct.* **8**:89.

Atkinson, P. H., Moyer, S. A., and Summers, D. F., 1976, Assembly of vesicular stomatitis virus glycoprotein and matrix protein into HeLa cell plasma membranes, *J. Mol. Biol.* **102**:613.

Barenholz, Y., Moore, N. F., and Wagner, R. R., 1976, Enveloped viruses as model membrane systems: Microviscosity of vesicular stomatitis virus and host cell membranes, *Biochemistry* **15**:3563.

Bates, S. R., and Rothblat, G. H., 1972, Incorporation of L cell sterols into vesicular stomatitis virus, *J. Virol.* **9**:883.

Bell, J. C., Brown, E. G., Takayesu, D., and Prevec, L., 1984, Protein kinase activity associated with immunoprecipitates of the vesicular stomatitis virus phosphoprotein NS, *Virology* 132:229.

Bergmann, J. E., Tokuyasu, K. T., and Singer, S. J., 1981, Passage of an integral membrane protein, the vesicular stomatitis virus glycoprotein, through the Golgi apparatus enroute to the plasma membrane, *Proc. Natl. Acad. Sci. U.S.A.* 78:1746.

Bishop, D. H. L., Repik, P., Obijeski, J. F., Moore, N. F., and Wagner, R. R., 1975, Restitution of infectivity to spikeless vesicular stomatitis virus by solubilized viral components, *J. Virol.* 16:75.

Bittman, R., Majuk, Z., Honig, D. J., Compans, R. W., and Lenard, J., 1976, Permeability properties of the membrane of vesicular stomatitis virions, *Biochim. Biophys. Acta* 433:63.

Blough, H. A., Tiffany, J. M., and Aaslestad, H. G., 1977, Lipids of rabies virus and BHK-21 cell membranes, *J. Virol.* 21:950.

Both, G. W., Moyer, S. A., and Banerjee, A. K., 1975, Transition and identification of the viral mRNA species isolated from subcellular fractions of vesicular stomatitis virus-infected cells, *J. Virol.* 15:1012.

Breindl, M., and Holland, J. J., 1976, Studies on the *in vitro* transcription and translation of vesicular stomatitis virus mRNA, *Virology* 73:106.

Brown, E. G., and Prevec, L., 1978, Proteins of vesicular stomatitis virus. IV. A comparison of the tryptic peptides of the vesicular stomatitis group of rhabdoviruses, *Virology* 89:7.

Brown, E. G., and Prevec, L., 1979, Comparative analyses of vesiculovirus proteins utilizing partial cleavage fragments at tryptophan residues, *Virology* 95:244.

Brown, E. G., and Prevec, L., 1982, Linear mapping of tryptophan residues in vesiculovirus M and N proteins by partial chemical cleavage, *J. Virol.* 43:311.

Bucher, D. J., Kharitonenkov, I. G., Zakomirdin, J. A., Grigoriev, V. B., Klimenko, S. M., and Davis, J. F., 1980, Incorporation of influenza virus M-protein into liposomes, *J. Virol.* 36:586.

Burge, B. W., and Huang, A. S., 1979, Conserved peptides in the proteins of vesicular stomatitis virus, *Virology* 95:445.

Capone, J., and Ghosh, H. P., 1984, Association of the nucleocapsid protein N of vesicular stomatitis virus with phospholipid vesicles containing the matrix protein M, *Can. J. Biochem. Cell. Biol.* 62:1174.

Capone, J., Toneguzzo, F., and Ghosh, H. P., 1982, Synthesis and assembly of membrane glycoproteins: Membrane anchoring COOH terminal domain of vesicular stomatitis virus envelope glycoprotein G contains fatty acids, *J. Biol. Chem.* 257:16.

Capone, J., Leblanc, P., Gerberg, G. E., and Ghosh, H. P., 1983, Localization of membrane proteins by the use of a photoreactive fatty acid incorporated *in vivo* into vesicular stomatitis virus, *J. Biol. Chem.* 258:1395.

Carroll, A. R., and Wagner, R. R., 1978, Reversal by certain polyanions of an endogenous inhibitor of the vesicular stomatitis virus-associated transcriptase, *J. Biol. Chem.* 253:3361.

Carroll, A. R., and Wagner, R. R., 1979, Role of the membrane (M) protein in endogenous inhibition of *in vitro* transcription by vesicular stomatitis virus, *J. Virol.* 29:134.

Cartwright, B., and Brown, F., 1972, Glycolipid nature of the complement-fixing host cell antigen of vesicular stomatitis virus, *J. Gen. Virol.* 15:243.

Cartwright, B., Smale, C. J., and Brown, F., 1969, Surface structure of vesicular stomatitis virus, *J. Gen. Virol.* 5:1.

Chatis, P. A., and Morrison, T. G., 1981, Mutational changes in the vesicular stomatitis virus glycoprotein affect the requirement of carbohydrate in morphogenesis, *J. Virol.* 37:307.

Chatis, P. A., and Morrison, T. G., 1983, Charcterization of the soluble glycoprotein released from vesicular stomatitis virus-infected cells, *J. Virol.* 45:80.

Clinton, G. M., Little, S. P., Hagen, F. S., and Huang, A. S., 1978, The matrix (M) protein of vesicular stomatitis virus regulates transcription, *Cell* 15:1455.

Clinton, G. M., Guerina, N. G., Guo, H. Y., and Huang, A. S., 1982, Host-dependent phos-phorylation and kinase activity associated with vesicular stomatitis virus, *J. Biol. Chem.* **257**:3313.

Combard, A., and Printz-Ané, C., 1979, Inhibition of vesicular stomatitis virus transcriptase complex by the virion envelope M protein, *Biochem. Biophys. Res. Commun.* **88**:117.

Compans, R. W., and Klenk, H.-D., 1979, Viral membranes, in: *Comprehensive Virology*, Vol. 13 (H. Fraenkel-Conrat and R. R. Wagner, eds.), pp. 293–407, Plenum Press, New York.

Crimmins, D. L., and Schlesinger, S., 1982, Physical properties of the glycoprotein of ve-sicular stomatitis virus measured by intrinsic fluorescence and aggregation, *Biochem-istry* **21**:3518.

David, A. E., 1977, Assembly of the VSV envelope: Transfer of viral polypepltides from polysomes to cellular membranes, *Virology* **76**:98.

De, B. P., Thornton, G. B., Luk, D., and Banerjee, A. K., 1982, Purified matrix protein of vesicular stomatitis virus blocks viral transcription *in vitro*, *Proc. Natl. Acad. Sci. U.S.A.* **79**:7137.

Dietzschold, B., 1977, Oligosaccharides of the glycoprotein of rabies virus, *J. Virol.* **23**:286.

Dille, B. J., and Johnson, T. C., 1982, Inhibition of vesicular stomatitis virus glycoprotein expression by chloroquine, *J. Gen. Virol.* **62**:91.

Doel, T. R., and Brown, F., 1978, Tryptic peptide analysis of the structural proteins of vesicular stomatitis virus, *J. Gen. Virol.* **38**:351.

Dubovi, E. J., and Wagner, R. R., 1977, Spatial relationships of the proteins of vesicular stomatitis virus: Induction of reversible oligomers by cleavable protein cross-linkers and oxidation, *J. Virol.* **22**:500.

Dunphy, W. G., Fries, E., Urbani, L., and Rothman, J. E., 1981, Early and late functions associated with the Golgi apparatus reside in distinct compartments, *Proc. Natl. Acad. Sci. U.S.A.* **78**:7453.

Eger, R., Compans, R. W., and Rifkin, D. B., 1975, The organization of the proteins of vesicular stomatitis virions: Labeling with pyridoxal phosphate, *Virology* **66**:610.

Eidelman, O., Schlegel, R., Tralka, T. S., and Blumenthal, R., 1984, pH Dependent fusion induced by vesicular stomatitis virus glycoprotein reconstituted into phospholipid ves-icles, *J. Biol. Chem.* **259**:4622.

Etchison, J. R., and Holland, J. J., 1974, Carbohydrate composition of the membrane gly-coprotein of vesicular stomatitis virus grown in four mammalian cell lines, *Proc. Natl. Acad. Sci. U.S.A.* **71**:4011.

Flamand, A., 1970, Etude génétique du virus de la stomatite vesiculaire: Classement de mutants thermosensibles spontanés en groupes de complementation, *J. Gen. Virol.* **8**:187.

Fong, B. S., and Brown, J. C., 1978, Asymmetric distribution of phosphatidylethanolamine fatty acyl chains in the membrane of vesicular stomatitis virus, *Biochim. Biophys. Acta* **510**:230.

Fong, B. S., Hunt, R. C., and Brown, J. C., 1976, Asymmetric distribution of phosphatidyl-ethanolamine in the membrane of vesicular stomatitis virus, *J. Virol.* **20**:658.

Friedman, R. M., and Pastan, I., 1969, Nature and function of the structural phospholipids of an arbovirus (SFV), *J. Mol. Biol.* **40**:107.

Fries, E., and Rothman, J. E., 1981, Transient activity of Golgi-like membranes as donors of vesicular stomatitis viral glycoprotein *in vitro*, *J. Cell. Biol.* **90**:697.

Fuller, S., von Bonsdorff, C. H., and Simons, K., 1984, Vesicular stomatitis virus infects and matures only through the basolateral surface of the polarized epithelial cell line, MDCK, *Cell* **38**:65.

Gahmberg, C. G., Simons, K., Renkonen, O., and Kääriäinen, L., 1972, Exposure of proteins and lipids in the Semliki Forest virus membrane, *Virology* **50**:259.

Gallione, C. J., and Rose, J. K., 1983, Nucleotide sequence of a cDNA clone encoding the entire glycoprotein from the New Jersey serotype of vesicular stomatitis virus, *J. Virol.* **46**:162.

Garreis-Wabnitz, C., and Kruppa, J., 1984, Intracellular appearance of a glycoprotein in VSV-infected BHK cells lacking the membrane-anchoring oligopeptide of the viral G-protein, Eur. Mol. Biol. Org. J. 3:1469.

Ghosh, H. P., 1980, Synthesis and maturation of glycoproteins of enveloped animal viruses, Rev. Infect. Dis. 2:26.

Gibson, R., Leavitt, R., Kornfeld, S., and Schlesinger, S., 1978, Synthesis and infectivity of vesicular stomatitis virus containing nonglycosylated G protein, Cell 13:671.

Gibson, R., Schlesinger, S., and Kornfeld, S., 1979, The nonglycosylated glycoprotein of vesicular stomatitis virus is temperature-sensitive and undergoes intracellular aggregation at elevated temperatures, J. Biol. Chem. 254:3600.

Gibson, R., Kornfeld, S., and Schlesinger, S., 1981, The effect of oligosaccharide chains of different sizes on the maturation and physical properties of the G protein of vesicular stomatitis virus, J. Biol. Chem. 256:456.

Gopalakrishna, Y., and Lenard, J., 1985, Sequence alterations in temperature-sensitive M protein mutants (complementation group III) of vesicular stomatitis virus, J. Virol. 56:655.

Green, R. F., Meiss, H. K., and Rodriguez-Boulan, E., 1981, Glycosylation does not determine segregation of viral envelope proteins in the plasma membrane of epithelial cells, J. Cell Biol. 89:230.

Gregoriades, A., 1980, Interaction of influenza M protein with viral lipid and phosphatidylcholine vesicles, J. Virol. 36:470.

Gregoriades, A., and Frangione, B., 1981, Insertion of influenza M protein into the lipid bilayer and localization of site of insertion, J. Virol. 40:323.

Heggeness, M. H., Scheid, A., and Choppin, P. W., 1980, Conformation of the helical nucleocapsids of paramyxoviruses and vesicular stomatitis virus: Reversible coiling and uncoiling induced by changes in salt concentration, Proc. Natl. Acad. Sci. U.S.A. 77:2631.

Hsu, M. C., Scheid, A., and Choppin, P. W., 1979, Reconstitution of membranes with individual paramyxovirus glycoproteins and phospholipid in cholate solution, Virology 95:476.

Hunt, L. A., Davidson, S. K., and Golemboski, D. B., 1983, Unusual heterogeneity in the glycosylation of the G protein of Hazelhurst strain of vesicular stomatitis virus, Arch. Biochem. Biophys. 226:347.

Imblum, R. L., and Wagner, R. R., 1974, Protein kinase and phosphoproteins of vesicular stomatitis virus, J. Virol. 13:113.

Irving, R. A., and Ghosh, H. P., 1982, Shedding of vesicular stomatitis virus soluble glycoprotein by removal of carboxy-terminal peptide, J. Virol. 42:322.

Jacobs, B. L., and Penhoet, E. E., 1982, Assembly of vesicular stomatitis virus: Distribution of the glycoprotein on the surface of infected cells, J. Virol. 44:1047.

Johnson, D. C., and Schlesinger, M. J., 1980, Vesicular stomatitis virus and Sindbis virus glycoprotein transport to the cell surface is inhibited by ionophores, Virology 103:407.

Johnson, D. C., Schlesinger, M. J., and Elson, E. L., 1981, Fluorescence photobleaching recovery measurements reveal differences in envelopment of Sindbis and vesicular stomatitis virus, Cell 23:423.

Kääriäinen, L., and Pesonen, M., 1982, Virus glycoproteins and glycolipids: Structure, biosynthesis, biological function and interaction with host, in: The Glycoconjugates, Vol. 4 (M. I. Horowitz, ed.), p. 191, Academic Press, New York.

Kang, C. Y., and Prevec, L., 1971, Proteins of vesicular stomatitis virus. III. Intracellular synthesis and extracellular appearance of virus-specific proteins, Virology 46:678.

Kelley, J. M., Emerson, S. U., and Wagner, R. R., 1972, The glycoprotein of vesicular stomatitis is the antigen that gives rise to and reacts with neutralizing antibody, J. Virol. 10:1231.

Kingsford, L., Emerson, S. U., and Kelley, J. M., 1980, Separation of cyanogen bromide cleaved peptides of the vesicular stomatitis virus glycoprotein and analysis of their carbohydrate content, J. Virol. 36:309.

Kiuchi, A., and Roy, P., 1984, Comparison of the primary sequence of spring viremia of carp virus M protein with that of vesicular stomatitis virus, *Virology* **134**:238.

Klenk, H.-D., and Choppin, P. W., 1971, Glycolipid content of vesicular stomatitis virus grown in baby hamster kidney cells, *J. Virol.* **7**:416.

Klenk, H.-D., and Rott, R., 1980, Cotranslational and post translational processing of viral glycoproteins, *Curr. Top. Microbiol. Immunol.* **90**:19.

Knipe, D. M., Baltimore, D., and Lodish, H. F., 1977a, Separate pathways of maturation of the major structural proteins of vesicular stomatitis virus, *J. Virol.* **21**:1128.

Knipe, D. M., Baltimore, D., and Lodish, H. F., 1977b, Maturation of viral proteins in cells infected with temperature sensitive mutants of vesicular stomatitis virus, *J. Virol.* **21**:1149.

Kotwal, G. J., and Ghosh, H. P., 1984, Role of fatty acid acylation of membrane glycoproteins: Absence of palmitic acid in glycoproteins of two serotypes of vesicular stomatitis virus, *J. Biol. Chem.* **259**:4699.

Kotwal, G. J., Capone, J., Irving, R. A., Rhee, S. H., Bilan, P., Toneguzzo, F., Hofmann, T., and Ghosh, H. P., 1983, Viral membrane glycoproteins: Comparison of the amino terminal amino acid sequences of the precursor and mature glycoproteins of three serotypes of vesicular stomatitis virus, *Virology* **129**:1.

Lafay, F., 1974, Envelope proteins of vesicular stomatitis virus: Effect of temperature-sensitive mutations in complementation groups III and V, *J. Virol.* **14**:1220.

Landsberger, F. R., and Compans, R. W., 1976, Effect of membrane protein on lipid bilayer structure: A spin-label electron spin resonance study of vesicular stomatitis virus, *Biochemistry* **15**:2356.

Leavitt, R., Schlesinger, S., and Kornfeld, S., 1977, Impaired intracellular migration and altered solubility of nonglycosylated glycoproteins of vesicular stomatitis virus and Sindbus virus, *J. Biol. Chem.* **252**:9018.

Lenard, J., 1978, Virus envelopes and plasma membranes, *Annu. Rev. Biophys. Bioeng.* **7**:139.

Lenard, J., and Compans, R. W., 1974, The membrane structure of lipid-containing viruses, *Biochim. Biophys. Acta* **344**:51.

Lenard, J., and Rothman, J. E., 1976, Transbilayer distribution and movement of cholesterol and phospholipid in the membrane of influenza virus, *Proc. Natl. Acad. Sci. U.S.A.* **73**:391.

Lingappa, V. R., Katz, F. N., Lodish, H. F., and Blobel, G., 1978, A signal sequence for the insertion of a transmembrane glycoprotein: Similarities to the signals of secretory proteins in primary structure and function, *J. Biol. Chem.* **253**:8667.

Little, S. P., and Huang, A. S., 1977, Synthesis and distribution of vesicular stomatitis virus-specific polypeptides in the absence of progeny production, *Virology* **81**:37.

Little, S. P., and Huang, A. S., 1978, Shedding of the glycoprotein from vesicular stomatitis virus-infected cells, *J. Virol.* **27**:330.

Lodish, H. F., and Kong, N., 1983, Reversible block in intracellular transport and budding of mutant vesicular stomatitis virus glycoproteins, *Virology* **125**:335.

Lodish, H. F., and Porter, M., 1980a, Specific incorporation of host cell surface proteins into budding vesicular stomatitis virus particles, *Cell* **19**:161.

Lodish, H. F., and Porter, M., 1980b, Heterogeneity of vesicular stomatitis virus particles: Implications for virion assembly, *J. Virol.* **33**:52.

Lodish, H. F., and Weiss, R. A., 1979, Selective isolation of mutants of vesicular stomatitis virus defective in production of the viral glycoprotein, *J. Virol.* **30**:177.

Lodish, H. F., Zilberstein, A., and Porter, M., 1981, Synthesis and assembly of transmembrane viral and cellular glycoproteins, *Methods Cell Biol.* **23**:5.

Maeda, M., Doi, O., and Akamatsu, Y., 1980, Behaviour of vesicular stomatitis virus glycoprotein in mouse LM cells with modified membrane-phospholipids, *Biochim. Biophys. Acta* **597**:552.

Magee, A. I., Koyama, A. H., Malfer, C., Wen, D., and Schlesinger, M. J., 1984, Release of fatty acids from virus glycoproteins by hydroxylamine, *Biochim. Biophys. Acta* **798**:156.

Majuk, Z., Bittman, R., Landsberger, F. R., and Compans, R. W., 1977, Effects of filipin on the structure and biological activity of enveloped viruses, *J. Virol.* **24**:883.

Mancarella, D. A., and Lenard, J., 1981, Interactions of wild-type and mutant M protein of vesicular stomatitis virus with viral nucleocapsid and envelope in intact virions: Evidence from [^{125}I]iodonaphthyl azide labeling and specific cross-linking, *Biochemistry* **20**:6872.

McSharry, J. J., 1977, The effect of chemical and physical treatments on the lactoperoxidase catalyzed iodination of vesicular stomatitis virus, *Virology* **83**:482.

McSharry, J. J., and Wagner, R. R., 1971a, Lipid composition of purified vesicular stomatitis virus, *J. Virol.* **7**:59.

McSharry, J. J., and Wagner, R. R., 1971b, Carbohydrate composition of vesicular stomatitis virus, *J. Virol.* **7**:412.

McSharry, J. J., Compans, R. W., and Choppin, P. W., 1971, Proteins of vesicular stomatitis virus and of phenotypically mixed vesicular stomatitis virus–simian virus 5 virions, *J. Virol.* **8**:722.

Miller, D. K., Feuer, B. I., Venderoef, R., and Lenard, J., 1980, Reconstituted G protein–lipid vesicles from vesicular stomatitis virus and their inhibition of VSV infection, *J. Cell Biol.* **84**:421.

Moore, N. F., Kelley, J. M., and Wagner, R. R., 1974, Envelope proteins of vesicular stomatitis virions: Accessibility to iodination, *Virology* **61**:292.

Moore, N. F., Barenholz, Y., McAllister, P. E., and Wagner, R. R., 1976, Comparative membrane microviscosity of fish and mammalian rhabdoviruses studied by fluorescence depolarization, *J. Virol.* **19**:275.

Moore, N. F., Patzer, E. J., Barenholz, Y., and Wagner, R. R., 1977a, Effect of phospholipase C and cholesterol oxidase on membrane integrity, microviscosity and infectivity of vesicular stomatitis virus, *Biochemistry* **16**:4708.

Moore, N. F., Patzer, E. J., Wagner, R. R., Yeagle, P. L., Hutton, W. C., and Martin, R. B., 1977b, The structure of vesicular stomatitis virus membrane: A phosphorus nuclear magnetic resonance approach. *Biochim. Biophys. Acta* **464**:234.

Moore, N. F., Patzer, E. J., Shaw, J. M., Thompson, T. E., and Wagner, R. R., 1978, Interaction of vesicular stomatitis virus with lipid vesicles: Depletion of cholesterol and effect on virion membrane fluidity and infectivity, *J. Virol.* **27**:320.

Morrison, T. G., and Lodish, H. F., 1975, Site of synthesis of membrane and non-membrane proteins of vesicular stomatitis virus, *J. Biol. Chem.* **250**:6955.

Morrison, T. G., and McQuain, C. O., 1978, Assembly of viral membranes: Nature of the association of vesicular stomatitis virus proteins to membranes, *J. Virol.* **26**:115.

Morrison, T. G., McQuain, C. O., and Simpson, D., 1978, Assembly of viral membranes: Maturation of vesicular stomatitis virus glycoprotein in the presence of tunicamycin, *J. Virol.* **28**:368.

Moyer, S. A., and Summers, D. F., 1974, Vesicular stomatitis virus envelope glycoprotein alterations induced by host cell transformation, *Cell* **2**:63.

Moyer, S. A., Tsang, J. M., Atkinson, P. H., and Summers, D. F., 1976, Oligosaccharide moieties of the glycoprotein of vesicular stomatitis virus, *J. Virol.* **18**:167.

Mudd, J. A., 1974, Glycoprotein fragment associated with vesicular stomatitis virus after proteolytic digestion, *Virology* **62**:573.

Newcomb, W. W., and Brown, J. C., 1981, Role of the vesicular stomatitis virus matrix protein in maintaining the viral nucleocapsid in the condensed form found in native virions, *J. Virol.* **39**:295.

Newcomb, W. W., Tobin, G. J., McGowan, J. J., and Brown, J. C., 1982, *In vitro* reassembly of vesicular stomatitis virus skeletons, *J. Virol.* **41**:1055.

Ogden, J. R., Pal, R., and Wagner, R. R., 1986, Mapping regions of the matrix (M) protein of vesicular stomatitis virus which binds ribonucleocapsids, liposomes, and monoclonal antibodies, *J. Virol.* **58**:860.

Pal, R., Barenholz, Y., and Wagner, R. R., 1980a, Effect of cholesterol concentration on organization of viral and vesicle membranes: Probed by accessibility to cholesterol oxidase, *J. Biol. Chem.* **255**:5802.

Pal, R., Petri, W. A., Jr., and Wagner, R. R., 1980b, Alteration of the membrane lipid composition and infectivity of vesicular stomatitis virus by growth in a Chinese hamster ovary cell sterol mutant and in lipid-supplemented baby hamster kidney clone 21 cells, *J. Biol. Chem.* **255**:7688.

Pal, R., Barenholz, Y., and Wagner, R. R., 1981a, Depletion and exchange of cholesterol from the membrane of vesicular stomatitis virus by interaction with serum lipoproteins or poly(vinylpyrrolidone) complexed with bovine serum albumin, *Biochemistry* **20**:530.

Pal, R., Barenholz, Y., and Wagner, R. R., 1981b, Transcription of vesicular stomatitis virus activated by paradaxin, a fish toxin that permeabilizes the virion membrane, *J. Virol.* **39**:641.

Pal, R., Barenholz, Y., and Wagner, R. R., 1981c, Paradaxin, a hydrophobic toxin of the Red Sea flat fish, disassembles the intact membrane of vesicular stomatitis virus, *J. Biol. Chem.* **256**:10209.

Pal, R., Wiener, J. R., Barenholz, Y., and Wagner, R. R., 1983, Influence of the membrane glycoprotein and cholesterol of vesicular stomatitis virus on the dynamics of viral and model membranes: Fluorescence studies, *Biochemistry* **22**:3624.

Pal, R., Grinnell, B. W., Snyder, R. M., Wiener, J. R., Volk, W. A., and Wagner, R. R., 1985b, Monoclonal antibodies to the M protein of vesicular stomatitis virus (Indiana serotype) and to a cDNA M-gene expression product, *J. Virol.* **85**:298.

Pal, R., Grinnell, B. W., Snyder, R. M., and Wagner, R. R., 1985c, Regulation of viral transcription by the matrix protein of vesicular stomatitis virus probed by monoclonal antibodies and temperature-sensitive mutants, *J. Virol.* **56**:386.

Patzer, E. J., Moore, N. F., Barenholz, Y., Shaw, J. M., and Wagner, R. R., 1978a, Lipid organization of the membrane of vesicular stomatitis virus, *J. Biol. Chem.* **253**:4544.

Patzer, E. J., Shaw, J. M., Moore, N. F., Thompson, T. E., and Wagner, R. R., 1978b, Transmembrane movement and distribution of cholesterol in the membrane of vesicular stomatitis virus, *Biochemistry* **17**:4192.

Patzer, E. J., Wagner, R. R., and Dubovi, E. J., 1979, Viral membranes: Model systems for studying biological membranes, *CRC Crit. Rev. Biochem.* **6**:165.

Pepinsky, R. B., and Vogt, V. M., 1979, Identification of retrovirus matrix proteins by lipid–protein cross-linking, *J. Mol. Biol.* **131**:819.

Perrault, J., and Kingsbury, D. T., 1974, Inhibition of vesicular stomatitis virus transcriptase in purified virions, *Nature (London)* **248**:45.

Pessin, J. E., and Glaser, M., 1980, Budding of Rous sarcoma virus and vesicular stomatitis virus from localized lipid regions in the plasma membrane of chicken embryo fibroblasts, *J. Biol. Chem.* **255**:9044.

Petri, W. A., Jr., and Wagner, R. R., 1979, Reconstitution into liposomes of the glycoprotein of vesicular stomatitis virus by detergent dialysis, *J. Biol. Chem.* **254**:4313.

Petri, W. A., Jr., and Wagner, R. R., 1980, Glycoprotein micelles isolated from vesicular stomatitis virus spontaneously partition into sonicated phosphatidylcholine vesicles, *Virology* **107**:543.

Petri, W. A., Jr., Estep, T. N., Pal, R., Thompson, T. E., Biltonen, R. L., and Wagner, R. R., 1980, Thermotropic behaviour of dipalmitoylphosphatidylcholine vesicles reconstituted with the glycoprotein of vesicular stomatitis virus, *Biochemistry* **19**:3088.

Petri, W. A., Jr., Pal, R., Barenholz, Y., and Wagner, R. R., 1981a, Fluorescence anisotropy of a fatty acid covalently linked *in vivo* to the glycoprotein of vesicular stomatitis virus, *J. Biol. Chem.* **256**:2625.

Petri, W. A., Jr., Pal, R., Barenholz, Y., and Wagner, R. R., 1981b, Fluorescence studies of dipalmitoylphosphatidylcholine vesicles reconstituted with the glycoprotein of vesicular stomatitis virus, *Biochemistry* **20**:2796.

Pinney, D. F., and Emerson, S. U., 1982, *In vitro* synthesis of triphosphate-initiated N-gene mRNA oligonucleotides is regulated by the matrix protein of vesicular stomatitis virus, *J. Virol.* **42**:897.

Portoukalian, J., Bugand, M., Zwingelstein, G., and Precausta, P., 1977, Comparison of the lipid composition of rabies virus propagated in NIL-2 cells maintained in monolayer versus spinner culture, *Biochim. Biophys. Acta* **489**:106.

Pringle, C. R., 1970, Genetic characterization of conditional lethal mutants of vesicular stomatitis virus induced by 5-fluorouracil, 5-azacytidine and ethylmethylsulfonate, *J. Virol.* **5**:559.

Reading, C. L., Penhoet, E. E., and Ballou, C. E., 1978, Carbohydrate structure of vesicular stomatitis virus glycoprotein, *J. Biol. Chem.* **253**:5600.

Reidler, J. A., Keller, P. M., Elson, E. L., and Lenard, J., 1981, A fluorescence photobleaching study of vesicular stomatitis virus infected BHK cells: Modulation of G protein mobility by M protein, *Biochemistry* **20**:1345.

Rindler, M. J., Ivanov, I. E., Plesken, H., Rodriguez-Boulan, E., and Sabatini, D. D., 1984, Viral glycoproteins destined for apical or basolateral plasma membrane domains traverse the same Golgi apparatus during their intracellular transport in doubly infected Madin–Darby canine kidney cells, *J. Cell Biol.* **98**:1304.

Robertson, J. S., Etchison, J. R., and Summers, D. F., 1976, Glycosylation sites of vesicular stomatitis virus glycoprotein, *J. Virol.* **19**:871.

Robertson, J. S., Etchison, J. R., and Summers, D. F., 1982, Comparison of the oligosaccharide structure of the glycoprotein of vesicular stomatitis virus and a thermolabile mutant (*tl*-17), *J. Gen. Virol.* **58**:13.

Robertson, M. A., Etchison, J. R., Robertson, J. S., Summers, D. F., and Stanley, P., 1978, Specific changes in the oligosaccharide moieties of VSV grown in different lectin-resistant CHO cells, *Cell* **13**:515.

Rodriguez-Boulan, E., 1983, Polarized assembly of enveloped viruses from cultured epithelial cells, *Methods Enzymol.* **98**:486.

Rodriguez-Boulan, E., and Pendergast, M., 1980, Polarized distribution of viral envelope proteins in the plasma membrane of infected epithelial cells, *Cell* **20**:45.

Rodriguez-Boulan, E., and Sabatini, D. D., 1978, Asymmetric budding of viruses in epithelial monolayers: A model system for study of epithelial polarity, *Proc. Natl. Acad. Sci. U.S.A.* **75**:5071.

Rose, J. K., and Bergmann, J. E., 1982, Expression from cloned cDNA of cell-surface secreted forms of the glycoprotein of vesicular stomatitis virus in eucaryotic cells, *Cell* **30**:753.

Rose, J. K., and Bergmann, J. E., 1983, Altered cytoplasmic domains affect intracellular transport of the vesicular stomatitis virus glycoprotein, *Cell* **34**:513.

Rose, J. K., and Gallione, C. J., 1981, Nucleotide sequences of the mRNA's encoding the vesicular stomatitis virus G and M proteins determined from cDNA clones containing the complete coding regions, *J. Virol.* **39**:519.

Rose, J. K., and Shafferman, A., 1981, Conditional expression of the vesicular stomatitis virus glycoprotein gene in *Escherichia coli*, *Proc. Natl. Acad. Sci. U.S.A.* **78**:6670.

Rose, J. K., Welch, W. J., Sefton, B. M., Esch, F. S., and Ling, N. C., 1980, Vesicular stomatitis virus glycoprotein is anchored in the viral membrane by a hydrophobic domain near the carboxy terminus, *Proc. Natl. Acad. Sci. U.S.A.* **77**:3884.

Rose, J. K., Doolittle, R. F., Anilionis, A., Curtis, P. J., and Wunner, W. H., 1982, Homology between the glycoproteins of vesicular stomatitis virus and rabies virus, *J. Virol.* **43**:361.

Rose, J. K., Adams, G. A., and Gallione, C. J., 1984, The presence of cysteine in the cytoplasmic domain of the vesicular stomatitis virus glycoprotein is required for palmitate addition, *Proc. Natl. Acad. Sci. U.S.A.* **81**:2050.

Rosen, C. A., Cohen, P.S., and Ennis, H. L., 1983, Identification of a new protein present in vesicular stomatitis virus-infected Chinese hamster ovary cells as a degradation product of viral M protein, *Virology* **130**:331.

Roth, M. G., and Compans, R. W., 1981, Delayed appearance of pseudotypes between vesicular stomatitis virus and influenza virus during mixed infection of MDCK cells, *J. Virol.* **40**:848.

Roth, M. G., Fitzpatrick, J. P., and Compans, R. W., 1979, Polarity of influenza and vesicular stomatitis virus maturation in MDCK cells: Lack of a requirement for glycosylation of viral glycoproteins, *Proc. Natl. Acad. Sci. U.S.A.* **76**:6430.

Rothman, J. E., and Fine, R., 1980, Coated vesicles transport newly synthesized membrane glycoproteins from endoplasmic reticulum to plasma membrane in two successive stages, *Proc. Natl. Acad. Sci. U.S.A.* **77**:780.

Rothman, J. E., and Fries, E., 1981, Transport of newly synthesized vesicular stomatitis viral glycoprotein to purified Golgi membrane, *J. Cell Biol.* **89**:162.

Rothman, J. E., and Lodish, H. F., 1977, Synchronized transmembrane insertion and glycosylation of a nascent membrane protein, *Nature (London)* **269**:775.

Rothman, J. E., Tsai, D. K., Dawidowicz, E. A., and Lenard, J., 1976, Transbilayer phospholipid asymmetry and its maintenance in the membrane of influenza virus, *Biochemistry* **15**:2361.

Rothman, J. E., Bursztyn-Pettegrew, H., and Fine, R. E., 1980, Transport of the membrane glycoprotein of vesicular stomatitis virus to the cell surface in two stages by clathrin-coated vesicles, *J. Cell Biol.* **86**:162.

Rott, R., and Klenk, H.-D., 1977, Structure and assembly of viral envelopes, in: *Virus Infection and the Cell Surface* (G. Poste and G. L. Nicholson, eds.), p. 47, North-Holland, Amsterdam.

Schlesinger, M. J., and Malfer, C., 1982, Cerulenin blocks fatty acid acylation of glycoproteins and inhibits vesicular stomatitis and Sindbis virus particle formation, *J. Biol. Chem.* **257**:9887.

Schlesinger, S., Malfer, C., and Schlesinger, M. J., 1984, The formation of vesicular stomatitis virus (San Juan strain) becomes temperature-sensitive when glucose residues are retained on the oligosaccharides of the glycoprotein, *J. Biol. Chem.* **259**:7597.

Schloemer, R. H., and Wagner, R. R., 1975, Association of vesicular stomatitis virus glycoprotein with virion membrane: Characterization of the lipophilic tail fragment, *J. Virol.* **16**:237.

Schmidt, M. F. G., and Schlesinger, M. J., 1979, Fatty acid binding to vesicular stomatitis virus glycoprotein: A new type of post-translational modification of the viral glycoprotein, *Cell* **17**:813.

Schmidt, M. F. G., and Schlesinger, M. J., 1980, Relation of fatty acid attachment to the translation and maturation of vesicular stomatitis virus and Sindbis virus membrane glycoproteins, *J. Biol. Chem.* **255**:3334.

Schnitzer, T. J., and Lodish, H. F., 1979, Noninfectious vesicular stomatitis virus particles deficient in the viral nucleocapsids, *J. Virol.* **29**:443.

Sefton, B. M., 1976, Virus-dependent glycosylation, *J. Virol.* **17**:85.

Sefton, B. M., and Gaffney, B. J., 1979, Complete exchange of viral cholesterol, *Biochemistry* **18**:436.

Shaw, J. M., Moore, N. F., Patzer, E. J., Correa-Freire, M. C., Wagner, R. R., and Thompson, T. E., 1979, Compositional asymmetry and transmembrane movement of phosphatidylcholine in vesicular stomatitis virus membranes, *Biochemistry* **18**:538.

Simpson, R. W., and Hauser, R. E., 1966, Influence of lipids on the viral phenotype. II. Interaction of myxoviruses and their lipid constituents with phospholipases, *Virology* **30**:684.

Stanley, P., Vivona, G., and Atkinson, P. H., 1984, ^{1}H NMR spectroscopy of carbohydrates from the G glycoprotein of vesicular stomatitis virus grown in parental and Lec4 Chinese hamster ovary cell, *Arch. Biochem. Biophys.* **230**:363.

Stoffel, W., and Bister, K., 1975, ^{13}C nuclear magnetic resonance studies on the lipid organization in enveloped virions (vesicular stomatitis virus), *Biochemistry* **14**:2841.

Stoffel, W., and Sorgo, W., 1976, Asymmetry of the lipid bilayer of Sindbis virus, *Chem. Phys. Lipids* **17**:324.

Stoffel, W., Anderson, R., and Stahl, J., 1975, Studies on the asymmetric arrangement of membrane-lipid-enveloped virions as a model system, *Hoppe-Seyler's Z. Physiol. Chem.* **356**:1123.

Stoffel, W., Bister, K., Schreiber, C., and Tunggal, B., 1976, ^{13}C-NMR studies of the membrane structure of enveloped virions, *Hoppe-Seyler's Z. Physiol. Chem.* **357**:905.

Stoffel, W., Schreiber, C., and Scheefers, H., 1978, Lipids with photosensitive groups as chemical probes for the structural analysis of biological membranes: On the localization of the G- and M-protein of vesicular stomatitis virus, *Hoppe-Seyler's Z. Physiol. Chem.* **359**:923.

Taube, S. E., and Rothfield, L. I., 1978, Isolation of the envelope of vesicular stomatitis virus, *J. Virol.* **26**:730.

Toneguzzo, F., and Ghosh, H. P., 1977, Synthesis and glycosylation *in vitro* of glycoprotein of vesicular stomatitis virus, *Proc. Natl. Acad. Sci. U.S.A.* **74**:1516.

Toneguzzo, F., and Ghosh, H. P., 1978, *In vitro* synthesis of vesicular stomatitis virus membrane glycoprotein and insertion into membrane, *Proc. Natl. Acad. Sci. U.S.A.* **75**:715.

Tsai, D. K., and Lenard, J., 1975, Asymmetry of influenza virus membrane bilayer demonstrated with phospholipase C, *Nature (London)* **253**:554.

Van Meer, G., and Simons, K., 1982, Viruses budding from either the apical or the basolateral plasma membrane domain of MDCK cells have unique phospholipid compositions, *Eur. Mol. Biol. Org. J.* **1**:847.

Van Meer, G., Simons, K., Op den Kamp, J. A. F., and van Deenen, L. L. M., 1981, Phospholipid asymmetry in Semliki Forest virus grown in baby hamster kidney (BHK-21) cells, *Biochemistry* **20**:1974.

Van Wyke, K. L., Yewdell, J. W., Reck, L. J., and Murphy, B. R., 1984, Antigenic characterization of influenza A virus matrix protein with monoclonal antibodies, *J. Virol.* **49**:248.

Wagner, R. R., 1975, Reproduction of rhabdoviruses, in: *Comprehensive Virology*, Vol. 4 (H. Fraenkel-Conrat and R. R. Wagner, eds.), pp. 1–93, Plenum, Press, New York.

Wagner, R. R., and Schnaitman, C. A., 1970, Proteins of vesicular stomatitis virus, in: *The Biology of Large RNA Viruses* (R. D. Barry and B. W. J. Mahy, eds.), p. 655, Academic Press, London.

Wagner, R. R., Schnaitman, T. C., Snyder, R. M., and Schnaitman, C. A., 1969, Protein composition of the structural components of vesicular stomatitis virus, *J. Virol.* **3**:611.

Wagner, R. R., Heine, J. W., Goldstein, G., and Schnaitman, C. A., 1971, Use of anti-viral–antiferritin hybrid antibody for localization of viral antigen in plasma membrane, *J. Virol.* **7**:274.

Wehland, J., Willingham, M. C., Gallo, M. G., and Pastan, I., 1982, The morphologic pathway of exocytosis of the vesicular stomatitis virus G protein in cultured fibroblasts, *Cell* **28**:831.

Weiss, R. A., and Bennett, P. L. P., 1980, Assembly of membrane glycoproteins studied by phenotypic mixing between mutants of vesicular stomatitis virus and retroviruses, *Virology* **100**:252.

Wiener, J. R., Pal, R., Barenholz, Y., and Wagner, R. R., 1983a, Influence of the peripheral matrix protein of vesicular stomatitis virus on the membrane dynamics of mixed phospholipid vesicles: Fluorescence studies, *Biochemistry* **22**:2162.

Wiener, J. R., Wagner, R. R., and Freire, E., 1983b, Thermotropic behavior of mixed phosphatidylcholine–phosphatidylglycerol vesicles reconstituted with the matrix protein of vesicular stomatitis virus, *Biochemistry* **22**:6117.

Wiener, J. R., Pal, R., Barenholz, Y., and Wagner, R. R., 1985, Effect of the vesicular stomatitis matrix protein on the lateral organization of lipid bilaiyers containing phospatidylglycerol: Use of fluorescent phospholipid analogs, *Biochemistry* **24**:7651.

Wilson, T., and Lenard, J., 1981, Interaction of wild-type and mutant M protein of vesicular stomatitis virus with nucleocapsids *in vitro*, *Biochemistry* **20**:1349.

Ye, Z., Pal, R., Ogden, J., Snyder, R. M., and Wagner, R. R., 1985, Monoclonal antibodies to the matrix protein of vesicular stomatitis virus (New Jersey serotype) and their effects on transcription, *Virology* **143**:657.

Yewdell, J. W., and Gerhard, W., 1981, Antigenic characterization of viruses by monoclonal antibodies, *Annu. Rev. Microbiol.* **35**:185.

Zakowski, J. J., and Wagner, R. R., 1980, Localization of membrane-associated proteins in vesicular stomatitis virus by use of hydrophobic membrane probes and cross-linking reagents, *J. Virol.* **36**:93.

Zakowski, J. J., Petri, W. A., Jr., and Wagner, R. R., 1981, Role of matrix protein in assembling the membrane of vesicular stomatitis virus: Reconstitution of matrix protein with negatively charged phospholipid vesicles, *Biochemistry* **20**:3902.

Zee, Y., Hackett, A. J., and Talens, L., 1970, Vesicular stomatitis virus maturation sites in six different host cells, *J. Gen. Virol.* **7**:95.

Zilberstein, A., Snider, M. D., Porter, M., and Lodish, H. F., 1980, Mutants of vesicular stomatitis virus blocked at different stages in maturation of the viral glycoprotein, *Cell* **21**:417.

Zilberstein, A., Snider, M. D., and Lodish, H. F., 1982, Synthesis and assembly of the vesicular stomatitis virus glycoprotein, *Cold Spring Harbor Symp. Quant. Biol.* **46**:785.

CHAPTER 4

Rhabdovirus Genomes and Their Products

JOHN ROSE AND MANFRED SCHUBERT

I. INTRODUCTION

Rhabdoviruses are relatively simple, membrane-enveloped viruses containing a single-stranded RNA genome. The genomic RNA is the negative sense—i.e., complementary to the messenger RNAs (mRNAs)—and is noninfectious. The virus particles must therefore contain an RNA-dependent RNA polymerase to generate the mRNAs (Baltimore *et al.*, 1970). Rhabdoviruses have a bacilliform, bullet-, or cone-shaped morphology and are known to infect vertebrates, invertebrates, and plants. The composition of various rhabdoviruses has been reviewed by McSharry (1979). The virus particles contain a helical, nucleocapsid core composed of the genomic RNA and protein. Generally, three proteins termed N (nucleocapsid), NS (originally indicating nonstructural), and L (large) are found to be associated with the nucleocapsid. An additional matrix (M) protein lies within the membrane envelope, perhaps interacting both with the membrane and the nucleocapsid core. A single glycoprotein (G) species spans the membrane and forms the spikes on the surface of the virus particle.

The purpose of this chapter is to review some of the recent work that contributes to our understanding of rhabdovirus genomes and the RNA and protein products encoded by them. Although the rhabdoviruses comprise a diverse group of enveloped viruses, we have decided to confine

JOHN ROSE • Departments of Pathology and Cell Biology, Yale University School of Medicine, New Haven, Connecticut 06510.   MANFRED SCHUBERT • Laboratory of Molecular Genetics, National Institute of Neurological and Communicative Disorders and Stroke, Bethesda, Maryland 20892.

our discussion largely to the two members of the group—vesicular stomatitis virus (VSV) and rabies virus—for which there is extensive characterization at the molecular level, including nucleotide-sequence information.

Nucleotide-sequence analysis of the VSV mRNAs and genomic RNA
from the Indiana serotype has been carried out over the past several years
largely from complementary DNA (cDNA) clones. From these published
data, we have been able to compile the complete sequence of the negative-
stranded RNA genome and the predicted amino acid sequences of the five
viral proteins. We review some of the interesting features of the genome
and the functions of the RNA and protein products encoded by it. In
addition, we discuss the comparison of sequences among the VSV serotypes and with a distant relative of VSV, rabies virus.

II. NUCLEOTIDE SEQUENCE AND ORGANIZATION
OF THE GENOME

A. Gene Order

Despite the wide spectrum of host organisms that can be infected
with rhabdoviruses, these viruses share a common strategy in gene expression and replication. This strategy is reflected in their genomic organization. The viral-specific proteins of many rhabdovirus species have been
identified and characterized. The genomic RNA and the organization of
the viral genes, however, have not been studied in great detail, with the
exception of vesicular stomatitis virus (VSV) and rabies virus, and very
recently with the fish rhabdovirus infectious hematopoietic necrosis virus
(discussed in Section III.C).

The transcriptional and translational mapping of the VSV genes by
UV irradiation revealed that transcription of the viral genes is carried out
sequentially. It was found that the UV-target sizes for the transcription
of each gene were independent of the size of each gene, but depended on
the position of the gene in the genome. The gene order for VSV-Indiana
could be established by this method (Abraham and Banerjee, 1976; Ball
and White, 1976). RNA duplex mapping, using genomic RNA and isolated
RNAs, confirmed the results (Herman *et al.*, 1978). Ultimately, the nucleotide sequence analyses of genomic RNA and mRNAs as well as of
cDNA clones established the complete primary structure of the VSV
genome. An almost identical gene order was established for rabies virus,

FIGURE 1. Nucleotide sequence of the VSV genome and predicted sequences of the VSV
proteins. The sequence of the VSV genome was compiled from the references described in
Table I and is shown as the (+)-strand complement (mRNA sense). The sequence is stored
in the Protein and Nucleic Acid Sequence Data Bases of the National Biomedical Research
Foundation, Washington, D.C.

```
                                                 M  S  V  T  V  K  R  I  I  D  N  T  V  I  V  P  K  L  P
ACGAAAGACAAACAAACCCAUUAUUAUCAUUAAAAAGGCUCAGGAGAAACUUCAACAGUAAUCAAAAGUGCUGUUACAGUCAGAGAAUCAUUGACAACACAGUCAUAGUUCCAAAACUUCCU
                                                                                                               120

    A  N  E  D  P  V  E  Y  P  A  D  Y  F  R  K  S  K  E  I  P  L  Y  I  N  T  T  K  S  L  S  D  L  R  Q  Y  V  Y  Q  Q  L
                                                                             50
GCAAAUGAGGAUCCAGUGGAAUACCCGGCAGAUUACUUCAGAAAGUCAAAGGAGAUUCCUCUUUACAUCAAUACAACAAAGUCUCUGAGUCUAAGGACUUAAGACGAUAUGUCUACCAACCUC
                                                                                                               240

    K  S  Q  N  V  S  I  I  H  V  N  S  Y  L  Y  Q  A  L  K  D  I  R  Q  K  L  D  K  D  W  S  S  F  Q  I  N  I  Q  K  A  Q
AAAUCCCAGAAAUGUAUCAAUCAUACAUGUCAACAGUCACUUGUAUUGGAGCAUUAAAGGAUAUCCGGGGUAAGUGAUAAAGAUUGGUCAAGUUCCGAAUUAAACAUCGGGAAAGCAGGG
100                                                                                                            360

    D  T  I  Q  I  F  D  L  V  S  L  K  A  L  D  Q  V  L  P  D  Q  V  S  D  A  S  R  T  S  A  D  D  K  W  L  P  L  Y  L  L
GAUACAAUCGGAAUAUUUGACCUUGUAUCCUUGAAAGCCUUGGACCAGGUACUUCCAGAUGGAGUAUCGGAUGCUUCCAGAACCAGCGCAGAUGACAAAUGGUUGCCUUUGUAUCUACUU
                                               150                                                             480

    Q  L  Y  R  V  Q  R  T  G  M  P  E  Y  R  K  K  L  M  D  Q  L  T  N  Q  C  K  M  I  N  E  Q  F  E  P  L  V  P  E  Q  R
GGCUUAAUACGAGUGCAGCGUACAGGCAUGCCUGAAUACAGAAAAAAGCUCAUGGAUCAGCUGACAAACCAAUGCAAAAUGAUCAAUGAACAGUUUGAACCUCUUGUUGCCGAAAGGCGU
                                                                             200                               600

    D  I  F  D  V  W  G  N  D  S  N  Y  T  K  I  V  A  A  V  D  M  F  F  H  M  F  K  K  H  E  C  A  S  F  R  Y  G  T  I  V
GACAUUUUUGAUGUAUGGGGAAAUGACAGUAAUUACACAAAAAUUGUCGCUGCAGUGGACAUGUUCUUCCACAUGUUCAAAAAACACGAGUGUGCUUCAUUCAGAUAUGGAACUAUAGUU
                                                                                                               720

    S  R  F  K  D  C  A  A  L  A  T  F  G  H  L  C  K  I  T  Q  M  S  T  E  D  V  T  T  W  I  L  N  R  E  V  A  D  E  M  V
UCCAGAUUCAAAGAUUGUGCUGCAUUGGCAACAUUUGGACACCUCUGCAAAAUAACCCAGAUGUCUACAGAAGAUGUAACCACCUGGAUUCUUAAUAGAGAGGUUGCAGAUGAAAUGGUC
                                                                             250                               840

    Q  M  M  L  P  Q  Q  E  I  D  K  A  D  S  Y  M  P  Y  L  I  D  F  G  L  S  S  K  S  P  Y  S  S  V  K  N  P  A  F  H  F
CAAAUGAUGCUUCCACAGCAAGAAAUUGAUAAGGCCGAUUCAUACAUGCCUUAUUUGAUCGACUUUGGAUUGUCUUCUAAAUCUCCAUAUUCUUCCGUCAAAAACCCUGCCUUCCACUUC
300                                                                                                            960

    W  G  Q  L  T  A  L  L  L  R  S  T  R  A  R  N  A  R  Q  P  D  D  I  E  Y  T  S  L  T  T  A  Q  L  L  Y  A  Y  A  V  G
UGGGGGCAAUUGACAGCUCUUCUGCUCAGAUCCACGAGAGCAAGGAAUGCCCGACAGCCUGAUGACAUUGAGUAUACAUCUCUUACUACAGCAGGUUUGUUAUACGCCUAUGCAGUAGGA
                                                                                                               1080

    S  S  A  D  L  A  Q  Q  F  C  V  G  D  N  K  Y  T  P  D  D  S  T  G  Q  L  T  T  N  A  P  P  Q  G  R  D  V  V  E  W  L
UCCUCUGCCGACAACUACGAGUUUUGUGUAGGAGAUAAACAAAUACACCAGAUGAUAGUACCCAGCAAGGCAGAGAUGUGGUGGUUGAAUGGGAUCGUUUGGUCUCUCCAAGGAUUGG
                                               350                                                             1200

    Q  W  F  E  D  Q  N  R  K  P  T  P  D  M  M  Q  Y  A  K  R  A  V  M  S  L  Q  G  L  R  E  K  T  I  G  K  Y  A  K  S  E
GGAUGGUUCGAAGAUCAAAACAGAAAACCGACUCCUGAUAUGAUGCAGUAUGCGAAAAGAGCUGUAAUGUCACUGCAGGGCCUAAGGGAGAAGACAAUUGGCAAGUAUGCAAAGUCAGAA
                                                                             400                               1320

    F  D  K  •                                                                                   M  D  N  L  T  K  V  R  E  Y  L  K  S  Y  S
UUUGGAUGACCCUAUAAUUCUCAGAUCACCUAUUAUAUAAUUAAUGCUACAUUAUGAAAAAAACUAACAGA................CAAAACAGAAUGGACAAUCUCACAAAAGUUCGUGAGUAUCUCAAGUCCUAUUCU
                                                                                                               1440

    R  L  D  Q  A  V  G  E  I  D  E  I  E  A  G  R  A  E  K  S  N  Y  E  L  F  G  E  D  Q  V  E  E  H  T  K  P  S  Y  F  Q
CGCUUGGAUCAGGCCGGUAGGGAGAGAUAGAGAUCGAGAUCGAAGCAGGCAGGGCGGAGAAGAGCAAUUAUGAGUUGUUCCAAGAGGAUGGGAUGGAAGAGCAUACUAAGCCCUCUUAUUUUCAG
                                                                             50                                1560

    A  A  D  D  S  D  T  E  S  E  P  E  I  E  D  N  G  L  Y  A  P  D  P  E  A  E  G  V  E  Q  F  I  G  Q  P  L  D  D  Y
GCAGCAGAUGAUAGCGACACAGAAUCUGAACCAGAAAUUGAAGAUAAUGGGCUUUAUGCACCAGAUCCAGAAGCUGAACAAGUUGAAGGCUUUAUACAGGGGGCCUUUAGAUGAUUAU
100                                                                                                            1680

    A  D  E  E  V  D  V  V  F  T  S  D  W  K  Q  P  E  L  E  S  D  E  H  G  K  T  L  R  L  T  S  P  E  Q  L  S  Q  E  G  K
GCAGAUGAGGAAGUGGAUGUGGUAUUUACUUCCGACUGGAAACAGCCUGAACUGGAAUCUGAUGAGCAUGGGAAAACUCUGAGGCUAACAUCUCCAGAGCAGUUGUCAAGGAGGGCAAGA
                                                                                                               1800

    S  G  W  L  S  T  I  K  A  V  V  G  S  A  K  Y  W  N  L  A  E  C  T  F  E  A  S  G  E  Q  V  I  M  K  E  R  G  I  T  P
UCCGGGUGGCUUUCGACGAUUAAAGCAGUCGUGCAAAGUGCUAAAUACUGGAAUCUGGCAGAGUGCACAUUUGAAGCAUCGGGAGAACAGGUUAUUAUGAAGGAGCGCCAGAUAACUCCG
                                               150                                                             1920

    D  V  Y  K  V  T  P  V  M  N  T  H  P  S  G  S  E  A  V  S  D  V  W  S  L  S  K  T  S  M  T  F  Q  P  K  K  A  S  L  Q
GAUGUAUAUAAGGUCACUCCAGUGAUGAACACACAUCCUUCUGGCUCAGAGGCAGUAUCAGAUGUUUGGUCUCUGUCAAAGACAUCCAUGACUUUCCAACCCAAGAAAGCAAGUCUUCAG
                                                                             200                               2040

    P  L  T  I  S  L  D  E  L  F  S  S  R  Q  E  F  I  S  V  G  Q  D  Q  R  M  S  H  K  E  A  I  L  L  G  L  R  Y  K  K  L
CCUCUCACCAUAUCCUUGGAUGAGCUCUUCUCAUCGAGACAGGAAUUCAUAUCUGUCGGACAGGAUCAGCGGAUGAGCCAUAAAGAAGCUAUUUUGCUUGGGCUUCGGUACAAGAAGUUG
                                               250                                                             2160

    Y  N  Q  A  R  V  K  Y  S  L  •                                                                   M  S  S  L  K  K  I  L  G  L  K
UACAACCAGGCGAGAGUCAAAUAUUCUCUGUAGACUAG..UAUGAAAAAAGAUAACAG..AUAUACGAUCUAAGUGUUAUCCCAAUCCUAUCAUCAUGAGUUCCUUAAAGAAGAUCCUCGGCCUUAAA
                                                                                                               2280

    Q  K  Q  K  K  S  K  K  L  G  I  A  P  P  P  Y  E  E  D  T  S  M  E  Y  A  P  S  A  P  I  D  K  S  Y  F  G  V  D  E  M
                                                                             50
AGGQAAAGQGUAAGAAAUCUAAGAAAUUGGGAUCGCACCACCCCCUUAUUGAAGAGGACACUAGCAUGGAGUAUGCUCCGAGCGCUCCAAUUGACAAAUCUUAUUUUGGAGUUGAUGAGAUGA
                                                                                                               2400

    D  T  Y  D  P  N  Q  L  R  Y  E  K  F  F  F  T  V  K  M  T  V  R  S  N  R  P  F  R  T  Y  S  D  V  A  A  A  V  S  H  W
UGGACACCUAUGAUCCGAAUCAAUUAAGAUAUGAGAAAUUCUUCUUUACAGUGAAAAUGACGGUUAGAUCCAACAGACCAUUCAGAACAUACUCAGAUGUGGCAGCCGCUGUAUCCCAUU
100                                                                                                            2520

    D  H  M  Y  I  G  M  A  G  K  R  P  F  Y  K  I  L  A  F  L  G  S  S  N  L  K  A  T  P  A  V  L  A  D  Q  G  Q  P  E  Y
GGGAUCACAUGUACAUCGGAAUGGCAGGGAAACGUCCUUCUACAAGAUCCUGGCGUUCUUGGGGUCAUCCAAUCUAAAGGCCACUCCAGCGGUAUUGGCAGAUCAAGGUCAACCAGAGU
                                                                             150                               2640

    H  T  H  C  E  G  R  A  Y  L  P  H  R  M  G  K  T  P  P  M  L  N  V  P  E  H  F  R  R  P  F  N  I  G  L  Y  K  G  T  I
AUCACACACUGUGAGGGUCGAGCAUAUUUACCACAUAGAAUGGGAAAGACCCUCCCCAUGUCAAUGGGUAUCCUACAGAGCACUUUCCAUGGGUUGCAGAAGGGAAGGGUUUAGACACGACUGA
                                                                                                               2760

    E  L  T  M  T  I  Y  D  D  E  S  L  E  A  A  P  M  I  W  D  H  F  N  S  S  K  F  S  D  F  R  E  K  A  L  M  F  G  L  I
UUGGAGCUCACAAUGACCAUCUACGAUGACGAGAGCUUGGAAGCAGCCCCUAUGAUCUGGGAUCAUUUCAACUCCUCCAAAUUUUCUGAUUUCAGAGAGAAAGCACUUAUGUUUGGCCUGA
                                                                             200                               2880

    V  E  K  K  A  S  Q  A  W  V  L  D  S  I  S  H  F  K  •
UUGAGAAGAAAGCAUCUCAGGCAUGGGUGCUGGAUUCUAUCUCUCAUUUCAAAUGACUAGUCUCUAACUUCUAGCUUCGAACAAUCCCCGGUUUACUACAGUCUCUCCUAAUUCCAGC
                                                                                                               3000

    M  K  C  L  L  Y  L  A  F  L  F  I  G  V  N
CUCUCGAACAACUAAUAUCCUGUCUUUUCUUUCCCUAUGAAAAAAACUAACAGAGAUGGAUCUGUUUCCUUGACACUAUGAAGUGCCUUUUGUACUUAGCCUUUUUAUUCAUUGGGUUAAUGUGGGGUGA
                                                                             50                                3120

    C  K  F  T  I  V  F  P  H  N  G  K  Q  N  W  K  N  V  P  S  N  Y  H  Y  C  P  S  S  S  D  L  N  W  H  N  D  L  I  G  T
AUUGGCAAGAUUCACCAUAGUGUUUCCCCACAAUGGGAAACAAAACUGGAAGAAUGUACCAUCCAAUUAUCAUUACUGCCCUUCAUCAUCUGAUCUAAAUUGGCAUAAUGAUUUAAUAGGCA
                                                                                                               3240

    A  I  G  V  K  M  P  K  S  H  K  A  I  Q  A  D  G  W  M  C  H  A  S  K  W  V  T  T  C  D  F  R  W  Y  G  P  K  Y  I  T
CAGGAUCCAUACAAGUGAAAAUGCCCAAGAGCCACAAGGCAAUCAUAGCAGGQGUGGAUGGGUGGAUGUGCCAUGCAAGCAAGUGGGUCACAACAUGUGAUUUCAGAUGGUAUGGCCCGAAGUAUAUAA
                                               100                                                             3360

    G  S  I  R  S  F  T  P  S  V  E  G  C  K  E  S  I  E  G  T  K  G  G  T  W  L  N  P  G  F  P  P  G  S  C  G  Y  A  T  V
CACAGUCCAUCCGAUCCUUCACUCCAUCUGUAGAAGGAUGCAAGGAGUCAAUCGAUGGAACAAAAGAAAAGGAACUUGGCUGAAUCCAGGCUUCCCUCCUGGAAGUUGUGGAUAUGCAACUGU
                                                                             150                               3480

    T  D  A  E  A  V  I  V  Q  V  T  P  H  H  V  L  V  D  E  Y  T  Q  E  W  V  D  S  G  F  I  N  G  K  C  S  N  Y  I  C  P
UGACGCUGGUGCCGAACAGUGGAUUGUUCCAGGUGACUCCUCACCAUGUGCUGGUUGAUGAAUACACACAGGAAUGGGUAGAUUCACAGUUCAUCAACGGGAAAUGCAGCAAUUACAUAUGUCCU
                                                                                                               3600

    T  V  H  N  S  T  T  W  H  S  D  Y  K  V  K  Q  L  C  D  S  N  L  I  S  M  D  I  T  F  F  S  E  D  G  E  L  S  S  L  G
CCACUGUCCAUAAACUCUACAACCUGGCAUUCUGAUUACAAAGUCAAAGGCUUAUGUGAUUCUAACCUCAUUUCCAUGGACAUCACCUUCUUCUCAGAGGACGGAGAGCUAUCUUCACUCCA
                                                                             250                               3720

    K  E  G  T  G  F  R  S  N  Y  F  A  Y  E  T  G  G  K  A  C  K  M  G  Y  C  K  H  W  G  V  R  L  P  S  G  V  W  F  E  M
GAAAGGAGGGACAGGGUUCAGAAGUAACUACUUUGCUUAUGAAACUGGAGGCAAGGCCUGCAAAAUGGGAUACUGCAAGCAUUGGGGAGUCAGACUCCCAUCAGGUGUUCUGGUUCGAAUGA
                                                                                                               3840
```

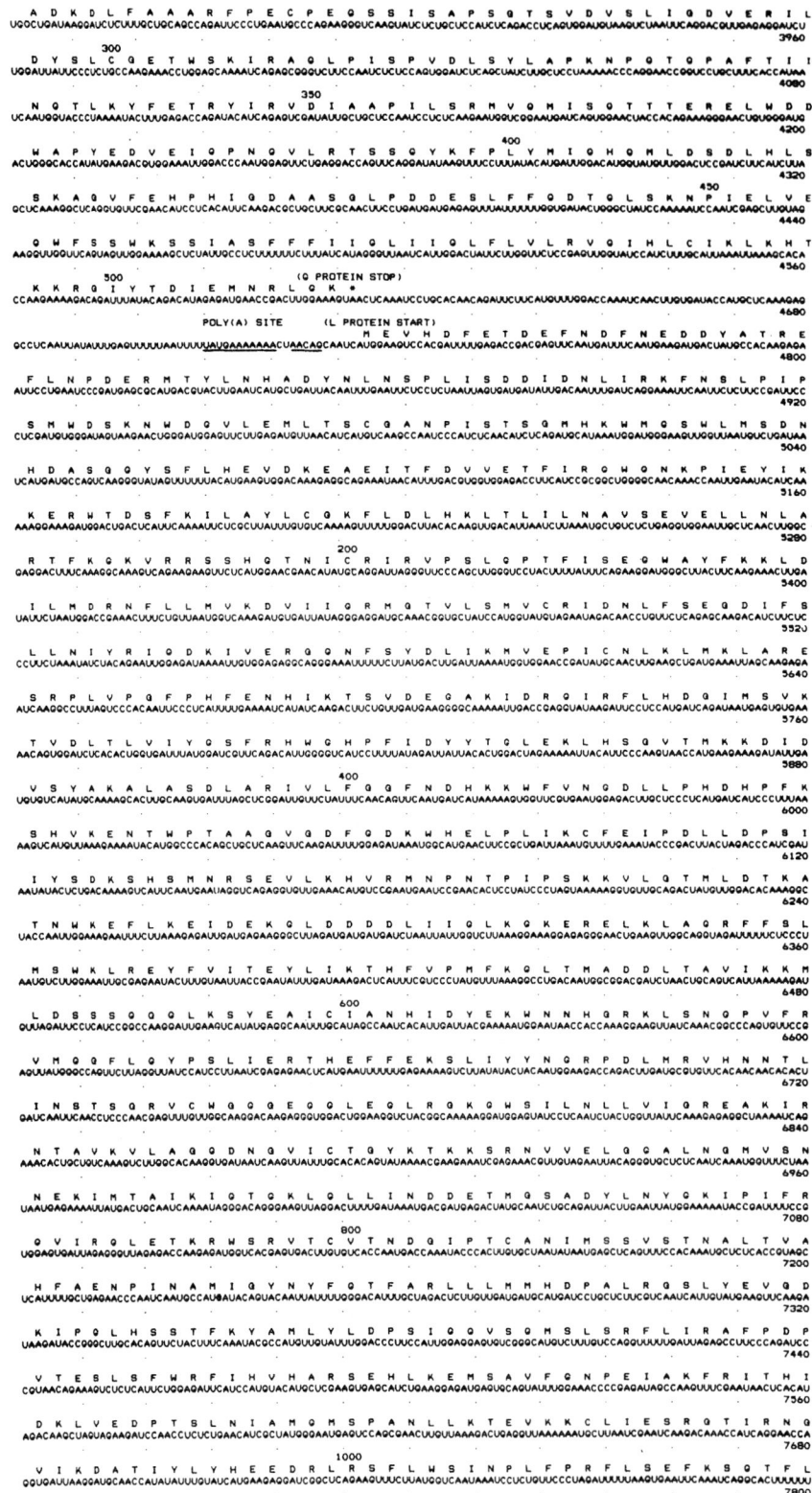

FIGURE 1. (Continued)

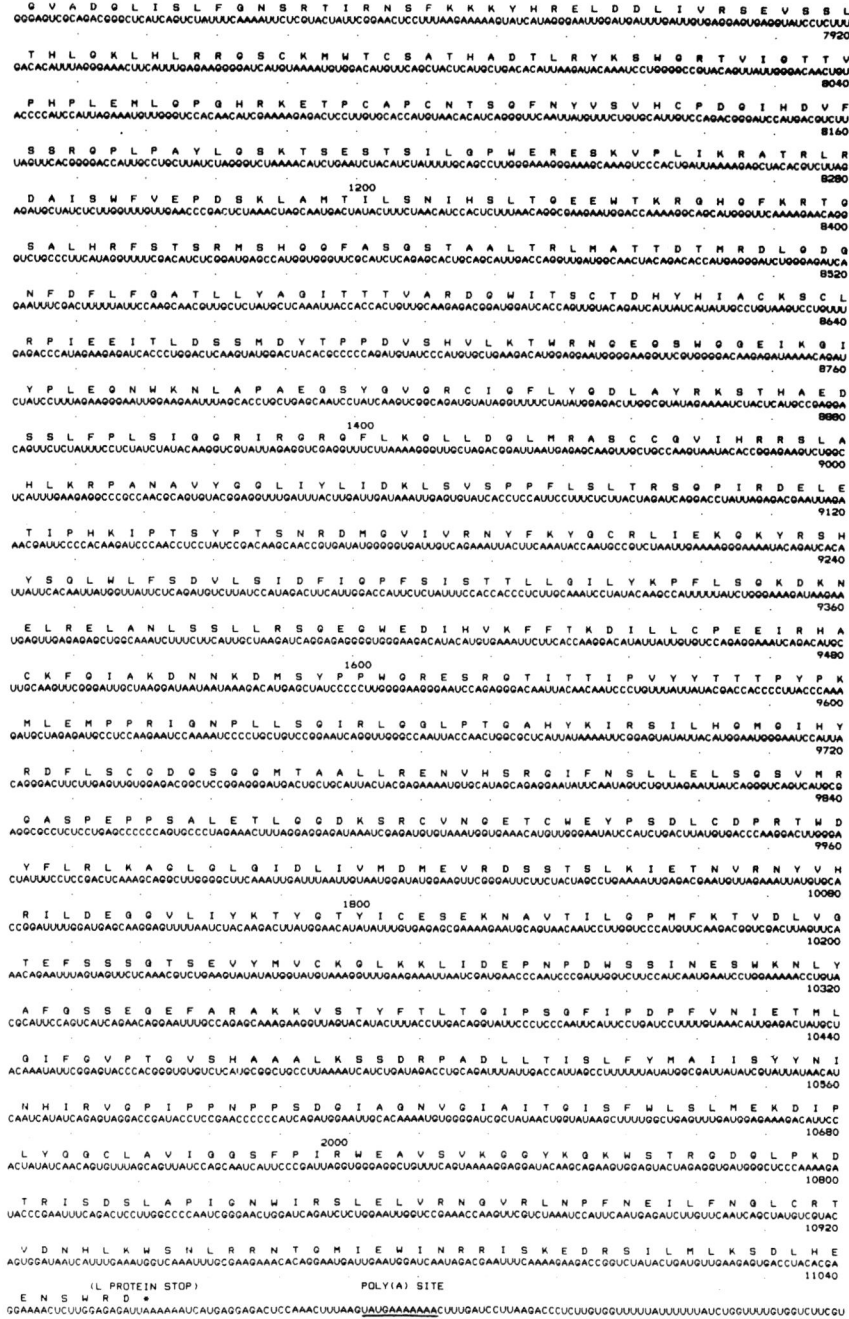

FIGURE 1. (Continued)

except that the nonstructural and matrix (M) proteins are replaced by the two matrix proteins (M_1, and M_2) of rabies virus (Flamand and Delagneau, 1978).

The genome size of approximately 11–12 kilobases (kb) as well as the gene order of rhabdoviruses are highly conserved. For example, the nucleocapsid (N) protein is always encoded by the first gene, while the large (L) polymerase gene is positioned at the 5' terminal end of the genome. The genes are transcribed sequentially, starting at the 3' end of the genome. Attenuation of transcription at each gene junction results in a decrease in the amounts of messages synthesized, depending on their distance form the 3'-terminal start site of transcription (Iverson and Rose, 1981). For example, the polymerase gene, L, which is positioned most distal from the polymerase entry site, is transcribed in low amounts compared to the N gene. This is consistent with the idea that the polymerase is needed in catalytic amounts, while the N protein is needed in high amounts for the encapsidation of the genomic minus- and plus-sense RNAs. Therefore, the organization of the genome together with the sequential and attenuated mode of transcription ensures that each gene product is synthesized in the ratios needed during the course of the infection, a characteristic of most, if not all, nonsegmented negative-strand viruses. Thus, the genome of rhabdoviruses represents a single, highly efficient transcriptional unit.

B. Nucleotide Sequence of the Vesicular Stomatitis Virus Genome

The sequence of the VSV genome is shown in Fig. 1. It is presented as the (+)-strand complement so that the regions corresponding to the mRNAs and the encoded proteins can be illustrated most easily. The majority of this 11,161-nucleotide sequence was derived from cDNA clones,

TABLE I. References from Which the Complete Sequence of the VSV Genome Was Assembled

Residue numbers	Strain	Reference
1–50	Glasgow	McGeoch and Dolan (1979)
51–1,376	San Juan	Gallione et al. (1981)
1,377–1,385, 2,200–2,208	San Juan	McGeoch (1979)
3,040–3,048, 4,714–4,722		Rose (1980)
1,386–2,199	San Juan	Gallione et al. (1981)
2,209–3,039, 3,049–4,713	San Juan	Rose and Gallione (1981)
4,723–11,102	Mudd–Summers	Schubert et al. (1984)
11,103–11,161	Mudd–Summers	Schubert et al. (1980)

and Table 1 lists the references from which the indicated portions of the sequence were compiled. Note that the sequence shown is derived almost entirely from two different isolates of VSV (San Juan and Mudd–Summers), both of the Indiana serotype. On the basis of our limited sequencing of the Mudd–Summers strain of VSV (unpublished data), we estimate that the Mudd–Summers and San Juan strains are likely to differ in about 2% of the nucleotides. About one fourth of these differences would be expected to affect the amino acids encoded in the VSV proteins (Gallione and Rose, 1985).

Figure 2 is a diagram of the genome drawn 3'–5' from left to right to be consistent with the sequence shown in Fig. 1. Transcription of the genome proceeds from left to right to generate a short, untranslated leader RNA of 47 nucleotides followed by five monocistronic mRNAs that encode the five viral proteins. Replication proceeds via a full-length positive-stranded copy of the genome. Transcription and replication are discussed in detail in Section IV.E.

The organization of the VSV genome is extremely compact. The complement of all but 70 nucleotides is found in the leader RNA and mRNAs. The nucleotides in the genome that are apparently not copied during transcription are illustrated in Fig. 3. There are two or three nucleotides between the end of the sequence complementary to the leader RNA and the beginning of the sequence complementary to the N protein mRNA and two nucleotides between the ends and starts of the other mRNAs. Also, there are 59 nucleotides after the L mRNA that are not transcribed. Of the 11,161 nucleotides in the VSV genome, all but 546 code for protein.

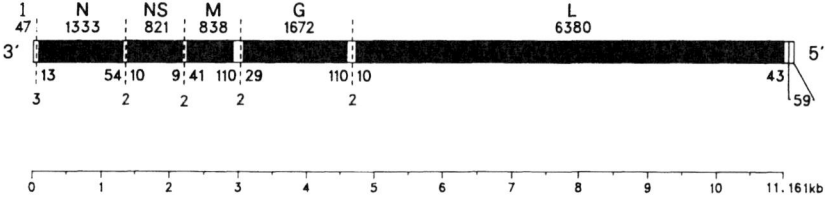

FIGURE 2. Map of the VSV genome. The genome is represented 3'–5' from left to right and shows the genes drawn approximately to scale. The numbers above the line indicate the lengths of the leader RNA and the mRNAs, including the first seven nucleotides of polyadenylate on each mRNA that is encoded by the genome. The solid regions represent the protein-coding portions of the genome, and the open portions represent the noncoding portions. The numbers immediately below the line indicate the lengths of the 5' and 3' untranslated portions of each mRNA. The dotted lines with numbers indicate the positions and lengths of the intergenic regions. The 59-nucleotide, nontranscribed, "trailer" region of the genome is indicated at the extreme 5' end of the genome.

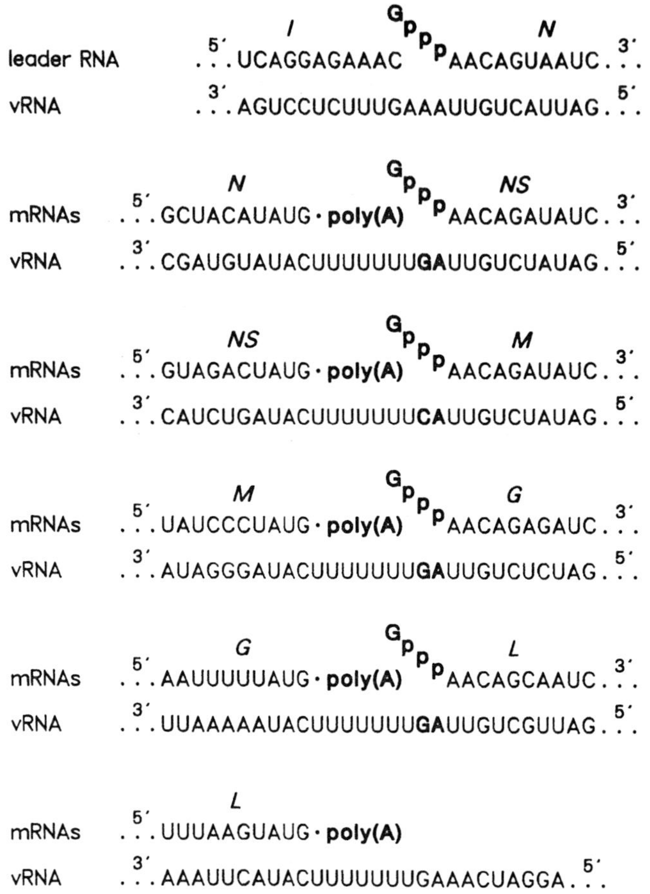

$$\text{leader RNA} \quad 5'...UCAGGAGAAAC \overset{G}{\underset{P_P}{P}} PAACAGUAAUC...3'$$

leader RNA   5´...UCAGGAGAAAC ^G_P_P^PAACAGUAAUC...3´ (*I* ... *N*)

vRNA   3´...AGUCCUCUUUGAAAUUGUCAUUAG...5´

mRNAs   5´...GCUACAUAUG · poly(A) ^G_P_P^PAACAGAUAUC...3´ (*N* ... *NS*)

vRNA   3´...CGAUGUAUACUUUUUUU**G**AUUGUCUAUAG...5´

mRNAs   5´...GUAGACUAUG · poly(A) ^G_P_P^PAACAGAUAUC...3´ (*NS* ... *M*)

vRNA   3´...CAUCUGAUACUUUUUUU**C**AUUGUCUAUAG...5´

mRNAs   5´...UAUCCCUAUG · poly(A) ^G_P_P^PAACAGAGAUC...3´ (*M* ... *G*)

vRNA   3´...AUAGGGAUACUUUUUUU**G**AUUGUCUCUAG...5´

mRNAs   5´...AAUUUUUAUG · poly(A) ^G_P_P^PAACAGCAAUC...3´ (*G* ... *L*)

vRNA   3´...UUAAAAAUACUUUUUUU**G**AUUGUCGUUAG...5´

mRNAs   5´...UUUAAGUAUG · poly(A) (*L*)

vRNA   3´...AAAUUCAUACUUUUUUUUGAAACUAGGA...5´

FIGURE 3. Gene junctions of VSV. The nucleotide sequences at the 3′ end of the leader RNA and the 5′ and 3′ ends of each mRNA are shown along with the corresponding genomic sequences (vRNA). The intergenic dinucleotides are indicated by bold letters.

C. Intergenic Structure and Transcription

Extensive homologies are found around the intergenic dinucleotides (Fig. 3). These regions have the common structure (3′)AUACUUUUUU UNAUUGUCNNUAG(5′), in which N indicates three variable positions and the intergenic dinucleotide is underlined. These dinucleotide spacers are GA, except at the NS–M junction, where the dinucleotide is CA. The first 11 nucleotides of the common sequence are complementary to the sequence (5′)...UAUGAAAAAAA...(3′) that occurs at the mRNA–polyadenylate [poly(A)] junction in each mRNA including L. Reiterative copying of the U residues by the VSV polymerase presumably generates the poly(A) tail on each mRNA (McGeoch, 1979; Rose, 1980; Schubert et al., 1980). The sequence complementary to the 5′ end of the mRNA follows the intergenic dinucleotide. The L mRNA also terminates with the sequence UAUG-poly(A) encoded by the sequence (3')AUACUUUUUUU and is presumably also polyadenylated by a polymerase "slippage" mechanism (Schubert et al., 1980; Schubert and Lazzarini, 1981).

A quantitative analysis of the extent of "readthrough" transcription (discussed in Section IV.E) that occurs at each of the intergenic regions in vivo has been carried out recently by Masters and Samuel (1984). The one dicistronic message that is found much more frequently than the others is the NS–M transcript, suggesting that the one nucleotide difference at this junction may affect the generation of these transcripts.

There are two general models that could explain the sequential nature of VSV transcription (Abraham and Banerjee, 1976; Ball and White, 1976), and these would involve either cleavage of a precursor or multiple, sequential transcription initiations. The cleavage model would propose that polymerase transcribes through the intergenic regions. Two cleavage events would then be required to remove the intergenic dinucleotide, and this would be followed by capping and methylation of the new 5'-terminus. Poly(A) would presumably be added before the cleavage by a polymerase "slippage" mechanism. Alternatively, poly(A) could be added after cleavage.

In the sequential initiation model, polymerase would terminate before copying the intergenic dinucleotide, presumably after repetitive copying of the U_7 sequence to generate about 200 nucleotides of poly(A). It is possible that the poly(A) could be added after termination. Synthesis of the next mRNA would require a separate initiation event. In either model, the conserved sequences around the intergenic dinucleotides would be critical signals affecting the activity of the polymerase or the cleavage enzymes. During replication, these signals would have to be ignored.

III. RNA PRODUCTS

A. Leader RNAs and Their Functions

1. Transcription

The synthesis of rhabdovirus messages is preceded by transcription of a short (+)-sense leader RNA that is presumably the first viral product made in an infected cell (Colonno and Banerjee, 1976). Sequence analyses revealed that (+)-sense leader RNAs of vesicular stomatitis virus (VSV) and rabies virus are between 47 and 58 nucleotides long, contain a 5'-terminal di- or triphosphate, and are initiated precisely at the 3' terminus of the (−)-sense genomic RNA (Colonno and Banerjee, 1978a,b; Kurilla et al., 1984; Keene et al., 1978; 1979; Rowlands, 1979).

UV inactivation of VSV transcription as well as the study of the kinetics of message synthesis demonstrated that the mRNAs are transcribed sequentially and that a single entry site for the polymerase complex exists at the exact 3' end of the genome (Abraham and Banerjee, 1976; Ball and White, 1976; Iverson and Rose, 1982; Emerson, 1982). Additional data suggest that the polymerase complex cannot move or freewheel along the template without simultaneous transcription (Keene *et al.*, 1981a; Emerson, 1982). It appears that transcription of (+)-sense leader RNA is necessary for the polymerase complex to reach the start site of the first gene, that for nucleocapsid (N) protein, two or three nucleotides downstream of the leader RNA termination site (Fig. 3). This distance is short and presumably enables the polymerase to recognize the start site of the N gene and proceed with transcription. Consistently, all intercistronic regions of VSV (Fig. 3) and those of paramyxoviruses analyzed to date are also only two or three nucleotides long (McGeoch, 1979; Rose, 1980; Gupta and Kingsbury, 1984).

2. Replication

Most defective interfering (DI) particles of VSV analyzed to date are unable to synthesize mRNAs. However, a low level of transcription of a short RNA product is observed from these particles (Reichmann *et al.*, 1974). This DI-product RNA was further characterized and sequenced (Emerson *et al.*, 1977; Schubert *et al.*, 1978; Semler *et al.*, 1978). With VSV, the DI-product RNA is 46 nucleotides long and is transcribed from the precise 3' terminus of the DI-particle genome. Hybridization and sequence analyses demonstrated that the 3'-terminal and 5'-terminal sequences of most DI-particle genomes are complementary for various lengths. In addition, the 3'-terminal sequences of these DI-particle RNAs are identical for at least 45 bases to the 3' ends of the (+)-sense genome of the parental virus (Schubert *et al.*, 1979; Kolakofsky, 1982; Perrault *et al.*, 1978).The structures and the generation of the DI-particle RNAs have been reviewed in detail earlier (Lazzarini *et al.*, 1981; Perrault, 1981; Meier *et al.*, 1984). Since the 3' ends of many DI-particle genomes are identical to the 3' end of the (+)-sense parental genome, the "DI-product RNA," referred to as (−)-sense leader RNA, is also found transcribed from (+)-sense VSV genomic RNA (Leppert *et al.*, 1979; Rao and Huang, 1979). In fact, its concentration in VSV-infected cells is 8 times higher than that of the (+)-sense leader RNA, suggesting that initiations by the polymerase occur more frequently on the (+)-sense genome than on the (−)-sense genome of the parental virus (Leppert and Kolakofsky, 1980). This is consistent with the fact that late in infection, (−)-sense genomes are approximately 4 times more abundant in infected cells than (+)-sense genomes (Simonsen *et al.*, 1979). In contrast, DI-particle (+)- and (−)-sense genomes, which contain identical 3'-terminal sequences, are present in approximately equimolar amounts (Moyer and Gatchell, 1979).

A comparison of the genomic (+)- and (−)-sense leader-region sequences from five different serotypes of VSV revealed a high degree of homology within the first 18 nucleotides of the 3' termini, and a consensus sequence can be deduced from the 5' termini of the leader RNAs: pppACGAANANNANNAAACA . . . (Giorgi et al., 1983). This sequence most likely contains the nucleation site for encapsidation of the leader and genomic RNAs (discussed in section IV.C). To a lesser degree, the termination sites for the various VSV leader RNAs are also conserved (Keene et al., 1980). The genomic sequences that control the preferred (−)-sense leader initiation, however, are most likely not present in the 3'-terminal genomic consensus sequence, since these sequences are identical for 21 bases in both the (+)- and (−)-sense genomes in the case of VSV-New Jersey (Keene et al., 1979). On the basis of in vitro methylation protection studies of nucleocapsid RNA in the presence or absence of soluble nonstructural (NS) and large (L) proteins it was proposed that the genomic leader regions—positions 16–30 from the 3' end of the VSV (−)-sense genome and positions 17–35 from the 3' end of the DI particle [or VSV (+)-sense] genome, respectively—may contain the binding site for the polymerase complex (Keene et al., 1981b; Isaac and Keene, 1982). The sequences of these regions differ between the (+) and (−) genomes and may differentially affect the affinity of the polymerase, thereby contributing to the preferred synthesis of (−)-sense leader and (−)-sense genomic RNA in vivo.

3. Cell Killing

An additional function for (+)-sense leader RNA that links its synthesis to the ability of VSV to kill the host cell has been proposed (Grinnell and Wagner, 1983). Unlike the case with rabies virus, cellular transcription and protein and DNA synthesis are rapidly inhibited on infection with VSV (Weck and Wagner, 1978; Wu and Lucas-Lenard, 1980; McGowan and Wagner, 1981; Wertz and Youngner, 1972). In contrast, no inhibition is observed when cells are infected with most DI particles of VSV (Weck and Wagner, 1979). The shutoff of cellular macromolecular synthesis by VSV is dependent on viral transcription, and the UV-target size for host shutoff is very small (Marcus and Sekellick, 1975; Weck et al., 1979; Wu and Lucas-Lenard, 1980), suggesting that only the (+)-sense leader RNA and possibly a short region of the N gene need to be transcribed. Thus, it appears that a newly synthesized viral protein product is not required. McGowan et al. (1982) demonstrated that the addition of purified (+)-sense, but not (−)-sense, leader RNA to soluble whole-cell extracts specifically inhibits DNA-dependent transcription by RNA polymerases II and III, using simian virus 40 (SV40) and adenovirus DNA templates. Synthetic oligonucleotides with sequences representing the middle portion of the leader RNA can also specifically inhibit in vitro RNA synthesis (Grinnell and Wagner, 1984). Kurilla et al. (1982) found that (+)-sense, but not (−)-sense, leader RNA accumulates in nuclei shortly after infection, where it associates with the cellular La protein. As yet it is the only VSV product detected in the nucleus. Interestingly, a slightly larger version of the VSV (−)-sense leader RNA (54 bases) was also found associated with the cellular La protein *in vivo* early during infection (Kurilla and Keene, 1983; Wilusz *et al.*, 1983). La protein usually interacts with precursor RNAs synthesized by RNA polymerase III. The significance of this interaction is unknown.

Although the inhibition of cellular transcription by (+)-sense leader RNA has never been demonstrated *in vivo*, the correlations among small UV-target size for cell killing, the *in vitro* inhibition of RNA polymerase II and III, and the accumulation of (+)-sense leader RNA in the nucleus are very suggestive that (+)-sense leader RNA may inhibit cellular macromolecular synthesis directly or by an as yet unknown mechanism. In contrast, (−)-sense leader RNA, which shares some sequence homology with (+)-sense leader RNA, does not inhibit RNA polymerases II and III and is not detected in the nuclei of infected cells. This is consistent with the absence of a shutoff of cellular macromolecular synthesis on infection by most DI particles. Interactions of the La protein with both the (+)- and the 54-nucleotide (−)-sense leader RNAs of VSV as well as with the rabies virus (+)-sense leader RNA seem to be a characteristic of rhabdovirus leader RNAs (Kurilla *et al.*, 1984). However, since rabies virus does not shut off the host cell, like VSV, it remains to be determined whether the formation of the La–leader complex plays a role in the host shutoff or whether interactions of the (+)-sense leader RNA of VSV with other, as yet unidentified, host proteins or nucleic acids are involved.

Despite their small size, the leader RNAs may encode or trigger many viral-specific functions. These may include polymerase binding and initiation; leading the polymerase to the start site for the first gene; N; encapsidation of the genomic RNA and thereby controlling replication vs. transcription through the availability of N protein; and, in the case of VSV, the shutoff of cellular macromolecular synthesis. The details of the mechanism of this VSV (+)-sense leader RNA interaction with the host cell remain an open question. Their study may not only contribute to our understanding of lytic and persistent infections, but also give insight into the mechanisms of macromolecular synthesis of the host cell.

B. Messenger RNA

VSV mRNAs contain a typical eukaryotic 5′-capping group $m^7G(5')ppp(5')$ and terminate with about 200 nucleotides of poly(A) (reviewed by Banerjee *et al.*, 1977; Banerjee, 1980) (discussed in Section IV.E). The first extensive sequences from the VSV mRNAs were obtained from the fragments of mRNA protected by ribosomes in a translation-initiation complex (Rose, 1977, 1978). In all five VSV mRNAs, the 40S or 80S ribosomes protected sequences of 30–45 nucleotides around the AUG closest to the 5' end of the mRNA, and this protection extended to the 5'-end in the N, NS, and L mRNAs. Also, low levels of ribosome binding to the second and third AUG codons in the N mRNA were detected. However, if translation initiated at these sites *in vivo*, only very short peptides of 23 and 30 amino acids would be encoded (Gallione *et al.*, 1981). These peptides have not yet been detected.

Complete sequences of the mRNAs that encode the N, NS, and matrix (M) proteins, and glycoprotein (G) were subsequently obtained from cDNA clones prepared from mRNA, and the complete sequence of the L mRNA was obtained from cDNA clones prepared from genomic RNA (Gallione *et al.*, 1981; Rose and Gallione, 1981; Schubert *et al.*, 1984). The lengths of the mRNAs and the proteins they encode are indicated in Table II. Each mRNA contains a single long open reading frame starting at the first AUG and terminating near the 3' end of the mRNA and relatively short noncoding regions at both ends (see Figs. 1 and 2). A table of codon usage in the VSV mRNAs' genome has been compiled in Fig. 4. It shows quite clearly that there is the typical eukaryotic bias against arginine condons of the form CGN and against any codon ending in CG. As noted previously (Rose and Gallione, 1981), there is also a strong bias against codon junctions of the type NNC–GNN throughout the VSV genome.

In the 815-nucleotide sequence reported previously (Gallione *et al.*, 1981) for the mRNA that encodes the VSV NS protein, we had noticed that there was a long second open reading frame that extended from nucleotide 531 to 806, while the termination codon for the NS protein was predicted to be at nucleotide 677. Because a single base deletion would result in fusion of the major open reading frame with this second frame, we decided to resequence the NS gene from a second cDNA clone. This second cDNA clone (pNS319) was sequenced completely and was indeed found to lack one T residue at nucleotide 575 (Hong and Rose, unpublished results). We then determined that the sequence of pNS319 (which

TABLE II. VSV Messenger RNAs and Proteins

Protein	Number of amino acids	Predicted mol.wt.[a]	Messenger RNA length
N	422	47,355	1,326
NS	265	29,878	814
M	229	26,064	831
G (nonglycosylated precursor)	511	57,416	1,665
L	2,109	240,707	6,373

[a] Molecular weights are calculated from the predicted protein sequences and include only the contribution from the amino acids. The two N-linked glycans on VSV G protein contribute approximately 6,000 to the apparent molecular weight. Sixteen N-terminal amino acids are cleaved from the G-protein precursor. All mRNAs listed are 7 nucleotides shorter than their corresponding genes shown in Fig. 2 because all A residues of the poly(A) tail were excluded, although the first 7 A residues of the poly(A) tail are complementary to the genome.

```
===============================================
F TTT 69.0  S TCT 84.0  Y TAT 77.0  C TGT 32.0
F TTC 89.0  S TCC 52.0  Y TAC 57.0  C TGC 24.0
L TTA 70.0  S TCA 67.0  * TAA  0.0  * TGA  0.0
L TTG 98.0  S TCG 16.0  * TAG  0.0  W TGG 71.0
===============================================
L CTT 43.0  P CCT 55.0  H CAT 72.0  R CGT 16.0
L CTC 37.0  P CCC 34.0  H CAC 25.0  R CGC  4.0
L CTA 41.0  P CCA 64.0  Q CAA 80.0  R CGA 25.0
L CTG 46.0  P CCG 20.0  Q CAG 41.0  R CGG 13.0
===============================================
I ATT 96.0  T ACT 62.0  N AAT 95.0  S AGT 57.0
I ATC 78.0  T ACC 50.0  N AAC 48.0  S AGC 24.0
I ATA 61.0  T ACA 73.0  K AAA134.0  R AGA 87.0
M ATG101.0  T ACG 19.0  K AAG 95.0  R AGG 33.0
===============================================
V GTT 48.0  A GCT 61.0  D GAT130.0  G GGT 44.0
V GTC 45.0  A GCC 34.0  D GAC 85.0  G GGC 36.0
V GTA 39.0  A GCA 71.0  E GAA112.0  G GGA100.0
V GTG 52.0  A GCG 12.0  E GAG 86.0  G GGG 46.0
===============================================
```

FIGURE 4. Codon usage in the VSV genome. The codon usage in the VSV mRNAs was compiled from the translated portions of each mRNA. The amino acids are indicated by the single-letter code beside each codon. The numbers indicate the number of times that each codon is used.

lacks all but 32 5'-nucleotides of the NS mRNA) was the correct one by sequencing the VSV genome in this region using a synthetic oligonucleotide primer. This correction changes the predicted C terminus of the NS protein and adds an additional 43 amino acids (Fig. 1). In addition to this frameshift error in the published sequence, there were also two clerical errors. The nucleotide sequence at positions 240 and 241 has been corrected to CA from the AG.

C. A Sixth Messenger RNA and Protein of Infectious Hematopoietic Necrosis Virus

Until recently, it was assumed that most rhabdovirus genomes would show the same organization as the two prototypes, VSV and rabies virus, both of which contain five genes. This assumption, however, has to be disregarded after the recent discovery of a sixth gene in the genome of infectious hematopoietic necrosis virus (IHNV). The gene order of IHNV was determined by R-loop mapping, using cDNA clones derived from polyadenylated viral mRNAs. The sixth gene, designated nonviral (NV), was tentatively mapped between the G and L genes: 3'-N-M_1-M_2-G-NV-L-5' (Kurath et al., 1985). The gene codes for a small NV protein of approximately 12 kilodaltons (kd). Poly (A) and hybrid-selected mRNA were translated into the 12-kd protein in vitro and corresponded to the

NV protein found in infected cells (Kurath and Leong, 1985). The function of the NV protein is unknown. It is not present in the virion, suggesting that it is not required for primary transcription. Since the discovery of the NV gene, it can now be expected that other rhabdovirus genomes may also encode additional gene products that have not been identified to date. Especially since the spectrum of host organisms that can be infected with rhabdoviruses is so broad, additional and possibly different gene products may be discovered with other rhabdoviruses in the future.

IV. PROTEINS

A. Glycoprotein and Its Functions

1. General Features and Biogenesis

Vesicular stomatitis virus (VSV) virions and the virions of other rhabdoviruses contain a single glycoprotein (G) species that forms spikes on the surface of the virions. The most recent estimate indicates that there are about 1200 molecules of VSV G protein per virus particle (Thomas et al., 1985), and these may be monomeric (Crimmins et al., 1983). The G-protein precursor contains 511 amino acids including a 16-amino-acid signal sequence (Lingappa et al., 1978; Chatis and Morrison, 1979; Irving et al., 1979; Rose and Gallione, 1981). Two asparagine-linked complex oligosaccharides are attached at amino acids 179 and 336 (Etchison et al., 1977; Reading et al., 1978; Rose and Gallione, 1981). The fatty acid palmitate is apparently esterified to cysteine 489 in the cytoplasmic domain (Magee et al., 1984; Rose et al., 1984).

VSV G protein is positioned in the virion such that the NH_2-terminal 95% of the polypeptide chain is external to the bilayer (Rose et al., 1980). G protein plays two roles in the virus life cycle: First, it is responsible for the binding of the virus to the host cell and for inducing uptake of the virus via fusion with the endosomal membrane (White et al., 1983). Second, during virus maturation, the interaction between the internal components of the virus and the cytoplasmic domain of the G protein presumably directs budding of the virus particle. Antibodies that neutralize virus infectivity are directed against the G protein (Kelley et al., 1972).

VSV G protein has been a major model system for the study of transport and processing of proteins that are directed to the plasma membrane (reviewed in Hubbard and Ivatt, 1981). G protein is inserted into the rough endoplasmic reticulum (RER) as a nascent chain (Rothman and Lodish, 1977), and both ends of the molecule play critical roles in this process. Like the majority of secreted and membrane proteins, the G-protein precursor contains a short hydrophobic NH_2-terminal signal (or leader peptide) that apparently initiates the vectorial discharge across the RER. The

G protein is not discharged completely across the RER, but becomes anchored stably in the membrane. This anchoring is accomplished by a hydrophobic domain of 20 amino acids that spans the bilayer (Rose *et al.*, 1980). The 29 COOH-terminal amino acids form the cytoplasmic domain of the molecule (Rose *et al.*, 1980; Rose and Gallione, 1981).

2. Expression and *in Vitro* Mutagenesis of the G-Protein Gene

Expression of the VSV G protein from cloned cDNA both in *Escherichia* coli (Rose and Shafferman, 1981) and in animal cells has been reported (Rose and Bergmann, 1982; Florkiewicz *et al.*, 1983; Riedel *et al.*, 1984). The expression in eukaryotic cells demonstrated that VSV G protein is transported to the cell surface in the absence of other VSV proteins (Rose and Bergmann, 1982). The availability of the systems for both stable and transient expression of G protein from cloned cDNA has allowed extensive analysis of the functions of the various domains and posttranslational modifications to the G protein.

Initial experiments illustrated the anchoring function of the transmembrane domain because G proteins lacking this domain (produced from a truncated G-protein cDNA) were secreted, albeit slowly (Rose and Bergmann, 1982). Further *in vitro* mutagenesis experiments demonstrated that the presence of the normal cytoplasmic domain of G protein was critical for rapid and efficient transfer of G protein from the RER to the Golgi apparatus (Rose and Bergmann, 1983). This domain may be responsible for directing the efficient incorporation of the G protein into vesicles budding from the RER and moving to the Golgi region. Other applications of these expression systems are described below.

3. Role of Glycosylation in Transport

The glycosylation of G protein has been studied in great detail and has been reviewed recently (Hubbard and Ivatt, 1981). Glycosylation of G protein follows the typical route involving addition of the high-mannose core sugars to the acceptor asparagine residues as the nascent polypeptide chain is transferred into the lumen of the RER. Subsequent trimming of these sugars occurs in the RER and Golgi apparatus. This trimming to a core structure is followed by addition of N-acetylglucosamine, galactose, fucose, and N-acetylneuraminic acid in the Golgi apparatus (Bergmann *et al.*, 1981).

The importance of the N-linked glycans in transport of the VSV G protein to the cell surface has been investigated using the antibiotic tunicamycin, which blocks all N-linked glycosylation. Virion production in VSV-infected cells treated with tunicamycin is almost completely inhibited as a result of the failure of nonglycosylated G protein to reach the cell surface (Leavitt *et al.*, 1977a; Morrison *et al.*, 1978). The G protein made in the presence of tunicamycin was less soluble in nonionic detergent and appeared to aggregate in the ER (Leavitt *et al.*, 1977b; Gibson *et al.*, 1979). Although the San Juan strain of the Indiana serotype appeared quite stringent in its requirement for carbohydrate, the Orsay strain appeared to have a less stringent requirement, especially when grown at lower temperature (Gibson *et al.*, 1979). These results suggested that the role of carbohydrate might be simply to maintain a certain conformation of the protein. There are ten amino acid differences between the G proteins of the Orsay and San Juan strains (Gallione and Rose, 1985), and one or more of these differences are presumably responsible for the differential responses to tunicamycin.

More recently, the role of glycosylation in transport of the G protein has been examined using a method that avoids all the potential side effects of tunicamycin. In these experiments, the procedure of oligo-nucleotide-directed mutagenesis was used to eliminate one or the other or both of the glycosylation sites in G protein. The results from these experiments (Machamer *et al.*, 1985) showed that G protein with either one of the two normal sites glycosylated was transported to the cell surface, while the double mutant lacking both sites was not transported beyond the Golgi apparatus. The simplest interpretation of these results is that the carbohydrate signals transport of the protein to the cell surface, although it is possible that glycosylation at either site is able to promote a conformational change required for transport. Work on creation of new glycosylation sites in G protein lacking the two normal sites is in progress. If creation of glycosylation sites at multiple sites in the protein is able to promote transport, then the argument that carbohydrate may play a direct role in transport will become more tenable.

4. Attachment of Fatty Acid

The G protein of VSV-Indiana is known to have 1–2 molecules of the fatty acid palmitate esterified to it. This modification is believed to occur just as G protein enters the Golgi apparatus (Schmidt and Schlesinger, 1979). Recent *in vitro* mutagenesis and expression experiments have focused on locating the site of and examining the function of fatty acid esterification in VSV G protein (Rose *et al.*, 1984). Deletion analysis suggested that the site of fatty acid addition might be at a cysteine in the cytoplasmic domain, and oligonucleotide-directed mutagenesis was then used to change this cysteine to a serine residue. G protein with this mutation does not have palmitate esterified to it, but is transported at a normal rate to the cell surface (Rose *et al.*, 1984). Although this experiment does not prove that the fatty acid is esterified to the cysteine residue, that it is esterified is very likely because the experiments of Magee *et al.* (1984) indicate that the fatty acid is attached to G protein via a thioester linkage.

The function of the fatty acid is unknown, but it appears not to have a role in G-protein transport to the cell surface. The experiments of Schlesinger and Malfer (1982) employing cerulenin to inhibit fatty acid acylation suggested that the fatty acid might play a role in VSV-particle formation. This result is complicated by the fact that the New Jersey serotype of VSV lacks both the cysteine and the fatty acid modification (Gallione and Rose, 1983), yet forms virus particles normally. Thus, there is certainly not a universal requirement for fatty acid acylation of the G protein in rhabdovirus-particle formation.

5. Fusion Activity

To cause an infection, enveloped viruses must release their nucleo-capsids into the cell cytoplasm. Experiments carried out over the past few years have provided strong evidence that this release occurs through fusion of the viral envelope with an internal cellular membrane (White *et al.*, 1983). By analogy with the fusion G proteins of the paramyxoviruses and myxoviruses, it seemed very likely that the single G protein of VSV would have such a fusion activity. Direct demonstration of this activity in G protein came from experiments in which a cDNA clone that encodes the G protein was expressed in animal cells (Florkiewicz and Rose, 1984; Riedel *et al.*, 1984). Cells that expressed the G protein were found to fuse if the pH of the medium was briefly reduced. These results provided clear evidence that the fusion activity resides in the G protein.

Although it is clear that the fusion activity resides in the G protein, the fusion domain within the polypeptide has not been clearly identified. Experiments have been reported in which a synthetic peptide corresponding to N-terminal amino acids of mature G protein caused hemolysis in a pH-dependent manner (Schlegel and Wade, 1984, 1985). Although it has been suggested from these experiments that the N terminus is the fusion domain, these experiments should probably be viewed with caution because of the very high peptide concentrations employed.

6. Sequence Comparisons of Rhabdovirus G Proteins

The predicted amino acid sequences of the G proteins of VSV and rabies virus were compared using a computer program that provides an optimal alignment and gives a statistical significance for the match (Doolittle 1981; Rose *et al.*, 1982). This computer-assisted analysis showed that the two G proteins are definitely related, since the alignment score was 12.6 standard deviations above the mean obtained from randomized comparisons. Since a score of 3.0 standard deviations above the mean is considered strong evidence of common ancestry, this analysis leaves little doubt that VSV and rabies share a common ancestor. Although the computer can easily discern this level of the relatedness, it should be pointed out that it is not readily detected without computer assistance. Indeed, although there is an overall identity of 20% between the two proteins in the optimal alignment, the matches are scattered and the longest identity is only four amino acids.

Sequence comparisons have also been done for VSV G proteins of the two serotypes, Indiana and New Jersey (Gallione and Rose, 1983). Here, the relatedness of the proteins is obvious without computer assistance and was found to be 51%.

7. Soluble Form of the G Protein

VSV- and rabies-virus-infected cells release a truncated, soluble form of the glycoprotein (G_s) into the medium. This protein lacks amino acids from the COOH terminus and is presumably generated by proteolytic degradation (Kang and Prevec, 1970; Little and Huang, 1977; Chatis and Morrison, 1982; Irving and Ghosh, 1982; Dietzschold et al., 1983). The COOH-terminal sequence of this form of the rabies virus G protein revealed that it was truncated within the transmembrane domain (Dietzschold et al., 1983).

Although the mechanism of generation of these and other soluble G protein forms is not known, recent evidence (Garreis-Wabnitz and Kruppa, 1984) shows that the VSV G_s protein is detected in very short metabolic labeling periods, suggesting that it might result from premature translation termination or rapid proteolysis immediately after or even during insertion into the endoplasmic reticulum. The function of this soluble glycoprotein is not known, but it has been suggested that it might play a role in viral pathogenesis (Little and Huang, 1977).

B. Matrix Protein

All rhabdoviruses contain a major structural protein called matrix (M) protein. In VSV, there are about 1800 molecules of M protein, making M the most abundant of the viral proteins (Thomas et al., 1985). The location of the M protein in the virion is not entirely certain, although it is clear that it is inside the viral envelope. Cross-linking experiments have suggested that M protein may form a bridge between the cytoplasmic domain of G protein and the nucleocapsid protein (Dubovi and Wagner, 1977; Zakowski and Wagner, 1980).

M protein initially appears as a soluble protein in the cytoplasm of VSV-infected cells and is then assembled into particles budding from the plasma membrane (Knipe et al., 1977a). However, it is not clear whether this M protein is associated with budding virus or whether there is a direct association between M protein and the plasma membrane. The M protein must play a critical role in virus assembly because mutations in M protein block the appearance of virus particles (Knipe et al., 1977b). M protein has been found to associate with membrane fractions from uninfected cells (Cohen *et al.*, 1971), suggesting that it might initiate the binding of nucleocapsid to the membrane. More recent experiments have suggested that M protein may interact with membrane phospholipids (Zakowski *et al.*, 1981).

The molecular weight of the VSV M protein calculated from the predicted protein sequence is 26,064 (Rose and Gallione, 1981). The predicted sequence of the M protein reveals some interesting features. First, the amino-terminal sequence of the protein contains 8 charged (lysine) residues out of the first 19 amino acids, and a triple proline sequence separates this basic, NH_2-terminal domain from the remainder of the protein. M Protein appears to play a negative role in regulating viral RNA synthesis (Clinton *et al.*, 1978b; Carroll and Wagner, 1979; Combard and Printz-Ané, 1979; Martinet *et al.*, 1979), and the extremely basic N-terminal domain of M protein is an obvious region that might interact with genomic RNA and regulate RNA synthesis. Another interesting feature of the M protein is the absence of any long hydrophobic stretches of amino acids that might interact with the lipid bilayer.

Two types of studies suggest that there may be an interaction between the M protein and the nucleocapsid. First, Newcomb and Brown (1981) have carried out a study in which they found that extraction of VSV virions with the detergent octylglucoside alone removed only the G protein and left a compact, viruslike, "skeleton" structure intact. However, extraction with octylglucoside in the presence of 0.5 M salt removed both the M and G proteins and produced a greatly extended nucleocapsid structure. It was concluded from this study that the M protein has a high affinity for the nucleocapsid and that it may play a role in holding the nucleocapsid together. De *et al.* (1982) have reported that addition of M protein to extended nucleocapsids causes condensation or clumping, again suggesting an important role for M protein in the structure of the nucleocapsid.

Like the nonstructural (NS) protein, the VSV M protein is known to be phosphorylated in VSV-infected cells (Imblum and Wagner, 1974; Moyer and Summers, 1974). The major phosphoamino acid found in the M protein was phosphoserine, although a low level of phosphotyrosine was also detected (Clinton and Huang, 1981). The functional significance of M-protein phosphorylation is unknown.

C. Nucleocapsid Protein

Rhabdovirus genomes are tightly associated with the viral nucleocapsid (N) protein. Approximately 1300 molecules of the N protein are associated with the VSV genome (Thomas *et al.*, 1985). The amino terminus of the N protein of VSV is blocked and has the structure *N*-acetyl-Ser-Val-Thr (Blumberg *et al.*, 1984). Two additional proteins, large (L) and

NS, are more loosely associated with the RNA–N protein complex to form the viral nucleocapsid structure. This nucleocapsid structure contains all the enzymatic activities required for the production of the viral mRNAs (discussed in Section IV.E). The complete sequences of cDNA clones that encode the N proteins from two VSV serotypes (Indiana and New Jersey) have been reported (Gallione *et al.*, 1981; Banerjee *et al.*, 1984). Important features of the N-protein structure have not been obvious from examination of the predicted sequence of the protein. The extent of sequence conservation during evolution is especially striking for the N proteins, for which the overall identity is 68% (Banerjee *et al.*, 1984). Presumably, the constraints on the sequence of the N protein are quite strong, because it must interact with the genome, other N-protein molecules, and the other VSV proteins. In contrast, the NS proteins of these two VSV serotypes show only 32% identity (Gill and Banerjee, 1985).

There is accumulating evidence that the N protein of VSV plays a crucial role in regulating the balance between viral transcription and replication in the cell. The (+)- and (−)-sense leader RNAs in infected cells are found encapsidated with the N protein, suggesting that they contain the nucleation site for encapsidation. This site most likely resides within the first 14 5'-terminal nucleotides, as demonstrated by *in vitro* encapsidation experiments (Blumberg and Kolakofsky, 1981; Blumberg *et al.*, 1983). Encapsidation by solubilized N protein seems to proceed unidirectionally from the 5' to the 3' end of the leader RNA. It is postulated that this association of N protein with the nascent transcript prevents the polymerase from recognizing termination sites on the genome, resulting in synthesis and encapsidation of full length (+)- and (−)-strand genomes. Thus, immediately after virus entry, transcription would be favored over replication because there would be no pool of N protein. Later, when excess N protein had accumulated, replication would be favored (Kingsbury, 1974; Blumberg *et al.*, 1981). In fact, the translation of purified N mRNA in an *in vitro* replication system appears to be the only requirement for the synthesis of (+)- and (−)-sense genomic RNAs of defective interfering particles (Patton *et al.*, 1984). *In vivo* and *in vitro* experiments employing a monoclonal antibody that recognizes free N protein also provide strong support for this model (Arnheiter *et al.*, 1985).

The N protein has been expressed from cloned cDNA in animal cells (Sprague *et al.*, 1983). The expression vector employed included the SV40 origin of replication. Transcription of the N gene was driven from the late promoter of SV40. Acute transfection of COS cells with this vector resulted in synthesis of abundant quantities of the N protein. Examination of the protein by indirect immunofluorescence microscopy showed that it was localized in the cytoplasm and had a granular appearance identical to that of N protein observed in VSV-infected cells. This pattern suggests that N protein may spontaneously aggregate or begin to form nucleocapsidlike structures in the cytoplasm.

D. NS Protein

1. Structure and Function

The NS phosphoprotein of VSV accumulates in the cytoplasm of infected cells, where it represents the most abundant soluble VSV protein. It migrates on denaturing sodium dodecyl sulfate (SDS)–polyacrylamide gels as a 40- to 60-kilodalton (kd) protein, depending on the bisacrylamide cross-linker concentration of the gel (Bell et al., 1984). The nucleotide sequence analysis of the NS gene of VSV-Indiana, from which the molecular weight of the NS was deduced, revealed that it is only 30 kd in size. Even complete phosphorylation of all serine and threonine residues could not account for this difference in molecular weights. The possibility of dimerization of the NS protein seems unlikely, since specific partial proteolytic cleavage of the NS protein results in the degradation of the expected fragments as deduced from the nucleotide sequence. Additional fragments that would be characteristic for dimer formation were not detected (Bell and Prevec, 1985). The anomalous electrophoretic mobility of NS may be explained by the presence of an unusual 44-amino-acid domain within the amino half of the protein that contains 18 negatively charged amino acids. This cluster of Asp and Glu may in part prevent the binding of SDS and thereby affect the mobility of the protein (Gallione et al., 1981).

A sequence comparison between the closely related San Juan and Mudd–Summers strains of Indiana serotype revealed 23 nucleotide differences, resulting in 10 amino acid changes (Hudson and Lazzarini, personal communication). The NS proteins of the Indiana and New Jersey serotypes, however, are only 32% homologous (Gill and Banerjee, 1985). The NS protein of the New Jersey serotype is 9 amino acids longer and has a molecular weight of 31.4 kd. Unlike the N and G proteins of the serotypes, which are, respectively, 68 and 50% homologous (Gallione and Rose, 1983; Banerjee et al., 1984), the NS protein surprisingly shows an extremely high degree of divergency. It appears that mutations in the NS gene can be tolerated more frequently than in the N and G genes. This is consistent with earlier genetic findings that the reversion rates of NS mutants are exceptionally high when compared to mutants of the other VSV genes (Pringle, 1975), suggesting that the function of a mutant NS protein can readily be restored by a second mutation at a different site. A short stretch of 21 amino acids near the carboxy terminus of the NS protein has greater than 90% homology and appears conserved compared to other regions of the protein. The significance of this conservation cannot be evaluated at this time. The least conserved region of the protein, with 34 mismatches in 39 amino acid residues, is found in the amino-terminal half of the protein. Nevertheless, the relative hydropathicity profile of the protein shows similarity in that a large number of hydrophilic amino acids are distributed throughout the amino half. It is possible that structural similarity exists between the NS proteins of both serotypes despite the enormous sequence divergency.

The NS protein, together with the L protein and the nucleocapsid template, are required during transcription (Emerson and Yu, 1975) (discussed in Section IV.E). In an *in vitro* transcription reaction, De and Banerjee (1984) exchanged the NS proteins of the VSV Indiana and New Jersey serotypes and found that the NS protein of Indiana could surprisingly substitute for the New Jersey NS protein using New Jersey ribonucleocapsid template and L protein. The NS protein from the New Jersey serotype, however, failed to substitute for the NS of Indiana in a reaction containing Indiana template and L protein. The fact that NS protein of the New Jersey serotype is functional in a heterologous transcription reaction suggests that although the homology between the primary structures of both proteins is very low, some similar structures may be conserved that allow specific interactions with the nucleocapsid template and the L protein.

The abundance of cytoplasmic NS protein suggests that NS may have additional functions besides during transcription. Complexes of the N and NS proteins have been detected in the cytoplasm of infected cells (Imblum and Wagner, 1975; Bell *et al.*, 1984; Peluso and Moyer, 1984). N protein has a strong tendency to self-aggregate under physiological conditions (Blumberg *et al.*, 1983; Sprague *et al.*, 1983). It is possible that NS protein may prevent self-aggregation of N proteins by forming N–NS complexes, thus masking the N–N binding sites used in nucleocapsid assembly (Arnheiter *et al.*, 1985). Further experimentation is needed to clarify the role of the N–NS complexes.

2. Phosphorylation of the NS Protein

There are more than 10 different phosphorylated subspecies of the NS protein that can be separated by two-dimensional gel electrophoresis (Hsu *et al.*, 1982). It is not known which of these phosphorylated forms of NS preferentially bind to nucleocapsids, interact with the L protein, are involved in transcription, or possibly even regulate transcription. Although differential effects of different phosphorylated forms of NS on transcription have been reported (Clinton *et al.*, 1978a; Kingsford and Emerson, 1980), it remains an unanswered question which of the multiple subspecies of the NS protein was the active component. Partial dephosphorylation of the NS protein by phosphatase does not change the pattern of the transcription products, but the amounts of the products are decreased (Kingsbury *et al.*, 1981). This suggests that a higher phosphorylated form of NS may possibly be needed during transcription *in vitro*.

The sites of phosphorylation of the NS protein have been mapped recently by Bell and Prevec (1985). Partial proteolytic cleavage of ^{32}P-labeled cytoplasmic NS protein revealed that the major phosphate residues are all located within the amino half of the NS protein and most probably between residues 35 and 106. Since only serine and threonine residues of the NS protein are phosphorylated (Clinton and Huang, 1981), there are 8 potential phosphorylation sites within this region. This could theoretically give rise to an enormous number of different phosphorylated forms of NS. This complexity may be simplified after future analysis of the major phosphorylation sites.

It is unknown which protein kinases phosphorylate the NS protein. Cellular protein kinases have been detected in enveloped viruses. A partially purified viral L-protein fraction is capable of phosphorylating the NS protein (Sanchez *et al.*, 1985). Protein kinase activity can also be found in immunoprecipitates from detergent-disrupted virus using antibodies against the NS protein (Bell *et al.*, 1984). In both cases, contamination with cellular protein kinases cannot be ruled out. It is a difficult task to dissect the functions of the multiple subspecies of NS, but it may be even more difficult to identify the protein that phosphorylates them.

E. Polymerase Protein

1. Polymerase Complex

All rhabdovirus as well as paramyxovirus genomes contain a large L gene that is commonly assumed to encode the RNA-dependent RNA polymerase. With VSV, as with other rhabdoviruses, the assignment of polymerase activity to a viral protein is complicated by the fact that transcription requires, besides the L protein, the small phosphoprotein NS and a nucleocapsid template, consisting of genomic RNA completely encapsidated by the viral N protein. Unencapsidated RNA does not serve as a template (Bishop and Roy, 1971; Emerson and Wagner, 1972). The requirement for L and NS has clearly been shown by Emerson and Yu (1975): Dissociation and reconstitution of the transcribing complex *in vitro* demonstrated that the individual components are transcriptionally inactive. Polymerase activity is restored only in the presence of both the L and NS proteins. There are three complementation groups of VSV mutants that can exert an RNA$^-$ phenotype (Pringle, 1982). Consistent with the *in vitro* studies, these conditional mutants contain defects in the L, NS, and N proteins, respectively. The L protein and the NS protein are therefore considered dissimilar subunits of a ribonucleocapsid-dependent RNA polymerase complex. Only because of its exceptionally large size of 241,000 d (Schubert *et al.*, 1984) and its presence in catalytic amounts of approximately 50 molecules per virion, compared to approximately 470 molecules of NS (Thomas *et al.*, 1985), is L suspected to carry the active center for polymerase activity.

The requirement for the L and NS proteins during *in vitro* transcription has been independently confirmed by a number of groups using either the Indiana or the New Jersey serotype of VSV (Naito and Ishihama, 1976; Hunt et al., 1984; De and Banerjee, 1984). It has also been shown that the binding of L protein to the template is dependent on the presence of NS protein (Mellon and Emerson, 1978). Ongradi et al. (1985), however, recently reported that in vitro transcription of the New Jersey serotype can be restored in the absence of NS protein, suggesting that the L protein together with the nucleocapsid template is sufficient for transcription. They conclude that the L protein is itself the transcriptase and that the NS protein may exert some control over transcription. This interesting observation is obviously in direct conflict with earlier results obtained by others. In both cases, the data are convincing, but how can both conclusions be right? An explanation for this discrepancy may be in the isolation procedures for the components, in a strain difference, in how the virus was propagated, or in how the activity of the transcriptase was assayed. We cannot distinguish among these possibilities at this time.

A large amount of data has been collected that describes the products of transcription and replication. In contrast, very little is known about the structure and functions of the active polymerase complex itself. On the basis of its L and NS content as well as its sedimentation coefficient of 11 S, Naito and Ishihama (1976) suggested that the polymerase complex of VSV contains L and NS in equimolar amounts, assuming that the L protein was about 4 times larger than the NS protein. Accurate molecular-weight determinations for both proteins, however, were not available at the time. The molecular weights of the L and NS proteins of the VSV Indiana serotype, as deduced from recent nucleotide sequence analyses, are 241,000 and 30,000 d, respectively (see Table II). Therefore, the L protein is 8 times larger than the NS protein instead of 4 times, as postulated earlier. This suggests that the amount of NS protein in the polymerase complex was underestimated by a factor of two. Consequently, the structure of the complex may possibly be $(L)_1(NS)_2$. This does not rule out that the active complex, when bound to the template, may contain more than one L protein. Intragroup complementation among VSV mutants of group I (L gene) may be indicative of intragenic complementation (Flamand, 1970), suggesting that the polymerase complex may contain two L-protein subunits, each carrying at least two different functional domains. However, if the L is a multifunctional protein, it is also possible that two mutant L proteins may perform these functions in a successive manner without being associated within the same complex. The precise composition of the polymerase complex is unknown, and its study is further complicated, since many different NS-protein subspecies that differ in their degree of phosphorylation have been identified by Hsu et al. (1982). It is not clear which of these species are associated with the active polymerase complex (Hsu et al., 1982; Kingsford and Emerson, 1980; Clinton et al., 1978a).

2. Is the Polymerase Complex Multifunctional?

The synthesis of functional mRNAs requires an impressive number of diverse activities. Besides specific binding to the template, initiation, elongation, and termination by the polymerase, the messages are capped by a guanylyltransferase, methylated by a (guanine-7)-methyltransferase and a 2'-0-methyltransferase, and polyadenylated by a poly(A) polymerase (Moyer et al., 1975; Rose, 1975; Ehrenfeld and Summers, 1972). The five monocistronic messages of VSV are sequentially transcribed (Abraham and Banerjee, 1976; Ball and White, 1976), and specific, highly conserved sequences at the beginning and end of each gene are recognized (see Fig. 3). These sequences, on the other hand, are ignored during replication, resulting in the synthesis of a full-length genomic RNA. Unlike transcription, replication requires ongoing protein synthesis (Huang and Manders, 1972; Wertz and Levine, 1973) and possibly the interaction of host factors (Pringle, 1978; Szilagyi et al., 1977; Obijeski and Simpson, 1974).

In the presence of the methyl donor S-adenosylmethionine, in vitro transcription by detergent-disrupted virus particles yields capped, methylated, and polyadenylated mRNAs (Abraham et al., 1975; Villarreal and Holland, 1973; Banerjee and Rhodes, 1973; Testa and Banerjee, 1977). Therefore, the virions must contain all the essential functions necessary for transcription and modification of functional messages. Whether all these functions are performed by viral proteins or are shared with cellular proteins packaged in the virion is unknown. If all these functions were carried out by the polymerase complex, it is obvious that there are more functions than can be accounted for by only the L and NS proteins, unless one or both of these proteins are multifunctional. Attempts to cap, methylate, or polyadenylate exogenous mRNAs during an in vitro transcription have not been successful. It appears not only that all these functions are carried out on nascent transcripts, but also that these functions are closely associated with the polymerase complex.

3. Polyadenylation of mRNAs

The synthesis of 100- to 200-nucleotide-long poly(A) tails during an in vitro transcription is dependent on the presence of all four nucleoside triphosphates, indicating a requirement for de novo message synthesis (Naito and Ishihama, 1976). Highly conserved polyadenylation signals at the end of each gene (see Fig. 3), which in the case of VSV contain a stretch of 7 U residues, suggest a chattering or slippage mechanism for polyadenylation by the polymerase complex (McGeoch, 1979; Schubert et al., 1980; Rose, 1980). It was proposed that repeated transcription of this short oligo (U) stretch results in the addition and covalent linkage of a poly(A) tail to its mRNA. Without apparently dissociating from the template, polyadenylation *in vitro* and *in vivo* is occasionally followed by faithful transcription of the short intercistronic region and the next gene. These dicistronic transcripts, containing two "messages" linked by intervening poly(A), indicate that the polymerase complex itself polyadenylates the messages (Herman *et al.*, 1978, 1980; Schubert and Lazzarini, 1981). Hunt (1983) identified a VSV L-gene mutant (*ts*G16) that aberrantly synthesizes unusually long poly(A) tails about 1 kb in length. By reconstitution of the polymerase complex after exchanging nucleocapsid templates and L and NS proteins from the mutant and wild-type virus, Hunt *et al.* (1984) were able to correlate the polyadenylation phenotype with a mutated L protein.

These data taken together strongly suggest that the L protein contains polymerase as well as poly(A) polymerase activity. In fact, polyadenylation must be considered to be a sequence-specified reiterated *transcription* of a short stretch of U residues. This is a general mechanism that has also been proposed with other nonsegmented as well as with segmented negative-strand RNA viruses (Gupta and Kingsbury, 1984; Robertson *et al.*, 1981). It differs significantly, however, from the posttranscriptional polyadenylation of host-cell messages. The switch from continuous to reiterated transcription during rhabdovirus polyadenylation involves the recognition of specific nucleotide sequences. How these sequences are recognized during transcription and how they are ignored during replication remain challenging questions.

4. Methylation of the Cap Structure

Horikami and Moyer (1982) studied the defects of some host-range mutants of VSV (Simpson and Obijeski, 1974; Obijeski and Simpson, 1974). *In vitro* transcription by the *hr* 1 mutant gives rise to normal capped and polyadenylated mRNAs. The cap structure, however, is not methylated, suggesting that one or both VSV-associated methyltransferase activities are virus-encoded. Because of its size, the L protein is a likely candidate for the methyltransferase. Is it only a coincidence that during *in vitro* transcription in the presence of the methyltransferase inhibitor *S*-adenosylhomocysteine, giant poly(A) tails are synthesized (Rose *et al.*, 1977)—as with the *ts*G16 mutant—or does this effect indicate colocalization of the functional domains for polymerization, polyadenylation, and methylations within the same protein, L? On the basis of the competitive inhibition of the VSV-associated methyltransferase by pyridoxal-5-phosphate (PLP) and the preferential binding of PLP to the L protein, as compared to N and NS, Morgan and Kingsbury (1981) concluded that the L protein may be the methyltransferase. Although there is evidence that one or even both methyltransferases may be of viral origin, the assignment to a viral protein clearly needs additional data.

5. Mechanism of Cap Formation

As yet, no conditional mutants of the capping function have been identified. It is still uncertain whether the guanylyltransferase is encoded by the virus genome or is a host-derived transferase that is packaged in the virion. Available data demonstrate that the mechanism of cap formation with rhabdovirus messages differs significantly from cellular mechanisms analyzed to date. Soluble cellular guanylyltransferases, isolated from nuclei, transfer guanosine monophosphate (GMP) to the 5′-terminal diphosphate of the message. In contrast, the messages of VSV and spring viremia of carp virus are capped by transfer of a guanosine diphosphate (GDP) to the 5′-terminal monophosphate of the message (Banerjee, 1980; Gupta and Roy, 1980). Although both reactions give rise to the same cap core structure, GpppN . . . , the mechanisms are quite different. In the presence of GTP, cellular guanylyltransferases form stable, covalent enzyme–GMP complexes that can readily be labeled using $[\alpha\text{-}^{32}P]$-GTP. These complexes are reactive intermediates in a capping reaction (Venkatesan and Moss, 1982; Wang *et al.*, 1982). Attempts to specifically identify the VSV-associated guanylyltransferase by this method were not successful. Although the viral phosphoproteins NS and M could be labeled in the presence of $[\alpha\text{-}^{32}P]$-GTP, $[\beta\text{-}^{32}P]$-GDP, or $[\alpha\text{-}^{32}P]$-dGTP, it is unlikely that the labeled proteins contained guanosine besides the phosphate residue because isotopic dilution experiments using unlabeled GTP, GDP, or dGTP did not, as expected, drastically decrease the labeling of both proteins. ATP, on the other hand, completely out-competed the labeling. It is interesting to note that dGTP is accepted by the guanylyltransferase and to some extent is also incorporated by the RNA polymerase (Schubert and Lazzarini, 1982). Unlike the case with cellular transferases, the NS-protein-bound phosphate was phosphatase- and alkali-sensitive (De and Banerjee, 1983; Schubert and Lazzarini, unpublished results). These data suggest that besides the difference in the mechanism of cap formation, the VSV-associated guanylyltransferase does not form detectable levels of stable enzyme–guanylyl intermediates that are a characteristic of cellular transferases. Despite these subtle differences between the viral and cellular guanylyl-transferases, it still remains to be determined whether capping of rhabdovirus messages is a viral function.

6. Expression of a Functional Recombinant L Protein

The polymerase of VSV and probably those of other rhabdoviruses and paramyxoviruses appear to be multifunctional complexes in which the L protein may perform many, if not all, of the functions described above. The 2109-amino-acid VSV L protein with a molecular weight of 241,000 d is a basic protein. Amino acid sequence analyses do not reveal any obvious characteristic domains with respect to hydrophobicity or the clustering of certain amino acids in particular regions that could be in-

dicative of structural or functional domains. A comparison of the L sequence with those of other nonsegmented negative-strand viruses awaits
sequence information from these viruses. Computer comparisons of the
L sequences to all sequences stored in data banks did not show any
homology to any of these sequences, including those of many proteins
that interact with nucleic acids, i.e., the polymerase proteins of influenza
virus (Schubert *et al.*, 1984).

As a new approach to identify and also to localize the multiple functional domains of the polymerase complex of VSV, Schubert *et al.* (1987)
have recently assembled the 6.4-kb L gene from partial, overlapping cDNA
clones. The gene was expressed in eukaryotic COS cells (Gluzman, 1981)
using an expression vector that contains the VSV L gene under control
of the SV40 late promoter. Rabbit antibodies were raised against synthetic
peptides that corresponded to the amino and carboxy termini of the L
protein. Both antibodies precipitate the L protein from the virion as well
as the recombinant L protein expressed from the expression vector (Fig.
5). The recombinant L protein (lanes 2 and 3) comigrates with the L protein
from the virion (lane 1). This demonstrates that the recombinant L protein
is identical in size to the virion L and that both termini as deduced from
the nucleotide-sequence analysis are conserved in the L protein.

If the recombinant L protein were functional, it should complement
and rescue conditional L-gene mutants of VSV at the restrictive temperature. The L-gene mutant *ts*G114 was chosen because it contains a mutated L protein that is unable to synthesize RNA at the restrictive temperature (40°C) (Hunt *et al.*, 1976). Cells were transfected with the expression
vector DNA with or without the L-gene insert. At 24 hr posttransfection,
the cells were infected with the L-gene mutant *ts*G114 and incubated at

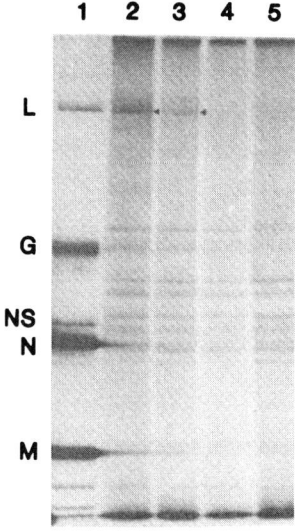

FIGURE 5. Immunoprecipitations of L protein of VSV.
COS cells were transfected with an SV40 expression
vector with (lanes 2 and 3) or without the VSV L-gene
insert (lanes 4 and 5). The expressed L protein was immunoprecipitated from [^{35}S]methionine-labeled cell extracts, using antibodies directed against the amino (lanes
2 and 4) or the carboxy terminus (lanes 3 and 5) of the
L protein. The precipitated proteins were separated on
a discontinuous SDS–polyacrylamide gel next to the
structural proteins of the virus (lane 1). From Schubert
et al. (1985).

40°C. The kinetics of virus release are shown in Fig. 6. Cells that were transfected with the vector that lacked the L-gene insert did not release any virus particles throughout the 24-hr time period. In contrast, cells that expressed the recombinant L protein released temperature-sensitive virus as early as 2 hr postinfection and reached a level of about 10^5 plaque-forming particles/ml.

These data show that the recombinant L protein is identical to the wild-type L protein with respect to size and function. It can transcribe and replicate and thereby complement and rescue the RNA minus L-gene mutant at the nonpermissive temperature. This unique system will, for the first time, allow the study of the organization of the functional domains of the L protein of a negative-strand virus through flexible recombinant DNA techniques. Site-specific mutagenesis will make it possible

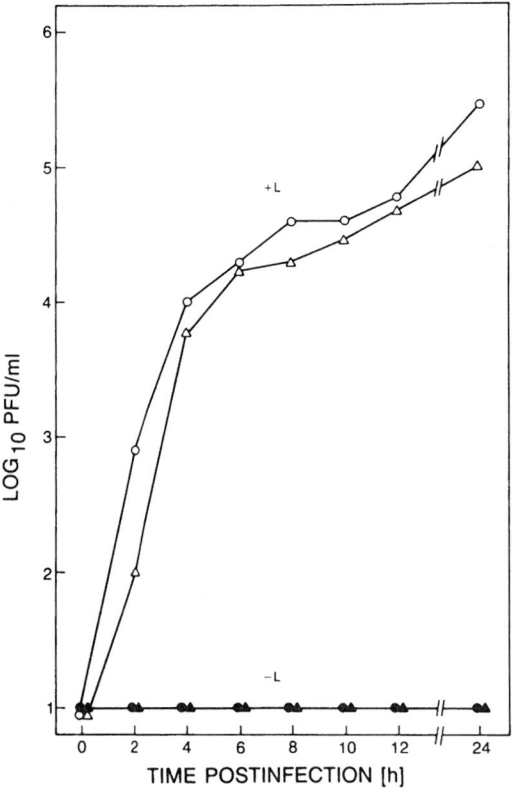

FIGURE 6. Rescue of a VSV L-gene mutant by expression of a recombinant DNA that encodes the L protein. COS cells were transfected with an SV40 expression vector with (○, △) or without (●, ▲) a VSV cDNA L-gene insert. The cells were incubated for 24 hr (○, ●) or 48 hr (△, ▲) at 37°C. After infection with the RNA⁻ VSV L-gene mutant tsG114 at 40°C, the nonpermissive temperature of the mutant, the titer of released virus was determined at the permissive temperature (32°C). From Schubert et al. (1985).

to study the effect of mutations on the performance of the multiple functions of the polymerase complex.

REFERENCES

Abraham, G., and Banerjee, A. K., 1976, Sequential transcription of the genes of vesicular stomatitis virus, *Proc. Natl. Acad. Sci. U.S.A.* **73**:1504.

Abraham, G., Rhodes, D. P., and Banerjee, A. K., 1975, The 5' terminal structure of the methylated mRNA synthesized *in vitro* by vesicular stomatitis virus, *Cell* **5**:51.

Arnheiter, H., Davis, N. L., Wertz, G., Schubert, M., and Lazzarini, R. A., 1985, Role of the nucleocapsid protein in regulating vesicular stomatitis virus RNA synthesis, *Cell* **41**:259.

Ball, L. A., and White, C. N., 1976, Order of transcription of genes of vesicular stomatitis virus, *Proc. Natl. Acad. Sci. U.S.A.* **73**:442.

Baltimore, D., Huang, A. S., and Stampfer, M., 1970, Ribonucleic acid synthesis of vesicual stomatitis virus. II. An RNA plymerase in the virion, *Proc. Natl. Acad. Sci. U.S.A.* **66**:572.

Banerjee, A. I., 1980, 5' Terminal cap structure in eucaryotic messenger ribonucleic acids, *Microbiol. Rev.* **44**:175.

Banerjee, A. K., and Rhodes, D. P., 1973, *In vitro* synthesis of RNA that contains polyadenylate by virion-associated RNA polymerase of vesicular stomatitis virus, *Proc. Natl. Acad. Sci. U.S.A.* **70**:3566.

Banerjee, A. K., Abraham, G., and Colonno, R. J., 1977, Vesicular stomatitis virus: Mode of transcription, *J. Gen. Virol.* **34**:1.

Banerjee, A. K., Rhodes, D. P., and Gill, D. S., 1984, Complete nucleotide sequence of the mRNA coding for the N protein of vesicular stomatitis virus (New Jersey serotype), *Virology* **137**:432.

Bell, J., and Prevec, L., 1985, Phosphorylation sites on phosphoprotein NS of vesicular stomatitis virus, *J. Virol.* **54**:697.

Bell, J., Brown, E. G., Takayesu, D., and Prevec, L., 1984, Protein kinase activity associated with immunoprecipitates of the vesicular stomatitis virus phosphoprotein NS, *Virology* **132**:229.

Bergmann, J. E., Tokuyasu, K. T., and Singer, S. J., 1981, Passage of an integral membrane protein, the vesicular stomatitis virus glycoprotein, through the Golgi apparatus en route to the plasma membrane, *Proc. Natl. Acad. Sci. U.S.A.* **78**:1746.

Bishop, D. H. L., and Roy, P., 1971, Kinetics of RNA synthesis by vesicular stomatitis virus particles, *J. Mol. Biol* **57**:513.

Blumberg, B. M., and Kolakofsky, D., 1981, Intracellular vesicular stomatitis virus leader RNAs are found in nucleocapsid structures, *J. Virol.* **40**:568.

Blumberg, B. M., Leppert, M., and Kolakofsky, D., 1981, Interaction of VSV leader RNA and nucleocapsid protein may control VSV genome replication, *Cell* **23**:837.

Blumberg, B. M., Giorgi, C., and Kolakofsky, D., 1983, N protein of vesicular stomatitis virus selectively encapsidates leader RNA *in vitro*, *Cell* **32**:559.

Blumberg, B. M., Giorgi, C., Rose, K., and Kolakofsky, D. 1984, Preparation and analysis of the nucleocapsid proteins of VSV and Sendai virus and analysis of the Sendai virus leader-np gene region, *J. Gen. Virol.* **65**:769–779.

Carroll, A., and Wagner, R., 1979, Role of the membrane (M) protein in endogenous inhibition of *in vitro* transcription by vesicular stomatitis virus, *J. Virol.* **29**:134.

Chatis, P. A., and Morrison, T. G., 1979, Vesicular stomatitis virus glycoprotein is anchored to intracellular membranes near its carboxyl end and is proteolytically cleaved at its amino terminus, *J. Virol.* **29**:957.

Chatis, P. A., and Morrison, T. G., 1982, Characterization of the soluble glycoprotein released from VSV infected cells, *J. Virol.* **45**:80.

Clinton, G. M., and Huang, A. S., 1981, Distribution of phosphoserine, phosphothreonine and phosphotyrosine in proteins of vesicular stomatitis virus, *Virology* **108**:510.

Clinton, G. M., Burge, B. W., and Huang, A. S., 1978a, Effects of phosphorylation and pH on the association of NS protein with vesicular stomatitis virus cores, *J. Virol.* **27**:340.

Clinton, G., Little, S., Hagen, F., and Huang, A., 1978b, The matrix protein of VSV regulates transcription, *Cell* **15**:1455.

Cohen, G., Atkinson, P. and Summers, D., 1971, Interaction of VSV structural proteins with HeLa cell plasma membranes, *Nature (London)*, **231**:121.

Colonno, R. J., and Banerjee, A. K., 1976, A unique RNA species involved in initiation of vesicular stomatitis virus RNA transcription *in vitro*, *Cell* **8**:197.

Colonno, R. J., and Banerjee, A. K., 1978a, Complete nucleotide sequence of the leader RNA synthesized *in vitro* by vesicular stomatitis virus, *Cell* **15**:93.

Colonno, R. J., and Banerjee, A. K., 1978b, Nucleotide sequence of the leader RNA of the New Jersey serotype of vesicular stomatitis virus, *Nucleic Acids Res.* **51**:4165.

Combard, A., and Printz-Ané, C., 1979, Inhibition of vesicular stomatitis virus transcriptase complex by the virion envelope protein, *Biochem. Biophys. Res. Commun.* **88**:117.

Crimmins, D. L., Mehard, W. B., and Schlesinger, S., 1983, Physical properties of a soluble form of the glycoprotein of vesicular stomatitis virus at neutral and acidic pH, *Biochemistry* **22**:5790.

De, B. P., and Banerjee, A. K., 1983, Specific binding of guanosine 5'-diphosphate with the NS protein of vesicular stomatitis virus, *Biochem. Biophys. Res. Commun.* **114**:138.

De, B. P., and Banerjee, A. K., 1984, Specific interactions of vesicular stomatitis virus L and NS proteins with heterologous genome ribonucleoprotein template lead to mRNA synthesis *in vitro*, *J. Virol* **51**:628.

De, B. P., Thornton, G. B., Luk, D., and Banerjee, A. K., 1982, Purified matrix protein of vesicular stomatitis virus blocks viral transcription *in vitro*, *Proc. Natl. Acad. Sci. U.S.A.* **79**:7137.

Dietzschold, B., Wiktor, T., Wunner, W., and Varrichio, A., 1983, Chemical and immunological analysis of the rabies virus soluble glycoprotein, *Virology* **124**:330.

Doolittle, R. F., 1981, Similar amino acid sequences: Chance or common ancestry?, *Science* **214**:149.

Dubovi, E. J., and Wagner, R. R., 1977, Spatial relationships of the proteins of vesicular stomatitis virus: Induction of reversible oligomers by cleavable protein cross-linkers and oxidation, *J. Virol.* **22**:500.

Ehrenfeld, E., and Summers, D. F., 1972, Adenylate-rich sequences in vesicular stomatitis virus messenger ribonucleic acids, *J. Virol.* **12**:683.

Emerson, S. U., 1982, Reconstitution studies detect a single polymerase entry site on the vesicular stomatitis virus genome, *Cell* **31**:635.

Emerson, S. U., and Wagner, R. R., 1972, Dissociation and reconstitution of the transcriptase and template activities of vesicular stomatitis B and T virions, *J. Virol.* **10**:297.

Emerson, S. U., and Yu, Y.-H., 1975, Both NS and L proteins are required for *in vitro* RNA synthesis by vesicular stomatitis virus, *J. Virol.* **15**:1348.

Emerson, S. U., Dierks, P. M., and Parsons, J. T., 1977, *In vitro* synthesis of a unique RNA species by a T particle of vesicular stomatitis virus, *J. Virol.* **23**:708.

Etchison, J., Robertson, J., and Summers, D., 1977, Partial structural analysis of the oligosaccharide moieties of the VSV glycoprotein by sequential and enzymatic degradation, *Virology* **78**:375.

Flamand, A., 1970, Etude génétique de la stomatite vésiculaire: Classement de mutants thermosensibles spontanées en groupes de complementation, *J. Gen. Virol.* **8**:187.

Flamand, A., and Delagneau, J. F., 1978, Transcriptional mapping of rabies virus *in vivo*, *J. Virol.* **28**:518.

Florkiewicz, R., and Rose, J., 1984, A cell line expressing the VSV glycoprotein fuses at low pH, *Science* **225**:721.

Florkiewicz, R., Smith, A., Bergmann, J. E., and Rose, J. K., 1983, Isolation of stable mouse cell lines that express cell surface and secreted forms of the vesicular stomatitis virus glycoprotein, *J. Cell Biol.* **97**:1381.

Gallione, C. J., and Rose, J. K., 1983, Nucleotide sequence of a cDNA clone encoding the entire glycoprotein from the New Jersey serotype of vesicular stomatitis virus, *J. Virol.* **46**:162.

Gallione, C. J., and Rose, J. K., 1985, A single amino acid substitution in a hydrophobic domain causes temperature-sensitive cell surface transport of a mutant viral glycoprotein, *J. Virol.* **54**:374–382.

Gallione, C. J., Greene, J. R., Iverson, L. E., and Rose, J. K., 1981, Nucleotide sequences of the mRNAs encoding the vesicular stomatitis virus N and NS proteins, *J. Virol.* **39**:529.

Garreis-Wabnitz, C., and Kruppa, J., 1984, Intracellular appearance of a glycoprotein in VSV-infected BHK cells lacking the membrane-anchoring oligopeptide of the viral G-protein, *Eur. Mol. Biol. Org. J.* **3**:1469.

Gibson, R. S., Schlesinger, and Kornfeld, S., 1979, The nonglycosylated glycoprotein of vesicular stomatitis virus is temperature-sensitive and undergoes intracellular aggregation at elevated temperatures, *J. Biol. Chem.* **254**:3600–3607.

Gill, D. S., and Banerjee, A. K., 1985, Vesicular stomatitis virus NS proteins: Structural similarity without extensive sequence homology, *J. Virol.* **55**:60.

Giorgi, C., Blumberg, B., and Kolakofsky, D., 1983, Sequence determination of the (+) leader RNA regions of the VSV Chandipura, Cocal and Piry serotype genomes, *J. Virol.* **46**:125.

Gluzman, Y., 1981, SV40-transformed simian cells support the replication of early SV40 mutants, *Cell* **23**:175.

Grinnell, B., and Wagner, R. R., 1983, Comparative inhibition of cellular transcription by vesicular stomatitis virus serotypes New Jersey and Indiana: Role of each viral leader RNA, *J. Virol.* **48**:88.

Grinnell, B., and Wagner, R. R., 1984, Nucleotide sequence and secondary structure of vesicular stomatitis leader RNA and homologous DNA involved in inhibition of DNA-dependent transcription, *Cell* **36**:533.

Gupta, K. C., and Kingsbury, D. W., 1984, Complete sequences of the intergenic and mRNA start signals in the Sendai virus genome: Homologies with the genome of vesicular stomatitis virus, *Nucleic Acids Res.* **12**:3829.

Gupta, K. C., and Roy, P., 1980, Alternate capping mechanisms for transcription of spring viremia of carp virus: Evidence for independent mRNA initiation, *J. Virol.* **33**:292.

Herman, R. C., Adler, S., Lazzarini, R. A., Colonno, R. J., Banerjee, A. K., and Westphal, H., 1978, Intervening polyadenylate sequences in RNA transcripts of vesicular stomatitis virus, *Cell* **15**:587.

Herman, R. C., Schubert, M., Keene, J. D., and Lazzarini, R. A., 1980, Polycistronic vesicular stomatitis virus RNA transcripts, *Proc. Natl. Acad. Sci. U.S.A.* **77**:4662.

Horikami, S. M., and Moyer, S. A., 1982, Host range mutants of vesicular stomatitis virus defective in *in vitro* RNA methylation, *Proc. Natl. Acad. Sci. U.S.A.* **79**:7694.

Hsu, C.-H., Morgan, E. M., and Kingsbury, D. W., 1982, Site-specific phosphorylation regulates the transcriptive activity of vesicular stomatitis virus NS protein, *J. Virol.* **43**:104.

Huang, A. S., and Manders, E. K., 1972, Ribonucleic acid synthesis of vesicular stomatitis virus. IV. Transcription by standard virus in the presence of defective interfering particles, *J. Virol.* **9**:909.

Hubbard, S. C., and Ivatt, R. J., 1981, Synthesis and processing of asparagine-linked oligosaccharides, *Annu. Rev. Biochem.* **50**:555.

Hunt, D. M., 1983, Vesicular stomatitis virus mutant with altered polyadenylic acid polymerase activity *in vitro*, *J. Virol.* **46**:788.

Hunt, D. M., Emerson, S. U., and Wagner, R. R., 1976, RNA temperature-sensitive mutants of vesicular stomatitis virus: L-protein thermosensitivity accounts for transcriptase restriction of group I mutants, *J. Virol.* **18**:596.

Hunt, D. M., Smith, E. F., and Buckley, D. W., 1984, Aberrant polyadenylation by a vesicular stomatitis virus mutant is due to an altered L protein, *J. Virol.* **52**:515.

Imblum, R. L., and Wagner, R. R., 1974, Protein kinase and phosphoproteins of vesicular stomatitis virus, *J. Virol.* **13**:113.

Imblum, R. L., and Wagner, R. R., 1975, Inhibition of viral transcriptase by immunoglobulin directed against the nucleocapsid NS protein of vesicular stomatitis virus, *J. Virol.* **15**:1357.

Irving, R. A., and Ghosh, H. P., 1982, Shedding of vesicular stomatitis virus soluble glycoprotein by removal of carboxyl-terminal peptide, *J. Virol.* **42**:322.

Irving, R. A., Toneguzzo, F., Rhee, S. H., Hofmann, T., and Ghosh, H. P., 1979, Synthesis and assembly of membrane glycoproteins: Presence of leader peptide in nonglycosylated precursor of membrane glycoprotein of vesicular stomatitis virus, *Proc. Natl. Acad. Sci. U.S.A.* **76**:570.

Isaac, C. L., and Keene, J. D., 1982, RNA polymerase associated interactions near template promoter sequences of defective interfering particles of vesicular stomatitis virus, *J. Virol.* **43**:241.

Iverson, L. E., and Rose, J. K., 1981, Localized attenuation and discontinuous synthesis during vesicular stomatitis virus transcription, *Cell* **23**:477.

Iverson, L., E., and Rose, J. K., 1982, Sequential synthesis of 5' proximal vesicular stomatitis virus mRNA sequences, *J. Virol.* **44**:356.

Kang, C. Y., and Prevec, L., 1970, Proteins of vesicular stomatitis virus. II. Immunological comparisons of viral antigens, *J. Virol.* **6**:20.

Keene, J. D., Schubert, M., Lazzarini, R. A., and Rosenberg, M., 1978, Nucleotide sequence homology at the 3' termini of RNA from vesicular stomatitis virus and its defective interfering particles, *Proc. Natl. Acad. Sci. U.S.A.* **75**:3225.

Keene, J. D., Schubert, M., and Lazzarini, R. A., 1979, Terminal sequences of vesicular stomatitis virus RNA are both complementary and conserved, *J. Virol.* **32**:167.

Keene, J. D., Schubert, M., and Lazzarini, R. A., 1980, Intervening sequences between the leader region and the nucleocapsid gene of vesicular stomatitis virus RNA, *J. Virol.* **33**:789.

Keene, J. D., Chien, I. M., and Lazzarini, R. A., 1981a, Vesicular stomatitis virus defective interfering particle containing a muted internal leader RNA gene, *Proc. Natl. Acad. Sci. U.S.A.* **78**:2090.

Keene, J. D., Thornton, B. J., and Emerson, S. U., 1981b, Sequence-specific contacts between the RNA polymerase of vesicular stomatitis virus and the leader RNA gene, *Proc. Natl. Acad. Sci. U.S.A.* **78**:6191.

Kelley, J. M., Emerson, S. U., and Wagner, R. R., 1972, The glycoprotein of vesicular stomatitis virus is the antigen that gives rise to and reacts with neutralizing antibody, *J. Virol.* **10**:1231.

Kingsbury, D. W., 1974, The molecular biology and paramyxoviruses, *Med. Microbiol. Immunol.* **160**:73.

Kingsbury, D. W., Hsu, C.-H., and Morgan, E. M., 1981, A role for NS protein phosphorylation in vesicular stomatitis virus transcription, in: *The Replication of Negative Strand Viruses* (D. H. Bishop and R. W. Compans, eds.), p. 821, Elsevier/North-Holland, New York.

Kingsford, L., and Emerson, S. U., 1980, Transcriptional activities of different phosphorylated species of NS protein purified from vesicular stomatitis virions and cytoplasm of infected cells, *J. Virol.* **33**:1097.

Knipe, D., Baltimore, D., and Lodish, H., 1977a, Separate pathways of maturation of the major structural proteins of vesicular stomatitis virus, *J. Virol.* **21**:1128.

Knipe, D., Baltimore, D., and Lodish, H., 1977b, Maturation of viral proteins in cells infected with temperature-sensitive mutants of vesicular stomatitis virus, *J. Virol.* **21**:1149.

Kolakofsky, D., 1982, Isolation of vesicular stomatitis virus defective interfering genomes with different amounts of 5' terminal complementarity, *J. Virol.* **41**:566.

Kurath, G., and Leong, J. C., 1985, Characterization of infectious hematopoietic necrosis virus mRNA species reveals a nonvirion rhabdovirus protein, *J. Virol.* **53**:462.

Kurath, G., Ahern, K. G., Pearson, G. C., and Leong, J. C., 1985, Molecular cloning of the six mRNA species of infectious hematopoietic necrosis virus, a fish rhabdovirus, and gene order determination by R-loop mapping, *J. Virol.* **53**:469.

Kurilla, M. G., and Keene, J. D., 1983, The leader RNA of vesicular stomatitis virus is bound by a cellular protein reactive with anti-La lupus antibodies, *Cell* **34**:837.

Kurilla, M. G., Pinwica-Worms, H., and Keene, J. D., 1982, Rapid and transient localization of the leader RNA of VSV in the nuclei of infected cells, *Proc. Natl. Acad. Sci. U.S.A.* **79**:5240.

Kurilla, M. G., Cabradilla, C. D., Holloway, B. P., and Keene, J. D., 1984, Nucleotide sequence and host La protein interactions of rabies virus leader RNA, *J. Virol.* **50**:773.

Lazzarini, R. A., Keene, J. D., and Schubert, M., 1981, The origins of defective interfering particles of the negative-strand RNA viruses, *Cell* **26**:145.

Leavitt, R., Schlesinger, S., and Kornfeld, S., 1977a, Tunicamycin inhibits glycosylation and multiplication of Sindbis and vesicular stomatitis viruses, *J. Virol.* **21**:375–385.

Leavitt, R., Schlesinger, S., and Kornfeld, S., 1977b, Impaired intracellular migration and altered solubility of nonglycosylated glycoproteins of vesicular stomatitis virus and Sindbis virus, *J. Biol. Chem.* **252**:9018–9023.

Leppert, M., and Kolakofsky, D., 1980, Effect of defective interfering particles on plus- and minus-strand leader RNAs in vesicular stomatitis virus-infected cells, *J. Virol.* **35**:704.

Leppert, M., Rittenhouse, L., Perrault, J., Summers, D. F., and Kolakofsky, D., 1979, Plus and minus strand leader RNAs in negative strand virus-infected cells, *Cell* **18**:235.

Lingappa, V. R., Katz, F. N., Lodish, H. F., and Blobel, G., 1978, A signal sequence for insertion of a transmembrane glycoprotein, *J. Biol. Chem.* **253**:8867.

Little, S. P., and Huang, A. S., 1977, Synthesis and distribution of vesicular stomatitis virus-specific polypeptides in the absence of progeny production, *Virology* **81**:37.

Machamer, C., Florkiewicz, R., and Rose, J., 1985, A single N-linked oligosaccharide at either of the two normal sites promotes cell surface transport of the vesicular stomatitis virus glycoprotein, *Mol. Cell. Biol.* **5**:3074.

Magee, A. I., Koyama, A. H., Malfer, C., Wen, D., and Schlesinger, M., 1984, Release of fatty acids from viral glycoproteins, *Biochim. Biophys. Acta* **798**:156–166.

Marcus, P. I., and Sekellick, M. J., 1975, Cell killing by viruses. II. Cell killing by vesicular stomatitis virus: A requirement for virion-derived transcription, *Virology*, **63**:176.

Martinet, C., Combard, A., Printz-Ané, C., and Printz, P., 1979, Envelope proteins and replication of vesicular stomatitis virus: *In vivo* effects of RNA⁺ temperature sensitive mutations on viral RNA synthesis, *J. Virol.* **29**:123.

Masters, P. S., and Samuel, C. E., 1984, Detection of *in vivo* synthesis of polycistronic mRNAs of vesicular stomatitis virus, *Virology* **134**:277.

McGeoch, D. J., 1979, Structure of the gene N:gene NS intercistronic junction in the genome of the vesicular stomatitis virus, *Cell* **17**:3199.

McGeoch, D. J., and Dolan, A., 1979, Sequence of 200 nucleotides at the 3' terminus of the RNA genome of vesicular stomatitis virus, *Nucleic Acids Res.* **6**:3199.

McGowan, J. J., and Wagner, R. R., 1981, Inhibition of cellular DNA synthesis by vesicular stomatitis virus, *J. Virol.* **38**:356.

McGowan, J. J., Emerson, S. U., and Wagner, R. R., 1982, The plus-strand leader RNA of VSV inhibits DNA-dependent transcription of adenovirus and SV40 genes in a soluble whole-cell extract, *Cell* **28**:325.

McSharry, J. J., 1979, The lipid envelope and chemical composition of Rhabdoviruses, in: *The Rhabdoviruses*, Vol. I (D. H. L. Bishop, ed.), p. 107, CRC Press, Boca Raton, Florida.

Meier, E., Harmison, G. G., Keene, J. D., and Schubert, M., 1984, Sites of copy choice replication involved in generation of vesicular stomatitis virus defective-interfering particle RNAs, *J. Virol.* **51**:515.

Mellon, M. G., and Emerson, S. U., 1978, Rebinding of transcriptase component (L and NS proteins) to the nucleocapsid template of vesicular stomatitis virus, *J. Virol.* **27**:560.

Morgan, E. M., and Kingsbury, D. W., 1981, Association of the transcriptase and RNA methyltransferase activities of vesicular stomatitis virus with the L protein, in: *The*

Replication of Negative Strand Viruses, (D. H. L. Bishop and R. W. Compans, eds.), pp. 815–820, Elsevier/North-Holland, New York.

Morrison, T. G., McQuain, C. O., and Simpson, D., 1978, Assembly of viral membranes: Maturation of the vesicular stomatitis virus glycoprotein in the presence of tunicamycin, *J. Virol.* **28**:368.

Moyer, S. A., and Gatchell, S. H., 1979, Intracellular events in the replication of defective interfering particles of vesicular stomatitis virus, *Virology* **92**:168.

Moyer, S. A., and Summers, D. F., 1974, Phosphorylation of VSV *in vivo* and *in vitro,* *J. Virol.* **13**:455.

Moyer, S. A., Abraham, G., Adler, R., and Banerjee, A. K., 1975, Methylated and blocked 5′ termini in vesicular stomatitis virus *in vivo* mRNAs, *Cell* **5**:59.

Naito, S., and Ishihima, A., 1976, Function and structure of RNA polymerase from vesicular stomatitis virus, *J. Biol. Chem.* **251**:4307.

Newcomb, W., and Brown, J., 1981, Role of the vesicular stomatitis virus matrix protein in maintaining the viral nucleocapsid in the condensed form found in native virions, *J. Virol.* **39**:295.

Obijeski, J. F., and Simpson, R. W., 1974, Conditional lethal mutants of vesicular stomatitis virus. II. Synthesis of virus-specific polypeptides in non-permissive cells infected with "RNA –" host restricted mutants, *Virology* **57**:369.

Ongradi, J., Cunningham, C., and Szilagyi, J. F., 1985, The role of polypeptides L and NS in the transcription process of vesicular stomatitis virus New Jersey using the temperature-sensitive mutant ts E1, *J. Gen. Virol.* **66**:1011.

Patton, J. T., Davis, N. L., and Wertz, G. W., 1984, N protein alone satisfies the requirement for protein synthesis during RNA replication of vesicular stomatitis virus, *J. Virol.* **49**:303.

Peluso, R. W., and Moyer, S. A., 1984, Vesicular stomatitis virus proteins required for the *in vitro* replication of defective interfering particle genome RNA, in: *Nonsegmented Negative Strand Viruses (Paramyxoviruses and Rhabdoviruses)* (D. H. L. Bishop and R. W. Compans, eds.), p. 153, Academic Press, San Francisco.

Perrault, J., 1981, Origin and replication of defective interfering particles, *Curr. Top. Microbiol. Immunol.* **93**:151–207.

Perrault, J., Semier, B. L., Leavitt, R. W., and Holland, J. J., 1978, Inverted complementary terminal sequences in defective interfering particle RNAs of vesicular stomatitis virus and their possible role in autointerference, in: *Negative Strand Viruses and the Host Cell* (B. W. J. Mahy and R. D. Barry, eds.), pp. 527–538, Academic Press, New York.

Pringle, C. R., 1975, Conditional lethal mutants of vesicular stomatitis virus, *Curr. Topics Microbiol. Immunol.* **69**:85.

Pringle, C. R., 1978, The tdCE and hrCE phenotypes: Host range mutants of vesicular stomatitis virus in which polymerase function is affected, *Cell* **15**:597.

Pringle, C. R., 1982, The genetics of vesiculoviruses, *Arch. Virol.* **72**:1.

Rao, D. D., and Huang, A. S., 1979, Synthesis of a small RNA in cells coinfected by standard and defective interfering particles of vesicular stomatitis virus, *Proc. Natl. Acad. Sci. U.S.A.* **76**:3472.

Reading, C. L., Penhoet, E., and Ballou, C. E., 1978, Carbohydrate structure of vesicular stomatitis virus glycoprotein, *J. Biol. Chem.* **253**:5600.

Reichmann, M. E., Villarreal, L. P., Kohne, D., Lesnam, J. A., and Holland, J. J., 1974, RNA polymerase activity and poly(A) synthesizing activity in defective T particles of vesicular stomatitis virus, *Virology* **58**:240.

Riedel, H., Kondor-Koch, C., and Garoff, H., 1984, Cell surface expression of fusogenic vesicular stomatitis virus G protein from cloned cDNA, *Eur. Mol. Biol. Org. J.* **3**:1477.

Robertson, J. S., Schubert, M., and Lazzarini, R. A., 1981, Polyadenylation sites of influenza virus mRNA, *J. Virol.* **38**:157.

Rose, J. K., 1975, Heterogeneous 5′ terminal structures occur on vesicular stomatitis virus mRNAs, *J. Biol. Chem.* **250**:8098.

Rose, J. K., 1977, Nucleotide sequences of ribosome recognition sites in messenger RNAs of vesicular stomatitis virus, *Proc. Natl. Acad. Sci. U.S.A.* **74:**3672.

Rose, J. K., 1978, Complete sequences of ribosome recognition sites on messenger RNAs of vesicular stomatitis virus, *Cell* **14:**345.

Rose, J. K., 1980, Complete intergenic and flanking gene sequences from the genome of vesicular stomatitis virus, *Cell* **19:**415.

Rose, J. K., and Bergmann, J. E., 1982, Expression from cloned cDNA of cell surface and secreted forms of the glycoprotein of vesicular stomatitis virus in eucaryotic cells, *Cell* **30:**753.

Rose, J. K., and Bergmann, J. E., 1983, Altered cytoplasmic domains affect intracellular transport of the vesicular stomatitis virus glycoprotein, *Cell* **34:**513.

Rose, J. K., and Gallione, C., 1981, Nucleotide sequences of the mRNAs encoding the VSV G and M proteins as determined from cDNA clones containing the complete coding regions, *J. Virol.* **39:**519.

Rose, J. K. and Shafferman, A., 1981, Conditional expression of the vesicular stomatitis virus glycoprotein in *Escherichia coli, Proc. Natl. Acad. Sci. U.S.A.* **78:**6670.

Rose, J. K., Lodish, H. F., and Brock, M., 1977, Giant heterogeneous polyadenylic acid on vesicular stomatitis virus mRNA synthesized *in vitro* in the presence of *S*-adenosylhomocysteine, *J. Virol.* **21:**683.

Rose, J. K., Welch, W. J., Sefton, B. M., Esch, F. S., and Ling, N. C., 1980. Vesicular stomatitis virus glycoprotein is anchored in the viral membrane by a hydrophobic domain near the COOH terminus, *Proc. Natl. Acad. Sci. U.S.A.* **77:**3884.

Rose, J. K., Doolittle, R. F., Anilionis, A., Curtis, P. J., and Wunner, W. H., 1982, Homology between the glycoproteins of vesicular stomatitis virus and rabies virus, *J. Virol.* **43:**361.

Rose, J. K., Adams, G., and Gallione, C., 1984, The presence of cysteine in the cytoplasmic domain of the vesicular stomatitis virus glycoprotein is required for palmitate addition, *Proc. Natl. Acad. Sci. U.S.A.* **81:**2050.

Rothman, J. E., and Lodish, H. F., 1977, Synchronized transmembrane insertion and glycosylation of a nascent membrane protein, *Nature (London)* **269:**775.

Rowlands, D. J., 1979, Sequences of vesicular stomatitis virus RNA in the region coding for leader RNA, N protein RNA and their junction, *Proc. Natl. Acad. Sci. U.S.A.* **76:**4793.

Sanchez, A., De, B. P., and Banerjee, A. K., 1985, *In vitro* phosphorylation of NS protein by the L protein of vesicular stomatitis virus, *J. Gen. Virol.* **66:**1025.

Schlegel, R., and Wade, M., 1984, A synthetic peptide corresponding to the NH$_2$-terminus of VSV glycoprotein is a pH-dependent hemolysin, *J. Biol. Chem.* **259:**4691.

Schlegel, R., and Wade, M., 1985, Biologically active peptides of the VSV glycoprotein, *J. Virol.* **53:**319.

Schlesinger, M. J., and Malfer, C., 1982, Cerulenin blocks fatty acid acylation of glycoproteins and inhibits VSV and Sindbis virus particle formation, *J. Biol. Chem.* **257:**9887.

Schmidt, M. F. G., and Schlesinger, M. J., 1979, Fatty acid binding to vesicular stomatitis virus glycoprotein: A new type of post-translational modification of the viral glycoprotein, *Cell* **17:**813–819.

Schubert, M., and Lazzarini, R. A., 1981, *In vivo* transcription of the 5'-terminal extracistronic region of vesicular stomatitis virus RNA, *J. Virol.* **38:**256.

Schubert, M., and Lazzarini, R. A., 1982, *In vitro* transcription of vesicular stomatitis virus: Incorporation of deoxyguanosine and deoxycytidine, and formation of deoxyguanosine caps, *J. Biol. Chem.* **257:**2968.

Schubert, M., Keene, J. D., Lazzarini, R. A., and Emerson, S. U., 1978, The complete sequence of a unique RNA species synthesized by a DI particle of VSV, *Cell* **15:**103.

Schubert, M., Keene, J. D., and Lazzarini, R. A., 1979, A specific internal RNA polymerase recognition site of VSV RNA is involved in the generation of DI particles, *Cell* **18:**749.

Schubert, M., Keene, J. D., Herman, R. C., and Lazzarini, R. A., 1980, Site on the vesicular stomatitis virus genome specifying polyadenylation and the end of the L gene mRNA, *J. Virol.* **34:**550.

Schubert, M., Harmison, G., and Meier, E., 1984, Primary structure of the vesicular stomatitis virus polymerase (L) gene: Evidence for a high frequency of mutations, *J. Virol.*
51:505.

Schubert, M., Harmison, G. G., Richardson, C. D., and Meier, E., 1985, Expression of a
cDNA encoding a functional, 241 kilodalton vesicular stomatitis virus RNA polymerase, *Proc. Natl. Acad. Sci. U.S.A.* **82**:7984.

Semler, B. L., Perrault, J., Abelson, J., and Holland, J. J., 1978, Sequence of a RNA templated
by the 3'-OH RNA terminus of defective interfering particles of vesicular stomatitis
virus, *Proc. Natl. Acad. Sci. U.S.A.* **75**:4704.

Simonsen, C. C., Batt-Hymphries, S., and Summers, D. F., 1979, RNA synthesis of vesicular
stomatitis virus-infected cells: *In vivo* regulation of replication, *J. Virol.* **31**:124.

Simpson, R. W., and Obijeski, J. F., 1974, Conditional lethal mutants of vesicular stomatitis
virus. I. Phenotypic characterization of single and double mutants exhibiting host restriction and temperature sensitivity, *Virology* **57**:357.

Sprague, J., Condra, J. H., Arnheiter, H., and Lazzarini, R. A., 1983, Expression of a recombinant DNA gene coding for the vesicular stomatitis virus nucleocapsid protein, *J.
Virol.* **45**:773.

Szilagyi, J. F., Pringle, C. R., and MacPherson, T. M., 1977, Temperature-dependent host
range mutation in vesicular stomatitis virus affecting polypeptide L, *J. Virol.* **22**:381.

Testa, D., and Banerjee, A. K., 1977, Two methyl transferase activities in the purified virions
of vesicular stomatitis virus, *J. Virol.* **24**:786.

Thomas, D., Newcomb, W. W., Brown, J. C., Wall, J. S., Hainfield, J. F., Trus, B. L., and
Alasdair, S. C., 1985, Mass and molecular composition of vesicular stomatitis virus: A
scanning transmission electron microscopy analysis, *J. Virol.* **54**:598.

Venkatesan, S., and Moss, B., 1982, Eucaryotic mRNA capping enzyme–guanylate covalent
intermediate, *Proc. Natl. Acad. Sci. U.S.A.* **79**:340.

Villarreal, L. P., and Holland, J. J., 1973, Synthesis of poly(A) *in vitro* by purified virions of
vesicular stomatitis virus, *Nature(London)* **245**:17.

Wang, D., Furuichi, Y., and Shatkin, A. J., 1982, Covalent guanylyl intermediate formed by
HeLa cell mRNA capping enzyme, *Mol. Cell. Biol.* **2**:993.

Weck, P. K., and Wagner, R. R., 1978, Inhibition of RNA synthesis in mouse myeloma cells
infected with vesicular stomatitis virus, *J. Virol.* **25**:720.

Weck, P. K., and Wagner, R. R., 1979, Transcription of vesicular stomatitis virus is required
to shut off cellular RNA synthesis, *J. Virol.* **30**:410.

Weck, P. K., Carroll, A. R., Shattuck, D. M., and Wagner, R. R., 1979, Use of U.V. irradiation
to identify the genetic information of vesicular stomatitis virus responsible for shutting
off cellular RNA synthesis, *J. Virol.* **30**:746.

Wertz, G. W., and Levine, M., 1973, RNA synthesis by vesicular stomatitis virus and a
small plaque mutant: Effects of cycloheximide, *J. Virol.* **12**:253.

Wertz, G. W., and Youngner, J. S., 1972, Inhibition of protein synthesis in L cells infected
with vesicular stomatitis virus, *J. Virol.* **9**:85.

White, J., Kielian, M., and Helenius, A., 1983, Viral fusion proteins, *Q. Rev. Biol. Phys.*
16:151.

Wilusz, J., Kurilla, M. G., and Keene, J. D., 1983, A host protein (La) binds to a unique
species of minus-sense leader RNA during replication of vesicular stomatitis virus,
Proc. Natl. Acad. Sci. U.S.A. **80**:5827.

Wu, F.-S., and Lucas-Lenard, J., 1980, Inhibition of RNA accumulation in mouse L cells
infected with VSV requires viral RNA transcription, *Biochemistry* **19**:804.

Zakowski, J. J., and Wagner, R. R., 1980, Localization of membrane associated proteins in
vesicular stomatitis virus by use of hydrophobic membrane probes and cross-linking
reagents, *J. Virol.* **36**:93.

Zakowski, J. J., Petri, W. A., and Wagner, R. R., 1981, Role of matrix protein in assembling
the membrane of vesicular stomatitis virus: Reconstitution of matrix protein with
negatively charged phospholipid vesicles, *Biochemistry* **23**:3902.

CHAPTER 5

Rhabdovirus Genetics

CRAIG R. PRINGLE

I. INTRODUCTION

A. Negative-Strand RNA Viruses

There are five families of RNA viruses in which the negative strand is sequestered in the extracellular virion. Viruses of two of these families, the Rhabdoviridae and the Paramyxoviridae, have unitary linear genomes, whereas viruses of the other three families, Arenaviridae, Bunyaviridae, and Orthomyxoviridae, have segmented genomes comprising, respectively, two, three, and seven or eight subunits. The informational macromolecules that comprise the genomes of rhabdoviruses and paramyxoviruses are among the largest functional RNA molecules and are exceeded in size only by those of the plus-strand coronaviruses. Reanney (1982, 1984) has calculated that the upper size limit for any RNA virus genome cannot be much in excess of 17,600 nucleotides (mol. wt. $\approx 5.7 \times 10^6$) as a consequence of the low copying fidelity of RNA polymerases. The segmentation of the genomes of the other negative-strand viruses may be a consequence of such constraints on molecular size or a device for decoupling the transcription of individual genes. Whatever the reason, the genetic properties of the segmented-genome viruses differ substantially from those of the unsegmented-genome viruses, because variation in the former is generated by reassortment of genome subunits as well as by mutation. Mutation is the sole mechanism of variation in unsegmented-genome viruses, since the intramolecular recombination observed with positive-

CRAIG R. PRINGLE • Department of Biological Sciences, University of Warwick, Coventry CV4 7AL, England.

strand RNA viruses does not seem to be permissible for any negative-strand RNA virus.

B. Rhabdoviruses

The family Rhabdoviridae comprises a diverse collection of viruses linked by a common bullet-shaped or bacilliform morphology. The host organisms include vertebrates (birds, fishes, amphibians, marsupials, and mammals), invertebrates (insects and arachnids), plants (monocotyledons and dicotyledons), and protists (amoeba). Several plant rhabdoviruses, such as lettuce necrotic yellows, bridge major taxonomic gaps, being capable of multiplying in both their plant host and their insect vector. Only a few of more than 75 named rhabdoviruses are included so far in the two recognized genera (Brown *et al.*, 1979).

The genus *Vesiculovirus* comprises nine viruses [the two serotypes of vesicular stomatitis virus (VSV), Chandipura virus, Isfahan virus, Piry virus, Jurona virus, Keuraliba virus, La Joya virus, and Yug Bugdanovac virus]. The Indiana serotype of VSV comprises the subtypes Indiana, Cocal, Argentina, and Brazil, and the New Jersey serotype the Concan and Hazelhurst subtypes. The vesiculoviruses mature and accumulate in the cytoplasm and bud from the plasma membrane. All five viral gene products—the large (L), glycosylated (G), matrix (M), nonstructural (NS), and nucleocapsid (N) proteins—are present in the virion. Four plant viruses (lettuce necrotic yellows virus, broccoli necrotic yellows virus, sonchus virus, and wheat striate mosaic virus) closely resemble the vesiculoviruses and have been designated the subgroup A plant rhabdoviruses.

The genus *Lyssavirus* comprises rabies virus and the five rabies-related viruses: Duvenhage virus, Kotonkan virus, Lagos bat virus, Mokola virus, and Obodhiang virus. Kotonkan virus is exceptional in being isolated from culicoides insects. Again, all five viral gene products [the L, G, M_2, M_1 (=NS), and N proteins] are present in the virion. These viruses mature and accumulate in the cytoplasm and bud predominantly from internal or *de novo* synthesized membranes. Four plant viruses (potato yellow dwarf virus, eggplant mottled dwarf virus, sonchus yellow net virus, and sowthistle yellow vein virus) closely resemble the lyssaviruses and have been designated the subgroup B plant rhabdoviruses. These viruses bud at the inner membrane of the nuclear membrane and accumulate perinuclearly. The plant rhabdoviruses, together with tomato spotted wilt virus (a possible bunyavirus) and carrot mottle virus (a possible togavirus), are unique among plant viruses in possessing outer envelopes, although specific receptors for attachment of any plant virus have not been identified. The ubiquitous presence of RNA-dependent RNA polymerase activity in normal plant tissue further highlights the evolutionary enigma of plant rhabdoviruses.

The remaining 37 named animal and 67 plant rhabdoviruses listed by Matthews (1982) are virtually uncharacterized, with the exception of Sigma virus, which had a long history of study as a hereditary factor of the insect *Drosophila* prior to its recognition as a rhabdovirus. It is possible that the genetic organization of some of these viruses may differ substantially from the vesiculovirus and lyssavirus patterns, since particle length (and presumably nucleic acid content) of the animal rhabdoviruses varies considerably, and some of the presumptive plant rhabdoviruses are known only as nucleus-associated unenveloped nucleocapsids. Indeed, Kurath and Leong (1985) have recently characterized the gene products of the fish rhabdovirus infectious hematopoietic necrosis virus (IHNV), which resembles rabies virus more than VSV in terms of its structural proteins. Six messenger RNAs (mRNAs) were detected by denaturing gel electrophoresis and their coding assignments determined by hybrid selection and *in vitro* translation experiments. Five mRNAs encoded the L, G, M_1, M_2, and N structural proteins, whereas the sixth encoded a nonstructural nonviral (NV) protein with a molecular weight of approximately 12,000. Kurath *et al.* (1985) determined the gene order by R-loop mapping using plasmids carrying complementary DNA (cDNA) copies of the six mRNAs of IHNV. The gene order was deduced to be (3')N-M_1-M_2-G-NV-L(5'), which is identical to that of rabies virus (Flamand and Delagneau, 1978) apart from the insertion of NV between G and L. The complete nucleotide sequence of the Pasteur strains of rabies virus has now been obtained confirming this gene order (Tordo *et al.*, 1986). In the rabies virus genome, however, there is a long intercistronic sequence between the G and L genes sufficient to accommodate an additional gene in place of the dinucleotide junction of the VSV genome. The additional sequence may represent either an evolving gene or a relict gene since all three reading frames are blocked. Thus, the vesiculovirus genome pattern may not be typical of the Rhabdoviridae as a whole.

The genetic properties of rhabdoviruses discussed in this chapter refer exclusively to the vesiculoviruses, the lyssaviruses, and Sigma virus of *Drosophila*. This survey of the genetics of rhabdoviruses covers developments up to the end of 1985. Earlier phases are described in reviews by Flamand (1980), Pringle (1975, 1977, 1982), and Pringle and Szilagyi (1980).

C. Comparison of the Rhabdovirus and Paramyxovirus Genomes

Comparison of the order of the genes in the genomes of three paramyxoviruses with the order of genes in VSV suggests that the rhabdovirus genome is the least complex of the unsegmented-genome negative-strand RNA viruses. Figure 1 indicates the order and possible functional homologies of the genes of VSV, Sendai virus, and respiratory syncytial virus. VSV lacks one of the two envelope G proteins of the parainfluenza viruses (Sendai virus), whereas the genome of the pneumovirus respiratory syn-

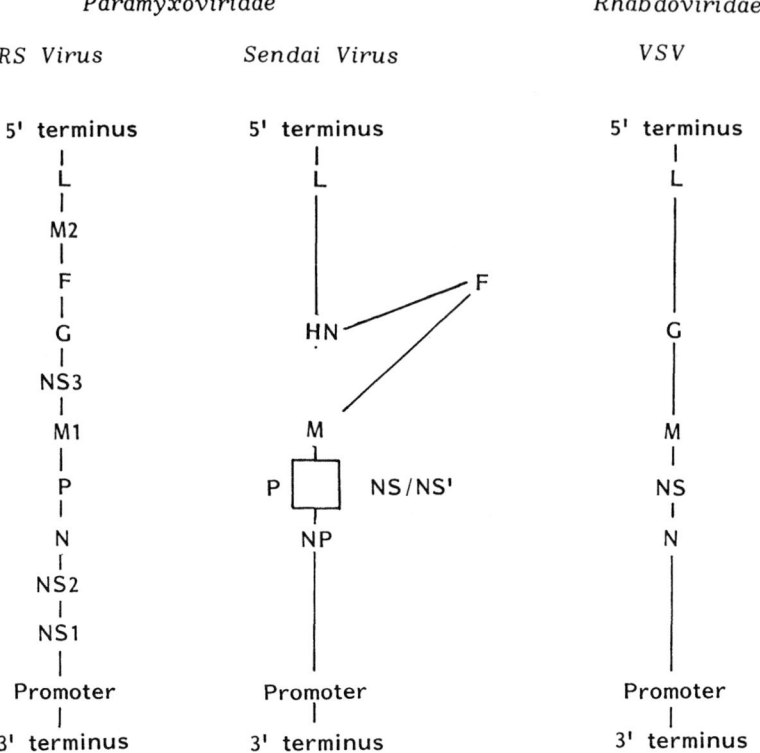

FIGURE 1. Genome structure of the unsegmented negative-strand RNA viruses. The diagram shows the order and presumed functional homologies of the genes of VSV, Sendai virus, and respiratory syncytial (RS) virus. (L) Large (polymerase) protein; (M) matrix protein; (F) fusion protein; (G) spike glycoprotein; (HN) hemagglutinin; (P) phosphoprotein; (N) nucleocapsid protein; (NP) nucleoprotein; (NS) nonstructural protein (in Sendai virus and RS virus) or polymerase-associated core protein (in VSV).

cytial virus is substantially more complex, encoding an additional unglycosylated membrane protein and two additional presumptive nonstructural proteins.

II. ORGANIZATION OF THE VESICULOVIRUS GENOME

A. General Features

The determination of the complete sequence of the L gene of the Mudd–Summers strain of the Indiana serotype of VSV (Schubert *et al.*, 1984) has closed the remaining gap in the sequence and established the basic features of the vesiculovirus genome. The nucleotide sequence of the L gene of the Mudd–Summers strain and the sequences of the leader region and the N, NS, M, and G genes of the San Juan strain of VSV

Indiana taken together establish the primary structure of the VSV genome. The genome, including the untranslated leader region, the five genes, the four intercistronic junctions, and the terminal tail region, is 11,162 nucleotides in length. The VSV genome is revealed as a remarkably simple entity; 99.37% of the nucleotides of the genome are transcribed, and 93.88% encode polypeptide-specifying information. The leader RNA and the five genes are arranged in a linear sequence without overlaps and with the minimum of untranscribed intervening and terminal sequences. The short 47-nucleotide untranslated leader sequence and the five monocistronic mRNAs are transcribed from the 3' terminus of the genome by a virion-associated RNA-dependent RNA polymerase, and the five mRNAs are present in infected cells in decreasing order of abundance in inverse relation to their proximity to the 3' terminus (Villareal et al., 1976). This divergence from equimolar synthesis has been termed "attenuation" to indicate partial termination of transcription occurring at or near the intercistronic boundaries (Iverson and Rose, 1981). The rate of transcription (elongation) in vitro appeared to be 3.7 nucleotides per second within all five genes and markedly slower at intergenic junctions. The order of the genes that determine the five structural proteins in the 5' to 3' direction is L, G, M, NS, and N; L denotes the gene that codes for the large viral polymerase; G, the gene that codes for the membrane glycoprotein that forms the virion spikes; M, the gene that codes for the unglycosylated matrix protein; NS, the gene that codes for the polymerase-associated phosphoprotein—the original designation NS, for nonstructural, being a misnomer [see Chapter 2 (Section II.B)]—and N, the gene that codes for the nucleocapsid protein, confirming earlier deductions from the kinetics of UV inactivation of gene function in vivo and in vitro with a coupled transcription–translation system. There is no apparent expansion of coding capacity by use of multiple reading frames and no splicing signals that might indicate processing of mRNA. The NS gene of the Mudd-Summers strain of VSV Indiana, however, contains a second open reading frame (Hudson et al., 1986) and a 7000 mol. wt. gene product has been identified by in vitro translation of a mixture of VSV polyadenylated mRNAs (Herman, 1986). Hybrid arrest translation showed that the VSV phosphoprotein (P) and the 7k protein were translated from the NS mRNA. The P and 7k proteins were both immunoprecipitated by an anti-P monoclonal antibody, indicating that the two products of the NS gene are encoded in the same reading frame. The results of hybrid arrest translation using mRNA fragments indicated that the 7k protein was produced by binding of ribosomes and initiation of translation internally at a site several hundred nucleotides downstream from the capped 5'-terminus. Herman (1986) has pointed out that small proteins encoded in the same open reading frame as the P protein are common in paramyxoviruses, and it would be logical to redesignate the NS gene as the P gene since the phosphoprotein is a virion component. The function of the 7k protein has not yet been established.

All the gene products apart from the leader RNA appear in the virion, the number of protein molecules per virion being in the following ranges: L, 20–30; G, 500–1500; M, 1600–4000; NS, 100–300; N, 1000–2000. The basic features of the organization of the VSV genome are summarized in Table I, together with the molecular weights of the viral proteins inferred from the sequence data.

B. Leader RNA

The leader RNA of 47 nucleotides is neither capped, polyadenylated, nor translated. Its function is undefined, but it is implicated in the inhibition of host RNA synthesis (McGowan *et al.*, 1982; Grinnell and Wagner, 1984). The nascent plus and minus leader RNA contains an encapsidation nucleation site perhaps represented by a five-times-repeated A residue at every third position from the 5′ end (Blumberg *et al.*, 1983). The leader region is separated from the first polypeptide-specifying gene by the sequence AAA (or AAAA in the New Jersey serotype) (Keene *et al.*, 1980; McGeoch *et al.*, 1980).

C. Intercistronic Junctions

The five genes are separated by junction regions of almost identical structure. The intergenic junctions consist of a unadecamer (3′)-AU-ACUUUUUUU-(5′) corresponding to the sequence (5′)-UAU-GAAAAAAA-(3′) at the mRNA–polyadenylic acid [poly(A)] junction to each message, which is presumed to be the signal for synthesis of the poly(A) tails of the mRNAs by reiterative copying ("slippage polymerization"). This unadecamer is followed by the dinucleotide GA (or CA in the case of the NS : M junction), which is untranscribed and marks the intercistronic junction. These dinucleotide spacers are followed immediately by sequences with the general form (5′)-AACAGNNAUC-(3′) complementary to the 5′ end of the mRNA, where N represents a variable site (McGeoch, 1979, 1981). At the 5′ end of the genome, there is an untranscribed region of 59 nucleotides beyond the 3′ terminus of the L gene. The function of this region is not known, but there is a polarity in the distribution of adenine and uridine residues at the 5′ end of the genome, such that 32 (53%) of the first 60 bases are adenines and 29 (48%) of the next 60 are uridines. The last 125 nucleotides of the genome can be arranged in a configuration with 62% of the bases paired. Schubert *et al.* (1980) have pointed out that this degree of secondary structure is greater than the self-complementarity of transfer RNA (tRNA), in which 55% of the bases are paired; consequently, it is likely that this potential secondary structure has some functional significance, even though N protein is normally associated with the 5′ terminus. There is, however,

TABLE I. Organization of the Genome of VSV-Indiana[a]

Regions	Sequences		Polypeptide-coding sequences (number of nucleotides)[b]	Gene product	Inferred molecular weight of polypeptides (number of amino acids)	Special features
	Untranscribed	Transcribed				
	Number of nucleotides					
3'-Terminal leader	—	47	—	(Leader RNA)	—	Function unknown, 48 nucleotides in some strains
Leader:N junction	3 (AAA)	—	—	—	—	Probably involved in inhibition of host RNA synthesis [(AAAA) in VSV-New Jersey]
N gene	—	1,333	(13) 1,266 [54]	Nucleocapsid protein (N)	47,355 (422)	Possible 30 and 23 amino acid coding regions within the N gene
N:NS junction	2 (GA)	—	—	—	—	
NS gene	—	822	(10) 666 [146]	Core protein (NS)	— 25,110 (222)	Phosphorylated domain of 18 negatively charged amino acids (responsible for anomalous PAGE mobility?)
NS:M junction	2 (CA)	—	—	—	—	
M gene	—	838	(41) 687 [103]	Unglycosylated matrix protein (M)	— 26,064 (229)	Most basic of VSV proteins

(continued)

TABLE I. (Continued)

Regions	Sequences		Polypeptide-coding sequences (number of nucleotides)[b]	Gene product	Inferred molecular weight of polypeptides (number of amino acids)	Special features
	Untranscribed	Transcribed				
	Number of nucleotides					
M:G junction	2 (GA)	—	—			—
G gene	—	1,672	(29) 1,533 [110]	Glycosylated membrane glycoprotein (G)	57,416 (511)	Signal sequence of 16 amino acids. Two glycosylation sites. Terminal fatty acid link (absent in VSV-New Jersey, VSV-Indiana, and VSV-Cocal)
G:L junction	2 (GA)	—	(10)	—	—	—
L gene	—	6,380	6,327 [43]	Core protein (L)	241,012 (2,109)	Short open reading frame in the negative strand; no known product
5'-Terminal tail	59	—	—	—	—	A-rich (32 of 59); 62% base-pairing possible with U-rich untranslated tail of L gene
Totals:	70	11,092				

Total number of nucleotides in genome: 11,162

[a] Compiled from McGeoch (1979, 1981), Gallione et al. (1981), Rose (1980), Rose (1979, 1981), Rose and Gallione (1981), and Schubert et al. (1984).

[b] Given as number of nucleotides in: (untranslated sequence); open reading frame; [untranslated region, including termination triplet and first seven As of poly(A) tail].

no indication of potential secondary structure at the intercistronic boundaries that could account for the phenomenon of attenuation.

D. L Gene

At 6380 nucleotides, the L gene accounts for 57.16% of the genome, approximating the size of the complete genome of tobacco mosaic virus (6395 nucleotides) and exceeding that of the entire genome of the ribophage MS2 (4569 nucleotides). The L protein has a molecular weight of 241,012, comprising 60.38% of the total coding capacity of the VSV genome. There are some short open reading frames in the genome-sense strand of the L gene, the largest of which is sufficient to encode a polypeptide with a molecular weight of about 10,000, but there is no evidence for the existence of a functional transcript. Consequently, there is no evidence in this rhabdovirus of the ambisense disposition of functional genetic information on both positive and negative strands that has been observed in the case of the S RNA of the bunyavirus Punta Tora (Akashi *et al.*, 1984) and the S RNA of the arenavirus Pichinde (Auperin *et al.*, 1984).

E. G Gene

The G gene, at 1672 nucleotides, is the next largest gene, encoding a polypeptide of 511 amino acids with a molecular weight of 57,416. The 511-amino-acid sequence includes an amino-terminal signal peptide of 16 amino acids that does not appear in the mature protein. There are two hydrophobic domains, one in the signal peptide and the other in the 20-amino-acid transmembrane segment close to the COOH terminus. There are two glycosylation sites, Asn-Ser-Thr at position 178–180 and Asn-Gly-Thr at position 335–337. One or two molecules of fatty acid are esterified to the G protein, probably as serine residues on the amino-terminal side of the hydrophobic transmembrane region, presumably stabilizing the association of this domain with the membrane (Rose and Gallione, 1981). Two of five serine residues are missing in this region of the G protein of VSV-New Jersey, which does not contain esterified fatty acid, and are probably the esterification sites (Gallione and Rose, 1983). Schlesinger *et al.* (1984) have reported that the San Juan strain but not the Orsay strain of VSV-Indiana became temperature-sensitive by inhibition of the removal of glucose in the initial stages of processing of the asparagine-linked oligosaccharides. The amino termini of the VSV and rabies virus G proteins are conserved and a peptide corresponding to the amino-terminal 25 amino acids of the VSV glycoprotein has been shown to be a pH-dependent hemolysin (Schlegel and Wade, 1985). It may be that the membrane-destabilizing properties of this domain are important for glycoprotein function.

F. M Gene

The M gene, with a size of 838 nucleotides, encodes a basic polypeptide of 229 amino acids with a molecular weight of 26,064. The amino terminus is highly basic, with 8 lysines in the first 19 amino acids. A triple proline sequence separates this basic domain from the rest of the molecule. This basic domain may interact with genomic RNA, since M-protein mutants and *in vitro* experiments have implicated M protein in the regulation of transcription (Rose and Gallione, 1981). Unlike the M protein of influenza virus, there are no long hydrophobic or nonpolar regions that might suggest a direct association of this protein with membranes.

G. NS Gene

The NS gene, with a size of 822 nucleotides, encodes a polypeptide of 222 amino acids with a molecular weight of 25,110. This protein runs anomalously in sodium dodecyl sulfate–polyacrylamide gel electrophoresis (SDS–PAGE) with an apparent molecular weight in the range 40,000–50,000. There is a domain with 18 negatively charged and no positively charged amino acids within a stretch of 44 in the amino-terminal half of the polypeptide. A deficiency of SDS binding in this region could explain the aberrant electrophoretic mobility; alternatively, the polypeptide could migrate as a dimer. The protein is phosphorylated, but dephosphorylation decreases the electrophoretic mobility and is not responsible for the aberrant mobility (Gallione et al., 1981). The NS protein is phosphorylated at multiple sites (6–10) and the phosphorylated sites on the NS protein of VSV-Indiana obtained either from infected cell cytoplasm (Bell and Prevec, 1985) or virions (Hsu and Kingsbury, 1985) have been mapped by chemical cleavage to the amino-terminal third of the molecule. Sites in a domain located to a peptide lying between amino acids 35 and 78 appear to be constitutively phosphorylated and five potential phosphorylation sites in this region are conserved in four different strains of VSV (Hudson et al., 1968; Rae and Elliott, 1986a). The results of Bell and Prevec (1985) were consistent with the existence of the NS polypeptide as a monomer, consequently the aberrant mobility of NS in polyacrylamide gel electrophoresis is unlikely to be a consequence of dimerization.

H. N Gene

Finally, the N gene, with a size of 1333 nucleotides, encodes a polypeptide of 422 amino acids with a molecular weight of 47,355. There are two additional AUG codons located in secondary ribosome-binding sites near the 5' end that could initiate synthesis of short polypeptides in a second reading frame (Gallione *et al.*, 1981). However, transcripts have not been identified, and although corresponding open reading frames are present in the Ogden strain of VSV-New Jersey (Banerjee *et al.*, 1984), they are shorter and the second is not present in the Missouri strain of VSV-New Jersey (McGeoch *et al.*, 1980). There is also an additional AUG codon in the Ogden sequence before the binding site, which is not present in the Missouri strain, and another before the third binding site. It is unlikely that these sites play any significant role in translation.

I. Gene Products

The five mRNAs are capped, methylated, and polyadenylated, whereas the leader RNA is not. The 3'-terminal adjacent region is conserved in at least five vesiculoviruses (McGeoch, 1981), although the polymerase-binding site located at position 16–30 in the middle of the leader region (Keene *et al.*, 1981b) does not appear to be. The precise function of the leader RNA is unclear, but plus-strand leader appears to be involved in the inhibition of cellular RNA synthesis, because *in vitro* VSV leader RNA or its cDNA inhibited transcription of adenovirus by polII and polIII (McGowan *et al.*, 1982; Grinnell and Wagner, 1984). It has been shown, however, using mutant *ts*G22(II), which is competent for primary transcription, that viral protein synthesis is required for inhibition of host macromolecular synthesis (Unger and Reichmann, 1973). Leader RNA has been shown to have an encapsidation nucleation site for N protein and to react specifically with La protein, an antigen that reacts with antibodies from patients with systemic lupus erythematosus (Kurilla and Keene, 1983). However, the plus-strand leader RNA of mutant virus isolated from persistently infected L cells, in which there is no inhibition of macromolecular synthesis, binds La protein as efficiently as wild-type virus. The La protein–leader interaction appears to be involved more in the control of replication rather than in direct inhibition of host macromolecular synthesis (Wilusz *et al.*, 1985). No base changes have been found in the leader sequence of several tdCE and hrCE host-range mutants of VSV-New Jersey; consequently, the putative host factor associated with these mutants (see Section VII) is unlikely to be the La protein (Keene, personal communication).

J. Codon Usage

There is nonrandom utilization of codons in all five mRNAs, with an overall deficiency of the dinucleotide CG affecting the codon usage for serine, proline, threonine, alanine, and arginine. The CG deficit is observed both within and between codons; consequently, it is unlikely to be an adaptation to a shortage of tRNA species that recognize the CG dinucleotide. The reason for the CG deficiency in viral RNA and vertebrate DNA in general remains unexplained.

III. COMPARATIVE SEQUENCE ANALYSIS

A. N-Gene Termini

McGeoch *et al.* (1980) compared the terminal sequences of the nucleocapsid (N)-protein gene mRNA of vesicular stomatitis virus (VSV)-Indiana (Glasgow strain) and VSV-New Jersey (Missouri strain, Hazelhurst subtype) to identify the positions of the initiation and termination of translation in each. About 200 residues adjacent to the 3'-terminal poly(A) tract of the N mRNA were determined. Taking the two regions together, there was 70.8% sequence homology between the two strains, compared with the 4–20% homology estimated from annealing experiments. Figure 2 is a graphic representation of the sequence homologies of the 3' terminus of the N genes of VSV-Indiana, VSV-New Jersey, VSV-Cocal, and Chandipura virus, showing the extent of sequence divergence and the regions of overall homology.

Figure 3 shows another example of this approach in which the 5' and 3' amino acid sequences derived from the nucleotide sequences of the N gene (McGeoch *et al.*, 1980; Banerjee *et al.*, 1984) of the Missouri and Ogden strains of VSV-New Jersey are compared. The Ogden and the Missouri strains represent the Concan and the Hazelhurst subtypes of VSV-New Jersey, respectively (Reichmann *et al.*, 1978). There is a considerable

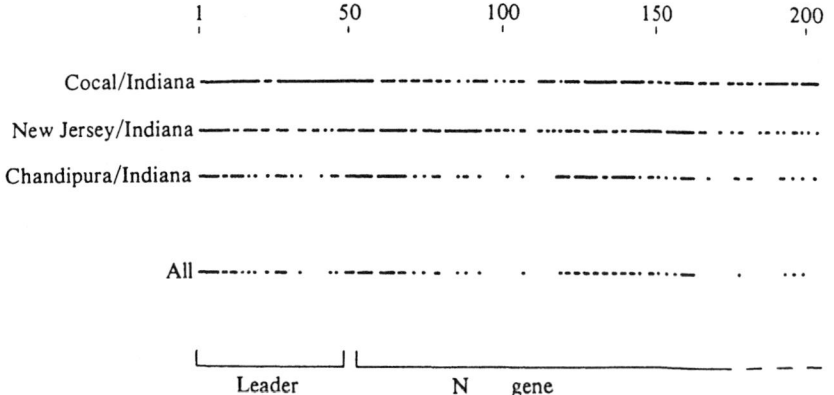

FIGURE 2. Comparison of the sequences at the 3' termini of four vesiculoviruses. For pairs of sequences, and for all four, regions of homology are represented by a heavy line. Where necessary, realignment has been made to allow for different leader and leader–N gene space lengths. Reproduced from McGeoch (1981).

A

```
          Met Ala Pro Thr Val Lys Arg Ile Ile Asn Asp Ser Ile Ile ***    15
AACAGATATCAAA ATG GCT CCT ACA GTT AAG AGA ATC ATT AAC GAC TCA ATT ATT XAG    58
************* *** *** *** *** *** *** *** *** *** ** *** ** ** *** **
AACAGATATCAAA ATG GCT CCT ACA GTT AAG AGA ATC ATT AAT GAC TCC ATA ATT CAG    58
          Met Ala Pro Thr Val Lys Arg Ile Ile Asn Asp Ser Ile Ile Gln    15
                                                        ---
```

```
Pro Arg Leu Pro Ala Asn Glu Asp Pro Val Glu Tyr Pro Ala Asp Tyr Phe Lys    33
CCC AGA TTA CCG GCC AAC GAG GAC CCG GTA GAA TAC CCG GCT GAT TAC TTC AAA    112
**  * *  ** *** *** **  *** **  *** *** **  **  **  *** *** *** *** ***
CCT AAA CTA CCG GCC AAT GAG GAT CCG GTA GAG TAT CCA GCT GAT TAC TTC AAA    112
Pro Lys Leu Pro Ala Asn Glu Asp Pro Val Glu Tyr Pro Ala Asp Tyr Phe Lys    33
    ---
```

```
Asn Asn Thr Asn Ile Val Leu Tyr Val Ser Thr Lys Val    46
AAT AAT ACC AAT ATA GTG TTA TAT GTG AGC ACC AAA GTA    151
*** **  *** **  *** **  *** *** *** ***  **  *** ***
AAT AAC ACC AAC ATA GTA TTA TAT GTG AGC ACT AAA GTA    151
Asn Asn Thr Asn Ile Val Leu Tyr Val Ser Thr Lys Val    46
```

B

```
    Lys Gly Arg Asn Val Val Asp Trp Leu Gly Trp Tyr Asp Asp Asn Gly Gly Lys    18
 CA AAA GGA AGA AAT GTC GTG GAC TGG CTT GGC TGG TAT GAT GAC AAC GGG GGA AAA    56
  *  *** *** *** *** **  **  *** *** **  **  *** *** *** *** **  **  **  ***
 CG AAA GGA AGA AAT GTT GTA GAC TGG CTC GGT TGG TAT GAT GAC AAT GGA GGG AAA    56
    Lys Gly Arg Asn Val Val Asp Trp Leu Gly Trp Tyr Asp Asp Asn Gly Gly Lys    18
```

```
Pro Thr Pro Asp Met Leu Asn Phe Ala Arg Arg Ala Val Asn Ser Leu Gln Ser    36
CCC ACA CCA GAT ATG CTC AAC TTC GCA AGA AGA GCA GTT AAC TCT CTG CAG TCA    110
*** *** *** *** *** *** *** *** *** ***  ** *** **  *   **  ** *** ***
CCC ACA CCG GAT ATG CTC AAC TTC GCA CGA AGA GCA GTC AGT TCG TTG CAG TCA    110
Pro Thr Pro Asp Met Leu Asn Phe Ala Arg Arg Ala Val Ser Ser Leu Gln Ser    36
                                            ---
```

```
Leu Arg Glu Lys Thr Ile Gly Lys Tyr Ala Lys Ala Glu Phe Asn Lys ---    53
CTT CGT GAG AAG ACA ATT GGC AAA TAT GCC AAG GCA GAG TTT AAC AAG TGA CAAGCC    167
**  **  *** **  *** *** *** *** **  **  **  *  * **  *** **  **  *** *  **
CTC CGC GAG AAA ACA ATT GGC AAA TAC GCA AAA GTA GAA TTT GAC AAA TGA CCAGGT    167
Leu Arg Glu Lys Thr Ile Gly Lys Tyr Ala Lys Val Glu Phe Asp Lys ---    53
                                        ---          ---
```

```
TTAAGATACCACTATCACTATTATATTCTATGTTATATATG    208
** * *** ** ***** *********** ** ********
TTGACATATCAATATCAATATTATATTCTGTGCTATATATG    208
```

FIGURE 3. A comparison of the 5'-end (A) and 3'-end (B) sequences of the N genes of the Missouri strain (McGeoch et al., 1980) and the Ogden strain (Banerjee et al., 1984) of VSV-New Jersey. The Missouri sequence is given above and the Ogden sequence below. Identical residues are indicated by asterisks, and coding changes are underlined.

degree of sequence divergence, perhaps not surprising since the Hazel-hurst subtype strain was isolated from a pig in Missouri in 1943 and the Concan subtype strain from a cow in Utah in 1949. Of the nucleotides at the 5' end, 10% are changed, and at the 3' end, 2.2% (or 14.2% overall). Of the 43 nucleotide changes, 33 are in the third codon position; therefore, the number of coding changes is low, one at the 5' and three at the 3' end (i.e., four overall). Thus, comparative nucleotide sequencing of a small region of the genome can be an accurate method of determining relationships within the family Rhabdoviridae, a method that is more sensitive and definitive than peptide mapping, annealing experiments, or oligonucleotide fingerprinting.

B. G Genes

Gallione and Rose (1983) have compared the nucleotide sequences of the glycoprotein (G) gene of VSV-Indiana and VSV-New Jersey (Ogden strain). The positions and sizes of the transmembrane domains, signal sequences, and glycosylation sites are identical in VSV-Indiana and VSV-New Jersey, but the lengths of the coding sequences differ, encoding 511 and 517 amino acids, respectively. The untranscribed regions of VSV-New Jersey are smaller, totaling 25 nucleotides compared with the 126 of VSV-Indiana.

The complete nucleotide sequence of the G gene of the ERA strain of rabies virus has been determined by Anilionis et al. (1981) and that of the CVS strain by Yelverton et al. (1983). The deduced amino acid sequences of the two G proteins show about 90% homology. The VSV-Indiana (San Juan strain) and rabies virus (ERA strain) G proteins are approximately the same size, with 511 and 524 amino acids, respectively. There are three putative glycosylation sites in the rabies virus G protein (Asn-X-Ser and Asn-X-Thr) and two in the G protein of VSV (Asn-X-Thr), the COOH-terminus-proximal glycosylation sites being aligned exactly. Comparison of the predicted amino acid sequences of the two glycoproteins has revealed significant homologies. Although the lengths of the two proteins differ, the signal and transmembrane sequences are aligned exactly. Nevertheless, the corresponding sequences are not highly conserved, which indicates that the only dominant requirement in these regions is hydrophobicity. In the optimum alignment, there is 20% identity of sequence (with higher values in limited regions, e.g., the 24 amino acids preceding and including the putative glycosylation sites nearest the COOH terminus). This value is less than that between Semliki Forest virus and Sindbis virus (47%) and about equal to that between the β-chain of human hemoglobin and human myoglobin or between the matrix (M) protein of VSV-Indiana and the M protein of influenza A virus. At the nucleotide level, there was much less apparent homology.

A remarkable amino acid homology has been reported between a segment of the rabies virus G protein and the entire sequence of the long (71 to 74-residue) curaremimetic neurotoxins of snake venom (Lentz et al., 1984). The greatest identity occurs in the region of the 20-amino-acid highly conserved "toxic" loop in the middle of the neurotoxin molecule, which is known to interact with the acetylcholine (ACh)-binding site of the ACh receptor. These observations and the fact that rabies virus binds to regions with abundant ACh receptors suggests that direct binding of rabies virus G protein to the ACh receptor may contribute to the neurotropism of this virus.

Kotwal et al. (1983) have compared the amino-terminal amino acid sequences of VSV-Indiana (Toronto HR strain), VSV-New Jersey (Concan subtype), and VSV-Cocal. There was little sequence homology in the 16-amino-acid (17-amino-acid in Cocal virus) signal-peptide region of the unglycosylated precursors, apart from a high content of hydrophobic amino acids. By contrast, the amino-terminal sequence of the mature envelope glycoproteins was highly conserved for at least 24 amino acids. The biological function of this highly conserved region remains to be elucidated.

C. NS Genes

The complete sequences of the NS genes of strains of both the Concan and Hazelhurst subtypes of the New Jersey serotype of VSV have been obtained by Gill and Banerjee (1985) and Rae and Elliott (1986a), respectively, and of the San Juan (Indiana subtype) and Mudd-Summers (Cocal subtype) of the Indiana serotype by Gallione et al. (1981) and Hudson et al. (1986), respectively. The mRNAs of the New Jersey strains are identical in length (856 nucleotides excluding poly A), as are those of the Indiana strains (814 nucleotides). There is about 50% homology between the NS genes of the two subtypes compared with 85% for the two New Jersey strains and 97% for the two Indiana strains. The greater divergence of the two New Jersey subtypes may reflect the marked difference in host range of the two subtypes (Schnitzlein and Reichmann, 1985).

Comparison of the predicted amino acid sequences of the NS proteins shows only 33% homology between the two subtypes, which is considerably less than the homology between the G (51%) and N (69%) proteins of these serotypes. Hudson et al. (1986) identified two domains: an acidic region (domain I) and a basic region (domain II). The conserved phosphorylation sites lie within domain I, and domain II is represented by the conserved carboxy-terminal region. Despite the lack of amino acid homology elsewhere in the molecule, the relative charge distribution and hydropathy are preserved. Hudson et al. (1986) suggested that domain II, by binding to a specific site on genome or antigenome RNA, is involved in the initiation of replication or transcription, whereas domain I, by mimicking the RNA template, binds N protein thereby facilitating polymerase entry. Eleven additional conserved potential phosphorylation sites are regularly spaced between domains I and II and phosphorylation at these sites would enhance the mimicry of RNA structure.

D. M Genes

The M genes of the San Juan strain (Rose and Gallione, 1981) and the Glasgow and Orsay strains (Gopalakrishna and Lenard, 1985) of VSV-Indiana have been sequenced. The coding region of the former strain differs from the latter pair in 13 nucleotides, causing 6 amino acid substitutions. The Glasgow and Orsay strains are more similar, differing by 8 nucleotides causing 4 amino acid changes.

IV. DELETION MUTANTS

A. Genome Structure

1. General Features

The defective interfering (DI) particles of vesicular stomatitis virus (VSV) and other rhabdoviruses that accumulate during multiplication at high multiplicities of infection (undiluted passage) contain only a portion of the viral genome and are formally equivalent to deletion mutants.

Characteristically, these incomplete genomes (in DI particles) interfere with the replication of the complete genome (standard virus). Nevertheless, they are themselves dependent on the polymerase of standard virus for their own multiplication, because DI particles invariably have lost a portion of the L gene and none can encode a functional polymerase. Since the complete sequence of the L gene (Schubert et al., 1984) and partial sequences of several DI particles have been determined, it is now possible to map precisely the extent of the deletions in DI-particle genomes and to characterize the sites of recombination of the deleted molecules (Meier et al., 1984).

Four types of DI-particle genomes can be distinguished, these being designated: (1) fusion, (2) panhandle, (3) snapback, and (4) compound (Lazzarini et al., 1981; Perrault, 1981). All four types of genomes have structural features in common; namely, variable portions of the L gene are deleted, the 5'-terminal regions of the deleted genomes are identical in sense and sequence to the 5' terminus of the minus strand of the complete genome, and the 3'-terminal regions are identical to the 5' terminus of either the minus or the plus strand of the complete genome. These features are diagrammed in Fig. 4.

2. Panhandle DI Particles

The snapback and panhandle (the most common type) DI particles have lost the entire N, NS, M, and G coding regions in addition to the 3' end of the L gene. These DI particles can neither transcribe functional mRNA nor complement any of the six groups of temperature-sensitive (ts) mutants. The extent of the deletions ranges from 90% for the smallest DI particles, which are all derived from ts mutants belonging to complementation group III VSV-Indiana, to 50% for the largest DI particles (Reichmann and Schnitzlein, 1979). These DI particles interfere specifically with the replication of homotypic standard virus, and less well or not at all with heterotypic viruses according to their degree of antigenic relatedness. The 3' termini of panhandle DI particles are identical in sense and sequence to the 3' terminus of the plus strand of standard virus and are hence complementary to the 5' terminus of the DI genome because of the inverted complementarity of the terminal regions of the standard

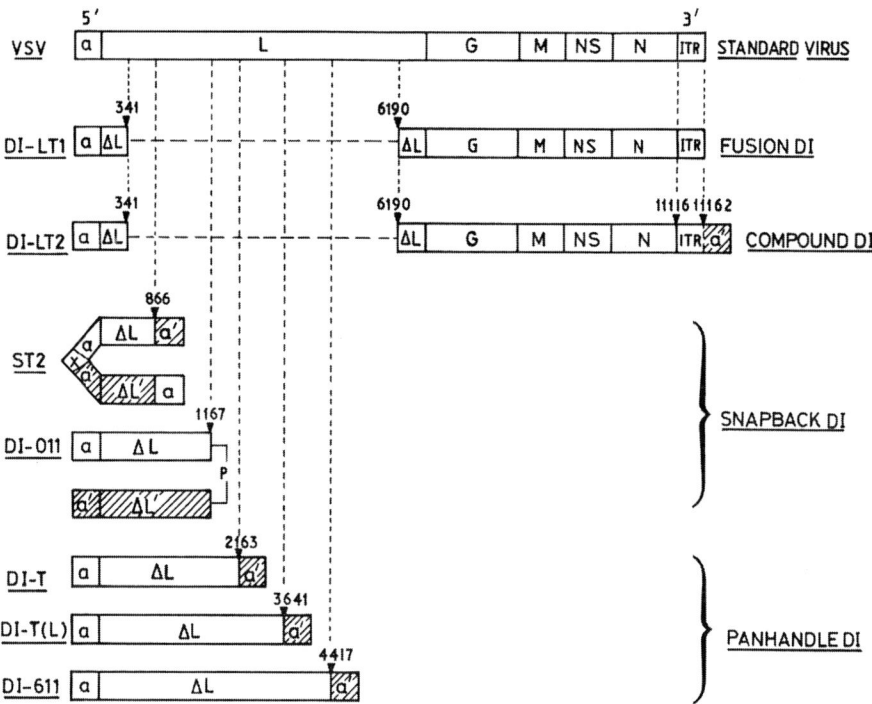

FIGURE 4. Structure of the genome of four types of deletion mutants of VSV. (a, a') Complementary terminal sequences; (ITR) inverted terminal repeat; (▨) complementary plus-sense sequences. Compiled from Meier *et al.* (1984), Lazzarini *et al.* (1981), and Nichol *et al.* (1984).

genome. The generation of this type of DI particle can be explained by disengagement of the polymerase during transcription of the 5' end of the plus-strand replicative template and reattachment at a site adjacent to the 5' terminus followed by copy-back synthesis to the 5' terminus. As a consequence, the 3' end of the DI-particle genome is identical to the 3' end of the plus strand.

Analysis of the sequences at the points of recombination in three panhandle DI particles of the Mudd–Summers strain by Meier *et al.* (1984) indicated that the boundary between the panhandle regions and the L gene was precisely defined. The added 3'-end sequences were 46 (or exceptionally 45) nucleotides long, suggesting that synthesis of the negative-strand leader RNA may have been the first step in the generation of these panhandle DI genomes. Presumably, the polymerase and the negative leader RNA detach from the plus-strand template and resume synthesis on a negative-strand template. The abundance of leader RNA transcripts and the excess of negative-strand templates over plus-strand templates in the infected cell favor a copy-choice model of generation of DI particles, rather than a copy-back model. O'Hara *et al.* (1984) have described DI

genomes with more complex configurations that appeared in the course
of prolonged undiluted lytic passages or from persistently infected cells,
but all can be interpreted by detachment of the polymerase and template
switching.

3. Snapback DI Particles

Snapback DI particles, which are less common, contain RNA with
extensive self-complementarity. Snapback DI011 has complementary sequences of 860 nucleotides, covalently linked by a single phosphate group.
The 5'-terminal linked sequence represents 7–8% of the standard genome.
A unique feature of this DI particle is that the parental and progeny strands
generated during replication have the same nucleotide sequence. Unlike
other DI particles, the covalently linked "turnaround" region shows no
homology to either the 5' or the 3' termini. The generation of DI particles
of this type is best explained by polymerase-template switching during
replicative synthesis on a plus-strand template, perhaps as a result of
"tailgating" behind a slower-preceding polymerase. If the switch to the
progeny strand occurs symmetrically across the replication fork, two completely complementary strands will result. Perrault and Leavitt (1977)
concluded on the basis of electron-microscope studies that snapback DI
particles derived from mutant tsG31(III) of the Indiana strain of VSV had
inverted complementary sequences at the plus–negative strand junction.
Sequence analysis of the termini has confirmed this structure (Nichol et
$al.$, 1984). It is apparent that other types of snapback particles exist in
addition to the DI011 type and are generated by a different mechanism.
The snapback DI particle designated ST2, which Nichol and colleagues
characterized, differs from DI011 in the possession of inverted complementary termini and reverse orientation of the hairpin. This is illustrated
in Fig. 4. It is likely that ST2 is derived directly from the panhandle ST1
DI particle present in the same material. Gillies and Stollar (1980) observed that snapback DI particles with 90–95% self-complementarity
were the predominant type of DI particles present after two passages of
standard VSV in $Aedes$ $albopictus$ cells, whereas snapback particles were
not detected among the DI particles generated by three passages in baby
hamster kidney (BHK) cells. The snapback particles generated in $A.$ $al-$
$bopictus$ cells induced interference in the mosquito cells only, unlike the
BHK-derived DI particles, which interfered in both cell types with similar
efficiency. A single passage in the heterologous culture reversed the specificity of interference.

4. Fusion DI Particles

The large DI particles may have panhandle-, fusion-, or compound-
type genomes. Large DI particles of the latter two types are derived predominantly from the Toronto HR strain of VSV-Indiana. Fusion-type DI
particles represent internal deletions of the L gene, leaving the terminal

regions intact. The HR LT1 DI has 5848 nucleotides deleted within the L gene. Compound-type DI particles resemble fusion DI particles in structure, except that the 3' terminus is a compound structure, as illustrated for DI LT2 in Fig. 4. The HR LT2 DI particle also has an L-gene deletion of similar extent and two initiation sites in tandem at the 3' end of the RNA. Unlike the HR LT1 DI particle, these particles were transcriptionally inactive,. and unlike the panhandle DI particles, which transcribed a 46-mer product, the internally located leader RNA region appeared to be muted (Keene *et al.*, 1981a). DI-particle genomes with parental ends predominated in early passages and were gradually replaced by genomes with compound complementary ends. It is unlikely that interfering ability is an attribute of a particular end structure, since Keene *et al.* (1979) have described a DI particle of VSV-New Jersey with a genome having parental ends that was able to induce interference. The LT DI particles can be transcribed *in vivo* and *in vitro* and complement *ts* mutants, because the G, M, NS, and N genes are undeleted. The G mRNA, however, yields a product larger than the viral G protein, presumably because of readthrough into the L gene. A distinguishing functional characteristic of the LT DI particles of VSV-Indiana is their ability to interfere heterotypically with VSV-New Jersey.

Analysis of the sequences around the recombination sites of fusion DI particles has ruled out the existence of a conserved signal sequence specifying termination or resumption of RNA synthesis. In addition, no stable secondary structures can be discerned in the regions adjacent to the site of recombination; therefore, secondary structure does not appear to be involved in the generation of DI particles. Nevertheless, certain sequence homologies adjacent to the points of termination and resumption of synthesis have been recognized. Two hexanucleotides, ATCTGA and GATTGG, separated by approximately the same number of intervening nucleotides occur in the genome four nucleotides before the beginning of the deletion and seven nucleotides before the end of the deletion. This suggests that weak base-pairing between the template and the nascent progeny strand occurs close to the site of resumption of synthesis, or perhaps there is a preferential association of the polymerase with these sequences. Synthesis of the compound HR LT2 DI particle can be explained best in terms of the copy-choice model described for panhandle DI particles (Section IV.A.2).

5. Induction

Perrault *et al.* (1981) have obtained a variant (Pol R1) of the Mudd–Summers strain of VSV-Indiana after several cycles of heat inactivation that appears to be an N-protein mutant (see Section VII.F.2). Subsequently, the Pol R1 variant generated DI particles that synthesized almost full-length copies of their RNA, unlike the wild-type panhandle DI particles, which produced a 46-mer transcript of the leader region only.

The mechanisms of induction of DI particles are obscure, but host factors are clearly involved. Suppression of host-cell macromolecular synthesis by treatment with actinomycin D prior to virus infection with VSV cloned and propagated in actinomycin-pretreated cells prevented the induction of DI particles. Actinomycin treatment had no effect on the replication of preexisting DI particles (Kang and Allen, 1978). These results are consistent with the view that DI particles play a role in the modulation of infection; i.e., induction of DI particles is a host-cell-dependent function that moderates the spread of virus in the host prior to the full development of the immune response.

B. Role of Host Factors

Holland *et al.* (1976) observed that a subline of Hela cells was incapable of generating DI particles of VSV. Kang *et al.* (1981) observed that DI particles of different sizes were generated by high-multiplicity passage in human and mouse cells and that none were induced in primary diploid fibroblasts (strain 544). Analysis of seven human–mouse L cell somatic-cell hybrids revealed that inability to generate DI particles was a dominant characteristic. Chromosome 16 was the only human chromosome common to all seven hybrids, and it was concluded that a factor on this chromosome is responsible for suppression of induction of DI particles. Both parents and all the hybrids, however, supported the replication of preexisting DI particles. Ability to generate DI particles was regained on loss of chromosome 16 from the hybrid clones. The mechanism of action of the suppressor function of chromosome 16 has not been determined.

V. MUTATION AND MUTABILITY

A. General Features

Mutation can be studied at two levels—the phenotypic level and the level of changes in the nucleotide sequence. Conditional lethal mutants, predominantly temperature-sensitive (*ts*) mutants, are suitable for the analysis of viral gene function, because this phenotype can be associated with mutation in any essential function of the viral genome.

The *ts* mutants of VSV-Indiana are designated conventionally as follows: phenotype (*ts*); origin B (for Bratislava), G (Glasgow), O (Orsay), L (Lodish), M (Massachusetts), T (Toronto), or W (Winnipeg); clone number; complementation group (in parentheses). For example: *ts*G11(I) is *ts* mutant 11 of the Glasgow strain classified in complementation group I.

Table II lists the phenotypes of the mutants and variants of rhabdoviruses that have been used in genetic analysis. The majority of these phenotypes represent single-step genetic mutations, but a few, such as the spontaneous variants of Sigma virus, probably represent multiple

TABLE II. Mutants of Rhabdoviruses

Virus	Serotype	Mutant designation	Phenotype	Mutagen	Reference
VSV	Indiana	L/s	Plaque-morphology	None, sp.	Wertz and Levine (1973)
	New Jersey	—	Plaque-morphology	UV	Schechmeister et al. (1967)
	Indiana (Orsay)	ts	Temperature-sensitive	None, sp.	Flamand (1970)
	Indiana (Winnipeg)	ts	Temperature-sensitive	5-FU, NA, EMS, P	Holloway et al. (1970)
	Indiana (Glasgow)	ts	Temperature-sensitive	5-FU, 5-AzaC, EMS	Pringle (1970)
	Indiana (Mass.)	ts	Temperature-sensitive	5-FU, NA, NTG	Rettenmier et al. (1975)
	New Jersey	ts	Temperature-sensitive	UV, NA	Pittman (1965)
	New Jersey [Hazlehurst]	ts	Temperature-sensitive	5-FU	Pringle et al. (1971)
	New Jersey (Concan)	ts	Temperature-sensitive	5-FU, 5-AzaC, NA, UV	Byrd et al. (1984)
	Cocal	ts	Temperature-sensitive	5-FU	Pringle and Wunner (1973)
	Indiana	tl	Thermolabile	5-FU	Zavada (1972)
	Indiana	hr	Host-range (restricted in HeLa cells)	5-FU, NTG	Simpson and Obijeski (1974)
	New Jersey	hrCE	Host-range (restricted in CE cells)	5-FU	Pringle (1978)
	Indiana	tdCE	Host-range (conditionally restricted in CE cells)	5-FU	Pringle (1978)
	New Jersey	tdCE	Host-range (conditionally restricted in CE cells)	5-FU	Szilagyi and Pringle (1975)
	Cocal	tdCE	Host-range (conditionally restricted in CE cells)	5-FU	Pringle (unpublished data)
	Indiana	Rif+	Rifampin sensitivity	None, sp.	Moreau (1974)
	Indiana	cd	Complementation-dependent	NA, NTG	Rettenmier et al. (1975)
	Indiana	P-	Host-protein-inhibition-defective	None	Francoeur et al. (1980)
	Indiana	P++	Host-protein-inhibition-enhanced	None	Stanners [personal communication]
	Indiana	PIF+	Plaque-interferon-positive	—	Francoeur et al. (1980)
	Indiana	Sdi-	DI-resistant	None, sp.	Horodyski et al. (1983)

(continued)

TABLE II. (*Continued*)

Virus	Serotype	Mutant designation	Phenotype	Mutagen	Reference
	Indiana	Pol R	Polyermase readthrough	None	Perrault et al. (1981)
	Indiana	am	+ RNA^w suppressor sensitive	5-FU	White and McGeoch (1985)
Chandipura	Nagpur	ts	Temperature-sensitive	5-FU	Pringle and Wunner (1973), Gadkari and Pringle (1980a)
		tdCE	Host-range (conditionally restricted in CE cells)	5-FU	Gadkari and Pringle (1980b), Rasool and Pringle (1986)
		PIF$^+$	Plaque-interferon-positive	5-FU	Rasool and Pringle (unpublished data)
Piry	—	agD	Affected growth in *Drosophila*	None	Brun (1981)
		rgD	Rapid growth in *Drosophila*	None	Brun (1984)
Rabies	CVS	ts	Temperature-sensitive	5-FU	Clark and Koprowski (1971), Bussereau and Flamand (1978)
		—	Plaque-size	None, sp.	Kondo (cited by Matsumoto, 1970)
		RV	Monoclonal-antibody-resistant	None	Lafon et al. (1983)
Sigma	—	ts	Temperature-sensitive	5-FU	Contamine (1973)
		haP	Host-adapted	—	Contamine (1981)
		Tr	Temperature-resistant	None, sp.	Ohanessian-Guillemain (1959)
		P$^+$/P$^-$	Host-range	None, sp.	Guillemain (1953)
		—	"Plaque-size"	None, sp.	Brun (1963)
		—	"Plaque-type"	None, sp.	Vigier (1966)
		rho	Replication-defective	None, sp.	Brun (1963)
		ultrarho	Replication-defective	None, sp.	Brun (1963)
		g$^+$/g$^-$	Germinal-transmission	None, sp.	Duhamel (1954)
		pr$^+$/pr$^-$	Egg-invasiveness	None, sp.	Duhamel (1954)
		V$^+$/V$^-$	Sperm-invasiveness	None, sp.	Goldstein (1949)

[a] (5-AzaC) 5-Azacytidine; (EMS) ethylmethane sulfonate; (5-FU) 5-fluorouracil; (NA) nitrous acid; (NTG) *N*-methyl-*N'*-nitro-*N*-nitrosoguanidine; (P) proflavine; (sp.) spontaneous; (UV) ultraviolet light.

genetic changes. In the case of the three *ts* mutants comprising complementation group E of vesicular stomatitis virus (VSV)-New Jersey and *ts*045(V) of VSV-Indiana, the inferred single-step nature of the change has been confirmed by oligonucleotide sequencing (see Section VII).

B. Mutation Frequency and Mutation Rate

In VSV, spontaneous mutation frequency for specific characteristics can be high even in the face of contrary selection (Flamand, 1980). Flamand (1980) calculated the rate of production of *ts* mutants in stocks of VSV-Indiana propagated in chick-embryo fibroblasts (CEF) as 1.1×10^3 using the formula: $m = 0.43f/\log N - \log N_o$, where m is the mutation rate, f is the frequency of mutants, N is the total population size, and N_o is the initial size of the population $(= 1$ in the case of a virus clone). In other words, the probability that a single base change resulting in a *ts* mutation will occur in a given replication event is approximately one in a thousand. Since about 90% of *ts* mutations in VSV are located in the polymerase gene, this figure of 10^3 approximates the mutation rate per gene. This value is similar to the intrinsic error rate for enzymatic nucleic-acid synthesis (Reanney, 1984). In the case of DNA synthesis, proofreading reduces the final error rate to 10^{-9} to 10^{-10}. Since there is no corrective mechanism in RNA synthesis, the error rate remains 10^5 to 10^6-fold greater than in DNA copying.

Because of the high mutation rate per nucleotide, any RNA genome can be described only in a probabilistic sense. For example, in the course of determining the sequence of the L gene of VSV-Indiana, each region of the genome was sequenced five times on average (Schubert *et al.*, 1984). At least 20 nucleotide differences were observed between different overlapping cDNA clones. Of these nucleotide differences, 15 represented base substitutions, 11 of which would produce an amino acid change in the gene product. The remaining 5 differences comprised 1 single-nucleotide addition and 4 single-nucleotide deletions. These changes would cause frameshifts with premature termination, producing truncated L-protein molecules. It is possible that some or all of these nucleotide changes were introduced during reverse transcription, but two of the deleted nucleotides were identical, and these changes at least were unlikely to have been generated during reverse transcription, since the enzyme would have to make the same error at the same site in independent transcripts. The final sequence established for the L gene, therefore, is a consensus sequence. A full length L-gene clone omitting the 20 nucleotides diverging from the consensus sequence was constructed and inserted into a simian virus 40 (SV40) expression vector. The vector was then transfected into COS cells, and the functional activity of the product of this L-gene construct was established by the complementation and rescue of a superinfecting nonrevertible polymerase-defective mutant *ts*G11(I) (Schubert and co-workers, personal communication), confirming the correctness of the sequence.

Flamand (1980) reported that the frequency of mutation of the Orsay stock of VSV was similar in different cell systems and that propagation of CEF-adapted virus in different cells did not result in genetic instability. Pringle and Wunner (1975) reported that the frequencies of *ts* mutants in stocks of VSV-Indiana, VSV-New Jersey, and VSV Cocal propagated under comparable conditions were similar and around 1%. Youngner *et al.* (1976) reported a frequency of 4.4% for their stock of VSV-Indiana. However, it is clear that the figure obtained will depend to some extent on the temperature arbitrarily chosen as the restrictive temperature and that comparison of data obtained in different laboratories is therefore not necessarily meaningful.

Flamand (1980) has shown that change in the temperature of incubation, on the other hand, resulted in a transient genetic instability. In one experiment, the frequency of *ts* mutants reached 8.3% at the second passage at 39.3°C compared with 2.6% at 30°C. It was concluded that passage at low multiplicity of infection under uniform temperature conditions did not result in change of the overall temperature sensitivity of the wild-type virus, unlike the situation commonly observed with picornaviruses, in which, for instance, in the case of poliovirus, nine passages at low temperature resulted in loss of ability to grow at high temperature.

Oligonucleotide mapping is the most sensitive method for detecting genetic change, short of complete sequencing, provided there is at least 90% sequence homology between viruses under comparison (Young *et al.*, 1981). In the case of VSV-Indiana, it has been shown by oligonucleotide fingerprinting that there are appreciable differences between strains isolated from different localities or at different times (Clewley *et al.*, 1977), whereas prolonged passage of any particular isolate in cultured cells produced little significant change (Clewley *et al.*, 1977; Holland *et al.*, 1980, 1982). Brand and Palese (1980) concluded from oligonucleotide mapping of influenza virus and VSV after 12 plaque-to-plaque passages in chick-embryo cells that VSV exhibited greater genetic stability than influenza virus. Nucleotide changes were detected in eight of eight subclones of influenza virus, but in only one of four VSV subclones at this passage level. On the other hand, Portner *et al.* (1980), using monoclonal antibodies, concluded that the mutation rates to antibody resistance for influenza virus, VSV, and Sendai virus were indistinguishable and approximately $10^{-4.5}$ per replication event.

The frequency of *ts* mutants reported in stocks of other rhabdoviruses varies over a wide range, from 0.3% to 12.5%, respectively, in the case of the lyssaviruses rabies virus and Lagos bat virus (Bussereau and Flamand, 1975; Clark and Wiktor, 1974). However, these comparisons are not meaningful in view of differences in restrictive temperature.

On the other hand, departures from this apparent genetic stability are common. An extreme example is the observation of Clark (1978) that several high-passage strains of rabies virus, previously regarded as genetically stable with no adult-mouse-killing potential, became neurovirulent after two to five passages in cultured human or mouse neuroblastoma cells, with a change in pathogenicity becoming evident even after a single passage. This newly acquired neurovirulence remained a stable genetic property during subsequent propagation in nonneuroblastoma cells.

It has also been observed that ts mutants accumulate during passage in the insect *Drosophila melanogaster* (Printz, 1970) and in *Drosophila* cells in culture (Mudd *et al.*, 1973). However, this phenomenon may be to a certain extent a function of the lower temperature of propagation.

C. Genome Evolution during Persistent or High-Multiplicity Infection

1. Persistent Infection

Extensive genetic changes occur in VSV maintained in persistently infected cells or passaged for a period at high multiplicities of infection. This phenomenon has been comprehensively studied at the molecular level by Holland and colleagues.

Holland and Villareal (1974) established a persistent infection in BHK-21 cells using tsG31(III), a ts mutant of complementation group III of VSV-Indiana, in the presence of its own homologous defective interfering (DI) particles at semipermissive temperature to restrict cytopathogenicity. A stable persistent infection was achieved after a few passages, and its subsequent maintenance was associated with the appearance of a new DI particle, rather than with temperature sensitivity of the initiating mutant or the small spherical DI particle so far uniquely associated with group III mutants (Reichmann *et al.*, 1971). The progress of this infection, designated Car4, has been monitored by T1-ribonuclease oligonucleotide mapping over the course of nine years. Continuous and progressive evolution was observed, with changes in the oligonucleotide maps accumulating sequentially. By 7.5 years, 24 changes had accumulated, which represents about 240 base changes, or 2% of the entire genome, since the large unique oligonucleotides represent only about 10% of the total sequence.

Comparison of the pattern of development of the T1-ribonuclease oligonucleotide-map changes in independently initiated persistent infections during sequential lytic cycles at high multiplicity with the pattern during persistent infection of *Aedes albopictus* cells indicates that the rapid genome evolution of VSV under these conditions is essentially random and unprogrammed (Spindler *et al.*, 1982). Factors involved in this process may include intergenomic complementation between lethal and semilethal mutants, *cis* and *trans* activity of a "mutator" gene, and selection mediated by DI particles.

Complementation-dependent mutations of VSV were identified by

Rettenmier *et al.* (1975) using a special screening procedure. It is clear, therefore, that such cryptic mutants exist and could contribute to the genetic variation that accumulates during persistent infection or in stocks maintained at high multiplicity.

A *trans*-acting "mutator" gene in VSV has been characterized by Pringle *et al.* (1981). The "mutator" phenotype is associated with mutant *ts*D1 of VSV-New Jersey, enhancing the mutability of homologous and heterologous VSV multiplying in the same cells. It was estimated that mutant *ts*D1 had accumulated base changes amounting to 0.5% of the genome, although it was only a few passages removed from the parental wild-type. It is clear that mutator-type genes occur in rhabdoviruses and could be responsible for rapid genome evolution.

2. Sdi⁻ Mutants

The selective effect of DI particles was first recognized by Kawai and Matsumoto (1977) when during persistent infection of BHK-21 cells by rabies virus there appeared small plaque mutants that were resistant to interference by the DI particles initially present in the culture. Subsequently, Horodyski and Holland (1980, 1981) observed the same phenomenon in their BHK persistently VSV-infected Car4 culture, and in addition they showed that there was a progressive replacement of these mutants. The appearance of resistant virus, the Sdi⁻ phenotype, was soon followed by the generation of new DI particles to which the virus was now fully susceptible, culminating in another cycle of mutation, escape, and susceptibility (Horodyski *et al.*, 1983). There is thus a simultaneous evolution of both virus and DI particles.

Complementation tests with these Sdi⁻ mutants indicated that at least two different viral functions were involved in the origin of Sdi⁻ mutants. Horodyski *et al.* (1983) also demonstrated, using chimeric DI particles, that the template RNA, rather than the envelope proteins of the DI particles, was responsible for the specific interference with the Sdi⁻ and Sdi⁺ mutants of the helper virus. Sequence analysis of the 5' and 3' termini of various Sdi⁻ mutants isolated sequentially during persistent infection or lytic infection at high multiplicity revealed a stepwise accumulation of stable base changes within the area of the replication-initiation site at their 5' end. Fewer mutations accumulated in the region of transcription initiation at their 3' terminus.

The most striking feature of the 5'-end sequence was the clustering of mutations in the sequence of 54 nucleotides adjacent to the terminus of the genome. There were 11 single site changes in the genomes of 9 different DI particles from the Car4 and Car6 independently derived persistent infections and from Car51 derived from Car4 after 5 years, whereas only 2 mutations were observed in the tail end of the L-gene mRNA. The 10 single site changes that were observed in Car4 and Car51 accumulated in a stepwise manner; i.e., once a base change appeared, it remained present in subsequent isolates, suggesting that selection operated in its favor. This stepwise accumulation of mutants was also observed in 11 Sdi⁻ mutants obtained by undiluted passage of mutant tsG31(III) itself; 5 changes accumulated sequentially and 4 coincided with 4 of the changes observed in the series of Sdi⁻ mutants obtained from the Car4 persistent infection. Interestingly, the Glasgow strain (from which the persistent infections were initiated) differed from the Mudd–Summers strain of VSV-Indiana at two sites within this 169-nucleotide region, but neither coincided with sites affected in Sdi⁻ mutants. The Sdi⁻ phenotype does not depend entirely on mutations in this region, however, since base substitutions were not observed in this region in 3 Sdi⁻ mutants of this series or in 2 derived from another persistent infection, Car21.

The 3'-terminal region of the same Sdi⁻ mutants was also determined from position 14 to positions 220–330 from the 3' terminus. This region does not differ between the two strains of VSV, and fewer changes were observed in the Sdi⁻ mutants at the 3' end; four mutations accumulated sequentially within the leader-RNA region, and another seven in the N-gene region (three of which were silent). However, several Sdi⁻ mutants exhibited no changes in this region. It is possible that the 3'-terminal region is more conserved than the 5' end because of its involvement in both replication and transcription control.

An invariable characteristic of all Sdi⁻ mutants is impaired virion polymerase activity. Presumably, since DI-particle-mediated selection affects replication, mutations involving the polymerase complex are also likely to affect its transcriptase function as well. It was concluded that they represented compensatory changes to accommodate Sdi⁻ mutations that affect replication- or encapsidation-gene products or both (O'Hara *et al.*, 1984).

Reconstruction experiments in BHK-21 cells confirmed that DI-particle-resistant virus was at a selective advantage over wild-type virus in the presence of DI particles and that this effect was observed even when the resistant mutant was added at low levels relative to wild-type virus in the inoculum (Horodyski and Holland, 1984). Consequently, recurrent DI-particle-mediated selection could be the driving force in the continual evolution of virus during persistent infection or passage at high multiplicity.

3. Temperature-Sensitive Mutants

Youngner and Quagliana (1976) reported that *ts* mutants mimicked DI particles and interfered with the multiplication of wild-type virus at both permissive and restrictive temperatures, when the *ts* mutants were present at 10-fold excess in the inoculum. Flamand (1980) and Spindler *et al.* (1982), however, did not observe interference using various *ts* mutants at equal multiplicity in the inoculum with wild-type virus. Nonetheless, two uncharacterized *ts* mutants isolated from the Car4 persistent infection at 60 and 76 months were able to interfere with wild-type VSV-Indiana and tsG31(III), the mutant used to initiate the Car4 persistent infection. Variants of this sort may augment the selective role of DI particles.

Persistent infections also initiated by mutant tsG31(III) have been established and maintained over a four-year period in neural (glioma and oligodendroglioma) cells and mouse L-929 cells (Huprikar et al., 1986). The virus derived from the nonneural L-929 cells and from oligodendroglioma cells remained highly neurovirulent in three-week-old Swiss mice, whereas virus from glioma cell cultures had lost virulence by the 20th passage and remained nonvirulent for a further 180 passages.

Neurovirulence rapidly returned, however, during intracerebral passage in mice, or more slowly during the course of passage in L-929 cells. The mechanism underlying the loss of neurovirulence in glioma cells but not in oligodendroglioma cells was not established, but passage in glioma cells was accompanied by selection of RNA -ve mutants. It is clear that alterations in pathogencis can occur by selection of mutants mediated by specific virus–host interactions.

4. DI Particles

The precise mechanism of interference remains to be elucidated, but it is clear that it operates at the level of RNA replication (reviewed by Reichmann and Schnitzlein, 1979; Lazzarini et al., 1981; Perrault, 1981), and DI-particle replication is required for interference to be expressed (Bay and Reichmann, 1979). The inverted terminal complementary structures common to the most frequent DI particles contain recognition sites for initiation of replication and encapsidation, giving the DI particles a competitive advantage.

O'Hara et al. (1984) also sequenced the 5' and 3' termini of 16 DI particles isolated from the same persistent infections or series of undiluted lytic passages of tsG31(III). All the 16 DI particles investigated were panhandle-type and derived from the 5' end of standard virus, displaying terminal complementarity for sequences extending from 54 to several hundred nucleotides. Of the 16 DI particles, 12 exhibited complex structural patterns with the sequences internal to the termini extensively rearranged. Some of these DI particles were derived from multiple intra- and interstrand recombinational events; nevertheless, the derivation of each can be explained by a polymerase-template switching model. Terminal complementarity with a minimum of 46 nucleotide pairs appears to be essential for the competitive survival of DI particles, but there appeared to be no defined sites for termination or restriction of replication. The degree of freedom accorded the internal rearrangements of the DI genome suggests that these sites are irrelevant in determining survival and interfering capacity. However, since some DI-particle genomes have the same replication-initiation sequence as resistant Sdi⁻ mutants, the internal sequences of the DI-particle genomes, despite their scrambled nature, must contain information that enables the Sdi⁻ polymerase to discriminate between the nondefective viral template and the defective DI genome. Neither overall size of the DI-particle genome nor size of the complementary region beyond the 46-nucleotide minimum appears to play any role in determining the competitive ability of DI particles, because large DI particles could supersede small DI particles during the evolution of a culture.

Base substitutions were also observed in the terminal regions of the DI genomes (some in the complementary region), and comparison of these base substitutions with the substitution present in the 5'-terminal region of the various Sdi⁻ mutants suggested that DI particles with specific terminal base substitutions were selected during virus evolution of persistent or acutely infected cultures. In fact, less than 1% of the bases in the region of DI genomes homologous or complementary to the first 170 bases of the 5' terminus of standard virus differed from those found at the same stage of persistent or acute infection in the homologous nondefective viral 5' terminus. In all cases, the 3' end of the DI-particle genomes had base substitutions complementary to those at the 5' end in the region of complementarity. It appears, therefore, that DI particles with specific arrays of terminal complementary base substitutions have selective advantage. The selected array of terminal base changes is generally similar to that in the nondefective virus that predominates at that stage in the evolution of the culture. One DI particle was exceptional in exhibiting a clustering of specific A → G (and complementary U → C) substitutions, possibly indicative of repetitive misincorporation by an error-prone polymerase. Indeed, the genetic properties of rhabdoviruses are dependent predominantly on the characteristics of the virion-associated polymerase.

Wilusz et al. (1985) have determined the nucleotide sequences of the terminal noncoding regions of VSV isolated after long-term persistent infection of mouse L cells. Several mutations were observed in the 3'-terminal region, but no 5'-end mutations were detected, in contrast to the hypermutability observed in the region of the genome of VSV propagated as a persistent infection of BHK cells (O'Hara et al., 1984). This difference may be a consequence of the apparent absence of DI particles from the L-cell carrier cultures. The role of DI particles in the generation of variation in unsegmented-genome viruses deserves further study.

VI. GENETIC INTERACTIONS

A. Absence of Genetic Recombination

Although the reality of intermolecular recombination has been fully established for three positive-strand RNA viruses, the picornaviruses, poliovirus (Agol et al., 1984), foot-and-mouth disease virus (King et al.,

1982), and the coronavirus mouse hepatitis virus (Lai *et al.*, 1985), there have been no reports of recombination in negative-strand RNA viruses apart from subunit reassortment in those viruses that have segmented genomes. Experiments with temperature-sensitive (*ts*) mutants indicate that the frequency of recombination, if it occurs, cannot be greater than 10^{-5}. Attempts to isolate recombinants of vesicular stomatitis virus (VSV)-New Jersey by selection of wild-type virus from mixed infections with *ts* mutants distinguishable by the differential electrophoretic mobility of four of the five virion polypeptides were no more successful (Pringle *et al.*, 1981).

B. Complementation Groups

Spontaneous and mutagen-induced *ts* mutants of the Glasgow, Massachusetts, Orsay, and Winnipeg strains of VSV-Indiana (Flamand, 1970; Holloway *et al.*, 1970; Pringle, 1970; Rettenmier *et al.*, 1975), VSV-Cocal (Pringle and Wunner, 1973), VSV-New Jersey (Hazelhurst subtype) (Pringle *et al.*, 1971), VSV-New Jersey (Concan subtype) (Byrd *et al.*, 1984), and Chandipura virus (Gadkari and Pringle, 1980a) have been classified into complementation groups (Tables IIIa–d). Temperature-sensitive mutants of rabies virus have also been isolated, but complementation has not been convincingly demonstrated even between mutants with distinct phenotypic properties, and these mutants remain ungrouped (Bussereau and Flamand, 1978). The homologies of the complementation groups of the three different strains of VSV-Indiana were established precisely (Flamand and Pringle, 1971; Cormack *et al.*, 1973).

Brun (1984) has described mutants of Piry virus that are temperature-sensitive (at 28°C) in *Drosophila* and has provisionally classified them into complementation groups. The assay system is complicated, making use of the CO_2 sensitivity of adult insects, and is beyond the scope of this chapter. Piry virus was chosen for this work because it most resembles Sigma virus in the extent of the CO_2 sensitivity it induces in *Drosophila*. Piry virus appears to be under the control of the same host regulatory genes as Sigma virus. Although Sigma virus is unable to multiply in mammalian cells, Brun has suggested that Sigma is a vesiculovirus that became trapped in a nonbiting dipteran and came to behave as a hereditary factor to ensure its survival.

Complementation between *ts* mutants derived from different viruses is not observed, except for limited complementation between RNA-positive mutants of the serologically related VSV-Indiana and VSV-Cocal (Pringle and Wunner, 1973). Recently, Byrd *et al.* (1984) demonstrated efficient complementation between the Concan and Hazelhurst subtypes of VSV-New Jersey, but not between the Concan subtype and VSV-Indiana. A closer relationship between the Concan subtype and VSV-Indiana was suspected because the HR LT1 defective interfering particle of the

TABLE IIIa. Complementation Groups of VSV-Indiana

Strain	Origin[a]	Complementation group						Unclassified	Total
		I	II	III	IV	V	VI		
Orsay	Spontaneous	58	2	2	5	3	1	0	71
Orsay	5-FU, NTG, NA	18	0	2	1	1	0	0	22
Glasgow	5-FU, 5-AzaC, EMS	177	2	3	22	0	0	6	210
Winnipeg	5-FU, NA, EMS, P	7	0	2	2	0	0	14	25
Massachusetts	5-FU, NA, NTG	—	0	3	0	2	2	1	8
Totals:		260	4	12	30	6	3	21	336
Gene assignments:		L	NS	M	N	G	(NS)		

[a] (5-AzaC) 5-Azacytidine; (EMS) ethylmethane sulfonate; (5-FU) 5-fluorouracil; (NA) nitrous acid; (NTG) N-methyl-N′-nitro-N-nitrosoguanidine; (P) proflavine; (UV) ultraviolet light.

TABLE IIIb. Complementation Groups of VSV-Cocal

Strain	Origin	Complementation group				Unclassified	Total
		α	β	γ	δ		
Glasgow	5-FU	29	2	3	6	0	40
Gene assignments:		—	—	G	—		

TABLE IIIc. Complementation Groups of VSV-New Jersey

Subtype	Strain	Origin	Complementation group						Unclassified	Total
			A	B	C	D	E	F		
Hazelhurst	Missouri	5-FU	17	21	4	1	3	2	1	49
Concan	Concan	5-FU, 5-AzaC, NA, UV	3	14	0	0	0	0	0	17
Totals:			20	35	4	1	3	2	1	66
Gene assignments:			N	L	M	(G?)	NS	L		

TABLE IIId. Complementation Groups of Chandipura Virus

Strain	Origin	Complementation group						Unclassified	Total
		ChI	ChII	ChIII	ChIV	ChV	ChVI		
Nagpur	5-FU	44	2	1	1	1	1	0	50
Gene assignments:		—	(G?)	—	—	(M?)	—		

latter virus cross-interferes with the Concan subtype but not with the Hazelhurst subtype, but the relationship was not reflected in an ability to cross-complement. It remains an anomaly that phenotypic mixing (see Section XI), which is a form of complementation, occurs between vesiculoviruses and a wide range of heterologous enveloped viruses, while complementation between ts mutants is restricted to serologically related viruses.

The occurrence of extensive intragenic complementation is responsible for the anomaly of six complementation groups and only five open reading frames in the genome. Analysis of a collection of 50 ts mutants of Chandipura virus resulted in identification of six groups (Gadkari and Pringle, 1980a) containing 44, 2, 1, 1, 1, and 1 mutant, respectively. Weak complementation was observed between individual members of the majority group (ChI) such that it could be subdivided into groups ChIa and ChIb. Intragroup complementation was most extensive in subgroup ChIb, and one mutant in this group complemented all but one (tsCh598) of the remaining mutants in group ChI. It is evident, therefore, that if mutant tsCh598 had not been isolated and included in the analysis, the number of complementation groups would have been increased to seven. Sporadic and limited intragenic complementation had been reported previously in analysis of the Orsay collection of spontaneous ts mutants (Flamand, 1970) and the induced Winnipeg mutants (Wong et al., 1972), but it was not observed in analysis of the larger collection of 5-fluorouracil-induced mutants of the Glasgow strain of VSV-Indiana. In circumstances in which intragenic complementation occurs, which depends mainly on the properties of the individual mutants, the number of complementation groups can be overestimated because intragenic and intergenic complementation cannot always be distinguished. The hypothesis of Gadkari and Pringle (1980a) that the sixth complementation group in three different rhabdoviruses (VSV-Indiana, VSV-New Jersey, and Chandipura virus) did not imply the existence of a sixth genome product has been confirmed by analysis of the functional properties of ts mutants of all three viruses and the complete sequencing of the genome of VSV-Indiana. The extent of intragenic complementation in group ChI of Chandipura virus suggests that the virion polymerase is a multimeric protein in its functional form.

Analysis of the functions of representative mutants of VSV-New Jersey suggested that mutants classified in complementation groups B and F were L-protein mutants, leaving the remaining four viral proteins to be assigned to groups A, C, D, and E. Conclusive gene assignments have been made in the case of three of the four (see Section VII). In VSV-Indiana, Deutsch et al. (1979) have argued from UV-rescue experiments that the solitary mutant (ts082) representing group VI in the Orsay collection of spontaneous ts mutants is an NS-protein mutant, thus reducing the number of intergenic complementation groups to five. The RNA-negative mutants comprising the sixth group of Rettenmier et al. (1975) have not been assigned, and it is not known whether they are the result of mutation in the L, NS, or another viral protein specifying a multifunctional or multimeric protein. In both these cases, the second product encoded in the NS gene of VSV-Indiana (Herman, 1986) may account for the enigmatic sixth complementation group tentatively identified by both Rettenmier et al. (1975) and Flamand (1980).

VII. GENE ASSIGNMENT

A. Homologies of the Complementation Groups

The gene assignments of the temperature-sensitive (ts) mutants of the complementation groups of vesicular stomatitis virus (VSV)-Indiana, VSV-Cocal, VSV-New Jersey, and Chandipura virus based on functional analyses have been reviewed comprehensively (Pringle and Szilagyi, 1980; Pringle, 1982). These data are summarized in Table IV, supplemented with more recent information that mainly concerns VSV-New Jersey. The increasing ease and speed of nucleotide-sequencing techniques promise to revolutionize this area.

The only assignment among the VSV-Indiana and VSV-New Jersey complementation groups that has defied functional analysis and limited sequencing is the single mutant that defines complementation group D of VSV-New Jersey. By exclusion, its temperature sensitivity should be the consequence of a lesion in the glycoprotein (G) gene, but there is no direct evidence of this. Mutant tsD1 uniquely exhibited atypical electrophoretic mobility of the G and nucleocapsid (N) virion polypeptides, and each of these characteristics could apparently revert to wild-type independently of the ts phenotype. Additional phenotypes, some involving the nonstructural (NS) polypeptide, appeared during sequential cloning of this mutant, indicating that mutations affecting polypeptide mobility were generated at high frequency during propagation of tsD1. Furthermore, mutations affecting the electrophoretic mobility of the G, N, NS, and matrix (M) polypeptides were induced in heterologous rhabdoviruses multiplying in the same cells. Recombintion and posttranslational modification of proteins were not involved in these phenotypes. Virions of tsD1 with complete or incomplete genomes appeared to be equally competent in this respect. It was concluded that tsD1 fortuitously carried a "mutator"-type mutation that could act both in cis and in trans.

Aberrant glycosylation could have accounted for the atypical mobility of some of the G-protein mutants, whereas initially the N-protein mutants exhibited faster mobilities in polyacrylamide gel, suggesting a smaller polypeptide. However, determination of the nucleotide sequence from each terminus of the N gene to about 200 nucleotides of one of these N-protein mutants, of a revertant, and of the wild-type parent did not reveal any changes compatible with synthesis of a shorter polypeptide, as a result of either premature termination or late initiation of translation. The mutant (and its revertant), however, differed from wild-type at two positions: a C → U change in the noncoding leader RNA and an A → G

TABLE IV. Gene Assignments

Virus	Group	RNA phenotype	Assignment	Critical evidence
VSV-Indiana	I	−	L	Thermolability of L in in vitro polymerase reconstitution (Hunt et al., 1976; Pringle and Szilagyi, 1980)
	II	− and +	NS	Partial proteolysis (Metzel and Reichmann, 1981) Tryptic peptide mapping (Lafay and Benejean, 1981)
	III	+	M	Partial proteolysis (Metzel and Reichmann, 1981) Tryptic peptide mapping (Lafay and Benejean, 1981) RNA heteroduplex mapping (Freeman and Huang, 1981) Nucleotide sequencing (Gopala krishna and Lenard, 1985)
	IV	−	N	Partial proteolysis (Metzel and Reichmann, 1981) Tryptic peptide mapping (Freeman and Huang, 1981)
	V	+	G	Rescue by pseudotype formation (Zavada, 1972) G protein in vivo instability (Knipe et al., 1977) Nucleotide sequencing and expression in vitro (Gallione and Rose, 1985)
	VI	+	(NS)	Phenotypic resemblance to group II in UV-inactivation experiments (Deutsch et al., 1979) Second ORF (Herman, 1986)
VSV-New Jersey	A	−	N	Thermolability, aberrant electrophoretic mobility (Byrd et al., 1984) Intracellular instability of N; thermolability of template activity in reconstituted N/ RNA complexes (Marks et al., 1985)
	B	−	L	In vitro polymerase reconstitution (Belle-Isle and Emereson, 1982; (Ongradi; et al., 1985b) Reversible thermolability in vitro (Pringle and Szilagyi, 1980)
	C	+	M	In vivo degradation, peptide mapping (Kennedy-Morrow and Lesnaw, 1984)

(continued)

TABLE IV. (*Continued*)

Virus	Group	RNA phenotype	Assignment	Critical evidence
	D	+	(G)	None. Assignment by exclusion only
	E	– and +	NS	Aberrant electrophoretic mobility (Evans *et al.*, 1979; Lesnaw *et al.*, 1979)
				In vitro polymerase reconstitution (Ongradi *et al.*, 1985a)
				Tryptic peptide maps (Maack and Penhoet, 1980)
				Nucleotide sequencing [McGeoch *et al.*, 1980 (Table VII); Rae and Elliott, 1986b (Fig. 7)]
	F	– and +	L	*In vitro* polymerase reconstitution (Belle-Isle and Emerson, 1982; Ongradi *et al.*, 1985)
VSV-Cocal	γ	+	G	Glycosylation defect (Buller, 1975; Kotwal *et al.*, 1986)
Chandipura	ChV	+	M	*In vivo* degradation (Gadkari and Pringle, 1980b)

change corresponding to an Asn → Asp substitution. The latter could not be responsible for the atypical phenotype, since it was present in both mutant and revertant.

The occurrence of two nucleotide changes within a stretch of approximately 400 extrapolates to 56 changes within a genome of 11,162 nucleotides, assuming random distribution. If this is compared with the estimate of Rowlands *et al.* (1980) of 200 changes accumulating in mutant *ts*G31(III) of VSV-Indiana maintained as a persistent infection over a period of 5 years, it is clear that rapid change in nucleotide sequence can occur during a few cycles of lytic infection.

Curiously, reversion of the G-protein aberrant mobility was never obtained without prior reversion of the N protein to a normal-mobility phenotype.

B. Large-Protein Mutants

1. Temperature-Sensitive Mutants

Temperature-sensitive mutants of complementation group I of VSV-Indiana, and groups B and F of VSV-New Jersey, have been unequivocally identified as large (L)-protein mutants (Table IV). In most cases, the assignment has been made in terms of reconstitution of *in vitro* preparations.

Hunt (1983) observed that the group I mutant *ts*G16(I) produced mRNA *in vitro* with longer poly(A) tracts than either wild-type virus or some other *ts* mutants including *ts*G13(I), another group I mutant. The overproduction of poly(A) tracts occurred at all temperatures. Homologous and heterologous reconstitution experiments showed that the atypical polyadenylation phenotype correlated with the presence of *ts*G16(I) L protein and not its NS or N protein (Hunt *et al.*, 1984). Poly(A) synthesis, measured as the ratio of AMP to UMP incorporation, was temperature-sensitive in reconstitution experiments whenever *ts*G16(I) L protein was present. Although a non-*ts* revertant of *ts*G16(I) was not included in this analysis, the evidence is consistent with the hypothesis that the L protein is associated with the atypical polyadenylation phenotype. Control experiments excluded any possibility that the phenotype might be due to contamination of the L-protein fraction by M protein, and the temperature sensitivity of the L protein in reconstitution experiments eliminated extraneous host components. These experiments demonstrate that the L protein of VSV affects the polyadenylation of mRNA, but do not prove that the VSV L protein is in fact the poly(A) polymerase.

2. Conditional Lethal Mutants

Many of the conditionally temperature-sensitive mutants of VSV and Chandipura virus that are unable to multiply in chick-embryo cells at 39°C, the *td*CE phenotype, exhibit thermosensitive *in vitro* polymerase activity. Mutant *td*CE3 of VSV-Indiana has been shown unequivocally by reconstitution experiments to be an L-protein mutant (Szilagyi *et al.*, 1977). These mutants, and also mutant *ts*B1 of VSV-New Jersey, which is also restricted in chick-embryo cells at 39°C, differ from the other conventional *ts* mutants because the thermosensitive defectiveness of *in vitro* virion polymerase activity is reversible, suggesting a conformational change, and is not the result of inactivation (Szilagyi and Pringle, 1979; J. F. Szilagyi, personal communication).

Frey and Youngner (1982) have described group I *ts* mutants of the Orsay strain of VSV-Indiana with a distinctive phenotype. These mutants originated from two independent cultures of persistently infected mouse L cells initiated by *ts*O23(III), a group III mutant. At permissive temperature, replicative RNA synthesis was enhanced and mRNA synthesis diminished in these mutants relative to wild-type or *ts*O23(III), although both were severely restricted at 40°C. Normal RNA synthesis was restored in non-*ts* revertants of these mutants. Paradoxically, protein synthesis was normal or even enhanced in mutant-infected cells despite the reduction in mRNA synthesis. It was suggested that these properties together with reduced inhibition of host macromolecular synthesis facilitate persistent infection. However, since mutants with this phenotype were subsequently isolated from stocks propagated by lytic infection, the evidence is not compelling.

3. Host-Range Mutants

The host-range (*hr*) mutants isolated by Simpson and Obijeski (1974), which multiply in chick-embryo cells but are restricted in human cells, were tentatively identified as polymerase mutants on the basis of their phenotypic properties. Horikami and Moyer (1982) have shown that one of these mutants, *hr*1, was defective for methylation of mRNA *in vitro*, although full-length polyadenylated mRNA with unmethylated guanylylated 5' termini were synthesized in normal amounts, and another mutant, *hr*8, was partially defective for methylation. The unmethylated mRNA synthesized by *hr*1 was translated less efficiently *in vitro*.

Horikami *et al.* (1984) found that synthesis of mRNA in restrictive Hep-2 cells infected by these two mutants was diminished and the residual guanylylated but unmethylated mRNA present was not translated *in vitro*. These results suggest that the methyltransferase activities associated with VSV are virus-coded and that there is a strict requirement for methylation of the guanosine residue of the cap structure. Revertants of these mutants have not been isolated and characterized. Horikami and colleagues also observed that coinfection of restrictive cells with an *hr* mutant and rabbitpox virus released the restriction and resulted in the synthesis of mRNA with fully methylated caps, presumably through the action of the poxvirus methyltransferase enzyme.

Two other *hr* mutants, *hr*5 and *hr*7, however, appeared to synthesize methylated mRNA in restricted cells and appeared to be blocked at the level of replication. These mutants were only poorly rescued by poxvirus coinfection. It is unlikely, therefore, that defective methylation is directly responsible for the host-range phenotype. Furthermore, permissive cells must be able to overcome the methylation defect, perhaps by activation of the viral enzyme. There is no evidence of a methylation defect associated with the *td*CE conditional host-range phenotype (Rasool and Pringle, 1986).

4. Suppressible Amber Mutants

White and McGeoch (1985) have described isolation of two putative amber mutants by screening mutagenized virus on cell lines possessing or lacking a functional amber suppressor to tRNA[tyr]. The mutants grow to high titer on su[+] cells but are restricted on su[-] cells. One mutant has been characterized and the mutational lesion localized to the L gene by complementation tests using *ts* mutants of the five complementation groups. No viral proteins are synthesized in su[-] cells, but in su[+] cells the 200-kd L protein is present in reduced amounts together with a novel 37-

kd protein which is possibly a truncated L protein. The properties of this mutant suggest that it has arisen by an amber mutation in the L gene.

C. Glycoprotein Mutants

1. Vesiculovirus Mutants

a. Glycosylation Defects

The *ts* mutants classified in group V of VSV-Indiana and group γ of VSV-Cocal are G-protein mutants on the basis of their phenotypic properties (Table IV). Other G-protein mutants include the *tl*17 mutant obtained by Zavada (1982), which shows extreme thermolability and was derived by cycles of mutagenesis and neutralization by specific antiserum. The *tl*17 mutant is a complex mutant, however, failing to complement *ts* mutants of both groups I and V. Robertson *et al.* (1982) found that the glycosylation sites and oligosaccharide structure of this mutant were unaltered, but the content of sialic acid and fucose was increased and small amounts of an additional branched side chain were present.

The *ts*γ1 mutant of VSV-Cocal, on the other hand, has a temperature-sensitive defect in glycosylation (Buller, 1975; Kotwal *et al.*, 1986). At permissive temperature only one of the two available sites is glycosylated. G protein containing only one oligosaccharide residue is still transported to the plasma membrane and is incorporated into infectious particles. At nonpermissive temperature neither of the two sites become glycosylated and the unglycosylated G protein is not transported to the cell surface. These results indicate that at least for VSV-Cocal glycosylation of the G protein is essential for its transport to the plasma membrane, and the presence of a single carbohydrate chain in sufficient for this purpose. The requirement for addition of carbohydrate for transport of the G protein from its site of synthesis on the endoplasmic reticulum to the cell membrane is strain-dependent. Transport was dependent on glycosylation at all temperatures in the San Juan strain of VSV-Indiana, but only at 39.8°C in the Orsay strain. Mutant *ts*011(V), one of two group V *ts* mutants derived from the Orsay strain, was not dependent on glycosylation at 39.8°C. The other, *ts*044(V), was dependent but became independent on reversion of the *ts* phenotype (Chatis and Morrison, 1981). These results indicate that glycosylation can be affected by simple genetic mutations and that the conformation of the polypeptide chain presumably determines whether addition of carbohydrates is necessary for migration of the G protein to the plasma membrane. Zilberstein *et al.* (1980) and Bergmann *et al.* (1981) analyzed the site of action of several group V *ts* mutants and concluded that three [*ts*045(V), *ts*L513(V), *ts*L501(V)] of four were not transported to the Golgi apparatus after synthesis on the endoplasmic reticulum. The fourth, *ts*L511(V), was unique among virus mutants in being blocked at a stage subsequent to migration to the Golgi apparatus.

b. Site of the ts045(V) Mutation

Gallione and Rose (1985) have determined the site of the ts045(V) mutational lesion responsible for conditional transport of G protein to the cell surface in a series of ingenious experiments with cloned cDNA copies of G-protein mRNA of ts045(V), a spontaneous non-ts revertant of ts045(V), and wild-type virus. The nucleotide sequence of each was determined with the results shown in Table Va. The mutant differed from wild-type at nine sites, producing five predicted amino acid substitutions. Surprisingly, in the revertant, three of the amino acid changes had reverted to wild-type. Since one or all of these changes could be responsible for the reversion of phenotype, Gallione and Rose constructed recombinant molecules *in vitro* and used an SV40-based vector to obtain expression

TABLE Va. Nucleotide Changes in mRNA and Predicted Amino Acid Substitutions in the G Protein for ts045(V) and a Non-ts Revertant[a]

	Residue no.				
Genotype	50	200	204	394	434
Wild-type (VSV-Indiana,	AAU	ACG	UUC	UUA	
Orsay strain)	Asn	Thr	Phe	Leu	—
ts045(V)	AAG	AUG	UCC	UCA	UUU
	Lys	Met	Ser	Ser	+ Phe[b]
Non-ts revertant	AAU	AUG	UUC	UCA	
	Asn	Met	Phe	Ser	—

[a]From Gallione and Rose (1985). [b](+ Phe) Insertion.

TABLE Vb. Temperature Sensitivity of Surface Antigen Expression at 39.8°C of COS-1 Cells Carrying Recombinant Constructs with Permutation of the Three Mutational Changes Associated with the ts045 Phenotype[a]

Predicted amino acid phenotype of recombinant			
Residue no.			
50	204	434	Temperature sensitivity of surface-antigen expression at 39.8°C
Lys	Ser	—	Sensitive
Asn	Phe	+ Phe[b]	Resistant
Lys	Phe	—	Resistant
Asn	Ser	+ Phe	Sensitive
Lys	Phe	+ Phe	Resistant
Asn	Ser	—	Sensitive

Conclusion: the Phe → Ser substitution at site 204 is responsible for the ts phenotype of mutant ts045(V).

[a] From Gallione and Rose (1985).[b] (+ Phe) Insertion.

of these recombinants in COS-1 cells. Expression was monitored by indirect immunofluoresence. Surface expression of G protein in COS-1 cells carrying the ts045(V) expression vector was temperature-sensitive, whereas it was not in the wild-type and revertant. Three sets of reciprocal recombinant molecules were constructed as indicated in Table Vb, and the temperature sensitivity of expression was assayed by immunofluorescence. The results show that a single substitution of phenylalanine by serine at site 204 is sufficient to prevent transport of G protein to the cell surface at 39.8°C. It is presumed that the co-reversion observed in the revertant reflects heterogeneity of the wild-type stock; however, other explanations are possible and the two apparently silent reverse mutations may affect some other important G-protein function. Previously, Rose and Bergmann (1983) had shown by in vitro mutagenesis that the 29-amino-acid cytoplasmic domain of the G protein was critical for efficient transport. Location of the ts045(V) mutation shows that other regions of the protein are also important.

Synthetic peptides corresponding to the amino-terminal region of the G protein have a pH-dependent hemolytic activity. However, oligonucleotide-directed mutagenesis has shown that this hemolytic activity is not synonymous with the membrane fusion function of the G protein. Woodgett and Rose (1986), by in vitro mutagenesis, produced a DNA-encoding G protein with a lysine–glutamic acid change at the amino terminus of the protein. The same amino acid substitution abolishes the hemolytic activity of the synthetic peptides, but the mutant protein expressed transiently in COS cells retained the pH-dependent fusion activity of wild-type virus.

It has been shown that at 39°C the G protein synthesized by mutant ts045(V) remains in an endoglycosidase H-sensitive state and is gradually degraded. Noninfectious virions lacking spikes are released to the exterior. Further characterization of the ts045(V) mutant has revealed that at 39°C the G protein is converted to the truncated soluble G_s form which is fully glycosylated but lacks the cytoplasmic and transmembrane portion of the molecule and which accumulates in the extracellular medium in the course of normal infection (Chen and Huang, 1986). Since G protein in the mutant is not transported to the surface, the G to G_s cleavage must occur intracellularly rather than at the surface as previously thought. There is some evidence that the other product of the cleavage reaction, the cytoplasmic tail and transmembrane region, is transported to the membrane and is involved in the budding of virions.

2. Lyssavirus Mutants

a. Monoclonal-Antibody-Resistant Mutants

Since the rhabdovirus G protein is the site of the antigen that induces neutralizing antibody, selection for resistance to neutralizing monoclonal antibodies has generated collections of G-protein mutants. Lafon et al.

(1983), using a battery of 24 neutralizing monoclonal antibodies to the CVS strain of rabies virus, have derived an epitope map of the rabies virus G protein, which has three functionally independent antigenic sites: I; IIA, IIB, IIC; and IIIA, IIIB.

Coulon et al. (1983) reported that nine of ten mutants of the CVS strain of rabies virus resistant to two monoclonal antibodies, 194-2 and 248-8, were either nonpathogenic or reduced in virulence for adult Swiss mice. These monoclonal antibodies were specific for the antigenic site III defined by Lafon et al. (1983). Dietzschold et al. (1983) confirmed and extended these observations and found that an amino acid substitution of arginine at position 333 in the G protein had occurred in all four neutralization-resistant nonpathogenic mutants examined by them. Arginine at position 333 was replaced by isoleucine in three mutants of the ERA strain and by glutamine in one mutant of the CVS strain.

b. Pathogenicity

Seif et al. (1985) have isolated a further 58 mutants of the CVS strain of rabies virus selected using four site-specific monoclonal antibodies to test the hypothesis that loss of pathogenicity is related to the amino acid substitution at position 333 of the G protein. These mutants were classified into five groups in terms of their pattern of resistance to the four monoclonal antibodies. Only mutants in one of the five groups (group II) were nonpathogenic or reduced in virulence. Eight of the nonpathogenic mutants were examined, and it was established by dideoxy sequencing on the RNA genome from a synthetic oligonucleotide primer that the arginine at position 333 was substituted by glutamine or glycine. In eight mutants that retained full pathogenicity, amino acid substitutions had occurred at positions 330 (lysine), 336 (asparagine), and 338 (isoleucine), but not at position 333. Thus, although amino acid substitutions in this region of the G protein alter neutralization, only substitution at position 333 (arginine) affects pathogenicity. The mechanism responsible for the loss of pathogenicity is not known.

3. Mutation and Secondary Structure

Wunner et al. (1985) have located the mutational lesion in two neutralization-resistant mutants representing epitopes IIA and IIIA by mapping of tryptic peptides and sequence analysis. The wild-type G protein exists in two forms, GI and GII, that differ only in glycosylation. Two of the four glycosylation sites from the nucleotide sequences predicted are used in GI and only one in GII. A single G-protein band corresponding to GI was present in one of the mutants, RV231-22. Sequence analysis of this mutant revealed a base mutation specifying an amino acid change six residues upstream from the predicted glycosylation site utilized in the GI form of G, but not in the GII form. Apparently, this amino acid substitution allows utilization of this glycosylation site in the GII form

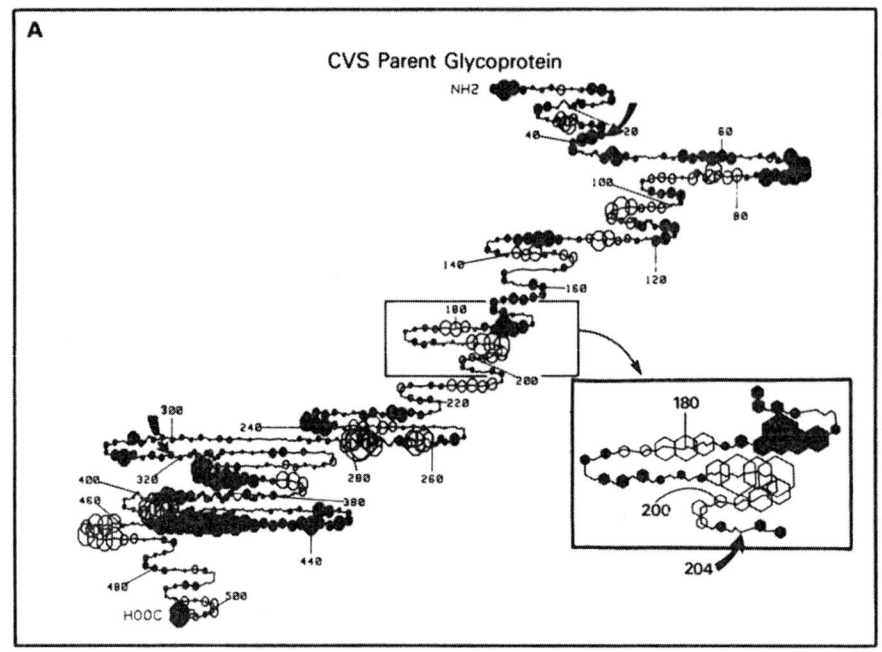

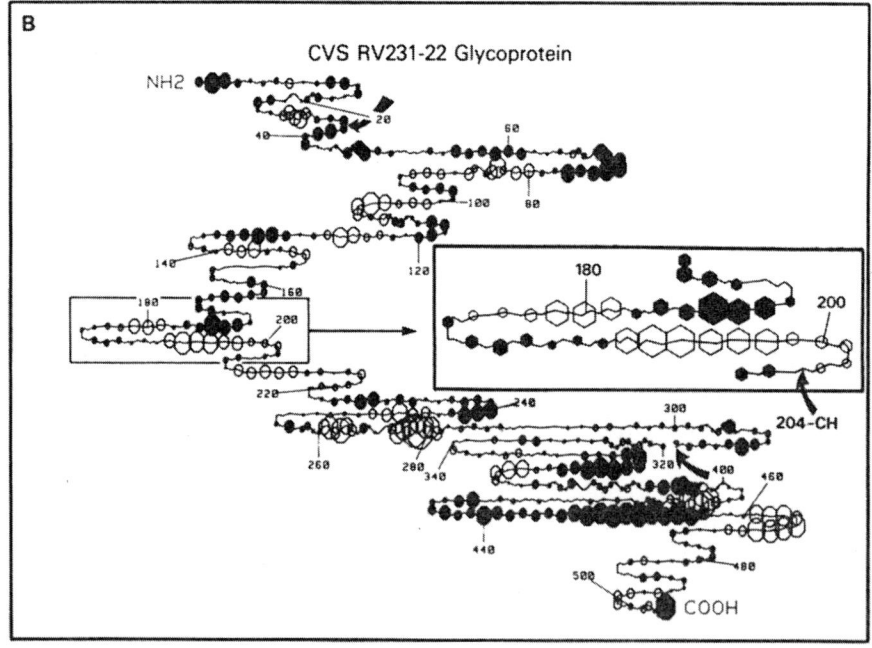

FIGURE 5. Predicted secondary structure and hydrophobicity of the G protein of the parental (CVS-11) strain of rabies virus (A) and the monoclonal-antibody-resistant mutant RV231-22 (B). (○) Hydrophilic regions; (●) hydrophobic regions. The radius of the circle over a residue site is proportional to the average hydrophilicity or hydrophobicity as calculated for that residue and the next five residues. The numbers correspond to amino acid residues from the NH$_2$-terminal lysine of the mature G protein. Heavy arrows indicate potential glycosylation sites at asparagine residues 37, 204, and 319. The enlarged insert shows the region of the mutation. Reproduced from Wunner *et al.* (1985).

of mutant RV231-22, resulting in comigration with GI and giving the observed single band in gel electrophoresis.

In the case of the other mutant, RV194-2, which represents antigenic site IIIA, both forms of G protein were present, although they differed slightly from wild-type in electrophoretic mobility. An extra glycosylation site was identified in both forms of G protein in this mutant. A single amino acid substitution of lysine by asparagine creates a new glycosylation site at residue 158 that is in fact utilized in this variant. However, this mutation was not observed in another variant, RV194-2(F2), from the same group, and it is unlikely to be solely responsible for the neutralization-resistant phenotype.

These results show clearly that not all predicted glycosylation sites are utilized and that the extent of glycosylation is probably dependent on the secondary structure of the molecule. A computer prediction of the conformational change in the region of the amino acid substitution in mutant RV231-22 is illustrated in Fig. 5. One of the four β-turns assigned to the predicted sequence of the wild-type G protein has been lost by the substitution of glutamic acid for lysine at residue 198, although local hydrophilicity and hydrophobicity are unchanged. The elimination of the β-turn in the vicinity of the glycosylation site apparently renders it more accessible for carbohydrate attachment. This analysis represented the first attempt to assess the precise effect of amino acid substitution on the conformational state of a rhabdovirus protein and a forerunner to the use of more advanced techniques in analysis of viral-gene function.

4. Enrichment Selection

The group V mutants initially were all derived from the Orsay strain of VSV-Indiana as spontaneous mutants, and none was found among 177 *ts* mutants of the Glasgow strain derived by mutagenesis (see Table IIIa). Later, other group V mutants were obtained by enrichment techniques. Lodish and Weiss (1979) have described a procedure for selective isolation of mutants defective in synthesis of G protein, which depends on rescue of pseudotypes of a group V *ts* mutant in cells preinfected with Rous-associated virus 1 (RAV-1) and incubated at restrictve temperature. This procedure also somewhat unexpectedly resulted in recovery of a slightly higher proportion of group IV N-protein mutants, which suggested that reduced synthesis of nucleocapsid resulted in a deficit of G protein at the surface.

D. Matrix-Protein Mutants

1. Temperature-Sensitive Mutants

The *ts* mutants of group III of VSV-Indiana, group C of VSV-New Jersey, and group ChV of Chandipura virus are M-protein mutants. Some of these mutants have been exploited experimentally in different situations. The M protein appears to be associated with both the plasma membrane and the viral nucleocapsid, and Lenard *et al.* (1981) reported that there was an enhanced association between the VSV-Indiana group III mutants and the plasma membrane in contradistinction to wild-type and non-*ts* revertants of these mutants. These observations suggest that the M-protein molecule has two independent binding sites and that one of these sites was preferentially affected in the mutants studied.

The M-protein mutants of all three viruses overproduce mRNA, and direct experimentation suggests that M protein is involved in control of transcription. The M protein of VSV plays a regulatory role in viral RNA synthesis in addition to its structural role. M protein from wild-type virus inhibits transcription *in vitro*, whereas M protein from group III mutants has no [*ts*023(III) and *ts*089(III)] or low [*ts*G31(III) and *ts*G33(III)] inhibitory activity. RNP-M protein interactions have been shown to be weaker for these four mutants (Carroll and Wagner, 1979; Wilson and Lenard, 1981). Pal *et al.* (1986) have reported that monoclonal antibody to epitope 1, but not antibody to epitopes 2 and 3, abolished the inhibitory activity of wild-type M protein. Monoclonal antibody to epitope 1 failed to bind to M protein from mutant *ts*023(III), whereas binding and inhibitory activity were restored by reversion of this mutant. These results indicate that exposure of epitope 1 on the surface of the M protein is necessary for inhibitory activity.

Experiments with *ts* mutants indicate that M protein plays a determining role in the polykaryocyte formation observed in certain host cells in place of the normal cytolytic response. The M-protein-defective mutant *ts*G31(III), unlike *ts* mutants representing the other four complementation groups, induced cell fusion at nonpermissive temperature (Storey and Yong Kang, 1985). Polykaryocyte formation can also be induced by inhibition of protein synthesis shortly after infection and it has been shown that the intracellular M-protein pool is the first to be depleted. It is likely, therefore, that cell fusion occurs when functional M protein falls below a critical level while G protein continues to be present in the surface membrane.

The nucleotide sequences of group III mutants, two spontaneous mutants (*ts*023 and *ts*089) derived from the Orsay wild-type and two 5-Fu-induced mutants (*ts*G31 and *ts*G33) derived from the Glasgow wild-type, have been published by Gopalakrishna and Lenard (1985). Each of the mutants differed from wild-type in one or two point mutations, all of which caused a change from or to a charged amino acid. The mutations were distributed throughout the coding sequence in contradistinction to the VSV-New Jersey NS gene mutations.

Analysis of a series of 25 non-*ts* revertants of these mutants has revealed greater complexity (J. Lenard, personal communication). Additional nucleotide changes have been found in each mutant such that *ts*023(III) carries three nucleotide changes producing three amino acid exchanges, *ts*089(III) carries two coding changes, *ts*G31(III) carries two coding and one noncoding change, and tsG33(III) carries three coding and one noncoding change. Twenty-four of the 25 revertants differ from their parent mutant by a single change only. In four cases only does this involve one of the mutated codons in the parental mutant; in the remainder the nucleotide changes are located at different sites producing a further coding change. It is evident that the M protein can tolerate mutation at many sites and that reversion occurs predominantly by intragenic suppression. It will be necessary to sequence low-passage material to determine unambiguously the primary site of mutation.

Small spherical defective interfering (DI) particles in which 90% of the genome is deleted are uniquely associated with the three 5-fluorouracil-induced group III mutants tsG31(III), tsG32(III), and tsG33(III). One of these mutants [tsG31(III)], in the presence of its homologous DI virus at semipermissive temperature to suppress cytopathogenicity, was used by Holland and Villareal (1974) to initiate long-term persistent infection of BHK-21 cells. This mutant and its DI particles were more manageable experimentally than other combinations of virus and DI particles.

Several ts mutants of VSV-Indiana induced unique forms of central nervous system (CNS) disease in mice (for a summary, see Pringle, 1982). Wild-type VSV inoculated intracerebrally induced fatal acute disease in 3 to 4-week-old mice. Mutant tsG31(III) and others induced a slow, progressive neurological disease characterised by hindlimb paralysis and a marked spongiform myelopathy in the spinal cord. Hughes and Johnson (1981) have shown by limited digestion with V8 protease that the M protein of tsG31(III) can be distinguished from wild-type M protein. A variant of tsG31(III) designated tsG31BP was isolated from the CNS and considered to be responsible for the CNS disease. This variant, which was a complex mutant unble to complement, had a wild-type-like M protein. By peptide mapping, its N protein was clearly different from that of either tsG31(III) or wild-type virus, and it exhibited abnormal nucleocapsid structure, as revealed by an atypical hyperchromatic shift on heating and other properties. This N-protein alteration and the reversion of the M protein may be associated, separately or together, with the unique spongiform myelopathy and hindlimb paralysis that follows intracerebral inoculation of tsG31(III).

2. P-Function Mutants

Stanners et $al.$ (1977) have identified mutations in a viral function, P, involved in the inhibition of host protein synthesis (see Section VIII.C). P⁻ mutants are defective in ability to inhibit host protein synthesis. There is circumstantial evidence that the P function may be located in the M gene, since the P⁻ tsT1026 and its non-tsP⁻ revertants exhibit a small change in the electrophoretic mobility of the virion M protein (Lodish and Porter, 1981); the mobility of the M polypeptide of these mutants is about 10% faster than that of the M protein of the HR wild-type from which they were derived. Stanners and colleagues (personal communication) have now characterized a further series of P⁻ mutants, one of which exhibits a similar M-protein electrophoretic-mobility shift.

E. Nonstructural Protein Mutants

Ths *ts* mutants of group II (and probably group VI) of VSV-Indiana, and group E of VSV-New Jersey, are NS-protein mutants. The group E mutants are remarkable in that the NS protein of each of the three mutants comprising this group has a distinctive electrophoretic mobility that reverts to wild-type mobility with reversion of the ts phenotype of *ts*E1 and *ts*E3; revertants of *ts*E2 have not been obtained. Therefore, these *ts* mutations appear to be directly responsible for the atypical mobility phenotype. Gross differences in phosphorylation were not responsible for the mobility differences (Evans *et al.*, 1979). Analysis of the temperature sensitivity of the *in vitro* polymerase activity of these *ts* mutants revealed that at 39°C, *ts*E1 was almost completely defective, *ts*E2 was partially defective, and *ts*E3 was not significantly different from wild-type. A non-*ts* revertant of *ts*E1 was restored to wild-type activity, indicating that the ts phenotype, the polymerase defect, and the NS-protein mobility difference were all correlated. The heat stabilities of the infectivity and polymerase activity of mutant *ts*E1 virions were equivalent to that of wild-type virions, the polymerase activity becoming heat-labile only after disruption of the virion and release of the core. Presumably, there is an interaction between the mutated polypeptide and an envelope protein in the virion that maintains the normal functional activity of the virion polymerase.

On the basis of these findings and the RNA phenotypes of the mutants, it has been proposed that the NS protein (like the other soluble component of the polymerase complex, the L protein) is a multifunctional protein that plays a role in mRNA transcription, genome replication, and virion maturation. The information on the properties of the RNA-negative *ts* mutants of VSV-Indiana and VSV-New Jersey is summarized in Table VI. The nature of the gene products transcribed under restrictive conditions by the nondefective mutants in these groups remains to be characterized.

About half the nucleotide sequence from the end of the NS gene of wild-type virus, the mutants *ts*E1, *ts*E2, and *ts*E3, and a non-*ts* revertant of *ts*E1 and *ts*E3 was determined by primer extension synthesis using a synthetic primer corresponding to a terminal sequence of the N gene (McGeoch *et al.*, unpublished data). A single base change producing an amino acid substitution in the predicted gene product was detected in all three group E mutants within a region of 18 nucleotides (positions 92–109 from the 5'-end of the mRNA). No other nucleotide changes were observed. The sequence of the two non-*ts* revertants was identical to wild-type, confirming the association of these nucleotide changes with the

TABLE VI. Phenotypes of Mutants of VSV in Complementation Groups Containing Mutants with Defects in RNA Synthesis

Virus	Complementation group	Gene assignment	Mutant	Phenotype at 39°C		
				Transcription	Replication	Heat stability
VSV-Indiana	I	L	tsG11(I)	Competent	Defective	Resistant
			tsG13(I)	Defective	Defective	Labile
	II	NS	tsG21(II)	Competent	Competent	Resistant
			tsG22(II)	Competent	Defective	Resistant
	IV	N	tsG41(IV)	Competent	Defective	Resistant
VSV-New Jersey	A	N	tsAl	Competent	Defective	Resistant
	B	L	tsBl	Defective	Defective	Labile but reversible
	E	NS	tsEl	Defective	Defective	Labile
			tsE2	Competent	Defective	Resistant
			tsE3	Competent	Competent	Resistant
	F	L	tsFl	Defective	Defective	Labile
			tsF2	Competent	Defective	Resistant

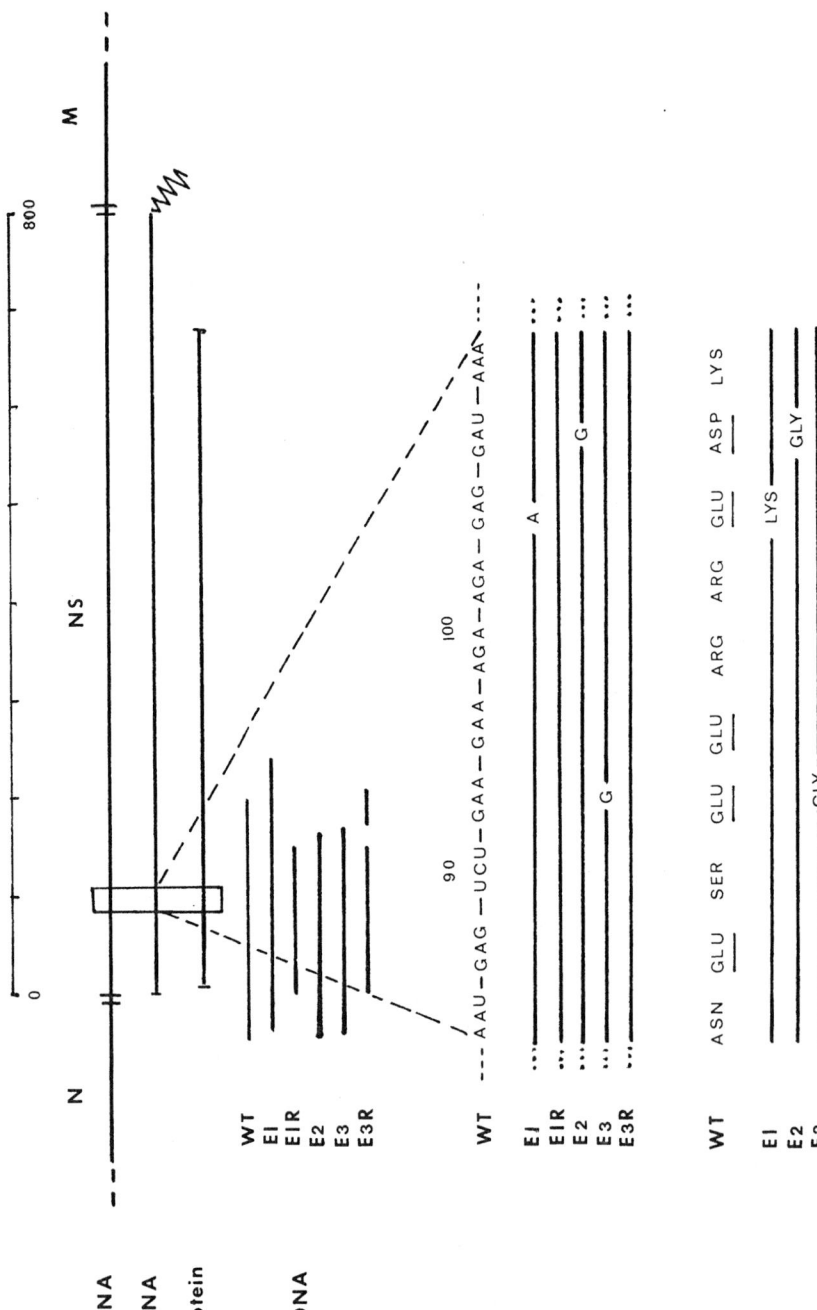

FIGURE 6. Location of single nucleotide changes in the NS gene of the three mutants that comprise complementation group E of VSV-New Jersey. (WT) Wild-type; (E1) *ts*E1; (E1R) a non-*ts* revertant of *ts*E1; (E2) *ts*E2; (E3) *ts*E3; (E3R) a non-*ts* revertant of *ts*E3. (This figure is based on unpublished data supplied by D. J. McGeoch and A. Dolan.)

phenotypic changes (Fig. 6). Although the sequence of these NS genes had still to be completed, it was reasonable to assume that the sites of the mutational lesions had been located to a region of six adjacent amino acids. Single amino acid substitution in this critical region drastically alters the functional activity of the NS protein.

Subsequently, full-length cDNA copies of all three mutant NS mRNAs have been obtained and their nucleotide sequence determined (Rae and Elliott, 1986b). No other nucleotide changes were detected in mutants tsE1 and tsE3, but a second change was detected at position 413 in mutant tsE2. The presence of a second mutation in tsE2 explains past failure to isolate non-ts revertants of this mutant. Rae and Elliott also cloned and sequenced full-length cDNA copies of the mRNA of non-ts revertants of tsE1 and tsE3. The revertant NS genes were identical to wild-type at the site of mutation. These results confirm that the mutations responsible for the three distinct phenotypes are clustered in a region of 18 nucleotides. Curiously, both non-ts revertants contained the same additional point mutation at position 513 producing a Met to Arg substitution. Figure 7 illustrates a prediction of the secondary structure of the wild-type and NS proteins according to the algorithm of Chou and Fasman (1978). The mutations responsible for the ts phenotype have marked effects on the

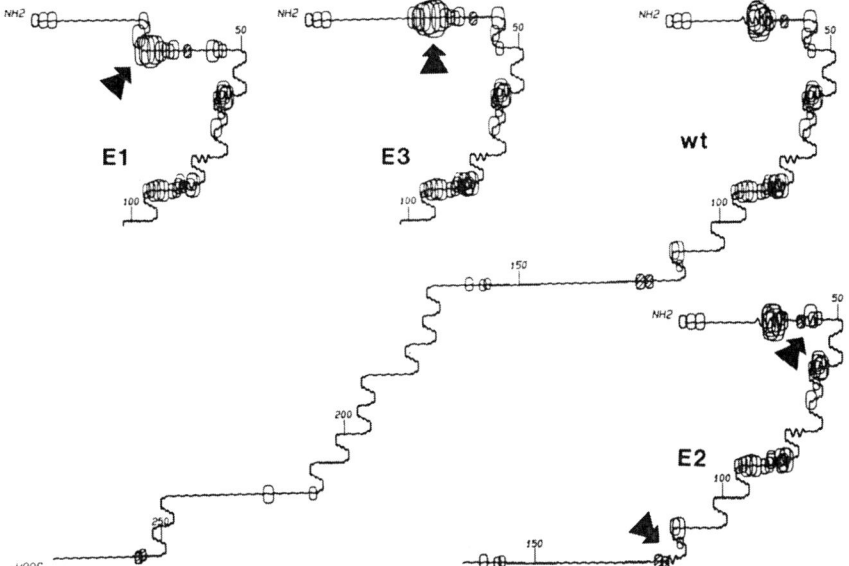

FIGURE 7. Secondary structures of the NS proteins of wild-type and complementation group E mutants of VSV-New Jersey predicted according to the method of Chou and Fasman (1978). The complete plot of wild-type (wt) NS protein is given, and the relevant portions of the plots of the group E mutants. The arrows indicate differences in the predicted secondary structures due to amino acid substitutions. ᴡ represents α helix, — represents β strand, — represents undefined, and a change of direction indicates β turn. The open ovals represent hydrophilic regions, and shaded ovals hydrophobic regions. Reproduced from Rae and Elliott (1986).

predicted secondary structure. The tsE1 mutation produces loss of a region of α-helix and changes in the position of a β-turn. The tsE3 mutation produces loss of the same α-helix without affecting the β-turn. The tsE2 mutation produces an additional region of α-helix and additional β-turns. (The second mutation in tsE2 at amino acid 114 and the two extraneous mutations in the revertants have little effect on the predicted secondary structure.) These effects on secondary structure may be responsible for the aberrant electrophoretic mobility of these mutants by affecting binding of sodium dodecyl sulfate.

Davis *et al.* (1986) have demonstrated that NS protein associated specifically with N protein in an *in vitro* replication system and that essentially all of the N protein was sequestered in N–NS complexes. This implies that these complexes are important in maintaining N protein in a functional state and in regulating the availability of N protein.

It has been proposed that different phosphorylated forms of NS protein may regulate RNA synthesis by their association with N protein rendering it active or inactive in replication. It is possible that the critical site identified by the complementation group E mutants plays a role in complex formation.

F. Nucleocapsid Protein Mutants

1. Temperature-Sensitive Mutants

The temperature-sensitive mutants classified in group IV of VSV-Indiana and group A of VSV-New Jersey are N-protein mutants. In addition to these mutants, some clones of mutant tsD1 of VSV-New Jersey have an N polypeptide that migrates in polyacrylamide gel faster than wild-type N, suggesting that the N polypeptide is around 10 amino acids shorter than normal. This characteristic is independent of the ts phenotype. Determination of the nucleotide sequence to about 200 acids from each terminus of the N gene did not reveal any changes compatible with synthesis of a shorter polypeptide by premature termination or late initiation of translation (Pringle *et al.*, 1981). Attacking this problem from a different direction, Brown and Prevec (1982), by specific chemical cleavages at tryptophan and methionine residues and enzymatic cleavage with carboxypeptidases A and B, concluded that residues near the carboxy terminus are responsible for the aberrant electrophoretic behavior of this mutant. No tryptic peptide differences were detected in this region, but a tryptic peptide difference was located outside this region. From these observations, it was concluded, first, that in agreement with the sequencing results, the missense point mutation responsible for the aberrant phenotype was located at least 40 residues from the carboxy terminus, and, second, that the altered electrophoretic mobility was the result of interaction between this site and a carboxy-terminal proximal region. Intramolecular interactions of this sort may be responsible for the majority of the electrophoretic variants generated by the tsD1 "mutator" polymerase.

2. Pol R Mutants

Perrault et al. (1981) have described mutants designated Pol R that generate unique DI particles. These DI particles read through the minus-strand leader termination signal in vitro, producing near full-length copies of the template RNA. These variants were obtained by a selection procedure involving several alternating cycles of heat inactivation and growth. The Pol R mutants were genetically stable following cloning by plaque isolation, but unexpectedly the virions were no more heat-resistant than wild-type. It was hypothesized that the Pol R viruses were mutated in a function involved in the switch from transcription to replication; in accordance with this hypothesis, it was later observed that the plus-strand leader–N gene junction was also read through efficiently. Perrault et al. (1983) have now shown by dissociation and reconstitution experiments that the mutation responsible for the readthrough phenotype is located in the N protein. Two-dimensional gel electrophoresis corroborated this conclusion, the magnitude of the pI shift observed with the N proteins of Pol R1 and Pol R2 being consistent with a single charge change. It is probable that the phenotype of Pol R1 and Pol R2 is the result of a similar single amino acid substitution in the N protein.

VIII. VIRUS–CELL INTERACTIONS

A. Cell-Killing Potential

Infection of host cells with vesiculoviruses generally results in a rapid cessation of host macromolecular synthesis. Marcus and Sekellick (1980) have used temperature-sensitive (ts) and deletion mutants to define the viral genes involved in the inhibition of cellular protein synthesis and the expression of cell killing. The cell-killing ability of (ts) mutants at permissive and restrictive temperatures was measured from survival curves of Vero cells infected at a range of multiplicities. Vesicular stomatitis virus (VSV) stocks on average contained 5 cell-killing units per plaque-forming unit. From UV-inactivation experiments, it was concluded that VSV virions are not toxic per se and that cell killing requires expression of only some of the five viral genes. Of 14 ts mutants examined, belonging to groups I, II, and IV, 6 did not induce cell killing at restrictive temperature, indicating that synthesis of functional core proteins is necessary for cell killing. Since the group II and group IV mutants tested were competent for primary transcription, it was concluded that primary transcription is necessary but not sufficient for cell killing. Loss of ability to inhibit cellular protein synthesis at restrictive temperatures correlated with loss of cell-killing ability and also with ability to induce interferon (Sekellick and Marcus, 1980), which suggested that the same gene function is involved in each case. Loss of ability of VSV to inhibit replication of pseudorabies virus, a nuclear DNA-containing virus, was also correlated with loss of ability to inhibit host protein synthesis.

From their mutant analysis, Marcus and Sekellick concluded that the proximate cell-killing factor was viral double-stranded RNA (dsRNA). Nilsen *et al.* (1981), although they could not detect viral dsRNA in wild-type infected HeLa cells using a cross-linking agent, were able to detect dsRNA by the same method in tsG114(I)- or tsG11(I)-infected cells following shift-up to restrictive temperature. These mutants were presumed to provide optimal conditions for detection of dsRNA, since mRNA synthesis but not replicative RNA synthesis is inhibited on shift-up (Wertz, 1978). These observations suggest that under such conditions, template RNA may not be complexed with protein and able to anneal with nascent complementary RNA.

Similarly highly purified preparations of nontranscribing defective interfering (DI) particles (i.e., 3'-end deletions) did not kill cells even at concentrations as high as 10^5 particles per cell (Doyle and Holland, 1973). It was found, as was originally predicted by Cooper and Bellett (1959) long before the discovery of DI particles, that suppression of viral replication by a transmissible interfering component did not provide protection against cell-killing particles.

B. Interferon Induction

Interferon treatment did not protect murine embryonal teratocarcinoma cells from infection with wild-type VSV. Nevertheless, Nilsen *et al.* (1981) observed that embryonal teratocarcinoma cells did exhibit interferon-induced resistance to infection by the mutants tsG11(I) and tsG114(I). The production of dsRNA by these two mutants under restrictive conditions appeared to be the common factor linking them with two picornaviruses that were also susceptible to the action of interferon in these cells. Interferon treatment of teratocarcinoma cells induces synthesis of the 2–5A oligonucleotide polymerase, but not the protein kinase activity observed in other interferon-treated cells, and this polymerase activity appears to be sufficient to protect these cells from viruses that induces dsRNA synthesis during infection.

Sekellick and Marcus (1980) observed that the mutants tsG11(I), tsG22(II), and tsG41(IV), which were defective for cell killing and inhibition of host protein synthesis, were also good inducers of interferon. In the case of tsG11(I) and the double-stranded DI particle (DI011), the maximal yield of interferon was obtained at a plaque-forming unit/cell ratio of 0.1–0.3, indicating that induction of interferon was probably a quantal effect and that interferon-inducing particles were present in greater numbers than infectious particles. Frey *et al.* (1981), on the other hand, comparing 12 DI particles with varying extents of double-strandedness, were unable to confirm this relationship. They did note, however, a correlation between interferon inducibility and the extent of virion contamination. It was concluded that ability to induce interferon is a determining factor in the initiation and maintenance of nonlytic persistent infection. The frequent association of *ts* mutants, predominantly of group I, or small-plaque mutants with persistent infections is consistent with this hypothesis.

C. Inhibition of Host Protein Synthesis

By systematic analysis of a particular temperature-sensitive mutant, *ts*T1026(I), and its non-*ts* revertants, Stanners *et al.* (1977) have identified a function designated the P function. This function is expressed by the normal VSV genome and is responsible for inhibition of protein synthesis in infected cells. Mutant *ts*T1026(I) is a group I *ts* mutant that is defective for inhibition of host protein synthesis. It is also a potent inducer of interferon and is able to initiate persistent infections in the absence of DI particles. Studies of non-*ts* revertants showed that *ts*T1026(I) is a complex mutant with a *ts* lesion in the L-protein gene and a non-*ts* lesion in the putative P function. These non-*ts* revertants, or R mutants, were defective for inhibition of host protein synthesis (P⁻), were efficient inducers of interferon, and failed to initiate persistent infections, unlike *ts*T1026(I) itself. Therefore, it was concluded that for the establishment of persistent infection, the P⁻ mutation is necessary to prevent inhibition of host-cell protein synthesis and the *ts* group I (L-protein) mutation to reduce transcription and cell-killing ability. Francoeur *et al.* (1980) later developed a procedure [the plaque interferon (PIF) assay] for selectively isolating P⁻ mutants that depends on the observation that the development of plaques initiated by the P⁻ mutants is arrested due to autoinduction of interferon, whereas plaques initiated by P⁺ virus increase in size linearly. One of these mutants has the same M-protein electrophoretic-mobility change observed in *ts*T1026(I) and its revertants. Lodish and Porter (1981), however, questioned the reality of the P function and argued that the rate of inhibition of host protein synthesis was dependent on the level of overall mRNA synthesis and not on a single gene product. Nevertheless, others have confirmed Stanners' findings (C. P. Stanners, personal communication), and the discrepancy may be due to the fact that the P function acts predominantly by inhibition of translational initiations and could have escaped detection in Lodish and Porter's experiments, which employed early exponential-phase L-cell cultures in which no free ribosomes were present and translational initiation was not rate-limiting. Recently, Stanners and colleagues (personal communication) have isolated P⁺⁺ mutants that enhance inhibition of host protein synthesis in the early exponential phase, providing additional support for the P function.

D. Inhibition of Host-Cell Transcription

Infection of vertebrate cells by VSV results in inhibition of cellular DNA and RNA synthesis, probably by different mechanisms. The transcriptase-defective mutant tsG114(I) has been employed to show that inhibition of both DNA and cellular RNA synthesis is dependent on at least limited transcription of the viral genome. Weck and Wagner (1979) showed that in MPC-11 cells, initiation of transcription may be due to diminished transcription of the cellular RNA polymerases, polII, more than of polI or polIII. On the basis of UV-inactivation studies, it was suggested that transcription of the leader sequence at the 3' terminus of the genome is required to initiate inhibition of cellular transcription (Weck et al., 1979). In support of these findings, Kurilla et al. (1982) demonstrated the presence of VSV leader RNA in the cell nucleus at the time of inhibition, and McGowan et al. (1982) found that in vitro transcription in HeLa-cell extracts of SV40 DNA by polII, and of adenovirus DNA by polII and polIII, was inhibited by addition of VSV leader RNA. VSV-New Jersey leader RNA was more effective than VSV-Indiana leader RNA in this respect (Grinnell and Wagner, 1983).

Several other lines of evidence, however, suggest that synthesis of more than leader RNA is required for inhibition of cellular RNA synthesis by VSV. In mouse L cells, Wu and Lucas-Lenard (1980) found that tsG41(IV), tsW10(IV), ts045(V), and ts011(V) did inhibit host RNA synthesis. They suggested that the group II function (the NS protein) is necessary for inhibition of host RNA synthesis as well as for limited transcription of the genome. Furthermore, UV-inactivation studies estimated that transcription of 17% of the genome (the target size of the N gene) was necessary for inhibition. Poirot et al. (1985) observed that mutant tsG22(II), which supports viral genome transcription but not replication at 30°C, also inhibited host RNA synthesis. This inhibition was abolished by protein-synthesis inhibitors, which do not affect transcription. These findings imply that viral protein synthesis is required for inhibition of host protein synthesis and that the inhibitory agent may be leader RNA complexed with either a viral protein (probably N) or perhaps a cellular protein, such as the La protein, that is in limiting supply or subject to rapid turnover (Kurilla and Keene, 1983). However, rabies virus does not inhibit host RNA synthesis, although its leader RNA interacts with La protein in the same manner as VSV (Kurilla et al., 1984).

Earlier, Genty (1975) had presented evidence indicating that in addition to a protein synthesis-dependent mechanism, inhibition of host RNA synthesis in VSV-infected chick-embryo cells is partially due to a reduced capacity to transport uridine. Infection of chick-embryo cells with *ts* mutants suggested that the M protein might be involved in modification of uridine transport.

IX. HOST RANGE

A. Permissive and Restrictive Cells

Vesicular stomatitis virus (VSV) infection produces cytopathic changes in a wide range of host cells. Mosquito cells, however, survive infection and continue to multiply, producing moderate amounts of VSV over long periods. Nevertheless, Sarvar and Stollar (1977) isolated a single clone of *Aedes albopictus* cells in which VSV produced a cytopathic response at 28 and 34°C. Other clones were obtained that were lysed by VSV at 34°C only. The host-cell genome plays an important role in expression of cell killing. A few types of mammalian cells are resistant to the cytopathic effects of VSV. These include rabbit kidney (DRK3) cells (Chen and Crouch, 1978), rabbit corneal (RC-60) cells (Thacore and Youngner, 1975), and human lymphoblastoid cells (Nowakowski *et al.*, 1973). Levinson *et al.* (1978) have also described a stock of VSV-Indiana that was restricted in duck-embryo cells, but not in chicken, quail, or pheasant cells.

Restriction of growth of VSV in DRK3 cells appeared to operate at the level of virus adsorption. Restriction of growth of VSV in rabbit corneal cells appeared to be due to inhibition of viral RNA replication, since transcription, translation, and posttranslational modification occurred normally. This restriction could be released by superinfection with wild-type rabbitpox virus or even certain host-range mutants defective in early functions. However, rabbitpox virus was unable to rescue VSV from persistently infected BHK-21 cells, which suggests that the mechanism of restriction of VSV multiplication in persistently infected cells is different from that in rabbit corneal cells (Hamilton *et al.*, 1980).

Lymphoblastoid cells of the T-cell lineage are permissive for VSV replication, whereas those of B-cell origin are restrictive (Nowakowski *et al.*, 1973). Most B-lymphoblastoid cells contain endogenous Epstein–Barr virus (EBV); e.g., Raji cells do not shed EBV, but the genome exists in an episomal form in multiple copies and some viral functions are expressed. Creager *et al.* (1981, 1982) showed that restriction of VSV in B cells is characterized by failure to inhibit host protein synthesis and by reduced VSV transcription and is dependent on the presence of endogenous EBV. Abolition of cell killing and the initiation of persistent infection were responses to the presence of the EBV genome in lymphoblastoid cells of B origin, since EBV-negative B cells were susceptible to VSV infection. Johnson and Herman (1984) observed that although VSV messages are abnormally modified in Raji cells, they were still functional and not rate-limiting. They concluded from analyses of the properties of the temperature-sensitive *ts*C1 matrix-protein mutant that mRNA synthesis in Raji cells is limited by the amount of available nucleocapsid and not by any defect in transcription. They concluded, as had Nowakowski *et al.* (1973) previously, that restriction of VSV multiplication in Raji cells might be at the level of replication. However, Piwnica-Worms and Keene (personal communication) have reported that the synthesis of minus- and plus-strand genome RNA is similar in permissive BHK-21 cells and restrictive Raji cells. At 16 hr postinfection, 12% of the genomic RNA molecules synthesized in BHK-21 cells were packaged and exported, as against 0.8% in Raji cells. Virions released from Raji cells had reduced infectivity and a glycoprotein (G) with a faster electrophoretic mobility than the G protein of the more infectious virions released from BHK-21 cells. It is probable, therefore, that there are multiple biochemical lesions in Raji cells that together mediate restriction of VSV multiplication.

B. Conditional Host-Range Mutants

1. Host Range and Methylation

Host-range (*hr*) mutants of VSV-Indiana have been isolated that were restricted in HeLa or Hep-2 cells and were simultaneously temperature-sensitive in chick-embryo cells (Obijeski and Simpson, 1974; Simpson and Obijeski, 1974). Primary transcription was not affected, but amplification of RNA synthesis was defective in restrictive Hep-2 cells (Morrongiello and Simpson, 1979; Moyer *et al.*, 1981). *In vitro* virion polymerase activity was diminished in some of these *hr–ts* mutants at 31°C, and *in vivo* weak (possibly intragenic) complementation was detected between some pairs of *hr–ts* mutants. A gradient of permissiveness was observed in a comparison of 30 cell lines, cells of human origin being the most restrictive (Simpson *et al.*, 1979). Furthermore, sublines of HeLa cells differed in restrictiveness, suggesting that host factors are concerned in viral replication. Simpson *et al.* (1979) considered these mutants to be large (L)-protein mutants.

Four of six of these *hr* mutants and a non-*ts* revertant of one of them could be rescued by superinfection with rabbitpox virus, which indicates that the ts and hr phenotypes of these mutants are independently determined (Moyer *et al.*, 1981). Horikami and Moyer (1982) reported that two of these *hr* mutants of VSV were defective in mRNA methylation *in vitro*, although guanylylation and polyadenylation were unaffected. Mutant *hr*1 was totally defective in methylation, although full-length messages were synthesized at the normal rate with 5' termini in the form GpppA. It is not known whether this mutant is defective only in the first or in both the first and second steps in the normal pathway of 5'-terminal methylation (i.e., GpppA → 7mGpppA → 7mGpppA^m), but the data suggest that at least one methyltransferase activity is encoded in the viral genome. The unmethylated mRNA produced by *hr*1 was poorly translated *in vitro*.

Mutant *hr*8 was partially defective for methylation, producing mRNA *in vitro* with predominantly GpppA and some GpppA^m 5' termini. Subsequently, Horikami *et al.* (1984) have shown that the mRNA synthesized *in vivo* in restrictive cells is also unmethylated and inactive in an *in vitro* reticulocyte translation system. The undermethylation of *hr*8, however, could be partially reversed *in vitro* by addition of large amounts of *S*-adenosylmethionine. The association of an mRNA methylation defect with a host-range phenotype was unexpected and poses the question of how the permissive host cell compensates for this defect. In this respect, it is interesting that permissive BHK-21 cells have 10-fold higher levels of endogenous *S*-adenosylmethionine than restrictive Rep-2 cells, and mRNA from BHK-21 cells is predominantly in the monomethylated (7mGpppA) form. Both *hr* mutants could be rescued from restriction by superinfection with a poxvirus, suggesting that the defective mRNA could be methylated by the poxvirus methyltransferases. However, two other *hr* mutants (*hr*5 and *hr*7) produced active capped and methylated mRNA in restrictive cells and appeared to be blocked at the level of replication. Consequently, host factors in general, rather than methylation deficiency in particular, appear to be implicated in the host-range phenotype. Methylation defects do not appear to be associated with the *td*CE mutants of VSV and Chandipura virus (Rasool and Pringle, 1986).

2. Host-Range and Polymerase Defectiveness

Conditional host-range mutants of VSV-New Jersey (designated *td*CE mutants), which multiply at 31 and 39°C in BHK-21 cells and at 31°C only in chick-embryo cells, have been characterized in some detail. These mutants were present in stocks of mutagenized virus at higher frequency than conventional *ts* mutants (Pringle, 1978). A reversible inhibition of *in vitro* polymerase activity at 39°C, suggesting conformational change rather than inactivation, that was exhibited by some *td*CE mutants was restored to normal activity on reversion of the *td*CE phenotype (Szilagyi and Pringle, 1975). The molecular site of the mutational lesion was shown to be the L polypeptide by dissociation and reconstitution experiments in the case of mutant *td*CE3 (Szilagyi *et al.*, 1977). Similar mutants have been isolated from mutangenized stocks of VSV-Indiana and Chandipura virus (Gadkari and Pringle, 1980b; Rasool and Pringle, 1986). The initial analysis of four VSV *td*CE mutants suggests that there might be discrete classes of polymerase defects, but comparison of the *in vitro* polymerase activity of 12 Chandipura virus *td*CE mutants at 39°C showed that each mutant was unique, the 12 ranging in polymerase activity from complete thermosensitivity to almost no thermosensitivity. The *td*CE phenotype is expressed irrespective of the cell type in which the virus is propagated. Therefore, the host-range component of the phenotype implies that a factor in the cell environment (absent in chicken-embryo cells and in the *in vitro* polymerase reaction mixture) is involved in the correct func-

tioning of the polymerase, maintaining it in an active configuration at
39°C (Szilagyi and Pringle, 1975). More than one host factor may be involved, since at least three types of *td*CE mutants of VSV-New Jersey
could be defined according to their ability to multiply in other types of
avian cells (Fig. 8). These *td*CE mutants tended to be conditionally temperature-sensitive in the embryonic cells of some species. The temperature sensitivity of *td*CE mutants in pluripotent murine embryonal carcinoma cells is no longer observed when the cultures undergo differentiation.
Several host factors may interact with the VSV polymerase, and the absence of these factors at certain stages of development may have a protective effect.

Temperature-shift experiments indicated that the host restriction
operated early in the growth cycle, and use of pseudotypes with the nucleocapsid of Chandipura virus and the envelope of VSV *td*CE mutants
confirmed that restriction occurs after adsorption (Pringle, 1978). The
putative host factors have not been identified, although some possible
mechanisms such as phosphorylation, methylation, interferon induction,
and heat shock or stress proteins have been excluded (Rasool and Pringle,
1986, and unpublished data).

HeLa cell extracts enhance *in vitro* VSV RNA transcription. Fractionation of these extracts indicated that stimulatory activity was associated mainly with fractions containing microtubules and microtubule-
associated-protein (MAP). There was no stimulation of influenza A virus

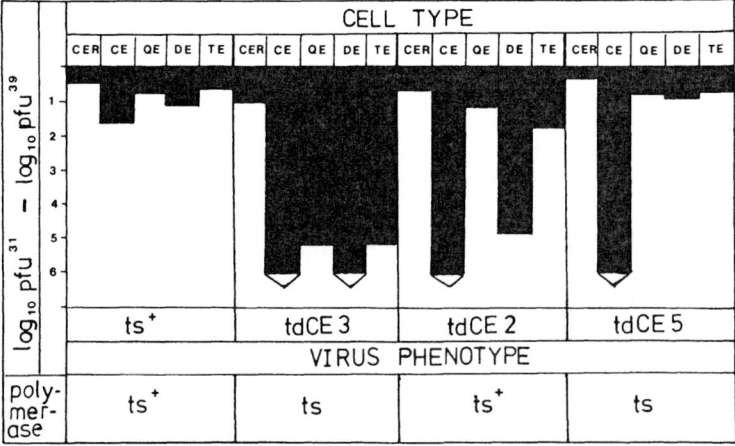

FIGURE 8. Efficiency of plating of three *td*CE mutants and wild-type virus in four avian
embryo cell types at 39°C. Mutants *td*CE2, *td*CE3, *td*CE5, and wild-type virus were assayed
on monolayers of hamster (CER), chick embryo (CE), Japanese quail embryo (QE), Pekin
duck embryo (DE), and turkey embryo (TE) cells. The mean values from five experiments
obtained from plaque counts after 48 hr of incubation at 31 or 39°C are plotted as histograms;
the longer the solid bar, the greater the temperature sensitivity. Data on the *in vitro* polymerase activity of these mutants are included. Reproduced from Pringle (1978).

transcription. Separation of MAP and tubulin indicated that stimulatory activity was associated with the MAP rather than tubulin (Hill *et al.*, 1986). It is conceivable that the MAP represents the specific host factors predicted by the genetic studies to be involved in VSV RNA replication. On the other hand, Moyer *et al.* (1986) found that tubulin stimulated *in vitro* VSV and Sendai virus transcription, and that addition of a monoclonal antibody to β-tubulin abolished all transcriptase activity. Direct association of tubulin and viral L protein was suggested because monoclonal and polyclonal antibodies to tubulin precipitated tubulin and L protein from the soluble fraction of infected cells. In view of the intimate association of L protein with the cytoskeleton throughout infection (Chatterjee *et al.*, 1984), and the presence of small amounts of tubulin in virions, it is likely that cytoskeletal components are the elusive host factors involved in replication. The specificity of these components remains to be explored in detail. Nonconditional *hr* mutants of VSV-New Jersey have also been isolated that are completely restricted in both chick-embryo and Mardin-Darby bovine kidney (MDBK) cells (Pringle, 1978). These mutants were present at low frequency only and so far have not been found in VSV-Indiana or Chandipura virus.

3. Piry Virus in *Drosophila*

Brun (1981) has described host-range (agD) mutants of Piry virus that have reduced ability to multiply and induce CO_2 sensitivity in adult *Drosophila melanogaster* (Paris strain). Many of these host-range mutants were also temperature-sensitive in chick-embryo cells. In the case of one of these mutants, it was shown that the restriction was under the control of a gene on chromosome III. The host gene that affects the multiplication of Piry virus was not one of the five genes known to be involved in the control of Sigma virus. Thus, either these two rhabdoviruses interact differently with the host organism or multiple host factors are concerned in the replication of rhabdoviruses in insects.

Most agD (affected growth in *Drosophila*) mutants are generally restricted in all strains of *Drosophila*. Other mutants, designated rgD (rapid growth in *Drosophila*), have been isolated from agD stocks or from wild-type by selection. These rgD mutants have lost general restriction in *Drosophila* strains but retain specific restriction in the Paris strain. Furthermore, they multiply in permissive strains with shorter incubation times and to higher yields than wild-type Piry virus (Brun, 1984). The agD and rgD mutants could be tentatively classified into complementation groups by a UV-inactivation–rescue procedure in which cells are infected with one mutant at low multiplicity in the presence of UV-inactivated wild-type virus or another mutant. This technique was developed by Deutsch (Deutsch, 1975, 1976; Deutsch *et al.*, 1977) for analysis of the function of *ts* mutants of VSV. The conclusion from preliminary analysis was that the majority of the agD and rgD mutants were L-protein mutants. Intracistronic complementation among L-protein (group I) mutants was frequent, resembling the pattern described for Chandipura virus by Gadkari and Pringle (1980a).

These observations are all consistent with the concept that the host-range properties of rhabdoviruses are determined primarily by specific interactions between host-cell factors and the viral polymerase.

X. PATHOGENESIS

A. Patterns of Disease and Protection

Altered patterns of disease have been associated with temperature-sensitive (ts) mutants of vesicular stomatitis virus (VSV) and Chandipura virus. In particular, changes in neurovirulence following intracerebral injection into adult mice or hamsters have been described, as have also the protective potential of ts mutants administered intranasally to adult mice. These observations have been summarized previously (Pringle, 1982) and have not been substantially extended.

Differences in the pathogenic properties of rabies virus ts mutants have been reported by Selimow and Nikotina (1970), Clark and Koprowski (1971), Aubert et al. (1980), and Bussereau et al. (1982). In the absence of characterization of these mutants by complementation analysis, direct comparison of these results is difficult.

B. Host Resistance

Fultz et al. (1981, 1982) used Syrian hamsters to study the protective effects of defective interfering particles, polyinosinic acid–polycytidylic acid, and interferon. VSV-Indiana wild-type produces a lethal disease in Syrian hamsters with 70% of the animals dying within 72 hr of injection of 10–100 plaque-forming units. The UTI strain, however, is relatively resistant, with 90% of animals surviving intraperitoneal injection. Resistance appeared to be multifactorial, with three independent loci interacting, two on autosomes and one on the X chromosome. Preliminary data suggested that the mechanism of resistance to VSV involves cells of the lymphohematopoietic system.

Murine resistance to intraperitoneal infection with street rabies virus is genetically determined. Certain strains of mice are fully susceptible, other strains do not exhibit signs of disease, and in still other strains, individual animals survive after clinical illness. Lodmel and Chesebro (1984) have shown that resistance to infection is controlled by two segregating genes and that the resistance genes in the SJL/J and CBA/J strains are allelic. These strains provide an opportunity to investigate how animals recover from infection and the role of the immune system in this process.

In a comparison of seven inbred strains of mice inoculated with i.p. street rabies virus, differences in susceptibility to CNS disease were correlated with restriction of virus multiplication within the CNS. Limitation of virus replication correlated with antibody response, resistant strains having high neutralizing antibody titers and susceptible strains low antibody levels (Lodmell and Ewalt, 1985). Resistant SJL/J mice were rendered susceptible by cyclophosphamide treatment, and resistance was restored by transfusion with immune spleen cells or serum. Other experiments indicated that antibody in the cerebrospinal fluid was not an important factor in resistance in mice which remained asymptomatic, but was important in survival of mice that developed clinical disease.

Templeton *et al.* (1986) have also studied the genetic control of serum-neutralizing antibody in mice. They showed that serum-neutralizing antibody response to vaccination and resistance to rabies virus challenge was controlled by separate unlinked genes. Hyperresponsiveness and hyporesponsiveness to rabies vaccination were controlled by dominant and recessive alleles, respectively. Survival after rabies virus challenge infection was dominant to nonsurvival. This result implies that even the most potent rabies vaccines may not induce protection in some genetically susceptible individuals. Host resistance to rabies infection is undoubtedly complex and cytotoxic T cells also play an essential role in protection.

XI. PHENOTYPIC MIXING AND PSEUDOTYPES

A. Prevalence of Phenotypic Mixing

Phenotypic mixing occurs widely between related and even unrelated enveloped viruses multiplying in the same cells. This contrasts with the pattern of genetic complementation, which in general is confined to closely related viruses. Phenotypic mixing between heterologous viruses involves the envelope glycoproteins only, and particles invested completely in the envelope antigens of a heterologous virus are described as "pseudotypes." Phenotype mixing of the internal proteins of heterologous viruses has not been observed.

Within the genus *Vesiculovirus*, heterologous complementation has been observed between the two subtypes (Concan and Hazelhurst) of vesticular stomatitis virus (VSV)-New Jersey and betweenVSV-Indiana and VSV-Cocal. In the latter, significant complementation involved only the two groups of mutants representing the envelope proteins. Complementation has not been observed in any of the other heterologous combinations tested, i.e., Indiana and New Jersey, New Jersey and Cocal, Cocal and Chandipura, Indiana and Chandipura, New Jersey and Chandipura. The level of discrimination of nongenetic interactions in pseudotype analysis is probably at least an order of magnitude better than that of complementation analysis using temperature-sensitive (ts) mutants.

The extent of phenotypic mixing between enveloped viruses is reviewed in Table VII. Phenotypic mixing in one direction at least has been recorded between rhabdoviruses and nine of the other ten families of enveloped viruses, the highest frequency (50%) of pseudotypes being recorded in coinfections with herpesvirus (Huang et al., 1974).

Preparations of pure pseudotypes can be obtained using deletion or conditional lethal mutants completely defective for glycoprotein (G) transport or synthesis; examples are BH-RSV (VSV) (Weiss et al., 1977) and VSV ts045(V) (MMTV) (Chan et al., 1982). Mutant ts045(V) of VSV-Indiana has been used extensively for production of pseudotype particles because the ts lesion renders the G protein thermolabile. This thermolability allows elimination of the non-phenotypically mixed VSV and facilitates resolution of the pseudotype particles enveloped in heterologous nonthermolabile protein. Zavada (1972) derived for the same purpose a more thermolabile (tl) nonleaky genetically stable G-protein multiple mutant tl17 by alternating 5-fluorouracil mutagenization and cycles of neutralization with specific antiserum. Weiss and Bennett (1980) demonstrated that pseudotypes of the form VSV(RSV) were obtained from RSV-infected cells superinfected with mutant ts045(V), whereas RSV(VSV) pseudotypes were produced in the same cells infected with the matrix (M)-protein mutant tsG31(III). Synthesis of M protein is essential for maturation of VSV, but not for the insertion of G protein into the plasma

TABLE VII. Phenotypic Mixing between Rhabdoviruses and Other
Enveloped Viruses[a]

Heterologous enveloped viruses		Rhabdovirus component of pseudotype[b]	
Family	Type	Nucleocapsid	Envelope
Togaviridae	RNA+	−	+
Coronaviridae	RNA-	NT	NT
Rhabdoviridae	RNA-	+	+
Paramyxoviridae	RNA-	NT	+
Arenaviridae	2 × RNA-	NT	+
Bunyaviridae	3 × RNA-	NT	+
Orthomyxoviridae	7–8 × RNA-	−	+
Retroviridae	2 × RNA+/ 1 × DNA	+	+
Herpesviridae	DNA (nuclear)	+	+
Poxviridae	DNA (cytoplasmic)	NT	+

[a] Adapted from Zavada (1982).   [b] (+) Identified; (−) not identified; (NT) not tested.

membrane. Consequently, no VSV particles are released from *ts*G31(III)-infected cells at restrictive temperature despite the presence of G protein in the plasma membrane.

B. Selective Incorporation of Host Antigens

Although cell-specific membrane proteins are generally excluded from the envelope of budding viruses during maturation, this exclusion is not absolute. Calafat *et al.* (1983) have demonstrated selective incorporation of murine cell-surface G proteins by immuno-electron microscopy with monoclonal antibodies; the Thy-1 G protein was present in the envelope of VSV and murine leukemia virus, but Pgp-1, T200, and the *H-2* antigens were excluded. Zavada *et al.* (1983b) demonstrated selective assembly of human cell-surface proteins into VSV envelopes using surface ^{125}I-labeled human tumor-cell lines and mutant *ts*045(V). At least three human tumor-specific antigens were detected in the VSV envelope. Little *et al.* (1983a,b) also found that a non-VSV antigen was present in the envelope of VSV grown in HeLa cells, since VSV propagated in HeLa cells was precipitated, although not neutralized, by antisera to HeLa-cell determinants. This non-VSV antigen appeared to correspond to a tumor-specific surface antigen of HeLa cells with a molecular weight of 75,000. Zavada (1982), in discussing the paradox of unrestricted phenotypic mixing between enveloped viruses and the almost complete exclusion of host proteins, suggested that there may be a common recognition mechanism for specific assembly of virus G proteins. Hence, the cell-surface proteins selected by VSV from human tumor-cell lines may be retrovirus-related gene products. Zavada and Huang (1984) have applied the same approach to analysis of proteins shed in particulate structures in large amounts from several human melanoma-cell lines. They found that the same sets of proteins were present in these structures as assembled into VSV particles propagated in these cell lines, and some of these proteins were antigenically cross-reactive with proteins from HeLa and T47D (breast carcinoma) cells. It was suggested that these proteins might be products of an endogenous human retrovirus.

Polakova *et al.* (1984) have produced a monoclonal antibody specific for one of these HeLa-cell antigens, using mutant *ts*045(V) propagated in HeLa cells at 39°C because particles devoid of the thermolabile G protein are produced under these conditions. This technique will facilitate characteristics of these HeLa-cell antigens that appear to be specifically associated with certain tumors or types of tissue. The data of Polakova and co-workers suggest that the tumor antigen recognized by their monoclonal antibody is present in low amounts and that VSV is highly selective and endowed with the ability to concentrate this antigen from the cell surface.

C. Pseudotypes as Biological Probes

Screening for pseudotypes with the genome of VSV and the envelope of a heterologous virus can be an extremely sensitive method for detection of cryptic or latent enveloped viruses (e.g., chick helper factor in avian cells) or as a rapid quantitation assay for slow or non cytopathogenic viruses. The sensitivity of the method can be increased using immunoprecipitation with *Staphylococcus aureus* rather than neutralization as the detection system (Zavada *et al.*, 1983a). Chandipura virus can be used in place of VSV in situations in which there are veterinary restrictions on use of VSV (Zavada *et al.*, 1979; Zavadova and Zavada, 1980). Analysis of pseudotype production has been employed as a quantitative procedure for detection and assay of bovine leukemia virus (Zavada *et al.*, 1979; Bruck *et al.*, 1982), mouse mammary tumor virus (Chan *et al.*, 1982), and human T-cell lymphoma virus types I and II (Clapham *et al.*, 1984).

XII. SIGMA VIRUS OF *DROSOPHILA*

A. General Features

Sigma virus of *Drosophila* has been identified as a rhabdovirus on the basis of size and bullet-shaped morphology. The complex biology, assay procedures, and germinal transmission of Sigma virus have been reviewed by Brun and Plus (1980) and Teninges *et al.* (1980) and will not be considered in detail here. The physical properties of Sigma virus remain uncharacterized, although Richard-Molard *et al.* (1984) have tentatively identified five virion proteins (p210, gp68, p57, p44, and p28) in partially purified virus from *Drosophila* tissue culture. The host range of Sigma virus is limited to *Drosophila*, whereas certain vertebrate rhabdoviruses (VSV, Piry virus, and Chandipura virus, but not rabies virus) can multiply in *Drosophila*. None of these viruses, Sigma virus included, produces pathogenic effects in insects or cytopathic effects in *Drosophila* cells *in vitro*. The vertebrate vesiculoviruses, however, produce the enhanced CO_2 sensitivity in adult insects that was earlier considered to be a unique property of Sigma virus. Sigma virus is not contagious and is transmitted germinally.

B. Temperature-Sensitive and Host-Range Mutants

Temperature-sensitive (*ts*) mutants and host-range variants have been used by Contamine (1984) to investigate the growth cycle of Sigma virus. Three classes of *ts* mutants have been defined: The first class (represented by *ts*4) has been designated early mutants (Contamine, 1973) and are mutants defective in hereditary transmission at restrictive temperature (32°C). These mutants are defective throughout the whole growth cycle, but maturation is not affected. The second class (represented by *ts*9) have been designated late mutants (Contamine, 1973) and are mutants defective in maturation. Maternal hereditary transmission is normal at elevated temperature. The third class (represented by haP7) are a second type of early mutants (Contamine, 1980, 1981) that exhibit an early temperature-sensitive phase only; hereditary transmission and maturation are not heat-sensitive. These mutants identify four stages in the growth cycle. At 20°C, there is a rapid adsorption–penetration phase (approximately 20 min), followed by a stage defined by the period of time corresponding to the temperature sensitivity of mutant haP7. This lasts from 0 to 7 hr. The next stage corresponds to the period of temperature sensitivity of *ts*4 and lasts from 4 hr to 60 hr, which covers the entire period of genome replication and can be subdivided into an initial phase from 4 to 11 hr, an intermediate period from 11 to about 60 hr during which the viral functions necessary to establish the stabilized state are expressed, and a final maturation phase defined by the temperature sensitivity of mutant *ts*9. The prolonged growth cycle is due to the presence of a particular restrictive allele [*ref (3)D*] of one of the host controlling genes.

At least five independent loci in *Drosophila* influence Sigma virus growth. The haP mutants of Sigma virus referred to above (many of which are also temperature-sensitive) were selected for specific resistance to gene *ref(2)P^p*. By this procedure, a range of haP mutants was obtained with a spectrum of host-restricted phenotypes. It was concluded that the haP phenotype represents mutation affecting a viral protein–genome complex, whereby virus interactions with any of the five independent host genes affecting Sigma virus replication may be modified (Coulon and Contamine, 1982; Contamine, 1984).

XIII. FUTURE PROSPECTS

Reverse mutagenesis and ability to express individual genes *in vitro* in different genetic environments will open up the study of the functional activity of rhabdovirus genes. The sequencing of variants selected for resistance to neutralization by monoclonal antibodies has initiated study of the effect of mutation at the level of secondary structure, a crucial stage in progress toward understanding of gene function.

The diversity of the family Rhabdoviridae awaits exploration.

ACKNOWLEDGMENTS. I am indebted to the many colleagues who supplied information prior to publication, and in particular to Duncan McGeoch, who made available unpublished results for inclusion in this chapter.

REFERENCES

Agol, V. I., Grachev, V. P., Drozdov, S. G., Kolesnikova, M. S., Kozlov, V. G., Ralph, N. M., Romanova, L. I., Tolskaya, E. A., Tyafanov, A. V., and Viktorova, E. G., 1984, Construction and properties of intertypic poliovirus recombinants: First approximation mapping of the major determinants of virulence, *Virology* **136**:41.

Akashi, H., Gay, M., Ihara, T., and Bishop, D. H. L., 1984, Localised conserved regions of the SRNA gene products of bunyaviruses are revealed by sequence analyses of the Simba serogroup Aino virus, *Virus Res.* **1**:51.

Anilionis, A., Wunner, W. H., and Curtis, P. J., 1981, Structure of the glycoprotein gene in rabies virus, *Nature (London)* **294**:275.

Aubert, M. F. A., Bussereau, F., and Blancon, J., 1980, Pathogenic, immunogenic and protective powers of ten temperature sensitive mutants of rabies virus in mice, *Ann. Virol. (Inst. Pasteur)* **131E**:217.

Auperin, D. D., Romanowski, V., Galinski, M., and Bishop, D. H. L., 1984, Sequencing studies of Pichinde arenavirus S RNA indicate a movel coding strategy, an ambisense viral S RNA, *J. Virol.* **52**:897.

Banerjee, A. K., Rhodes, D. P., and Gill, D. S., 1984, Complete sequence of the mRNA coding for the N protein of vesicular stomatitis virus (New Jersey serotype), *Virology* **137**:432.

Bay, P. H. S., and Reichmann, M. E., 1979, UV inactivation of the biological activity of defective interfering particles generated by vesicular stomatitis virus, *J. Virol.* **32**:876.

Bell, J. C., and Prevec, L., 1985, Phosphorylation sites on phosphoprotein NS of vesicular stomatitis virus. *J. Virol.* **54**:697.

Belle-Isle, H. D., and Emerson, S. U., 1982, Use of a hybrid infectivity assay to analyse primary transcription of temperature-sensitive mutants of the New Jersey serotype of vesicular stomatitis virus, *J. Virol.* **43**:37.

Bergmann, J. E., Tokuyasu, K. T., and Singer, S. J., 1981, Passage of an integral membrane protein, the vesicular stomatitis virus glycoprotein, through the Golgi apparatus en route to the plasma membrane, *Proc. Natl. Acad. Sci. U.S.A.* **78**:1746.

Blumberg, B. M., Giorgio, C., and Kolakofsky, D., 1983, N protein of vesicular stomatitis virus selectively encapsidates leader RNA *in vitro*, *Cell* **32**:559.

Brand, C., and Palese, P., 1980, Sequential passage of influenza virus in embryonated egg or tissue culture: Emergence of mutants, *Virology* **107**:424.

Brown, E., and Prevec, L., 1982, Characterization of the electrophoretic mobility mutation in the N protein of the *ts* D1 mutant of vesicular stomatitis virus New Jersey serotype, *Can. J. Biochem.* **60**:1065.

Brown, F., Bishop, D. H. L., Crick, J., Francki, R. I. B., Holland, J. J., Hall, R., Johnson, K., Martelli, G., Murphy, F. A., Obijeski, J. F., Peters, D., Pringle, C. R., Reichmann, M. E., Schneider, L. G., Shope, R. E., Simpson, D. I. H., Summers, D. F., and Wagner, R. R., 1979, Rhabdoviridae, *Intervirology* **12**:1.

Bruck, C., Portelle, D., Burny, A., and Zavada, J., 1982, Topographical analysis by monoclonal antibodies of BLV-gp 51 epitopes involved in viral functions, *Virology* **122**:353.

Brun, G., 1963, Etude d'une association du virus et de son hôte la drosophile: l'Etat stabilisée, *Thèse Biol. Exp.*

Brun, G., 1981, Are the *Drosophila ref* genes for Piry and Sigma rhabdoviruses identical?, in: *The Replication of Negative Strand Viruses* (D. H. L. Bishop and R. W. Compans, eds.), pp. 921–928, Elsevier/North-Holland, New York.

Brun, G., 1984, Host-range mutants of Piry virus: A new type of mutant in *Drosophila*, in: *Negative Strand Viruses: Paramyxoviruses and Rhabdoviruses* (D. H. L. Bishop and R. W. Compans, eds.), pp. 413–420, Academic Press, Orlando, Florida.

Brun, G., and Plus, N., 1980, The viruses of Drosophila, in: *The Genetics and Biology of Drosophila* (M. Ashburner and T. R. F. Wright, eds.), pp. 625–693, Academic Press, London.

Buller, R. M. L., 1975, Biological and biochemical characterization of vesicular stomatitis virus and temperature-sensitive maturation mutants, Ph.D. thesis, University of Glasgow.

Bussereau, F., and Flamand, A., 1978, Isolation and preliminary characterization of *ts* mutants of rabies virus, in: *Negative Strand Viruses and the Host Cell* (B. W. J. Mahy and R. D. Barry, eds.), pp. 701–708, Academic Press, New York.

Bussereau, F., Benejean, J., and Saghi, N., 1982, Isolation and study of temperature-sensitive mutants of rabies virus, *J. Gen. Virol.* **60**:153.

Byrd, A. D., Kennedy-Morrow, J., Marks, M. D., and Lesnaw, J. A., 1984, Functional relationships within the New Jersey serotype of vesicular stomatitis virus: Genetic and physiological comparisons of the Hazelhurst and Concan subtypes, *J. Gen. Virol.* **65**:1769.

Calafat, J., Janssen, H., Demant, P., Helgers, J., and Zavada, J., 1983, Specific selection of host cell glycoproteins during assembly of murine leukaemia virus and vesicular stomatitis virus: Presence of Thy-1 glycoprotein and absence of H-2, Pgp-1 and T-200 glycoproteins on the envelopes of these virus particles, *J. Gen. Virol.* **64**:1241.

Carroll, A. R., and Wagner, R. R., 1979, Role of the membrane protein in endogenous inhibition of *in vitro* transcription by vesicular stomatitis virus, *J. Virol.* **29**:134.

Chan, J. C., East, J. L., Bowen, J. M., Massey, R., and Schochetman, G., 1982, Monoclonal and polyclonal antibody studies of VSV (hr MMTV) pseudotypes, *Virology* **120**:54.

Chatis, P. A., and Morrison, T. G., 1981, Mutational changes in the VSV glycoprotein affect the requirement of carbohydrate in morphogenesis, *J. Virol.* **37**:307.

Chatterjee, P. K., Cervera, M. M., and Penman, S., 1984, Formation of vesicular stomatitis virus nucleocapsid from cytoskeletal framework-bound N protein: Possible model for structure assembly, *Mol. Cell. Biol.* **4**:2231.

Chen, C.-Y., and Crouch, N. A., 1978, Shope fibroma virus-induced facilitation of vesicular stomatitis virus adsorption and replication in nonpermissive cells, *Virology* **85**:43.

Chen, S. S.-L., and Huang, A. S., 1986, Further characterization of the vesicular stomatitis virus temperature-sensitive 045 mutant: Intracellular conversion of the glycoprotein to a soluble form, *J. Virol.* **59**:210.

Chou, P. Y., and Fasman, G. D., 1978, Prediction of the secondary structure of proteins from their amino acid sequence, *Adv. Enzymol.* **47**:45.

Clapham, P., Nagy, K., and Weiss, R. A., 1984, Pseudotypes of human T-cell leukemia virus types 1 and 2: Neutralization by patients' sera, *Proc. Natl. Acad. Sci. U.S.A.* **81**:2886.

Clark, H. F., 1978, Rabies viruses increase in virulence when propagated in neuroblastoma cell culture, *Science* **199**:1072.

Clark, H. F., and Koprowski, H., 1971, Isolation of *ts* conditional lethal mutants of "fixed" rabies virus, *J. Virol.* **7**:295.

Clark, H. F., and Wiktor, T. J., 1974, Plasticity of phenotypic characters of rabies related viruses: Spontaneous variation in plaque morphology, virulence and temperature-sensitivity characters of serially propagated Lagos bat and Mokola viruses, *J. Infect. Dis.* **130**:608.

Clewley, J. P., Bishop, D. H. L., Kang, C. Y., Coffin, J., Schnitzlein, W. M., Reichmann, M. E., and Shope, R. E., 1977, Oligonucleotide fingerprints of RNA species obtained from rhabdoviruses belonging to the vesicular stomatitis virus subgroup, *J Virol.* **23**:152.

Contamine, D., 1973, Etude de mutants thermosensibles du virus Sigma, *Mol. Gen. Genet.* **124**:233.

Contamine, D., 1980, Two types of early mutants among temperature-sensitive mutants of *Drosophila* Sigma virus, *Ann. Virol. (Inst. Pasteur)* **131E**:113.

Contamine, D., 1981, Role of the *Drosophila* genome in Sigma virus multiplication. I. Role of the *ref (2) P* gene; selection of host-adapted mutants at the non-permissive allele P^p, *Virology* **114**:474.

Contamine, D., 1984, The late functions of *Drosophila* Sigma virus, *Arch. Virol.* **82**:31.

Cormack, D. V., Holloway, A. F., and Pringle, C. R., 1973, Temperature-sensitive mutants of vesicular stomatitis virus: Homology and nomenclature, *J. Gen. Virol.* **19**:295.

Cooper, P. D., and Bellett, A. J. D., 1959, A transmissible interfering component of vesicular stomatitis virus preparations, *J. Gen. Microbiol.* **21**:485.

Coulon, P., and Contamine, D., 1982, Role of the *Drosophila* genome in sigma virus multiplication. II. Host spectrum variants among haP mutants, *Virology* **123**:381.

Coulon, P., Rollin, P. E., and Flamand, A., 1983, Molecular basis of rabies virus virulence. II. Identification of a site on the CVS glycoprotein associated with virulence, *J. Gen. Virol.*, **64**:693.

Creager, R. S., Cardamone, J. J., and Youngner, J. S., 1981, Human lymphoblastoid cell lines of B- and T cell origin: Different responses to infection with vesicular stomatitis virus, *Virology* **111**:211.

Creager, R. S., Whitaker-Dowling, P., Frey, T. K., and Youngner, J. S., 1982, Varied response of human B-lymphoblastoid cell lines to infection with vesicular stomatitis virus, *Virology* **121**:414.

Davis, N. L., Arnheiter, H., and Wertz, G. W., 1986, Vesicular stomatitis virus N and NS proteins form multiple complexes, *J. Virol.* **59**:751.

Deutsch, V., 1975, Nongenetic complementation of group V temperature-sensitive mutants of vesicular stomatitis virus by UV-irradiated virus, *J. Virol.* **15**:788.

Deutsch, V., 1976, Parental G protein reincorporation by a vesicular stomatitis virus temperature-sensitive mutants of complementation group V at nonpermissive temperature, *Virology* **69**:607.

Deutsch, V., Muel, B., and Brun, G., 1977, Action spectra for the rescue of temperature-sensitive mutants of vesicular stomatitis virus by ultraviolet-irradiated virions at nonpermissive temperature, *Virology* **77**:294.

Deutsch, V., Muel, B., and Brun, G., 1979, Temperature-sensitive mutant *ts* 082 of vesicular stomatitis virus. 1. Rescue at non-permissive temperature by UV-irradiated virus, *Virology* **93**:286–290.

Dietzschold, B., Wunner, W. H., Wiktor, T. J., Lopes, A. D., Lafon, M., Smith, C. L., and Koprowski, H., 1983, Characterization of an antigenic determinant of the glycoprotein that correlates with pathogenesis of rabies virus, *Proc. Natl. Acad. Sci. U.S.A.* **80**:70.

Doyle, M., and Holland, J. J., 1973, Prophylaxis and immunization in mice by use of virus-free defective T particles to protect against intracerebral infection by vesicular stomatitis virus, *Proc. Natl. Acad. Sci. USA* **70**:2105.

Duhamel, C., 1954, Etude de la sensibilité héréditaire a l'anhydride carbonique chez la Drosophile: Description de quelques variants du virus, *C. R. Acad. Sci.* **239**:1157.

Evans, D., Pringle, C. R., and Szilagyi, J. J., 1979, Temperature-sensitive mutants of complementation group E vesicular stomatitis virus New Jersey serotype possess altered NS polypeptides, *J. Virol.* **31**:325.

Flamand, A., 1970, Etude génétique du virus de la stomatite vesiculaire: Classement de mutants thermosensibles spontanées en groupes de complementation, *J. Gen. Virol.* **8**:187.

Flamand, A., 1980, Rhabdovirus genetics, in: *Rhabdoviruses*, Vol. II (D. H. L. Bishop, ed.), pp. 115–140, CRC Press, Boca Raton, Florida.

Flamand, A., and Delagneau, J. F., 1978, Transcriptional mapping of rabies virus *in vivo, J. Virol.* **28**:518.

Flamand, A., and Pringle, C. R., 1971, The homologies of spontaneous and induced temperature-sensitive mutants of vesicular stomatitis virus isolated in chick embryo and BHK-21 cells, *J. Gen. Virol.* **11**:81.

Francoeur, A. M., Lam, T., and Stanners, C. P., 1980, PIF, a highly sensitive plaque assay for induction of interferon, *Virology* **105**:526.

Freeman, G. J., and Huang, A. S., 1981, Mapping temperature-sensitive mutants of vesicular stomatitis virus by RNA heteroduplex formation, *J. Gen. Virol.* **57**:103.

Frey, T. K., and Youngner, J. S., 1982, Novel phenotype of RNA synthesis expressed by vesicular stomatitis virus isolated from persistent infections, *J. Virol.* **44**:167.

Frey, T. K., Frielle, D. W., and Youngner, J. S., 1981, Standard vesicular stomatitis virus is required for interferon induction in L cells by defective interfering particles, in: *The*

Replication of Negative Strand Viruses (D. H. L. Bishop and R. W. Compans, eds.), pp. 901–907, Elsevier/North-Holland, New York.

Fultz, P. N., Shadduck, J. A., Kang, C. Y., and Streilein, J. W., 1981, Genetic analysis of resistance to lethal infections of vesicular stomatitis virus in Syrian hamsters, *Infect. Immun.* **32:**1007.

Fultz, P. N., Shadduck, J. A., Kang, C. Y., and Streilein, J. W., 1982, Mediators of protection against lethal systemic vesicular stomatitis virus infection in hamsters: Defective interfering particles, polyinosinate–polycytidylate, and interferon, *Infect. Immun.* **37:**679.

Gadkari, D. A., and Pringle, C. R., 1980a, Temperature-sensitive mutants of Chandipura virus. I. Inter- and intra-group complementation, *J. Virol.* **33:**100.

Gadkari, D. A., and Pringle, C. R., 1980b, Temperature-sensitive mutants of Chandipura virus. II. Phenotype characteristics of the six complementation groups. *J. Virol.* **32:**107.

Gallione, C. J., and Rose, J. K., 1983, Nucleotide sequence of a cDNA clone encoding the entire glycoprotein from the New Jersey serotype of vesicular stomatitis virus, *J. Virol.* **46:**162.

Gallione, C. J., and Rose, J. K., 1985, A single amino acid substitution in a hydrophobic domain causes temperature sensitive cell-surface transport of a mutant viral glycoprotein, *J. Virol.* **54:**374.

Gallione, C. J., Greene, J. R., Iverson, L. E., and Rose, J. K., 1981, Nucleotide sequences of the mRNA's encoding the vesicular stomatitis virus N and NS proteins, *J. Virol.* **39:**529.

Genty, N., 1975, Analysis of uridine incorporation in chicken embryo cells infected by vesicular stomatitis virus and its temperature sensitive mutants: Uridine transport, *J. Virol.* **15:**8.

Gill, D. S., and Banerjee, A. K., 1985, Vesicular stomatitis virus NS proteins: Structural similarity without extensive sequence homology, *J. Virol.* **55:**60.

Gillies, S., and Stollar, V., 1980, Generation of defective interfering particles of vesicular stomatitis virus in *Aedes albopictus* cells, *Virology* **107:**497.

Goldstein, L., 1949, Contribution a l'étude de la sensibilité hereditaire au gaz carbonique chez la Drosophile: Mise en évidence d'une forme nouvelle du génoide, *Bull. Biol. Fr. Belg.* **83:**177.

Gopalakrishna, Y., and Lenard, J., 1985, Sequence alterations in the temperature-sensitive M-protein mutants (complementation group III) of vesicular stomatitis virus, *J. Virol.* **56:**655.

Grinell, B., and Wagner, R. R., 1983, Comparative inhibition of cellular transcription by vesicular stomatitis virus serotypes New Jersey and Indiana: Role of each viral leader RNA, *J. Virol.* **48:**88.

Grinnell, B. W., and Wagner, R. R., 1984, Nucleotide sequence and secondary structure of VSV leader RNA and homologous DNA involved in inhibition of DNA-dependent transcription, *Cell* **36:**533.

Guillemain, A., 1953, Découverte et localization d'une gene empêchant le multiplication du virus de la sensibilité hereditaire au CO_2 chez D. M., *C. R. Acad. Sci.* **236:**1085.

Hamilton, D. H., Moyer R. W., and Moyer, S. A., 1980, Characterization of the non-permissive infection of rabbit cornea cells by vesicular stomatitis virus, *J. Gen. Virol.* **49:**273.

Herman, R. C., 1986, Internal initiation of translation on the vesicular stomatitis virus phosphoprotein mRNA yields a second protein, *J. Virol.* **58:**797.

Hill, V. M., Harmon, S. A., and Summers, D. F., 1986, Stimulation of vesicular stomatitis virus *in vitro* RNA synthesis by microtubule-associated protein, *Proc. Natl. Acad. Sci. U.S.A.* **83:**5410.

Holland, J. J., and Villareal, L. P., 1974, Persistent noncytocidal vesicular stomatitis virus infections mediated by defective T particles that suppress virion transcriptase, *Proc. Natl. Acad. Sci. U.S.A.* **71:**2956.

Holland, J. J., Villareal, L. P., and Breindl, M., 1976, Factors involved in the generation and replication of rhabdovirus defective T particles, *J. Virol.* **17:**805.

Holland, J. J., Kennedy, S. I. T., Semler, B. L., Jones, C. L., Roux, L., and Grabau, E. A., 1980, Defective interfering RNA viruses and the host cell response, in: *Comprehensive Virology*, Vol. 16 (H. Fraenkel-Conrat and R. R. Wagner, eds.), pp. 137–192, Plenum Press, New York.

Holland, J. J., Spindler, K. R., Horodyski, F. M., Grubau, E. A., Nichol, S. T., and VandePol, S., 1982, Rapid evolution of RNA genomes, *Science* **215**:1577.

Holloway, A. F., Wong, P. K. Y., and Cormack, D. V., 1970, Isolation and characterization of temperature-sensitive mutants of vesicular stomatitis virus, *Virology* **42**:917.

Horikami, S. M., and Moyer, S. A., 1982, Host range mutants of vesicular stomatitis virus defective in *in vitro* RNA methylation, *Proc. Natl. Acad. Sci. U.S.A.* **79**:7694.

Horikami, S. M., De Ferra, F., and Moyer, S. A., 1984, Characterization of the infections of permissive and non-permissive cells by host range mutants of vesicular stomatitis virus defective in RNA methylation, *Virology* **138**:1.

Horodyski, F. M., and Holland, J. J., 1980, Virus isolated from cells persistently infected with vesicular stomatitis virus show altered interations with defective interfering particles, *J. Virol.* **36**:627.

Horodyski, F. M., and Holland, J. J., 1981, Continuing evolution of virus–DI particle interaction during VSV persistent infection, in: *The Replication of Negative Strand Viruses* (D. H. L. Bishop and R. W. Compans, eds.), pp. 887–892, Elsevier/North-Holland, New York.

Horodyski, F. M., and Holland, J. J., 1984, Reconstruction experiments demonstrating selective effects of defective interfering particles on mixed populations of vesicular stomatitis virus, *J. Gen. Virol.* **65**:819.

Horodyski, F. M., Nichol, S. T., Spindler, K. R., and Holland, J. J., 1983, Properties of DI particle-resistant mutants of vesicular stomatitis virus isolated from persistent infections and from undiluted passages, *Cell* **33**:801.

Hsu, C-H., and Kingsbury, D. W., 1985, Constitutively phosphorylated residues in the NS protein of vesicular stomatitis virus, *J. Biol. Chem.* **260**:8990.

Huang, A. S., Palma, E. L., Hewlett, M., and Roizman, B., 1974, Pseudotype formation between enveloped RNA and DNA viruses, *Nature (London)* **252**:743.

Hudson, L. D., Condra, C., and Lazzarini, R. A., 1986, Cloning and expression of a viral phosphoprotein: Structure suggests vesicular stomatitis virus NS may function by mimicking an RNA template, *J. Gen. Virol.* **67**:1571.

Hughes, J. V., and Johnson, T. C., 1981, Alteration in peptide structure of vesicular stomatitis virus mutant and its central nervous system isolate, *J. Gen. Virol.* **53**:309.

Hunt, D. M., 1983, Vesicular stomatitis virus mutant with altered polyadenylic acid polymerase activity *in vitro*, *J. Virol.* **46**:788.

Hunt, D. M., Emerson, S. U., and Wagner, R. R., 1976, RNA-negative temperature-sensitive mutants of vesicular stomatitis virus: L protein thermosensitivity accounts for transcriptase restriction of group I mutants, *J. Virol.* **18**:596.

Hunt, D. M., Smith, E. F., and Buckley, D. W., 1984, Aberrant polyadenylation by a vesicular stomatitis virus mutant is due to an altered L protein, *J. Virol.* **52**:515.

Huprikar, J., Rabinowitz, S. G., Dal Canto, M. C., and Rundell, M. K., 1986, Persistent infection of a temperature-sensitive G31 vesicular stomatitis virus mutant in neural and nonneural cells; biological and virological characteristics, *J. Virol.* **58**:493.

Iverson, L. E., and Rose, J. K., 1981, Localized attenuation and discontinuous synthesis during vesicular stomatitis virus transcription, *Cell* **23**:477.

Johnson, G. P., and Herman, R. C., 1984, Non-permissive infection of lymphoblastoid cells by vesicular stomatitis virus. I. Synthesis and function of the viral transcripts, *Virus Res.* **1**:259.

Kang, C. Y., and Allen, R., 1978, Host function dependent induction of defective interfering particles of vesicular stomatitis virus, *J. Virol.* **25**:202.

Kang, C. Y., Weide, L. G., and Tischfield, J. A., 1981, Suppression of vesicular stomatitis virus defective interfering particle generation by a function(s) associated with human chromosome 16, *J. Virol.* **40**:946.

Kawai, A., and Matsumoto, S., 1977, Interfering and noninterfering defective particles generated by a rabies small plaque variant virus, *Virology* **76**:60.

Keene, J. D., Schubert, M., and Lazzarini, R. A., 1979, Terminal sequences of vesicular stomatitis virus RNA are both complementary and conserved, *J. Virol.* **32**:167.

Keene, J. D., Schubert, M., and Lazzarini, R. A., 1980, Intervening sequence between the leader region and the nucleocapsid gene of vesicular stomatitis virus RNA, *J. Virol.* **33**:789.

Keene, J. D., Chien, I. M., and Lazzarini, R. A., 1981a, Vesicular stomatitis virus defective particle contains a muted internal leader RNA gene, *Proc. Natl. Acad. Sci. U.S.A.* **78**:2090.

Keene, J. D., Thornton, B. T., and Emerson, S. U., 1981b, Sequence-specific contacts between the RNA polymerase of vesicular stomatitis virus and the leader RNA gene, *Proc. Natl. Acad. Sci. U.S.A.* **78**:6191.

Kennedy-Morrow, J., and Lesnaw, J. A., 1984, Structural and functional characterization of the RNA-positive complementation groups, C and D, of the New Jersey serotype of vesicular stomatitis virus: Assignment of the M gene to the C complementation group, *Virology* **132**:38.

King, A. M. Q., McCahon, D., Slade, W. R., and Newman, J. W. I., 1982, Recombination in RNA, *Cell* **29**:921.

Knipe, D., Lodish, H. F., and Baltimore, D., 1977, Analysis of the defects of temperature-sensitive mutants of vesicular stomatitis virus: Intracellular degradation of specific viral proteins, *J. Virol.* **21**:1140.

Kotwal, G. J., Capone J., Irving, R., Rhee, S. H., Bilan, P., Toneguzzo, F., Hotmann, T., and Ghosh, H. P., 1983, Viral membrane glycoproteins: Comparison of the amino terminal amino acid sequences of the precursor and mature glycoproteins of three serotypes of vesicular stomatitis virus, *Virology* **129**:1.

Kotwal, G. J., Buller, R. M. L., Wunner, W. H., Pringle, C. R., and Ghosh, H. P., 1986, Role of glycosylation in transport of vesicular stomatitis virus envelope glycoprotein. A new class of mutant defection in glycosylation and transport of G protein. *J. Biol. Chem.* **261**;8936.

Kurath, G., and Leong, J. C., 1985, Characterization of infectious hematopoietic necrosis virus mRNA species reveals a nonvirion rhabdovirus protein, *J. Virol.* **53**:462.

Kurath, G., Ahern, K. G., Pearson, G. D., and Leong, J. C., 1985, Molecular cloning of the six mRNA species of infectious hematopoietic necrosis virus, a fish rhabdovirus, and gene order determination by R loop mapping, *J. Virol.* **53**:469.

Kurilla, M. G., and Keene, J. D., 1983, The leader RNA of vesicular stomatitis virus is bound by a cellular protein reactive with anti-La lupus antibodies, *Cell* **34**:837.

Kurilla, M. G., Piwnica-Worms, H., and Keene, J. D., 1982, Rapid and transient localization of the leader RNA of VSV in the nuclei of infected cells, *Proc. Natl. Acad. Sci. U.S.A.* **79**:5240.

Kurilla, M. G., Cabradilla, C. D., Holloway, B. P., and Keene, J. D., 1984, Nucleotide sequence and host La protein interactions of rabies virus leader RNA, *J. Virol.* **50**:773.

Lafay, F., and Benejean, J., 1981, Temperature-sensitive mutants of vesicular stomatitis virus: Tryptic peptide maps of the proteins modified in complementation groups II and IV, *Virology* **111**:93.

Lafon, M., Wiktor, T. J., and Macfarlan, R. I., 1983, Antigenic sites on the CVS rabies virus glycoprotein: Analysis with monoclonal antibodies, *J. Gen. Virol.* **64**:843.

Lai, M. M. C., Baric, R. S., Makino, S., Keck, J. G., Egbert, J., Leibowitz, J. L., and Stohlman, S. A., 1985, Recombination between nonsegmented RNA genomes of murine coronaviruses, *J. Virol.* **56**:449.

Lazzarini, R. A., Keene, J. D., and Schubert, M., 1981, The origin of defective interfering particles of the negative-strand RNA viruses, *Cell* **26**:145.

Lenard, J., Wilson, T., Mancarella, D., Reidler, J., Keller, P., and Elson, E., 1981, Interaction of mutant and wild type M protein of vesicular stomatitis virus with nucleocapsids and membranes, in: *The Replication of Negative Strand Viruses* (D. H. L. Bishop and R. W. Compans, eds.), pp. 855–863, Elsevier/North-Holland, New York.

Lentz, T. L., Wilson, P. T., Hawrot, E., and Speicher, D. W., 1984, Amino acid sequence similarity between rabies virus glycoprotein and snake venom curaremimetic neurotoxins, *Science* **226:**847.

Lesnaw, J. A., Dickson, L. R., and Curry, R. H., 1979, Proposed replicative role of the NS polypeptide of vesicular stomatitis virus: Structural analysis of an electrophoretic variant, *J. Virol.* **31:**8.

Levinson, W., Oppermann, H., Rubinstein, P., and Jackson, L., 1978, Host range restrictions of vesicular stomatitis virus on duck embryo cells, *Virology* **85:**612.

Little, L. M., Lanman, G., and Huang, A. S., 1983a, Immunoprecipitating human antigens associated with vesicular stomatitis virus grown in HeLa cells, *Virology* **129:**127.

Little, L. M., Zavada, J., Der, C. J., and Huang, A. S., 1983b, Identity of HeLa cell determinants acquired by vesicular stomatitis virus with a tumor antigen, *Science* **220:**1069.

Lodish, H. F., and Porter, M., 1981, Vesicular stomatitis virus mRNA and inhibition of translation of cellular mRNA—is there a P function in vesicular stomatitis virus?, *J. Virol.* **38:**504.

Lodish, H. F., and Weiss, R. A., 1979, Selective isolation of mutants of vesicular stomatitis virus defective in production of the viral glycoprotein, *J. Virol.* **30:**177.

Lodmell, D. L., and Chesebro, B., 1984, Murine resistance to street rabies virus: Genetic analysis by testing second-backcross progeny and verification of allelic resistance genes in SJL/J and CBA/J mice, *J. Virol.* **50:**359.

Lodmell, D. L., and Ewalt, L. C., 1985, Pathogenesis of street rabies virus infections in resistant and susceptible strains of mice, *J. Virol.* **55:**788.

Maack, C. A., and Penhoet, E. E., 1980, Biochemical characterization of the *ts*El mutant of vesicular stomatitis virus (New Jersey), *J. Biol. Chem.* **255:**9249.

Marcus, P., and Sekellick, M. J., 1980, Cell-killing by vesicular stomatitis virus: The prototype rhabdovirus, in: *Rhabdoviruses*, Vol. III (D. H. L. Bishop, ed.), pp. 13–50, CRC Press, Boca Raton, Florida.

Marks, D. M., Kennedy-Morrow, J., and Lesnaw, J. A., 1985, Assignment of the temperature-sensitive lesion in the replication mutant *ts* A1 of vesicular stomatitis virus to the N gene, *J. Virol.* **53:**44.

Matsumoto, S., 1970, Rabies virus, *Adv. Virus Res.* **16:**257.

Matthews, R. E. F., 1982, Classification and nomenclature of viruses: Fourth report of the International Committee on Taxonomy of Viruses, *Intervirology* **17:**1.

McGeoch, D. J., 1979, Structure of the gene N : gene NS intercistronic junction in the genome of VSV, *Cell* **17:**673.

McGeoch, D. J., 1981, Structural analysis of animal virus genomes, *J. Gen. Virol.* **55:**1.

McGeoch, D. J., Dolan, A., and Pringle, C. R., 1980, Comparison of nucleotide sequences in the genomes of the New Jersey and Indiana serotypes of vesicular stomatitis virus, *J. Virol.* **33:**69.

McGowan, J. J., Emerson, S. U., and Wagner, R. R., 1982, The plus strand leader RNA of vesicular stomatitis virus inhibits DNA-dependent transcription of adenovirus and SV40 genes in a soluble whole cell extract, *Cell* **28:**325.

Meier, E., Harmison, G. G., Keene, J. D., and Schubert, M., 1984, Sites of copy choice replication involved in generation of vesicular stomatitis virus defective interfering particle RNAs, *J. Virol.* **51:**515.

Metzel, P. S., and Reichmann, M. E., 1981, Characterization of vesicular stomatitis virus mutants by partial proteolysis, *J. Virol.* **37:**248.

Moreau, M.-C., 1974, Inhibition of a vesicular stomatitis virus mutant by rifampin, *J. Virol.* **14:**517.

Morrongiello, M. P., and Simpson, R. W., 1979, Conditional lethal mutants of vesicular stomatitis virus. 4. RNA species detected in non-permissive cells infected with host restricted mutants, *Virology* **93:**506.

Moyer, S. A., Horikami, S. M., and Moyer, R. W., 1981, The effect of the host cell and heterologous viruses on VSV production, in: *The Replication of Negative Strand Viruses* (D. H. L. Bishop and R. W. Compans, eds.), pp. 965–970, Elsevier/North-Holland, New York.

Moyer, S. A., Baker, S. C., and Lessard, J. L., 1986, Tubulin: A factor necessary for the synthesis of both Sendai virus and vesicular stomatitis virus RNAs, *Proc. Natl. Acad. Sci. U.S.A.* **83**:5405.

Mudd, J. A., Leavitt, R. W., Kingsbury, D. T., and Holland, J. J., 1973, Natural selection of mutants of vesicular stomatitis virus by cultured cells of *Drosophila melanogaster, J. Gen. Virol.* **20**:341.

Nichol, S. T., O'Hara, P. J., Holland, J. J., and Perrault, J., 1984, Structure and origin of a novel class of defective interfering particle of vesicular stomatitis virus, *Nucleic Acids Res.* **12**:2775.

Nilsen, T. W., Wood, D. L., and Baglioni, C., 1981, Cross-linking of viral RNA by 4'-aminomethyl-4,5',8-trimethylpsoralen in HeLa cells infected with encephalomyocarditis virus and the *ts*G114 mutant of vesicular stomatitis virus, *Virology* **109**:82.

Nowakowski, M., Bloom, B. R., Ehrenfeld, E., and Summers, D. F., 1973, Restricted replication of vesicular stomatitis virus in human lymphoblastoid cells, *J. Virol.* **12**:1272.

Obijeski, J. F., and Simpson, R. W., 1974, Conditional lethal mutants of vesicular stomatitis virus. II. Synthesis of virus-specific polypeptides in non-permissive cells infected with "RNA –" host restricted mutants, *Virology* **57**:369.

Ohanessian-Guillemain, A., 1959, Etude génétique du virus hereditaire de la Drosophile (σ); Mutations et recombination génétique, *Ann. Genet.* **1**:59.

O'Hara, P. J., Nichol, S. T., Horodyski, F. M., and Holland, J. J., 1984, Vesicular stomatitis virus defective interfering particles can contain extensive genomic sequence rearrangements and base substitutions, *Cell* **36**:915.

Ongradi, J., Cunningham, C., and Szilagyi, J. F., 1985a, The role of polypeptides L and NS in the transcription process of vesicular stomatitis virus New Jersey using the temperature-sensitive mutant *ts*E1, *J. Gen. Virol.* **66**:1011.

Ongradi, J., Cunningham, C., and Szilagyi, J. F., 1985b, Temperature sensitivity of the transcriptase of mutants *ts*B1 and *ts*F1 of vesicular stomatitis virus New Jersey is a consequence of mutation affecting polypeptide L, *J. Gen. Virol.* **66**:1507.

Pal, R., Grinnell, B. W., Snyder, R. M., and Wagner, R. R., 1986, Regulation of viral transcription by the matrix protein of vesicular stomatitis virus probed by monoclonal antibodies and temperature-sensitive mutants, *J. Virol.* **56**:386.

Perrault, J., 1981, Origin and replication of defective interfering particles, *Curr. Top. Microbiol. Immunol.* **93**:151.

Perrault, J., and Leavitt, R. W., 1977, Inverted complementary terminal sequences in single-stranded RNAs and snap-back RNAs from vesicular stomatitis virus defective interfering particles, *J. Gen. Virol.* **38**:35.

Perrault, J., Lane, J. L., and McClure, M. A., 1981, *In vitro* transcription alterations in a vesicular stomatitis virus variant, in: *The Replication of Negative Strand Viruses* (D. H. L. Bishop and R. W. Compans, eds.), pp. 829–836, Elsevier/North-Holland, New York.

Perrault, J., Clinton, G. M., and McClure, M. A., 1983, RNP template of vesicular stomatitis virus regulates transcription and replication functions, *Cell* **35**:175.

Pittman, D., 1965, Temperature-sensitive mutants of a rod-shaped RNA animal virus, *Genetics* **52**:468.

Poirot, M. K., Schnitzlein, W. N., and Reichmann, M. E., 1985, The requirement of protein synthesis and VSV inhibition of host cell RNA synthesis, *Virology* **140**:91.

Polakova, K., Zavadova, Z., Zavada, J., and Russ, G., 1984, Monoclonal antibody against an antigen selectively assembled into vesicular stomatitis virus virions from HeLa cells, *Int. J. Cancer* **34**:91.

Portner, A., Webster, R. G., and Bean, W. H., 1980, Similar frequencies of antigenic variation in Sendai virus, vesicular stomatitis virus and influenza A virus, *Virology* **104**:235.

Pringle, C. R., 1970, Genetic characteristics of conditional lethal mutants of vesicular stomatitis virus induced by 5-fluorouracil, 5-azacytidine and ethyl methane sulphinate, *J. Virol.* **5**:559.

Pringle, C. R., 1975, Conditional lethal mutants of vesicular stomatitis virus, *Curr. Top. Microbiol. Immunol.* **69**:85.

Pringle, C. R., 1977, Genetics of rhabdoviruses, in: *Comprehensive Virology*, Vol. 9 (H. Fraenkel-Conrat and R. R. Wagner, eds.), pp. 239–290, Plenum Press, New York.

Pringle, C. R., 1978, The *td*CE and *hr*CE phenotypes: Host range mutants of vesicular stomatitis virus in which polymerase function is affected, *Cell* 15:597.

Pringle, C. R., 1982, The genetics of vesiculoviruses, *Arch. Virol.* 72:1.

Pringle, C. R., and Wunner, W. H., 1973, Genetic and physiological properties of temperature-sensitive mutants of Cocal virus, *J. Virol.* 12:677.

Pringle, C. R., and Wunner, W. H., 1975, A comparative study of the structure and function of the VSV genome, in: *Negative Strand Viruses*, Vol. 2 (B. W. J. Mahy and R. D. Barry, eds.), pp. 707–723, Academic Press, New York.

Pringle, C. R., and Szilagyi, J. F., 1980, Gene assignment and complementation group, in: *Rhabdoviruses*, Vol. II (D. H. L. Bishop, ed.), pp. 141–161, CRC Press, Boca Raton, Florida.

Pringle, C. R., Duncan, I. B., and Stevenson, M., 1971, Isolation and characterization of temperature-sensitive mutants of vesicular stomatitis virus, New Jersey serotype, *J. Virol.* 8:836.

Pringle, C. R., Devine, V., Wilkie, M., Preston, C. M., Dolan, A., and McGeoch, D. J., 1981, Enhanced mutability associated with a temperature-sensitive mutant of vesicular stomatitis virus, *J. Virol.* 39:377.

Printz, P., 1970, Adaptation du virus de la stomatite vesiculaire à *Drosophila melanogaster*, *Ann. Inst. Pasteur Paris* 119:520.

Rae, B. P., and Elliott, R. M., 1986a, Conservation of potential phosphorylation sites in the NS proteins of the New Jersey and Indiana serotypes of vesicular stomatitis virus, *J. Gen. Virol.* 67:1351.

Rae, B. P., and Elliott, R. M., 1986b, Characterization of the mutations reponsible for the electrophoretic mobility differences in the NS proteins of vesicular stomatitis virus-New Jersey complementation group E mutants, *J. Gen. Virol.* 67:2635.

Rasool, N., and Pringle, C. R., 1986, *In vitro* transcriptase deficiency of temperature-dependent host range mutants of Chandipura virus, *J. Gen. Virol.* 67:851.

Reanney, D. C., 1982, The evolution of RNA viruses, *Annu. Rev. Microbiol.* 36:47.

Reanney, D. C., 1984, The molecular evolution of viruses, in: *The Microbe 1984: I. Viruses* (B. W. J. Mahy and J. R. Pattison, eds.), pp. 175–196, Cambridge University Press.

Reichmann, M. E., and Schnitzlein, W. M., 1979, Defective interfering particles of rhabdoviruses, *Curr. Top. Microbiol. Immunol.* 86:123.

Reichmann, M. E., Pringle, C. R., and Follett, E. A. C., 1971, Defective particles in BHK cells infected with temperature-sensitive mutants of vesicular stomatitis virus, *J. Virol.* 8:154.

Reichmann, M. E., Schnitzlein, W. M., Bishop, D. H. L., Lazzarini, R. A., Beatrice, S. T., and Wagner, R. R., 1978, Classification of the New Jersey serotype of vesicular stomatitis virus into two subtypes, *J. Virol.* 25:446.

Rettenmier, C. W., Dumont, R., and Baltimore, D., 1975, Screening procedure for complementation-dependent mutants of vesicular stomatitis virus, *J. Virol.* 15:41.

Richard-Molard, C., Blondel, D., Wyers, F., and Dezelee, S., 1984, Sigma virus: Growth in *Drosophila melanogaster* cell culture; purification; protein composition and localization, *J. Gen. Virol.* 65:91.

Robertson, J. S., Etchison, J. R., and Summers, D. F., 1982, Comparison of the oligosaccharide structure of the glycoprotein of vesicular stomatitis virus and a thermolabile mutant *tl*17, *J. Gen. Virol.* 58:13.

Rose, J. K., 1980, Complete intergenic and flanking gene sequences from the genome of vesicular stomatitis virus, *Cell* 19:415.

Rose, J. K., and Bergmann, J. E., 1983, Altered cytoplasmic domains affect intracellular transport of the vesicular stomatitis virus glycoprotein, *Cell* 34:513.

Rose, J. K., and Gallione, C. J., 1981, Nucleotide sequences of the mRNA's encoding the vesicular stomatitis virus G and M proteins determined from cDNA clones containing the complete coding regions, *J. Virol.* 39:519.

Rowlands, D., Grabau, E., Spindler, K., Jones, C., Semler, B., and Holland, J., 1980, Virus protein changes and RNA termini alterations evolving during persistent infection, *Cell* **19**:871.

Sarvar, N., and Stollar, V., 1977, Sindbis virus-induced cytopathic effect in clones of *Aedes albopictus* (Singh) cells, *Virology* **80**:390.

Schechmeister, I. L., Streckfuss, J., and St. John, R., 1967, Comparative pathogenicity of vesicular stomatitis virus and its plaque type mutants, *Arch. Gesamte Virusforsch.* **39**:203.

Schlegel, R., and Wade, M., 1985, Biologically active peptides of the vesicular stomatitis virus glycoprotein, *J. Virol.* **53**:319.

Schlesinger, S., Malfer, C., and Schlesinger, M. J., 1984, The formation of vesicular stomatitis virus (San Juan strain) becomes temperature-sensitive when glucose residues are retained as the oligosaccharides of the glycoproteins, *J. Biol. Chem.* **259**:7597.

Schnitzlein, W. M., and Reichmann, M. E., 1985, Characterization of New Jersey vesicular stomatitis virus isolates from horses and black flies during the 1982 outbreak in Colorado, *Virology* **142**:426.

Schubert, M., Keene, J. D., Herman, R. C., and Lazzarini, R. A., 1980, Site of the vesicular stomatitis virus genome specifying polyadenylation and the end of the L gene mRNA, *J. Virol.* **34**:550.

Schubert, M., Harmison, G. G., and Meier, E., 1984, Primary structure of the vesicular stomatitis virus polymerase (L) gene: Evidence for a high frequency of mutations, *J. Virol.* **51**:505.

Seif, I., Coulon, P., Rollin, P. E., and Flamand, A., 1985, Rabies virulence: Effect on pathogenicity and sequence characterization of rabies virus mutations affecting antigenic site III of the glycoprotein, *J. Virol.* **53**:926.

Sekellick, M. J., and Marcus, P., 1980, Persistent infection of rhabdoviruses, in: *Rhabdoviruses*, Vol. III (D. H. L. Bishop, ed.), pp. 67–98, CRC Press, Boca Raton, Florida.

Selimow, M. A., and Nikotina, L. F., 1970, The "rct 40" marker of fixed rabies virus, *Vopr. Virusol.* **15**:161.

Simpson, R. W., and Obijeski, J. F., 1974, Conditional lethal mutants of vesicular stomatitis virus. I. Phenotypic characterization of single and double mutants exhibiting host restriction and temperature sensitivity, *Virology* **57**:357.

Simpson, R. W., Obijeski, J. F., and Morrongiello, M. P., 1979, Conditional lethal mutants of vesicular stomatitis virus. 3. Host range properties, interfering capacity and complementation patterns of specific hr mutants, *Virology* **93**:493.

Spindler, K. R., Horodyski, F. M., and Holland, J. J., 1982, High multiplicities of infection favor rapid and random evolution of vesicular stomatitis virus, *Virology* **119**:96.

Stanners, C. P., Francoeur, A. M., and Lam, T., 1977, Analysis of a VSV mutant with attenuated cytopathogenicity: Mutation in viral function, P, for inhibition of protein synthesis, *Cell* **11**:273.

Storey, D. G., and Yong Kang, C., 1985, Vesicular stomatitis virus-infected cells from which the intracellular pool of functional M proteins is reduced in the presence of G protein, *J. Virol.* **53**:374.

Szilagyi, J. F., and Pringle, C. R., 1975, Virion transcriptase activity differences in host range mutants of vesicular stomatitis virus, *J. Virol.* **16**:927.

Szilagyi, J. F., and Pringle, C. R., 1979, Effect of temperature-sensitive mutation on the RNA transcriptase activity of vesicular stomatitis virus New Jersey, *J. Virol.* **30**:692.

Szilagyi, J. F., Pringle, C. R., and Macpherson, T. M., 1977, Temperature-dependent host range mutation in vesicular stomatitis virus affecting polypeptide L, *J. Virol.* **22**:381.

Templeton, J. W., Holmberg, C., Garber, T., and Sharp, R. M., 1986, Genetic control of serum-neutralizing-antibody response to rabies vaccination and survival after a rabies challenge infection in mice, *J. Virol.* **59**:98.

Teninges, D., Contamine, D., and Brun, G., 1980, *Drosophila* Sigma virus, in: *Rhabdoviruses*, Vol. III (D. H. L. Bishop, ed.), pp. 113–134, CRC Press, Boca Raton, Florida.

Thacore, H. R., and Youngner, J. S., 1975, Abortive infection of a rabbit cornea cell line by vesicular stomatitis virus: Conversion to productive infection by superinfection with vaccinia virus, *J. Virol.* **16:**322.

Tordo, N., Poch, O., Ermine, A., Keith, G., and Rougeon, F., 1986, Walking along the rabies virus genome: Is the large G–L intergenic region a remnant gene? *Proc. Natl. Acad. Sci. U.S.A.* **83:**3914.

Unger, J. T., and Reichmann, M. E., 1973, RNA synthesis in temperature sensitive mutants of vesicular stomatitis virus, *J. Virol.* **12:**570.

Vigier, P., 1966, Contribution a l'étude de l'mutabilité génétique du virus de la Drosophile, *Ann. Genet.* **9:**5.

Villareal, L. P., Breindl, M., and Holland, J. J., 1976, Determination of molar ratios of vesicular stomatitis virus induced RNA species in BHK-21 cells, *Biochemistry* **15:**1663.

Weck, P. K., and Wagner, R. R., 1979, Inhibition of RNA synthesis in mouse myeloma cells infected with vesicular stomatitis virus, *J. Virol.* **25:**770.

Weck, P. K., Carroll, A. R., Shattuck, D. M. and Wagner, R. R., 1979, Use of UV irradiation to identify the genetic information of vesicular stomatitis virus responsible for shutting off cellular RNA synthesis, *J. Virol.* **30:**746.

Weiss, R. A., and Bennett, P. L. P., 1980, Assembly of membrane glycoproteins studied by phenotypic mixing between mutants of vesicular stomatitis virus and retroviruses, *Virology* **100:**252.

Weiss, R. A., Boettiger, D., and Murphy, H. M., 1977, Pseudotypes of avian sarcoma viruses with the envelope properties of vesicular stomatitis virus, *Virology* **76:**808.

Wertz, G. W., 1978, Isolation of possible replicative intermediate structures from vesicular stomatitis virus infected cells, *Virology* **85:**271.

Wertz, G. W., and Levine, M., 1973, RNA synthesis of vesicular stomatitis virus and a small plaque mutant: Effects of cycloheximide, *J. Virol.* **12:**253.

White, B. T., and McGeoch, D. J., 1985, Suppressible amber mutants of vesicular stomatitis virus Indiana serotype, *Virus Res.* (Suppl. 1):27.

Wilson, T., and Lenard, J., 1981, Interaction of wild type and mutant M protein of vesicular stomatitis virus with nucleocapside *in vitro*, *Biochemistry* **20:**1349.

Wilusz, J., Youngner, J. S., and Keene, J. D., 1985, Base mutations in the terminal noncoding regions of the genome of vesicular stomatitis virus isolated from persistent infections of L cells, *Virology* **140:**249.

Wong, P. K. Y., Holloway, A. F., and Cormack, D. V., 1972, Characterization of three complementation groups of vesicular stomatitis virus, *Virology* **50:**829.

Woodgett, C., and Rose, J. K., 1986, Amino-terminal mutation of the vesicular stomatitis virus glycoprotein does not affect its fusion activity, *J. Virol.* **59:**486.

Wu, F. S., and Lucas-Lenard, J. M., 1980, Inhibition of RNA accumulation in mouse L cells infected with vesicular stomatitis virus requires viral ribonucleic acid transcription, *Biochemistry* **19:**804.

Wunner, W. H., Dietzschold, B., Smith, C. L., Lafon, M., and Golub, E., 1985, Antigenic variants of CVS rabies virus with altered glycosylation sites, *Virology* **140:**1.

Yelverton, E., Norton, S., Obijeski, J. F., and Goeddel, D. V., 1983, Rabies virus glycoprotein analogs: Biosynthesis in *Escherichia coli*, *Science* **219:**614

Young, J. F., Taussig, R., Aaronson, R. P., and Palese, P., 1981, Advantages and limitations of the oligonucleotide mapping technique for the analysis of viral RNAs, in: *Replication of Negative Strand Viruses* (D. H. L. Bishop and R. W. Compans, eds.), pp. 209–219, Elsevier/North-Holland, New York.

Youngner, J. S., and Quagliana, D. O., 1976, Temperature-sensitive mutants of vesicular stomatitis virus are conditionally defective particles that interfere with and are rescued by wild-type virus, *J. Virol.* **19:**102.

Youngner, J. S., Dubovi, E. J., Quagliana, D. O., Kelly, M., and Preble, O. T., 1976, Role of temperature-sensitive mutants in persistent infections initiated with vesicular stomatitis virus, *J. Virol.* **19:**90.

Zavada, J., 1972, VSV pseudotype particles with the coat of avian myeloblastosis virus, *Nature (London) New Biol.* **240:**122.

Zavada, J., 1982, The pseudotype paradox, *J. Gen. Virol.* **63**:15.

Zavada, J., and Huang, A. S., 1984, Further characterization of proteins assembled by vesicular stomatitis virus from human tumour cells, *Virology* **138**:16.

Zavada, J., Cerny, L., Zavadova, Z., Bozonova, J., and Altstein, A. D., 1979, A rapid neutralization test for antibodies to bovine leukemia virus, with the use of rhabdovirus pseudotypes, *J. Natl. Cancer Inst.* **62**:95.

Zavada, J., Russ, G., Zavadova, Z., and Sabo, A., 1983a, Vesicular stomatitis virus phenotypically mixed with retroviruses: An efficient detection method, *Acta Virol.* **27**:110.

Zavada, J., Zavadova, Z., Russ, G., Polakova, K., Rajcani, J., Stend, J., and Loksa, J., 1983b, Human cell surface proteins selectively asembled into vesicular stomatitis virus virions, *Virology* **127**:345.

Zavadova, Z., and Zavada, J., 1980, Pseudotypes of vesicular stomatitis virus with coat antigen of bovine leukaemia virus—VSV(BLV): Antigenic surface mosaic and the roles of precipitating antibodies and polycations, *Acta Virol.* **24**:166.

Zilberstein, A., Snider, M. D., Porter, M., and Lodish, H. F., 1980, Mutants of vesicular stomatitis virus blocked at different stages in maturation of the viral glycoprotein, *Cell* **21**:417.

CHAPTER 6

Transcription of Vesicular Stomatitis Virus

SUZANNE URJIL EMERSON

I. INTRODUCTION

Up until 1970, it was not understood why the deproteinized genomic RNAs of positive-strand viruses such as poliovirus were infectious while those of negative-strand viruses such as vesicular stomatitis virus (VSV) were not. An explanation of this difference was obtained when Baltimore *et al.* (1970) incubated highly purified virions of VSV with nonionic detergent, salts, and nucleoside triphosphates and demonstrated that RNA was synthesized *in vitro*. There was an RNA-dependent RNA polymerase packaged in the virion! Since the RNA synthesized *in vitro* was complementary to the genomic RNA, it was thought to be messenger RNA (mRNA), and the enzyme responsible for its synthesis was named the "transcriptase." It is now clear that the obligatory first biosynthetic step in the infectious cycles of negative-strand viruses is transcription. Since the host cells appear to lack enzymes capable of utilizing RNA as templates, the infecting virus particle must carry its own supply of transcriptase into the cell to initiate the viral reproductive cycle. It is generally assumed that the viral polymerase is totally or mostly coded for by viral genes.

Because of its historical significance as the first negative-strand virus shown to carry an RNA polymerase, as well as for technical reasons of ease of growth and purification, VSV has been one of the most intensively studied RNA viruses in regard to transcription. Although genetic, bio-

SUZANNE URJIL EMERSON • Department of Microbiology, University of Virginia School of Medicine, Charlottesville, Virginia 22908.

chemical, and immunological methods have been exploited, the large body of evidence accumulated in the last decade and a half is still contradictory and incomplete, and although the overall process has been defined, a precise understanding of VSV transcription has yet to be obtained.

VSV has five genes that code for five proteins, designated nucleocapsid (N), nonstructural (NS)—which is actually a misnomer [see Chapter 2 (Section II.B)]—large (L), glycoprotein (G), and matrix (M), and all five proteins are components of the infectious virion (Wagner, 1975). Since there are only five viral proteins, the original assumption was that it should be fairly straightforward to determine which viral proteins were required for transcription and exactly which function(s) each protein performed. However, due to the structural simplicity of the virion compared to the biochemical complexity of transcription, each protein must be multifunctional and must act in conjunction with other proteins, so that it has proved to be extremely difficult to uncouple and study the multiple steps involved in transcription.

II. GENOME ORGANIZATION

The VSV genome consists of a single piece of RNA, 11,162 nucleotides long (Schubert *et al.*, 1984), that during transcription serves as the template for the synthesis of six separate and nonoverlapping RNAs (Fig. 1). VSV transcribes five capped and polyadenylated mRNAs that code for the five viral proteins and a small 47-nucleotide-long leader RNA that is not translated (Banerjee *et al.*, 1985). Early studies demonstrated that the mRNAs were not transcribed in equimolar amounts (Villarreal *et al.*, 1976). This unequal transcription was later found to correlate with gene order. UV-inactivation studies of gene target size showed that only the N gene had a target size corresponding to the actual gene size, while the

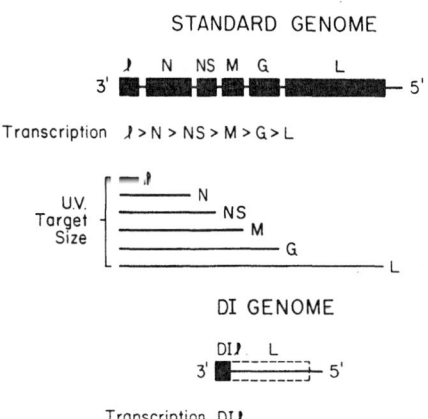

FIGURE 1. Schematic representation of the genomes of the standard and of the most common type of defective interfering (DI) particle of VSV.

other four structural genes had target sizes much larger than predicted (Abraham and Banerjee, 1976; Ball and White, 1976). These data indicated that the genes are transcribed sequentially, in a polar manner, in the order (3')-N-NS-M-G-L-(5'). The N mRNA is transcribed most frequently and the L mRNA least. Annealing of leader RNA to 3'-labeled genome indicated that the leader gene starts at the exact 3' end of the genome and is located 3' to all the other genes (Colonno and Banerjee, 1977). With the cloning and sequencing of the L gene now completed, the entire sequence of the wild-type genome is known (Schubert et al., 1984). The intergenic regions consists of 4 or 5 bases between the leader–N-gene junction and only 2 bases between each two of the other genes. At the 5' terminus, 59 bases are not normally transcribed. Therefore, the transcriptase copies 99.4% of the standard genome into complementary RNA.

In contrast to the standard virions, defective interfering (DI) particles of VSV have a modified structure and a different transcription pattern. The 3' ends of DI particles code for a 46-base-long DI leader RNA that is analogous but not identical to the standard leader RNA (Emerson et al., 1977; Schubert et al., 1978). Adjacent to the leader gene, most DI particles contain sequences that originate within the L gene and continue to the normal 5' terminus of the standard virus genome (Lazzarrini et al., 1981). The great majority of the DI particles transcribe only the DI leader, which means transcription is restricted to a very small portion of the DI genome.

III. PROTEINS INVOLVED IN TRANSCRIPTION

Nonsegmented RNA viruses of the negative sense do not undergo genetic recombination, so this potent tool cannot be used to analyze VSV transcription. Genetic complementation assays have been useful, however. Because the virus apparently lacks a mechanism for proofreading and correcting polymerase mistakes, there is a high mutation rate estimated to be at least 10^{-3} (Flamand, 1980; Pringle, 1982). These mutations often result in the production of temperature-sensitive (ts) mutants that can be placed into genetic complementation groups and the phenotype of each mutant determined. Since each complementation group should consist of mutants altered in one particular protein, the phenotype of the mutants can be used to identify the function of that protein. The ts mutants of the two serotypes of VSV, New Jersey serotype and Indiana serotype, have been classified into five clear-cut complementation groups and a borderline sixth group (Pringle, 1982). With each serotype, ts mutants from two of the groups representing the M and G proteins synthesized normal amounts of both RNA and replicative RNA at nonpermissive temperature. Each of the other three (or four) complementation groups contains ts mutants that displayed an RNA⁻ phenotype at the nonper-

missive temperature. Therefore, three or four viral proteins were implicated as components of the viral RNA synthetic machinery.

Identification of the viral proteins involved in transcription was first
obtained through partial disruption of virions and *in vitro* transcription
analyses (Fig. 2). Treatment of virions with nonionic detergent at low
ionic strength solubilizes the virion envelope and releases the M and G
proteins, leaving a particulate ribonucleocapsid protein (RNP) core that
can be purified by centrifugation (Bishop and Roy, 1972; Emerson and
Wagner, 1972; Szilagyi and Uryvayev, 1983). The RNP core contains the
viral RNA genome, a major structural N protein, and two minor proteins,
L and NS (Wagner, 1975). Since the RNP (depleted of G and M proteins)
retained all the transcriptional activity of the virions, the three RNP
proteins N, NS, and L were logical candidates for the proteins implicated
in transcription by the complementation assays. This assumption was
verified by *in vitro* reconstitution assays.

Disruption of virions with nonionic detergent at high ionic strength
released the L and NS as well as the G and M proteins from the RNA–N

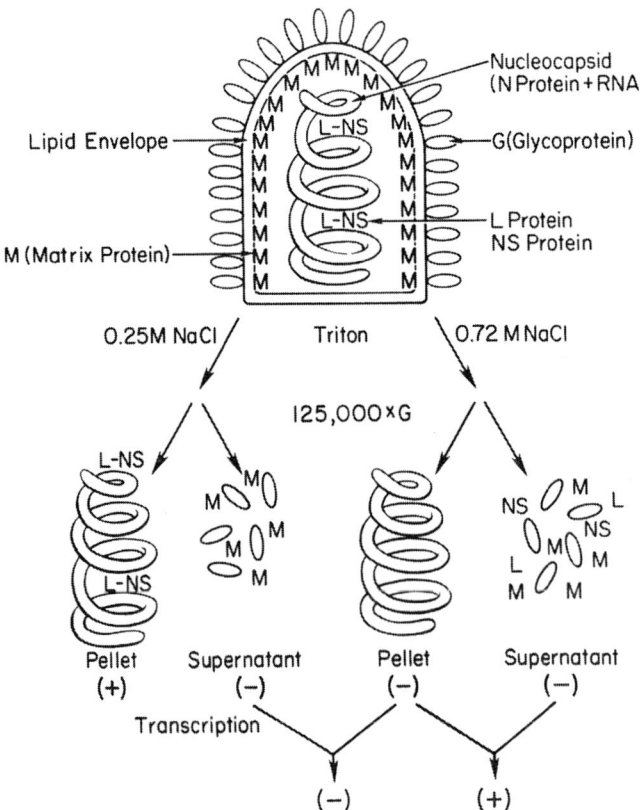

FIGURE 2. Diagram for dissociation of VSV virions under conditions that allow recovery
of transcription activity.

protein complex. The disrupted virions were partitioned by centrifugation into a pellet fraction (RNA–N protein) and a supernatant fraction (soluble L, NS, G, and M proteins) (Emerson and Wagner, 1972). Neither the pellet nor the supernatant fraction alone could synthesize RNA *in vitro*. Recombination of the two fractions, however, reconstituted the active transcription complex. Removal of the N protein from the RNA eliminated transcription. To date, no substrate other than VSV genomic RNA encapsidated with N protein has served as a template for the virion polymerase. It can be concluded, therefore, that N protein is required for transcription. Purification of the solubilized proteins by ion-exchange chromatography and their selective readdition to transcription reactions demonstrated that both L and NS proteins are required for reconstitution of transcription *in vitro* (Fig. 3) (Emerson and Wagner, 1973; Emerson and Yu, 1975).

Antibody experiments have also shown that L, NS, and N are integral components of the active transcription complex. Monospecific antibodies to the L, NS, and N proteins have all been shown to inhibit *in vitro* transcription, while certain hybridoma antibodies to NS and N proteins are also inhibitory (Imblum and Wagner, 1974a; Carroll and Wagner, 1978; Harmon and Summers, 1982; De *et al.*, 1982; Williams and Emerson, unpublished data). There is abundant evidence, then, to indicate that these three viral proteins are required for transcription *in vitro*.

Genetic analyses have convincingly demonstrated that L protein is required for transcription *in vivo* also. Certain *ts* mutants of complementation group I of the Indiana serotype are unable to transcribe at

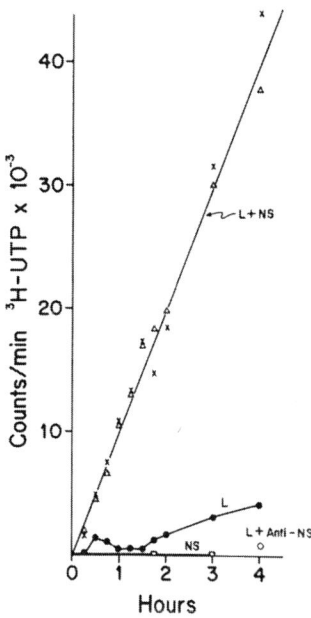

FIGURE 3. *In vitro* reconstitution experiment showing that both L and NS proteins are required for transcription. L and NS proteins were purified by ion-exchange column chromatography and recombined with purified template consisting of the N protein–genomic RNA complex. RNA synthesis was measured by incorporation of [³H]-UTP into trichloroacetic-acid-insoluble material. The template or L fraction often contains a small amount of residual NS protein as demonstrated by the 77% reduction in RNA synthesis by the L–template combination in the presence of antibody to NS protein. Redrawn from Emerson and Yu (1975).

nonpermissive temperatures (Pringle, 1982). *In vitro* reconstitution studies have shown that the *ts* phenotype of group I mutants was correlated with an altered L protein (Hunt *et al.*, 1976). Mixing experiments using different combinations of wild-type (*wt*) and *ts* mutant proteins demonstrated that the template (RNA–N protein) and NS proteins of *ts*114 (group I) were able to reconstitute an active transcription complex at nonpermissive temperatures when *wt* L protein was added. Unfortunately, it was not possible to purify an L protein from *ts*114 that was active at either permissive or nonpermissive temperatures, apparently because the mutant L protein was so labile. However, *ts*114 virions could transcribe at nonpermissive temperatures if L protein purified from wild-type virions was included in the reaction. It was concluded therefore, that the *ts* mutants of group I that have a transcription defect have a temperature-sensitive L protein.

Recently, the monumental task of sequencing and reconstructing a DNA clone of the entire wild-type L gene has been accomplished (Schubert *et al.*, 1984). The cloned L gene has been expressed in COS cell lines, and the resulting L protein is able to complement coinfecting *ts*114 virions at nonpermissive temperatures and rescue the mutant virus (Schubert *et al.*, 1985). These rescue results prove conclusively that the L protein is required for transcription *in vivo*.

In vitro transcription analyses of the New Jersey *ts* mutants have been inconclusive. However, hybrid infectivity assays have confirmed that L protein is required for transcription (Belle-Isle and Emerson, 1982). In these experiments, L and NS proteins purified from *ts* mutant and *wt* viruses were recombined in all possible combinations with a *wt* template, and the reconstituted RNP was used to infect cells. If proper transcription occurred, *wt* mRNAs would be synthesized and the infection would proceed normally. Quantitation of virus yields at permissive vs. nonpermissive temperatures was used to identify the *ts* protein. The data from these experiments showed that the two New Jersey transcription-negative complementation groups both had *ts* L proteins. One of these mutants represented the putative sixth complementation group. Therefore, the low complementation indices obtained when these two mutants were used in mixture to infect cells appears to reflect intragenic complementation. These results suggest that two different L functions, represented by the two mutants, are required for *in vivo* transcription.

Genetic evidence for the *in vivo* requirement for NS and N proteins in transcription has not been obtained, since the complementation groups corresponding to these proteins currently contain only mutants competent for primary transcription (Pringle, 1982). The transcription capacity of these mutants at nonpermissive temperatures probably means that these proteins are not thermolabile if they are synthesized and folded at permissive temperatures, but rather display their *ts* characteristics only when they are both synthesized and analyzed at nonpermissive temperatures.

The *in vivo* evidence that NS protein is required for transcription is indirect. RNP infectivity assays showed that there was over a 200-fold increase in infectious virus released if NS protein, as well as L protein, was included on the infecting RNP (Belle-Isle and Emerson, 1982). The small amount of virus produced when exogenous NS protein was omitted from the complex most likely resulted from trace contamination by endogenous NS protein.

IV. STOICHIOMETRY OF THE TRANSCRIPTION COMPLEX

The ratios of large (L), nonstructural (NS), and nucleocapsid (N) proteins required to form a functional transcription complex are not yet resolved, since there is a lack of agreement between various physical and functional measurements. The N protein is the major protein of the ribonucleocapsid protein (RNP), completely encapsidating the genomic RNA and protecting it from nuclease digestion (Wagner, 1975). There is no evidence for displacement or movement of N protein, and its role in transcription is believed to be strictly structural. The estimated number of N-protein molecules per wild-type genome ranges from 1000 to 2300. From electron-microscopic analyses of the RNP, Nakai and Howatson (1968) calculated that there were 1000 N-protein molecules, while Bishop and Roy (1972) based their calculation of 2300 N molecules per genome on polyacrylamide-gel and biochemical analyses. Using high-resolution scanning transmission electron microscopy (STEM). Thomas et al. (1985) estimated 1258 N molecules per genome, or approximately 1 N molecule per 9 bases. In these STEM images, N-protein subunits appeared as wedge-shaped, bi-lobed structures. Although the absolute number of N molecules is not definite, it appears from electron micrographs that the protein is arranged evenly along the entire length of the genome.

Since the L and NS proteins are both required for transcription and there is evidence for their association in solution (Naito and Ishihama, 1976), it is generally assumed that they function as a complex that carries out the enzymatic steps of transcription. Polyacrylamide-gel analyses of purified virions indicated that there were about 96-protein L and 401 NS-protein molecules per 2000 N-protein molecules (numbers based on corrected molecular weights of 240,000, 25,000, and 47,000, respectively), or 4 times as many NS as L molecules per virion (Mellon and Emerson, 1978). Rebinding assays showed that at saturation, the numbers of L and NS molecules that were re-bound to purified template equaled those found in intact virions, suggesting that all the L and NS in virions is also bound to template. STEM analyses of virions, on the other hand, suggest that there are 50 copies of L protein to 466 of NS, for a ratio of 1 : 9.3 (Thomas et al., 1985). A different physical measurement, ultrastructural electron-microscopic analysis of RNPs tagged with antibodies to L and NS proteins, yielded numbers of 30–35 L molecules and 60–70 NS molecules per template, for a ratio of 1 : 2 (Harmon et al., 1985). All the physical measurements presented above are unable to distinguish between polymerase enzymes bound to promoters and those merely packaged in the virion. Since there are only six genes and a maximum of six promoters, whereas there are many more L and NS molecules, most of the L and NS detected in these measurements may reflect packaging requirements, and the stoichiometry of the transcriptionally active complex could be quite different.

Functional assays have also produced disagreement on the ratio of L to NS. De and Banerjee (1985) titrated L and NS proteins against template and found that optimal transcription occurred at a ratio of 1 L to 70 NS molecules, to give 2800 NS molecules per template. Because this number of NS molecules is greater than the number of N molecules, the conclusion was that NS protein is required in stoichiometric rather than catalytic amounts. This requirement for high levels of NS protein is difficult to reconcile with the rebinding data that show that the template is saturated with NS protein when approximately 400 molecules are bound. The differences reported could be due to a significant difference between an actively transcribing complex and a quiescent one. However, an alternative explanation is suggested by the results of Naito and Ishihama (1976). They purified various L–NS complexes by column chromatography and plotted the L/NS protein ratio vs. transcription activity *in vitro*. Using the corrected molecular weights for the proteins, they reported maximal transcription with an L/NS ratio of 1 : 2. These numbers agree reasonably well with the ratios determined by quantitating all the proteins in the virions. However, they found additionally that if they titrated complexes containing suboptimal NS protein, more NS protein than predicted needed to be added in order to obtain optimum transcription. These results suggest that the polymerase complex, once dissociated, may reform inefficiently *in vitro* and an excess of NS protein may be required to drive the formation of an active complex.

All these studies do agree that there are more NS proteins than L proteins involved in transcription, and there are more polymerases present than could be bound to the limited number of potential promoters. Additionally, there is evidence that the transcriptase is a heteromultimeric enzyme. The evidence for *in vitro* complementation of transcription by two different temperature-sensitive (*ts*) L mutants suggests that there are a minimum of 2 L molecules per complex (Belle-Isle and Emerson, 1982; Pringle, 1982). Other experiments demonstrating rescue of *ts* mutants by UV-irradiated viruses indicate that intragenic complementation of NS occurs, which suggests that there are also 2 or more NS molecules per enzyme complex (Deutsch *et al.*, 1979).

V. TRANSCRIPTION MODELS

It is well accepted that there is a gradient of transcription such that each viral gene is transcribed less frequently than the gene 3′ to it. Therefore, all models of transcription must account for this polarity as well as consider why the mRNAs are all capped, methylated, and polyadenylated, while the leader RNA is neither capped, methylated, nor polyadenylated.

Originally, a model of precursor cleavage followed by posttranscriptional modification was proposed (Colonno and Banerjee, 1976). Supporting evidence for this model included the discoveries of putative precursor RNAs in the form of covalently linked transcripts derived from adjacent genes. In the case of mRNAs, some of these transcripts were covalently linked by polyadenylic acid [poly(A)] sequences (Herman *et al.*, 1978). In addition, leader sequences attached 5' to the N mRNA and trailer sequences attached 3' to the L mRNA have been found (Chinchar *et al.*, 1982; Schubert and Lazzarini, 1981). These linked RNAs could represent processing intermediates. The structure of the cap also lends credence to the processing model. In a processing model, a maximum of one phosphate group could be donated to the cap structure by the first A residue. *In vitro* labeling experiments have indeed shown that only the α-phosphate of labeled ATP is donated to the cap structure, while the capping GTP donates both an α- and a β-phosphate. These results are consistent with, but do not prove, a precursor model. Evidence against this model includes the discovery of short, triphosphate-initiated RNAs corresponding to the 5' ends of mRNAs as well as identification of a low percentage of full-length mRNAs containing 5' triphosphate termini. Because these RNAs have multiple phosphates on their 5' ends, they could not have arisen by processing and must reflect initiation events. It has not been possible to process the putative precursor RNAs *in vitro*, so there is no evidence that they are normal intermediates. In addition, they are found at a very low frequency. Although the processing model is really out of favor at this time, it is still theoretically possible.

A second model, which accounts just as well for the polarity of transcription, is the stop–start model (Banerjee *et al.*, 1977). According to this model, the polymerase initiates and terminates RNA chain synthesis at each gene junction, but all 3' proximal sequences must be transcribed before the 5' adjacent gene can be completely transcribed. Indirect evidence for a stop–start mechanism is provided by defective interfering (DI) particles. The majority of DI particles transcribe only the DI leader. If RNA processing were involved in generating leader RNAs, some RNA transcribed from the remainder of the DI genome should be found. However, the DI particles effectively terminate transcription at the end of the DI-particle leader gene and no other RNAs are synthesized (Emerson *et al.*, 1977). Since most of the DI particles have deletions covering the start of the L gene, it is probable that the DI-particle transcriptase synthesizes the DI-particle leader, terminates synthesis at the end of the leader gene, and is unable to reinitiate transcription because the initiation region of the L gene is absent. Direct evidence that the transcriptase can indeed initiate at the beginning of internal genes derives from the identification of numerous small RNAs containing triphosphate termini. Under conditions of matrix (M)-protein inhibition of transcription, a set of small oligonucleotides corresponding to the first 11–14 bases of the N gene are actually synthesized in molar excess over leader RNA (Pinney and Emerson, 1982a,b). These RNAs have triphosphate termini and are faithfully initiated at the exact end of the gene in large amounts, so the transcriptase is clearly able to initiate efficiently at discrete internal sequences. Other, longer, triphosphate-initiated oligonucleotides corresponding to the 5' ends of the N- NS- and M-gene mRNAs have also been identified (Testa et al., 1980a). Although these uncapped oligonucleotides are found in greater abundance than the uncleaved putative precursor RNAs, like the possible precursor RNAs, they have not been chased into mature mRNAs or even into capped oligonucleotides. Therefore, it is unclear whether they represent true transcription intermediates or simply common artifacts.

The stop–start model is currently favored over the processing model. However, there are two conflicting variations of the stop–start model, and there are data to support both of them. One variation of this model proposes that there is a single promoter or polymerase entry site on the very 3' end of the genome. In this model, sequential synthesis is obligatory, because the only way for the transcriptase to reach internal genes is to transcribe the 3' proximal sequences first (Banerjee et al., 1977). The second variation relies on a cascade mechanism to explain polarity (Testa et al., 1980a). In this variation, each gene has its own promoter or polymerase entry site, but although initiations occur simultaneously at each gene, internal transcriptases are unable to completely traverse an internal gene until the adjacent 3' sequences are copied.

Evidence for both models has been obtained by partial reaction experiments. The leader RNA has the sequence 5' ACG, which is easily distinguished from the shared 5' AACAG sequence of the mRNAs. Therefore, if a transcription reaction is performed with ATP and CTP as the only available nucleotides, synthesis from the leader gene should result in an AC dimer, while that from any internal gene should yield AACA. Similarly if ATP, CTP, and GTP are included and UTP is omitted, the predicted products of the leader and internal genes can be distinguished. When the products of partial transcription by solubilized virions were characterized, triphosphate-initiated oligonucleotides derived from both leader and internal genes were identified (Chanda and Banerjee, 1981; Naeve and Summers, 1981). These results demonstrate that internal initiations can occur concomitantly with leader initiations and are consistent with there being a separate polymerase entry site or promoter for each gene. However, reconstitution studies yielded quite different results. Transcription in the presence of only ATP and CTP by a reconstituted complex, as opposed to solubilized virions, produced only AC: leader RNAs, but no mRNAs, were being initiated in the reconstituted system (Fig. 4) (Emerson, 1982). If the reconstituted complexes were allowed to transcribe in the presence of all four ribonucleoside triphosphates for

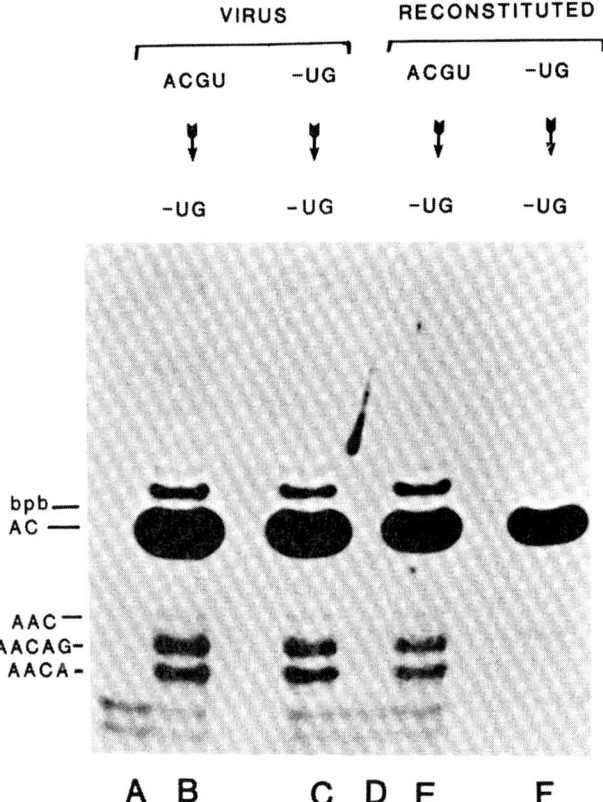

FIGURE 4. Two-step transcription reactions indicating that there is a single polymerase entry site on the VSV template. Virions or reconstituted complexes were incubated in a transcription mixture containing either all four ribonucleoside triphosphates or only ATP and CTP. After 45 min, the samples were centrifuged through a Sephadex G50 column to remove all nucleotides, and the effluent containing the viral components was incubated with [α-³²P]-CTP and ATP for 3 hr. The RNA products were purified, treated with alkaline phosphatase, and analyzed on a 20% polyacrylamide gel. The nucleotides present in the first and second incubation are shown above and below the arrows, respectively. (A, D) Isotope controls; (B, C) virions; (E, F) reconstituted complexes; (bpb) Bromophenol blue dye; (AC) initiating oligonucleotide from leader; (AAC, AACAG, AACA) initiating oligonucleotides from mRNA. Reprinted from Emerson (1982).

sufficient time to begin synthesizing mRNAs and UTP and GTP were then removed, both leader and mRNA oligonucleotides were synthesized. These data suggest that there is a single polymerase entry site, that it is located at the 3' end of the genome, and that polymerases can reach internal sites only by transcribing the 3' proximal sequences. In addition, these experiments indicate that once a polymerase has reached the start of an internal gene and is stopped there, either by removal of nucleotides or by packaging of the RNP into virions, the polymerase is frozen in position but can resume synthesis at that site when transcription is again permitted (Fig. 5). This conclusion is supported by the observation that lowering the NaCl concentration from 0.1 to 0.05 M increases leader AC synthesis 5-fold but does not affect mRNA starts. The interpretation of these results is that lowering the salt increases *de novo* polymerase initiations, and since the polymerases can bind directly only to the 3' end, only AC synthesis is increased. However, when Thornton *et al.* (1984) repeated the reconstitution experiments with an AC reaction, they found both leader and mRNA starts, although 60% of the label in the "mRNA" oligonucleotide band was in sequences corresponding to neither leader nor mRNA.

Until further experiments are done, there is no convincing way to reconcile the aforediscussed data, and it may be useful to review numerous caveats. First of all, these are all *in vitro* experiments, and the results may not accurately reflect what occurs *in vivo*. Also, the partial

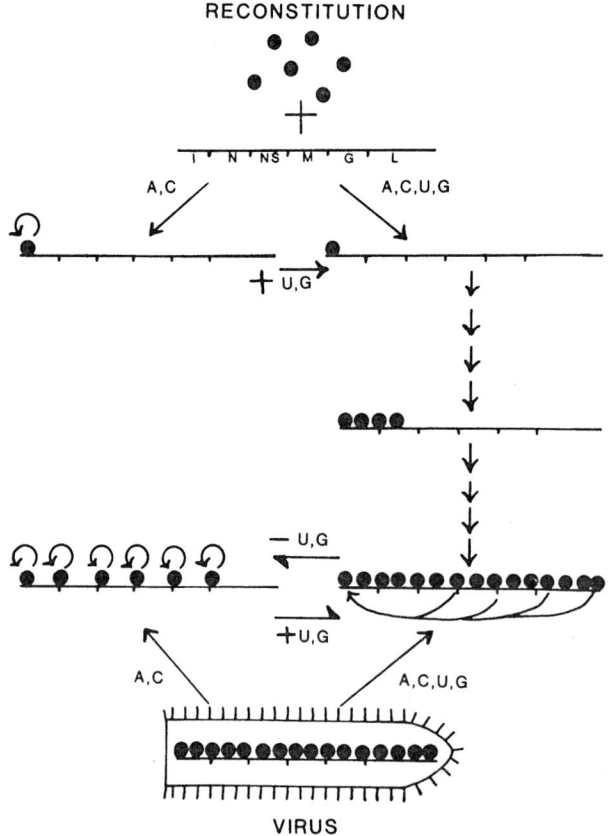

FIGURE 5. Model interpreting two-step incubation results. (●) Polymerases; (—) template, with the 3' terminus on the left; (A, C, U, G) nucleoside triphosphates present in each reaction. The curved arrows denote reiterative transcription. Reprinted from Emerson (1982).

reaction products as well as most of the small RNAs described previously all have triphosphate ends rather than the normal caps and cannot be chased into larger products, so they may not represent bona fide precursors. Additionally, NS protein has been implicated in L-protein bindings, and NS protein is extremely difficult to remove totally from the template. If only a small percentage of the original NS remained bound to template at internal sites, it could possibly direct L there during reconstitution, whereas a native polymerase complex or soluble NS might enter only at the 3' end. Also, these experiments all generate short RNAs, and the question of reiterative transcription complicates the issue. Reiterative transcription has been postulated to account for poly(A) synthesis and the synthesis of short oligonucleotides corresponding to N mRNA starts. With reiterative transcription, a polymerase would have to initiate binding only once to transcribe a sequence many times, while synthesis of a longer product would require a rate-limiting initiation step for each round of synthesis; therefore, quantitative comparisons between synthesis of short and synthesis of long RNAs may be very misleading. Finally, small differences in reaction conditions or virus stocks could lead to different results. The salt concentration has been shown to quantitatively alter the RNA product distribution, and the M protein, which can affect transcription, is normally present *in vitro* in much higher amounts than *in vivo* (Pinney and Emerson, 1982b).

Since normally the mRNAs are capped, analyses of the synthesis of capped RNAs might reflect *in vivo* steps more accurately. However, the *in vitro* analysis of capped RNAs has also produced conflicting results. Iverson and Rose (1982) analyzed T1 digests of transcribed RNA to determine the kinetics of synthesis. Unique T1-ribonuclease oligonucleotides derived from within the first 50 bases of the leader, N, NS, M, and G RNAs can be separated and identified on two-dimensional fingerprints. Quantitation of these unique oligonucleotides synthesized under conditions conducive to normal mRNA transcription showed sequential transcription of the 5' ends of the RNAs. The results were deemed inconsistent with a simultaneous-initiation model. Piwnica-Worms and Keene (1983) also concluded that there is a single entry site on the basis of kinetics of synthesis of small capped RNAs *in vitro*. Short capped RNAs from the 5' termini of N and NS genes were found to appear sequentially during transcription. Since synthesis of these short oligonucleotides had UV-inactivation kinetics similar to those of the mRNAs, the oligonucleotides appeared to be normal precursors derived by sequential initiation events. A different approach was used by Talib and Hearst (1983) to obtain data supporting multiple promoters. They found that aurintricarboxylic acid or vanadyl ribonucleoside complexes inhibited mRNA and leader synthesis, but allowed synthesis of a small capped RNA approximately 68 bases long. Although the 5' sequence of this oligonucleotide was not determined, the last 19 bases correspond to N-gene sequences. Their conclusion was that there are actually different types of polymerase complexes, one for leader synthesis and one for mRNA synthesis. It is not clear why these inhibitors would prohibit full-length mRNA synthesis while allowing only N-mRNA initiations that prematurely terminated at a precise site within the N gene.

VI. FUNCTIONS OF TRANSCRIPTION PROTEINS

A. Nucleocapsid Protein

The N protein encapsidates the viral genomic RNA to generate the helical ribonucleocapsid (Wagner, 1975). The interaction of N protein with the RNA is resistant to high salt concentrations or 8 M urea but can be disrupted with very severe protein denaturants such as sodium dodecyl sulfate (SDS) or anidinium hydrochloride (Blumberg et al., 1984). As far as is known, the N protein serves strictly in a structural role in transcription, although it may be involved in regulating replication (see Chapter 7). The N protein is absolutely required for transcription, since neither deproteinized genomic RNA nor various synthetic ribopolymers will serve as a template for the viral transcriptase (Emerson and Wagner, 1972). Because some regions of the genome are theoretically capable of extensive base-pairing and some defective interfering (DI)-particle genomes can actually form perfect hairpin duplexes, one function of N protein is probably to keep the genome RNA in an extended form suitable for transcription. The N protein also efficiently protects the genome from nucleases. In fact, the polymerase can actually transcribe effectively in the presence of high levels of pancreatic nuclease. Therefore, the N protein alone or the N protein in combination with the polymerase must interact with virtually every nucleotide in the genome. Footprinting studies have shown that the methylation patterns of deproteinized and N-encapsidated genome RNA are identical (Keene et al., 1981) Therefore, the N protein does not interact directly with the bases, but apparently binds to the phosphate backbone of the RNA.

It seems likely that N protein also serves to concentrate the polymerase on the template. Although the actual number of polymerase molecules per template is uncertain, it is clearly more than six, the maximum number of promoters. The DI particles carry a comparable number of L and NS molecules per template unit length, as do standard particles, yet only a small fraction of the DI genome is transcribed (Emerson and Wagner, 1972; Schubert et al., 1978). Therefore, clearly, at least in DI particles, the vast majority of polymerases are not simultaneously involved in transcription, but are merely strongly associated with the template. Since complexes of soluble N and NS proteins can be isolated (Bell et al., 1984), it may be that N–NS interactions are responsible for packaging the polymerase onto the template.

B. Large and Nonstructural Proteins

L and NS proteins are both required for transcription. If, as currently suspected, no host proteins are involved, these two viral proteins together are responsible for proper initiation, elongation, capping, two methylations, polyadenylation, and either termination or processing. The central question is how these two proteins carry out all these functions and in addition are regulated so that under other circumstances they function to replicate rather than transcribe the genome.

The L and NS proteins can bind to each other in solution (Naito and Ishihama, 1976), but whether they exist as a complex on the template has not been directly shown. Rebinding experiments using purified templates, L, and NS proteins demonstrated that NS by itself could bind to saturation on the template, while L protein was unable to bind to template unless NS protein was also present (Mellon and Emerson, 1978). When both proteins were present at saturating levels, the amounts of re-bound L and NS per template were the same as those found in mature virions. Thus, it appears that one function of NS is to promote L interaction with the template and, by so doing, to determine the amount of L on the template. The simplest interpretation of these results is that the two proteins form a complex on the template. This binding of NS to template could represent packaging via N-protein interactions or specific binding to RNA sequences or both.

Footprinting studies suggest that NS may be the protein that recognizes the promoter. The 3' end of the genome is extremely important in that both transcription and replication begin here, while N encapsidation is initiated on the complementary sequences (Blumberg et al., 1983). The importance of this region is emphasized by the sequence conservation within the first 18 bases for the Indiana, New Jersey, Cocal, Piry, and Chandipura minus-strand genomes and for the Indiana and New Jersey plus-strand genomes (Giorgi et al., 1983; Keene et al., 1979). When the 3' sequences of the standard and DI genomes were compared, they were found to diverge after base 18. Using dimethyl sulfate methylation patterns to study protein interaction with the genome RNA, Keene et al. (1981) found that NS protein changed the base methylation pattern of bases 15–30 at the 3' end of the genome. When DI-particle genomes were similarly studied, methylations of bases 17–37 were altered by NS binding (Isaac and Keene, 1982). Since intimate association of protein and RNA can alter the susceptibility of bases to methylation, these results demonstrate that NS binds tightly and specifically to these regions. Since NS is required for L binding, NS probably functions by recognizing these sequences and thereby directing L to the neighboring initiation sequences at the 3' end. The fact that NS binds at the region where the DI-particle and standard genomes diverge may have implications for how DI particles compete with standard particles during replication (see Chapter 8).

The L protein has long been assumed to be the enzymatic unit of the transcriptase. This assumption is based not on data, but rather on the intuitive logic that since transcription involves so many steps, a very large protein must be required to carry out the multitude of functions. With the one exception discussed below, RNA synthesis and modification occur only when all three transcription proteins (N, NS, and L) are present simultaneously. In addition, modification occurs only on nascent transcripts; it has not been possible to use viral proteins to modify exogenous RNAs. This inability to uncouple the various steps in transcription has made it extremely difficult to assign specific functions.

There is general agreement that both L and NS proteins are required for mRNA synthesis. However, De and Banerjee (1984) reported that the L protein of either vesicular stomatitis virus (VSV)-Indiana or VSV-New Jersey was able to synthesize AC (leader) and AACA (mRNA) oligonucleotides during partial reactions in the absence of NS protein. They concluded that L protein was sufficient for initiation, but that NS was required for elongation. However, synthesis of both AC and AACA includes elongation as well as initiation steps, so it is not clear what length transcript could be synthesized before NS protein was required. Monoclonal antibodies to NS protein inhibit both AC transcription and mRNA synthesis, indicating that in solubilized virions at least, NS protein is normally intimately associated with the initiation step of leader synthesis (Williams and Emerson, unpublished data). The report that L alone can synthesize small oligonucleotides is important because it is the first direct evidence that L is the actual catalytic unit of the polymerase. In cases like this, however, where transcription is probably reiterative, a very low level of NS contamination of either template or the L protein might support oligonucleotide synthesis but not mRNA production. Therefore, because NS is so difficult to remove completely and because it binds very strongly to the 3' initiation region of the template where the AC reaction occurs (Keene et al., 1981), the conclusion that L alone can synthesize oligonucleotides must be accepted with some reservation.

There is reasonable, but not unequivocal, evidence that the RNA modification steps of capping, methylation, and polyadenylation are carried out by viral proteins. The origin of phosphates in the cap structure is unique, and the mechanism of cap formation suggests that a viral rather than a cellular enzyme is involved (Banerjee et al., 1977). Neither the capping protein nor the minimum RNA substrate for capping is known. Many small uncapped RNAs that represent the 5' termini of mRNAs have been identified. If capping depends solely on sequence recognition, any of these small RNAs should be substrates, because they all contain the first five bases common to all the mRNAs. Unfortunately, capping requirements appear to be more complex than this. There are data that suggest that there is a minimum size limit that must be attained before capping occurs. The shortest capped transcript identified thus far is 23 bases long (Schubert et al., 1982), and Piwnica-Worms and Keene (1983) described oligonucleotides from the 5′ end of the N mRNA that were not capped when 35 bases long, but were capped when 37 bases long. It may be that in the polymerase complex, the capping site is distant from the elongation site and only transcripts over a certain length are able to reach the capping site. If this were so, it would seem that addition of exogenous RNAs to soluble enzyme would overcome this length requirement. However, triphosphate-initiated oligonucleotides corresponding to the first 11–14 bases at the 5′ end of the N gene are not capped *in vitro* when mixed with solubilized viral proteins and GTP in the presence or absence of ATP (Green and Emerson, unpublished data). The failure to cap these RNAs could be due to, among other things, the wrong number of phosphates on the 5′ end, the inability of the capping protein(s) to interact with soluble RNA, or the wrong length or conformation of RNA. In the absence of an *in vitro* system that will cap exogenous RNA, it will be very difficult to determine the protein(s) involved and the correct substrate.

VSV carries out two methylation reactions. First, the mRNAs are capped, then methylated on the first A residue, and finally methylated on the cap G (Testa and Banerjee, 1977). Both methylation reactions appear to be carried out by viral proteins. Two host-range mutants of VSV have been studied that synthesize capped but nonmethylated RNA *in vitro* and in certain cell lines, whereas other host-range mutants make normally methylated RNA under the same conditions. Since at least two of the host-range mutants synthesize unmethylated RNAs in the same cells in which other mutants make normally methylated RNAs, the viruses and not the cells must have the altered methylase activities (Horikami and Moyer, 1982; Horikami *et al.*, 1984). Although the viral protein or proteins responsible for the methylations have not yet been identified, mixed reconstitution experiments using wild-type and mutant virus proteins should answer the question.

Polyadenylation is the best understood of the RNA-modification reactions. Originally, Schubert *et al.* (1980) postulated that a sequence of seven U residues at the 5′ terminus of the L gene controls polyadenylation. This conclusion was confirmed when the same sequence was found at the terminus of each of the other four protein-coding genes but not at the end of the leader gene (see Chapter 4). Because the seven U residues at the polyadenylation site are not enough to code directly for the observed poly(A) tracts of 100 or more A residues, a chattering or reiterative transcription of the stretch of U residues was proposed (Schubert *et al.*, 1980). Identification of tandem mRNAs linked by long poly(A) tracts confirmed that poly(A) synthesis occurs via transcription, rather than by posttranscriptional addition after RNA release from the template (Herman *et al.*, 1978).

Polyadenylation appears to be a function of the L protein. A group I temperature-sensitive (*ts*) mutant of the Indiana serotype synthesizes abnormally long poly(A) tracts (Hunt, 1983). However, since the aberrant polyadenylation was not related to the ts phenotype, there was no genetic evidence that the *ts* L mutation was responsible. *In vitro* reconstitution assays mixing wild-type and mutant proteins demonstrated directly that the L protein of the mutant was responsible for the synthesis of the long poly(A) (Hunt *et al.*, 1984). Since polyadenylation involves chain elongation, these results are consistent with the supposition that L is also responsible for overall RNA chain elongation.

S-adenosyl methionine (SAM) is the methyl donor for VSV RNA methylations, and *S*-adenosylhomocysteine (SAH) is a competitive inhibitor of the methylation reactions. Rose *et al.* (1977) found that addition of SAH to *in vitro* transcriptions causes the production of abnormally long poly(A) tracts. Originally, it was suggested that there was a relationship between methylation at the 5' end of the RNA and polyadenylation at the 3' end. However, in the absence of both SAH and SAM, where there is also a very low level of methylation, the poly(A) is of normal length. Therefore, it may be that L protein contains the methylase activities, as well as the polyadenylase activity, and the interaction of SAH with L protein results in a subtle change that causes L to synthesize longer poly(A).

As evidence accumulates, it seems more and more likely that the L protein carries out most, if not all, enzymatic steps connected with transcription. A major question is still how L functions are regulated and what part NS protein plays.

VII. ROLE OF ATP IN TRANSCRIPTION

The K_m for ATP in vesicular stomatitis virus (VSV) transcription is 10-fold higher than for any of the other nucleoside triphosphates (Testa and Banerjee, 1979). High concentrations of ATP could be required for a specific step or for one or more of a multitude of steps including initiations, polyadenylations, or protein phosphorylation. The extensive phosphorylation of nonstructural (NS) protein is intriguing. NS protein is phosphorylated *in vivo*, and virion-associated kinases phosphorylate it *in vitro*. The DNA sequence of the NS gene indicates that the NS protein contains 33 potential phosphorylation sites at serine and threonine residues (Gallione *et al.*, 1981; Marnell and Summers, 1984). Peptide mapping analyses showed that at least 21 of the sites are phosphorylated (Hsu *et al.*, 1982). When NS protein is examined by O'Farrell gels, multiple species are resolved, so phosphorylation leads to a heterogeneous population of NS (Hsu and Kingsbury, 1982). Whether the different forms of NS have different functions or are functionally identical has not been resolved.

One approach to determining the effect of phosphorylation on NS activity is to purify the different phosphorylated forms and test them for their ability to support transcription in an *in vitro* reconstitution system.

DEAE column chromatography was used to separate the NS protein from virions into two populations displaying different average levels of phosphorylation (Kingsford and Emerson, 1980). Only the more highly phosphorylated population was able to reconstitute transcription *in vitro*. However, at saturation of activity, reconstitution by the active population gave only 50% as much transcription as did unfractionated NS protein. Addition of the underphosphorylated population, which was inactive by itself, stimulated transcription by the more phosphorylated population. Both DEAE-purified populations were equally effective at mediating large (L)-protein binding to template, so the underphosphorylated species can react with both template and L protein (Williams and Emerson, unpublished data). These results are consistent with the view that phosphorylation can affect some critical NS function in transcription, but they neither prove it nor provide insight into the mechanism.

Clinton *et al.* (1978) identified two differently phosphorylated populations of NS protein separable on acid urea gels. Ribonucleocapsid proteins (RNPs) from either virions or the cytoplasm of infected cells contained the less phosphorylated form, while the more phosphorylated form was free in the cytoplasm and not bound to RNPs. Bands possibly corresponding to these were seen on SDS–urea gels by Kingsford and Emerson (1980) for both the transcriptionally active and the inactive populations of NS purified from virions. Because different gel systems, cell lines, and virus stocks are used by different investigators and because the NS population is so complex, results from different laboratories are almost impossible to compare. It seems that either pure species of NS, rather than heterogeneous populations, or specific antibodies to NS will be required to give definitive results.

Phosphatase treatment of NS protein has also been employed to study the role of phosphorylation in NS function. Bacterial alkaline phosphatase treatment of RNPs removes phosphates from NS and decreases subsequent transcription by 80% (Kingsbury *et al.*, 1981). This phosphatase-treated NS protein could be rephosphorylated *in vitro* by virion-associated kinases, but the lost transcription ability was not restored. Sinacore and Lucas-Lenard (1982) carried out similar phosphatase experiments and showed that along with decreased transcription, most of the NS and L proteins were released from the template. There is an apparent correlation between the phosphate content of NS protein and transcriptional efficiency. It remains to be shown whether the role of phosphates is to increase the charge of the protein in a gross way or whether there are specific sites that must be phosphorylated to activate NS function.

There are protein kinases in VSV virions that phosphorylate NS *in vitro* mainly at serine and threonine residues. The major kinase activity in the virion derives from a cellular, rather than a viral, protein (Imblum and Wagner, 1974b; Harmon *et al.*, 1983). The cellular kinase pp60*src*, which phosphorylates tyrosine residues, is also found in VSV virions (Clinton *et al.*, 1982). Bell *et al.* (1984) found that antibodies to NS protein precipitated protein complexes containing kinase activity including c-*src*, so kinases have a high affinity for the RNP complex or the viral proteins in it. Therefore, there is more than one kinase in purified virions, and the amount is substantial. Sinacore and Lucas-Lenard (1982) reported that a 60-fold decrease in endogenous kinase still left nearly saturating amounts of kinase activity with respect to endogenous protein phosphorylation.

Purified L protein is also reported to have a serine kinase activity that phosphorylates NS protein *in vitro* (Sanchez *et al.*, 1985). 8-Azido-ATP was shown to label purified L protein, and cross-linking of the ATP to L protein inhibited transcription. Since the 8-azido-ATP at concentrations way below the K_m for transcription could both bind to L and be used to phosphorylate NS protein, the conclusion was that L protein is a kinase for NS protein and that this phosphorylation is required for transcription. Whether or not L protein serves to phosphorylate NS, there must be other requirements as well, since phosphatase-treated NS protein is not reactivated in the presence of L protein and virion-associated cellular kinases, and the inactive population of NS from DEAE columns remains inactive under reconstitution conditions in which phosphorylation should occur.

The high K_m for ATP could reflect initiation requirements. Commonly, the K_m for initiation of RNA synthesis is higher than that for elongation, and the leader RNA and all five mRNAs have A as their first residue. The ATP analogue β-γ-imido-ATP has been used to analyze the requirement for ATP in transcription. Since the β–γ phosphate bond cannot be cleaved but the α–β phosphate bond can be, this analogue is not a substrate for kinases but can be incorporated into RNA. In the presence of β-γ-imido-ATP, VSV transcription was drastically reduced, leading Testa and Banerjee (1979) to propose that cleavage of a β–γ phosphate bond is required for initiation. They also reported that preincubation with ATP and CTP, followed by a chase in the presence of all four nucleoside triphosphates with β-γ-imido-ATP substituted for ATP, allowed full-length RNA synthesis representative of replication rather than transcription (Testa *et al.*, 1980b). In contrast, we found that although β-γ-imido-ATP greatly reduced overall RNA synthesis, it supported synthesis of both standard and defective-interfering-particle leader RNAs as well as synthesis of the initiating oligonucleotides AC from leader and the 14-mer from the nucleocapsid (N)-protein gene. The same results were obtained using ATP concentrations 10-fold below the K_m (Green and Emerson, 1984). These results suggest that initiation *per se* is not inhibited by β-γ-imido ATP, but that efficient mRNA synthesis requires a hydrolyzable bond in this position and high concentrations of ATP. The effects of β-γ-imido ATP on RNA synthesis cannot be readily explained, but they emphasize that a critical step must occur at the junction of the leader and the N gene or that there is a great difference in the synthesis of leader RNA as compared to that of authentic mRNA.

VIII. CONCLUSIONS

Vesicular stomatitis virus (VSV) transcription appears to be carried out solely by virally encoded proteins. Three of the five viral proteins—nucleocapsid (N), large (L), and nonstructural (NS)—are absolutely required for mRNA synthesis, while the remaining two viral proteins—glycoprotein (G) and matrix (M)—do not participate. The N protein plays a passive role in transcription and is probably required to keep the genomic RNA in an extended form; the N protein may also interact directly with the NS protein. The transcriptase itself is probably multimeric, with at least two molecules of both L and NS per holoenzyme. The NS protein is highly and heterogeneously phosphorylated, so many different species of NS are found in the virion. Although phosphorylation of NS is important for its function, it is at present unclear whether there are specific sites that must be phosphorylated or whether differences in phosphorylation reflect different functions of NS. NS protein most likely serves as the promoter-recognition protein and is required for L-protein binding to template. The L protein is probably responsible for the entire spectrum of enzymatic activities, including initiation, elongation, methylation, capping, and polyadenylation. However, the evidence linking these catalytic functions with L protein is generally indirect, and many more data are required to confirm this conclusion.

The actual mechanism of VSV transcription is unique. Full-length mRNAs are synthesized by initiation and termination events at the termini of each gene. There is still controversy over whether there is a single promoter located at the 3' end of the genome or whether each gene has its own promoter with the requirement that sequential transcription be controlled by a cascade mechanism. None of the viral RNA modification steps occurs on exogenously added RNAs, so methylation, capping, and polyadenylation are all tightly coupled to transcription. If all these modification reactions as well as initiation and elongation are carried out by the L protein, it will be extremely interesting to determine the spatial distribution of the active sites on the L molecule.

The most interesting questions about VSV transcription have yet to be answered: Why is the leader not capped? Is the sequence conservation at the 5' end of each mRNA required for capping, or does it serve some other function? If polyadenylation results from reiterative transcription of oligouridylate, what makes the enzyme recycle rather than terminate? What controls termination: RNA sequence? RNA secondary structure? protein modifications? Exactly what happens at the intercistronic junctions? Does the high K_m for ATP reflect steps other than initiation? How is RNA synthesis regulated such that the same template can be used for synthesis of either mRNAs or unit-length replicative RNA? These are only a few of the questions that remain to be answered. VSV, a structurally simple virus, has solved the complex problems of RNA synthesis, modification, and regulation using only three proteins. When the final story of VSV transcription is complete, it should provide fascinating insights into protein–protein and protein–RNA interactions.

REFERENCES

Abraham, G., and Banerjee, A. K., 1976, Sequential transcription of the genes of vesicular stomatitis virus, *Proc. Natl. Acad. Sci. U.S.A.* **73:**1504–1508.

Ball, L. A., and White, C. N., 1976, Order of transcription of genes on vesicular stomatitis virus, *Proc. Natl. Acad. Sci. U.S.A.* **73:**442–446.

Baltimore, D., Huang, A. S., and Stampfer, M., 1970, Ribonucleic acid synthesis of vesicular stomatitis virus. II. An RNA polymerase in the virion, *Proc. Natl. Acad. Sci. U.S.A.* **66:**572–576.

Banerjee, A. K., Abraham, G., and Colonno, R. J., 1977, Vesicular stomatitis virus: Mode of transcription, *J. Gen. Virol.* **34:**1–8.

Banerjee, A. K., De, B. P., and Sanchez, A., 1985, Transcription of vesicular stomatitis virus genome RNA, *Viral Messenger RNA* (Y. Becker, ed.), pp. 197–224, Martinus Nijhoff Publishing, Boston.

Bell, J. C., Brown E. G., Takayesu, D., and Prevec, L., 1984, Protein kinase activity associated with immunoprecipitates of the vesicular stomatitis virus phosphoprotein NS, *Virology* **132:**229–238.

Belle-Isle, H. D., and Emerson, S. U., 1982, Use of a hybrid infectivity assay to analyze primary transcription of temperature-sensitive mutants of the New Jersey serotype of vesicular stomatitis virus, *J. Virol.* **43:**37–40.

Bishop, D. H. L., and Roy, P., 1972, Dissociation of vesicular stomatitis virus and relation of the virion proteins to the viral transcriptase, *J. Virol.* **10:**234–243.

Blumberg, B. M., Giorgi, C., and Kolakofsky, D., 1983, N protein of vesicular stomatitis virus selectively encapsidates leader RNA *in vitro, Cell* **32:**559–567.

Blumberg, B. M., Giorgi, C., Rose, K., and Kolakofsky, D., 1984, Preparation and analysis of the nucleocapsid proteins of vesicular stomatitis virus and Sendai virus, and analysis of the Sendai virus leader–NP gene region, *J. Gen. Virol.* **65:**769–779.

Carroll, A. R., and Wagner, R. R. 1978, Inhibition of transcription by immunoglobulins directed against the ribonucleoprotein of homotypic and heterotypic vesicular stomatitis virus. *J. Virol.* **25:**675–684.

Chanda, P. K., and Banerjee, A. K., 1981, Identification of promoter-proximal oligonucleotides, and a unique oligonucleotide pppGpC from *in vitro* transcription products of vesicular stomatitis virus, *J. Virol.* **39:**93–103.

Chinchar, V. G., Amesse, L. S., and Portner, A., 1982, Linked transcripts of the genes for leader and N message are synthesized *in vitro* by vesicular stomatitis virus, *Biochem. Biophys. Res. Commun.* **105:**1296–1302.

Clinton, G. M., Burge, B. W., and Huang, A. S., 1978, Effects of phosphorylation and pH on the association of NS protein with vesicular stomatitis virus cores, *J. Virol.* **27:**340–346.

Clinton, G. M., Guerina, N. G., Guo, H., and Huang, A. S., 1982, Host dependent phosphorylation and kinase activity associated with vesicular stomatitis virus. *J. Biol. Chem.* **257:**3313–3319.

Colonno, R. J., and Banerjee, A. K., 1976, A unique RNA species involved in initiation of vesicular stomatitis virus RNA transcription *in vitro, Cell* **8:**197–204.

Colonno, R. J., and Banerjee, A. K., 1977, Mapping and initiation studies on the leader RNA of vesicular stomatitis virus, *Virology* **77:**260–268.

De, B. P., and Banerjee, A. K., 1984, Specific interactions of vesicular stomatitis virus L and NS proteins with heterologous genome ribonucleoprotein template lead to mRNA synthesis *in vitro, J. Virol.* **51:**628–634.

De, B. P., and Banerjee, A. K., 1985, Requirements and functions of vesicular stomatitis virus L and NS proteins in the transcription process in vitro, Biochem. Biophys. Res. Commun. **126**:40–49.

De, B. P., Tahara, S. M., and Banerjee, A. K., 1982, Production and characterization of monoclonal antibody to the N protein of vesicular stomatitis virus (Indiana serotype), Virology **122**:510–514.

Deutsch V., Muel, B., and Brun, G., 1979, Temperature-sensitive mutant ts 082 of vesicular stomatitis virus. 1. Rescue at non-permissive temperature by UV-irradiated virus, Virology **93**:286–290.

Emerson, S. U., 1982, Reconstitution studies detect a single polymerase entry site on the vesicular stomatitis virus genome, Cell **31**:635–642.

Emerson, S. U., and Wagner, R. R., 1972, Dissociation and reconstitution of the transcriptase and template activities of vesicular stomatitis B and T virions, J. Virol. **10**:297–309.

Emerson, S. U., and Wagner, R. R., 1973, L protein requirement for in vitro RNA synthesis by vesicular stomatitis virus, J. Virol. **12**:1325–1335.

Emerson, S. U., and Yu, Y. H., 1975, Both NS and L proteins are required for in vitro RNA synthesis by vesicular stomatitis virus, J. Virol. **15**:1348–1356.

Emerson, S. U., Dierks, P. M., and Parsons, J. T., 1977, In vitro synthesis of a unique RNA species by a T particle of vesicular stomatitis virus, J. Virol. **23**:708–716.

Flamand, A., 1980, Rhabdovirus genetics, in: Rhabdoviruses, Vol. II (D. H. L. Bishop, ed.), pp. 115–140, CRC Press, Boca Raton, Florida.

Gallione, C. J., Greene, J. R., Iverson, L. E., and Rose, J. K., 1981, Nucleotide sequences of the mRNAs encoding the vesicular stomatitis virus N and NS proteins, J. Virol. **39**:529–535.

Giorgi, C., Blumberg, B., and Kolakofsky, D., 1983, Sequence determination of the (+) leader RNA regions of the vesicular stomatitis virus Chandipura, Cocal, and Piry serotype genomes, J. Virol. **46**:125–130.

Green, T. L., and Emerson, S. U., 1984, Effect of the beta–gamma phosphate bond of ATP on synthesis of leader RNA and mRNAs of vesicular stomatitis virus, J. Virol. **50**:255–257.

Harmon, S. A., and Summers, D. F., 1982, Characterization of monospecific antisera against all five vesicular stomatitis virus-specific proteins: Anti-L and anti-NS inhibit transcription in vitro, Virology **120**:194–204.

Harmon, S. A., Marnell, L. L., and Summers, D. F., 1983, The major ribonucleoprotein associated protein kinase of vesicular stomatitis virus is a host cell protein, J. Biol Chem. **258**:15,283–15,290.

Harmon, S. A., Robinson, E. N., Jr., and Summers, D. F., 1985, Ultrastructural localization of L and NS enzyme subunits on vesicular stomatitis virus RNPs using gold sphere–staphylococcal protein A–monospecific IgG conjugates, Virology **142**:406–410.

Herman, R. C., Adler, S., Lazzarini, R. A., Colonno, R. J. Banerjee, A. K., and Westphal, H., 1978, Intervening polyadenylate sequences in RNA transcripts of vesicular stomatitis virus, Cell **15**:587–596.

Horikami, S. M., and Moyer, S. A., 1982, Host range mutants of vesicular stomatitis virus defective in in vitro RNA methylation, Proc. Natl. Acad. Sci. U.S.A. **79**:7694–7698.

Horikami, S. M., DeFerra, F., and Moyer, S. A., 1984, Characterization of the infections of permissive and nonpermissive cells by host range mutants of vesicular stomatitis virus defective in RNA methylation, Virology **138**:1–15.

Hsu, C. H., and Kingsbury, D. W., 1982, NS phosphoprotein of vesicular stomatitis virus: Subspecies separated by electrophoresis and isoelectric focusing, J. Virol. **42**:342–345.

Hsu, C. H., Morgan, E. M., and Kingsbury, D. W., 1982, Site-specific phosphorylation regulates the transcriptive activity of vesicular stomatitis virus NS protein, J. Virol. **43**:104–112.

Hunt, D. M., 1983, Vesicular stomatitis virus mutant with altered polyadenylic acid polymerase activity in vitro, J. Virol. **46**:788–799.

Hunt, D. M., Emerson, S. U., and Wagner, R. R., 1976, RNA-temperature-sensitive mutants of vesicular stomatitis virus: L protein thermosensitivity accounts for transcriptase restriction of group I mutants, *J. Virol.* **18**:596–603.

Hunt, D. M., Smith, E. F., and Buckley, D. W., 1984, Aberrant polyadenylation by a vesicular stomatitis virus mutant is due to an altered L protein, *J. Virol.* **52**:515–521.

Imblum, R. L., and Wagner, R. R., 1974a, Inhibition of viral transcriptase by immunoglobulin directed against the nucleocapsid NS protein of vesicular stomatitis virus, *J. Virol.* **15**:1357–1366.

Imblum, R. L., and Wagner, R. R., 1974b, Protein kinase and phosphoproteins of vesicular stomatitis virus, *J. Virol.* **13**:113–124.

Isaac, C. L., and Keene, J. D., 1982, RNA polymerase-associated interactions near template promoter sequences of defective interfering particles of vesicular stomatitis virus, *J. Virol.* **43**:241–249.

Iverson, L. E., and Rose, J. K., 1982, Sequential synthesis of 5'-proximal vesicular stomatitis virus mRNA sequences, *J. Virol.* **44**:356–365.

Keene, J. D., Schubert, M., and Lazzarini, R. A., 1979, Terminal sequences of vesicular stomatitis virus RNA are both complementary and conserved, *J. Virol.* **32**:167–174.

Keene, J. D., Thornton, B. J., and Emerson, S. U. 1981, Sequence-specific contacts between the RNA polymerase of vesicular stomatitis virus and the leader RNA gene, *Proc. Natl. Acad. Sci. U.S.A.* **78**:6191–6195.

Kingsbury, D. W., Hsu, C. H., and Morgan, E. M., 1981, A role for NS-protein phosphorylation in vesicular stomatitis virus transcription, in: *The Replication of Negative Strand Viruses* (D. H. L. Bishop and R. W. Compans, eds.), pp. 821–827, Elsevier/North-Holland, Amsterdam.

Kingsford, L., and Emerson, S. U., 1980, Transcriptional activities of different phosphorylated species of NS protein purified from vesicular stomatitis virus and cytoplasm of infected cells, *J. Virol.* **33**:1097–1105.

Lazzarini, R. A., Keene, J. D., and Schubert, M., 1981, The origins of defective interfering particles of the negative-strand RNA viruses, *Cell* **26**:145–154.

Marnell, L. L., and Summers, D. F., 1984, Characterization of the phosphorylated small enzyme subunit, NS, of the vesicular stomatitis virus RNA polymerase, *J. Biol. Chem.* **259**:13,518–13,524.

Mellon, M. G., and Emerson, S. U., 1978, Rebinding of transcriptase components (L and NS proteins) the nucleocapsid template of vesicular stomatitis virus, *J. Virol.* **27**:560–567.

Naeve, C. E., and Summers, D. F., 1981, Initiation of transcription by vesicular stomatitis virus occurs at multiple sites, in: *The Replication of Negative Strand Viruses* (D. H. L. Bishop and R. W. Compans, eds.), pp. 769–779, Elsevier/North-Holland, Amsterdam.

Naito, S., and Ishihama, A., 1976, Function and structure of RNA polymerase from vesicular stomatitis virus, *J. Biol. Chem.* **251**:4307–4314.

Nakai, T., and Howatson, A. F., 1968, The fine structure of vesicular stomatitis virus, *Virology* **35**:268–281.

Pinney, D. F., and Emerson, S. U., 1982a, Identification and characterization of a group of discrete initiated oligonucleotides transcribed *in vitro* from the 3' terminus of the N-gene of vesicular stomatitis virus, *J. Virol.* **42**:889–896.

Pinney, D. F., and Emerson, S. U., 1982b, *In vitro* synthesis of triphosphate-initiated N-gene mRNA oligonucleotides is regulated by the matrix protein of vesicular stomatitis virus, *J. Virol.* **42**:897–904.

Piwnica-Worms, H., and Keene, J. D., 1983, Sequential synthesis of small capped RNA transcripts *in vitro* by vesicular stomatitis virus, *Virology* **125**:206–218.

Pringle, C. R., 1982, The genetics of vesiculoviruses, *Arch. Virol.* **72**:1–34.

Rose, J. K., Lodish, H. F., and Brock, M. L., 1977, Giant heterogeneous polyadenylic acid on vesicular stomatitis virus mRNA synthesized *in vitro* in the presence of S-adenosylhomocysteine, *J. Virol.* **21**:683–693.

Sanchez, A., De, B. P., and Banerjee, A. K., 1985, *In vitro* phosphorylation of NS by the L protein of vesicular stomatitis virus, *J. Gen. Virol.* **66**:1025–1036.

Schubert, M., and Lazzarini, R. A., 1981, In vivo transcription of the 5'-terminal extracistronic regions of vesicular stomatitis virus RNA, *J. Virol.* **38**:256–262.

Schubert, M., Keene, J. D., Lazzarini, R. A., and Emerson, S. U., 1978, The complete sequence of a unique RNA species synthesized by a DI particle of VSV, *Cell* **15**:103–112.

Schubert, M., Keene, J. D., Herman, R. C., and Lazzarini, R. A., 1980, Site on the vesicular stomatitis virus genome specifying polyadenylation and the end of the L gene mRNA, *J. Virol.* **34**:550–559.

Schubert, M., Harmison, G. G., Sprague, J., Condra, C. S., and Lazzarini, R. A., 1982, In vitro transcription of vesicular stomatitis virus: Initiation with GTP at a specific site within the N cistron, *J. Virol.* **43**:166–173.

Schubert, M., Harmison, G. G., and Meier, E., 1984, Primary structure of vesicular stomatitis virus polymerase (L) gene: Evidence for a high frequency of mutations, *J. Virol.* **51**:505–514.

Schubert, M., Harmison, G. G., and Richardson, C. D., and Meier, E., 1985, Expression of a cDNA encoding a functional, 241 kilodalton vesicular stomatitis virus RNA polymerase, *Proc. Natl. Acad. Sci. U.S.A.* **82**:7984–7988.

Sinacore, M. S., and Lucas-Lenard, J., 1982, The effect of the vesicular stomatitis virus-associated protein kinase on viral mRNA transcription in vitro, *Virology* **121**:404–413.

Szilagyi, J. F., and Uryvayev, L., 1973, Isolation of an infectious ribonucleoprotein from VSV containing an active RNA transcriptase, *J. Virol.* **11**:279–286.

Talib, S., and Hearst, J. E., 1983, Initiation of RNA synthesis in vitro by vesicular stomatitis virus: Single internal initiation in the presence of aurintricarboxylic acid and vanadyl ribonucleoside complexes, *Nucleic Acids Res.* **11**:7031–7042.

Testa, D., and Banerjee, A. K., 1977, Two methyltransferase activities in the purified virions of vesicular stomatitis virus, *J. Virol.* **24**:786–793.

Testa, D., and Banerjee, A. K., 1979, Initiation of RNA synthesis in vitro by vesicular stomatitis virus, *J. Biol. Chem.* **254**:2053–2058.

Testa, D., Chanda, P. K., and Banerjee, A. K., 1980a, Unique mode of transcription in vitro by vesicular stomatitis virus, *Cell* **21**:267–275.

Testa, D., Chanda, P. K., and Banerjee, A. K., 1980b, In vitro synthesis of the full-length complement of the negative-strand genome RNA of vesicular stomatitis virus, *Proc. Natl. Acad. Sci. U.S.A.* **77**:294–298.

Thomas, D., Newcomb, W. W., Brown, J. C., Wall, J. S., Hainfeld, J. F., Trus, B. L., and Steven, A. C., 1985, Mass and molecular composition of vesicular stomatitis virus: A STEM analysis, *J. Virol.* **54**:598–607.

Thornton, G. B., De, B. P., and Banerjee, A. K., 1984, Interaction of L and NS proteins of vesicular stomatitis virus with its template ribonucleoprotein during RNA synthesis in vitro, *J. Gen. Virol.* **65**:663–668.

Villarreal, L. P., Briendl, M., and Holland, J. J., 1976, Determination of molar ratios of vesicular stomatitis virus induced RNA species in BHK-21 cells, *Biochemistry* **15**:1663–1667.

Wagner, R., 1975, Reproduction of rhabdoviruses, in: *Comprehensive Virology*, Vol. 4 (H. Fraenkel-Conrat and R. Wagner, eds.), pp. 1–93, Plenum Press, New York.

CHAPTER 7

The Role of Proteins in Vesicular Stomatitis Virus RNA Replication

GAIL W. WERTZ, NANCY L. DAVIS, AND JOHN PATTON

I. INTRODUCTION

The negative-strand RNA genome of vesicular stomatitis virus (VSV) serves as template for two types of RNA-synthesis reactions, transcription and replication. A major distinction between the two RNA-synthesis processes is a requirement for protein synthesis. Transcription, the synthesis of the leader RNA and five discrete monocistronic messenger RNAs (mRNAs), does not require protein synthesis. Replication, the production of full-length copies of the viral RNA, requires continuous synthesis of viral proteins. This chapter will focus on the question of what newly synthesized protein or proteins are required to effect and maintain the transition from transcription to replication of the negative-strand RNA. Reviews of negative-strand virus RNA replication, including discussions of the nature of the template and specific sequences involved in replication, were presented by Wertz (1980), Lazzarini et al. (1981), and Ball and Wertz (1981). This chapter will concentrate on work that has been carried out since these previous reviews.

GAIL W. WERTZ, NANCY L. DAVIS, AND JOHN PATTON • Department of Microbiology and Immunology, School of Medicine, The University of North Carolina, Chapel Hill, North Carolina 27514. Present address of G.W.W.: Department of Microbiology, University of Alabama, Birmingham, Alabama 35294. Present address of N.L.D.: Department of Microbiology, North Carolina State University, Raleigh, North Carolina 27650. Present address of J.P.: Department of Biology, University of Southern Florida, Tampa, Florida 33620.

II. VESICULAR STOMATITIS VIRUS RNA-SYNTHETIC REACTIONS

The genome of VSV is expressed to yield two different types of RNA products: either five discrete capped, methylated, and polyadenylated mRNAs and a leader RNA or a complete, all-inclusive positive-strand copy of the genome that serves as template to replicate the negative-strand genome.

Transcription initiates at the 3' end of the genome and occurs in an obligatory sequential fashion 3' to 5', which reflects the physical arrangement of the five viral genes (Fig. 1) (for further details, see Chapter 6). Transcription does not result in products that contain a complete copy of the genome. Four or five bases at the leader–N junction (Keene *et al.*, 1980), two bases at each intercistronic junction (McGeoch, 1979; Rose, 1980), and the trailer RNA (Schubert and Lazzarini, 1981) are not represented in the transcripts. The exact mechanism that produces these discrete transcripts is unknown. Transcription has been reproduced in highly purified *in vitro* systems. Reconstitution experiments have shown that the active template for transcription is not naked RNA but RNA associated with nucleocapsid (N) protein in a nucleocapsid structure and containing the nonstructural (NS) and large (L) proteins, which are the enzymatic components of the transcription complex (Emerson and Wagner, 1972; Emerson and Yu, 1975; Naito and Ishihama, 1976). This nucleocapsid structure is capable of initiating infection in cells; however, only

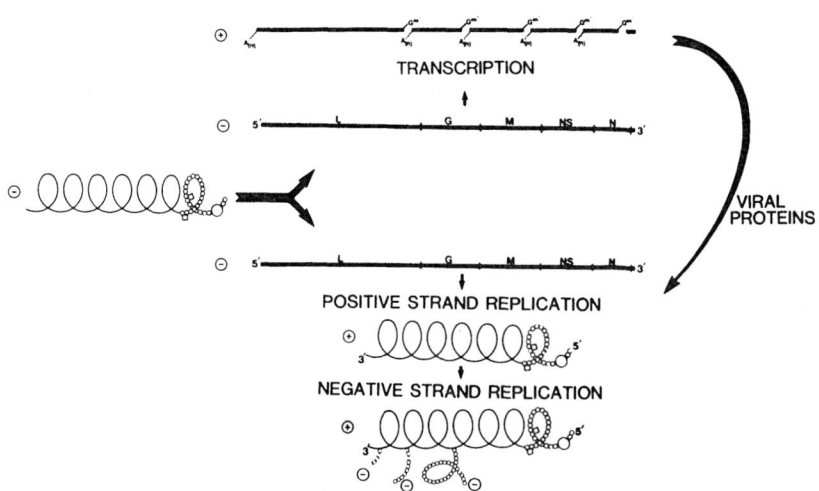

FIGURE 1. VSV RNA-synthetic reactions.

mRNAs and leader RNA are synthesized in the absence of viral protein synthesis.

As diagrammed in Fig. 1, a full-length, all-inclusive positive-strand copy of the genome, the first step in replication, is produced from the infecting nucleocapsid templates consisting of RNA and N, NS, and L proteins only when viral protein synthesis occurs. Inhibition of viral protein synthesis, even at late times in infection, when replication is ongoing, results in rapid cessation of replication, indicating a constant need for protein to support replication (Wertz and Levine, 1973; Perlman and Huang, 1973). Unlike mRNAs, the full-length positive strand is neither capped nor polyadenylated. Furthermore, both the full-length positive-strand and the negative-strand genomic RNA are found only in the form of N-coated nucleocapsid structures, never as naked RNAs. In this form, they are resistant to ribonucleases, whereas the unencapsidated mRNAs are not (Soria et al., 1974).

What proteins need to be made to change the RNA-synthesis activity of the nucleocapsid templates (which contain the N-coated RNA template and the NS and L proteins) from transcription of discrete mRNAs to replication of full-length RNA? In considering the nature of the requirement for protein synthesis, it has been postulated by numerous workers that because the products of replication are N-protein-coated, the requirement for protein synthesis may be for a continual supply of the nucleocapsid protein. Moreover, since the major difference between encapsidated RNAs and nonencapsidated mRNAs is the presence of a leader sequence, it has been proposed that N protein may play a crucial role in the transition from transcription to replication by binding to a site in the leader RNA that may be the nucleation site for encapsidation. It is postulated that the binding of N protein at this point would serve to begin encapsidation and at the same time to attenuate a termination signal that would normally lead to production of discrete mRNAs (Leppert et al., 1979). This event would be dependent on the availability of N protein and would determine the balance between replication and transcription.

It is also possible that other proteins may be required to promote the change from transcriptive to replicative activity of the nucleocapsid templates. For example, it is possible that a newly synthesized L molecule may be required, although previous evidence suggests that the same enzyme is responsible for both mRNA transcription and genome synthesis (Perlman and Huang, 1973). Alternatively, an L molecule modified by association with a newly synthesized form of the phosphoprotein, NS, which has been shown to exist in different phosphorylation forms, may be required (Clinton et al., 1979; Kingsford and Emerson, 1980). Finally, it is also possible that a host factor may play a role in the transition (Pringle, 1978) or in regulation of the transition (Keene et al., 1984).

A. Considerations in Analyzing the Requirements for Replication

Several aspects of VSV replication constituted specific constraints in the approaches that could be employed to study the requirement for protein synthesis in replication. First, the active template for replication was not naked RNA, but RNA in a nucleocapsid structure (Emerson and Wagner, 1972). Second, of the five VSV proteins, the one that was postulated to be important to replication, the nucleocapsid (N) protein, was not soluble under most isolation conditions (Blumberg *et al.*, 1984). Third, continuous protein synthesis was known to be required for replication *in vivo* (Wertz and Levine, 1973; Perlman and Huang, 1973). These constraints ruled out development of a simple *in vitro* system in which the various proteins could be purified and added to the RNA template individually or in combinations to test their efficacy in promoting replication. As an alternative, several approaches using permeabilized cells and infected-cell or uninfected-cell lysates were explored in the effort to develop pliant replication systems. These are described below.

III. EARLY REPLICATION SYSTEMS

Condra and Lazzarini (1980) developed a permeable-cell system from VSV-infected cells that was capable of carrying out transcription as well as synthesis of genome-length RNA and its subsequent assembly into nucleocapsids. Synthesis of full-length positive- and negative-strand RNA was also reported in extracts of VSV-infected (Hill *et al.*, 1981) and uninfected (Hill and Summers, 1982) HeLa-cell S-10 extracts using intracellular ribonucleoprotein particles as template. Although viral protein synthesis was observed in these extracts, full-length RNA synthesis was not sensitive to inhibition of protein synthesis in the infected HeLa S-10 extracts. Presumably, this was due to the high concentration of viral proteins present in the extract at the time it was made. In contrast, full-length RNA synthesis was sensitive to protein-synthesis inhibition in uninfected-cell extracts in which mRNA made by transcribing templates was the source of viral proteins. In the latter system, *de novo* synthesis of genome-length RNA was not detected. Ghosh and Ghosh (1982) also reported synthesis of 42S RNA of both positive and negative polarity in a coupled transcription–translation system using RNP particles from VSV-infected HeLa cell extracts and a ribosomal extract from uninfected HeLa cells. The sensitivity of 42S RNA synthesis to protein synthesis inhibition was not tested.

A cytoplasmic extract prepared by lysolecithin treatment of VSV-infected BHK cells also supported both VSV mRNA synthesis and synthesis of genome-length RNA or the RNA of a defective interfering (DI) particle (Peluso and Moyer, 1983). Since the DI RNA was smaller than standard genomic RNA and had correct initiation sites for replication (Lazzarini *et al.*, 1981), it was an efficient template. Synthesis of genome-length RNA was shown to be initiated in this system by use of virion-derived DI templates. Virion-derived nucleocapsids, in contrast to nucleocapsids isolated from infected cells, have been shown to have no nascent chains attached (Chanda and Banerjee, 1981). Therefore, their ability to be replicated in the system showed that both initiation and elongation were occurring. However, in this system, as in the infected-cell extracts described above, RNA replication and nucleocapsid formation were not dependent on *de novo* protein synthesis, but used the preformed soluble proteins present in the cell at the time the extract was prepared. Thus, in this system, the protein-synthesis requirement for replication could not be studied. As described in Section V, monoclonal antibodies were used to investigate the role of individual proteins in replication in this system.

A third approach was the development of a reconstituted *in vitro* system consisting of (1) an mRNA-dependent reticulocyte lysate to carry out protein synthesis, (2) purified VSV mRNAs to direct the synthesis of VSV proteins, and (3) nucleocapsids containing both positive- and negative-strand VSV genomes (Davis and Wertz, 1982). This system supported synthesis of leader and the five mRNAs as well as genome-length RNA. Negative-strand RNA products represented 2–5% of the total RNA synthesized. The genome-length RNA was encapsidated in structures indistinguishable from authentic viral nucleocapsids (Patton *et al.*, 1983a). RNA replication in this system was shown to be sensitive to inhibition of protein synthesis. Furthermore, the level of replication was a function of the amount of concurrent viral protein synthesis. It was also shown that a purified VSV DI particle (DI-T) could be used as template in this system and that as a function of protein synthesis, the DI nucleocapsid template could be switched from synthesis of only the 46-base DI leader RNA to replication of genome-length RNA of both positive and negative polarity (Wertz, 1983).

Each of these three types of systems was capable of synthesizing genome-length RNA. The systems were clearly distinguished, however, by the fact that in those made from infected-cell extracts, replication was not sensitive to the inhibition of protein synthesis, whereas it was in systems prepared from uninfected-cell lysates or in those that used the reticulocyte lysate to program synthesis of proteins directed by exogenously added mRNAs.

IV. DISSECTING THE REQUIREMENTS FOR REPLICATION *IN VITRO*

None of the systems described above was ideal for studying replication. To dissect the protein-synthesis requirement in replication, a system was needed in which replication was sensitive to protein synthesis and in which the synthesis of individual viral proteins or their availability to purified nucleocapsid templates could be controlled at will. The development of such a system was reported by Patton et al. (1984a). This system consisted of three major components, as shown in Table I: (1) purified nucleocapsid templates containing genome-length RNA coated with N protein and having the NS and L polymerase proteins attached, (2) an mRNA-dependent protein-synthesizing reticulocyte lysate, and (3) individual VSV mRNAs isolated by hybridization–selection with individual cDNA clones. Thus, in this system, the synthesis of individual proteins could be controlled by the addition of one or another of the purified mRNAs to program synthesis of the desired protein.

A second important aspect of this system was the templates. In a system in which one wishes to link protein synthesis to replication and strictly control the availability of mRNA in the system, it is not possible to use standard VSV nucleocapsids as templates. These templates will transcribe as well as replicate in the system and hence rapidly make all five mRNAs available to the protein-synthesis system. For these reasons, the DI particle called DI-T (VSVI-DI0.25) was selected as template. As shown in Fig. 2, this DI particle lacks genetic information for the N, NS, M, and G genes and contains only partial information for the L gene (Leamnson and Reichmann, 1974; Stamminger and Lazzarini, 1979). Therefore, this DI particle does not direct mRNA synthesis; in the absence of protein synthesis, it synthesizes only the 46-base DI-leader RNA (Emerson et al., 1977; Schubert et al., 1978; Semler et al., 1978). However, in infected cells, in the presence of helper viral protein synthesis, both positive- and negative-strand genome-length RNAs are synthesized in addition to leader (Leppert and Kolakofsky, 1980; Rao and Huang, 1979, 1980). Another important feature of this DI-particle template was that the 3′ terminus of the positive strand (the initiation site for negative-

TABLE I. VSV in Vitro RNA-Synthesis System

Components
1. Messenger-RNA-dependent rabbit reticulocyte protein-synthesis lysate
2. Individual hybrid-selected VSV mRNAs
3. Intracellular or virion-derived nucleocapsids

Specifics[a]			
50	mM HEPES, pH7.6	0.05	mM 20 amino acids
10	mM creatine phosphate	2.0	mM magnesium acetate
1	mM ATP	2.0	mM DTE
0.6	mM GTP	66	mM NH₄Cl
0.6	mM CTP	14	mM potassium acetate
0.1	mM UTP		

[a] (HEPES) N-2-Hydroxyethylpiperazine-N-2-ethane-sulfonic acid; (DTE) dithioerythritol.

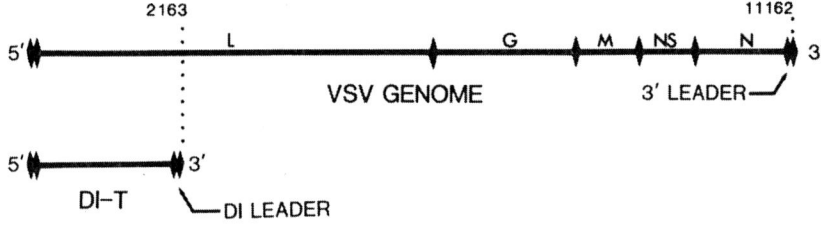

FIGURE 2. Map of VSV and VSV DI-T RNA.

strand replication) was identical to the 3' terminus of standard VSV positive strands (Keene *et al.*, 1978; Lazzarini *et al.*, 1981). The 5' and 3' termini of the DI are also complementary (Keene *et al.*, 1979; Perrault and Leavitt, 1976). Therefore, the 3'-terminal sequences of both positive and negative RNA strands, which are the initiation sites for replication, are identical.

Thus, this DI particle was an appropriate template to use to investigate the requirements for protein synthesis in replication: It was small and did not direct mRNA synthesis, yet had the correct sites for initiation of replication and was known to replicate efficiently in cells coinfected with standard VSV (Leppert and Kolakofsky, 1980; Rao and Huang, 1979, 1980). The suitability of this DI nucleocapsid as a template to investigate

FIGURE 3. Agarose–urea gel electrophoresis of RNA products synthesized by DI-particle nucleocapsids *in vitro.* RNA products were labeled with [³H]-UTP in the cell-free system programmed with the following components: (1) no DI templates; (2) DI templates, no VSV mRNA; (3) DI templates, VSV mRNA; (4) DI templates, VSV mRNA, and cycloheximide (50 μg/ml); (5) marker DI-template RNA. Incubation was for 180 min at 30°C. Reprinted from Wertz (1983) with permission.

the requirements for RNA replication was demonstrated directly: The DI template was shown to switch from synthesis of only the 46-base DI-leader RNA to replication of both genome-length positive- and negative-strand RNA in this *in vitro* system as a function of viral protein synthesis (Fig. 3) (Wertz, 1983).

A. Nucleocapsid Protein Alone Fulfills the Requirement for Continuous Protein Synthesis in Replication

The protein requirement for replication was examined by programming the *in vitro* system described above and summarized in Table I with the individual hybrid-selected N, NS, or M mRNAs (Patton *et al.*, 1984a). The aim of these experiments was to determine whether the N, NS, or matrix (M) protein, when translated individually in the presence of enzymatically competent nucleocapsid templates containing the N, NS, and L proteins, could promote the synthesis of genome-length RNA. As shown in Fig. 4, synthesis of both viral RNA and proteins was examined simultaneously by double-labeling with [^{3}H]-UTP and [^{35}S]methionine. The amount of replication in the presence of varying concentrations of individually synthesized proteins was compared with the level of replication in the presence of all the VSV proteins. These experiments showed that synthesis of N protein alone was sufficient for replication of genome-length DI-particle RNA of both positive and negative polarity (Fig. 4, lanes 6–8). No replication occurred in the absence of protein synthesis (Fig. 4, lane 1), and neither the NS nor the M protein by itself was able to support replication (Fig. 4, lanes 9, 10). Furthermore, when the levels of genome synthesis and N-protein expression were compared for reactions containing either total mRNA or N mRNA alone, the amount of genome synthesis supported by a given amount of N-protein synthesis was identical in both reactions (Fig. 5) (Patton *et al.*, 1984a). Therefore, the presence of the other viral proteins did not affect the efficiency of N protein in supporting replication. It was also shown in this system that synthesis of N protein alone was able to support replication when nucleocapsid templates were derived from budded DI virions (Patton *et al.*, 1984a). This finding implies that N protein fulfills the protein-synthesis requirement for *de novo* RNA replication, because no evidence exists for the presence of initiated RNAs on budded virions (Chanda and Banerjee, 1981). This result also ruled out the possibility that newly made NS or L protein was required to begin replication.

B. Presynthesized Nucleocapsid Protein Can Support Replication, Inefficiently

The requirement for concurrent synthesis in replication was tested in the defined *in vitro* system (Patton *et al.*, 1984a). It would be predicted

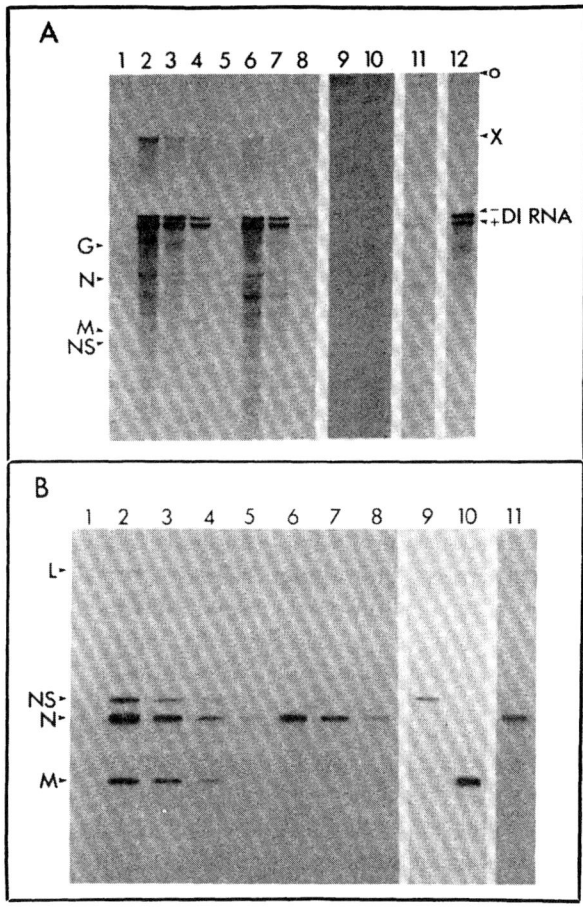

FIGURE 4. Synthesis of DI genome-length RNA and viral proteins *in vitro* in reactions containing intracellular nucleocapsids and programmed with total VSV mRNA or individual mRNAs. Reactions contained both [^{3}H]-UTP and [^{35}S]methionine and were incubated for 3 hr at 30°C. (B) Samples (1 μl) were analyzed by polyacrylamide gel electrophoresis for ^{35}S-labeled proteins. (A) The remainder after deproteinization was analyzed for ^{3}H-labeled RNAs by electrophoresis on agarose–urea gels. RNA and protein products are shown from a reaction containing no added VSV mRNA (1); from reactions containing 0.74 μg (2), 0.185 μg (3), 0.037 μg (4), and 0.007 μg (5) of total polyadenylated VSV mRNA; from reactions containing 5 μl (6), 1.25 μl (7), and 0.3 μl (8) of hybrid-selected N mRNA; from a reaction containing 4 μl of hybrid-selected NS mRNA (9); and from a reaction containing 4 μl of hybrid-selected M mRNA (10). (11) Products from a reaction in which 4 μl of hybrid-selected N mRNA was translated for 45 min and then nucleocapsids and anisomycin were added for 2 hr and 15 min. (12) RNA purified from DI intracellular nucleocapsids labeled *in vivo* with [^{3}H]uridine. The positions of ^{3}H-labeled VSV mRNAs run in parallel lanes are indicated. Reprinted from Patton *et al.* (1984a) with permission.

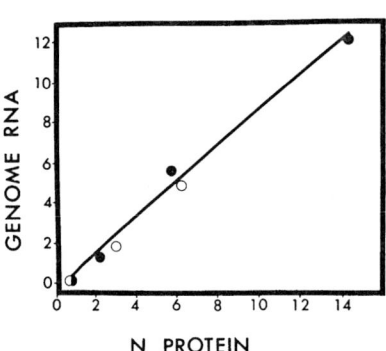

FIGURE 5. Correlation of genome-length RNA synthesis and N-protein expression. The fluorograms shown in Fig. 4 were analyzed by densitometry. Exposure times were chosen to give band intensities in the linear range. The relative absorbance units of ^{35}S-labeled N protein for a given reaction are plotted vs. the total amount of genome-length RNA (plus and minus strands) synthesized in that reaction. Results are shown from reactions to which varying amounts of total VSV mRNA (●) or hybrid-selected N mRNA (○) were added. The line has been fitted with a correlation coefficient of 0.995 to the results from reactions containing total VSV mRNA, using a least-squares analysis. Reprinted from Patton *et al.* (1984a) with permission.

that a preexisting pool of N protein could support RNA replication, since full-length RNA synthesis and encapsidation were observed in infected-cell extracts in the presence of cycloheximide (Hill and Summers, 1982; Peluso and Moyer, 1983). The ability of N protein synthesized before the addition of nucleocapsid templates and anisomycin (presynthesized N protein) to support replication was examined using purified N mRNA and the *in vitro* system (Patton *et al.*, 1984a). It was found that presynthesized N protein could support DI-RNA replication, but at a greatly reduced efficiency (see Fig. 4, lane 11). The amount of genome-length RNA synthesized per unit of N protein in reactions consisting of only presynthesized N protein was not more than 25% of that in reactions in which N protein and RNA were synthesized simultaneously. The lowered efficiency of replication per unit of pretranslated N protein was dependent on N-protein concentration. As the concentration of N protein increased above a certain level, replication decreased rather than increased as was shown with concurrent synthesis of N protein (Howard *et al.*, 1986). N protein translated alone, in the absence of genomic RNA replication to draw from the pool of N protein, appears to aggregate when a critical concentration is reached and is less active in supporting replication (Howard *et al.*, 1986). Infected-cell extracts, in contrast, may contain a factor that maintains the presynthesized N protein in a functional state. This is discussed in Section VI in relation to a model for N-protein solubility.

C. Nucleocapsids Are Formed in the Presence of Nucleocapsid Protein Alone

The ability of newly replicated RNA to form nucleocapsid structures in the presence of N protein alone was investigated. The products of replication in the presence of hybrid-selected N mRNA were analyzed for nuclease resistance and buoyant density in cesium chloride gradients. It was shown (Patton *et al.*, 1984a) that newly synthesized N protein was assembled into nucleocapsid structures that banded in cesium chloride gradients at the same position as authentic nucleocapsids. The genome-length RNA in nucleocapsids synthesized in the presence of N protein alone was also resistant to digestion by ribonuclease. Therefore, on the basis of these experiments, it was concluded that assembly of stable, nuclease-resistant nucleocapsids is dependent solely on a supply of N protein. Furthermore, no ribonuclease-sensitive (unencapsidated) genome-length RNA was detected. This finding shows that replication and encapsidation with N protein are closely coupled, and it explains why replication is dependent on a constant supply of protein.

V. ANTIBODIES AS PROBES FOR PROTEIN FUNCTION IN REPLICATION

A. *In Vitro* Studies

In vitro systems that contain infected-cell extracts or transcribing nucleocapsid templates, or both, also contain, of necessity, all five of the viral proteins. It is not possible in these systems to test separately the effects of individual viral proteins on replication. However, antibodies against specific viral proteins have been used in an effort to deduce the role of these proteins in replication. Monospecific antibodies against the N and NS proteins caused marked inhibition of replication *in vitro* (Hill and Summers, 1982). Inhibition would be expected if these antibodies bound to N and NS proteins on the nucleocapsid template and impeded the movement of the polymerase. It was not shown in these experiments whether or not the anti-N and anti-NS antibodies bound to nucleocapsids. However, the fact that the same preparation of anti-N antibody did not inhibit mRNA synthesis raised the possibility that it inhibited replication by another mechanism. The presence of anti-L antibody unexpectedly stimulated full-length RNA synthesis *in vitro;* the reason for this effect is not known.

This approach is limited by the fact that monospecific, polyclonal antisera to the N, NS, and L proteins usually react with the nucleocapsid structure to interrupt replication. They are not generally useful probes for studying the activity of these proteins in their soluble form. The isolation of monoclonal antibodies was a way to overcome this limitation by allowing selection of antibodies specific for soluble forms of the N and NS proteins that did not react with the nucleocapsid structure.

Several monoclonal antibodies have been tested for their effects on RNA replication. Most anti-N monoclonal antibodies inhibited both transcription and replication, presumably by binding to the nucleocapsid template (Peluso and Moyer, 1984). Recently, however, an anti-N monoclonal antibody has been isolated and shown to react with free N protein, but not with nucleocapsid-bound N protein. This antibody was used in the following studies to provide a unique link between *in vitro* and *in vivo* approaches to viral RNA replication.

Two monoclonal antibodies directed against distinct regions of the N protein were used as probes for N-protein function both *in vitro* and *in vivo* (Arnheiter et al., 1984, 1985). Antibody 1 bound to nucleocapsids as well as to soluble N protein, while antibody 2 reacted only with un-assembled N protein. Transcription *in vitro* was inhibited by antibody 1, but not by antibody 2. When tested in an *in vitro* replication system that includes enzymatically active DI nucleocapsids and N protein translated from hybrid-selected N mRNA, both antibodies inhibited the synthesis of full-length RNA. Antibody 1 prevented all nucleocapsid-directed RNA synthesis by binding to the template. Antibody 2, however, inhibited only replication, because it bound to and inactivated the pool of N protein required for this process. The result obtained with antibody 2 is analogous to that observed in reactions from which a source of N protein is omitted (Section IV. A).

B. Infected-Cell Studies

The effect of these antibodies on virus-directed processes within the infected cell was tested using the technique of microinjection of tissue-culture cells (Arnheiter et al., 1984, 1985). Microinjection of antibody 2 led to effects that would be predicted if N-protein function and the synthesis of full-length RNA synthesis were linked *in vivo* as they are *in vitro*. The presence of antibody 2 in a cell subsequently infected with VSV resulted in the low levels of viral mRNA and protein synthesis produced by primary transcription and translation. The reduced amount of N protein that was made was not assembled into nucleocapsid structures even at 4 hr after infection. If antibody 2 were inhibiting replication *in vivo* in the same way as it did *in vitro*, i.e., by binding to soluble N protein and rendering it incapable of supporting replication, it would be predicted that increasing amounts of N protein resulting from translation of primary transcripts would eventually titrate the antibody. After a delay, the production of nucleocapsids and progeny virus would proceed. This is exactly the result that is seen; virus production by cells previously injected with antibody 2 reaches normal levels, but is delayed 3–6 hr. Although the synthesis of full-length RNA could not be assayed directly in these experiments, these results are consistent with those obtained *in vitro*, showing that a source of functional N protein is required for VSV-genome replication *in vivo* as well as *in vitro*.

VI. ROLE OF PROTEIN COMPLEXES IN REPLICATION

A. N and NS Proteins Form Complexes Both *in Vivo* and *in Vitro*

By their very nature, reconstituted *in vitro* systems like the ones described above are simplified in order to define minimal requirements. In the case described above (Patton *et al.*, 1984a), the minimal requirement for protein species that must be synthesized to promote genome RNA synthesis was identified as the N protein alone. It would not be surprising to find that the replication process as it occurs within the infected cell, with all the viral genes being actively expressed, might include factors other than the simple interaction of N protein and active templates. In fact, there are several lines of evidence that suggest that N protein may function not as a monomeric protein subunit, but as part of a protein complex that includes NS protein.

Complexes of N and NS proteins in the soluble cytoplasmic fraction of VSV-infected cells were first described by Bell *et al.* (1984). They used monospecific antisera prepared against denatured viral proteins to immunoprecipitate native proteins from infected-cell cytoplasm. The soluble fraction remaining after removal of the intracellular nucleocapsids by centrifugation contained N and NS proteins that could be coprecipitated with anti-NS antibody; N protein appeared in the immunoprecipitate because it was associated with NS protein.

Peluso and Moyer (1984) extended these studies using several monoclonal antibodies specific for N protein. They separated the soluble N protein from infected-cell cytoplasm into two fractions by centrifugation. One, containing 40% of the total free N protein, sedimented as a high-molecular-weight (>300,000) species. The other, lower-molecular-weight N-containing fraction contained N–NS complexes that could be immunoprecipitated with anti-N monoclonal antibody. These N–NS complexes involved about 10% of the total NS protein.

Similar studies of viral proteins synthesized *in vitro* also demonstrated the association of N protein with NS protein (Davis *et al.*, 1986). The reticulocyte-lysate translation system programmed with all five VSV mRNAs contained complexes of N and NS proteins that were immunoprecipitated both with anti-N monoclonal antibody and with anti-NS monoclonal antibody. Separation by velocity sedimentation of viral proteins synthesized *in vitro* showed that under these conditions, essentially all the N protein was associated with NS protein. In further experiments, it was possible, by adjusting the relative amounts of purified N and NS mRNA, to produce an excess of either uncomplexed N or NS protein that appeared as slower-sedimenting species. Thus, the complex between these two proteins *in vitro* appears to be specific.

1. Does the N–NS Complex Play a Role in Genome Replication?

The existence of a complex containing N and NS proteins both *in vivo* and *in vitro* is well established (see above). Experiments have been carried out to test the role of this complex in the process of genome replication. The addition of gradient fractions containing N–NS complexes isolated from infected-cell cytoplasm to active nucleocapsid templates led to synthesis of genome-length RNA (Peluso and Moyer, 1984). Additional evidence comes from *in vitro* studies. Reaction mixtures programmed with total polyadenylated viral RNA are able to support RNA-genome replication (Wertz, 1983); essentially all the free N protein made in the reactions was complexed with NS protein (Davis *et al.*, 1986). Therefore, a portion of the N protein associated with NS must be active in RNA replication.

If the requirement for new protein in RNA replication can be met by N protein alone, as shown in the *in vitro* replication system (Patton *et al.*, 1984a), what additional function is provided by the associated NS protein? Under what conditions is this additional function required? We propose that a necessary role is played by soluble NS protein in replication in the infected cell and that this additional function is required only as N protein achieves high concentration. Although a precise role for soluble NS protein in replication has not been proven, we will discuss the most reasonable possibilities in light of present findings.

2. NS May Function to Prevent Aggregation of N

Several lines of evidence show that N protein has a marked tendency to self-associate. It has been suggested that this characteristic may be related to self-assembly of N protein during encapsidation of viral RNA (Blumberg *et al.*, 1983). N protein synthesized in reticulocyte lysates programmed with high concentrations of hybrid-selected N mRNA formed high-molecular-weight aggregates (Davis, Howard, and Wertz, unpublished results), N protein purified from virion nucleocapsids formed large aggregates *in vitro* (Blumberg *et al.*, 1983), and 40% of the non-nucleocapsid-bound N protein in cytoplasmic extracts of infected cells sedimented as large aggregates (Peluso and Moyer, 1984). A further demonstration of N-protein aggregation comes from work using a eukaryotic expression vector. When N protein was expressed from a simian virus 40 vector in COS cells in the absence of any other VSV proteins, the majority of the expressed N protein was found in aggregates in the transfected cells (Sprague *et al.*, 1983). Taken together, these observations indicate that N protein self-associates to a significant degree. Furthermore, depending on concentration, the majority of the N protein can appear in an aggregated state, e.g., in *in vitro* reactions programmed with large amounts of purified N mRNA or in transfected cells in which only N protein is being made.

Available evidence indicates that high-molecular-weight aggregates of N protein are not active in RNA replication (Peluso and Moyer, 1984; Patton *et al.*, 1984a). A high-molecular-weight form of N protein isolated from the soluble fraction of infected cells was found to have much lower activity than the lower-molecular-weight N- and NS-containing fractions in supporting replication (Peluso and Moyer, 1984). In addition, indirect data came from experiments with cells transfected with an N-protein expression vector and subsequently infected with temperature-sensitive N-protein mutants. The N protein expressed from the transfected gene could not rescue the mutants, possibly due to its aggregated state (Sprague, personal communication).

These findings are consistent with the idea that for N protein to function in replication, its natural tendency to aggregate or self-assemble must be inhibited. The association of N protein with NS protein in the soluble pool, which is probably superseded by the N–N association during encapsidation, may perform this function (see Fig. 6). Alternatively, in conditions under which N aggregation does not occur readily, e.g., in *in vitro* systems programmed with hybrid-selected mRNA in which N protein is present in low concentrations and is rapidly sequestered into growing nucleocapsids, NS protein is not required to maintain N protein in an unaggregated functional state. To test this possibility, Howard *et al.* (1986) compared the efficiency of increasing concentrations of presynthesized N protein alone to support replication *in vitro* with the ability of high concentrations of N protein presynthesized in the presence of NS protein to support replication. Presynthesized N in the presence of NS protein at a molar ratio of NS to N of 1 : 2 was approximately four times

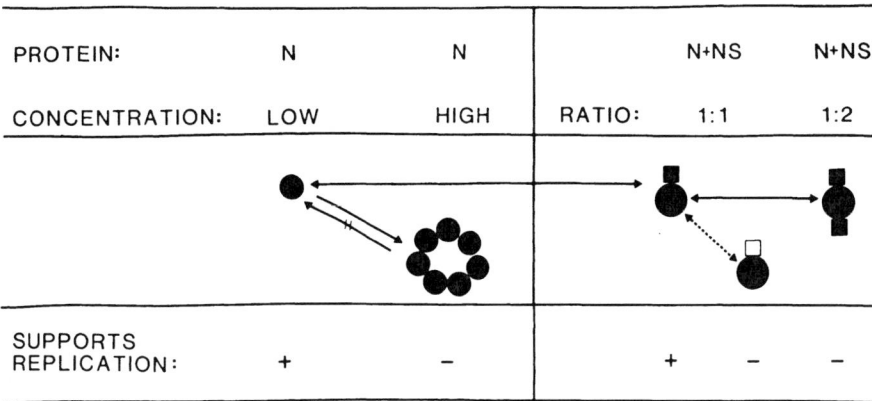

FIGURE 6. A model for interactions of N and NS proteins in the soluble pool. N-protein monomers can support replication; aggregates formed at high concentrations of N are inactive in replication. Association of N with NS at a ratio of 1 : 1 prevents aggregation and maintains N in a functional state. Complexes of N and NS with increased ratios of NS to N do not support replication. Alternatively, an alteration in the NS protein may render the N–NS complex unable to support replication. (●) N; (■) NS; (□) altered NS.

more efficient in supporting RNA replication than the same amount of presynthesized N protein alone. Furthermore, it was demonstrated that under conditions of N-protein aggregation (high concentrations) the presence of presynthesized NS protein prevented aggregation of N and allowed it to support replication (Howard et al., 1986; Howard and Wertz, manuscript in preparation).

3. N–NS Complexes May Be Involved in Maintaining the Balance between Transcription and Replication

NS protein, because it is a phosphoprotein that is found in different phosphorylated forms (Clinton et al., 1979; Kingsford and Emerson, 1980; Hsu et al., 1982), is an attractive candidate for a viral protein involved in regulation, i.e., in maintaining the balance among viral processes. More than one type of evidence suggests that the activity of NS protein in transcription varies with its state of phosphorylation (Kingsford and Emerson, 1980; Hsu et al., 1982; Witt and Summers, 1980). Experiments with temperature-sensitive mutants point to roles for NS protein in replication as well as in transcription (Lesnaw et al., 1979; Szyilagyi and Pringle, 1979). It may be that the balance between replication and transcription is determined not only by the total level of N protein in the cell, but also bv the availability of *functional* N protein. Whether a given N-protein molecule is functional may be determined by the form of NS protein with which it is complexed. In this scheme, the different phosphorylated forms of NS protein regulate viral RNA synthesis by their association with N protein, rendering it either active or inactive in replication (Fig. 6). Two predictions of this model are that (1) under appropriate conditions NS protein can act as an inhibitor of RNA replication and (2) more than one type of N–NS complex can be formed. The following observations made with the reticulocyte-lysate *in vitro* replication system support both these predictions.

In experiments using this *in vitro* replication system and hybrid-selected viral mRNAs, it was found that increasing ratios of purified NS to N mRNA led to decreasing levels of genome-length-RNA synthesis (Patton et al., 1984b). This inhibition by NS protein synthesized *in vitro* was not seen in *in vitro* assays for viral transcription (Howard and Wertz, unpublished observations). Inhibition of RNA replication was also observed when hybrid-selected NS mRNA was added to total polyadenylated viral RNA, thereby increasing the ratio of NS protein to N protein synthesized (Patton et al., 1984b). It is not clear yet whether the inhibition of replication observed *in vitro* is the result of an unusual phosphorylation pattern of the NS protein that is made or whether an abnormal ratio of NS protein to N protein alone is sufficient to inhibit replication under these conditions. The inhibitory effect of NS protein *in vitro* may reflect, perhaps in an exaggerated way, a normal function of NS protein in maintaining the balance between viral RNA-synthesis processes.

In further experiments with the reticulocyte *in vitro* replication system, we observed more than one species of N–NS complex in lysates programmed with total polyadenylated viral RNA (Davis *et al.*, 1986). The N–NS complexes had different ratios of NS protein to N protein and similar but distinguishable sedimentation rates (Fig. 7). It is not yet clear how many distinct complexes are present in these mixtures or how their relative activities in replication compare. Another unanswered question is whether different N–NS complexes contain different phosphorylated forms of NS protein. Answers to these questions may help to determine whether NS protein regulates RNA synthesis by forming active and inactive complexes with N protein.

Bell *et al.* (1984) have observed that immune precipitates of N–NS complexes include a tightly associated protein kinase activity that phosphorylates serine residues of NS protein *in vitro*. It is interesting to speculate that this complex-associated enzymatic activity may be involved in the interconversion of active and inactive forms of the N–NS complex.

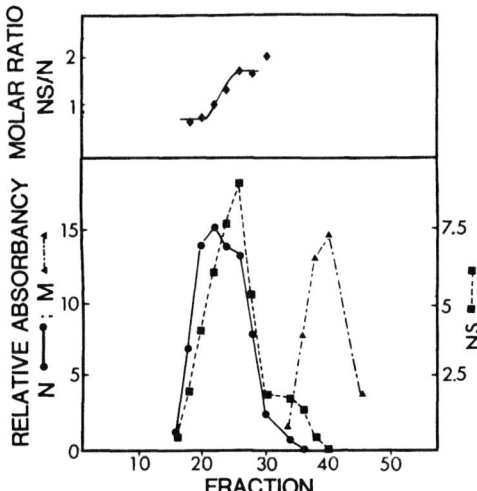

FIGURE 7. Rate velocity sedimentation of VSV proteins synthesized *in vitro*. Total polyadenylated RNA from VSV-infected cells was translated for 3 hr in a micrococcal nuclease-treated reticulocyte lysate containing [³⁵S]methionine. A portion of the reaction mixture was layered onto a 10–30% (wt./wt.) glycerol gradient that contained 0.01 M Tris, pH 7.4, 0.001 M EDTA, and 0.15 M NaCl and centrifuged for 20 hr at 234,000 g. Fractions of the gradient containing ³⁵S-labeled proteins were identified and analyzed by sodium dodecyl sulfate–polyacrylamide gel electrophoresis. Densitometric analysis of the gel autoradiogram was used to quantitate each viral protein in each gradient fraction. The resulting sedimentation profile of the N, NS, and M proteins is shown in the bottom panel. The molar ratio of NS to N in each fraction, shown in the top panel, was calculated using the relative incorporation of [³⁵S]methionine into NS and N and the number of methionine residues in the two proteins. Sedimentation is from right to left.

VII. DO TEMPLATE PROTEINS HAVE A ROLE IN REGULATING REPLICATION?

In the preceding sections, we have considered ways in which the availability of N protein for encapsidation to support replication could be controlled. Another point at which replication could be regulated is at the level of the template: either (1) by modification of template proteins or (2) by affecting enzymatic activity of proteins on the template.

A. Modification of Template Proteins

Perrault et al. (1983) have identified mutants of VSV, designated Pol R1 and R2, that have a modification of the N protein of the viral template. In vitro, in the absence of protein synthesis, these mutants have a high frequency of reading through, rather than terminating, at the leader boundary. This has some but not all of the characteristics of a replicative event. The products of this synthesis are small—usually 300–800 nucleotides long. The phenotype of these mutants raises the possibility that an alteration in one or more of the template N proteins, perhaps by posttranslational modification, may affect read through at intercistronic boundaries and hence be a factor involved in the transition to replication.

B. Exchange of Template Proteins

Work discussed in Sections IV and V has led to the conclusion that only a supply of N protein is required for replication and that de novo synthesis of NS or L protein is not required. However, a change in the activity of a protein on the template could be involved in regulation of replication. In considering how a change in protein activity could occur, we asked the question, can proteins exchange on and off nucleocapsids? Is this a means by which viral proteins could gain access to nucleocapsid templates in order to regulate their RNA-synthesis capabilities? Studies were carried out to determine whether newly made viral proteins could associate with nucleocapsids in the absence of RNA replication. If proteins did associate with templates in the absence of RNA synthesis, it could be concluded that they are able to exchange with, or bind in addition to, existing nucleocapsid proteins. An inhibitor of VSV RNA synthesis was used in vitro to determine whether a direct correlation existed between the association of N and NS proteins with nucleocapsids and the level of RNA synthesis. The inhibitor used was 2',3'-dideoxycytidine triphosphate (ddCTP), a compound that equally inhibited VSV RNA replication and transcription in vitro, but had little or no effect on translation of viral mRNAs (Patton et al., 1983b). A comparison of reactions containing the usual CTP or varying amounts of ddCTP showed that the association of newly made N protein with nucleocapsids was directly dependent on the level of RNA synthesis. However, the ratio of newly made NS protein to N protein in the nucleocapsid fraction increased 4.5-fold when the replication level was reduced by ddCTP. This result is consistent with the idea that NS protein, in contrast to N protein, may associate with parental nucleocapsids during replication (Patton et al., 1983a).

The possible association of newly made NS protein with parental templates was tested more directly by measuring the ability of ^{35}S-labeled NS protein alone to associate with nucleocapsids under conditions in which no RNA replication was occurring. In these experiments, reactions contained DI-T intracellular nucleocapsids and were programmed with only NS mRNA. Since N mRNA was not added to the system to support replication and could not be synthesized by the DI-T nucleocapsids, any observed association of newly made NS protein with nucleocapsids would have to be with parental nucleocapsids. The results are shown in Fig. 8. Electrophoretic analysis of DI-T nucleocapsids recovered from reactions by centrifugation on sucrose gradients showed that newly synthesized NS protein can associate with parental nucleocapsids, and this association does not require replication.

FIGURE 8. Association of viral proteins with DI nucleocapsids in reactions programmed with VSV total or NS mRNA. After incubation at 30°C for 90 min, reactions containing VSV (poly A$^+$) total mRNA (1, 2) or hybrid-selected NS mRNA (3, 4) were diluted to 2 ml with buffer (10 mM HEPES, 2 mM MgCl$_2$, 2 mM DTE, 66 mM NH$_4$Cl, 14 mM potassium acetate, 0.5% Nonidet P-40, pH 7.6) and centrifuged through 10-ml gradients of 15–30% sucrose in a Beckman SW40 rotor at 38,500 rpm for 2 hr at 4°C. After fractionation, portions of fractions corresponding to the position reached by marker nucleocapsids in these gradients were analyzed by electrophoresis on 10% polyacrylamide gels. (1,) ^{35}S-labeled proteins sedimenting to the nucleocapsid position from reactions with (1) or without (2) DI nucleocapsids and programmed with VSV poly(A$^+$) total mRNA; (3, 4) ^{35}S-labeled proteins sedimentating to the nucleocapsid position from reactions with (3) or without (4) DI nucleocapsids and programmed with hybrid-selected NS mRNA.

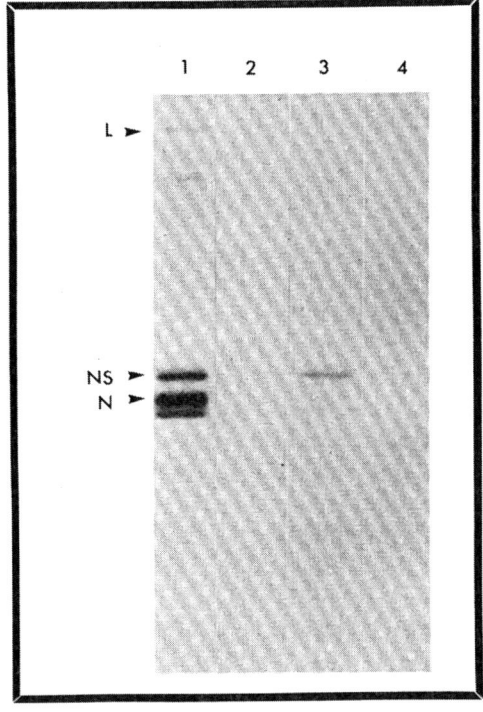

Hypothetically, the association of newly made NS protein with parental nucleocapsids must require the displacement of bound NS protein from these templates, because they are presumably saturated with NS protein. To test this possibility, we examined the ability of parental nucleocapsid templates to release NS protein under various reaction conditions. Wild-type nucleocapsids labeled *in vivo* with [^{35}S]methionine were added to reactions containing total mRNA and 1 mm of either ddCTP or CTP (control), but no [^{35}S]methionine. In both the control reaction and the reaction containing ddCTP, significant amounts of prelabeled NS protein (and L protein) were released from the input nucleocapsids. Further, this release was not dependent on RNA synthesis.

To determine whether the release of NS protein was dependent on protein synthesis, cycloheximide was added to reactions containing either CTP or ddCTP. The amount of NS protein released from parental nucleocapsids in reactions in which protein synthesis was blocked was several times less than that observed in reactions in which protein synthesis occurred (Patton, Davis, and Wertz, unpublished data). Indeed, in reactions containing cycloheximide, the amount of NS protein released was the same as that released by nonincubated parental nucleocapsid templates. In summary, these results indicate that concurrent with the synthesis of viral protein and the binding of newly made NS protein to parental nucleocapsids, bound NS protein is released from parental nucleocapsids. The fact that the release of NS protein is protein-synthesis-dependent suggests that newly synthesized viral proteins catalyze this release. However, this exchange of NS protein on parental nucleocapsids is not dependent on RNA synthesis.

It is not clear what actually causes the release of NS protein from nucleocapsids. Is the bound NS protein being modified by the newly synthesized proteins, perhaps kinases, or phosphatases so that its affinity for nucleocapsids is decreased? Results from several studies suggest that the degree of phosphorylation of the NS protein can affect its ability to bind to VSV nucleocapsids. Sinacore and Lucas-Lenard (1982) have shown that treatment of transcriptionally active nucleocapsids with bacterial alkaline phosphatase results in not only the partial dephosphorylation of bound NS protein, but also the release of the majority of the bound NS protein from the nucleocapsid. Witt and Summers (1980) have shown that further phosphorylation of NS protein bound to nucleocapsids in *in vitro* kinase reactions also causes a significant proportion of the NS protein to be released. However, these investigators found that the release did not affect the level of transcription by the nucleocapsids and thus proposed that the released NS protein represented excess NS protein packaged during virion assembly. Clinton *et al.* (1979) have shown that the phosphatases necessary to cause the partial dephosphorylation of NS protein are endogenous to the cytoplasmic extracts of infected and uninfected cells. In addition, these investigators have shown that virion-associated kinases have the capacity to convert NS1 protein, a less phosphorylated species, to NS2, a more highly phosphorylated species. Thus, in infected cells, it is apparent that enzymes are present that can cause the interconversion of forms of NS protein resulting in the production of distinct forms of NS protein that could differ in their binding efficiencies for the VSV nucleocapsid. The exchange of bound and released NS protein may provide an avenue by which NS proteins with different functional activities could interact with nucleocapsid templates and thereby possibly regulate RNA synthesis. This proposal is hypothetical at present. The data presented above are included in order to point out that exchange of proteins on the nucleocapsids does occur and to suggest that the functional significance of this exchange needs to be evaluated in relation to RNA-synthesis events.

VIII. DISCUSSION AND SUMMARY

The work described in this chapter has shown by several approaches that N protein is required for replication (Patton et al., 1984a; Arnheiter et al., 1985) and that N protein alone can satisfy the requirement for protein synthesis to support the transition from transcription to replication by nucleocapsid templates containing the L and NS proteins (Patton et al., 1984a). N protein alone is also capable of binding to nascent RNA to form nucleocapsid structures; newly replicated RNA is found only in nucleocapsid structures (Patton et al., 1984a). These findings also indicate that the NS and L proteins present on the nucleocapsid templates are capable of catalyzing replication and that de novo synthesis of these proteins is not required for replication.

The finding that N protein alone supports replication is consistent with, although it does not prove, a model in which binding of N protein to nascent leader RNA functions to promote read through of the termination signal at the end of leader and to begin nucleocapsid assembly (Leppert et al., 1979). Thus, the availability of N protein would determine the balance between replication and transcription. This model would explain why genome-length RNAs are encapsidated whereas mRNAs are not: The mRNAs lack a leader sequence that contains the proposed nucleation site. Leader RNAs have been found encapsidated with N protein in infected cells, and N has been shown to preferentially assemble with leader over other viral transcripts in vitro (Blumberg et al., 1981, 1983).

Thus, it would seem that N protein is necessary, and can be envisioned as sufficient, for replication of nucleocapsid templates bearing the NS and L proteins: A suitable model exists whereby availability of N protein could determine the balance between replication and transcription. An important question to consider at this point is whether N protein alone is optimal for replication. Are there other factors operating to regulate replication, perhaps at (1) the level of availability of N protein or (2) the level of template activity?

A. Availability of Nucleocapsid Protein

It has been observed that N protein self-aggregates (Blumberg et al., 1983; Sprague et al., 1983; Peluso and Moyer, 1984). In in vitro replication systems, it has been shown that N protein by itself self-aggregates as a function of increased concentration and that in the aggregated state it is less able to support replication (Peluso and Moyer, 1984; Howard et al., 1986). In conjunction with this, complexes between N protein and NS protein have been shown to exist both in vivo and in vitro and to be able to function in replication (Bell et al., 1984; Peluso and Moyer, 1984; Davis et al., 1986). The most compelling evidence for a role of NS in maintaining the availability of N for replication comes from experiments showing that NS can rescue the ability of N to support replication. High concentrations of N presynthesized alone aggregate and are unable to support replication. The same concentration of N presynthesized in the presence of NS protein, however, is able to support replication efficiently (Howard et al., 1986). These observations lead one to conclude that while N protein is sufficient for replication, other factors are involved in maintaining the availability of N protein. For example, it appears that binding of NS protein to N protein to form NS–N complexes may be a mechanism by which N protein is prevented from self-aggregation and thereby maintained in a form capable of supporting replication. This proposal assumes that the NS protein would be displaced during the encapsidation reaction.

Another possible function of an N–NS complex may be to modulate the availability of N protein for replication by creating pools of N–NS complexes containing different forms of NS protein, some of which are inactive in replication. The inactive complexes, which could be activated by the phosphorylation or dephosphorylation of the NS phosphoprotein, would represent a reserve pool (subset) of N protein, ready to be tapped when the supply of active N protein dropped too low. Therefore, there are several ways postulated in which NS protein could play a role in determining the level of free N protein available to support replication.

B. Availability of Leader to Nucleocapsid Protein

Another mechanism by which the N–leader interaction may be regulated has been proposed by Keene et al. (1984). Complexes of cellular La protein with leader have been observed at times early in infection, preceding the appearance of leader–N complexes (Kurilla and Keene, 1983; Wilusz et al., 1983). This has led to the proposal that the leader–La interaction may provide a mechanism by which the availability of leader to interact with N protein may be regulated.

Evidence for the two types of complexes, the N–NS complexes and the leader–La complexes, has been documented as described above. Both these complexes have been postulated to play a role in regulating the interaction of N protein with the nascent RNA to promote replication. It has been shown that N protein is required for replication; now it would appear that the task in further elucidating the process of replication is to determine whether these other proposed factors do in fact regulate the replication process.

C. Other Factors

The possibility that changes in the template proteins may play a role in regulating the transition to replication has also been presented. The Pol R mutants display a high proportion of readthrough at the leader–N boundary and have templates containing an altered N protein (Perrault et al., 1983). Thus, this presents the possibility that an altered N protein, perhaps even a single altered N protein that is present at a crucial site on the template, may effect the decision to transcribe or replicate.

Another consideration is regulation of the activities of template proteins. The functions of template proteins may be regulated by interaction with other newly synthesized viral proteins. Relative to this, it was shown that NS protein, a protein often proposed as a regulatory factor in VSV RNA synthesis because of its existence in different phosphorylated forms, is able to exchange on and off nucleocapsid templates. This exchange of NS protein may be a mechanism by which the enzymatic activity of template proteins is regulated.

D. Summary

In summary, major advances have been made in the past five years in elucidating the requirements for replication of the negative-strand RNA virus VSV. It has been demonstrated that N protein fulfills the requirement for protein synthesis in replication. Satisfactory models whereby this protein could regulate replication have been presented. While N protein alone can support replication, however, several lines of investigation have suggested that regulation of replication and, indeed, optimal levels of replication, may involve the function of other factors. We have reviewed these possibilities and suggest that the next task in investigating replication is to elucidate whether these factors do function in the regulation of replication in the cell and, if they do, how.

ACKNOWLEDGMENTS. The authors thank their colleagues for encouragement and constructive comments. The authors' research was supported by Public Health Service Grants AI12464 and AI15134 from the National Institute of Allergy and Infectious Disease.

REFERENCES

Arnheiter, H., Dubois-Dalcq, M., Schubert, M., Davis, N., Patton, J., and Lazzarini, R., 1984, Microinjection of monoclonal antibodies to vesicular stomatitis virus nucleocapsid protein into host cells: Effect on virus replication, in: *Nonsegmented Negative Strand Viruses* (D. H. L. Bishop and R. W. Compans, eds.), pp. 393–398, Academic Press, New York.

Arnheiter, H., Davis, N. L., Wertz, G. W., Schubert, M., and Lazzarini, R. A., 1985, Role of the nucleocapsid protein in regulating vesicular stomatitis virus RNA synthesis, *Cell* **14:**259–267.

Ball, L. A., and Wertz, G. W., 1981, VSV RNA synthesis: How can you be positive?, *Cell* **26:**143.

Bell, J. C., Brown, E. G., Takayesu, D., and Prevec, L., 1984, Protein kinase activity associated with immunoprecipitates of the vesicular stomatitis virus phosphoprotein NS, *Virology* **132:**229.

Blumberg, B. M., and Kolakofsky, D., 1981, Intracellular vesicular stomatitis virus leader RNAs are found in nucleocapsid structures, *J. Virol.* **40:**568.

Blumberg, B. M., Leppert, M., and Kolakofsky, D., 1981, Interaction of VSV leader RNA and nucleocapsid protein may control VSV genome replication, *Cell* **23:**837.

Blumberg, B. M., Giorgi, C., and Kolakofsky, D., 1983, N protein of vesicular stomatitis selectively encapsidates leader RNA *in vitro*, *Cell* **32:**559.

Blumberg, B., Colomba, G., Rose, K., and Kolakofsky, P., 1984, Preparation and analysis of the nucleocapsid proteins of vesicular stomatitis virus and Sendai virus, and analysis of the Sendai virus leader–NP gene region, *J. Gen. Virol.* **65:**769.

Chanda, P. K., and Banerjee, A. K., 1981, Identification of promoter-proximal oligonucleotides and a unique dinucleotide, pppGpC, from *in vitro* transcription products of vesicular stomatitis virus, *J. Virol.* **39:**93.

Clinton, G. M., Burge, B. W., and Huang, A. S., 1979, Phosphoproteins of vesicular stomatitis virus: Identity and interconversion of phosphorylated forms, *Virology* **99:**84.

Condra, J., and Lazzarini, R., 1980, Replicative RNA synthesis and nucleocapsid assembly in vesicular stomatitis virus-infected permeable cells, *J. Virol.* **36:**796.

Davis, N. L., and Wertz, G. W., 1982, Synthesis of vesicular stomatitis virus negative-strand RNA *in vitro*: Dependence on viral protein synthesis, *J. Virol.* **41:**821.

Davis, N. L., Arnheiter, H., and Wertz, G. W., 1986, Vesicular stomatitis virus N and NS proteins from multiple complexes, *J. Virol.* **59:**751.

Emerson, S., and Wagner, R., 1972, Dissociation and reconstitution of the transcriptase and template activities of vesicular stomatitis B and T virions, *J. Virol.* **10:**297.

Emerson, S. U., and Yu, Y.-H., 1975, Both NS and L proteins are required for *in vitro* RNA synthesis by vesicular stomatitis virus, *J. Virol.* **15:**1348.

Emerson, S., Dierks, P., and Parsons, J., 1977, *In vitro* synthesis of a unique RNA species by a T particle of vesicular stomatitis virus, *J. Virol.* **23:**708.

Ghosh, K., and Ghosh, H. P., 1982, Synthesis *in vitro* of full length genomic RNA and assembly of the nucleocapsid of vesicular stomatitis virus in a coupled transcription–translation system, *Nucleic Acids Res.* **20:**6341.

Hill, V. M., and Summers, D. F., 1982, Synthesis of VSV RNPs *in vitro* by cellular VSV RNPs added to uninfected HeLa cell extracts: VSV protein requirements for replication *in vitro*, *Virology* **123:**407.

Hill, V., Marnell, L., and Summers, D., 1981, *In vitro* replication and assembly of vesicular stomatitis virus nucleocapsids, *Virology* **113:**109.

Howard, M., Davis, N., Patton, J., and Wertz, G., 1986, Roles of vesicular stomatitis virus N and NS proteins in viral RNA replication, in: *The Biology of Negative Strand Viruses* (B. Mahy and D. Kolakofsky, eds.), pp. 134–140, Elsevier Press, New York.

Hsu, C.-H., Morgan, E. M., and Kingsbury, D. W., 1982, Site-specific phosphorylation regulates the transcriptive activity of vesicular stomatitis virus NS protein, *J. Virol.* **43:**104.

Keene, J., Schubert, M., Lazzarini, R., and Rosenberg, M., 1978, Nucleotide sequence ho-
    mology at the 3' termini of RNA from VSV and its defective interfering particles, *Proc.
    Natl. Acad. Sci. U.S.A.* **75**:3225.
Keene, J., Schubert, M., and Lazzarini, R., 1979, Terminal sequences of VSV RNA are both
    complementary and conserved, *J. Virol.* **32**:167.
Keene, J. D., Schubert, M., and Lazzarini, R., 1980, Intervening sequence between the leader
    region and the nucleocapsid gene of vesicular stomatitis virus RNA, *J. Virol.* **33**:789.
Keene, J. D., Kurilla, M., Wilusz, J., and Chambers, J., 1984, Interactions between cellular
    La protein and leader RNA, in: *Nonsegmented Negative Strand Viruses* (D. H. L. Bishop
    and R. W. Compans, eds.), pp. 103–108, Academic Press, New York.
Kingsford, L., and Emerson, S. U., 1980, Transcriptional activities of different phosphoryl-
    ated species of NS protein purified from vesicular stomatitis virions and cytoplasm of
    infected cells, *J. Virol.* **33**:1097.
Kurilla, M., and Keene, J., 1983, The leader RNA of VSV is bound by a cellular protein
    reactive with anti-La lupus antibodies, *Cell* **34**:937.
Lazzarini, R., Keene, J., and Schubert, M., 1981, The origins of defective interfering particles
    of the negative-strand RNA viruses, *Cell* **26**:145.
Leamnson, R., and Reichmann, M., 1974, The RNA of defective vesicular stomatitis virus
    particles in relation to viral cistrons, *J. Mol. Biol.* **85**:551.
Leppert, M., and Kolakofsky, D., 1980, Effect of defective interfering particles on plus and
    minus-strand leader RNAs in VSV infected cells, *J. Virol.* **35**:704.
Leppert, M., Rittenhouse, L., Perrault, J., Summers, D., and Kolakofsky, D., 1979, Plus and
    minus strand leader RNAs in negative strand virus-infected cells, *Cell* **18**:735.
Lesnaw, J. A., Dickson, L. R., and Curry, R. H., 1979, Proposed replicative role of the NS
    polypeptide of VSV: Structural analysis of an electrophoretic variant, *J. Virol.* **31**:8.
McGeoch, D. J., 1979, Structure of the gene N: gene NS intercistronic junction in the
    genome of vesicular stomatitis virus, *Cell* **17**:673.
Naito, S., and Ishihama, A., 1976, Function and structure of RNA polymerase from vesicular
    stomatitis virus, *J. Biol. Chem.* **251**:4307.
Patton, J. T., Davis, N. L., and Wertz, G. W., 1983a, Cell-free synthesis and assembly of
    vesicular stomatitis virus nucleocapsids, *J. Virol.* **45**:155.
Patton, J. T., Davis, N. L., and Wertz, G. W., 1983b, Inhibition of VSV RNA synthesis by
    2',3'-dideoxycytidine 5'-triphosphate, *J. Gen. Virol.* **64**:743.
Patton, J. T., Davis, N. L., and Wertz, G. W., 1984a, N protein alone satisfies the requirement
    for protein synthesis during RNA replication of VSV, *J. Virol.* **49**:303.
Patton, J. T., Davis, N. L., and Wertz, G. W., 1984b, Role of VSV proteins in RNA replication,
    in: *Nonsegmented Negative Strand Viruses* (D. H. L. Bishop and R. W. Compans, eds.),
    pp. 147–152, Academic Press, New York.
Peluso, R. W., and Moyer, S., 1983, Initiation and replication of vesicular stomatitis virus
    genome RNA in a free-cell system, *Proc. Natl. Acad. Sci. U.S.A.* **80**:3198.
Peluso, R. W., and Moyer, S. A., 1984, Vesicular stomatitis virus proteins required for the
    in vitro replication of defective interfering particle genome RNA, in: *Nonsegmented
    Negative Strand Viruses* (D. H. L. Bishop and R. W. Compans, eds.), pp. 153–160,
    Academic Press, New York.
Perlman S. M., and Huang, A. S., 1973, RNA synthesis of VSV. V. Interaction between
    transcription and replication, *J. Virol.* **12**:1395.
Perrault, J., and Leavitt, R., 1976, Inverted complementary terminal sequences in single-
    stranded RNAs and snap-back RNAs from VSV defective interfering particles, *J. Gen.
    Virol.* **38**:35.
Perrault, J., Clinton, G., and McClure, M., 1983, RNP template of vesicular stomatitis virus
    regulates transcription and replication functions, *Cell* **35**:175.
Pringle, C., 1978, The tdCE and hrCE phenotypes: Host range mutants of vesicular sto-
    matitis virus in which polymerase function is affected, *Cell* **15**:597.
Rao, D., and Huang, A., 1979, Synthesis of a small RNA in cells coinfected by standard and
    defective interfering particles of VSV, *Proc. Natl. Acad. Sci. U.S.A.* **76**:3742.

Rao, D., and Huang, A. S., 1980, RNA synthesis by vesicular stomatitis virus. X. Transcription and replication by defective interfering particles, *J. Virol.* **36**:756.

Rose, J. K., 1980, Complete intergenic and flanking gene sequences from the genome of vesicular stomatitis virus, *Cell* **19**:415.

Schubert, M., and Lazzarini, R., 1981, *In vivo* transcription of the 5'-terminal extracistronic region of vesicular stomatitis virus RNA, *J. Virol.* **38**:256.

Schubert, M., Keene, J., Lazzarini, R., and Emerson, S., 1978, The complete sequence of a unique RNA species synthesized by a DI particle of VSV, *Cell* **15**:103.

Semler, B., Perrault, J., Abelson, J., and Holland, J., 1978, Sequence of RNA templated by the 3'-OH RNA terminus of defective interfering particles of VSV, *Proc. Natl. Acad. Sci. U.S.A.* **75**:4704.

Sinacore, M. S., and Lucas-Lenard, J., 1982, The effect of the vesicular stomatitis virus-associated protein kinase on viral mRNA transcription *in vitro*, *Virology* **121**:404.

Soria, M., Little, S., and Huang, A., 1974, Characterization of vesicular stomatitis virus nucleocapsids. I. Complementary 40S RNA molecules in nucleocapsids, *Virology* **61**:270.

Sprague, J., Condra, J. H., Arnheiter, H., and Lazzarini, R. A., 1983, Expression of a recombinant DNA gene coding for the vesicular stomatitis virus nucleocapsid protein, *J. Virol.* **45**:773.

Stamminger, G., and Lazzarini, R., 1979, Analysis of the RNA of defective VSV particles, *Cell* **3**:85.

Szilagyi, J. F., and Pringle, C. R., 1979, Effect of temperature-sensitive mutation on activity of the RNA transcriptase of VSV New Jersey, *J. Virol.* **30**:692.

Wertz, G., 1980, RNA replication, in: *Rhabdoviruses*, Vol. II (D. H. L. Bishop, ed.), pp. 75–93, CRC Press, Boca Raton, Florida.

Wertz, G. W., 1983, Replication of vesicular stomatitis virus defective interfering particle RNA *in vitro:* Transition from synthesis of defective interfering leader RNA to synthesis of full-length defective interfering RNA, *J. Virol.* **46**:513.

Wertz, G. W., and Levine, M., 1973, RNA synthesis by vesicular stomatitis virus and a small plaque mutant: Effects of cycloheximide, *J. Virol.* **12**:253.

Wilusz, J., Kurilla, M., and Keene, J., 1983, A host protein (La) binds to a unique species of minus-sense leader RNA during replication of VSV, *Proc. Natl. Acad. Sci. U.S.A.* **80**:5827.

Witt, D. J., and Summers, D. F., 1980, Relationship between virion-associated kinase-effected phosphorylation and transcription activity of vesicular stomatitis virus, *Virology* **107**:34.

CHAPTER 8

Defective Interfering Rhabdoviruses

JOHN J. HOLLAND

I. INTRODUCTION AND BACKGROUND

This review focuses on biological and molecular properties of defective interfering (DI) particles of the rhabdoviruses, but coverage of DI particles of other RNA viruses is frequently included to emphasize common principles, significant difference, or phenomena that have not yet been observed or examined among the rhabdoviruses. Nearly all definitive studies of rhabdovirus DI particles have been carried out with the prototype rhabdovirus, vesicular stomatitis virus (VSV), and VSV DI particles are probably the most studied and best understood of all DI particles. Space limitations prevent complete coverage of the DI-particle field, and the reader is referred to earlier reviews for supplemental background information (Huang and Baltimore, 1977; Reichmann and Schnitzlein, 1979, 1980; Holland *et al.*, 1980; Perrault, 1981; Lazzarini *et al.*, 1981; Blumberg and Kolakofsky, 1983).*

DI particles were first clearly defined by Huang and Baltimore (1970) when they recognized the ubiquity and the similarities of homologous interfering particles among divergent groups of RNA and DNA viruses. To encompass all these interfering viruses, they proposed the general term "defective interfering particles," or "DI particles." The definition of DI

---

* Following completion of this review, others have recently completed a number of outstanding reviews regarding DI particles *in vitro* and *in vivo* (Dimmock and Barrett, 1986; Nayak *et al.*, 1985; Huang, 1987; Schlesinger, 1987).

JOHN J. HOLLAND • Department of Biology, University of California at San Diego, La Jolla, California 92093.

particles by Huang and Baltimore in 1970 has proved appropriate for all groups of viruses, and their description of the nature of DI particles still applies without modification more than 15 years later. DI particles are subgenomic virus particles (lacking greater or lesser percentages of the virus genome). They contain virus structural proteins and antigens. They require homologous parental virus for replication. They replicate preferentially at the expense of helper virus, thereby causing interference. As will be seen, the last-named property does not hold for all DI particles.

DI-particle interference was first described by Henle and Henle (1943) as lowered infectivity for mice of influenza viruses passaged undiluted in embryonated chicken eggs. This phenomenon was investigated extensively by von Magnus (1951, 1954). He demonstrated that serial undiluted passage of influenza virus *in ovo* led to strong interference after several passages and that this interference was due to replicating defective virus particles that he called "incomplete virus particles." These "incomplete viruses" exhibited hemagglutinating activity, but lacked infectivity. He also demonstrated partial protection of mice against influenza virus challenge through interference by incomplete virus. Mims (1956) showed that Rift Valley fever virus (a phlebovirus) produces an analogous phenomenon in mice. Serial undiluted passage of virus in mice produced incomplete virus that interfered with virus growth, prolonged virus incubation periods, and led to survival of some mice, whereas dilute virus passage led to high virus titers and rapid disease and death.

Cooper and Bellett (1959) and Bellett and Cooper (1959) analyzed this phenomenon in some detail using VSV in cell culture. They proved that the interfering virus consisted of sedimentable particles that accumulated after serial undiluted passages in cell culture. These particles were shown to be transmissible (i.e., to replicate with helper virus) and to sediment more slowly than helper virus, but inexplicably, antiviral antiserum failed to neutralize their interfering particles and interferon could not be ruled out. Hackett (1964) obtained electron micrographs showing that VSV DI particles are shorter than the long rod of infectious VSV. Crick *et al.* (1966), Huang *et al.* (1966), and Hackett *et al.* (1967) demonstrated that discrete bands of DI particles of VSV could be separated from infectious virus by velocity sedimentation in sucrose gradients, and this led to the first definitive biological and biochemical characterization of purified DI particles of VSV (Huang and Wagner, 1966a,b). Since Huang and Baltimore (1970) defined the general nature and widespread occurrence of DI particles, they have been observed and characterized in nearly every class of RNA and DNA viruses (Huang and Baltimore, 1977, Stollar, 1980, Nayak, 1980; Holland *et al.*, 1980: Perrault, 1981; Lazzarini *et al.*, 1981). Despite the ubiquity of DI particles, they have been characterized only rather poorly for most groups of viruses, because of the difficulty in purifying them free of infectious virus and vice versa. Conversely, rhabdovirus DI particles are more extensively characterized than those of any other viruses because of the relative ease of separating rhabdovirus DI particles (and nucleocapsids) from each other and from infectious virus (and virus nucleocapsids).

Because rhabdovirus DI particles are so widely employed in basic studies, and because there are so many strains of rhabdoviruses and each can generate numerous different DI particles, a uniform system of nomenclature was proposed (Reichmann *et al.*, 1980) to deal with this problem. It was proposed to designate the serotype (e.g., VSI = VSV-Indiana), the isolate initial (and mutant number if any) (e.g., VSI *ts*G12 = VSV-Indiana Glasgow isolate, temperature-sensitive mutant number 2 of complementation group I), and the DI-particle fractional length with genome-end derivation and percentage self-annealing in parenthesis. Thus, VSI *ts*G12 DI 0.27(5′,24%) designates a DI particle derived from VSV-Indiana Glasgow strain complementation group I mutant number 2. The DI-particle genome is 27% of the length of the virus genome RNA, it is derived from the 5′ end of the virus genome, and it self-anneals 24% of its RNA. After 5 years, it now appears that this terminology has not been, and probably will not be, widely adopted. Investigators often employ numerous different DI particles, many of which are not characterized regarding fractional length and self-annealing percentages. Those that are so characterized have more often been referred to by initials such as DI-T, MST, DI-LT, DI-LT$_2$, DI-ST, or some other simple designation. Regardless of the nomenclature employed, it is important for investigators to identify the viral serotype, strain, and mutant origin of their DI particles and to provide a name for each that will distinguish it from other DI particles generated by the same parent virus and from DI particles widely employed in other laboratories.

II. GENERAL BIOLOGICAL CHARACTERISTICS OF DEFECTIVE INTERFERING PARTICLES

A. Size, Shape, and Nucleocapsid Content of Defective Interfering Particles

DI particles generally exhibit the same shape and symmetry characteristics as parental virus. DI particles of rhabdoviruses are usually smaller than parental virus because the deleted genome RNA assembles into a shorter nucleocapsid (which determines virus length as it becomes enveloped at the cell surface). Figure 1 shows electron-microscopic (EM) comparison of a vesicular stomatitis virus (VSV)-Indiana large DI particle with its parental virus. The flexible helical nucleocapsid rods are seen regularly coiled within each, but there are fewer coils in DI particles—the number being related to the total ribonucleoprotein (RNP) length. The careful EM studies of Naeve *et al.* (1980) showed that the mean length of VSV viral RNP is 4,095 nm in its uncoiled (10-nm-diameter)

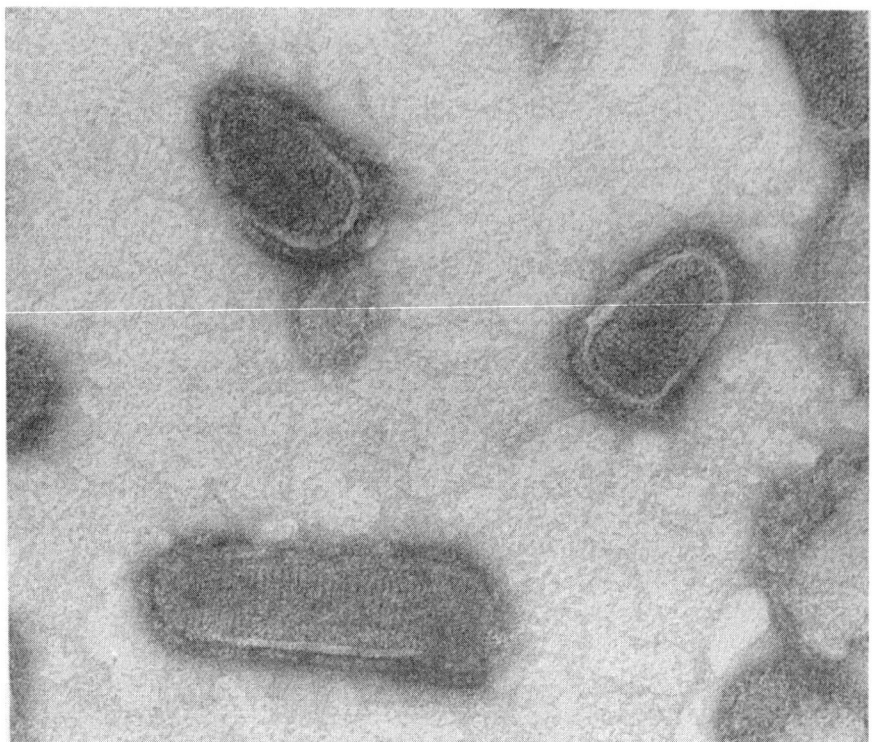

FIGURE 1. Electron micrograph of an infectious virion and associated long DI particles of VSV-Indiana serotype.

long threadlike form. This is coiled in the cytoplasm (in the presence of divalent cations) into 20 × 700 nm structures that undergo a final conformational change during virus maturation to the 50 × 175 nm form observed in virions. Obviously, the lengths of each form are shortened progressively in increasingly smaller DI particles. The smallest DI particles of VSV observed are only about 10% of viral genome length, and the largest are about 50–60%. It is likely that DI nucleocapsids smaller than 0.1 fractional length are able to be formed and replicated, and to interfere, but that they cannot be matured and enveloped efficiently at the cell surface. It can be seen from Fig. 1 that a DI nucleocapsid of fractional length 0.05 or less could not coil more than a few times within an envelope and would be unlikely to form stable particles of a uniform size unless several nucleocapsids were included. Rhabdoviruses might sometimes form DI particles containing two or more DI nucleocapsids, but if they do, such "tandem" particles are rare and fail to "breed true" or to compete successfully with "nontandem" DI genomes.

A different situation obtains in icosahedral viruses in which packaging constraints may require a minimal amount of RNA for stable virions to be matured. DI particles of some icosahedral viruses contain multiple copies of subgenomic-size RNA. This is true of alphaviruses such as Sindbis virus, in which multiple copies of DI RNA are encapsidated in sufficient numbers to provide a near-normal content of viral RNA (S.I.T. Kennedy et al., 1976; Guild et al., 1977). Alternatively, some icosahedral viruses such as poliovirus may exhibit packaging constraints on the extent of deletion of viral RNA. Apparently, in poliovirus, only minimally deleted DI genomes can be packaged and serially propagated (Cole and Baltimore, 1973a–c; Cole, 1975; Cole et al., 1971; Lundquist et al., 1979; Nomoto et al., 1979; McClure et al., 1980). The deletions of known poliovirus DI genomes range from about 4 to 16% of the viral genome, probably because DI genomes much below 85% of virion size cannot be stably encapsidated in the icosahedral shell. In any case, the helical symmetry of rhabdovirus RNA nucleocapsids generally leads to single DI RNA species within single, shorter nucleocapsids that mature to smaller DI virions at the cell surface. The situation is not so clear with the other helical-symmetry negative-strand RNA viruses such as myxoviruses, paramyxoviruses, arenaviruses, bunyaviruses, and others in which cell-surface maturation of short DI nucleocapsids does not produce DI virions of uniformly smaller size than those of standard virus. It is likely that multiple DI genomes are frequently matured within the variable-size envelopes of these viruses. And because polyploid virions have been reported for some of these, it is also likely that DI nucleocapsids and helper virus nucleocapsids might often be incorporated together within individual virions of these viruses. Interestingly, Barrett et al. (1984a) and R.E. Johnson et al. (1975) found Sindbis virus particles to be slightly larger than their homologous DI particles.

B. Rhabdovirus Defective Interfering Particles Interfere Mainly at the Level of Virus Replication

As is discussed in Section V, rhabdoviruses can form many different sizes of DI particles by one or more "RNA recombinational events" at a variety of different genomic sites. By far the most common, and probably the most competitive, class of VSV DI particles are those derived from the 5' half of the genome [i.e., those that contain portions of the large (L)-protein gene] (Reichmann and Schnitzlein, 1979). This most commonly encountered class of DI particles is genetically inert in vivo and in vitro (Huang and Manders, 1972; Perrault and Holland, 1972b), directing neither transcription nor translation. In cells doubly infected with helper virus plus DI particles, the following characteristics are observed: (1) Primary transcription (from input virus templates) proceeds normally. (2) Virion-size RNA replication is strongly inhibited, and, therefore, secondry transcription (from newly replicated viral templates) is also strongly suppressed. (3) The reduced amounts of viral replication, encapsidation, and maturation proteins that result from these reduced levels of viral transcription and translation become devoted largely to replication of DI RNA genomes and to their maturation into DI particles (Huang and Manders, 1972; Perrault and Holland, 1972b; Khan and Lazzarini, 1977).

The studies cited above led to the conclusion that this majority class of DI particles are replicative, transcriptionally inert particles that exert their interfering effects by outcompeting standard virus at the level of genome replication. The reasons for DI-particle replicative advantages are not known, but possible mechanisms are discussed in Sections V and VI. Previously, lack of an *in vitro* replication system has hindered biochemical studies of viral and DI replication. However, VSV *in vitro* replication systems have recently been developed (Batt-Humphries *et al.*, 1979; Hill *et al.*, 1981; Davis and Wertz, 1982; Patton *et al.*, 1983; Hill and Summers, 1982; Peluso and Moyer, 1983). These systems have now been shown to support replication of DI nucleocapsids as well as viral nucleocapsids (Peluso and Moyer, 1983; Wertz, 1983) (see also Chapter 7). This might eventually allow molecular dissection of the host and viral components, contributing to the selective replication–encapsidation of DI genomes during interference. As is discussed in Sections V and VI, certain less common VSV DI particles direct both transcription and translation of VSV genes, and their interference effects can operate at levels other than, or in addition to, replication (Johnson and Lazzarini, 1977, 1978; Bay and Reichmann, 1979, 1981). However, such transcribing DI particles are very rare among rhabdoviruses (Schnitzlein and Reichmann, 1976, 1977a,b; Adler and Banerjee, 1976), probably because they compete poorly with the common DI particles (Perrault and Semler, 1979).

In contrast to rhabdoviruses, transcribing DI particles of this internal deletion type are the most common and competitive of influenza DI particles (Nayak *et al.*, 1982; Jennings *et al.*, 1983; Akkina *et al.*, 1984). They are also very common DI-particle types of Sendai virus (Amesse *et al.*, 1982; Kailash *et al.*, 1983), but nontranscribing "stem" DI-particle types are also very common in Sendai virus (Kolakofsky, 1976). Finally, there are indications that DI particles of Sendai virus can affect the cell-surface expression and turnover of viral glycoproteins (G proteins), so there are subtle specific effects (as yet unexplained) of DI particles in addition to replicative interference (Roux *et al.*, 1984, 1985). Paralta *et al.* (1981) reported an aberrant G protein replacing the viral G proteins of lymphocytic choriomeningitis virus, so membrane effects of DI particles may occur frequently. Weiss *et al.* (1974) and Guild and Stollar (1975) have reported unusual proteins synthesized (or abnormally processed) in Sindbis-virus-infected cells during DI-particle interference. In rhabdovirus-infected cells, DI-particle interference does not usually cause selective inhibition of any of the VSV proteins (Little and Huang, 1977), but measles virus DI particles appear to inhibit selectively the synthesis of all viral proteins except the nucleocapsid (N) protein (Rima and Martin, 1979). Such "phenotypic" effects of DI particles might sometimes result simply from cumulative kinetic effects of replicative interference on viral protein synthesis, processing, maturation, turnover, and other functions. Regardless of the mechanisms, such effects could obviously have profound biological consequences in infected cells, tissues, and organs.

C. Host-Cell Type Profoundly Influences Defective-Interfering-Particle Replication or Interference or Both

There are numerous reports showing that DI-particle replication or interference or both can vary greatly depending on the host-cell type infected. This has been shown for VSV (Huang and Baltimore, 1970; Perrault and Holland, 1972a; Holland et al., 1976a; Kang et al., 1981; Youngner et al., 1981), for influenza virus (Choppin, 1969; De and Nayak, 1980; Crumpton et al., 1981), for Sendai virus (Kingsbury and Portner, 1970), for alphaviruses (Stark and Kennedy, 1978; Levin et al., 1973; Eaton, 1975; Igarashi and Stollar, 1976; C.-C. King et al., 1979; Steacie and Eaton, 1984), for flaviviruses (Darnell and Koprowski, 1974; Brinton, 1983; Brinton and Fernandez, 1983; Brinton et al., 1984), for mengovirus (McClure et al., 1980), for encephalomyocarditis virus (Radlof and Young, 1983), and for a number of other viruses. The molecular basis for these "high-interference" vs. "low-interference" cell influences is unclear, but insights may be obtained using cell-free replication systems. In some "low-interference cell lines," DI-particle generation might be faulty, but most often, exogenously supplied DI particles either are replicated efficiently and fail to interfere well [as was seen with VSV DI particles in MDCK, MDBK, and PK-15 cells by Perrault and Holland (1972a)] or replicate poorly [as was observed by (Holland et al. (1976a) in a HeLa cell line (which nevertheless exhibited DI-particle interference with parental VSV)]. Host-cell factors for replication–encapsidation and/or other viral processes must be involved, but their nature and site or sites of action are unknown.

Kang et al. (1981) concluded from studies of DI-particle replication in mouse–human hybrid cells that human chromosome 16 encodes a factor or factors able to suppress de novo generation of VSV DI particles, but not the replication of added DI particles. A remarkable series of studies (Brinton, 1983; Brinton and Fernandez, 1983; Brinton et al., 1984) related control of DI-particle replication to an autosomal dominant allele of mice that controls resistance to flavivirus encephalitis. This resistance allele is not common in inbred laboratory mouse strains, but it is present in wild mouse populations in California and Maryland (Brinton and Nathanson, 1981). The presence of this resistance gene leads to rapid accumulation of DI particles in mouse cells in culture and in the brains of infected mice in vivo (Brinton et al., 1984).

A very striking effect of cell origin was reported by C.-C. King et al. (1979) and investigated further by Gillies and Stollar (1980). C.-C. King *et al.* (1979) showed that Sindbis virus DI particles generated in chick cells did not interfere in *Aedes albopictus* mosquito cells and that those generated in mosquito cells failed to interfere in chick cells. Gillies and Stollar (1980) observed the identical phenomenon with VSV DI particles grown in baby hamster kidney (BHK)-21 cells or in *A. albopictus* cells. They further demonstrated that a single passage of DI particles from either cell source into the heterologous cell type completely reversed the pattern of interference specificity. This surprising finding was analogous to the serotype-specificity patterns reported for VSV DI particles by Schnitzlein and Reichmann (1977b), who found that VSV-Indiana DI-particle RNA genomes interfered very poorly with VSV-New Jersey infectious virus. However, if pseudotype DI particles that contained VSV-Indiana RNA within VSV-New Jersey N, envelope, and other proteins were prepared, these particles interfered strongly with VSV-New Jersey. Again in this case, a single further passage of these VSV-Indiana RNA genomes within cells infected by VSV-Indiana completely abolished the enhanced interference with VSV-New Jersey. Obviously, either viral proteins or host-cell phenotypic modification (or selection) can change the replicative specificity of DI-particle interference.

The molecular nature of all the aforementioned host-cell influences will be determined only with difficulty, but they are clearly of great importance biologically. It is virtually certain that the ability of DI particles to protect cells and tissues *in vivo* (see Section VII) will frequently be determined by the cell type in which the virus is replicating (and possibly by the cell type in which an infecting virus or DI particle or both were replicated previously).

D. The Amount of Interference Exerted by a Given Defective-Interfering-Particle Type in Any Cell Can Vary Enormously

The early studies of Bellett and Cooper (1959) showed that increasing the input multiplicities of infection (MOIs) of VSV DI particles led to a progressive decrease in yields of infectious virus. Their dose–response curves (DI input MOI vs. infectious virus yield curves) fit the Poisson distribution well for a one DI hit = zero cell yield response. Bellett and Cooper (1959) concluded that DI-particle interference was generally all-or-none, with infected cells yielding either a normal virus yield ($P = 0$ class for DI-particle hits) or zero infectious virus yield ($P \geq 1$ for DI particles). However, this study was complicated by the possibility of interferon effects, since antiviral antiserum failed to neutralize DI particles and there was evidence that interferon might be involved (as discussed in Section II.F, some DI particles strongly induce or enhance interferon production). Sekellick and Marcus (1980) carefully tested this hypothesis using high virus-input multiplicities and DI-particle multiplicities of 1 or 16 followed by yield analysis from many individual cells "fished out" into micropipettes. They showed convincingly that single cells capable of yielding over 5000 plaque-forming units (PFU) generally yield 0 PFU of infectious virus if simultaneously infected with virus plus a single DI particle. They employed GMK-Vero cells unable to respond to interferon, so the direct interfering ability of DI particles rather than interferon was involved.

The all-or-none effect of DI particles is probably operative in most situations involving "wild-type" VSVs and their homologous DI particles. As discussed in Section II.C, this is obviously not true in "low-interference cell lines." Furthermore, it is not necessarily true when DI particles coinfect cells with heterotypic helper virus. When New Jersey serotype VSV helper virus was coinfected with DI particles of VSV-Indiana serotype, the heterologous DI particles replicated well, but did not interfere significantly (Schnitzlein and Reichmann, 1977a,b; Khan and Lazzarini, 1977; Adachi and Lazzarini, 1978). As mentioned in Section II, C, however, pseudotype VSV-Indiana DI-particle genomes previously encapsidated and matured within VSV-New Jersey proteins did interfere strongly (Schnitzlein and Reichmann, 1977b).

In a more natural homotypic interference system, Stampfer et al. (1969) observed greatly reduced DI-particle interference (but significant DI-particle replication) if VSV-Indiana DI particles were added to cells 2.5 hr after infection by homologous helper virus.

Last, the continuous evolution or "coevolution" of virus mutants and of their associated DI particles can profoundly affect the degree of DI-particle interference with homotypic helper virus. This is true of rabies virus (Kawai and Matsumoto, 1977), or VSV (Horodyski and Holland, 1980, 1981, 1983), of lymphocytic choriomeningitis virus (Jacobsen and Pfau, 1980), of Sindbis virus (Weiss and Schlessinger, 1981), of RNA phage f1 (Enea and Zinder, 1982), and of West Nile virus, a flavivirus (Brinton and Fernandez, 1983). In all of these viruses, mutants arose (following DI-particle selective pressures) that were very resistant to the preexisting DI particles. Quantitative assays showed that widely varying degrees of resistance can be obtained (Horodyski and Holland, 1980, 1981, 1983). The degree of mutant-virus "escape" from DI-particle interference ranges from very slight to complete resistance. This phenomenon is reviewed in detail in Section IX. Finally, Winship and Thacore (1979) reported that Shope fibroma virus gene expression without replication can prevent VSV DI-particle replication, but not VSV replication, in monkey-kidney cells.

DI-particle interference in nature may frequently be all-or-none, as observed by Bellett and Cooper (1959) and Sekellick and Marcus (1980). At times, however, the levels of DI-particle interference can vary enormously depending on cell type, virus serotype, virus mutant phenotype, DI-particle type, and relative times of entry into the cell of virus and of DI particles.

E. Defective Interfering Particles Can Allow Survival of Cells Infected by Otherwise Lethal Viruses

Wagner *et al.* (1963) first demonstrated that VSV could establish persistent infections in cell culture. They demonstrated that small-plaque mutants were much more likely than large-plaque virus to initiate persistence in L cells. Holland and Villarreal (1974) showed that purified DI particles regularly led to establishment of persistent BHK-21-cell infections *in vitro* when coinfecting with standard infectious VSV-Indiana. This was particularly effective if a *ts*G31 mutant, a matrix (M)-protein mutant of Pringle (1970), was used with the DI particles. Since then, there have been numerous reports of DI-particle facilitation of the establishment or maintenance, or both, of persistent infections of a variety of cells with a large variety of viruses. This literature is reviewed in Section VIII. Holland *et al.* (1976b) demonstrated that although nearly all cells are persistently infected and carrying VSV and DI-particle genomes, they can regularly be "cured" by cloning the persistently infected cells under antiserum. Clearly, DI-particle effects can lead to two biologically important results: (1) persistent infection of cells that otherwise would have been destroyed by viral cytopathology and (2) complete recovery of cells from virus infection that would have been lethal in the absence of DI particles. DI-particle sparing effects on cells, tissues, organs, and animals are covered in detail in Sections VII and VIII.

F. Defective Interfering Particles Can Facilitate Production of Interferon

Marcus and Sekellick (1977) reported that certain DI particles that contain totally self-complementary or "snapback" RNA are excellent inducers of interferon. They found that a single molecule of DI double-stranded RNA (dsRNA) fully induced a maximum yield of interferon in "aged" chick-embryo cells. Other, "nonsnapback" structural types of DI particles did not induce interferon as efficiently. However, Frey *et al.* (1979) also found this DI particle to be a potent inducer of interferon in L cells, but other DI particles of differing degrees of self-complementarity were also effective inducers. They observed no correlation between interferon-inducing capacity and degree of self-complementarity of DI-genome RNA. Frey *et al.* (1981) later reported that contaminating infectious-virus levels were correlated with the interferon-inducing capacity of DI particles. However, Sekellick and Marcus (1982) later ruled out significant infectious-virus contribution to the interferon induction caused by the snapback DI particles, and they showed that nearly all DI particles induced interferon in "aged" chick cells, but only about 1 in 400 did so in the L cells employed by Frey *et al.* (1981).

The situation is complicated by the fact that infectious VSV mutants can also induce interferon by forming dsRNA (Nilsen *et al.*, 1981). Also nonmutant wild-type VSV can inhibit cell protein synthesis by inducing dsRNA (Thomas and Wagner, 1982). Furthermore, VSV infection strongly inhibits cell transcription, at least partly via small leader RNA (Grinnell and Wagner, 1984). Since the leader regions of both virus and DI particles may be involved as switches in VSV transcription–replication and in DI-particle replication (Blumberg *et al.*, 1983; Kurilla and Keene, 1983), any interpretation of DI-particle induction of interferon when contaminating viruses are present in varying numbers becomes exceedingly complex. However, it is clear that DI particles can induce and enhance interferon production, and this might frequently be a significant biological function of DI particles *in vivo*. In fact, Sendai virus, a very widely used inducer of interferon, requires DI particles along with virus for its high-level induction of interferon (M.D. Johnston, 1981).

G. Defective Interfering Particles and Helper Viruses Exhibit Cyclical Interactions

Palma and Huang (1974) demonstrated regular cyclical patterns of VSV-Indiana virus yields that were inversely correlated with homologous DI-particle yields when uninfected cells were added regularly to infected cultures. After virus cyclic replication to high levels, DI particles soon replicated to high levels (using infectious virus as helper). At the maximum level of DI-particle yields, infectious-virus replication regularly became depressed until relatively few helper viruses remained to support DI-particle replication. Then when DI particle levels dropped, virus yields once again cycled up to maximum and the cycle was repeated. Identical smooth, cyclic virus–DI particle interactions were observed by Kawai *et al.* (1975) in BHK-21 cells persistently infected by rabies virus. Infectious-virus shedding by the carrier rose and fell in a cyclic, out-of-phase manner with shedding of DI particles. Grabau and Holland (1982) disrupted cells persistently infected for many years by either VSV or rabies virus, isolated intracellular viral nucleocapsids, and analyzed them on sucrose gradients. Analysis of both VSV and rabies carrier cells at intervals for many weeks demonstrated that the ratio of virus-size nucleocapsids to DI-size nucleocapsids fluctuated dramatically and rather regularly. Whenever the ratio of rabies DI nucleocapsids to virus nucleocapsids was maximum, incorporation of [^{3}H]uridine into total intracellular nucleocapsids was minimal, and vice versa. This indicates clearly that viral and DI nucleocapsids were undergoing regular cyclical interactions intracellularly and that DI nucleocapsids were cyclically regulating viral synthesis in these persistently infected cells even at times when little or no infectious mature virus was being shed.

Cave *et al.* (1984) recently obtained intriguing results in mice infected intranasally with 10^8 PFU VSV-Indiana together with different amounts of homologous DI particles. Lower levels of DI particles were more protective than higher input levels, and the lower levels led to detection of DI-size RNA in individual mouse brains. This strongly suggests that cyclical virus–DI particle interactions were occurring *in vivo*. This phenomenon is discussed in more detail in Section VII, but these results raise the intriguing possibility that cyclical virus–DI particle interaction patterns may affect the outcome of virus infections *in vivo*. If so, such cyclic bursts of virus and DI synthesis could greatly influence interferon production, natural killer (NK)-cell responses, and specific immunocyte responses in very complex ways. In an early study before DI particles were well characterized, Mims (1956) observed profound cyclic variations in blood levels of Rift Valley fever virus in pooled mice during serial undiluted intravenous passages. This observation confirms that cycling can occur *in vivo*.

H. Segmented-Genome RNA Viruses Exhibit Reassortment with Their Defective Interfering Particles

Although it is not directly relevant to rhabdoviruses, it should be noted that DI particles of segmented-genome viruses generally carry unaltered genome segments in addition to deleted or rearranged segments (Nayak, 1980; Nonoyama *et al.*, 1970; Schuerch *et al.*, 1974; Ahmed and Graham, 1977), so they can be subject to gene reassortment with standard virus. For example, Ahmed and Fields (1981) demonstrated reassortment between DI particles of reovirus and either wild-type or temperature-sensitive mutant reovirus. Because of interference due to DI particles, the reassortment occurred at a low level, but they were able to rescue mutations from nondeleted genes. This type of segment rescue is not possible with the nonsegmented rhabdoviruses, of course, but covalent RNA recombination does infrequently occur between segments of influenza virus (S. Fields and Winter, 1982; Jennings *et al.*, 1983) and might *very infrequently* recombine information between DI particles and infectious rhabdoviruses. RNA recombination is discussed in detail in Sections V and VI.

III. ASSAY SYSTEMS FOR DEFECTIVE INTERFERING PARTICLES

A number of different methods have been employed to detect the presence of and to quantitate rhabdovirus DI particles.

A. Electron Microscopy

Because rhabdovirus DI particles are truncated as compared to infectious standard virus, they can be distinguished from virus in electron micrographs (Hackett, 1964). A number of investigators have used EM to assay for the presence of DI particles in virus preparations or to estimate the level of infectious-virus contamination of purified DI-particle preparations. Although tedious, this method is effective and reasonably quantitative if the methods used to concentrate or purify virions and DI particles do not differentially enrich for one or the other, if aggregation (particularly differential aggregation) is minimized, and if statistically significant numbers of standard virions and DI particles are counted to justify estimates of relative proportions of each. It is clear from an inspection of Fig. 2 that DI particles outnumber virus particles in this field. If many more such fields were counted, an estimate of relative proportions of each in this preparation could be obtained. EM on sections of cells, tissues, and organs is not a good method to estimate relative numbers of

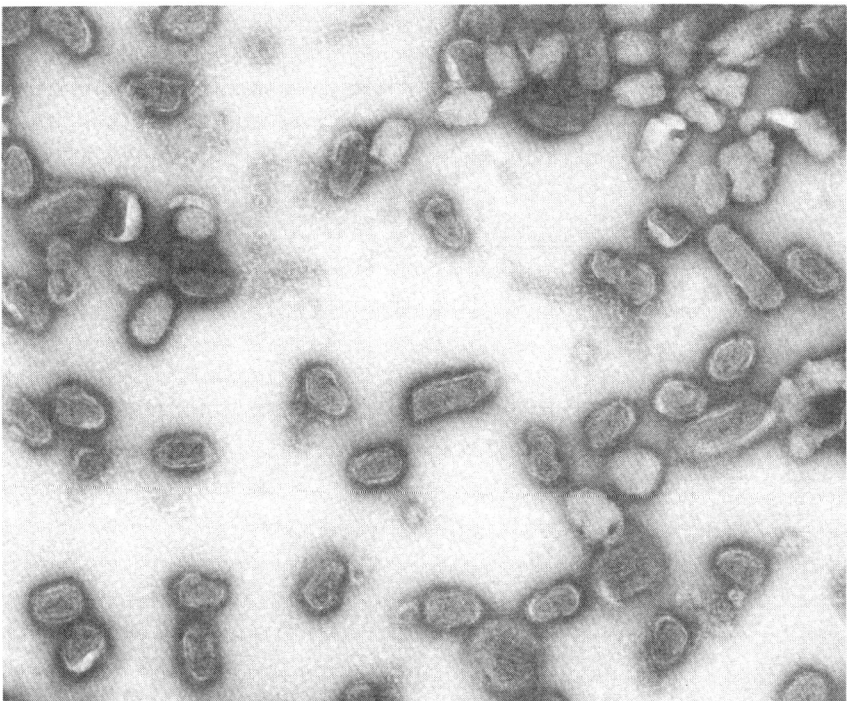

FIGURE 2. Electron micrograph demonstrating how relative proportions of infectious VSV and its DI particles can be estimated in electron micrographs. It is obvious that DI particles greatly outnumber virus particles in this pool, but many more such fields would have to be counted to obtain an accurate estimate of their relative numbers.

virus particles and DI particles being shed from cells, since tangential sections of infectious-virus particles often resemble DI particles.

B. Velocity-Gradient Sedimentation

Huang *et al.* (1966), Crick *et al.* (1966), and Hackett *et al.* (1967) demonstrated that rhabdovirus DI particles can be separated from infectious-virus particles on sucrose gradients. This method is employed as a step in purification, and sharp bands of virus and various sizes of DI particles can be delineated. Relative numbers of virus particles, and of each DI particle, can readily be determined from optical-density measurements of the particles recovered from each band. This is the most direct and widely employed method for recovery and quantitation of virus and DI particles of rhabdoviruses. However, it is not sufficiently sensitive to detect very small proportions of DI particles. By repeated purification of DI particles using sucrose gradients, it is possible to achieve more than 10^9-fold purification of rhabdovirus DI particles and to free them completely from infectious virus for biological studies (Doyle and Holland, 1973). This degree of purification is difficult to achieve, and most of the starting DI particles are lost, but virus-free preparations are necessary for certain applications. It is difficult to verify that a concentrated DI-particle preparation is free of infectious virus because of strong interference with a minimal number of infectious-virus particles. This can be done, however, by serial blind passages of the DI preparation, using as a control an aliquot of the same DI-particle preparation to which has been added several plaque-forming units of infectious virus before the first passage on sensitive cell cultures (Doyle and Holland, 1973). It should be noted that it is not feasible to purify infectious virions of rhabdoviruses free of DI-particle contamination using sucrose gradients. DI particles (and virus particles) always aggregate to some extent, and aggregates below the main bands of DI particles can cosediment with the main virus band. Methods to eliminate or minimize DI particles in virus pools are discussed in Section IV.

C. Biological-Amplification Assay

Direct velocity-gradient sedimentation assay as described above is an excellent method to purify and quantitate the virions and DI particles present in a virus pool, but it cannot detect very low levels of DI particles present along with large amounts of infectious virus. However, this method can be made extremely sensitive by carrying out one or several biological-amplification cycles in cell culture prior to velocity-gradient analysis (Holland and Villarreal, 1975). This is done simply by adsorbing a rhabdovirus inoculum that is a possible source of DI particles to sensitive cells, following which infectious helper virus is added at a multiplicity of 100 to assure support for nearly all DI particles present and to assure that only one cycle of virus and DI-particle replication ensues. The first cycle yield is then analyzed on sucrose gradients. If no DI particles are detected, a portion of the yield from the first cycle is adsorbed to sensitive cells and analyzed in a second cycle. If necessary, a third cycle can be employed. A control series is included in which only helper virus is used to initiate the first cycle, and after four or five cycles, DI particles that were present or generated in the first-cycle helper-virus pool will finally be detected. The number of passages before this occurs will, of course, vary with virus strain, assay cell type, and other factors, as well as with DI numbers originally present, and must be checked for each system.

Holland and Villarreal (1975) quantitated the amplification factor for DI particles and found that vesicular stomatitis virus (VSV)-Indiana DI particles were amplified about 20,000-fold per single cycle, while HEP rabies virus DI particles were amplified about 4000- to 5000-fold per cycle at *low DI-particle input multiplicities*. Of course, at DI-particle inputs exceeding 1 per cell, amplification factors should diminish, and such diminution was observed. Holland *et al.* (1976b) also showed that this assay can be employed to detect nonmatured DI nucleocapsids within acutely and persistently infected cells. The nucleocapsids released from disrupted cells were introduced into first-cycle assay cells using DEAE–dextran-facilitated uptake as originally described by Brown *et al.*(1967).

This type of amplification assay requires careful standardization for each virus–cell system prior to use in order to determine amplification factors, number of cycles allowable before helper-virus-derived DI particles appear, and other variables. Once standardized, the assay is sensitive and reproducible and has the advantage that large amounts of the DI particles assayed are available for biological and biochemical characterizations. Its main disadvantage is that it is effective for rhabdovirus DI particles, but not for most other virus groups. A similar sucrose-gradient analysis of viral- and DI-nucleocapsid RNA sizes can allow its employment for Sendai virus amplification assays and probably for many other negative-strand viruses in which DI-particle sizes are not reliably different from standard virion size (Kolakofsky, 1976, 1979; Roux and Holland, 1979). Amplification assays simplify detection, isolation, and characterization of rhabdovirus DI particles under conditions in which direct isolation is possible, but expensive and impractical. For example, VSV DI particles produced in newborn-mouse brains could be detected directly only when dozens of mouse brains were pooled, whereas a single amplification assay easily quantitated and isolated the same DI particles from individual mouse brains (Holland and Villarreal, 1975). One cannot be certain, however, that all DI particles from all cells *in vivo* (or from cells of other species) will be amplified (or interfere) in the assay cells employed.

D. Yield-Reduction Assays

Yield-reduction assays for rhabdoviruses were first developed by Bellett and Cooper (1959) to measure biological activity (interference) due to DI particles added to populations of cells. They employed Poisson distribution analysis of the decrease in virus yield with increasing DI-particle input to show an all-or-none effect on virus yields as described in Section II.D. This approach, or variations of it, have been employed extensively in a number of virus–cell systems to measure biologically active DI particles in sensitive cells (Kowal and Stollar, 1980; Fuller and Marcus, 1980; Treuhaft and Beem, 1982). Sekellick and Marcus (1980) used this approach to analyze yields from individual cells as described previously. Another variation (infectious-center reduction by DI particles added at varying dilutions to sensitive cell populations) has been effective for measuring interfering units of influenza virus DI particles (Janda *et al.*, 1979), of lymphocytic choriomeningitis virus (Welsh *et al.*, 1972), and of Sindbis virus (R.E. Johnston *et al.*, 1975).

E. Reduction of Virus-Specified RNA Synthesis

A somewhat different method was employed by Barrett *et al.* (1981) to estimate numbers of Semliki Forest virus DI particles. Instead of yield reduction, they measured suppression of virus-directed [^{3}H]uridine incorporation into RNA of infected cells in the presence of actinomycin D. Incorporation of [^{3}H]uridine was depressed to approximately the low level of uninfected control cells when large numbers of DI particles were added to infected-cell monolayers, but with increasing DI-particle dilution, it increased sigmoidally to the level of virus-alone controls. Barrett and coworkers could easily estimate numbers of biologically active DI particles from the dilution giving 50% inhibition because the response at intermediate levels of DI particles was linear. This assay was reproducible, and it should be applicable to a variety of virus–cell systems. It is much simpler and more rapid than yield-reduction assays, particularly for viruses with prolonged plaque-assay times.

F. Focus-Forming (Negative-Plaque) Assays

It was clearly established using VSV (Holland and Villarreal, 1974) and rabies virus (Kawai *et al.*, 1975) that DI particles can protect cells infected with otherwise lethal virus and thereby allow indefinite cell survival and prolonged persistent infection (see Section II.E). Several DI-particle assay systems are based on this cell-sparing effect of DI particles.

Popescu *et al.* (1976) developed a sensitive, powerful technique for the assay and isolation of biologically active DI particles of lymphocytic choriomeningitis virus. Monolayers of sensitive cells are infected with low multiplicities of infectious helper virus; then the putative source of DI particles is added at various dilutions. Attachment of a single biologically active DI particle to a cell will protect that cell and nearby cells from virus cytopathology because the DI particles are amplified when virus infects the same cell, and the DI particles produced spread to adjacent cells and protect them from virus killing. This results in irregular islands of protected, attached cells against a background of dead and dying detached cells. Some of the DI-particle-protected cells probably multiply to help form the protected cell colony. This assay has also been applied to VSV DI particles (Winship and Thacor, 1979) and to rabies virus DI particles (Kawai and Matsumoto, 1982).

Wherever it can be employed, this type of assay system is very sensitive. However, the system is difficult to establish with most viruses, and when it does work well, many of the foci (negative plaques) are often so small and irregular that counting of foci can require frequent decisions regarding whether a small colony of cells is a DI-particle focus or a clump of dying cells. Kawai and Matsumoto (1982) improved the definition of their rabies DI-particle foci by a temperature shift up to 39–41°C after an initial period at 34–36°C. This caused increased cytopathic effect on unprotected cells and arrested virus replication.

Because low-input virus multiplicities must be employed, many or most DI particles may not encounter helper virus before they decay, so establishment of focus assays may often require careful balancing of cell type, virus strain, virus input multiplicity, temperature control, and other factors. Helper virus must be free or nearly free of contaminating DI particles. Kawai and Matsumoto (1982) estimated that approximately 235 input physical DI particles of rabies virus were required to form each scored focus in their assay (as compared to about 57 virus physical particles/PFU), so this can be a quite sensitive assay system when difficulties are surmounted.

G. Protected-Cell Adherence Assay

Macdonald and Yamamoto (1978) employed a much simpler method to detect cells protected by DI particles of infectious pancreatic necrosis virus of fish. They measured the ratio of reattaching to nonreattaching cells after infected-cell monolayers were exposed to dilutions of DI particles, trypsinized, and then replated on plastic surfaces. The number of cells that reattached gave good quantitation of the numbers of biologically active DI particles. This simpler cell-protection method might be useful in many other virus–DI particle systems.

H. Protected-Cell Dye-Uptake Assay

Treuhaft (1983) modified a colorimetric assay for interferon. The assay is based on the uptake of neutral red dye by living cells. Cell monolayers protected from respiratory syncytial virus cell-killing by homologous DI particles took up decreasing amounts of dye in linear proportion to the dilution of input DI particles. The dye was extracted with 50% ethanol and measured spectrophometrically at 540 nm. The dilution that yielded 63% protection of control dye uptake was used to estimate the biologically active DI-particle MOI of 1. This simple method might also have wide applicability to other virus systems.

I. Assays Based on Ratios of Viral-Size and Defective-Interfering-Particle-Size RNAs in Cells, Tissues, and Organs

Direct measurement of the sizes of radiolabeled viral RNA in infected cells has been widely employed to determine ratios of virus to DI particles in virus pools. See, for example, Weiss *et al.* (1974), Guild and Stollar (1975), and Stollar (1980), who employed such assays for alphaviruses. These indirect methods need not be used for rhabdoviruses because their DI particles and DI nucleocapsids are so easily distinguished in other ways, as outlined above. However, it will often be informative to measure directly the ratios of virus-size and DI-size RNA in tissues and organs of infected animals. Cave *et al.* (1984) have chosen the utility of Northern hybridization for detection of VSV DI-particle RNA in the brains of individual mice. They employed nick-translated plasmids containing molecularly cloned complementary DNA (cDNA) of VSV DI particles and were able to detect about 150 pg or more of viral RNA. They obtained clear evidence for replication of DI particles in individual mouse brains with this technique. Dubensky *et al.* (1984) used a whole-mouse section hybridization technique to localize VSV RNA synthesis (and polyoma virus synthesis) within tissues and organs. Combinations of these approaches should be very useful in viral pathogenesis studies.

J. Assays for the *Relative* Replicating and Interfering Capacities of Defective Interfering Particles with Different Helper Viruses

All the assays described above are useful for detecting and quantitating DI particles, or their interfering and replicating activities, or both. However, DI particles and their helper viruses can coevolve with each other in very complex ways in which DI-particle interference patterns are generally not all-or-none, but change constantly (Horodyski and Holland, 1983). This phenomenon is discussed in Section IX. Khan and Lazzarini (1977) fractionated sucrose gradients continuously through an absorbance monitor to quantitate yields of DI particles and virus. Under conditions in which interference and DI-particle replication are uncoupled (Adachi and Lazzarini, 1978), this is an ideal assay system. This absorbance monitoring was employed by Horodyski and Holland (1980, 1981, 1983) in concert with widely varying DI-input levels and different mutant VSV helper viruses to quantitate relative interference interactions of evolving virus and DI particles. To do this, it was necessary to prestandardize one biological unit of *each* DI-particle type as the quantity of that DI particle that gave 37% yield of parental, sensitive standard virus. When the DI particle/standard virus particle ratios in the yields were determined for varying DI-unit inputs, it was possible to quantitate relative interfering activities of a DI particle against numerous virus mutants. Although it is tedious, this type of assay can provide accurate determination of relative interfering abilities over more than a 100,000-fold range.

Finally, it must be emphasized that *none* of the assay systems discussed above is sensitive enough to detect DI particles at very low, seed levels (which could be very important biologically.) However, the stochastic nature of DI-particle generation can allow their detection (see Section IV.A).

IV. RATES OF GENERATION OF DEFECTIVE INTERFERING PARTICLES AND THEIR CONTAMINATION OF VIRUS POOLS

Since the pioneering work of von Magnus (1954), virologists have been aware that undiluted passaging of virus pools might introduce undesireable levels of DI particles. A widespread misconseption has been an unfortunate by-product of this awareness. The scientific literature frequently contains statements or unstated assumptions suggesting that virus pools are "free" of DI particles or that DI particles are not involved in certain virus activities because the virus pool was prepared by dilute passage series, or the virus was cloned prior to preparation of the virus pool, or "sensitive" assay systems failed to detect DI particles. Although these statements and assumptions may be correct in certain cases, they are probably most often incorrect. The rates of DI-particle generation are unknown for most viruses, and they apparently can vary greatly depending on the virus strain, the host-cell type, and other factors. For example, picornaviruses amplify DI particles at a very low rate as compared to other RNA viruses, and dozens or hundreds of undiluted passages may be required to obtain detectable amounts (Cole *et al.*, 1971; McClure *et al.*, 1980; Radloff and Young, 1983). In contrast, rabies virus (HEP) Flury strain can generate detectable DI particles in the first undiluted passage from a clonal pool (Holland *et al.*, 1976a). Only virus particles were observed at 4 days postinfection, but an array of DI particles was visible as bands in sucrose gradients at days 8 and 11. In employment of the sensitive focus-forming assay for rabies virus DI particles, Kawai and Matsumoto (1982) pointed out that the cloned helper virus, which inevitably contains a few DI particles, should ideally give fewer than 5 "background foci" due to contaminating DI particles.

Stampfer *et al.* (1971) developed the best method available to eliminate or greatly reduce the numbers of DI particles present in a vesicular stomatitis virus (VSV) pool. They serially cloned VSV and prepared virus pools from very dilute cloned virus—so dilute that multiple plaques were the source of the virus in the working pool. Holland *et al.* (1976a) confirmed that this technique can give DI-free virus pools, but only if the virus-pool size is kept very small. They estimated that the rate of VSV DI-particle generation is about 10^{-7} to 10^{-8} per infectious virus replication in BHK-21 cells, since expansion of clones much beyond 10^8 PFU regularly gives rise to DI-particle contamination. Inevitably, nearly all high-titered VSV pools exceeding 10^9 PFU/ml will be contaminated. DI-particle rates for other negative-strand RNA viruses have not been determined, but are probably as high as for rhabdoviruses. It is of interest in this regard that high levels of DI particles are required along with infectious Sendai virus for efficient induction of interferon (M.D. Johnston, 1981), as mentioned in Section II.F. That Sendai virus had been the most widely employed inducer of interferon for many years before this fact was discovered suggests that the Sendai virus pools employed had generally been unknowingly (and rather heavily) contaminated with DI particles. Barrett *et al.* (1981) reported that Semliki Forest virus DI particles are generated at a very high frequency.

Even when the rate of DI-particle generation has been determined for a virus–cell system, it cannot be assumed that this rate will hold true for different virus strains or mutants or that it will be constant in different cell types. Depolo and Holland (1986) have recently isolated mutants of VSV-Indiana infectious virus (after more than 10 years of persistent infection) that generate new DI particles at much higher rates than do parental wild-type strains of VSV.

A. "Randomness Assay" for the Presence of Very Low Proportions of Defective Interfering Particles in Virus Pools

Holland *et al.* (1976a) described a method to verify the presence or absence of extremely small seed quantities of DI particles in any virus pool, and this method has been applied to other viruses. The assay is based on the fact that the generation of DI particles is a stochastic process in which many different sizes and genome types of DI particles are generated randomly (see Section V). If a cloned VSV pool contains even tiny seed quantities of newly generated DI particles, these particles will be greatly amplified during subsequent undiluted passages, and the most competitive will always predominate as major DI bands in sucrose gradients following independent undiluted passage series in a given cell line. Therefore, if a contaminated virus pool is used to initiate five independent passage series (of three or four or more undiluted passages each) in BHK-21 cells, the resulting DI-particle bands in all five sucrose-gradient analyses of each series will be identical. If, however, the initiating virus pool is free of DI particles, there is no seed inoculum of DI particles to be amplified. Therefore, the DI-particle bands observed following each independent passage series will be randomly different because each band pattern will represent DI particles newly generated during the first undiluted passage of each (Holland et al., 1976a).

This provides a simple, sensitive way to verify that a cloned pool of virus is free (of seed quantities) of DI particles. It also provides a means for the generation and repeated production of new DI-particle types. This randomness of new DI-particle generation has allowed the method to be employed for Sendai virus DI particles (Kolakofsky, 1979), for influenza virus DI particles (Janda et al., 1979), and for simian virus 40 (SV40) DI particles (Norkin and Tirrell, 1982). Interestingly, Norkin and Tirrell (1982) found that SV40 virus pools verified to be free of DI particles in this way showed no tendency to generate any particular DI-particle class or to have picked up any particular segment(s) of host-cell DNA. Earlier workers had concluded that a cloned pool of SV40 would regularly integrate near, and pick up, certain host-cell DNA sequences, and these would regularly appear in DI particles. This is another example of the confounding effects of seed quantities of DI particles in cloned virus pools that are believed (but not verified) to be free of DI particles. In the absence of tests for randomness of generation, statements that virus pools are free of "significant numbers" of DI particles can have no scientific credibility.

B. Cloning of Defective Interfering Particles

Because DI-particle generation is random and occurs at high frequencies, it may frequently be useful to investigators to clone a particular DI particle. The most straightforward way to do this is to employ a focus-forming assay (see Section III.E) for DI particles and to pick and amplify the DI particles present in a well-isolated focus. Alternatively, single cells infected with high multiplicities of virus and low multiplicities of DI particles, and scoring as zero-virus-yielders following isolation by micromanipulation (Sekellick and Marcus, 1980), can be used for seed quantities of cloned DI particles to be amplified. Neutral-red-stained cells from the DI-particle assay of Treuhaft (1983) or adherent cells from the adherence DI-particle assay of Macdonald and Yamamoto (1978) might also serve as single-cell sources of cloned DI particles for amplification (see Section III.C) if low-multiplicity inputs of DI particles were used.

It must be emphasized that cloning of DI particles can provide only clonally derived DI particles. As soon as such clonally derived DI particles are expanded to large numbers for molecular studies, they must inevitably become contaminated with newly generated DI particles. However, the contaminating new DI particles will generally represent only a small percentage of the total. This percentage, of course, depends on the replicative and competitive efficiency of the cloned DI particles, but their earlier amplification to large numbers usually will allow them to predominate for many passages.

V. GENOME STRUCTURES AND MECHANISMS OF GENERATION OF DEFECTIVE INTERFERING PARTICLES

There have been two well-organized analytical reviews covering molecular structures and generation mechanisms of RNA virus DI particles

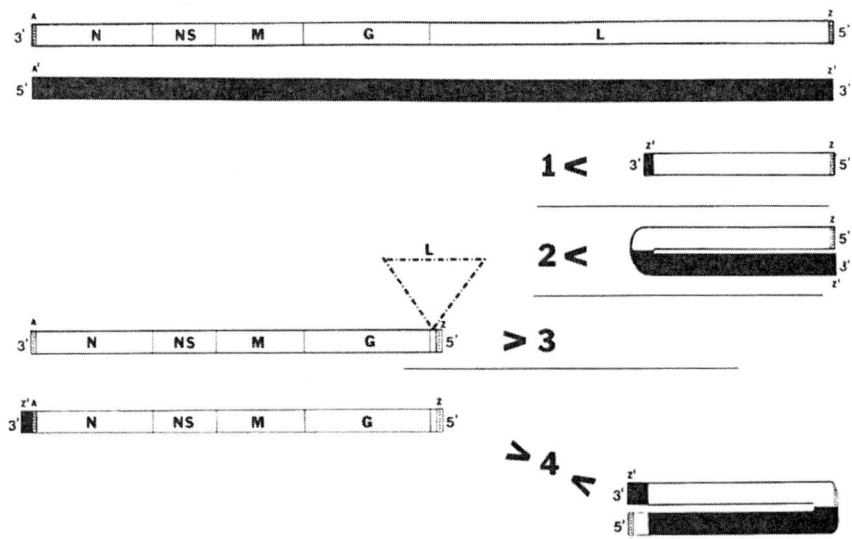

FIGURE 3. Schematic representation of the four major classes of VSV DI-particle genome structures and their relationship to the VSV parental genome. These structures are derived from numerous references cited in the text. Most of the sequence data on virus structure were derived by Rose and Gallione (1981) and Gallione et al. (1981) for the nucleocapsid (N), nonstructural (NS), matrix (M), and glycoprotein (G) genes and by Schubert et al. (1984) for the large (L) protein gene. Most of the extensive sequence information regarding VSV DI particles was derived by Lazzarini, Keene, Schubert, and colleagues (Lazzarini et al., 1981), and many other pertinent references are cited in the text. Note that the VSV negative-strand (top) genome 3' terminus (A) is a site of initiation for both transcription and replication–encapsidation, whereas the 3' terminus (Z') of the antigenome plus strand (in black, below) is involved only in replication–encapsidation (see Chapter 7).

Class 1. Stem or panhandle DI particles: This common class of VSV nontranscribing DI particle contains no complete VSV gene. These contain varying portions of the L gene and its adjacent 5' viral terminus, with a "stem" or "panhandle" antigenome Z' terminus

(Lazzarini *et al.*, 1981; Perrault, 1981), so this section will be kept brief and restricted mainly to rhabdovirus [vesicular stomatitis virus(VSV)] DI particles. The four basic classes of VSV DI genomes are schematized in Fig. 3 and discussed below.

A. 5'-End-Derived Stem or Panhandle Class (Class 1)

This is the simplest structural class of VSV DI genomes, and they are, in all cell systems examined so far, the most abundant and competitive class (along with the snapback type). As can be seen in Fig. 3, they consist of the 5' end of the standard virus and varying portions of the adjacent large (L)-protein gene. This type of DI particle was fist detected in Sendai virus DI particles (Kolakofsky, 1976; Leppert *et al.*, 1977) and later in VSV (Perrault, 1976; Perrault and Leavitt, 1977; Perrault *et al.*, 1978). All DI particles of VSV examined so far contain the virus RNA 5'

---

recombined to form a new 3' terminus attached to the truncated L-gene segment. No transcription-initiation site exists in this DI genome or in its antigenomic "plus-strand" complement, so they are replicative templates when interacting with helper virus gene products.

Class 2. Snapback or hairpin DI particles: These simple snapback DI particles contain genomic minus-sense RNA and covalently linked antigenomic plus-sense RNA that are complementary to the minus-sense RNA over its entire length. The minus strand is co-linear with a portion of the virus L gene and its adjacent 5' terminus. A single phosphate links the plus- and minus-sense RNAs of these DI particles, so there are no separate plus and minus strands to act as templates for each other. Each time these DI particle genomes replicate, they make an exact copy of themselves. These common DI particles are non-transcribing because they contain only the Z' replication-initiation site at their 3' terminus.

Class 3. Simple internal-deletion DI particles: This transcribing class of DI particles is very rare in rhabdoviruses because they compete poorly with other classes of DI particles. Note that a simple internal deletion removes most of the L-protein gene, but all other genes are retained together with normal virus 5' and 3' termini. The 3' terminus (A) initiates both transcription and replication at the 3'-leader region. After leader synthesis, polar transcription of N, NS, M, and G genes can proceed in the usual manner (see Chapter 6).

Class 4. mosaic rearrangement DI particles: This class of complex DI particles is not a single structural type, but includes all DI particles with extensively rearranged genomes. Only two rather simple mosaic-class genomes are illustrated. The DI genome on the lower left is identical to (and possibly is derived from) the transcribing DI particle illustrated for class 3, except that a short Z stem has been recombined onto the 3' end. This simple stem addition silences the transcription-initiation site (presumably by internalizing it) and enables the resulting nontranscribing DI particle to compete more effectively with other DI particles. The complex DI-particle genome illustrated on the lower right is a mosaic snapback DI particle that apparently arose by hairpin copy-back from a class 1 nonsnapback DI-particle genome. It contains the same 54-base Z and Z' stems at its termini that the class 1 parental DI genome contains, but the turnaround region linking the plus-sense RNA and minus-sense RNA also contains internalized Z' and Z stems that are inverted complements of the termini stems. Many other bizarre structures of mosaic DI particles have been observed, including deletions, repetitions, extensive base substitutions, and others. These mosaic-type DI particles are common in persistent infections, repeated high-multiplicity passages, and other situations in which earlier-generated DI-particle genomes can evolve and rearrange.

end (in the negative-strand sense), but this class of DI particles has lost the virus 3' sequences and replaced them with a complementary copy of the virion 5' end (i.e., a 3' end identical in sequence to the 3' terminus of the viral antigenomic plus strand). The "stem" or "panhandle" self-complementarity of these DI particles varies from 45 base pairs to about 150 base pairs in length. The termini of VSV-Indiana and of many VSV-Indiana "stem"- or "panhandle"-type DI particles have now been se-quenced (Keene *et al.*, 1978; Schubert *et al.*, 1978, 1979, 1980; Semler *et al.*, 1978, 1979; McGeoch and Dolan, 1979; Rowlands, 1979; Yang and Lazzarini, 1983; Nichol *et al.*, 1984; O'Hara *et al.*, 1984a,b; Meier *et al.*, 1984). In all these DI particles, the inverted complementary termini ex-hibit a 5' terminus that is homologous in sequence to the 5' end of the standard virus genome for at least the stem length. The 3' stem in all these DI particles is complementary to the virus 5' end (and its own 5' end) for 45 to about 150 bases.

DI particles of this class 1 stem type are therefore mainly negative-sense portions of the L gene and the 5' terminus, plus a small positive-sense (5'-complementary) segment comprising the 3'-stem terminus. Un-like virus, these DI particles sometimes form mature particles containing positive-strand RNAs along with the predominant minus-strand-polarity DI particles (Roy *et al.*, 1973; Perrault and Leavitt, 1977). The percentages of plus- and minus-strand DI genomes in mature particles vary with different stem or panhandle DI particles, but the negative-sense RNA genomes are generally predominant.

B. Snapback or Hairpin Class (Class 2)

As can be seen in Fig. 3, and as the name implies, this type of DI particle is self-complementary over its entire length. This DI structure was discovered because on deproteinization the genome "snaps back" into a stable hairpin duplex molecule (Lazzarini *et al.*, 1975; Perrault, 1976; Perrault and Leavitt, 1977). Sequence analysis of this type of DI particle (Johnson and Lazarini, 1977; Keene *et al.*, 1977, 1979; Schubert *et al.*, 1979) shows that half of each genome is homologous to the 5' end of the virus negative strand and a portion of the adjacent L gene. The other half is the 3' complement of the 5' half. Schubert and Lazzarini (1981) sequenced the joining region that links the 5' half of the snapback DI011 to its 3' complement and found that this genome is a perfect duplex joined by a single phosphate. This interesting structure negates the con-cept of plus and minus strands in these DI genomes, since each replication of one of these DI particles gives rise to a DI genome identical in sequence to the template genome. This type of DI genome is commonly encoun-tered in VSV virus pools. Recently, the end sequences and joining se-quences of another snapback DI particle (ST2) have been determined, and this is a more complex, multiply rearranged DI genome that apparently arose by rearrangement of progenitor (nonsnapback ST1) DI particle (Nichol et al., 1984). These multiple rearrangements place this snapback DI particle into the mosaic rearrangement class (see Section V.D).

C. Simple Internal-Deletion Class (Class 3)

The simple internal-deletion type of DI particle is the least complex DI-genome rearrangement. In the prototype HRLT DI particle (Fig. 3), the internal deletion excised most of the L-protein gene, leaving both viral termini and the other four genes intact (Perrault and Semler, 1979; Epstein et al., 1980; Yang and Lazzarini, 1983). This is an extremely rare type of VSV DI particle. It was originally found to be unique because it mapped to the 3' half of the genome in annealing studies (Reichmann and Schnitzlein 1979, 1980). It is also unique in being able to transcribe the four nondeleted (N, NS, M, G) cistrons (Chow et al., 1977; Colonno et al., 1977; Johnson and Lazzarini, 1977, 1978; Johnson et al., 1979). Presumably due to its transcribing activity, this DI particle has the unique ability to kill cells when infecting alone (Marcus et al., 1977) and to interfere heterotypically with New Jersey Concan subgroup VSV (but not with New Jersey Hazelhurst subgroup VSV) (Prevec and Kang, 1970; Schnitzlein and Reichmann, 1976, 1977a,b; Adachi and Lazzarini, 1978; Reichmann et al., 1978). Transcribing VSV DI particles are probably very rare because they do not compete effectively with other, nontranscribing DI particles and are quickly displaced by them (Perrault and Semler, 1979).

In contrast to rhabdoviruses, some other RNA viruses frequently generate DI particles of simple internal-deletion type, and some of these DI particles transcribe actively. For example, poliovirus DI particles contain only small internal deletions, and the remainder of the genome is transcribed and translated (Cole and Baltimore, 1973a–c; McClure et al., 1980; Lundquist et al., 1979; Phillips et al., 1980). Most influenza virus DI particles are also internal-deletion genomes retaining both the 3' and 5' ends, and many are able to transcribe and direct aberrant translation products (Nayak and Sivasubramanian, 1983; Jennings et al., 1983; S. Fields and Winter, 1982; Akkina et al., 1984). Sendai virus generates stem-type DI particles similar to those of VSV (Kolakofsky, 1976, 1979; Leppert et al., 1977) and also can generate internal-deletion DI particles at high frequency, and these particles retain the 3' and 5' ends of the virus genome plus a portion of the L gene (Amesse et al., 1982; Kailash et al., 1983). The alphaviruses also generate a variety of classes of DI particles, with 5', 3' internal-deletion types being common (Stollar, 1979, 1980; Pettersson, 1981; Lehtovaara et al., 1981). The rhabdoviruses might prove to be somewhat unique in selecting against DI particles of the internally deleted type.

D. Mosaic Rearrangement Types (Class 4)

This last class of DI genome structure is not a fixed structural type, but a catchall class including many bizarre, extensively rearranged DI RNAs. Previously, it has been assumed that the most commonly isolated VSV DI particles (of classes 1 and 2 above) are by far the most abundant. However, previous studies have been done mainly on DI particles isolated following only a relatively low number of undiluted passages. O'Hara *et al.* (1984a) recently sequenced the termini of 16 VSV DI particles isolated at intervals from persistent infections and from undiluted lytic passages carried out for prolonged periods. These particles all contained 5' stem- or panhandle-type self-complementarity, but the sequences immediately internal to these stems were extensively rearranged in a variety of ways. It appears that although the simpler class 1 and class 2 DI-particle types may predominate in virus populations following a low number of undiluted passages, very long-term passages and persistent infections provide the time and conditions required for generation and selective replication of the mosaic types of DI particles.

Extremely bizarre, multiply rearranged DI particles had been found earlier for influenza virus (S. Fields and Winter, 1982) and for Semliki Forest virus (Pettersson, 1981; Lehtovaara *et al.*, 1981; Kääriänen *et al.*, 1981), so multiple genome rearrangements in DI particles may be widespread. The most bizarre DI particles yet encountered are those of Sindbis virus that have been described by Monroe and Schlesinger (1983, 1984). They obtained two different isolates of Sindbis virus DI particles in which the 5' terminus of the virus had been replaced by sequences that are virtually identical to sequences 10–75 of a rat transfer RNA[Asp]. Only two base differences were seen within this sequence. The DI-particle populations containing these sequences were cloned three consecutive times between one DI particle generation and the other, so they arose twice independently and must therefore have a strong probability of generation and amplification, at least in chicken cells, in which they are commonly observed. Great insight has recently been obtained regarding the termini sequences essential for replication of Sindbis virus DI particles and even for replication of foreign genes placed within these termini by recombinant DNA approaches (Tsiang *et al.*, 1985; Schlesinger, 1987).

Figure 3 illustrates two of the mosaic-type DI-particle genomes characterized for VSV. The top one is HRLT-2, which was present in the same virus pool with the transcribing HRLT and which might have been derived from it (Perrault and Semler, 1979; Epstein *et al.*, 1980). It differs from the transcribing HRLT only in containing an added stem sequence (Perrault and Semler, 1979; Epstein *et al.*, 1980), and this stem addition apparently completely silences the transcription-initiation activity of the viral 3' end. The viral 3' end is internal to this added stem by only 72 nucleotides from the DI 3' end (Keene *et al.*, 1981a), but this internalization apparently inactivates it. Furthermore, this stem addition renders this DI particle much more competitive than its transcribing counterpart lacking the 3' stem (Perrault and Semler, 1979).

The second mosaic-type DI particle schematized in Fig. 3 is the genome of ST2 DI particles. This is a snapback particle derived from the same virus pool as the nonsnapback ST1 DI particle (which has a stem or panhandle 54 bases long) (Nichol et al., 1984). ST2, however, is about twice the length of ST1 genomes and contains the same 54-base stem at its 3' end linked to the same portion of the L gene and 5' end, but it is covalently linked to its complementary positive strand at the equivalent of the ST1 5' end. Therefore, this snapback mosaic DI particle almost certainly arose from an ST1 DI particle in the virus pool. A variety of mosaic snapback DI genomes and mosaic stem DI genomes of VSV have now been characterized from persistent infections and prolonged-passage series (O'Hara et al., 1984a). They contain deletions, repeated sequences, clustered repetitive base substitutions and other rearrangements, so it is clear that rhabdovirus DI particles can become extensively evolved and rearranged and that mosaic DI particles differ from the other types only in the extent to which they are rearranged from the parental virus sequences. No cellular RNA sequences have yet been detected in rhabdovirus DI particles, but only a relative few have been examined thoroughly to be sure that they contain only virus sequences.

E. Mechanisms of Generation of Defective Interfering Particles

Both Leppert et al. (1977) and Huang (1977) proposed that the structural features of stem-type DI particles were easily explained by their having been generated by viral replicase "template-switching" or "copyback," in which the polymerase leaves its template, carrying the 3' end of the nascent chain with it, then resumes nascent-chain elongation after binding near the 3' end of the nascent-chain ribonucleoprotein. This is equivalent to older copy-choice models for DNA recombination, and this basic replicase-error model for generation of stem-type DI-particles can be generalized to explain the generation of all four classes of VSV DI particles, (Lazzarini et al., 1981; Perrault, 1981), by using interstrand or intrastrand "replicase leaps," or both. This model is clearly expounded in the two review articles cited, so it will not be illustrated again here. All structural data and sequence data obtained so far for negative-strand (and positive-strand) RNA viruses are compatible with this model. See, for example, Yang and Lazzarini (1983), Schubert et al. (1984), Keene et al. (1981a), O'Hara et al., (1984a), Nichol et al. (1984), and Meier et al. (1984) for VSV DI-genome structures, Amesse et al. (1982) for Sendai virus DI structures, and S. Fields and Winter (1982), Nayak and Sivasubramanian (1983), and Jennings et al. (1983) for influenza virus DI-particle structures. The influenza virus DI particles sequenced by S. Fields and Winter (1982) and Jennings et al. (1983) provide very strong evidence that strand-switching (both intra- and inter-strand) of viral replicase was the likely mode of generation.

In none of the aforementioned studies was there any evidence for involvement of cellular splicing enzymes, since splicing consensus sequences have not been observed at junctions. In most of the sequences examined so far, there is no evidence for sequence specificity either at the sites of replicase premature termination or at sites of resumption of synthesis (Lazzarini *et al.*, 1981; Perrault, 1981; Yang and Lazzarini, 1983; De and Perrault, 1982; Schubert *et al.*, 1984; O'Hara *et al.*, 1984a; S. Fields and Winter, 1982; Jennings *et al.*, 1983; Nayak and Sivasubramanian, 1983). Although some very weak, imprecise sequence recognition might be operative in a few examples (Schubert *et al.*, 1979; Keene *et al.*, 1981a; Meier *et al.*, 1984), the nonspecific "generalized replicase-leap" model of Perrault (1981) appears to generally correct.

It must be concluded that most RNA-virus DI particles probably arise by polymerase errors as originally proposed. This does not negate the possibility that rare cases of DI-particle generation might involve cellular splicing enzymes, or cellular nucleases and ligases, or other mechanisms. The obvious question arises as to the reason for the great frequency of these copy-choice RNA recombinations in generating DI particles of negative-strand viruses, whereas recombination of the helper negative-strand viruses has not been detectable (Granoff, 1959a, 1962, 1964; Wong *et al.*, 1971; Pringle and Wunner, 1975; Holland *et al.*, 1982). In contrast, recombination of the picornaviruses occurs at a quite high level (Cooper, 1968, 1977; A.M.Q. King *et al.*, 1982; Agol *et al.*, 1984). It must be assumed that the "replicase leaps" of picornaviruses (and of other RNA viruses that recombine at high levels) frequently result in resumption of synthesis at a specific site on the new template. Presumably, the prematurely terminated nascent chain directs the leaping replicase to the homologous site on the new template by nascent primer–new template base-pairing hybridization. This is much more likely to occur with positive-strand viruses, in which the templates and nascent-strand RNAs are naked, and much more improbable with negative-strand viruses, in which the template and nascent RNAs are encapsidated (Hill *et al.*, 1979).

Specificity of replicase leaps in positive-strand viruses could explain why DI-particle generation is much more rare in these viruses. In contrast, a very low specificity of replicase leaps in negative-strand viruses can generate rich assortments of DI particles, but only extremely rarely would the improbably precise replicase-switching occur that could produce viable recombinants of infectious virus.

VI. MECHANISMS OF DEFECTIVE-INTERFERING-PARTICLE INTERFERENCE

The mechanisms of DI-particle interference are not known for rhabdoviruses or for any other RNA virus. Perrault (1981) has cogently reviewed all the data bearing on interference mechanisms, and the reader is referred to his coverage for greater detail than space permits here. Because the mechanisms are clearly different, nontranscribing and transcribing DI particles will be treated separately here.

A. Nontranscribing Defective Interfering Particles

The early studies of Huang and Manders (1972) and Perrault and Holland (1972b) demonstrated that most DI-particle interference—i.e., by classes 1, 2, and 4 (see Section V) of nontranscribing DI particles—probably occurs at the level of replication of viral and DI genomes. These DI particles are unable to transcribe themselves; they do not inhibit primary transcription by the virus; they rapidly and strongly suppress viral genome replication, while DI-particle genomes replicate efficiently; and finally, they strongly suppress secondary transcription—an inevitable outcome of the reduction of synthesis of new viral templates. These findings were extended by Palma and Huang (1974) and Stamminger and Lazzarini (1977). Since the reduction of secondary transcription leads to a relative shortage of viral replicase and encapsidation proteins, and since the reduced numbers of virus templates are devoted mainly to transcription, while DI nucleocapsids are devoted only to replication, then nearly all the limited viral replication–encapsidation polypeptides should be shunted to DI-particle replication at the expense of viral replication.

The structures of classes 1, 2, and 4 DI-particle genomes reviewed above (see Fig. 3) make it clear why they do not transcribe: They are rendered transcriptionally silent by a stem addition even when they possess viral 3'-end transcription initiation sites and downstream viral cistrons. The latter are almost always lacking in vesicular stomatitis virus (VSV) DI particles in any case. The major remaining question regards the reason viral template nucleocapsids confine themselves almost exclusively to transcription in the presence of replicating DI nucleocapsids. In fact, there are at least three situations in which they carry out both viral transcription and replication efficiently while also replicating large quantities of DI particles:

1. In low-interference cell lines (see Section II.C).
2. In heterotypic interference involving DI-particle replication with helper virus of a different VSV serotype (Schnitzlein and Reichmann, 1977b; Khan and Lazzarini, 1977; Adachi and Lazzarini, 1978).
3. Whenever DI particles are added to cells several hours or more following virus infection (Stampfer et al., 1969).

These results suggest that some regulatory signal controls transcription-to-replication switching, that this signal can involve host-cell factors, that it is serotype-specific, and that it is less restrictive later in infection. The predominance of stem- or panhandle-type DI particles and snapback DI particles suggests that the 3'-stem sequence favors both DI-particle negative-strand (and positive-strand) replication, just as it favors virus negative-strand replication over virus positive-strand replication.

Rao and Huang (1980) observed that the total number of viral templates participating in RNA replication is rather constant and is determined early during infection. This would explain why DI particles added late do not interfere. But what is this host-cell-determined, early switch that regulates replication? It could operate at the level of viral RNA encapsidation or at the level of replicase action, or both.

Replication of negative-strand viruses utilizes nucleocapsid templates, and the nascent replicase products are also nucleocapsidated (Soria et al., 1974; Batt-Humphries et al., 1979, Hill et al., 1979, 1981; Davis and Wertz, 1982; Peluso and Moyer, 1983; Patton et al., 1983; Wertz, 1983) (see also Chapter 7). As can be seen from the aforecited references, VSV replication is complex and its requirements are only now being studied in cell-free systems, but it is now obvious that replication is obligatorily coupled with encapsidation. This had led to the intriguing suggestion that the switch that controls the transcription-vs.-replication decision might be the availability of nucleocapsid (N) protein for encapsidation (Leppert et al., 1979; Leppert and Kolakofsky, 1980). Leppert and colleagues suggested that the requirement for N protein resides at the level of chain termination of "leader RNA" synthesized off the first 50 nucleotides from the 3' end of the viral genome. Termination leads to free leader RNA and to initiation of transcription of downstream genes. Conversely, lack of termination (due to N-protein encapsidation beginning at encapsidation–nucleation sites in the leader sequence) leads to replication–encapsidation of the entire genome-size RNA. This suggestion has now been confirmed (Patton et al., 1984).

The behavior of small leader RNAs (Colonno and Banerjee, 1978; Rao and Huang, 1979, 1980; Leppert et al., 1977, 1979; Leppert and Kolakofsky, 1980; Blumberg et al., 1981) is generally consistent with the major postulates of this model, and in vitro self-assembly of N protein to selectively encapsidate leader RNA strongly supports specific leader RNA–N protein interactions as a major determinant of encapsidation specificity (Blumberg et al., 1983; Blumberg and Kolakofsky, 1981, 1983). Blumberg et al. (1983) suggest that the recognition site for initiation of VSV encapsidation is a five-times-repeated A residue at every third position from the 5' end of the leader chain. This is consistent with sequences observed so far in most serotypes, strains, and mutants of VSV, but rabies virus leader sequences deviate slightly from this rule (Kurilla et al., 1984).

Perrault et al. (1980, 1981, 1983) studied a class of VSV mutants designated Pol R mutants in which termination of transcription at leader sites is specifically suppressed. These mutants give rise to a much higher proportion of "readthrough transcripts" in vitro (from both virus and DI particles) then do wild-type viruses. Reconstitution experiments suggested that the N protein of template nucleocapsids is responsible, and it was proposed that regulation of termination at leader sites could be due to modification of the N protein.

Banerjee and his colleagues have shown another way in which the coupling of replication with translation (or N-protein availability) can be bypassed (Testa *et al.*, 1980; Chanda *et al.*, 1980, 1983). They used imido analogues of ATP in which the β–γ bond is nonhydrolyzable and found that addition of analogue before initiation of transcription inhibited VSV transcription *in vitro*, but addition after initiation allowed "readthrough synthesis" of full-length complements of viral RNA at low ATP concentrations. High ATP levels ($K_m \approx 500$ μM) are required to initiate VSV transcription, whereas K_m's for GTP, UTP, and CTP are low (22–33 μM) (Testa and Banerjee, 1979). Green and Emerson (1984) suggested that imido analogues of ATP do not inhibit initiation of leader RNA synthesis, but rather inhibit subsequent downstream mRNA synthesis, whereas Perrault and McLear (1984) suggested that transcription initiation rather than termination at leader sites is affected by β-γ-imido ATP. They also found that Pol R mutants were much less sensitive to replacement of ATP by imido-ATP and suggested that ATP hydrolysis may be important in transcription–replication switching. Clearly, more information is required regarding molecular aspects of both transcription and replication of rhabdoviruses before detailed molecular aspects of DI-particle interference can be elucidated (see Chapter 6 and 7).

Overall, the majority of experimental evidence points to replication–encapsidation as the level at which nontranscribing DI nucleocapsids interfere with virus nucleocapsids. The nature of the transcription–replication switch is not known, but encapsidation is intimately involved in replication initiation. *In vitro* experiments with antisera to viral proteins are consistent with the thesis that N-protein availability limits replication (Hill and Summers, 1982). Anti-NS sera inhibited both transcription and replication VSV *in vitro*, while anti-N inhibited replication and not transcription, and anti-M and anti-G affected neither. Strangely, anti-L inhibited transcription, but *increased* replication up to 2-fold! Also, N-protein pool sizes failed to correlate well with *in vitro* replication, and stabilities of transcription complexes differed from those of replication complexes, suggesting that enzyme-complex differences rather than a simple "switch" may be operative (Hill and Summers, 1982; Hill *et al.*, 1981). Only further experimentation *in vitro* can determine whether altered polymerase complexes, altered templates, N-protein availability (or perhaps a combination of these) regulates the balance between transcription and replication and the manner in which DI particles co-opt replication–encapsidation functions.

B. Transcribing Defective Interfering Particles

Transcribing DI particles (class 3 in Fig. 3) clearly interfere in a different manner than the majority of VSV DI particles. They interfere with other serotypes (Prevec and Kang, 1970; Schnitzlein and Riechmann, 1976). They can replicate with heterotypic helper virus without causing interference (Schnitzlein and Reichmann, 1977b; Khan and Lazzarini, 1977; Adachi and Lazzarini, 1978). The addition of a stem sequence at the 3' end (class 4 DI genomes in Fig. 3) silences their transcriptive ability and changes their interference behavior (Perrault and Semler, 1979; Epstein *et al.*, 1980; Keene *et al.*, 1981a). Their UV-target size is about 42% that of the DI genome (Bay and Reichmann, 1979), whereas other DI particles have a target size equal to genome size. Finally, Bay and Reichmann (1981, 1982) showed that their transcribing ability is required for interference. Furthermore, they interfere with primary transcription of both homotypic and heterotypic helper virus, a phenomenon that does not occur with nontranscribing DI particles.

Using chimeric DI particles with thermolabile polymerase, Bay and Reichmann (1981, 1982) observed that at nonpermissive temperature, these unique DI particles could not self-transcribe, and concomitantly they lost the ability to inhibit virus primary transcription and to interfere heterotypically (but they still were replicated). They suggested a model for interference proposing that there are more frequent dissociation events during transcription of the virus nucleocapsids due to their longer lengths. This mechanism would also make these transcribing DI particles unique, since smaller DI genome size is not a factor in the interference exerted by nontranscribing DI particles.

Among other RNA viruses, transcribing DI particles are often the majority class. Influenza virus DI particles usually contain internal-deletion mutants of one of the large polymerase genes, and some of these DI particles can be transcribed and translated to produce aberrant polypeptides (Chanda *et al.*, 1983; Akkina *et al.*, 1984a). However, subgenomic DI particles of other influenza virus genomic segments can also be found (Jennings *et al.*, 1983). Akkina *et al.* (1984b) have proposed that influenza virus transcribing DI RNA subgenomic fragments may interfere by selective amplification due to smaller size. This, together with packaging exclusion of full-size viral polymerase genes whenever a subgenomic DI polymerase nucleocapsid is matures (Smith and Hay, 1982; Akkina *et al.*, 1984b), was proposed to explain interference with influenza virus.

Sendai virus is another RNA virus in which internal-deletion DI genomes with conserved 3' and 5' termini may be commonly observed (Amesse *et al.*, 1982; Kailash *et al.*, 1983). These DI genomes contain the leader transcription-initiation sequence, but 3' sequences of the NP gene are fused to 5' sequences of the L gene, erasing all viral transcription-termination signals. Therefore, Amesse *et al.* (1982) propose that these "fusion DI RNAs" act exclusively as templates for replication and are functionally equivalent to stem or panhandle copy-back DI particles, despite their "transcription"-initiation ability.

It appears that there are probably at least as many different DI-particle interference mechanisms as there are different replication strategies among

RNA viruses, but the ability to generate DI genomes by replicative error is ubiquitous, and each virus eventually selects and amplifies the more competitive of the subgenomic structures generated.

VII. DEFECTIVE INTERFERING PARTICLES REPLICATE AND INTERFERE *IN VIVO* AND CAN PROTECT ANIMALS

The very first evidence for DI particles, influenza virus harvested from embryonated eggs (Henle and Henle, 1943), involved protection of mice with undiluted virus pools containing DI particles (although "inactive virus" was believed responsible). The early characterization of influenza "incomplete virus" by von Magnus (1951, 1954) also showed that DI particles reduced virus multiplication in mice and provided some protection against lethality. Mims (1956) passaged Rift Valley fever virus serially in mice by the intravenous route and observed profound cyclic variations in virus levels due to DI particle–virus cycling (Palma and Huang, 1974). At the point in cycling where virus titers were minimal and "incomplete" virus maximal, Mims (1956) observed prolonged incubation periods and survival of some mice.

Doyle and Holland (1973) employed DI particles completely freed of infectious virus by extensive purification, and they observed that large numbers of these particles could provide 100% protection of adult mice from vesicular stomatitis virus (VSV) lethality when injected together with low (but otherwise lethal) doses of infectious virus. DI particles provided less protection as the virus challenge dose was increased, but symptoms were different and deaths were reduced and greatly delayed. Crick and Brown (1977) suggested that some of this protection might be due to immunizing ability of DI particles. However, they immunized 2 days *before* challenge with chemically inactivated virus to obtain strong protection and got lesser protection (against homologous and heterotypic VSV) when inactivated virus was given together with challenge virus. Jones and Holland (1980), however, carefully tested this possibility. They used UV-inactivated virus and UV-inactivated DI particles given simultaneously (or up to 3 days prior to challenge) and obtained no protection from death (although noninactivated DI particles gave strong protection). In agreement with Crick and Brown (1977), however, they did observe prolongation of time till death when UV-inactivated "snapback" DI particles were given at the time of challenge. Since the "snapback" type of DI particle is common in virus pools and is a strong interferon inducer (Sekellick and Marcus, 1978), interferon might have been involved in the protection observed by Crick and Brown. Rabinowitz et al. (1977) also observed that VSV DI particles prolonged time to death in mice lethally infected with VSV, and they studied the pathology of the slowly ascending hindlimb paralytic disease that ensued after dual infection. They observed unusual pathological changes that were not caused by wild-type virus or by temperature-sensitive mutants of VSV. These changes included parenchymal necrosis of spinal cord, secondary demyelination, spongiform degeneration in gray matter of neuropil, and inflammatory changes. Faulkner et al. (1979) proved that VSV DI particles replicate and cause true homologous interference with VSV in central nervous system (CNS) neurons cultured in vitro, so DI particles can function effectively in differentiated neurons.

Although adult mice are protected by large doses of VSV DI particles, they appeared to replicate DI particles rather inefficiently during high-multiplicity intracerebral passages (Holland and Villarreal, 1975), whereas newborn mice generated and replicated VSV DI particles very efficiently. In fact, VSV DI particles could be purified directly from the pooled brains of newborn mice, and they were shown to be as active biologically as are DI particles grown in vitro (Holland and Villarreal, 1974). Also, HEP Flury rabies virus generated an array of DI particles on the first intracerebral passage in newborn mice in the aforecited study.

It is not clear why newborn mice replicate DI particles so much more efficiently, but there are two possible explanations. First, this could be a cell effect, because Huang (1977) found that DI-particle interference was stronger in mouse fibroblasts in vitro when they were derived from young mice as compared to old mice.

Second, the recent studies of Cave et al. (1984) suggest that dosage, route of innoculation, and timing may be critical in the replication and interference (and detection) of VSV DI particles in adult mice. They used Northern hybridization with cloned DNA probes to assay for virus-size and DI-size genomic RNA in infected mouse brains. They found only virus RNA in brains of mice infected intranasally with 10^8 virus PFU alone or with 10^8 PFU plus an equivalent amount of DI particles, and most of the mice died. However, when the same amount of virus was administered intranasally along with 1000-fold fewer DI particles, significant protection of mice occurred and DI-size RNA was present in their brains in large amounts (together with virus-size RNA) (Cave et al., 1984). There are a number of important implications in this unexpected finding: (1) VSV DI particles can replicate and interfere well in adult mice even when initially present as a very small proportion of a virus inoculum. (2) Cycling of DI particles and virus apparently can occur in individual infected animals. (3) Protection by DI particles depends on route of infection, relative dose and timing of input virus and DI particles, cells and organs involved, and other factors. DI-particle interference is a much more complex phenomenon in vivo than was previously believed. (4) Attempts to detect DI-particle replication in vivo (and, of course, to detect virus in vivo) may succeed at one time and fail at another, depending on the dynamics of cycling, tissues assayed, route and dosages of infection, virus strain, DI-particle type, and other factors. If such cycling occurs in humans (infected with influenza virus, for example), it seems likely that chance variations in dose, stochastic variations in cycling dynamics, immune-response timing, and other variables will often constitute the crucial difference between mild, moderate, and severe, or even fatal, disease. These findings can have far-reaching significance in virology and deserve extensive follow-up investigation.

Fultz and her colleagues have carried out thorough studies of hamster infection by VSV and of protection by DI particles (Fultz et al., 1981, 1982a,b, 1984). Intraperitoneal injection of only 100 PFU was lethal for 90% of Syrian hamsters, but most could be protected by administration of DI particles or poly(I)–poly(C). Homologous interference was probably involved in the DI-particle protection, but a role for interferon cannot be ruled out because interferon is induced with or without added DI particles in infected hamsters. Interestingly, most DI-particle-protected hamsters appeared to become persistently infected by VSV, although infectious virus was only rarely recoverable (see Section VII). This VSV–hamster system is an interesting one because primary virus multiplication is mainly in spleen, liver, and other visceral sites, with the CNS being only secondarily infected.

A number of nonrhabdovirus systems have also shown DI-particle interference in vivo, but they cannot be reviewed in detail here. Spandidos and Graham (1976) and Graham (1977) showed that DI particles of dsRNA viruses can strongly affect virus replication in vivo. Injection of newborn rats at 2 days of age with reovirus led to death of all animals, but if DI particles were injected together with infectious reovirus, over 60% survived. These protected rats developed a severe runting syndrome in which DI particles and infectious virus were shed chronically.

Dimmock, Barrett, and colleagues have carried out thorough studies of the effects of Semliki Forest virus DI particles on virus infection by the intracerebral, intranasal, and intraperitoneal routes in mice. They observed significant protection that was demonstrated to be due, not to immunizing effects of the DI particles, but to their interfering effects (Dimmock and Kennedy, 1978; Crouch et al., 1982; Barrett and Dimmock, 1984a–d; Barrett et al., 1984a,b). They found that different DI-particle types either (1) protected mice and left them immune to challenge (2) protected them but left them susceptible to later challenge, or (3) failed to protect them despite strong interfering ability in cell culture. This is another indication that DI-particle types may vary widely in protective ability in vivo. Protection was related to their ability to reduce virus levels in vivo, the immunizing DI particle–virus interactions exhibiting less suppression of virus replication in organs and tissues than did the nonimmunizing interactions. DI particles were not detected in mouse tissues. However, if cycling interactions occur, detection would likely be dependent on dose, tissue examined, timing, and other factors (Cave et al., 1984).

Rabinowitz and Huprikar (1979) showed protective effects DI particles on the pathogenesis of influenza virus infection of mice inoculated intranasally. Popescu and Lehmann-Grube (1977), using a focus-forming assay, found DI particles in organs of adult mice acutely infected with lymphocytic choriomeningitis virus and in persistently infected mice that had been infected as neonates. The relative numbers of virus plaque-forming units and DI-particle focus-forming units varied in different organs, as might be expected if cell type were influencing DI-particle replication–interference. Mims (1956) showed clear cycling interference during serial passage of Rift Valley fever (phlebovirus) in mice as alluded to above. Also, as mentioned in Section II.C, Brinton and colleagues (Brinton, 1983; Brinton and Fernandez, 1983; Brinton et al., 1984) have shown a remarkable relationship between a resistance allele in mice that confers resistance to flavivirus encephalitis and the rapid accumulation of West Nile virus DI particles in the brains of mice carrying this resistance gene (and in their cells in culture).

No investigators have yet demonstrated the presence of DI particles in tissue or body fluids of infected animals or humans in nature. This is probably because few attempts have been made to do so and because technical problems in detecting DI particles or proving their presence would make such demonstration difficult with most systems. There seems little doubt that lymphocytic choriomeningitis virus DI particles would be detectable in persistently infected mice trapped in the wild (Popescu and Lehmann-Grube, 1977), and it would be quite feasible and interesting to attempt detection of VSV DI particles in animals and insect vectors in nature. However, logistic problem and the disparate goals of viral field epidemiologists and of laboratory investigators tend to prevent this being accomplished. In any case, nothing that has been observed in cells or in laboratory animals would suggest that DI particles are laboratory curiosities that are somehow not generated or not important in nature. Replicative-error levels in all RNA virus replication should assure that DI particles will be generated at a significant level during virus spread in nature, but their abilities to replicate efficiently and to interfere should vary in different cell types, with different virus strains and DI types, as occurs in cultured cells and in laboratory animal studies.

After this chapter was completed, Bean et al. (1985) and Webster et al. (1986) implicated poor production (or nonproduction) of DI particles along with viral mutation in the extreme virulence of the severe outbreak of influenza A virus in chickens in the eastern United States during 1983–1984. Likewise, Huang and colleagues obtained further evidence for the ability of DI particle cycling to influence virus infections and host immune responses and disease outcomes in an unpredictable manner (Cave et al., 1985; Huang, 1987). Furthermore, the behavior of field isolates of VSV with regard to DI particle generation and interference also suggested DI particle involvement in nature (Huang et al., 1986). Finally, Dimmock and colleagues carried out extensive studies of DI particle effects in vivo. They showed (Dimmock et al., 1986) that influenza virus DI particles ameliorated virus pathology, apparently by modulating immunopathology and antibody responses in the lungs of infected mice. They also demonstrated (Atkinson *et al.*, 1986) that a small percentage of mice protected by DI particles showed prolonged persistence of virulent Semliki Forest virus in brain cells, and that mice protected from Semliki Forest virus lethality by DI particles can show persisting abnormalities of central nervous system neurotransmitters (Barrett *et al.*, 1986).

VIII. DEFECTIVE INTERFERING PARTICLES IN PERSISTENT INFECTIONS

Holland *et al.* (1980) have reviewed in some detail the involvement of DI particles in persistent infections, and the reader is referred to this earlier review for background information that cannot be included here. Youngner and Preble (1980) have also reviewed persistent infections by RNA viruses with an emphasis on systems that involve temperature-sensitive (*ts*) mutants, interferon, and factors other than DI particles.

A. Rhabdovirus Persistent Infections

Wagner *et al.* (1963) were the first to report persistent infections by rhabdoviruses *in vitro*. They demonstrated that small-plaque mutants of vesicular stomatitis virus (VSV) caused chronic infections in which cells continued to replicate while shedding virus. Fernandes *et al.* (1964) showed that rabies virus, which is usually less cytopathic than VSV, regularly and rather easily establishes persistent "endosymbiotic relationships" with cultured cells. Wiktor and Clark (1972) observed cyclic variations in hamster cells and other cells persistently infected by rabies virus. They suggested that interferon might be the factor that causes these cyclic changes *in vitro*. Huang and Baltimore (1970) speculated that DI particles might exert a virus-regulatory role in both acute and persistent virus diseases. Holland and Villarreal (1974) used multiply cloned *ts* mutants of VSV together with highly purified DI particles. They demonstrated that the DI particles were necessary to establish persistence in BHK-21 cells. They found that the *ts*G31 mutant of Pringle (1970) regularly established persistence, whereas wild-type virus or other *ts* mutants were not regularly successful in doing so, even with Di particles present (unless a particularly effective long DI particle was used). DI particles recovered from carrier cells quickly suppressed *in vitro* transcription levels of virus grown in their presence (or selected by their interference effects).

Early during DI-particle-mediated persistence, cytopathology crises were frequent, but after 90 days, cells resumed normal growth rates, despite over 99% of cells being infected (i.e., being infectious centers shedding both virus and DI particles). These cells have remained persistently infected for over 11 years, but infectious virus is very difficult to isolate after 7 years, even though most cells exhibit virus antigens. Kawai *et al.* (1975) established persistent rabies virus infections in BHK-21 cells (without deliberate addition of DI particles, but effects of high- vs. low-multiplicity infection indicated their likely presence). Following the establishment of persistent infection, regular cyclical interactions between virus shedding and DI-particle shedding were observed. Small-plaque mutants of rabies virus replaced the original large-plaque virus after a number of subcultures. Later characterization of these small-plaque mutants showed that they had become resistant to interference caused by the DI particles generated by the original large-plaque virus, but not to DI particles that they generated themselves (Kawai and Matsumoto, 1977). This was the first description of such mutants (termed Sdi⁻) in any virus system, and this phenomenon is described in detail in Section IX.B.

Youngner *et al.* (1976, 1978), Youngner and Preble (1980), and Frey and Youngner (1982) have extensively characterized mouse L cells persistently infected by VSV. Although DI particles were needed initially to establish persistence, they were not required for maintenance of persistence. Only a small percentage of L cells shed virus or expressed virus antigens; *ts* mutants arose and quickly predominated in the carrier cells, and these mutants could establish persistent infection in the absence of DI particles when infecting at low multiplicities. Interferon played a role in these persistent infections because antiserum to interferon caused increased virus shedding and destruction of the carrier cultures. Ramseur and Friedman (1977) also implicated *ts* mutants and interferon in establishing and maintaining VSV persistent infections in L cells, as did Nishiyama (1977), who in addition implicated DI particles in establishment. Nishiyama *et al.* (1978) showed that *ts* small-plaque mutants of VSV recovered from persistent infections of L cells are better interferon inducers and can initiate persistence alone at low multiplicities without added interferon.

Stanners *et al.* (1977) characterized a *ts* mutant of VSV that regularly establishes persistence and found that it is a double mutant that affects viral RNA polymerase and viral shutoff of protein synthesis. Frey and Youngner (1982) also found altered polymerase and shutoff phenotypes of VSV recovered from L-cell persistent infections. These *ts* mutants showed reduced transcription of RNA and enhanced replication in cells at 34 or 37°C. Holland and Villarreal (1974) observed that infectious VSV replicated (and selected?) in the presence of DI particles from persistence exhibits greatly reduced *in vitro* virion transcriptase activity. Villarreal and Holland (1976) showed that both VSV and rabies persistent infections exhibit greatly reduced levels of virus transcription *in vivo* and strong suppression of viral RNA replication. O'Hara *et al.* (1984b) also showed that VSV Sdi⁻ mutants selected during repeated high-multiplicity passages in the presence of DI particles exhibit greatly reduced levels of virion transcriptase *in vitro*. It appears that DI-particle interference and persistent infection of cells (with or without DI particles) tend to select polymerase mutants. Most VSV *ts* mutants selected in L-cell persistence belong to complementation group I which affects large (L) (polymerase) protein (Youngner and Preble, 1980; Frey and Youngner, 1982).

The long-term carrier cultures of BHK-21 cells persistently infected with VSV and its DI particles (designated Car4) have been extensively characterized for the past 10 years. Most of the cells exhibit virus antigens despite very low levels of virus shedding, the cells contain intracellular virus and DI nucleocapsids, and prolonged exposure of light cultures to antiviral antiserum can cure the cells of virus persistence (Holland *et al.*, 1976b). The genome of the original, cloned *ts*G31 mutant used to establish the carrier state has undergone extensive and continuing mutational evolution (Holland *et al.*, 1979; Rowlands *et al.*, 1980; O'Hara *et al.*, 1984a), and the associated DI particles continually change and evolve (Horodyski and Holland, 1983; O'Hara *et al.*, 1984b). Many of these mutational changes are probably involved in the greatly reduced virus shedding (and frequent lack of virus shedding) observed after years of persistence. In fact, the infrequent infectious-virus isolates obtained after 10 years of persistence are probably multiply mutated as compared to the majority of intracellular viral genomes that never mature to form infectious virus. All virus recovered late is *ts*, weakly cytopathic, and avirulent.

Youngner and Quagliana (1976) reported that *ts* mutants of VSV interfere with replication of wild-type VSV at permissive and nonpermissive temperatures. However, they employed "interfering" *ts* virus at a multiplicity of 1, while wild-type virus was added at a 10-fold lower multiplicity of 0.1. These results do not prove interfering ability, because wild-type virus had no opportunity to compete equally in most infected cells. In fact, Flamand (1980) could not detect *ts*-mutant interference with wild-type virus when both infected at equal multiplicity, and Spindler and Holland (1982) repeated this experiment and observed no interference by *ts* mutants at equal multiplicities. Likewise, Spindler and Holland (1982) found that most VSV mutants obtained after years of persistent infection did not interfere significantly with wild-type virus, but one mutant recovered after 76 months of persistence showed strong homologous interference (but not heterotypic interference with VSV-New Jersey). Thus, this mutant behaved as a nondefective homologous interfering particle, and the participation of DI particles was ruled out by nucleocapsid analysis. Such interference might be important at times in the evolution of competing VSV genomes during persistent infections, but its molecular basis has not been explored. Because the genome of VSV evolves continuously and rapidly during persistent infection, numerous other interesting mutants can be isolated.

One of the mutant types of VSV obtained during persistent infection is a phenotype that escapes surveillance by NK cells. When persistently infected tumor cells are injected into nude mice, the persistently infected cells are recognized as virus-infected and are killed by NK cells (Reid *et al.*, 1979; Minato *et al.*, 1979; Jones *et al.*, 1980). Interferon activation of

NK-cell responses is important in this process, because antiinterferon antibody prevents this rejection of virus-infected tumor cells (Reid *et al.*, 1981). However, if large numbers of tumor cells persistently infected with VSV are injected, tumors will arise in some nude mice after many months. When VSV recovered from such "escaped" tumors is cloned, then used to start new persistent infections in BHK-21 tumor cells, these cells will regularly form tumors and metastases that shed large amounts of this mutant virus and its DI particles.

Cells infected with this virus are able to activate NK cells very efficiently, but are not killed by them (Minato *et al.*, 1979; Jones *et al.*, 1980). Genome evolution of these VSV mutant types occurs at an extremely high rate in tumors in nude mice, presumably due to selective pressures of immunocytes, antibody, and other factors *in vivo* (Spindler *et al.*, 1982). Approximately 1% VSV-genome sequence divergence occurred *in vivo* in only 12–15 weeks. Recently, several independent VSV mutants selected in nude mice for ability to escape NK cells (and allow tumor formation) have been isolated and sequenced (VandePol and Holland, 1986). These mutants show clustered mutations in the glycoprotein (G), but not in the matrix (M) protein, indicating a role of VSV G protein in NK-cell targeting to infected cells.

Another interesting mutant-type VSV isolated from persistent infection was a *ts* small-plaque mutant isolated after 70 months (Spindler *et al.*, 1982). After initial isolation and cloning, it gave a yield of only 2.3 $\times$ 10^{-5} PFU per BHK-21 cell! However, during five low-multiplicity passages, its yields gradually increased over 10^7-fold. Such phenotypes probably predominate intracellularly after years of persistent infection, during which the ability to produce mature infectious virus is probably not frequently or strongly selected.

Andzhaparidze *et al.* (1981) showed that the host-cell type in which rabies virus persisted influenced both the character of virus–DI particle interactions and the virulence phenotype of released virus. They established persistent infection with the Pasteur strain of rabies virus both in BHK-21 cells and in human epithelial cells (Hep-2 cells). Both persistently infected cells released DI particles, but cyclical fluctuations in virus release and in percentage of antigen-positive cells were seen in the BHK-21 cells, but not in Hep-2 cells (in which virus release and numbers of antigen-positive cells remained constantly rather high). Virus recovered from BHK-21 carrier cells early during persistence caused rabies with prolonged incubation period (12–14 days vs. 6–7 days). However, virus released from the BHK-21 cells after 50 passages (5–7 days, passage time) had gradually lost all virulence, whereas virus from Hep-2 cells retained its mouse-killing capacity for more than 3 years. Clearly, the host cells in which a virus persists can exert differing and biologically important selective pressures in addition to the constant selective pressures exerted by associated DI particles (Andzhaparidze *et al.*, 1981).

Finally, infectious hematopoietic necrosis virus, a rhabdovirus that causes severe diseases of trout and salmon, produces persistent infections in salmon cells in which *ts* mutants, small-plaque mutants, and DI particles are all apparently involved (Engleking and Leong, 1981). In conclusion, it is clear that DI particles can play important roles in either or both the establishment and the maintenance of rhabdovirus persistent infections, but small-plaque mutants, *ts* mutants, and mutations that affect transcription, replication, host-cell shutoff, maturation, surface protein, and other factors are also involved.

B. Defective-Interfering-Particle Involvement in Persistent Infections by Other Viruses

The paramyxovirus Sendai virus is noteworthy for its ability to establish persistence very readily in the presence of homologous DI particles. Kimura *et al.* (1975) and Nishiyama *et al.* (1976) demonstrated that *ts* mutants of Sendai virus could easily establish persistent infection of BHK-21 cells and that *ts* mutants are readily derived from persistent Sendai virus infections. In these early studies, they did not look for DI-particle involvement. Roux and Holland (1979, 1980) showed that DI particles protect cells strongly from the lethal effects of wild-type Sendai virus and that persistent infection immediately ensues. Roux and his colleagues also characterized important modulating effects of DI particles (and of persistent infection) on cell-surface expression of the viral hemagglutinin–neuraminidase (HN) glycoprotein and on its turnover (Roux and Waldvogel, 1981, 1982, 1983; Roux and Beffy, 1984; Roux *et al.*, 1984, 1985).

It was found that virus alone kills all BHK-21 cells within several days, whereas addition of DI particles causes nearly all the cells to survive indefinitely as virus carrier cultures that show only early, occasional crises. Yoshida *et al.* (1982) have found that their BHK-21–Sendai virus carrier cells maintained over 10 years continue to produce *ts*-virus and DI-particle RNAs. The *ts* virus recovered from these carrier cells is only weakly cytopathic (except virus recovered during crises), and it regularly establishes persistent infections in LLKMK$_2$ cells without the presence of DI particles during initiation or maintenance.

Presumably, the lack of appearance of DI particles during maintenance of persistence in these cells is a sign that LLCMK$_2$ is a low-interference cell line for Sendai virus, since nearly 100% of the cells are infected, and it seems very unlikely that there are no polymerase errors generating DI particles during maintenance. In any case, this is another example of the evolution of *ts*, weakly cytopathic virus during long persistence, and this *ts* virus interfered with wild-type virus. Also, mixtures of this mutant virus and wild-type virus readily established persistent infections without added DI particles.

Roux and Waldvogel (1983) showed, using cell-surface immune precipitation of iodinated proteins, that in persistently infected cells and during Sendai virus coinfection with DI particles, HN glycoprotein of the virus is not inserted as efficiently into the cell membrane, and inserted HN is less accessible to antibody. Using monoclonal antibodies to HN or fusion (F_o) glycoprotein of Sendai virus, Roux et al. (1984) showed that HN was restricted in expression at the cell surface relative to F_o during persistence and during DI-particle interference, and an increased HN turnover rate was detected using pulse–chase experiments. Increased turnover of HN correlated with decreased maturation of HN (and M protein) into virus particles.

Roux et al. (1985) compared HN behavior in three different types of infection: In lytic virus infection, HN accumulated stably at the cell surface (half-life over 10 hr). When HN was less highly expressed due to DI-particle interference, the half-life was only 2 hr, and this rapid turnover involved reinternalization. When HN was expressed very poorly at the cell surface (in long-term persistent infection), HN did not reach the cell surface but was degraded beforehand. F_o did not show these changes in turnover rate. Such changes in persistently infected, or DI-particle-infected, cells in vivo could obviously affect interactions with the immune system. The authors suggest that M protein regulates this cell-surface modulation of HN and propose that survival of cells in the presence of DI particles is due to the cell-surface modulation. Finally, Roux and Beffy (1984) showed that replication of persisting Sendai-virus genomes does not require host-cell division.

Welsh and Oldstone (1977) and Huang et al. (1978) have previously shown that DI particles of lymphocytic choriomeningitis virus and of VSV caused reduced virus-antigen expression on cell surfaces. Defects of M-protein synthesis of function or both probably cause measles G-protein modulation in subacute sclerosing panencephalitis brain cells (Hall and Choppin, 1979; Machamer et al., 1981; Carter et al., 1983; ter Meulen et al., 1983). Modulating effects on virus surface G-proteins are also exerted by antibody (Fujinami and Oldstone, 1984). Together with modulating effects of DI particles or virus mutations or both, antibody modulation can aid in escape from immune surveillance. Why this should occur only rarely in certain diseases is not clear. Perhaps it occurs commonly, but is manifest as disease only when it progresses to involve large numbers of cells. Finally, it is interesting that VSV causes cell fusion when functional M protein is reduced in cells in the presence of G protein (Storey and Kang, 1985), since cell-to-cell fusion can allow enveloped virus genomes to spread slowly and persist without need for mature-virus synthesis. Failure of persisting RNA viral genomes to form mature virus is probably common and important in animals and in human diseases (Koch et al., 1984; Goswami et al., 1984; Mills et al., 1984).

DI particles have been associated with persistent infections by many different viruses, including lymphocytic choriomeningitis virus (Popescu and Lehmann-Grube, 1977); Japanese encephalitis virus(Schmaljohn and

Blair, 1977); Sindbis virus (Weiss *et al.*, 1980; Igarashi *et al.*, 1977); measles virus (Rima *et al.*, 1977); mumps virus (McCarthy *et al.*, 1981; Andzha-paridze *et al.*, 1982); reovirus (Ahmed and Graham, 1977); infectious pancreatic necrosis virus, a fish dsRNA virus (J.C. Kennedy and Macdon-ald, 1982); Tacaribe virus (Gimenez and Compans, 1980); Junin virus (Damonte *et al.*, 1983); Toscana virus (Verani *et al.*, 1984); herpesviruses (Dauenhauer *et al.*, 1982; Koschka-Dierich *et al.*, 1982); a baculovirus (Burand *et al.*, 1983); and others.

A somewhat disturbing reminder of the inapparent nature of stable RNA virus persistence was provided by van der Zeijst *et al.* (1983a,b), who found lymphocytic choriomeningitis virus persistently infecting some Syrian hamster tumor-cell lines (which had not knowingly been infected), and DI-size RNAs were seen in these carrier cells. An unusual SV40 persistent infection studied by O'Neill *et al.* (1982) contained two com-pleting subgenomic DI RNA circles—one that encoded early functions and the other late functions. These subgenomic DNA circles could pro-ductively infect cells when added together. No such complementing DI genomes of RNA viruses have been found in persistent infections, al-though it is theoretically possible for them to form.

In conclusion, DI particles clearly can be involved in either or both the establishment and the maintenance of many types of persistent in-fections by rhabdoviruses and many other viruses. But persistent infec-tions are very complex host–virus relationships involving many other factors, such as immune-system interactions and rapid evolution of nu-merous mutant virus phenotypes. Even the host cells may coevolve with virus mutants, as was observed by Ahmed *et al.* (1981) during persistent infections of L cells by reovirus.

IX. DEFECTIVE INTERFERING PARTICLES CAN INFLUENCE VIRUS EVOLUTION

A. High Rates of Mutation and Evolution of RNA Viruses

Mutation frequencies of RNA viruses have been known for decades to be extremely high (Granoff, 1959a,b, 1964; B.N. Fields and Joklik, 1969; Pringle, 1970; Domingo *et al.*, 1976, 1978). In addition, the evolution of RNA genomes can proceed at rates that exceed 10^6-fold the rates of ev-olution of their eukaryotic hosts. This subject has been reviewed by Hol-land *et al.* (1982). The lack of proofreading enzymes in RNA replicases plus inherently high error levels in RNA replication account for the ex-treme mutation frequencies. Base misincorporation frequencies can ex-ceed 10^{-4} base substitutions per base incorporated, both in RNA phages (Domingo *et al.*, 1976, 1978) and in RNA animal viruses (Steinhauer and Holland, 1986). These extreme error rates assure that the majority of progeny of replicating genomes are mutant in one or more sites and that even clonal virus pools are exceedingly diverse mixtures of genomes. Only strong biological selection for the most fit, competitive members of this mixture can allow the "consensus sequence" that is obtained when "a virus genome" is sequenced. Domingo et al. (1976, 1978) showed that the genome of wild-type QB phage outgrew that mutants with replication rates of 0.8–0.9 and came to a predictable equilibrium with wild-type predominating. They proposed that while mutants arose at a high rate, they were strongly selected against, so that a dynamic equilibrium of viral genomes ensued. We review below how selective pressures of DI particles (or of any continuing selective influence) can upset dynamic equilibria and promote the extreme rates of evolution of which RNA genomes are capable.

There are many conditions of virus passage in which vesicular stomatitis virus (VSV) genomes exhibit reasonable stability as measured by oligonucleotide mapping (Clewley et al., 1977; Freeman et al., 1978; Brand and :palese, 1980; Rowlands et al., 1980; Holland et al., 1980b; Spindler et al., 1982). Diluted passage in cell culture generally was observed to allow genome stability in the aforecited studies, although with influenza virus (but not VSV), Brand and Palese (1980) observed rapid genome evolution during diluted passages. In contrast (Holland et al., 1979; Semler et al., 1979; Rowlands et al., 1980), VSV persistent infection (in which DI-particle genomes are constantly present) leads to continuous rapid VSV-genome evolution. Furthermore, rapid VSV-genome evolution was observed during serial undiluted passages in BHK-21 cells (but not during diluted passage series in which DI-particle interference is minimized) (Holland et al., 1980b). Youngner et al. (1981) reported that VSV temperature-sensitive (ts) mutants appeared and accumulated more rapidly during 10 undiluted passages in BHK-21 cells only if the BHK-21 cells were pretreated with actinomycin D. They postulated a host-cell "fidelity factor" suppressed by actinomycin D pretreatment, because L cells (treated or untreated) showed high ts-mutant accumulation, whereas MDCK cells always showed low ts-mutant accumulation. However, it is clear that even without actinomycin pretreatment, genome changes evolve rapidly during serial undiluted passages in BHK-21 cells (Holland et al., 1980b; Spindler et al., 1982; O'Hara et al., 1984a). Ahmed et al. (1980) also observed rapid appearance of ts mutants during serial undiluted passage of reovirus in L cells. Pringle et al. (1981) characterized a mutant of VSV with a mutator phenotype. It will be interesting to determine whether mutator phenotypes are common during persistent infections and undiluted passages of VSV. Schubert et al. (1984), in sequencing multiple cDNA clones of the large (L) (polymerase) protein gene of a cloned VSV pool, observed 16 mutations among overlapping clones, and at least 2 of these were not due to reverse transcriptase error, because they contained the identical mutation at a single site. Schubert et al. (1984) suggested that some L-gene point mutations might give rise to full-genome-length DI particles of VSV and also that further erosion of polymerase fidelity might contribute to events that occur during persistent infection. Sobrino *et al.* (1983) showed by oligonucleotide mapping a foot-and-mouth disease clones derived from a common pool that most genomes in a virus population probably differ from each other by 2–8 mutations. Obviously, a selective pressures of any kind will influence the "consensus sequence" of viral genomes present in a population.

B. Defective-Interfering-Particle-Resistant Mutants of RNA Viruses

As mentioned in Section VII.A, Kawai and Matsumoto (1977) were the first to observe the appearance of DI-particle-resistant mutants (which they designated Sdi⁻ mutants) in cells persistently infected with rabies virus. The original large-plaque virus was soon replaced by small-plaque mutant virus, and these small-plaque mutants were found to be resistant to interference by the DI particles originally present early during persistent infection of BHK-21 cells. However, when the small-plaque Sdi⁻ mutants generated new DI particles, these particles interfered with the Sdi⁻ parental virus (and with the original large-plaque rabies virus). The same phenomenon has now been observed to occur during persistent infections with a number of other RNA viruses, including VSV (Horodyski and Holland, 1980), lymphocytic choriomeningitis virus (Jacobsen and Pfau, 1980), Sindbis virus (Weiss and Schlesinger, 1981), and West Nile virus (Brinton and Fernandez, 1983). The fact that Sdi⁻ mutants have been isolated from persistent infections by such different virus types suggests strongly that DI particles were exerting selective pressures during these persistent infections prior to their isolation. Horodyski and Holland (1983) also showed that Sdi⁻ mutants of VSV appear during serial undiluted passages in BHK-21 cells, and Enea and Zinder (1982) isolated DI-particle-resistant mutants of the filamentous, male-specific DNA coliphage f₁ by serial transfer with excess DI particles.

Horodyski and Holland (1980, 1981) developed a quantitative assay for relative interfering ability based on the ratio of DI particles to virus in yields following various DI-particle inputs (see Section III.J). With the use of this assay system, it was shown that numerous Sdi⁻ mutants of VSV are selected continuously in a stepwise manner during persistent infection and during serial undiluted passages. Throughout this evolution of various Sdi⁻ mutant viruses, there is a constant appearance and disappearance of new DI-particle types that also exhibit altered interference properties with the various Sdi⁻ mutants (Horodyski and Holland, 1983). Thus, during persistence or undiluted passages, there is a constant dynamic "coevolution" of virus and associated DI particles. Complementation tests with Sdi⁻ mutants indicated that mutation in at least two different virus factors can produce VSV Sdi⁻ mutants, and studies with chimeric DI particles showed that DI-genome RNA rather than encapsidation–maturation proteins on the Di particles determines phenotype of the particles in interactions with various Sdi⁻ mutants of VSV.

The genes involved in Sdi⁻ mutations are not yet identified, but presumably they involve replication–encapsidation proteins (N, NS, L). In fact, Enea and Zinder (1982) used heteroduplex mapping to identify two sites in the gene involved in their IR (Sdi⁻) phage mutants and found that two mutations in gene II (a replicase-associated protein-encoding gene) are required. Sequencing of the two identified sites showed that one mutation occurred in the untranslated region 20 bases before the first AUG codon for gene II, and the other was a $G \rightarrow A$ transition within gene II that caused a Thr $\rightarrow$ Ile change at amino acid 183 (of the 410-amino-acid gene II protein). The molecular nature of DI-particle interference in this phage is not known, and it is not yet clear how these Sdi⁻ mutations in gene II suppress interference, but gene II protein is a site-specific strand-specific nicking enzyme essential for f1 replication in association with host-cell enzymes.

O'Hara et al. (1984a) sequenced the 3' and 5' termini of a number of Sdi⁻ mutants of VSV from serial undiluted passages and from persistent infections. They observed the stepwise accumulation of stable 5'-terminal mutations and of fewer 3'-terminal mutations in the viral genomes. Once these mutations appeared, they were stable thereafter during passages or persistent infection, suggesting strong biological selection to maintain them.

Isaac and Keene (1982) and Keene et al. (1981b) have identified binding sites for VSV-encoded transcription–replication proteins within VSV termini, and Blumberg et al. (1983) and Giorgi et al. (1983) identified sites for initiation of encapsidation within the termini (see Section VI.A). Therefore, the evolving termini mutations that occur in the presence of DI particles are probably compensatory changes that give better fit to altered replication–encapsidation proteins of the Sdi⁻ mutant viruses. In support of this possibility is the finding of Wilusz et al. (1985) that VSV mutants obtained from prolonged persistent infections of L cells (in which DI particles are seldom present) do not accumulate 5'-terminal base substitutions. This finding suggests that the 5' mutations that accumulate in the presence of DI particles probably were selected by DI interfering activities. However, Wilusz et al. (1985) did observe base substitutions in the 3' termini of virus mutants from persistently infected L cells. One mutation at position 21 from the 3' end was observed in both mutants they sequenced, and the same $U \rightarrow G$ or A transversion at position 21 was observed by O'Hara et al. (1984a) in mutants from persistently infected BHK-21 cells after 5 years. Wilusz et al. (1985) suggested that these changes at the 3' end of the genome may affect transcription, the transcription–translation balance (see above), or the ability to shut off host macromolecular synthesis (Grinell and Wagner, 1984). They suggested that such changes could be important in virus–host cell interactions during persistent infections.

O'Hara *et al.* (1984b) sequenced the termini of 16 different DI particles isolated at intervals from persistent infection and undiluted passages and found that their termini generally reflected the base substitutions observed at the same time in homologous infectious virus recovered at the same point in coevolution. Of a total of over 3300 DI-terminal bases sequenced, less than 1% differed from virus at the same stage of passage or persistence. This is strong evidence that the termini of both virus and DI genomes are subjected to similar selective pressures while coevolving in complex competing mixtures of virus and DI genomes. Horodyski and Holland (1984) performed reconstruction experiments with DI particles as mixtures of wild-type and Sdi⁻ mutant viruses and demonstrated that, in fact, DI particles strongly select Sdi⁻ mutants over wild-type virus and when the Sdi⁻ mutant is added initially to the mixed population at very low levels relative to wild-type. Clearly, DI particles can upset a viral population equilibrium that otherwise favors wild-type predominance.

Horodyski and Holland (1983) showed that resistance to an original DI particle continues to increase in a stepwise manner through repeated undiluted passages as new Sdi⁻ mutants appear, and that resistance to that original DI particle disappears as new Sdi⁻ mutants "escape" from more recently generated DI particles. This kind of Sdi⁻ mutant "escape" might continue indefinitely as new Sdi⁻ mutants and their newly generated DI particles appear, compete, and disappear, or there might be a limit to the number of new DI particles against which new Sdi⁻ mutants can mutate to resistance. If only a few cycles of "escape" are possible, then "nonescaping" virus or "inescapable" DI particles or both might appear after continuous competition with Sdi⁻ mutants, but this does not appear to occur (DePolo *et al.*, 1987). Each Sdi⁻ mutant is soon replaced by another that is in turn better able to escape interference by those DI particles that suppressed the previous Sdi⁻ mutant. Some of these later-arising mutants are remarkably resistant (over 200,000-fold) to formerly effective DI particles. Obviously, this kind of continuing selective pressure intracelluarly could constantly disrupt viral-population equilibria. This should drive rapid viral evolution during persistent infections particularly, but also during acute infections whenever virus and DI particles accumulate locally.

Whenever virus-population equilibrium is upset (by DI particles, antibody, or any selective force), mutational change at many genomic sites can be promoted simply by suppression of the dominant viral genome that carries the previously most competitive consensus sequence. Thus, DI particles are capable of driving viral-genome evolution intracellularly to produce unforseeable phenotypes. Also, albeit extremely infrequently, DI-particle-genome sequences might be transferred, by replicative error, back to viral genomic RNA to give viable recombinant virus. No matter how rare, such events could contribute significantly to viral-genome evolution, because mutational constraints on DI genomes (except for termini) are clearly much less stringent than on infectious virus genomes

(O'Hara *et al.*, 1984a,b). DI genomes would then represent an evolving pool of extensively mutated and rearranged viral-genome subsets that only very infrequently might contribute highly altered sequences back to virus genomes. If this takes place in an environment in which the recombinant virus is not only viable but also selectively replicated, then DI genomes participate in the modular evolution of their parental virus genomes, just as integrated defective modules of DNA phages contribute to the modular evolution of their parental viruses (and hosts) (Botstein and Susskind, 1974; Botstein, 1980). Ahmed and Fields (1981) have clearly demonstrated reassortment between nondeleted segments of reovirus DI particles and infectious reovirus. Covalent recombination between RNA-virus genomes and their DI-particle genomes probably also occurs, but at a much lower frequency.

X. SUMMARY

RNA virus DI particles are subgenomic deletion mutants that arise during replicative polymerase errors. Although rates of generation of DI particles can vary, their generation should occur at some level in all cells, since polymerase error is inevitable in any cell able to replicate an RNA virus. Once DI particles are generated, their ability to replicate and interfere varies enormously depending on DI-particle genome structure, helper-virus strain and phenotype, host-cell type, time of entry into infected cells, relative competitive ability vs. other DI particles, and other factors. Rhabdovirus DI particles fall into one of several different genome-structure classes, with one or both of the infectious virus genome termini being retained in all of them (since these termini are required for replication–encapsidation functions). DI-genome sequences within the conserved termini can be extensively rearranged, mutated, deleted, or reiterated, or some combination thereof, especially in DI particles from persistent infections and high passage levels.

Most rhabdovirus DI particles interfere mainly at the level of virus replication–encapsidation, but transcribing classes of DI particles (which are rare in rhabdoviruses, but common in some viruses) appear to exert effects that result from their transcription–translation products. DI-particle interference can lead to survival of cells infected by otherwise lethal viruses, thereby leading to their being persistently infected or "cured" of virus infection or both. DI particles can be generated and replicated and interfere in animals *in vivo*, although their presence in infections of animals or humans in nature has never been demonstrated (or ruled out) in field studies. DI particles apparently undergo cyclical interactions with their helper viruses in animal tissues (just as they do in cell culture), and this could have far-reaching significance for the outcome of natural virus infections, since it would be a complex stochastic process in tissues and organs. DI particles can also enhance interferon production, modulate surface expression of viral proteins, affect their transport, processing, and turnover, and alter the timing and basic pathology of a virus infection *in vivo*.

DI particles constitute a powerful intracellular selective force that can upset viral-genome population equilibria and help drive rapid viral evolution. Since (with the exception of their termini) there are far fewer constraints to prevent very rapid evolution of DI RNA genomes, these genomes can constitute a source of extensively rearranged viral RNA that only *very* infrequently might become recombined back into virus genomes to cause profound evolutionary change (when environmental selective forces permit or favor such change).

Most assay systems for DI particles are useful for detecting and quantitating rather large numbers of DI particles, but are inadequate to detect very low (but biologically significant) levels of DI particles contaminating cloned virus pools. Without employment of a very sensitive "randomness of generation assay" for minuscule seed quantities of DI particles, claims that cloned virus pools are free of "significant amounts" of DI particles have no validity.

REFERENCES

Adachi, T., and Lazzarini, R. A., 1978, Elementary aspects of autointerference and the replication of defective interfering virus particles, *Virology* **87**:152.

Adler, R., and Banerjee, A. K., 1976, Analysis of the RNA species isolated from defective particles of vesicular stomatitis virus, *J. Gen. Virol.* **33**:51.

Agol, V., Grachev, V., Drozdov, S., Kolesnikova, M., Kozlov, V., Ralph, N., Romanova, L., Tolskaya, E., Tzufanov, A., and Victorova, E., 1984, Construction and properties of intertypic poliovirus recombinants: First approximation mapping of the major determinants of neurovirulence, *Virology* **136**:41.

Ahmed, R., and Fields, B. N., 1981, Reassortment of genome segments between reovirus defective interfering particles and infectious virus: Construction of temperature-sensitive and attenuated virus by rescue of mutations from DI particles, *Virology* **111**:351.

Ahmed, R., and Graham, A. F., 1977, Persistent infections in L cells with temperature sensitive mutants of reovirus, *J. Virol.* **23**:250.

Ahmed, R., Chakraborty, P. R., and Fields, B. N., 1980, Genetic variation during lytic reovirus infection: High-passage stocks of wild-type reovirus contain temperature-sensitive mutants, *J. Virol.* **334**:285.

Ahmed, R., Canning, W. M., Kaufmann, R. S., Sharpe, A. H., Hallum, J. V., and Fields, B. W., 1981, Role of the host cell in persistent viral infection: Coevolution of L cells and reovirus during persistent infection, *Cell* **25**:325.

Akkina, R. K., Chambers, T. M., and Nayak, D. P., 1984a, Expression of defective-interfering influenza virus-specific transcripts and polypeptides in infected cells, *J. Virol.* **51**:395.

Akkina, R. K., Chambers, T. M., and Nayak, D. P., 1984b, Mechanism of interference by defective interfering particles of influenza virus: Differential reduction of intracellular synthesis of specific polymerase proteins, *Virus Res.* **1**:687.

Amesse, L. S., Pridgen, C. L., and Kingsbury, D. W., 1982, Sendai virus DI RNA species with conserved virus genome termini and extensive internal deletions, *Virology* **118**:17.

Andzhaparidze, O. G., Bogomolova, N. N., Boriskin, Y. S., Bektermirova, M. S., and Drynov, I. D., 1981, Comparative study of rabies virus persistence in human and hamster cell lines, *J. Virol.* **37**:1.

Andzhaparidze, O.G., Boriskin, Y. S., Bogomolova, N. N., and Drynov, I. D., 1982, Mumps virus-persistently infected cell cultures release defective interfering virus particles, *J. Gen. Virol.* **63**:499.

Atkinson, T., Barrett, A., Mackenzie, A., and Dimmock, N., 1986, Persistence of virulent Semliki Forest virus in mouse brain following coinoculation with defective interfering particles, *J. Gen. Virol.* **67**:1189.

Barrett, A. D. T., and Dimmock, N. J., 1984a, Variation in homotypic and heterotypic interference by defective interfering viruses derived from different strains of Semliki Forest virus and from Sindbis virus, *J. Gen. Virol.* **65**:1119.

Barrett, A. D. T., and Dimmock, N. J., 1984b, Properties of host and virus which influence defective interfering virus-mediated protection of mice against Semliki Forest virus lethal encephalitis, *Arch. Virol.* **81**:185.

Barrett, A. D. T., and Dimmock, N. J., 1984c, Modulation of Semliki Forest virus-induced infection of mice by defective interfering virus, *J. Infect. Dis.* **150**:98.

Barrett, A. D. T., and Dimmock, N. J., 1984d, Modulation of a systemic Semliki Forest virus infection in mice by defective interfering virus, *J. Gen. Virol.* **65**:1827.

Barrett, A. D. T., Crouch, C. F., and Dimmock, N. J., 1981, Assay of defective-interfering Semliki Forest virus by the inhibition of synthesis of virus-specified RNAs, *J. Gen. Virol.* **54**:273.

Barrett, A. D. T., Cubitt, W. D., and Dimmock, N. J., 1984a, Defective interfering particles of Semliki Forest virus are smaller than particles of standard virus, *J. Gen. Virol.* **65**:2265.

Barrett, A. D. T., Guest, A. R., Mackenzie, A., and Dimmock, N.J., 1984b, Protection of mice infected with a lethal dose of Semliki Forest virus by defective interfering virus: Modulation of virus multiplication, *J. Gen. Virol.* **65**:1909.

Barrett, A. D. T., Cross, A., Crow, T., Johnson, J., Guest, A., and Dimmock, N. J., 1986, Subclinical infections in mice resulting from the modulation of a lethal dose of Semliki Forest virus with defective interfering viruses: Neurochemical abnormalities in the central nervous system, *J. Gen. Virol.* **67**:1727.

Batt-Humphries, S., Simonsen, C. C., and Ehrenfeld, E., 1979, Full-length viral RNA synthesized *in vitro* by vesicular stomatitis virus-infected HeLa cell extracts, *Virology* **96**:88.

Bay, P. H. S., and Reichmann, M. E., 1979, UV inactivation of the biological activity of defective interfering particles generated by vesicular stomatitis virus, *J. Virol.* **32**:876.

Bay, P. H. S., and Reichmann, M. E., 1981, *In vivo* inhibition of primary transcription of vesicular stomatitis virus by a defective interfering particle, in: *The Replication of Negative Strand Viruses* (D. H. L. Bishop and R. W. Compans, eds.), pp. 879–885, Elsevier/North-Holland, New York.

Bay, P. H. S., and Reichmann, M. E., 1982, *In vitro* inhibition of primary transcription of vesicular stomatitis virus by a defective interfering particle, *J. Virol.* **41**:172.

Bean, W., Kawaoka, Y., Wood, J., Pearson, J., and Webster, R., 1985, Characterization of virulent and avirulent A/Chicken/Pennsylvania/83 influenza A viruses: Potential role of defective interfering RNAs in nature, *J. Virol.* **54**:151.

Bellett, A. J. D., and Cooper, P. D., 1959, Some properties of the transmissible interfering component of vesicular stomatitis virus preparations, *J. Gen. Microbiol.* **21**:498.

Blumberg, B. M., and Kolakofsky, D., 1981, Intracellular vesicular stomatitis virus leader RNAs are found in nucleocapsid structures, *J. Virol.* **40**:568.

Blumberg, B. M., and Kolakofsky, D., 1983, An analytical review of defective infections of vesicular stomatitis virus, *J. Gen. Virol.* **64**:1839.

Blumberg, B. M., Leppert, M., and Kolakofsky, D., 1981, Interaction of VSV leader RNA and nucleocapsid protein may control VSV genome replication, *Cell* **23**:831.

Blumberg, B. M., Giorgi, C., and Kolakofsky, D., 1983, N. protein of vesicular stomatitis virus selectively encapsidates leader RNA *in vitro*, *Cell* **32**:559.

Botstein, D., 1980, A modular theory of virus evolution, in: *Animal Virus Genetics* (B. N. Fields, R. Jaenisch, and C. F. Fox, eds.), pp. 11–20, Academic Press, New York.

Botstein, D., and Susskind, M. M., 1974, Regulation of lysogeny and the evolution of temperate bacterial viruses, in: *Mechanisms of Viral Disease* (W. S. Robinson and C. F. Fox, eds.), pp. 363–384, W. A. Benjamin, Menlo Park, California.

Brand, C., and Palese, P., 1980, Sequential passage of influenza virus in embryonated eggs or tissue culture: Emergence of mutants, *Virology* **107**:424.

Brinton, M. A., 1983, Analysis of West Nile virus (WNV) particles produced by cell cultures from genetically resistant and susceptible mice indicates enhanced amplification of DI particles by resistant cultures, *J. Virol.* **46**:860.

Brinton, M. A., and Fernandez, A. V., 1983, A replication efficient mutant of West Nile virus is insensitive to DI particle interference, *Virology* **129**:107.

Brinton, M. A., and Nathanson, N., 1981, Genetic determinants of virus susceptibility: Epidemiologic implications of murine models, *Epidemiol. Rev.* **3**:115.

Brinton, M. A., Blank, K. J., and Nathanson, N., 1984, Host genes that influence susceptibility to viral disease, in: *Concepts in Viral Pathogenesis* (A. L. Notkins and M. B. Oldstone, eds.), pp. 71–78, Sringer-Verlag, New York.

Brown, F., Cartwright, B., Crick, J., and Smale, C. J., 1967, Infective substructure from vesicular stomatitis virus, *J. Virol.* **1**:368.

Burand, J. P., Wood, H. A., and Summers, M. D., 1983, Defective particles from a persistent baculovirus infection in *Trichoplusia ni* tissue culture cells, *J. Gen. Virol.* **64**:391.

Carter, M. J., Willcocks, M. M., and ter Meulen, V., 1983, Defective translation of measles virus matrix protein in a subacute sclerosing panencephalitis cell line, *Nature (London)* **305**:153.

Cave, D. R., Hagen, F. S., Palma, E. L., and Huang, A. S., 1984, Detection of vesicular stomatitis virus RNA and its defective interfering particles in individual mouse brains, *J. Virol.* **50**:86.

Cave, D., Hendrickson, F., and Huang, A., 1985, Defective interfering virus particles modulate virulence, *J. Virol.* **55**:366.

Chanda, P. K., Yong Kang, C., and Banerjee, A. K., 1980, Synthesis *in vitro* of the full-length complement of defective interfering particle RNA of vesicular stomatitis virus, *Proc. Natl. Acad. Sci. U.S.A.* **77**:3927.

Chanda, P. K., Chambers, T. M., and Nayak, D. P., 1983, *In vitro* transcription of DI particles of influenza virus produced polyadenylic acid-containing complementary RNAs, *J. Virol.* **45**:55.

Choppin, P. W., 1969, Replication of influenza virus in a continuous cell line: High yield of infective virus from cells inoculated at high multiplicity, *Virology* **39**:130.

Chow, J. M., Schnitzlein, W. M., and Reichmann, M. E., 1977, Expression of genetic information contained in the RNA of a defective interfering particle of vesicular stomatitis virus, *Virology* **77**:579.

Clewley, J. P., Bishop, D. H. L., Kang, C. Y., Coffin, J., Schnitzlein, W. M., Reichmann, M. E., and Shope, R. E., 1977, Oligonucleotide fingerprints of RNA species obtained from rhabdoviruses belonging to the vesicular stomatitis virus subgroup, *J. Virol.* **23**:152.

Cole, C. N., 1975, Defective interfering (DI) particles of poliovirus, *Prog. Med. Virol.* **20**:180.

Cole, C. N., and Baltimore, D., 1973a, Defective interfering particles of poliovirus, 2. Nature of the defect. *J. Mol. Biol.* **76**:325.

Cole, C. N., and Baltimore, D., 1973b, Defective interfering particles of poliovirus. 3. Interference and enrichment, *J. Mol. Biol.* **76**:345.

Cole, C. N., and Baltimore, D., 1973a, Defective interfering particles of poliovirus, 4. Mechanisms of enrichment, *J. Virol.* **12**:1414.

Cole, C. N., Smoler, D., Wimmer E., and Baltimore, D., 1971, Defective interfering particles of poliovirus, 1. Isolation and physical properties, *J. Virol.* **7**:478.

Colonno, R. J., and Banerjee, A. K., 1978, Complete nucleotide sequence of the leder RNA synthesized in vitro by vesicular stomatitis virus, *Cell* **15**:93.

Colonno, R. J., Lazzarini, R. A., Keene, J. D., and Banerjee, A. K., 1977, *In vitro* synthesis of messenger RNA by a defective interfering particle of vesicular stomatitis virus, *Proc. Natl. Acad. Sci. U.S.A.* **74**:1884.

Cooper, P. D., 1968, A genetic map of poliovirus temperature sensitive mutants, *Virology* **35**:584.

Cooper, P. D., 1977, Genetics of picornaviruses, in: *Comprehensive Virology*, Vol. 9 (H. Fraenkel-Contrat, and R. R. Wagner eds.), pp. 133–207, Plenum Press, New York.

Cooper, P. D., and Bellett, A. J. D., 1959, A transmissible interfering component of vesicular stomatitis virus preparations, *J. Gen. Microbiol.* **21**:485.

Crick, J., and Brown, F., 1977, *In vivo* interference in vesicular stomatitis virus infection, *Infect. Immun.* **15**:354.

Crick, J., Cartwright, B., and Brown, F., 1966, Interfering components of vesicular stomatitis virus, *Nature (London)* **211**:1204.

Crouch, C. F., Mackenzie, A., and Dimmock, N. J., 1982, The effect of defective interfering Semliki Forest virus on the histopathology of infection with virulent Semliki Forest virus in mice, *J. Infect. Dis.* **146**:411.

Crumpton, W. M., Avery, R. J., and Dimmock, N. J., 1981, Influence of the host cell on the genomic and subgenomic RNA content of defective-interfering influenza virus, *J. Gen. Virol.* **53**:173.

Damonte, E. B., Merrich, S. E., and Coto, C. E., 1983, Response of cells persistently infected with arenaviruses to superinfection with homotypic and heterotypic viruses, *Virology* **129**:474.

Darnell, M. B., and Koprowski, H., 1974, Genetically determined resistance to infection with group B arboviruses. II. Increased production of interfering particles in cell cultures from resistant mice, *J. Infect. Disc.* **129**:248.

Dauenhauer, S. A., Robinson, R. A., and O'Callahan, D. J., 1982, Chronic production of defective interfering particles by hamster embryo cultures of herpesvirus persistently infected and oncogenically transformed cells, *J. Gen. Virol.* **60**:1.

Davis. N. L., and Wertz, G. W., 1982, Synthesis of vesicular stomatitis virus negative-strand RNA *in vitro*: Dependence on viral protein synthesis, *J. Virol.* **41**:821.

De, B. K., and Nayak, D. P., 1980, Defective interfering influenza viruses and host cells: Establishment and maintenance of persistent influenza virus infection in MDBK in HeLa cells, *J. Virol.* **36**:847.

De, B. K., and Perrault, J., 1982, Signal sequence involved in the generation of an internal deletion DI RNA for VSV, *Nucleic Acids. Res.* **10**:6919.

DePolo, N. H., and Holland, J. J., 1986, Very rapid generation/amplification of defective interfering particles by vesicular stomatitis virus variants isolated from persistent infection, *J. Gen. Virol.* **67**:1195.

DePolo, N. H., Giachetti, C., and Holland, J. J., 1987, Continuing coevolution of virus and DI particles during undiluted passages: Virus mutants exhibiting nearly complete resistance to formerly dominant DI particles, *J. Virol.* **61**:454.

Dimmock, N. J., and Barrett, A., 1986, Defective viruses in diseases, *Curr. Topics Microbiol. Immunol.* **128**:55.

Dimmock, N. J., and Kennedy, S. I. T., 1978, Protection of lethally infected mice with low doses of defective-interfering Semliki Forest virus, *J. Gen. Virol.* **39**:231.

Dimmock, N. J., Beck, S., and McLain, L., 1986, Protection of mice from lethal influenza: Evidence that defective interfering virus modulates the immune response and not virus multiplication, *J. Gen. Virol.* **67**:839.

Domingo, E., Flavell, R. A., and Weissman, C., 1976, *In vitro* site-directed mutagenesis: Generation and properties of an infectious extracistronic mutant of bacteriophage QB, *Gene* **1**:3.

Domingo, E., Sabo, D., Taniguchi, T., and Weissman, C., 1978, Nucleotide sequence heterogeneity of an RNA phage population, *Cell* **13**:735.

Dolye, M., and Holland, J. J., 1973, Prophylaxis and immunization in mice by use of virus-free defective T particles to protect against intracerebral infection by vesicular stomatitis virus, *Proc. Natl. Acad. Sci. U.S.A.* **70**:2105.

Dubensky, T. W., Murphy, F. A., and Villarreal, L. P., 1984, Detection of DNA and RNA virus genomes in organ systems of whole mice: Patterns of mouse organ infection by polyomavirus, *J. Virol.* **50**:779.

Eaton, B. T., 1975, Defective interfering particles of Semliki Forest virus do not interfere with viral RNA synthesis in *Aedes albopictus* cells, *Virology* **68:**534.

Enea, V., and Zinder, N. D., 1982, Interference resistant mutants of phage f1, *Virology* **122:**222.

Engleking, H. M., and Leong, J. C., 1981, IHNV persistently infects chinook salmon embryo cells, *Virology* **109:**47.

Epstein, D. A., Herman, R. C., Chien, I., and Lazzarini, R. A., 1980, Defective interfering particle generated by internal deletion of the vesicular stomatitis virus genome, *J. Virol.* **33:**818.

Faulkner, G., Dubois-Dalcq, M., Hoogke-Peters, E., McFarland, H. F., and Lazzarini, R. A., 1979, Defective interfering particles modulate VSV infection of dissociated neuron cultures, *Cell* **17:**979.

Fernandes, M. V., Wiktor, T. J., and Koprowski, H., 1964, Endosymbiotic relationship between animal viruses and their host cells, *J. Exp. Med.* **120:**1099.

Fields, B. N., and Joklik, W. K., 1969, Isolation and preliminary genetic and biochemical characterization of temperature-sensitive mutants of reovirus, *Virology* **37:**335.

Fields, S., and Winter, G., 1982, Nucleotide sequences of influenza virus segments 1 and 3 reveal mosaic structure of a small viral RNA segment, *Cell* **28:**303.

Flamand, A., 1980, Rhabdovirus genetics, in: *Rhabdoviruses*, Vol. II (D. H. L. Bishop, ed.), pp. 115–140, CRC Press, Boca Raton, Florida.

Freeman, G. J., Rao, D. D., and Huang, A. S., 1978, Genome organization of vesicular stomatitis virus: Mapping *ts*G41 and the defective interfering T particle, in: *Negative Strand Viruses and the Host Cell*, (B. W. J. Mahy and R. D. Barry eds.), pp. 261–270, Academic Press, London.

Frey, T. K., and Youngner, J. S., 1982, Novel phenotype of RNA synthesis expressed by vesicular stomatitis virus isolated from persistent infection, *J. Virol.* **44:**167.

Frey, T. K., Jones, E. V., Cardamone, Jr., J. J., and Youngner, J. S., 1979, Induction of interferon in L cells by defective interfering (DI) particles of vesicular stomatitis virus: Lack of correlation with content of [±] snapback RNA, *Virology* **99:**95.

Frey, T. K., Frielle, D. W., and Youngner, J. S., 1981, Standard vesicular stomatitis virus is required for interferon induction in L cells by defective interfering particles, in: *The Replication of Negative Strand Viruses* (D. H. L. Bishop and R. W. Compans, eds.), pp. 901–907, Elsevier/North-Holland, New York.

Fujinami, R. S., and Oldstone, M. B. A., 1984, Antibody initiates virus persistence: Immune modulation of measles virus infection, in: *Concepts in Viral Pathogenesis* (A. L. Notkins and M. B. A. Oldstone, eds.), pp. 187–193, Springer-Verlad, New York.

Fuller, F. J., and Marcus, P. I., 1980, Interferon induction by viruses—Sindbis virus: Defective interfering particles temperature-sensitive for interferon induction, *J. Gen. Virol.* **48:**391.

Fultz, P. N., Shadduck, J. A., Kang, C.-Y., and Streilein, J. W., 1981, Genetic analysis of resistance to lethal infections of vesicular stomatitis virus in Syrian hamsters, *Infect. Immun.* **32:**1007.

Fultz, P. N., Shadduck, J. A., Kang, C.-Y., and Streilein, J. W., 1982a, Vesicular stomatitis virus can establish persistent infections in Syrian hamsters, *J. Gen. Virol.* **63:**493.

Fultz, P. N., Shadduck, J. A., Kang, C.-Y., and Streilein, J. W., 1982b, Mediators of protection against lethal systemic VSV infection in hamsters: Defective interfering particles, polyinosinate–polycylidylate, and interferon, *Infect. Immun.* **37:**679.

Fultz, P. N., Holland, J. J., Knobler, R., and Oldstone, M. B. A., 1984, Long-term persistence by vesicular stomatitis virus in hamsters, in: *Nonsegmented Negative Strand Viruses* (B. Mahy and R. Barry, eds.), pp. 489–496, Academic Press, New York.

Gallione, C. J., Greene, J. R., Iverson, L. E., and Rose, J. K., 1981, Nucleotide sequences of the mRNA's encoding the vesicular stomatitis virus N and NS proteins, *J. Virol.* **39:**529.

Gillies, S., and Stollar, V., 1980, The production of high yields of infectious vesicular stomatitis virus in *A. albopictus* cells and comparisons with replication in BHK-21 cells, *Virology* **107:**509.

Gimenez, H. B., and Compans, R. W., 1980, Defective interfering Tacaribe virus and persistently infected cells, *Virology* **107:**229.

Giorgi, C., Blumberg, B., and Kolakofsky, D., 1983, Sequence determination of the (+) leader RNA regions of the vesicular stomatitis virus Chandipura, Cocal and Piry serotype genomes. *J. Virol.* **46**:125.

Goswami, K. K. A., Cameron, K. R., Russell, W. C., Lange, L. S., and Mitchell, D. N., 1984, Evidence for the persistence of paramyxoviruses in human bone marrows, *J. Gen. Virol.* **65**:1881.

Grabau, E. A., and Holland, J. J., 1982, Analysis of viral and defective-interfering nucleocapsids in acute and persistent infection by rhabdoviruses, *J. Gen. Virol.* **60**:87.

Graham, A. F., 1977, Possible role of defective virus in persistent infection in: *Microbiology 1977* (D. Schlessinger, ed.), pp. 445–450, American Society for Microbiolgy, Washington, D. C.

Granoff, A., 1959a, Studies on mixed infection with Newcastle disease virus. I. Isolation of Newcastle disease virus mutants and tests for genetic recombination between them, *Virology* **9**:636.

Granoff, A., 1959b, Studies on mixed infection with Newcastle disease virus. II. The occurrence of NDV heterozygotes and study of phenotypic mixing involving serotype and thermal stability, *Virology* **9**:649.

Granoff, A., 1962, Heterozygous and phenotypic mixing with Newcastle disease virus, *Cold Spring Harbor Symp. Quant. Biol.* **27**:319.

Granoff, A., 1964, Nature of the Newcastle disease virus population, in: *Newcastle Disease Virus, an Evolving Pathogen,* (R. P. Hanson, ed.), pp. 107–118, University of Wisconsin Press, Madison.

Green, T. L., and Emerson, S. U., 1984, Effect of the β-γ phosphate bond of ATP on synthesis of leader RNA and mRNAs of vesicular stomatitis virus, *J. Virol.* **50**:255.

Grinnell, B. W., and Wagner, R. R., 1984, Nucleotide sequence and secondary structure of VSV leader RNA and homologous DNA involved in inhibition of DNA-dependent transcription, *Cell* **36**:533.

Guild, G. M., and Stollar, V., 1975, Defective interfering particles of Sindbis virus. 3. Intracellular viral RNA species in chick embryo cell cultures, *Virology* **67**:24.

Guild, G. M., Flores, L., and Stollar, V., 1977, Defective interfering particles of Sindbis virus. 4. Virion RNA species and molecular weight determination of defective double-stranded RNA, *Virology* **77**:158.

Hackett, A. J., 1964, A possible morphological basis for the autointerference phenomenon in vesicular stomatitis virus, *Virology* **24**:51.

Hackett, A. J., Schaffer, F. L., and Madin, S. H., 1967, The separation of infectious and autointerfering particles in vesicular stomatitis virus preparations, *Virology* **31**:114.

Hall, W. W., and Choppin, P. W., 1979, Evidence for lack of synthesis of the M polypeptide of measles virus in brain cells in subacute sclerosing panencephalitis, *Virology* **99**:443.

Henle, W., and Henle, G., 1943, Interference of inactive virus with the propagation of virus of influenza, *Science* **98**:87.

Hill, V. M., and Summers, D. F., 1982, Synthesis of VSV RNAs *in vitro* by cellular extracts: VSV protein requirements for replication *in vitro, Virology* **123**:407.

Hill, V. M., Simonsen, C. C., and Summers, D. F., 1979, Characterization of vesicular stomatitis virus replicating complexes isolated in renografin gradients, *Virology* **99**:97.

Hill, V. M., Marnell, L., and Summers, D. F., 1981, *In vitro* replication and assembly of vesicular stomatitis virus nucleocapsids, *Virology* **113**:109.

Holland, J. J., and Villarreal, L. P., 1974, Persistent noncytocidal vesicular stomatitis virus infections mediated by defective T particles that suppress virion transcriptase, *Proc. Natl. Acad. Sci. U.S.A.* **71**:2956.

Holland, J. J., and Villarreal, L. P., 1975, Purification of defective interfering T particles of vesicular stomatitis and rabies viruses generated *in vivo* in brains of newborn mice, *Virology* **67**:438.

Holland, J. J., Villarreal, L. P., and Breindl, M., 1976a, Factors involved in the generation and replication of rhabdovirus defective T particles, *J. Virol.* **17**:805.

Holland, J. J., Villarreal, L. P., Welsh, R. M., Oldstone, M. B. A., Kohne, D., Lazzarini, R., and Scolnick, E., 1976b, Long-term persistent vesicular stomatitis virus and rabies virus infection of cells *in vitro*, *J. Gen. Virol.* **33**:193.

Holland, J. J., Grabau, E., Jones, C. L., and Semler, B. L., 1979, Evolution of multiple genome mutations during long term persistent infection by vesicular stomatitis virus, *Cell* **16**:495.

Holland, J. J., Kennedy, S. I. T., Semler, B. L., Jones, C. L., Roux, L., and Grabau, E. A., 1980a, Defective interfering RNA viruses and the host cell response, in: *Comprehensive Virology*, Vol. 16 (H. Fraenkel-Contrat and R. R. Wagner, eds.), pp. 137–192, Plenum Press, New York.

Holland, J. J., Spindler, K., Grabau, E., Semler, B., Jones, C., Horodyski, F., Rowlandi, D., Janis, B., Reid, L., Minato, N., and Bloom, B., 1980b, in: *Animal Virus Genetics* (B. Fields, R. Jaenisch, and C. F. Fox, eds.), pp. 695–709, Academic Press, New York.

Holland, J., Spindler, K., Horodyski, F., Grabau, E., Nichol, S., and Van-dePol, S., 1982, Rapid evolution of RNA genomes, *Science* **215**:1577.

Horodyski, F. M., and Holland, J. J., 1980, Viruses isolated from cells persistently infected with vesicular stomatitis virus show altered interactions with defective interfering particles, *J. Virol.* **36**:627.

Horodyski, F. M., and Holland, J. J., 1981, Continuing evolution of virus–DI particle interaction resulting during VSV persistent infection, in: *Replication of Negative Strand Viruses* (D. H. L. Bishop and R. Compans, eds.), pp. 887–892, Elsevier, Amsterdam.

Horodyski, F., and Holland, J., 1983, Properties of DI particle-resistant mutants of vesicular stomatitis virus isolated from persistent infection and from undiluted passages, *Cell* **33**:801.

Horodyski, F. M., and Holland, J. J., 1984, Reconstruction experiments demonstrating selective effects of defective interfering particles on mixed population of vesicular stomatitis virus, *J. Gen. Virol.* **65**:819.

Huang, A. S., 1977, Viral pathogenesis and molecular biology, *Bacteriol. Rev.* **41**:811.

Huang, A. S., 1987, Modulation of viral disease processes by defective interfering particles, in: *RNA Genetics* (E. Domingo, P. Ahlguist, and J. Holland, eds.), CRC Press, Boca Raton, Florida (in press).

Huang, A. S., and Baltimore, D., 1970, Defective viral particles and viral disease processes, *Nature (London)* **226**:325.

Huang, A. S., and Baltimore, D., 1977, Defective interfering animal viruses, in: *Comprehensive Virology*, Vol. 10 (H. Fraenkel-Conrat and R. R. Wagner, eds.), pp. 73–116, Plenum Press, New York.

Huang, A. S., and Manders, E. K., 1972, Ribonucleic acid synthesis of vesicular stomatitis virus. 4. Transcription by standard virus in the presence of defective interfering virus particles, *J. Virol.* **9**:909.

Huang, A. S., and Wagner, R. R., 1966a, Defective T particles of vesicular stomatitis virus. 2. Biologic role in homologous interference, *Virology* **30**:173.

Huang, A. S., and Wagner, R. R., 1966b, Comparative sedimentation coefficients of RNA extracted from plaque-forming and defective particles of vesicular stomatitis virus, *J. Mol. Biol.* **22**:381.

Huang, A. S., Greenwalt, J. W., and Wagner, R. R., 1966, Defective T particles of vesicular stomatitis virus. 1. Preparation, morphology, and some biologic properties, *Virology* **30**:161.

Huang, A. S., Little, S. P., Oldstone, M. B. A., and Rao, D., 1978, Defective interfering particles: Their effect on gene expression and replication of vesicular stomatitis virus, in: *Persistent Viruses* (J. Stevens, G. Todaro, and C. F. Fox, eds.), pp. 399–408, Academic Press, New York.

Huang, A., Wu, T., Yilma, T., and Lanman, G., 1986, Characterization of virulent isolates of vesicular stomatitis virus in relation to interference by defective particles, *Microb. Pathogenesis* **1**:206.

Igarashi, A., and Stollar, V., 1976, Failure of defective interfering particles of Sindbis virus produced in BHK or chicken cells to affect viral replication in *Aedes albopictus* cells, *J. Virol.* **19**:398.

Igarashi, A., Koo, R., and Stollar, V., 1977, Evolution and properties of *Aedes albopictus* cell cultures persistently infected with Sindbis virus, *Virology* **82**:69.

Isaac, C. L., and Keene, J. D., 1982, RNA polymerase-associated interactions near template-promoter sequences of defective interfering particles of VSV, *J. Virol.* **43**:241.

Jacobsen, S., and Pfau, C. J., 1980, Viral pathogenesis and resistance to defective interfering particles, *Nature (London)* **283**:311.

Janda, J. M., Davis, A. R., Nayak, D. P., and De, B. K., 1979, Diversity and generation of defective intefering influenza virus particles, *Virology* **95**:45.

Jennings, P. A., Finch, J. T., Winter, G., and Robertson, J. S., 1983, Does higher order structure of influenza virus ribonucleoprotein guide sequence rearrangements in influenza viral RNA? *Cell* **34**:619.

Johnson, L. D., and Lazzarini, R. A., 1977, Replication of viral RNA by a defective interfering vesicular stomatitis virus particle in the absence of helper virus, *Proc. Natl. Acad. Sci. U.S.A.* **74**:4387.

Johnson, L. D., and Lazzarini, R. A., 1978, Gene expression by a defective interfering particle of vesicular stomatitis virus, in: *Persistent Virus: ICN–UCLA Symposia on Molecular and Cellular Biology*, Vol. XI (J. G. Stevens, G. J. Todaro, and C. F. Fox, eds.), pp. 409–416, Academic Press, New York.

Johnson, L. D., Binder, M., and Lazzarini, R. A., 1979, A defective interfering vesicular stomatitis virus particle that directs the synthesis of functional proteins in the absence of helper virus, *Virology* **99**:203.

Johnston, M. D., 1981, The characteristics required for a Sendai virus preparation to induce high levels of interferon in human lymphoblastoid cells, *J. Gen. Virol.* **56**:175.

Johnston, R. E., Jovell, D. R., Brown, D. T., and Falkner, P., 1975, Interfering passages of Sindbis virus: Concomitant appearance of interference, morphological variants and truncated viral RNA, *J. Virol.* **16**:951.

Jones, C. L., and Holland, J. J., 1980, Requirements for DI particle prophylaxis against VSV infection *in vivo*, *J. Gen. Virol.* **49**:215.

Jones, C. L., Spindler, K. R., and Holland, J. J., 1980, Studies on tumorigenicity of cells persistently infected with vesicular stomatitis virus, *Virology* **103**:158.

Kääriänen, L., Pettersson, R. F., Keränen, S., Lehtovaara, P., Söderlund, H., and Ukkonen, P., 1981, Multiple structurally related defective-interfering RNAs formed during undiluted passages of Semliki Forest virus, *Virology* **113**:686.

Kailash, G. G., Gupta, K. C., and Kingsbury, D. W., 1983, Genomic and copy-back 3' terminiy in Sendai virus defective interfering RNA species, *J. Virol.* **45**:659.

Kang, C. Y., Weide, L. G., and Tischfield, J. A., 1981, Suppression of vesicular stomatitis virus defective interfering particle generation by a function(s) associated with human chromosome 16, *J. Virol.* **40**:946.

Kawai, A., and Matsumoto, S., 1977, Interfering and noninterfering defective particles generated by a rabies small plaque variant virus, *Virology* **76**:60.

Kawai, A., and Matsumoto, S., 1982, A sensitive bioassay for detecting defective interfering particles of rabies virus, *Virology* **122**:98.

Kawai, A., Matsumoto, S., and Tanabe, K., 1975, Characterization of rabies viruses recovered from persistently infected BHK cells, *Virology* **67**:520.

Keene, J. D., Rosenberg, M., and Lazzarini, R. A., 1977, Characterization of the 3' terminus of RNA isolated from vesicular stomatitis virus and from its defective interfering particles, *Proc. Natl. Acad. Sci. U.S.A.* **74**:1353.

Keene, J. D., Schubert, M., Lazzarini, R. A., and Rosenberg, M., 1978, Nucleotide sequence homology at the 3' termini of RNA from vesicular stomatitis virus and its defective interfering particles, *Proc. Natl. Acad. Sci. U.S.A.* **75**:3225.

Keene, J. D., Schubert, M., and Lazzarini, R. A., 1979, Terminal sequences of vesicular stomatitis virus RNA are both complementary and conserved, *J. Virol.* **32**:167.

Keene, J. D., Chien, I. M., and Lazzarini, R. A., 1981a, Vesicular stomatitis defective interfering particle containing a muted, internal leader RNA gene, *Proc. Natl. Acad. Sci. U.S.A.* **18**:2090.

Keene, J. D., Thornton, B. J., and Emerson, S. U., 1981b, Sequence-specific contacts between the RNA polymerase of vesicular stomatitis virus and the leader RNA gene, *Proc. Natl. Acad. Sci. U.S.A.* **78**:6191.

Kennedy, J. C., and Macdonald, R. D., 1982, Persistent infection with infectious pancreatic necrosis virus mediated by defective-interfering (DI) virus particles in a cell line showing strong interference but little DI replication, *J. Gen. Virol.* **58**:361.

Kennedy, S. I. T., Bruton, C. J., Weiss, B., and Schlesinger, S., 1976, Defective interfering passages of Sindbis virus: Nature of the defective virion RNA, *J. Virol.* **19**:1034.

Khan, S. R., and Lazzarini, R. A., 1977, The relationship between autointerference and the replication of a defective interfering particle, *Virology* **77**:189.

Kimura, Y., Ito, Y., Shimokata, K., Nishiyama, Y., Nagata, I., and Kitoh, J., 1975, Temperature sensitive virus derived from BHK cells persistently infected with HVJ, *J. Virol.* **15**:55.

King, A. M. Q., McCahon, D., Slade, W. R., and Newman, J. W. I., 1982, Recombination in RNA, *Cell* **29**:921.

King, C.-C., King, M. W., Garry, R. F., Wan, K. M.-M., Ulug, E. T., and Waite, M. R. F., 1979, Effect of incubation time on the generation of defective-interfering particles during undiluted serial passage of Sindbis virus in *Aedes albopictus* and chick cells, *Virology* **96**:229.

Kingsbury, D. W., and Portner, A., 1970, On the genesis of incomplete Sendai virions, *Virology* **42**:872.

Koch, E. M., Neubert, W. J., and Hofschneider, P. H., 1984, Lifelong persistence of paramyxovirus Sendai-6194 in C129 mice: Detection of a latent viral RNA by hybridization with a cloned genomic cDNA probe, *Virology* **136**:78.

Kolakofsky, D., 1976, Isolation and characterization of Sendai virus DI RNAs, *Cell* **8**:547.

Kolakofsky, D., 1979, Studies on the generation and amplification of Sendai virus DI genomes, *Virology* **93**:589.

Koschka-Dierich, C., Wermer, F. J., Bauer, I., and Fleckenstein, B., 1982, Structure of nonintegrated, circular *Herpesvirus saimiri*, and *Herpesvirus ateles* genomes in tumor cell lines in *in vitro* transformed cells, *J. Virol.* **44**:295.

Kowal, K. J., and Stollar, V., 1980, Differential sensitivity of infectious and defective-interfering particles of Sindbus virus to ultraviolet irradiation, *Virology* **103**:149.

Kurilla, M. G., and Keene, J. D., 1983, The leader RNA of vesicular stomatitis virus is bound by a cellular protein reactive with anti-La lupus antibodies, *Cell* **34**:837.

Kurilla, M. G., Cabradilla, C. D., Holloway, B. P., and Keene, J. D., 1984, Nucleotide sequence and host La protein interactions of rabies virus leader RNA, *J. Virol.* **50**:773.

Lazzarini, R. A., Weber, G. H., Johnson, L. D., and Stamminger, G. M., 1975, Covalently linked message and anti-message (genomic) RNA from a defective vesicular stomatitis virus particle, *J. Mol. Biol.* **97**:289.

Lazzarini, R. A., Keene, J. D., and Schubert, M., 1981, The origins of defective interfering particles of the negative-strand RNA viruses, *Cell* **26**:145.

Lehtovaara, P., Söderlund, H., Keränen, S., Pettersson, R., and Kääriäninen, L., 1981, 18S defective interfering RNA of Semliki Forest virus contains a triplicated repeat, *Proc. Natl. Acad. Sci. U.S.A.* **78**:5353.

Leppert, M., and Kolakofsky, D., 1980, Effect of defective interfering particles on plus and minus strand leader RNAs in vesicular stomatitis virus-infected cells, *J. Virol.* **35**:704.

Leppert, M., Kort, L., and Kolakofsky, D., 1977, Further characterization of Sendai virus DI RNAs: A model for their generation, *Cell* **12**:539.

Leppert, M., Rittenhouse, L., Perrault, J., Summers, D. F., and Kolakofsky, D., 1979, Plus and minus strand leader RNAs in negative-strand virus-infected cells, *Cell* **18**:735.

Levin, J. G., Ramseur, J. M., and Grimley, P. M., 1973, Host effect on arbovirus replication: Appearance of defective interfering particles in murine cells, *J. Virol.* **12**:1401.

Little, S. P., and Huang, A. S., 1977, Synthesis and distribution of VSV specific polypeptides in the absence of progeny production, *Virology* **81**:37.

Lundquist, R. E., Sullivan, M., and Maizel, J. V., 1979, Characterization of a new isolate of poliovirus defective interfering particles, *Cell* **18**:79.

Macdonald, R. D., and Yamamoto, T., 1978, Quantitative analysis of defective interfering particles in infectious pancreatic necrosis virus preparations, *Arch. Virol.* **57**:77.

Machamer, C. E., Stephenson, J. R., and Zweerink, H. J., 1981, Cells infected with a cell-associated subacute sclerosing panencephalitis virus do not express M protein, *Virology* **108**:515.

Marcus, P. I., and Sekellick, M. J., 1977, Defective interfering particles with covalently linked [±] RNA induce inteferon, *Nature (London)* **266**:815.

Marcus, P. I., Sekellick, M. J., Johnson, L. D., and Lazzarini, R. A., 1977, Cell killing by viruses. V. Transcribing defective interfering particles of VSV function as cell killing particles, *Virology* **82**:242.

McCarthy, M., Wolinsky, J. S., and Lazzarini, R. A., 1981, A persistent infection of Vero cells by egg-adapted mumps virus, *Virology* **114**:343.

McClure, M. A., Holland, J. J., and Perrault, J., 1980, Generation of defective interfering particles in picornaviruses, *Virology* **100**:408.

McGeoch, D. J., and Dolan, A., 1979, Sequence of 200 nucleotides at the 3' terminus of the genome RNA of vesicular stomatitis virus, *Cell* **6**:3199.

Meier, E., Harmison, G. G., Keene, J. D., and Schubert, M., 1984, Sites of copy choice replication involved in generation of vesicular stomatitis virus defective-interfering particle RNAs, *J. Virol.* **51**:515.

Mills, B. G., Singer, F. S., Weiner, L. P., Suffin, S. C., Stabile, E., and Holst, B. A., 1984, Evidence for both respiratory syncytial virus and measles virus antigens in the osteoclasts of patients with Paget's disease of bone, *Clin. Orthopaed.* **183**:303.

Mims, C. A. C., 1956, Rift Valley fever in mice. 4. Incomplete virus: Its production and properties, *Br. J. Exp. Pathol.* **37**:129.

Minato, N., Bloom, B. R., Jones, C., Holland, J. J., and Reid, L. M., 1979, Mechanism of rejection of virus persistently infected tumor cells by athymic nude mice, *J. Exp. Med.* **149**:1117.

Monroe, S. S., and Schlesinger, S., 1983, RNAs from two independently isolated defective interfering particles of Sindbis virus contain a cellular tRNA sequence at their 5' ends, *Proc. Natl. Acad. Sci. U.S.A.* **80**:3279.

Monroe, S. S., and Schlesinger, S., 1984, Common and distinct regions of defective-interfering RNAs of Sindbis virus, *J. Virol.* **49**:865.

Naeve, C. W., Kolakofsky, C. M., and Summers, D. F., 1980, Comparison of vesicular stomatitis virus intracellular and virion ribonucleoproteins, *J. Virol.* **33**:856.

Nayak, D. P., 1980, Defective interfering influenza viruses, *Annu. Rev. Microbiol.* **34**:619.

Nayak, D. P., and Sivasubramanian, N., 1983, The structure of the influenza defective interfering (DI) RNAs and their progenitor genes, in: *Genetics of Influenza Viruses* (P. Palese and D. W. Kingsbury, eds.), pp. 255–279, Springer-Verlag, Vienna.

Nayak, D. P., Sivasubramanian, N., Davis, A. R., Cortini, R., and Sung, J., 1982, Complete sequence analyses show that two defective interfering influenza viral RNAs contain a single internal deletion of polymerase genes, *Proc. Natl. Acad. Sci. U.S.A.* **79**:2216.

Nichol, S. T., O'Hara, P. J., Holland, J. J., and Perrault, J., 1984, Structure and origin of a novel class of defective interfering particle of vesicular stomatitis virus, *Nucleic Acids Res.* **12**:2775.

Nilsen, T. W., Wood, D. L., and Baglioni, C., 1981, Cross-linking of viral RNA by 4'-aminomethyl-4,5',8-trimethylpsoralen in HeLa cells infected with encephalomyocarditis virus and the *ts*G114 mutant of vesicular stomatitis virus, *Virology* **109**:82.

Nishiyama, Y., 1977, Studies of L cells persistently infected with VSV: Factors involved in the regulation of persistent infection, *J. Gen. Virol.* **35**:265.

Nishiyama, Y., Ito, Y., Shimokata, K., Kimura, Y., and Nagata, J., 1976, Relationship between establishment of persistent infection of hemagglutinating virus of Japan and the properties of the virus, *J. Gen. Virol.* **32**:73.

Nishiyama, Y., Ito, Y., and Shimokata, K., 1978, Properties of viruses selected during persistent infection of L cells with VSV, *J. Gen. Virol.* **40**:481.

Nomoto, A., Jacobson, A., Lee, Y. F., Dunn, J., and Wimmer, E., 1979, Defective interfering particles of poliovirus: Mapping of the deletion and evidence that the deletion in the genome of DI (1), (2), and (3) is located in the same region, *J. Mol. Biol.* **128**:179.

Nonoyama, M., Watanabe, Y., and Graham, A. F., 1970, Defective virions of reovirus, *J. Virol.* **6**:693.

Norkin, L. C., and Tirrell, S. M., 1982, Emergence of simian virus 40 variants during serial passage of plaque isolates, *J. Virol.* **42**:730.

Nayak, D., Chambers, T., and Akkina, R., 1985, Defective interfering (DI) RNAs of influenza viruses: Origin, structure, expression, and interference, *Curr. Topics Microbiol. Immunol.* **114**:103.

O'Hara, P. J., Nichol, S. T., Horodyski, F. M., and Holland, J. J., 1984a, Vesicular stomatitis virus defective interfering particles can contain extensive genomic sequence rearrangements and base substitutions, *Cell* **36**:915.

O'Hara, P. J., Horodyski, F. M., Nichol, S. T., and Holland, J. J., 1984b, Vesicular stomatitis virus mutants resistant to defective interfering particles accumulate stable 5'-terminal and fewer 3'-terminal mutations in a stepwise manner, *J. Virol.* **49**:793.

O'Neill, F. J., Maryon, E. B., and Carrol, D., 1982, Isolation and charcterization of defective SV$_{40}$ genomes which complement for infectivity, *J. Virol.* **43**:18.

Palma, E. L., and Huang, A. S., 1974, Cyclic production of vesicular stomatitis virus caused by defective interfering particles, *J. Infect. Dis.* **126**:402.

Patton, J. T., Davis, N. L., and Wertz, G. W., 1983, Cell-free synthesis and assembly of vesicular stomatitis virus nucleocapsids, *J. Virol.* **45**:155.

Patton, J., Davis, N., and Wertz, G., 1984, N protein alone satisfies the requirement for protein synthesis during RNA replication of vesicular stomatitis virus, *J. Virol.* **49**:303.

Peluso, R. W., and Moyer, S. A., 1983, Initiation and replication of vesicular stomatitis virus genome RNA in a cell-free system, *Proc. Natl. Acad. Sci. U.S.A.* **80**:3198.

Peralta, L. M., Bruns, M., and Lehmann-Grube, F., 1981, Biochemical composition of lymphocytic choriomeningitis virus interfering particles, *J. Gen. Virol.* **55**:475.

Perrault, J., 1976, Cross-linked double-stranded RNA from a defective vesicular stomatitis virus particle, *Virology* **70**:360.

Perrault, J., 1981, Origin and replication of defective interfering particles, *Curr. Top. Microbiol. Immunol.* **93**:151.

Perrault, J., and Holland, J. J., 1972a, Variability of vesicular stomatitis virus autointerference with different host cells and virus serotypes, *Virology* **50**:148.

Perrault, J., and Holland, J. J., 1972b, Absence of transcriptase activity or transcription-inhibiting activity in defective interfering particles of vesicular stomatitis virus, *Virology* **50**:159.

Perrault, J., and Leavitt, R. W., 1977, Characterization of snap-back RNAs in vesicular stomatitis defective interfering virus particles, *J. Gen. Virol.* **38**:21.

Perrault, J., and McLear, P. W., 1984, ATP dependence of vesicular stomatitis virus transcription initiation and modulation by mutation in the nucleocapsid protein, *J. Virol.* **51**:635.

Perrault, J., and Semler, B. L., 1979, Internal genome deletions in two distinct classes of defective interfering particles of vesicular stomatitis virus, *Proc. Natl. Acad. Sci. U.S.A.* **76**:6191.

Perrault, J., Semler, B. L., Leavitt, R. W., and Holland, J. J., 1978, Inverted complementary sequences in defective interfering particle RNAs of vesicular stomatitis virus and their possible role in autointerference, in: *Negative Strand Viruses and the Host Cell* (B. W. J. Mahy and R. D. Barry, eds.), pp. 527–538, Academic Press, New York.

Perrault, J., Lane, J. L., and McClure, M. A., 1980, A variant VSV generates defective interfering particles with replicase-like activity *in vitro*, in: *Animal Virus Genetics: ICN–UCLA Symposia on Molecular and Cellular Biology*, Vol. XVIII (B. Fields, R. Jaenisch, and C. F. Fox, eds.), pp. 379–390, Academic Press, New York.

Perrault, J., Lane, J. L., and McClure, M. A., 1981, *In vitro* transcription alterations in a vesicular stomatitis virus variant, in: *The Replication of Negative Strand Viruses* (D. H. L. Bishop and R. W. Compans, eds.), pp. 829–836, Elsevier/North-Holland, New York.

Perrault, J., Clinton, G. M., and McClure, M. A., 1983, RNA template of vesicular stomatitis virus regulates transcription and replication functions, *Cell* **35**:175.

Pettersson, R., 1981, 5'-Terminal nucleotide sequence of Semliki Forest virus 18S defective interfering RNA in heterogeneous and different from the genomic 42S RNA, *Proc. Natl. Acad. Sci. U.S.A.* **78**:115.

Phillips, B. A., Lundquist, R. E., and Maizel, J. V., Jr., 1980, Absence of subviral particles and assembly activity in HeLa cells infected with defective-interfering (DI) particles of poliovirus, *Virology* **100**:116.

Popescu, M., and Lehmann-Grube, F., 1977, Defective interfering particles in mice infected with lymphocytic choriomeningitis virus, *Virology* **77**:78.

Popescu, M., Schaefer, H., and Lehmann-Grube, F., 1976, Homologous interference of lymphocytic choriomeningitis virus: Detection and measurement of interference foci-forming units, *J. Virol.* **20**:1.

Prevec, L., and Kang, C. Y., 1970, Homotypic and heterotypic interference by defective particles of vesicular stomatitis virus, *Nature (London)* **228**:25.

Pringle, C. R., 1970, Genetic characteristics of conditional lethal mutants of vesicular stomatitis virus induced by 5-fluorouracil, 5-azacytidine and ethylmethane sulfonate, *J. Virol.* **5**:559.

Pringle, C. R., and Wunner, W. H., 1975, A comparative study of the structure and function of the vesicular stomatitis virus genome, in: *Negative Strand Viruses* (B. W. J. Mahy and R. D. Barry, eds.), pp. 707–723, Academic Press, London.

Pringle, C. R., Devine, V., Wilkie, M., Preston, C. M., Dolan, A., and McGeogh, D. J., 1981, Enhanced mutability associated with a temperature-sensitive mutant of vesicular stomatitis virus, *J. Virol.* **39**:377.

Rabinowitz, S. G., and Huprikar, J., 1979, The influence of defective interfering particles of the PR-8 strain of influenza A virus on the pathogenesis of pulmonary infection of mice, *J. Infect. Dis.* **140**:305.

Rabinowitz, S. G., Dal Canto, M. C., and Johnson, T. C., 1977, Infection of the central nervous system produced by mixtures of defective-interfering particles and wild type vesicular stomatitis virus in mice, *J. Infect. Dis.* **136**:59.

Radlof, R. J., and Young, S. A., 1983, Defective interfering particles of encephalomyocarditis virus, *J. Gen. Virol.* **64**:1637.

Ramseur, J. M., and Friedman, R. M., 1977, Prolonged infection of interferon-treated cells by VSV: Possible role of temperature sensitive mutants and interferon, *J. Gen. Virol.* **37**:523.

Rao, D. D., and Huang, A. S., 1979, Synthesis of a small RNA in cells coinfected by standard and defective interfering particles of vesicular stomatitis virus, *Proc. Natl. Acad. Sci. U.S.A.* **76**:3742.

Rao, D. D., and Huang, A. S., 1980, RNA synthesis of vesicular stomatitis virus. X. Transcription and replication by defective interfering particles, *J. Virol.* **36**:756.

Reichmann, M. E., and Schnitzlein, W. M., 1979, Defective interfering particles of rhabdoviruses, *Curr. Top. Microbiol. Immunol.* **86**:124.

Reichmann, M. E., and Schnitzlein, W. M., 1980, Rhabdovirus defective particles: Origin and genome assignments, in: *Rhabdoviruses*, Vol. II (D. H. L. Bishop, ed.), pp. 189–200, CRC Press, Boca Raton, Florida.

Reichmann, M. E., Schnitzlein, W. M., Bishop, D. H. L., Lazzarini, R. A., Beatrice, S. T., and Wagner, R. R., 1978, Classification of the New Jersey serotype of vesicular stomatitis virus into two subtypes, *J. Virol.* **25**:446.

Reichmann, M. E., Bishop, D. H. L., Brown, F., Crick, J., Holland, J. J., Kang, C.-Y., Lazzarini, R., Moyer, S., Perrault, J., Prevec, L., Pringle, C. R., Wagner, R. R., Youngner, J. S., and Huang, A. S., 1980, Proposal for a uniform nomenclature for defective interfering viruses of vesicular stomatitis virus, *J. Virol.* **34**:792.

Reid, L. M., Jones, C. L., and Holland, J., 1979, Virus carrier state suppresses tumorigenicity of tumor cells in athymic (nude) mice, *J. Gen. Virol.* **42**:609.

Reid, L. M., Minato, N., Gresser, J., Holland, J. J., Kadish, A., and Bloom, B. R., 1981, Influence of anti-mouse interferon serum on the growth and metastasis of tumor cells persistently infected with virus and of human prostatic tumors in athymic nude mice, *Proc. Natl. Acad. Sci. U.S.A.* **78**:1171.

Rima, B. K., and Martin, S. J., 1979, Effect of undiluted passage on the polypeptides of measles viruses, *J. Gen. Virol.* **44**:135.

Rima, B. K., Davidson, W. B., and Martin, S. J., 1977, The role of defective interfering particles in persistent infection of Vero cells by measles, *J. Gen. Virol.* **35**:89.

Rose, J. K., and Gallione, C. J., 1981, Nucleotide sequence of the mRNA's encoding the vesicular stomatitis virus G and M proteins determined from cDNA clones containing the complete coding regions, *J. Virol.* **39**:519.

Roux, L., and Beffy, P., 1984, Cell division does not affect Sendai virus genome replication in persistently infected BHK cells, *J. Gen. Virol.* **65**:2055.

Roux, L., and Holland, J. J., 1979, Role of defective interfering particles of Sendai virus in persistent infections, *Virology* **93**:91.

Roux, L., and Holland, J. J., 1980, Viral genome synthesis in BHK21 cells persistently infected with Sendai virus, *Virology* **100**:53.

Roux, L., and Waldvogel, F. A., 1981, Establishment of Sendai virus persistent infection: Biochemical analysis of the early phase of a standard plus defective interfering virus infection of BHK cells, *Virology* **112**:400.

Roux, L., and Waldvogel, F. A., 1982, Instability of viral M protein in BHK-21 cells persistently infected with Sendai virus, *Cell* **28**:293.

Roux, L., and Waldvogel, F. A., 1983, Defective interfering particles of Sendai virus modulate HN expression at the surface of infected BHK cells, *Virology* **130**:91.

Roux, L., Beffy, P., and Portner, A., 1984, Restriction of cell surface expression of Sendai virus hemagglutinin-neuraminidase glycoprotein correlates with its higher instability in persistently and standard plus defective interfering virus infected BHK-21 cells, *Virology* **138**:118.

Roux, L., Beffy, P., and Portner, A., 1985, Three variations in the cell surface expression of the hemagglutinin-neuraminidase glycoprotein of Sendai virus, *J. Gen. Virol.* **66**:987.

Rowlands, D. J., 1979, Sequences of VSV RNA in the region coding for leader RNA, N mRNA and their function, *Proc. Natl. Acad. Sci. U.S.A.* **76**:4793.

Rowlands, D. J., Grabau, E., Spindler, K., Jones, C., Semler, B., and Holland, J. J., 1980, Virus protein changes and RNA termini alterations evolving during persistent infection, *Cell* **19**:871.

Roy, P., Repik, P., Hefti, E., and Bishop, D. H. L., 1973, Complementary RNA species isolated from vesicular stomatitis (HR) strain defective virions, *J. Virol.* **11**:915.

Schlesinger, S., 1987, The generation and amplification of defective interfering RNAs, in: *RNA Genetics* (E. Domingo, P. Ahlquist, and J. Holland, eds.), CRC Press, Boca Raton, Florida (in press).

Schmaljohn, C., and Blair, C. D., 1977, Persistent infection of cultured mammalian cells by Japanese encephalitis virus, *J. Virol.* **24**:580.

Schnitzlein, W. M., and Reichmann, M. E., 1976, The size and cistronic origin of defective vesicular stomatitis virus particle RNAs in relation to homotypic and heterotypic interference, *J. Mol. Biol.* **101**:307.

Schnitzlein, W. M., and Reichmann, M. E., 1977a, Interference and RNA homologies of New Jersey serotype isolates of vesicular stomatitis virus and their defective particles, *Virology* **77**:490.

Schnitzlein, W. M., and Reichmann, M. E., 1977b, A possible effect of viral proteins on the specificity of interference by defective vesicular stomatitis virus particles, *Virology* **80**:275.

Schubert, M., and Lazzarini, R. A., 1981, Studies on the structure and origin of a snapback DI particle of vesicular stomatitis virus, *J. Virol.* **37**:661.

Schubert, M., Keene, J. D., Lazzarini, R. A., and Emerson, S. U., 1978, The complete sequence of a unique RNA species synthesized by a DI particle of VSV, *Cell* **15**:103.

Schubert, M., Keene, J. D., and Lazzarini, R. A., 1979, A specific internal RNA polymerase recognition site of VSV RNA is involved in the generation of DI particles, *Cell* **18**:749.

Schubert, M., Keene, J. D., Herman, R. C., and Lazzarini, R. A., 1980, The site on the VSV genome specifying polyadenylation and the end of the L gene mRNA, *J. Virol.* **34**:550.

Schubert, M., Harmison, G. G., and Meier, E., 1984, Primary structure of the vesicular stomatitis virus polymerase (L) gene: Evidence for a high frequency of mutations, *J. Virol.* **51**:505.

Schuerch, A. R., Matsuhisa, T., and Joklik, W. K., 1974, Temperature-sensitive mutants of reovirus. 6. Mutant *ts*447 and *ts*556 particles that lack either one or two genome segments, *Intervirology* **3**:36.

Sekellick, M. J., and Marcus, P. I., 1978, Persistent infection. I. Interferon-inducing defective-interfering particles as mediators of cell sparing: Possible role in persistent infection by vesicular stomatitis virus, *Virology* **85**:175.

Sekellick, M. J., and Marcus, P. I., 1980, Viral interference by defective particles of vesicular stomatitis virus measured in individual cells, *Virology* **104**:247.

Sekellick, M. J., and Marcus, P. I., 1982, Inteferon induction by viruses. VIII. Vesicular stomatitis virus: ($\pm$) DI-011 particles induce interferon in the absence of standard virions, *Virology* **117**:280.

Semler, B. L., Perrault, J., Abelson, J., and Holland, J. J., 1978, Sequence of a RNA templated by the 3'-OH RNA terminus of defective interfering particles of vesicular stomatitis virus, *Proc. Natl. Acad. Sci. U.S.A.* **75**:4704.

Semler, B. L., Perrault, J., and Holland, J. J., 1979, The nucleotide sequence of the 5' terminus of vesicular stomatitis virus RNA, *Nucleic Acids Res.* **6**:3923.

Smith, G. L., and Hay, A. J., 1982, Replication of the influenza virus genome, *Virology* **118**:96.

Sobrino, F., Dovila, M., Ortin, J., and Domingo, E., 1983, Multiple genetic variants arise in the course of replication of foot-and-mouth disease virus in cell culture, *Virology* **128**:310.

Soria, M., Little, S. P., and Huang, A. S., 1974, Characterization of vesicular stomatitis virus nucleocapsids. I. Complementary 40S RNA molecules in nucleocapsids, *Virology* **61**:270.

Spandidos, D. A., and Graham, A. F., 1976, Generation of defective virus after infection of newborn rats with reovirus, *J. Virol.* **20**:234.

Spindler, K. R., and Holland, J. J., 1982, A mutant standard virus isolated from vesicular stomatitis virus persistent infection interferes specifically with wild type virus replication, *J. Gen. Virol.* **62**:363.

Spindler, K. R., Horodyski, F. M., and Holland, J. J., 1982, High multiplicities of infection favor rapid and random evolution of vesicular stomatitis virus, *Virology* **119**:98.

Stamminger, G. M., and Lazzarini, R. A., 1977, RNA synthesis in standard and autointerfered vesicular stomatitis virus infections, *Virology* **77**:202.

Stampfer, M., Baltimore, D., and Huang, A. S., 1969, Ribonucleic acid synthesis of vesicular stomatitis virus. I. Species of ribonucleic acid found in Chinese hamster ovary cells infected with plaque-forming and defective particles, *J. Virol.* **4**:154.

Stampfer, M., Baltimore, D., and Huang, A. S., 1971, Absence of interference during high-multiplicity infection by clonally purified vesicular stomatitis virus, *J. Virol.* **7**:409.

Stanners, C. P., Francoeur, A. M., and Larn, T., 1977, Analysis of VSV mutant with attenuated cytopathogenicity: Mutation in viral function P for inhibition of protein synthesis, *Cell* **11**:273.

Stark, C., and Kennedy, S. I. T., 1978, The generation and propagation of defective interfering particles of Semliki Forest virus in different cell types, *Virology* **89**:285.

Steacie, A. D., and Eaton, B. T., 1984, Properties of defective interfering particles of Sindbis virus generated in vertebrate and mosquito cells, *J. Gen. Virol.* **65**:333.

Steinhauer, D. S., and Holland, J. J., 1986, Direct method for quantitation of extreme polymerase error frequencies at selected single base sites in viral RNA, *J. Virol.* **57**:219.

Stollar, V., 1979, Defective interfering particles of togaviruses, *Curr. Top. Microbiol. Immunol.* **86**:35.

Stollar, V., 1980, Defective interfering alphaviruses, in: *The Togaviruses: Biology, Structure, Replication* (R. W. Schlesinger, ed.), pp. 427–455, Academic Press, New York.

Storey, D. G., and Kang, C. Y., 1985, Vesicular stomatitis virus-infected cells fail when the intracellular pool of functional M protein is reduced in the presence of G protein, *J. Virol.* **53**:374.

Ter Meulen, V., Stephenson, J. R., and Kieth, H. W., 1983, Subacute sclerosing panencephalitis, in: *Comprehensive Virology*, Vol. 18 (H. Fraenkel-Conrat and R. R. Wagner, eds.), pp. 105–159, Plenum Press, New York.

Testa, D., and Banerjee, A. K., 1979, Initiation of RNA synthesis *in vitro* by vesicular stomatitis virus: Role of ATP, *J. Biol. Chem.* **254**:2053.

Testa, D., Chanda, P. K., and Banerjee, A. K., 1980, In vitro synthesis of the full-length complement of the negative-strand genome RNA of vesicular stomatitis virus, *Proc. Natl. Acad. Sci. U.S.A.* **77**:294.

Thomas, J. R., and Wagner, R. R., 1982, Evidence that vesicular stomatitis virus produces double-stranded RNA that inhibits protein synthesis in a reticulocyte lysate, *J. Virol.* **44**:189.

Treuhaft, M. W., 1983, A colorimetric assay for quantification of defective interfering particles of respiratory syncytial virus, *J. Gen. Virol.* **64**:1301.

Treuhaft, M. W., and Beem, M. O., 1982, Defective interfering particles of respiratory syncytial virus, *Infect. Immun.* **37**:439.

Tsiang, M., Monroe, S., and Schlesinger, S., 1985, Studies of defective interfering RNAs of Sindbis virus with and without tRNA^asp sequences at their 5' termini, *J. Virol.* **54**:38.

Vandepol, S. B., and Holland, J. J., 1986, Evolution of vesicular stomatitis virus in athymic nude mice: Mutations associated with natural killer selections, *J. Gen. Virol.* **67**:441.

Van der Zeijst, B. A. M., Noyes, B. E., Mirault, M., Parker, B., Osterhaus, A. D. M. E., Swyayd, E. A., Bleumink, N., Horzinek, M. C., and Stark, G. R., 1983a, Persistent infection of some standard cell lines by lymphocytic choriomeningitis virus: Transmission of infection by an intracellular agent, *J. Virol.* **48**:249.

Van der Zeijst, B. A. M., Bleumink, N., Crawford, L. V., Swyryd, E. A., and Stark, G. R., 1983b, Viral proteins and RNAs in BHK cells persistently infected by lymphocytic choriomeningitis virus, *J. Virol.* **48**:262.

Verani, P., Nicoletti, L., and Marchi, A., 1984, Establishment and maintenance of persistent infection by the *phlebovirus* Toscana in Vero cells, *J. Gen. Virol.* **65**:367.

Villarreal, L. P., and Holland, J. J., 1976, RNA synthesis in BHK$_{21}$ cells persistently infected with vesicular stomatitis virus and rabies virus, *J. Gen. Virol.* **33**:213.

Von Magnus, P., 1951, Propagation of the PR 8 strain of influenza virus in chick embryos. 3. Properties of the incomplete virus produced in serial passages of undiluted virus, *Acta Pathol. Microbiol. Scand.* **29**:156.

Von Magnus, P., 1954, Incomplete form of influenza virus, *Adv. Virus Res.* **2**:59.

Wagner, R. R., Levy, A., Snyder, R., Ratcliff, G., and Hyatt, D., 1963, Biologic properties of two plaque variants of vesicular stomatitis virus (Indiana serotype), *J. Immunol.* **91**:112.

Webster, R., Kawaoka, Y., and Bean, W., Jr., 1986, Molecular Changes in A/Chicken/Pennsylvania/83 (H5N2) influenza virus associated with acquisition of virulence, *Virology* **149**:165.

Weiss, B., and Schlesinger, S., 1981, Defective interfering particles of Sindbis virus do not interfere with the homologous virus obtained from persistently infected BHK cells but do interfere with Semliki Forest virus, *J. Virol.* **37**:840.

Weiss, B., Goran, D., Cancedda, R., and Schlesinger, S., 1974, Defective interfering passages of Sindbis virus: Nature of the intracellular defective viral RNA, *J. Virol.* **14**:1189.

Weiss, B., Rosenthal, R., and Schlesinger, S., 1980, Establishment and maintenance of persistent infection by Sindbis virus in BHK cells, *J. Virol.* **33**:463.

Welsh, R. M., and Oldstone, M. B. A., 1977, Inhibition of immunologic injury of cultured cells infected with lymphocytic choriomeningitis virus: Role of defective interfering virus in regulating viral antigenic expression, *J. Exp. Med.* **145**:1449.

Welsh, R. M., O'Connell, C. M., and Pfau, C. J., 1972, Properties of defective lymphocytic choriomeningitis virus, *J. Gen. Virol.* **17**:355.

Wertz, G. W., 1983, Replication of vesicular stomatitis virus defective interfering particle RNA *in vitro:* Transition from synthesis of defective interfering leader RNA to synthesis of full length defective interfering RNA, *J. Virol.* **46:**513.

Wiktor, T. J., and Clark, H. F., 1972, Chronic rabies virus infection of cell cultures, *Infect. Immun.* **6:**988.

Wilusz, J., Youngner, J. S., and Keene, J. D., 1985, Base mutations in the terminal noncoding regions of the genome of vesicular stomatitis virus isolated from persistent infections of L cells, *Virology* **140:**249.

Winship, T. R., and Thacore, H. R., 1979, Inhibition of vesicular stomatitis virus-defective interfering particle synthesis by Shope fibroma virus, *Virology* **93:**515.

Wong, P. K. Y., Holloway, A. F., and Cormack, D. V., 1971, A search for recombination between temperature-sensitive mutants of vesicular stomatitis virus, *J. Gen. Virol.* **13:**477.

Yang, F., and Lazzarini, R. A., 1983, Analysis of the recombination event generating a vesicular stomatitis virus deletion defective interfering particle, *J. Virol.* **45:**766.

Yoshida, T., Hamaguchi, M., Naruse, H., and Nagai, Y., 1982, Persistent infection by a temperature-sensitive mutant isolated from a Sendai virus (HVJ) carrier culture: Its initiation and maintenance without aid of defective interfering particles, *Virology* **120:**329.

Youngner, J. S., and Preble, O. T., 1980, Viral persistence: Evolution of viral populations, *Compr. Virol.* **16:**73.

Youngner, J. S., and Quagliana, D. O., 1976, Temperature sensitive mutants of vesicular stomatitis virus are conditionally defective particles that interfere with and are rescued by wild type virus, *J. Virol.* **19:**102.

Youngner, J., Dubovi, E. J., Quagliana, D. O., Kelly, M., and Preble, O. T., 1976, Role of temperature sensitive mutants in persistent infections initiated with vesicular stomatitis virus, *J. Virol.* **19:**90.

Youngner, J. S., Preble, O. T., and Jones, E. V., 1978, Persistent infection of L cells with vesicular stomatitis virus: Evolution of virus populations, *J. Virol.* **28:**6.

Youngner, J. S., Jones, E. V., Kelley, M., and Frielle, D. W., 1981, Generation and amplification of temperature-sensitive mutants during serial undiluted passages of vesicular stomatitis virus, *Virology* **108:**87.

CHAPTER 9

Rabies Viruses—Pathogenesis and Immunity

WILLIAM H. WUNNER

I. INTRODUCTION

The outcome of rabies virus infection in all warm-blooded hosts, including humans, is generally fatal. For this reason, rabies virus is by far the most important rhabdovirus (member of the genus *Lyssavirus*) for study with regard to understanding its mode of pathogenesis and immunity. The long history of research in rabies pathogenesis beginning early in the 19th century is clear testimony of the importance continually placed on the need to better understand the nature of the disease caused by rabies virus, which in the recorded history of rabies was possibly recognized more than 4000 years ago (Steele, 1975; Miller and Nathanson, 1977), and to determine its course of development in hope of finding a means of prevention.

A. A Historical Perspective

Rabies virus produces an acute central nervous system (CNS) infection that is clinically manifested by alternating symptoms of depression and agitation leading to extreme types of violent behavior. Long after the Greeks and Romans (ca. 500 B.C.–100 A.D.) recognized that signs of aggressiveness typifying clinical rabies were hallmarks of an infection that could be transmitted by the inoculation of saliva of a "mad" dog by biting, Zinke, in 1804, performed the first experiment that demonstrated,

WILLIAM H. WUNNER • The Wistar Institute of Anatomy and Biology, Philadelphia, Pennsylvania 19104.

by painting saliva from a rabid dog into incisions he had made on the foreleg of a healthy dog, that rabies could be transmitted in saliva (Steele, 1975). The experiments of Zinke (which had to be repeated many times) and others that followed, including those of Galtier (1881), later convinced authorities that rabies was an infectious disease associated primarily with dogs and that dog-control regulations such as quarantine, elimination of stray dogs, and muzzling might lead to its elimination (Steele, 1975; Brown and Crick, 1979). At the time Galtier (1881) was conducting his experiments with dogs and rabbits, Pasteur et al. (1881, 1882) were using infected saliva to study transmission of rabies from humans to animals and animals to animals. They published their observations on the role of the CNS in rabies, concluding that "the central nervous system and especially the bulb which joins the spinal cord to the brain are particularly concerned and active in the development of the disease." Steele (1975), in his recounting of the history of experimental pathogenesis of rabies, points out that Pasteur also reported "success in producing rabies by the injection of central nervous system material and spinal fluid" from infected animals, indicating that Pasteur at this time was aware that the virus was not solely in the saliva. Since Pasteur was bothered by the experimental difficulties of the long incubation period following inoculation of infected saliva, Steele relates, "he soon found that by injecting brain material from rabid animals directly into the brains of dogs, the incubation period was shortened to 1 to 2 weeks, or at most 3 weeks. This was an important advance in experimental rabies studies."

It is interesting that the findings from experiments with rabbits, guinea pigs, and dogs performed almost a century ago by DiVestea and Zagari, Fermi, Babes, Roux, Bardach, Cantani, Bartarelli, and others (cited by Baer, 1975; Steele, 1975) established the basic tenet of rabies pathogenesis. They demonstrated that virus inoculated into peripheral nerves produced rabies and that if the sciatic nerve was cut or cauterized after peripheral injection, the virus was unable to spread to different tissues of the body, including the salivary gland. Moreover, they had established, as others who followed had concluded (Baer, 1975), that virus was not only transported to the brain along nerves, but also reached the salivary gland from the brain in rabid dogs by traveling along nerves.

The study of pathological changes in tissues of animals with rabies began at the turn of the century as well. In 1903, Negri (1903) reported on his investigations of nerve cells and his finding of the inclusion bodies (which bear his name) in large numbers in the neurons of rabid animals. Regarding "Negri bodies" as typical of the infection that produces rabies, naturally or experimentally, Negri is credited with establishing the first diagnostic test for rabies. Also at this time, Remlinger (1903) was looking for the etiological agent of rabies as he described the production of rabies in rabbits by inoculating filtrates of brain material from rabid dogs and rabbits. Remlinger did not realize that he had established the viral etiology of rabies. He argued, rather, that the etiological agent of rabies was not a microorganism or a parasite as Negri proposed from his observations. Remlinger thought that he may have transmitted a nonviral toxic substance.

Protective immunity to rabies infection was also a primary concern at the end of the 19th century, and Pasteur considered the possibility that persons exposed to rabies could be immunized using attenuated virus. In the early 1880s, Pasteur *et al.* (1881) found that if one passed rabies from dog to monkey and then from monkey to monkey, the virulence of the virus fell off at each passage. What was particularly striking to Pasteur was that if the virus was used to inoculate other animals, it not only remained attenuated, but also protected those animals from developing rabies on subsequent challenge with virulent virus. Although the method originally was not very reliable and its practical application was doubtful because the degree of attenuation could not be controlled, Pasteur proceeded to try to make dogs refractory to rabies by subcutaneously injecting suspensions of fragments of rabies-virus-infected rabbit spinal cords, beginning with attenuated preparations and using increasingly more virulent material until he finally inoculated more than 50 dogs with a cord suspension that was fully virulent. All the dogs resisted rabies infection when injected intracerebrally with the virulent material. Thus, the first method of prophylaxis was begun, which instigated a long succession of treatments for rabies (reviewed by Wiktor, 1985a) and experimental studies to understand the mechanisms of the host's immune defense against rabies infection.

Two further important developments that were heralded as technological breakthroughs for the study of rabies pathogenesis and immunity to infection should be mentioned; both have to do with the propagation of rabies virus in more manageable systems within the laboratory. One of these developments was the successful propagation of virus in developing chick embryos in 1940 (Dawson, 1939; Bernkopf and Kligler, 1940), which led to the production of a chick-embryo-origin vaccine (Koprowski and Cox, 1948). Concurrently, vaccines were also being derived from rabies-virus-infected newborn and adult mouse brains (Habel, 1940). These vaccines of tissue origin were particularly useful in studies of the immune response to vaccines in general. Growth of virus in these nervous tissues also encouraged more intensive and extensive laboratory studies of rabies virus and its biological, immunological, and pathogenic properties in the ensuing years. The other significant development was the growth of rabies virus in cell culture, a development that occurred over the next 20 years (reviewed by Wiktor and Clark, 1975; Clark, 1979) and provided a more efficient method for the propagation of rabies virus. Cell cultures for growing virus not only led to production of more potent and economical vaccines capable of protecting man and animals against infection, but also opened up all the possible uses of modern biotechnology and immunology that can be applied advantageously today. These developments were crucial to studies of virus morphology, structure, and replication that are mentioned in this chapter in the context of rabies virus pathogenesis and immunity.

Rabies virus, the etiological agent of rabies disease, was first described from visualizations in the electron microscope in 1962 using crude preparations obtained from infected hamster kidney-cell-culture and negative-staining procedures (Almeida *et al.*, 1962) as well as thin sections of animal brain tissue infected with a street rabies virus (Matsumoto, 1962, 1963). (The term "street" is commonly used to refer to viruses that have not been serially passaged in animals or tissue culture.) A year later, the characteristic morphological and structural features of rabies virus were described and related to those of vesicular stomatitis virus (Davies *et al.*, 1963; Atanasiu *et al.*, 1963a; Pinteric *et al.*, 1963). The morphological aspects of rabies virus and the characteristic intraneuronal matrices seen within the CNS have been noted by practically all investigators studying the pathogenesis of rabies virus. Studies were soon focused on precisely how rabies virus moves from the inoculation site to peripheral nerves and the CNS, in the hope of defining the course of infection and thereby establishing new strategies to successfully prevent virus from reaching those parts of the body where it irreversibly causes death.

B. Perspective for the Future

The global problem of rabies control and eventual eradication is still a major concern for research investigators, vaccine producers, and public health officials alike, despite eminent efforts (and many have been truly effective) to understand rabies as a disease, to treat infections, and to implement proper controls for the prevention of virus spread. Most of the enzootic rabies in the world today is in the Third World countries, where canine rabies is largely uncontrolled and human rabies remains a sizeable clinical problem. But rabies is also enzootic in several species and epizootically active in other areas of the world as well, including Europe and the Americas (Kaplan, 1983). Therefore, preventive treatment of rabies virus infection in animals, both wild and domestic, as well as postexposure treatment in humans throughout the world continues to be a major priority. To meet the challenges of the next few decades, new approaches must be taken not only to increase public awareness of the rabies problem but also to control the disease in wild animals. It is also hoped that as a result of further investigations, many of the remaining unresolved aspects of rabies pathogenesis and immunity will soon be understood. Certainly, application of the rapidly expanding modern techniques in molecular biology and immunology will continue to provide new approaches that could add substantial depth to the study of rabies and help to turn exciting new ideas and fresh insights into more practical solutions. The intention in this chapter is to review the salient features of rabies pathogenesis and immunity that have been experimentally defined and to emphasize for the student and experienced investigator alike that future efforts in the study of rabies will require strong multidisciplinary connections in both laboratory and clinical investigations.

II. PATHOGENESIS

A. Early Rabies Virus Infection

1. Preneural Rabies Pathogenesis

The incubation period of rabies is extremely variable and often quite prolonged, with the result that some cases of human rabies are either unsuspectingly misdiagnosed or not linked to a transmission event at the time the individual presents clinical signs of the disease. The incubation period in humans is generally between 15 days and 1 year (R. T. Johnson, 1982). In dogs, foxes, skunks, raccoons, bats, and rodents, the incubation period is also extremely varied, and although variation is often explained in terms that merely reflect on the relative susceptibility of these species (Tierkel, 1975), there are a number of factors (including host species and virus substrain) that conceivably contribute to this variability in the early stages of rabies infection. One such factor is the manner in which virus-laden saliva is introduced into bite or scratch wounds inflicted by rabid animals, the most common route of entry. Virus deposited in subcutaneous and muscle tissues of the extremities (furthest away from the head) is more likely to result in longer incubation periods unless the introduction of rabid saliva is deep into torn muscle and virus is given immediate access to endings of motor- and sensory-nerve fibers (Murphy, 1977). Experimental studies of the variability of rabies pathogenesis (production of disease) in different mammalian vector hosts in relation to the area of the body exposed suggest that wounds on the hindlegs and feet are generally less lethal than those directed to the head and neck (Baer, 1975). Oral and intranasal routes of virus entry have also been described both in animal colonies (Winkler *et al.*, 1972) and in laboratory accidents involving humans (Winkler *et al.*, 1973; *Morbidity and Mortality Weekly Report*, 1977) and have been particularly implicated in the transmission of virus among bats clustered in high-density cave dwellings (Constantine, 1962; Winkler, 1968). Little is known about the precise mechanism of viral entry once virus contacts nasal mucosal cells (although efforts to define cellular membrane receptors for rabies virus attachment have been initiated), and the significance of these routes for virus spread and maintenance of natural infection remains obscure (Atanasiu, 1965; Hronovsky and Benda, 1969; Afshar, 1979; Fischman and Schaeffer, 1971).

The duration of the initial period in which virus is still recoverable from tissue is relatively short (R. T. Johnson, 1965, Baer *et al.*, 1968). Thereafter, the virus goes into an eclipse phase in which the virus either remains localized or proceeds on its pathway to the CNS. At no time, however, does the virus become viremic in the natural course of infection (Schindler, 1961; R. T. Johnson, 1971). The apparent local sequestering of virus at the site of inoculation for periods of days (Murphy and Bauer, 1974), or possibly weeks in experimental animals (Baer and Cleary, 1972), or even months in humans (Hattwick and Gregg, 1975), is a puzzling aspect of rabies virus infection. Baer and Cleary (1972), studying the pathogenesis of street virus infection in mice by footpad inoculation, observed that a marked reduction in mortality could be obtained by amputating the foot as long as 18 days after injection. This implies that the virus remains at the local wound site for some time before coursing through the peripheral nerves into the CNS.

Murphy and colleagues have shed the most light on this question of where the virus is retained following inoculation, through their immunofluorescence and electron-microscopic studies of experimental rabies virus infection in hamsters (Murphy et al., 1973a; Murphy and Bauer, 1974; Murphy, 1977; Harrison and Murphy, 1978). These and related studies have been recently reviewed by Murphy (1982, 1985). They were the first to observe viral antigen accumulating in striated muscle following intramuscular inoculation of a number of rabies viruses or rabieslike viruses (Makola and Lagos bat viruses) into the hindleg of young hamsters (Fig. 1) and no involvement of other tissues that might account for a slow progression of virus spread. Infection of muscle at the site of inoculation, which was evident by the number of localized immunofluorescent foci that appeared on day 3 after inoculation and essentially doubled by day 6, represents an initial potential site for virus replication. Electron-microscopic examination of serially collected inoculated tissue confirmed the productive infection of these muscle fibers by showing virus shedding into nearby extracellular spaces (Harrison and Murphy, 1978). Muscle infection has also been observed in young skunks infected intramuscularly with street rabies virus (Charlton and Casey, 1979) and in dogs (Fekadu and Shaddock, 1984). While these studies collectively establish that rabies virus can infect muscle cells and replicate in them prior to its invasion of the peripheral nervous system (PNS) and CNS, the infection is nevertheless confined to a small number of myocytes close to the inoculation site and produces few infectious virus particles prior to neurological infection.

2. Neural Pathogenesis—Transit of Virus to the Central Nervous System

A second site to become exposed to virus after muscle-cell infection (although infection of muscle cells may not be necessary) is the unmyelinated nerve endings deep in muscle that wrap around modified muscle cells or tendons and form neuromuscular (sensory stretch proprioceptors) or neurotendinal spindles. Exposed sensory-nerve endings are important

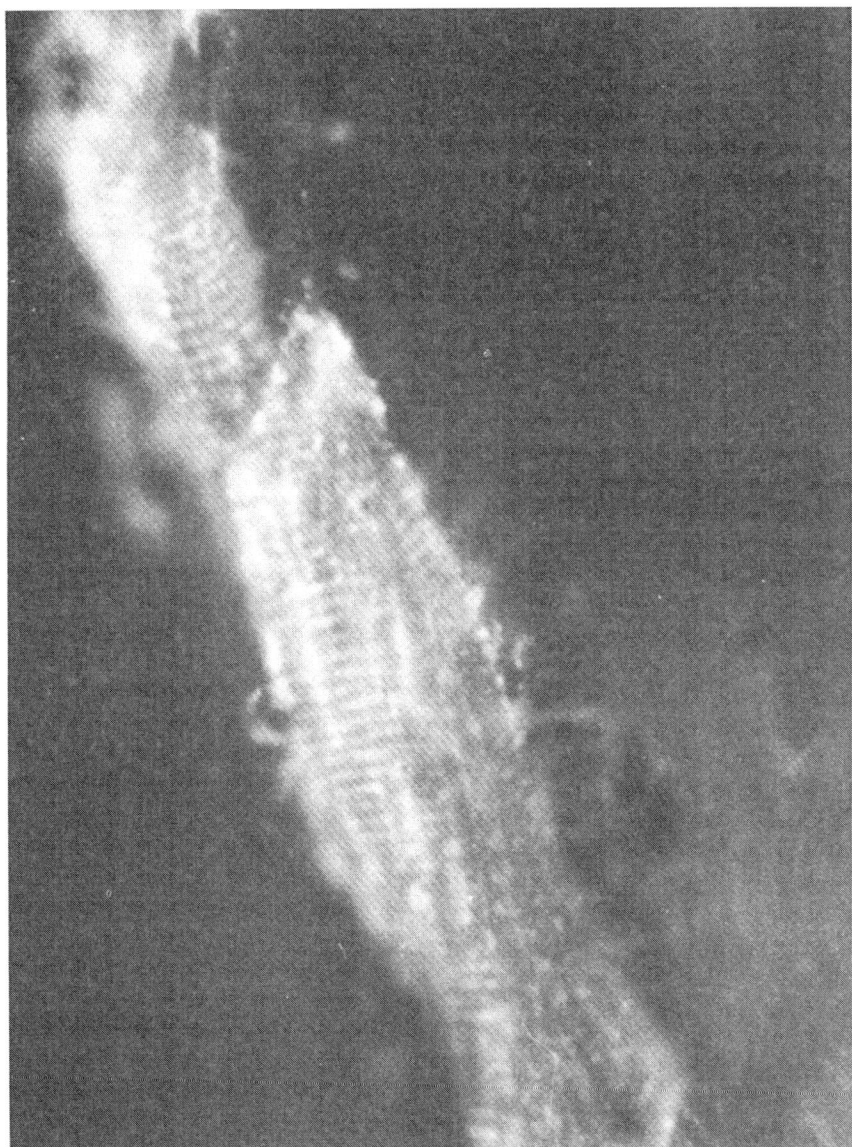

FIGURE 1. Experimental rabies virus infection of a hamster muscle cell. Viral antigen is visualized in the cytoplasm by immunnofluorescence. Frozen section. × 1200. Supplied by F. A. Murphy and A. K. Harrison.

sites of virus entry into the PNS and might actually be the primary site of virus attachment leading directly to neuronal infection in badly torn muscle of a bite wound (Murphy *et al.*, 1973a). Similarly, the many sensory-nerve endings of epithelial and subepithelial tissues exposed by superficial abrasions of skin and mucous membranes may also serve as primary sites for virus entry. Rabies virus antigen has been detected by immunofluorescence in neuromuscular and neurotendinal spindles in the form of fine particulate or aggregate granules dispersed throughout the cells (Fig. 2). Later in the infection, Murphy and colleagues detected antigen in the fine nerves within muscles, tendons, and adjoining connective

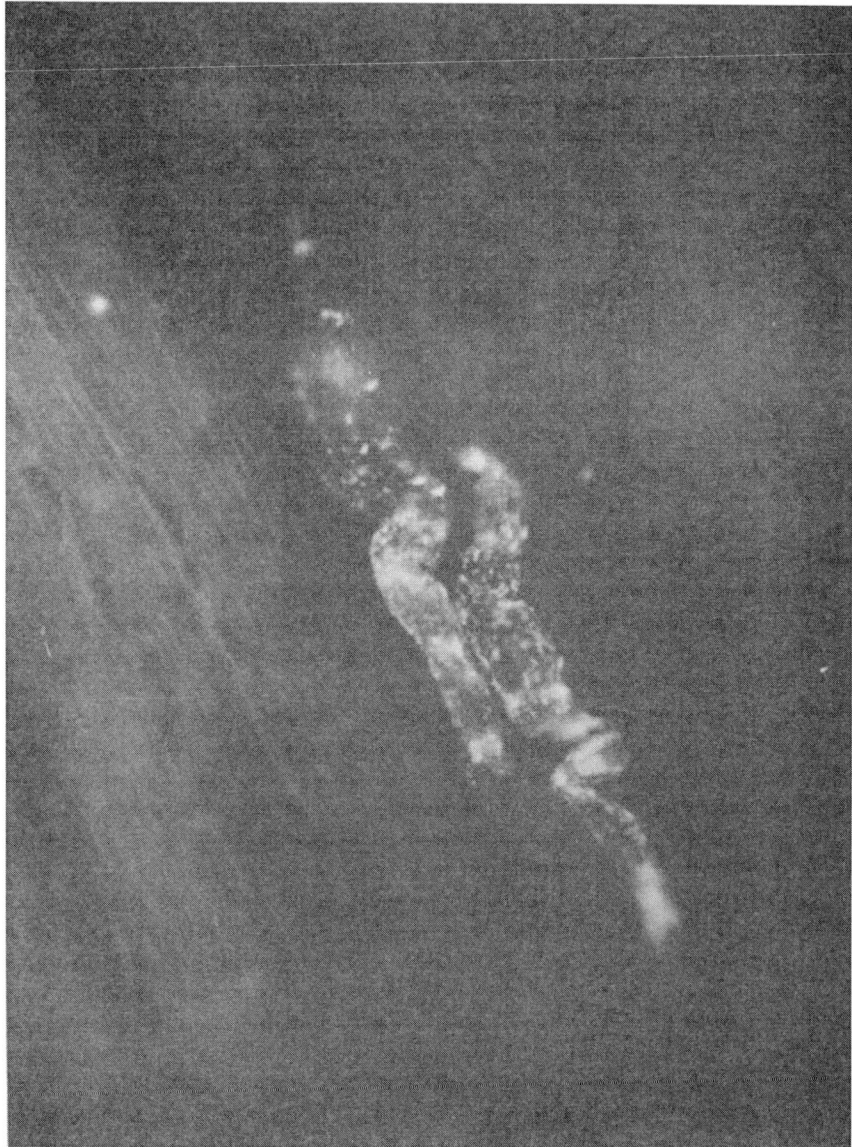

FIGURE 2. Rabies virus localized in two cells of a neuromuscular spindle within muscle of an experimentally infected hamster. Viral antigen is visualized by fluorescent-antibody staining. Frozen section. ×500. Supplied by F. A. Murphy and A. K. Harrison.

tissue leading from spindles to large nerve trunks. This was the evidence needed to support the earlier contention by Dean *et al.* (1963) and others (reviewed by Baer, 1975) that virus spreads from the site of inoculation to the spinal cord by way of the nerves. In these earlier studies, a variety of techniques including the cutting of sciatic and saphenous nerves either before or after footpad inoculation of fixed virus was used to establish that virus travels along nerve pathways at a rate of approximately 3 mm/hr (Dean *et al.*, 1963). Baer and co-workers (Baer *et al.*, 1968; Baer and Cleary, 1972) observed that when a street virus isolate was used, the rate of progression was much slower, and more variable, such that neurectomy or amputation of the hindlimb above the inoculation site many days after inoculation was able to prevent the animal from becoming rabid.

Another site deep in muscles where rabies virus has been detected by electron microscopy (Harrison and Murphy, 1978), as shown in masseter muscle of baby hamster (Fig. 3) and by immunofluorescence (Watson *et al.*, 1981), is in the region of more sparsely distributed motor-nerve endings (motor endplates). Watson *et al.* (1981) showed in mice inoculated with rabies virus (CVS strainlike) in the hindlimb by the footpad route, or in both the muscle and footpad, that as early as 1 hr postinoculation, viral antigen fluorescence was concentrated at motor endplate sites within the muscle. These sites could also be correlated in double-stained preparations with cholinesterase-positive sites signifying the localization of virus to high-density nicotinic acetylcholine receptors (Burrage and Lentz, 1981). The motor-axon terminal at the neuromuscular junction has recently been investigated using isolated mouse diaphragms with attached phrenic nerves as a possible site for specific attachment of rabies virus (Lentz *et al.*, 1982). The question of whether the acetylcholine receptors in these and other tissues participate in rabies virus infection is discussed in Section II.D.1, but the observation of rabies virus binding at neuromuscular junctions offers a satisfactory explanation for viral uptake by the PNS at the site of inoculation. Uptake of rabies virus at motor-nerve endings was confirmed by Watson *et al.* (1981) by detection of viral antigen in ventral-horn cells in the spinal cord and dorsal-root ganglion cells at 20 hr after virus inoculation and early detection at cholinesterase-positive sites.

The entry of rabies virus into peripheral nerves begins the neuronal phase of the infection, which is characterized by the centripetal (toward the center), movement of virus within axons to the CNS. This centripetal transit of virus (or viral nucleocapsid) that courses first to the spinal cord and then to the brain is suspected to occur by passive migration in the axoplasm of peripheral nerves (Baer *et al.*, 1965, 1968; Kristensson and Olson, 1973), where rabies virions and viral matrices have been detected (Jensen *et al.*, 1969; Murphy *et al.*, 1973a). The best evidence of rabies virus (or viral nucleocapsid) transport by axonal flow in peripheral nerves comes from experiments in which colchicine and vinblastine, two alkaloids capable of inhibiting axonal transport in either direction, were

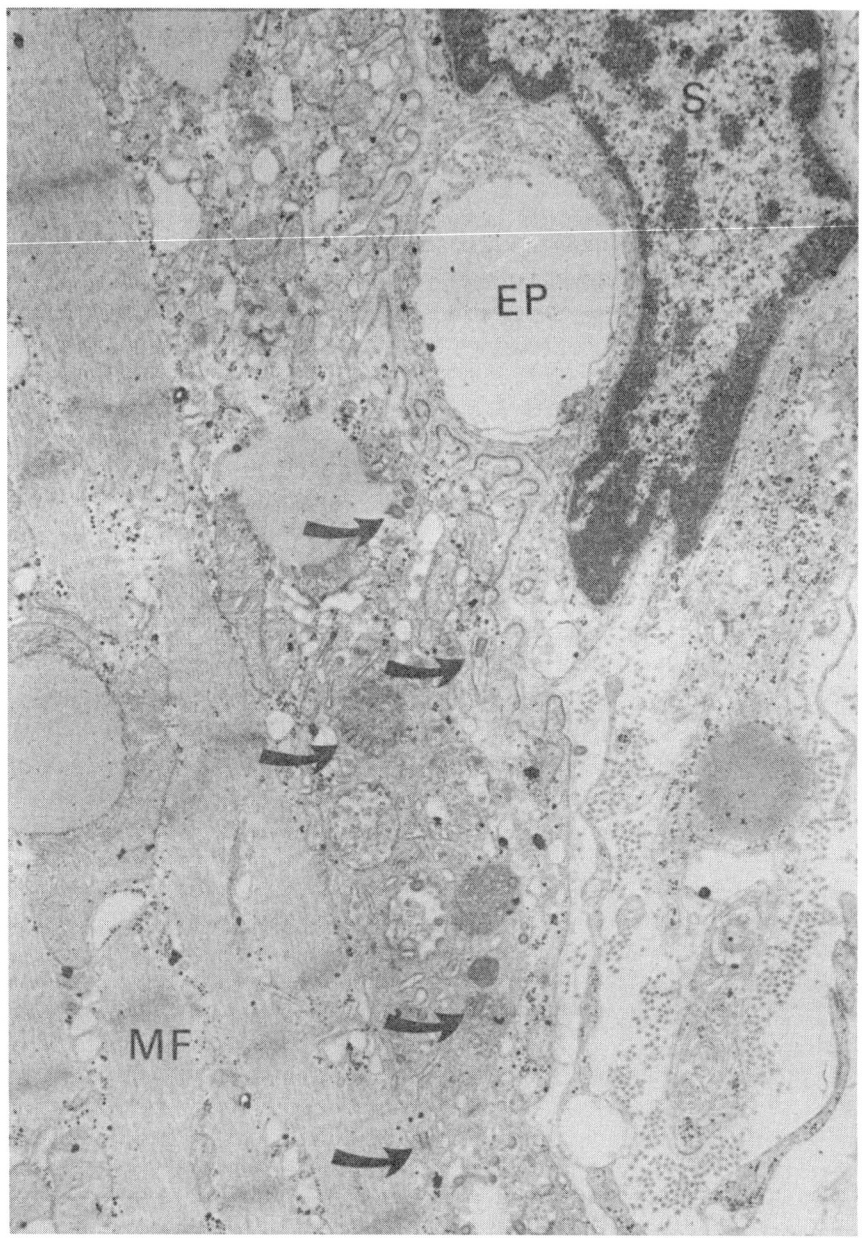

FIGURE 3. Area of muscle cell adjacent to motor endplate (EP) heavily infected with rabies
virus. Virus particles (arrows) can be found budding from sarcoplasmic reticulum in area
between masseter muscle fiber (MF) and Schwann cell (S). Electron micrograph. ×29,400.
Supplied by F. A. Murphy and A. K. Harrison.

applied locally to sciatic nerves by means of small pieces of blotting paper at different times before (Heaney et al., 1976) and after (Bijlenga and Heaney, 1978) footpad inoculation of rabies virus. Colchicine was most effective in stopping rabies virus infection when applied 3 days before intraplantar inoculation with the CVS strain of rabies virus, but also arrested progression of infectious substance when applied up to 6 hr postinfection. By 8 hr, the infection could not be prevented by either colchicine or vinblastine; presumably, the "virus" had already passed through the local axons of the nerve. Infections with sylvatic rabies virus (isolated from fox salivary glands) were more sensitive to the action of colchicine (Tsiang, 1978) and vinblastine (Bijlenga and Heaney, 1978), presumably because the virus persists longer at the inoculation site and the drug has a greater chance of blocking virus spread. Although rabies virus appears to move along nerve trunks by passive flow, other mechanisms of virus transport may facilitate its spread, such as movement of virus across cell-to-cell junctions, including synaptic junctions. Harrison and Murphy (1978) provided evidence of virus budding on axonal membranes by electron microscopy and observed intraaxonal accumulation of virus particles and viral inclusions at nodes of Ranvier as infection progressed to the brain. Budded virus particles were also detected "trapped" in spaces between axons and their myelin sheaths, but very little virus was ever observed in intercellular compartments, where it could become the target for specific host defense mechanisms. Virus budding requires virus maturation at plasma membranes, and the implication of such observations is that virus may be capable of spreading from cell to cell before reaching the brain as it appears to do within the brain in rabies virus infection.

B. Rabies Virus Infection of the Central Nervous System

1. Spinal Cord

The spinal cord is the usual entry route for virus that is destined to reach the brain. This is clear from the ascending paralysis seen in many experimental animals and in some human cases and has been confirmed by sequential immunofluorescence studies reviewed by Baer (1975) and Murphy (1977). Neuronal cell bodies in dorsal-root ganglia of the spinal cord are the first to become heavily infected by virus as it ascends by axoplasmic flow toward the brain. Virus replication in the perikaryon of ganglionic neurons is so active that massive amounts of viral antigen rapidly accumulate (Murphy et al., 1973a). Most neuronal processes within ganglia and dendritic nerve trunks leading to the spinal cord become heavily infected, but nonneuronal satellite cells of ganglia are spared (Murphy et al., 1973a). What is not clear is whether virus replication in spinal-root ganglia is necessary for continued centripetal spread of infection (R. T. Johnson, 1971), since the virus can move directly through the ganglion to the cord (Murphy, 1977). Because of the marked accumulation of viral antigen in spinal ganglion cells, many of the earlier rabies studies using immunofluorescence techniques heralded the spinal ganglia and the spinal cord itself as the initial sites for virus detection (R. T. Johnson, 1965; Baer et al., 1965; reviewed by Schneider, 1975). Kligler and Bernkopf (1943) detected virus 72 hr after intraperitoneal or subcutaneous inoculation of mice with "fixed" virus, first in the upper cord and 24 hr later in the entire CNS. If inoculation was by hind footpad, the virus was first detected in the lumbar and thoracic spinal-cord segments at 72 hr, in the cervical cord at 96 hr, and in the brain at 120 hr. Clearly, as these and other studies indicate, the distribution of infectious virus in the cord depends on the various routes of virus inoculation, although distribution may also be a reflection of strain properties and possibly other factors, including host species (Schneider, 1975).

In the studies by Murphy et al. (1973b) that followed, hamsters inoculated intramuscularly with rabies virus (CVS strain or vampire bat street strain) or rabieslike viruses (Mokola and Lagos bat) produced virus that was detectable by immunofluorescence in spinal-cord neurons that initially appeared as a fairly particulate pattern of viral antigen in the cytoplasm (Fig. 4) and rapidly developed into massive amounts in aggregate form. By 4–5 days after inoculation, the "antigenic masses were spread evenly throughout the cord, obliterating the earlier delineation of grey and white matter," according to Murphy et al. (1973b); "Late in infection, immunofluorescence was contiguous from spinal nerves to their junctions in the spinal cord at all levels, lumbar to cervical."..."None of the supportive structures of the spinal cord, including meninges or central canal ependyma, however, ever contained antigen." The results were the same, whether virus was inoculated intramuscularly or by the intraplantar route. When virus was administered orally or intranasally, spinal-cord involvement occurred later in infection, presumably as a result of the centrifugal (outward from the brain) neural spread of virus.

Inflammatory lesions due to perivascular infiltration are the most common and frequently noted histological signs of change in the spinal cord and brainstem of animals and humans inflicted with rabies (Dupont and Earle, 1965; Jubb and Kennedy, 1963; Perl, 1975; Murphy, 1977). The infiltrate is composed primarily of lymphocytes and macrophages and occasionally polymorphonuclear cells (reviewed by Perl, 1975). When inflammatory lesions are intense, polymorphonuclear cells will predominate particularly in the brainstem region, and more so in the young than in adults (Tangchai et al., 1970). These ganglionic changes, though not exclusively confined to rabies (Jubb and Kennedy, 1963), could serve as diagnostic lesions for rabies, since they are not encountered in diseases that must be considered in the differential diagnosis of rabies (Herzog, 1965). The intensity of the inflammatory lesions in rabies encephalitis tends to be affected by the host species infected, the virus strain, and the course of the disease (Perl, 1975). The most severe inflammatory changes

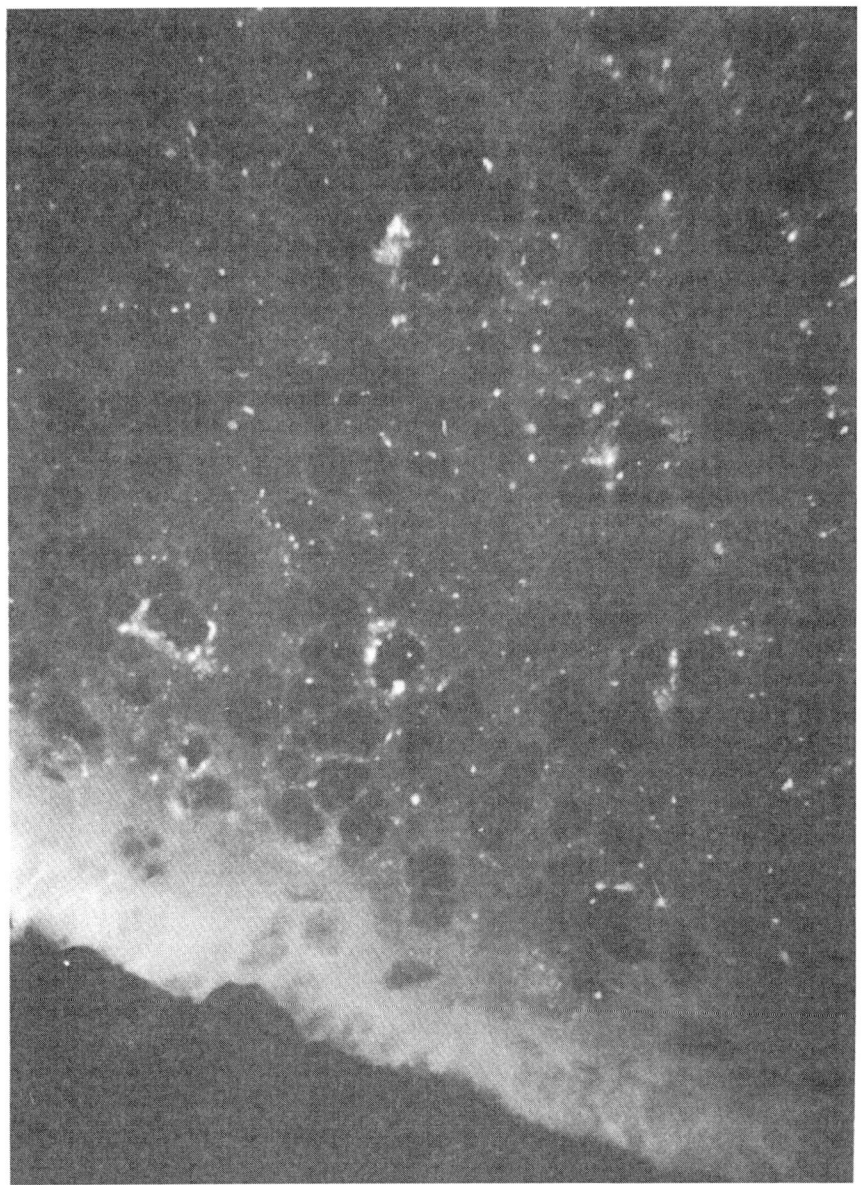

FIGURE 4. Rabies virus CVS strain infection of cervical cord 3 days after intramuscular inoculation of hindlimb of a hamster. ×500. Supplied by F. A. Murphy and A. K. Harrison.

have been associated with rabies virus isolated from Mexican freetail bats, with Flury LEP (low egg passage) rabies vaccine virus that occasionally induced rabies in vaccinated dogs, and with the rabieslike Mokola and Lagos bat viruses (reviewed by Clark and Prabhakar, 1985). Chronic (prolonged) rabies virus infections are also known to produce intense encephalitic lesions (Murphy *et al.*, 1980).

The lesions that tend to have more significance in rabies diagnosis are the eosinophilic masses measuring 1 to approximately 30 μm in diameter that appear to be the main morphologically distinct lesions of rabies localized in virus-infected neurons. These are the inclusion bodies of rabies, first described by Negri (1903), that form a matrix characteristically recognized by a basophilic inner body that is surrounded by a more acidophilic peripheral region (reviewed by Clark and Prabhakar, 1985) and under the electron microscope appears to have a homogeneous ground substance composed of relatively electron-dense filamentous aggregates (Goodpasteur, 1925; Matsumoto, 1962, 1963; Hummeler *et al.*, 1967) (see Figs. 13 and 14). Some investigators, however, consider them unsuitable for diagnosis of rabies, since small eosinophilic inclusions may also appear within the neuronal cytoplasm of normal animals of many species, notably the cat (Szlachta and Habel, 1953; Tierkel, 1959; Innes and Saunders, 1962). The true nature and exact composition of the Negri neuronal cytoplasmic inclusion bodies have been described by Miyamoto and Matsumoto (1965). They showed that the Negri body was a matrix composed predominantly of viral ribonucleoprotein (nucleocapsid) as the eosinophilic substance. The internal structure of the matrix was revealed by light and electron microscopy as being comprised of strands or "granules of chromatin-like material." The history of the Negri body and an account of its significance for virus growth in nerve cells are reviewed by Matsumoto (1970). From numerous observations of Negri bodies in undamaged virus-infected nerve cells, it is clear that there is considerable variation in size, morphology, and location of the cytoplasmic inclusions with the nerve cell. Also, the capacity of rabies viruses to form inclusion bodies is gradually lost after serial passage through experimental animals (Perl, 1975). Negri bodies are particularly found within neurons of peripheral and basal ganglia of the cord, in pons and medulla of the brainstem, and in ganglion cells of Ammon's horn of the hippocampus, as well as in other areas of the brain, including cerebral cortex and thalamus and Purkinje cells of cerebellum. The inclusions are seen most prominently in the largest neurons, such as the large ganglion cells of Ammon's horn and the Purkinje cells of the cerebellum (Fig. 5). These histopathological features of rabies encephalitis caused by street virus are essentially similar in wild, domestic, and laboratory animals infected naturally or artificially and in man (Clark and Prabhakar, 1985). Because not all cases of rabies show the typical pathological picture (Dupont and Earle, 1965; Perl, 1975), pathological examination alone may not lead to a certain diagnosis of

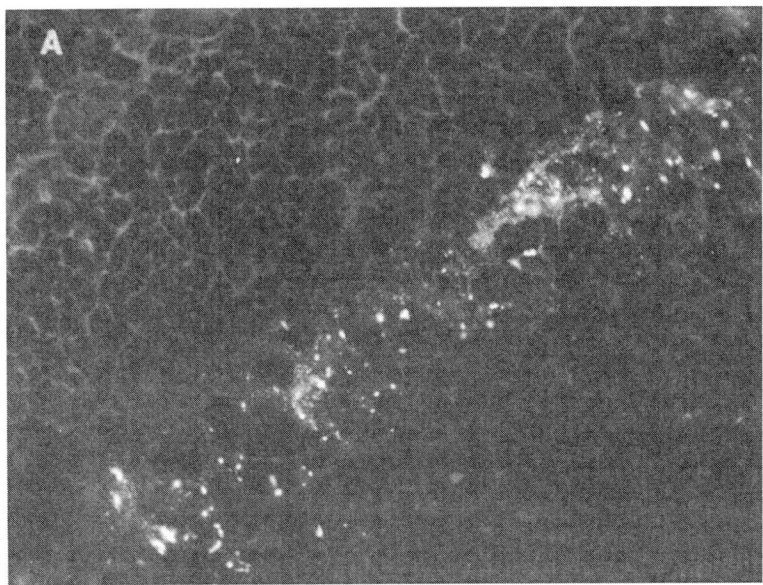

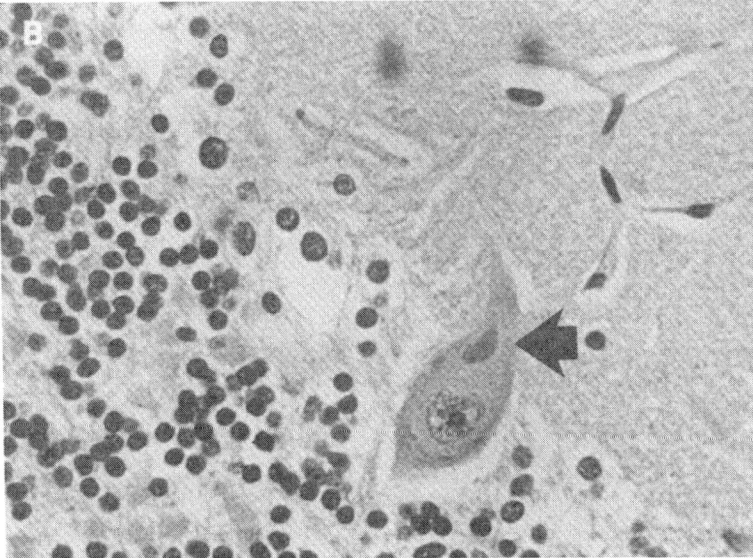

FIGURE 5. (A) Localization of fixed rabies virus CVS strain in the Purkinje-cell layer of the cerebellum of a young hamster at 5 days after intramuscular inoculation. ×400. (B) Prominent Negri body located in a Purkinje cell of the cerebellum of a mouse infected with street rabies virus. Paraffin section, Lendrum's stain. ×400. Supplied by F. A. Murphy and A. K. Harrison.

rabies. Diagnosis of rabies infection in the absence of any pathological evidence, however, can still be reliably made by a fluorescent-antibody test or mouse-inoculation test.

2. Brain

Rabies virus moves rapidly to the brain once it reaches the spinal cord. That it does has been demonstrated many times in mice, rats, and hamsters by an increase of immunofluorescent antigen within hours after virus has reached the spinal cord (Dean *et al.*, 1963; Baer *et al.*, 1965; Murphy *et al.*, 1973a,b). By the time virus reaches the brain, the infection in animals and humans is irreversible, and the outcome either is fatal or results in permanent brain damage. Murphy (1977) considers that the "precise and consistent localization of rabies infection in the brain is the most likely explanation for the specific signs and symptoms at various stages of clinical illness in man and animals." On reaching the brain, the virus continues to spread rapidly, so that eventually the infection involves "virtually every neuron" of the brain, to the exclusion, on the whole, of all supporting cells in the brain (Murphy *et al.*, 1973b). The occasionally noted exceptions to exclusive neuronal infection in the brain have been the infections of astrocytes and other glial elements (reviewed by Schneider, 1975). However, passive transport of "virus genome" within the nerve tracts could account for the most rapid spread of infection in the brain, whereas the spread of virus by alternative pathways involving astrocytes and glia would theoretically take much longer.

It was considered for a long time that virus maturation in infected brain occurred within the cytoplasm of host cells and that lack of virus budding from the host-cell plasma membrane was characteristic of rabies virus replication in the brain. Although this was thought to be the case in CNS tissue, virus budding was frequently observed from various extraneural tissue cells of rabid animals (Dierks *et al.*, 1969; Murphy *et al.*, 1973a) and from cell-surface membrane in a variety of tissue-culture systems (Atanasiu *et al.*, 1963b; Davies *et al.*, 1963; Hummeler *et al.*, 1967; Matsumoto and Kawai, 1969; Iwasaki *et al.*, 1973; Matsumoto *et al.*, 1974; Tsiang *et al.*, 1983). Cytoplasmic maturation within infected neuronal cells of brain commonly appeared to occur by budding on the intracytoplasmic membranes of the endoplasmic reticulum; the outer lamella of the nuclear envelope and Golgi membranes were occasionally involved (Murphy *et al.*, 1973b; Matsumoto, 1975). More recently, rabies virus budding was observed on the plasma membrane of neuronal cells, altering the traditional concept that virus maturation involves only intracellular membrane systems. Currently, the view is that budding of rabies virus from neuronal-cell plasma membrane not only is a general phenomenon, but also does not seem to be restricted to particular strains of the virus or to certain animal species (Iwasaki *et al.*, 1975; Iwasaki and Clark, 1975; Charlton and Casey, 1979) or to human rabies (Iwasaki *et al.*, 1985), in contrast to earlier beliefs (reviewed by Matsumoto, 1975).

In the budding process, viral nucleocapsid [genome RNA plus nucleoprotein (N), phosphoprotein (NS), and transcriptase (L) protein] is contributed from the cytoplasm; sometimes excess amounts of nucleocapsid are present in the form of cytoplasmic inclusions (Negri bodies). Viral envelope is contributed by morphologically modified host-cell membrane [with viral glycoprotein (G) inserted] (Wagner *et al.*, 1971). The actual mechanism that initiates the outfolding of a patch of modified cellular membrane that contains the viral components and results in the emergence of the bullet-shaped structure of the mature virus particle (Morrison, 1980) remains unclear at this time.

3. Street vs. Fixed Rabies Virus Infection of the Spinal Cord and Brain

The pathogenic patterns of street and fixed rabies viruses in the brain of experimentally infected rodents are quite distinct with regard to occurrence of cytopathic changes in infected neurons. Typically, street viruses leave nearly all neurons in the brain and cord of infected animals intact with very few abnormalities in organelle structure, whereas fixed viruses cause widespread and severe damage to infected neurons in the cord and brain (Murphy, 1977; Miyamoto and Matsumoto, 1967; Iwasaki and Clark, 1975). Murphy (1977) considers that "this difference must relate to the mechanism of neuronal dysfunction, the mode of virus spread in the brain, and the nature of the stimulus of inflammation infiltration." Iwasaki and Clark (1975), in a comparative chronological ultrastructural study of rabies virus-infected suckling and adult mice, examined specimens from the cerebral cortex, hippocampus, cerebellar cortex, and pons after intracerebral inoculation with fixed and street viruses. They observed in the fixed virus-infected mice that the first ultrastructural changes appeared on day 2 in the perikarya and in various cellular processes of neuronal cells at the site of inoculation, as evidenced by the development of small aggregates of a "granulofibrillary substance" that resembled matrix (Fig. 6). Two or more matrices were frequently seen per neuron, and most were less than 2 μm in diameter. At 24-hr after matrix development, the first signs of virus maturation were detectable within the perikaryon (Fig. 7) (including budding from the perikarya) and in various cellular dendritic processes (Figs. 8 and 9). On days 3 and 5, extracellular virus particles were observed in intercellular spaces (Figs. 8, 9, and 10); intracellular virus particles in fixed virus-infected brains (Fig. 11) were seen only in mice surviving when the experiment was terminated.

These conspicuous features of fixed virus infection in suckling mice were more severe in the adult mouse brain. Extensive degeneration of neurons and marked proliferation of glial cells (mostly microglia) were seen in the pyramidal-cell layer (particularly of Ammon's horn) of the

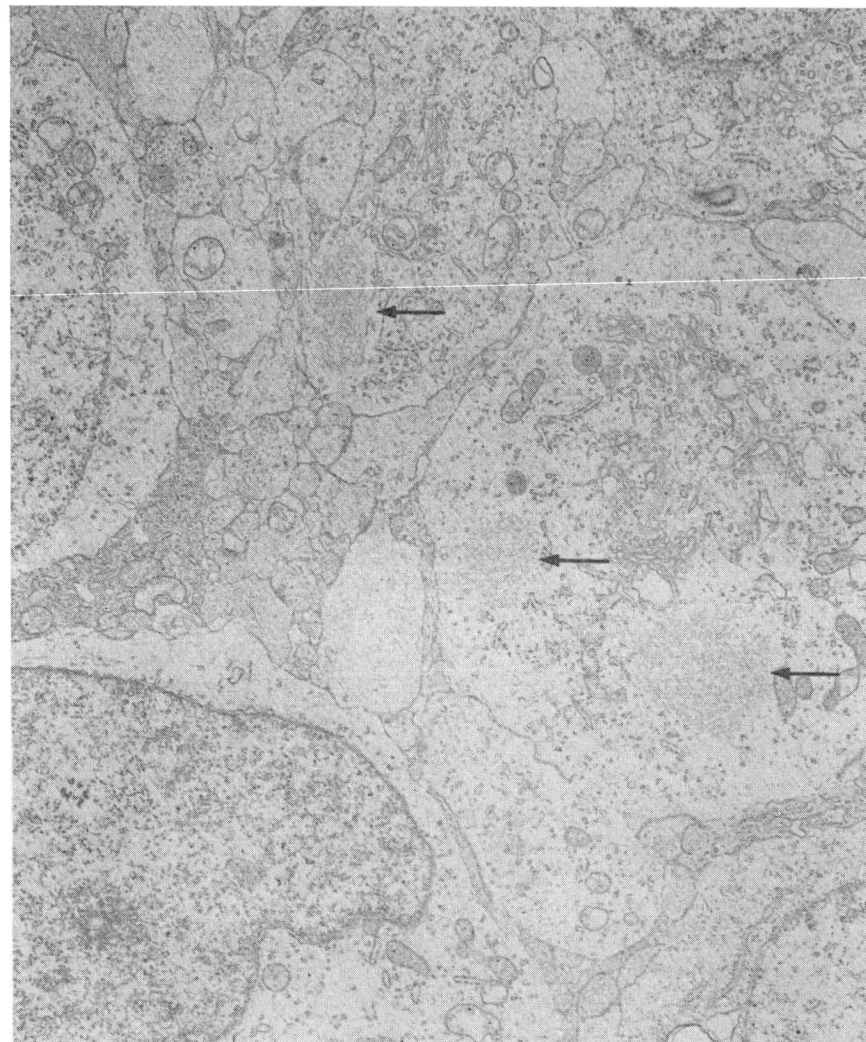

FIGURE 6. Formation of small matrices (arrows) in a neuron of the hippocampus of a suckling mouse 48 hr after intracerebral inoculation with the ERA strain of rabies virus. × 10,200. Supplied by Y. Iwasaki.

hippocampus, in thalamus (especially the ventrolateral portion), and in Purkinje cells, according to Iwasaki and Clark (1975). Neuron degeneration was also seen, but less regularly, in the cerebral cortex, midbrain, and pontine tegmentum. The reaction of glial cells was not as conspicuous in the cerebellum or pons as it was in other regions of the brain examined. Infiltration of monocluear cells in the perivascular space of parenchymal vessels and in subarachnoidal space was mild. The appearance of virus budding from plasma membrane of neurons in fixed virus-infected adult

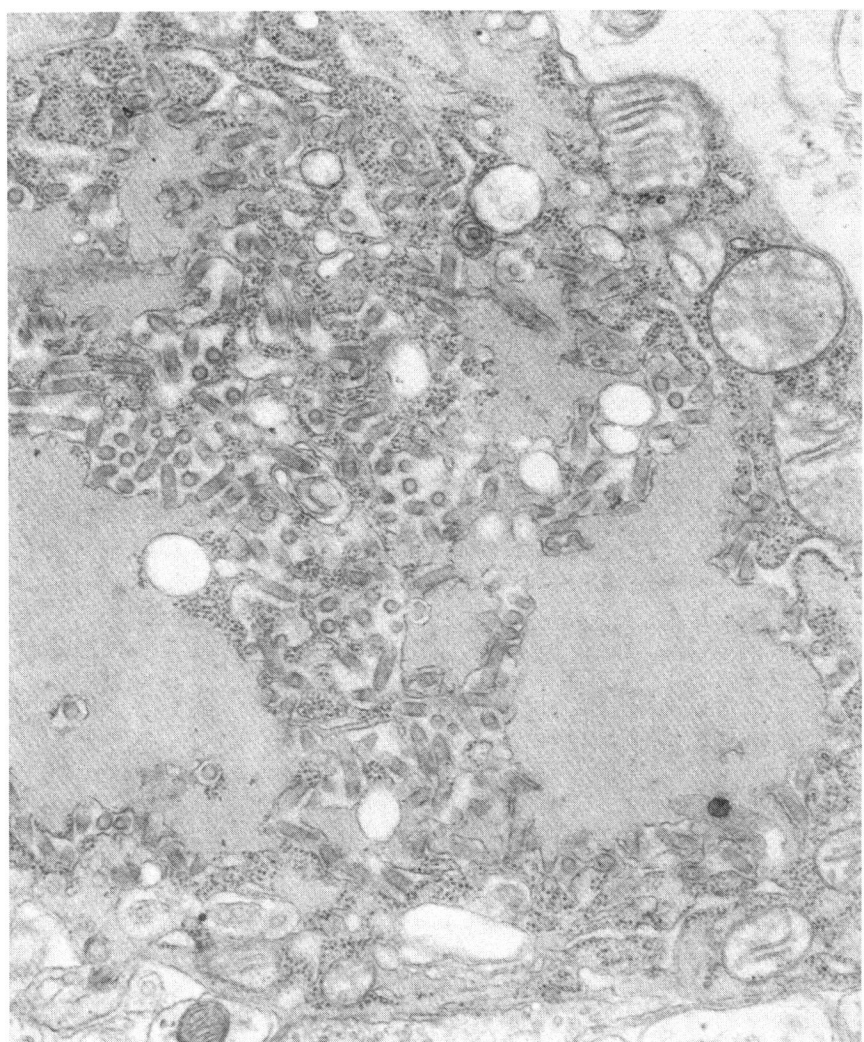

FIGURE 7. Rabies virions within a mouse neuron. Virus particles and viral nucleocapsid inclusions fill the cytoplasm of the intact neuron. Infection of mouse is with human rabies virus isolate. Plastic section for electron microscopy. ×50,000. Supplied by F. A. Murphy and A. K. Harrison.

mouse brain was similar to that seen in suckling mice, but in the adult mouse, virus budding was more frequent from the dendritic and axonal processes than from the perikaryon. The presence of virions in intracellular spaces was no different from that observed in fixed virus-infected suckling mouse brain, and occasionally virions were seen at synaptic termini in the adult mouse (Fig. 12).

The main cytopathic differences between street and fixed virus infections are the striking histopathological appearance of large intracy-

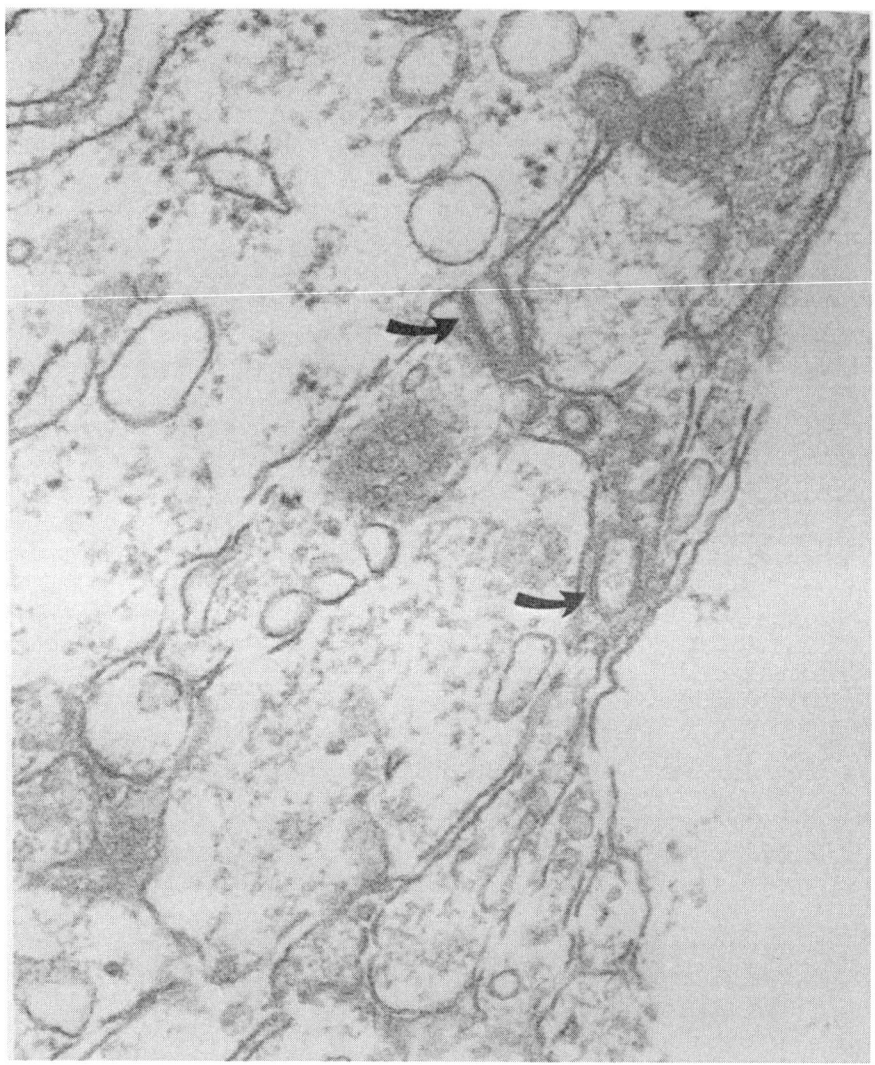

FIGURE 8. Rabies virions (arrows) budding from the perikaryon of the hippocampus and in the intercellular spaces of dendritic processes 120 hr after infection of a suckling mouse with ERA strain rabies virus. ×84,000. Supplied by Y. Iwasaki.

toplasmic eosinophilic inclusion bodies within neurons of the street virus-infected mice (Iwasaki and Clark, 1975) (Fig. 13) and the observed intimacy between viral matrices and virus-particle production (Fig. 14) (Iwasaki and Clark, 1975; Murphy, 1975). The eosin-stained inclusion bodies in the street virus-infected mouse brains appeared to be outside the perikaryon in various cellular processes. In fixed virus-infected specimens, inclusion bodies were not discernible in hematoxylin/eosin-stained preparations (Iwasaki and Clark, 1975). In street virus infections, virus

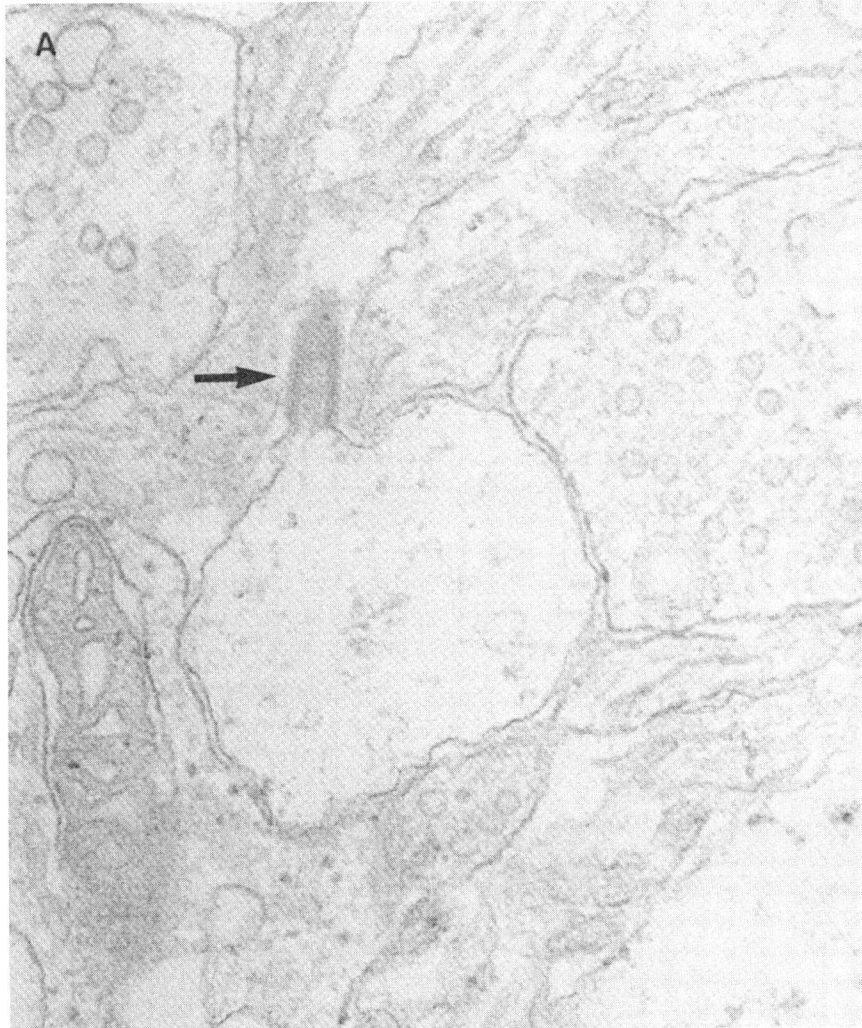

FIGURE 9. Rabies virion (arrows) budding from a dendritic process in the hippocampus of a suckling mouse 72 hr (A) and 120 hr (B) after intracerebral inoculation with ERA strain rabies virus. (A) ×111,000; (B) ×105,000. Supplied by Y. Iwasaki.

particles are generally associated with cytoplasmic membranes adjacent to nearly all inclusions (Fig. 15). In fixed virus infections, virus particles are not associated with the inclusions seen in infected cells (Miyamoto and Matsumoto, 1967; Murphy, 1975; Iwasaki and Clark, 1975). Nevertheless, the distribution of neurons bearing inclusion bodies in street virus infections of suckling and adult mice was similar to the distribution of degenerative neurons in fixed virus-infected mouse brains, suggesting that the pathogenic pathways of the two virus types are comparable.

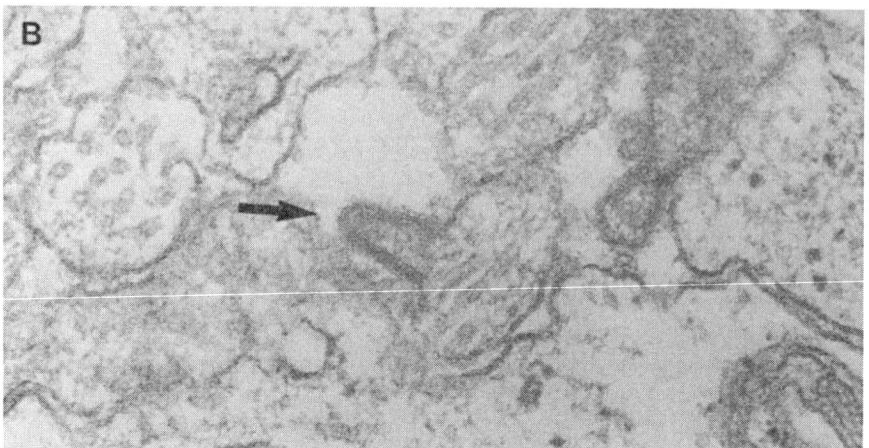

FIGURE 9. (Continued)

Iwasaki and Clark (1975) found inclusion bodies and virions "in almost all pyramidal cells in Ammon's horn and in the neurons in the thalamus, less frequently in Purkinje cells, neurons of the cerebral cortex, hypothalamus, substantia nigra, nucleus oculomotorius, nucleus medianus cerebellum, nucleus vestibule medianus, nucleus reticularis tegmenti pontis, nucleus septium, nucleus amygdaloideus, and very infrequently in nucleus caudatus and putamen" of street virus-infected suckling mouse brain. They also observed that "the thalamic nuclei were inevitably involved in all cases and in severe cases, groups of neurons underwent necrosis that was often accompanied by neuronophagia. Necrotic neurons were also sporadically encountered in other regions among those neurons bearing inclusion bodies. Perivascular and subarachnoidal mononuclear-cell infiltration was seen rarely and only in mild form" (Iwasaki and Clark, 1975). The inclusions seen in street virus-infected adult mice, unlike those seen in suckling mice, had a basophilic inner body (see Figs. 13 and 14) typical of Negri bodies and an intense mononuclear-cell infiltration within the perivascular and subarachnoidal spaces. The proliferation of microglia in street virus-infected adult mice was more severe than that in street virus-infected suckling mice, but much less extensive than that in fixed virus-infected adult mice (Iwasaki and Clark, 1975).

The ultrastructural features of virus replication in street virus-infected adult mice were similar to those seen in suckling mice. As well as virus particles being found near matrices in the cytoplasm of neurons and their processes, virus was also observed budding from infected neuronal plasma membrane (Fig. 16).

Similar pathogenic features of localized rabies virus infection have been detected in the brains of other infected animal species. Neurons and their processes in nearly all regions of the brain of striped skunks inoculated intramuscularly in the hindlimb with street virus isolated from

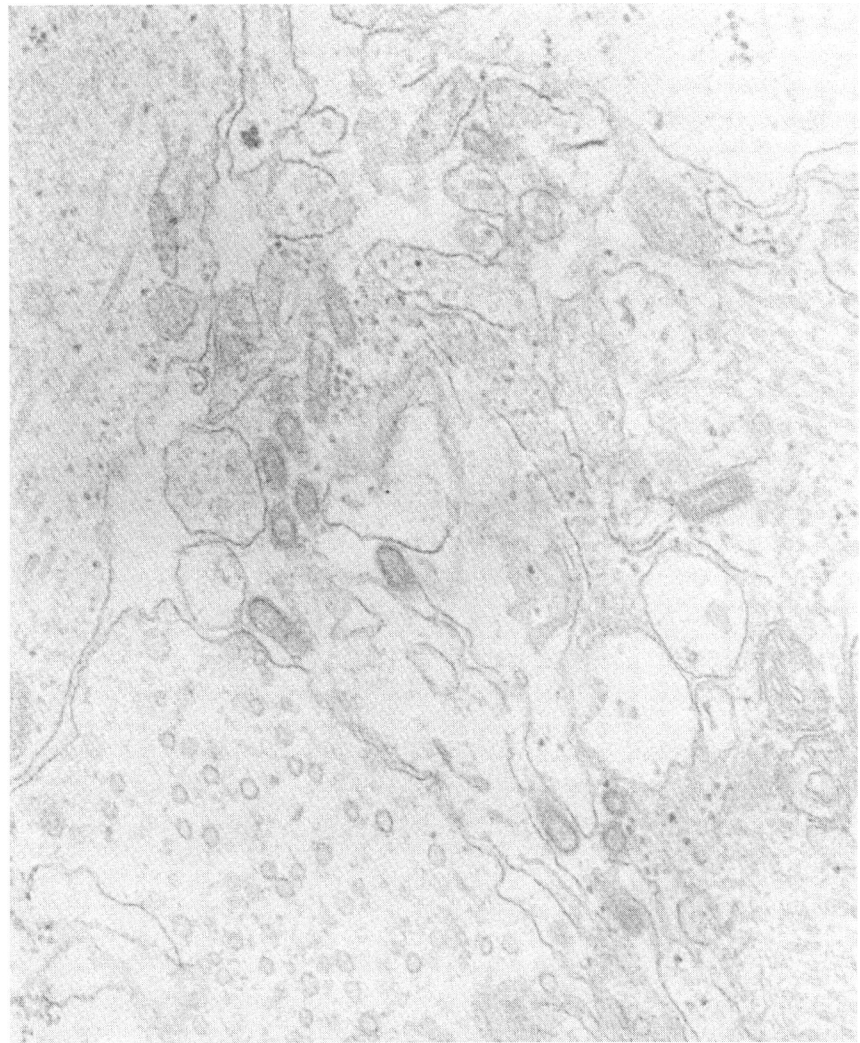

FIGURE 10. Rabies virions in the intercellular space within the hippocampus of a suckling mouse 72 hr after intracerebral inoculation with the CVS strain of rabies virus. ×68,000. Supplied by Y. Iwasaki.

rabid skunk salivary gland contained viral antigen in the terminal stages of the disease (Charlton and Casey, 1979). Not only was virus observed budding on the plasma membrane of dendrites and perikarya of neurons in the brain, but also evidence of budding virions on the postsynaptic membrane and adjacent plasma membrane of axon terminals with simultaneous viropexis by the presynaptic axon terminal indicates a transneuronal transfer of infection. These findings support the observations (discussed further below) of Iwasaki and Clark (1975), who found evidence

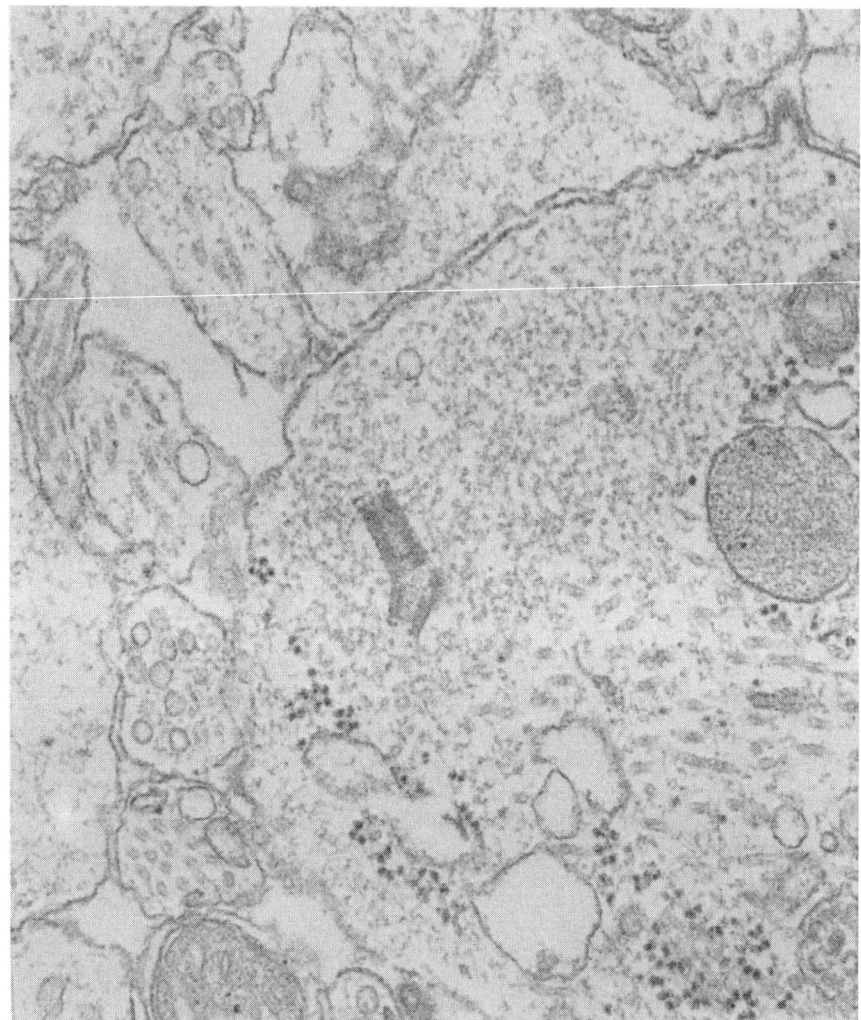

FIGURE 11. Intracytoplasmic maturation of virus particles 120 hr after intracerebral inoculation of a suckling mouse with ERA strain of rabies virus. ×66,000. Supplied by Y. Iwasaki.

of direct cell-to-cell transfer of rabies virus in the brains of mice as an alternative mechanism of virus transfer to the spread of virus by movement from neuronal plasma membrane into intercellular spaces or via intracellular movement of virus as described in studies with hamsters (Murphy *et al.*, 1973b).

In human rabies, many of the observations attributed to street virus infections in animals are generally found in neural perikarya and cellular processes of the human brain (Morecki and Zimmerman, 1969; Leech, 1971; DeBrito *et al.*, 1973; Sung *et al.*, 1976; Iwasaki *et al.*, 1985), with

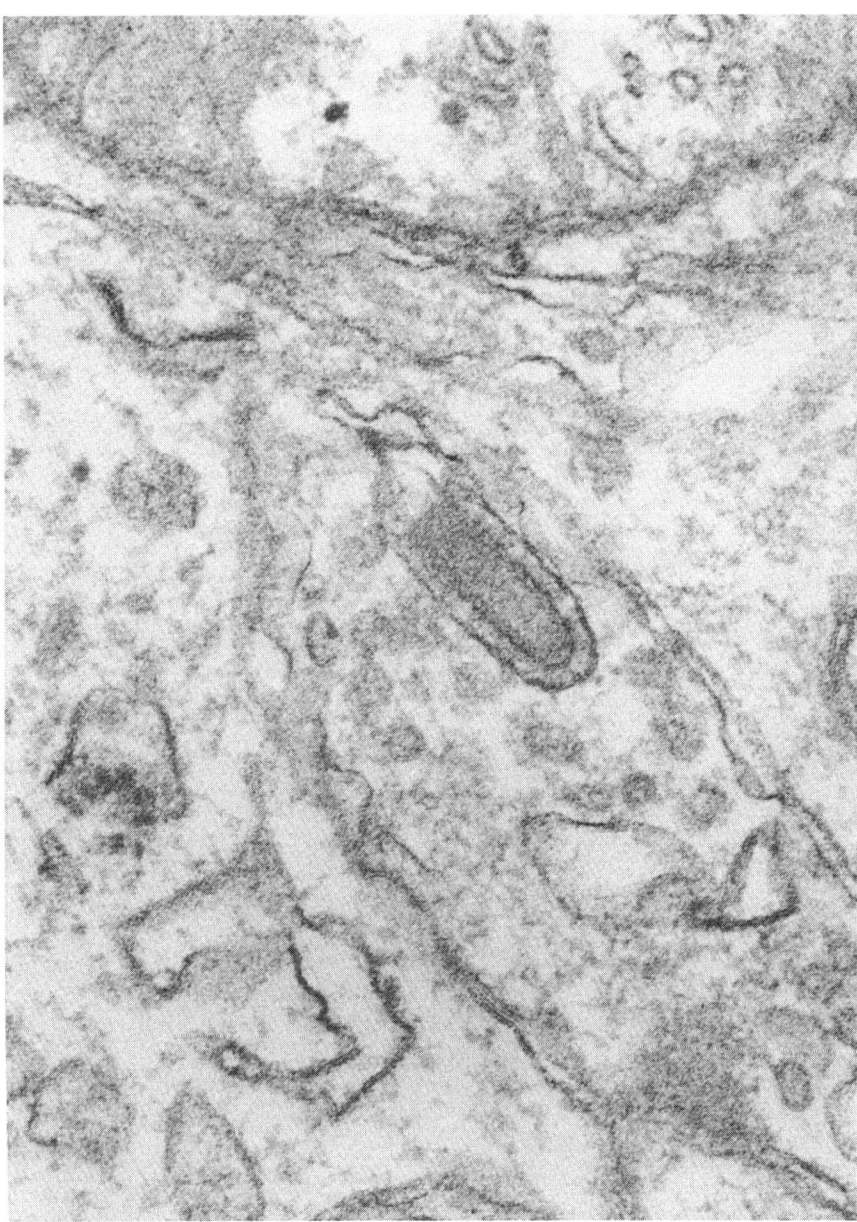

FIGURE 12. Rabies virion at a synaptic terminal in the cerebral cortex of an adult mouse 144 hr after intracerebral inoculation with CVS rabies virus. × 155,400. Supplied by Y. Iwasaki.

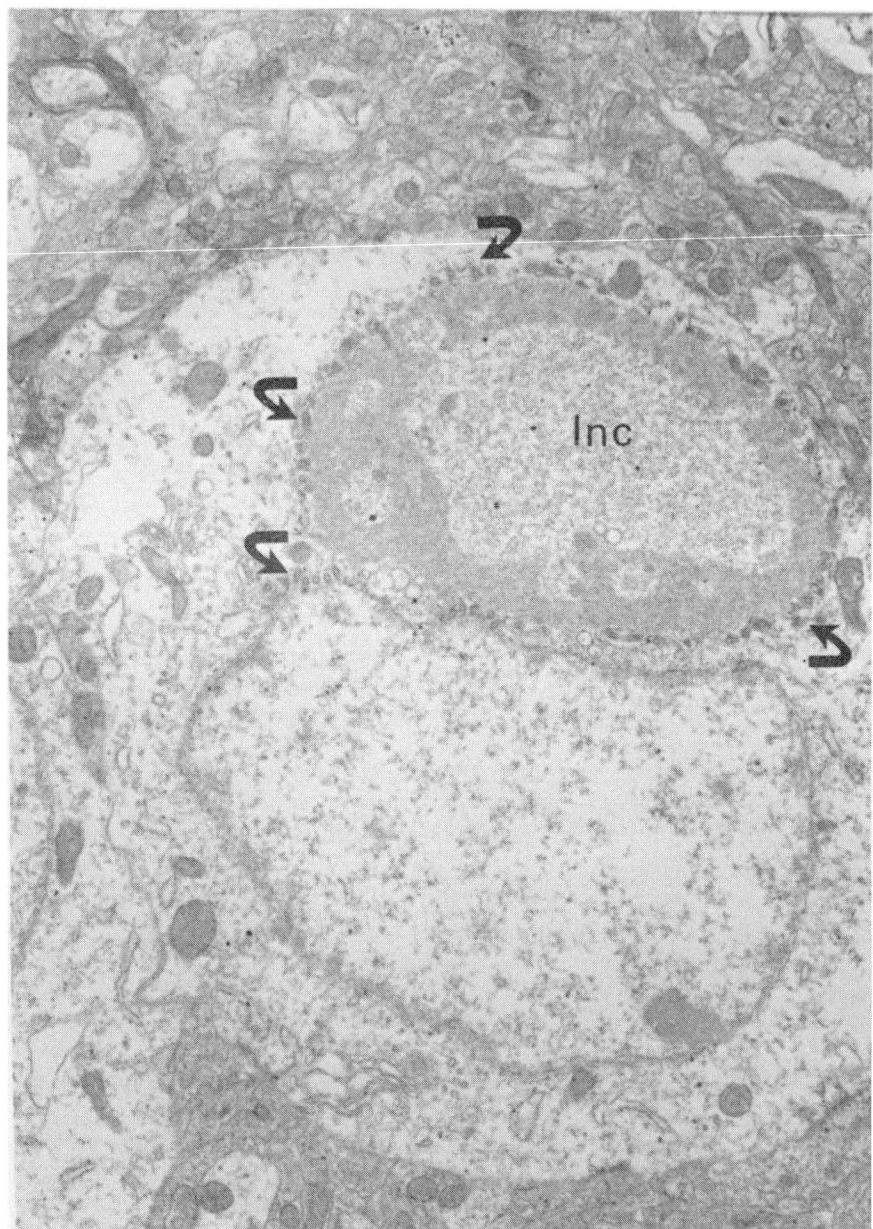

FIGURE 13. A large cytoplasmic inclusion body (Inc) in a pyramidal cell of the hippocampus in a street rabies virus-infected adult mouse, 7 days after intracerebral inoculation. Virus particles (arrows) associate with the large inclusion body. × 13,900. Supplied by Y. Iwasaki.

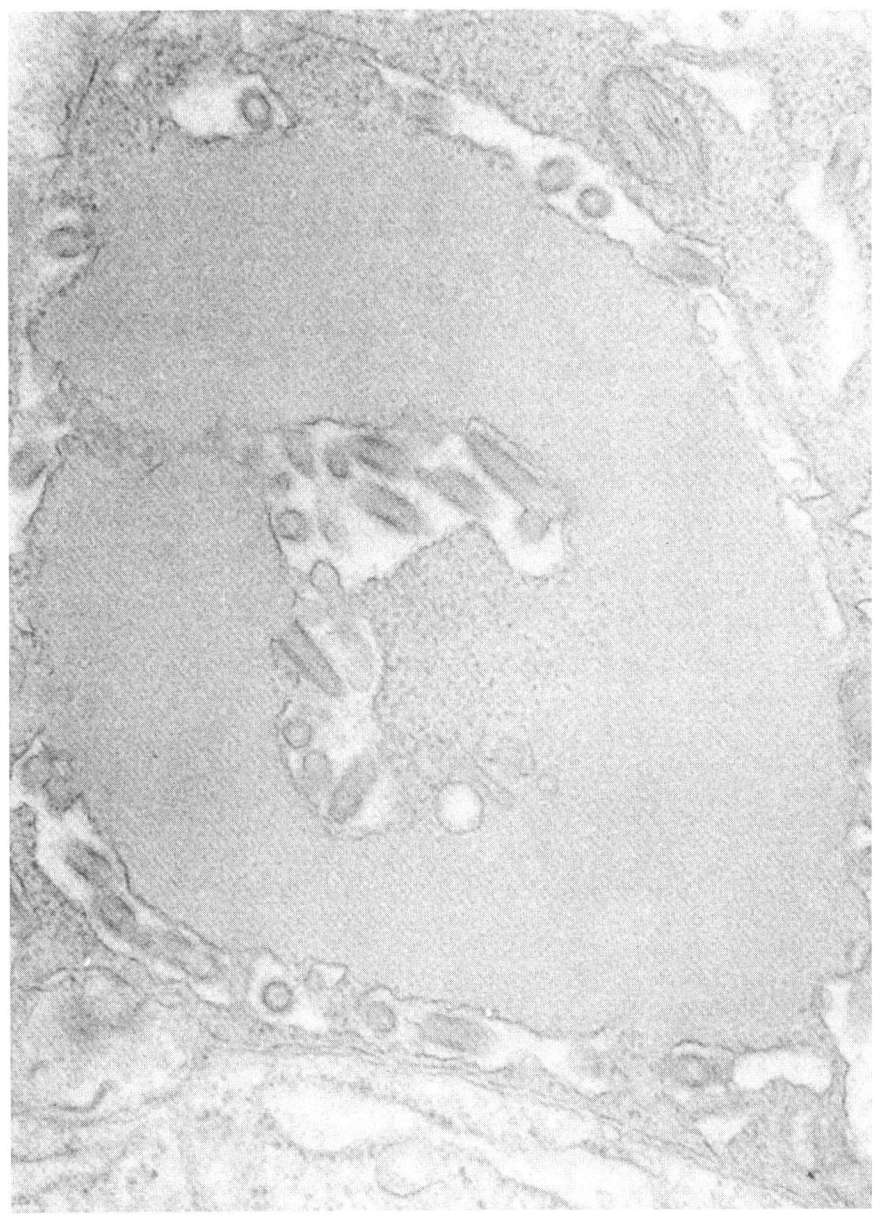

FIGURE 14. Higher magnification of matrix and rabies virions in close association. Hippocampus of street virus-infected adult mouse. ×69,300. Supplied by Y. Iwasaki.

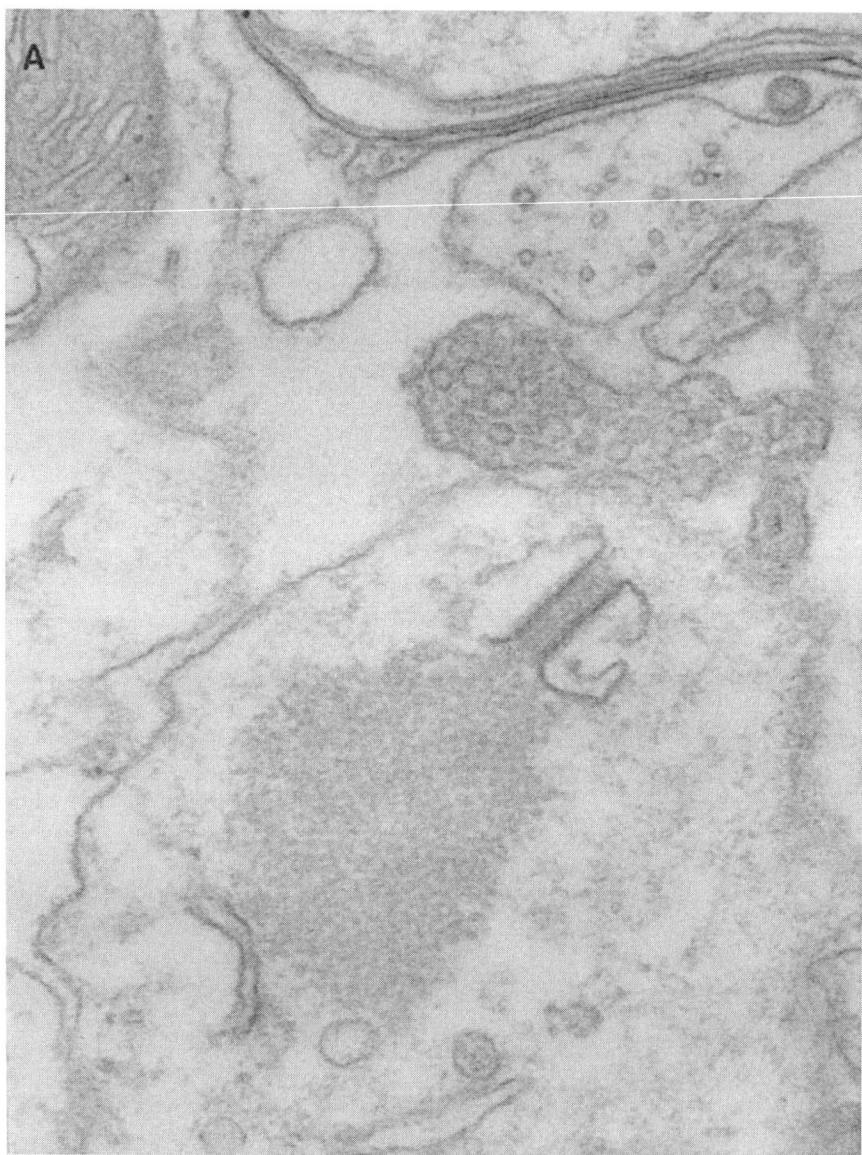

FIGURE 15. Virus-particle formation in association with cytoplasmic membrane adjacent to matrix inclusion. Street virus infections of adult mouse hippocampus (A) 7 days and suckling mouse hippocampus (B, C) 11 days after intracerebral inoculation. (A, C) ×96,000; (B) ×105,000. Supplied by Y. Iwasaki.

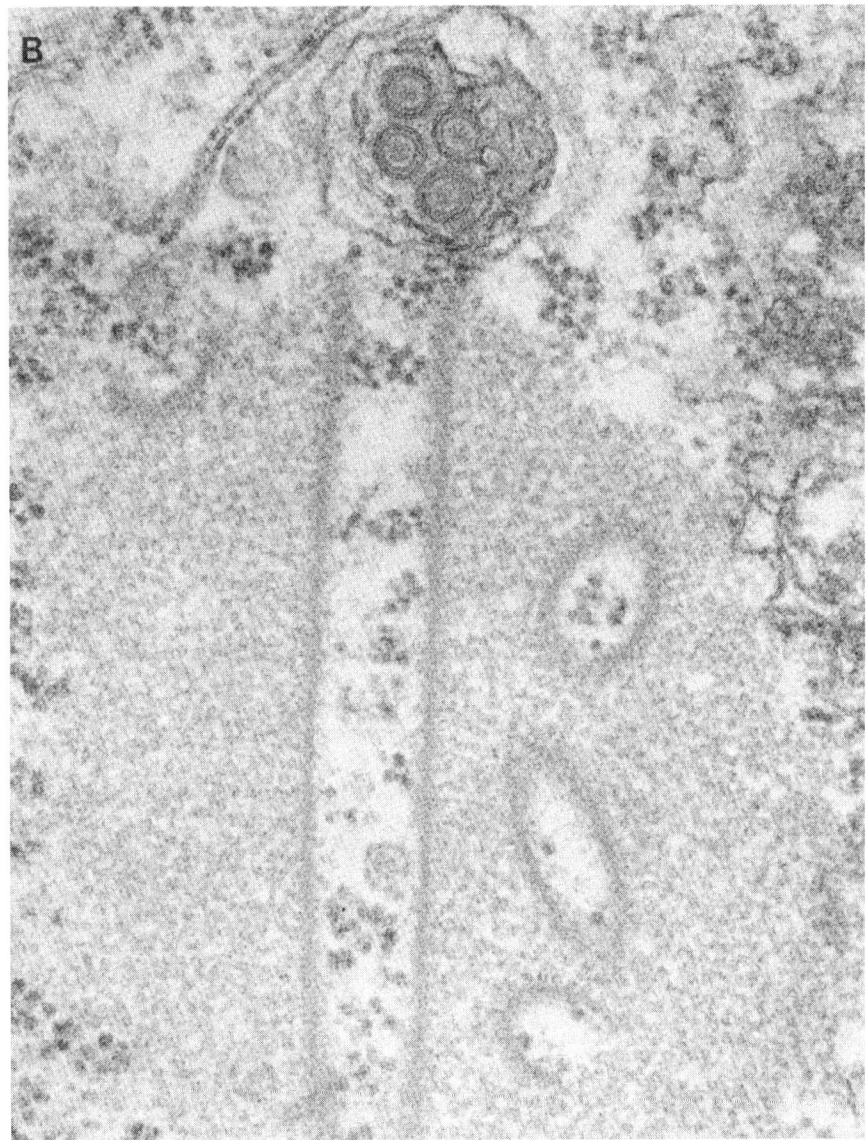

FIGURE 15. *(Continued)*

the exception in most cases reported (excluding those of Iwasaki *et al.*, 1985) that virus budding from the plasma membrane of infected neuronal cells was not detected. The recent ultrastructural and immunohisto-chemical studies of Iwasaki *et al.* (1985) on the brains of two cases of human rabies dramatically change what has been a traditional concept. Not only were virions found budding into intercellular spaces (Fig. 17), but also intracellular virions were sometimes observed in the absence of

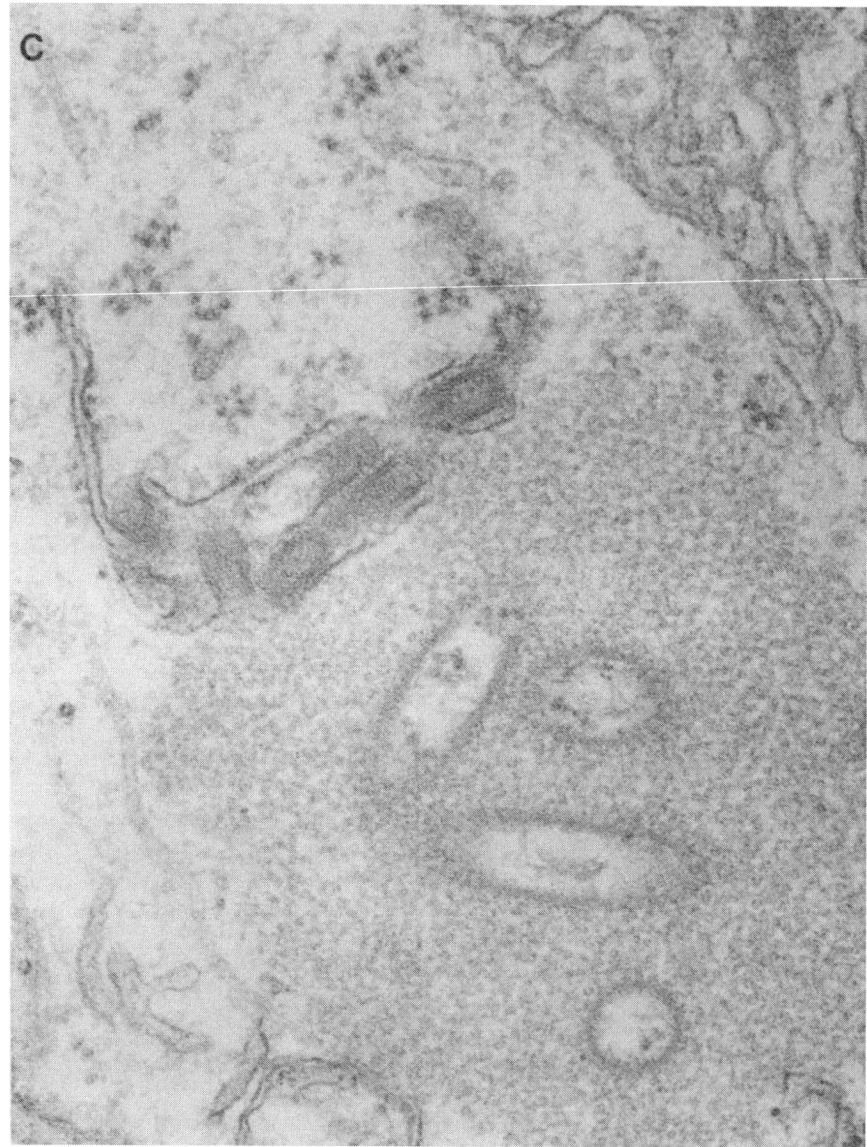

FIGURE 15. (Continued)

or remote from intracellular viral inclusions (Fig. 18). This latter observation tends to indicate that the virus can mature without inclusion
formation (Iwasaki et al., 1985). If this possibility is real, it would explain
why Negri bodies were occasionally not found in virologically confirmed
human rabies cases (Dupont and Earle, 1965; Duenas et al., 1973; Gold-

wasser and Kissling, 1958). It should also be noted that the longer the duration of the disease, the more chance there is of finding inclusions.

4. Mechanisms of Virus Spread within the Brain

The spread of rabies virus in the brain appears to occur by several pathways. There is evidence of *active transport* of the virus by multipli-

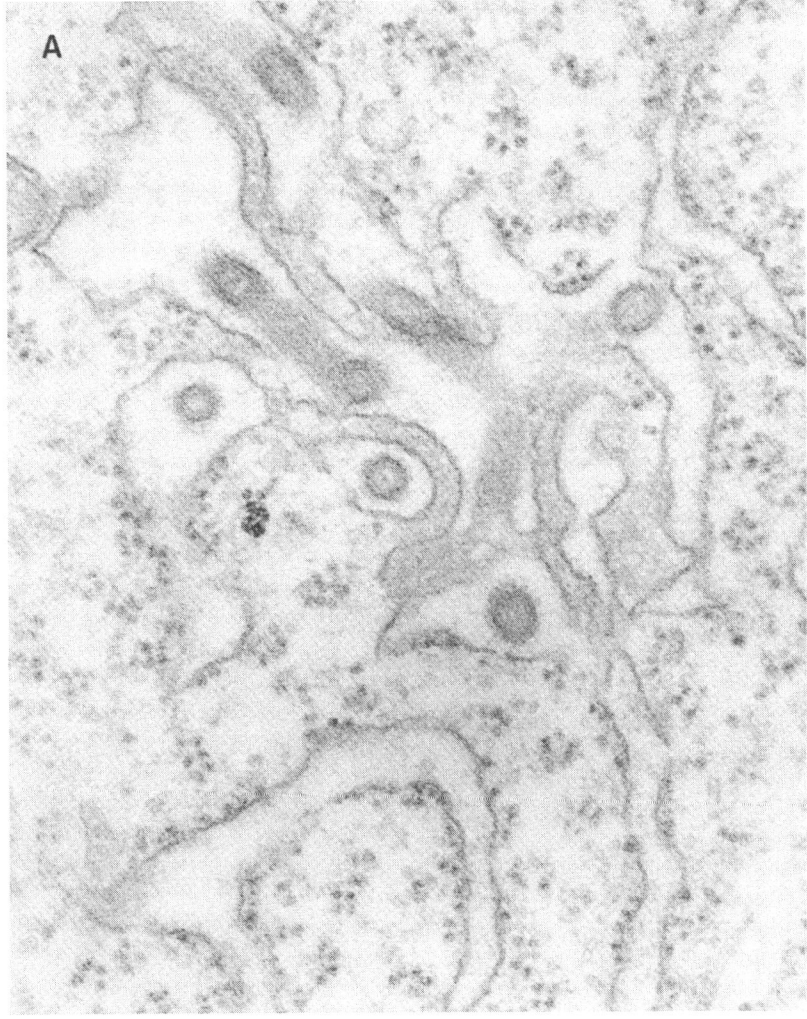

FIGURE 16. Rabies virus maturation in the perikaryon of adult mouse hippocampus (A) and in transit at a synapse within the hippocampus (B) 7 days after inoculation with street virus. (A) ×96,000; (B) ×138,600. Supplied by Y. Iwasaki.

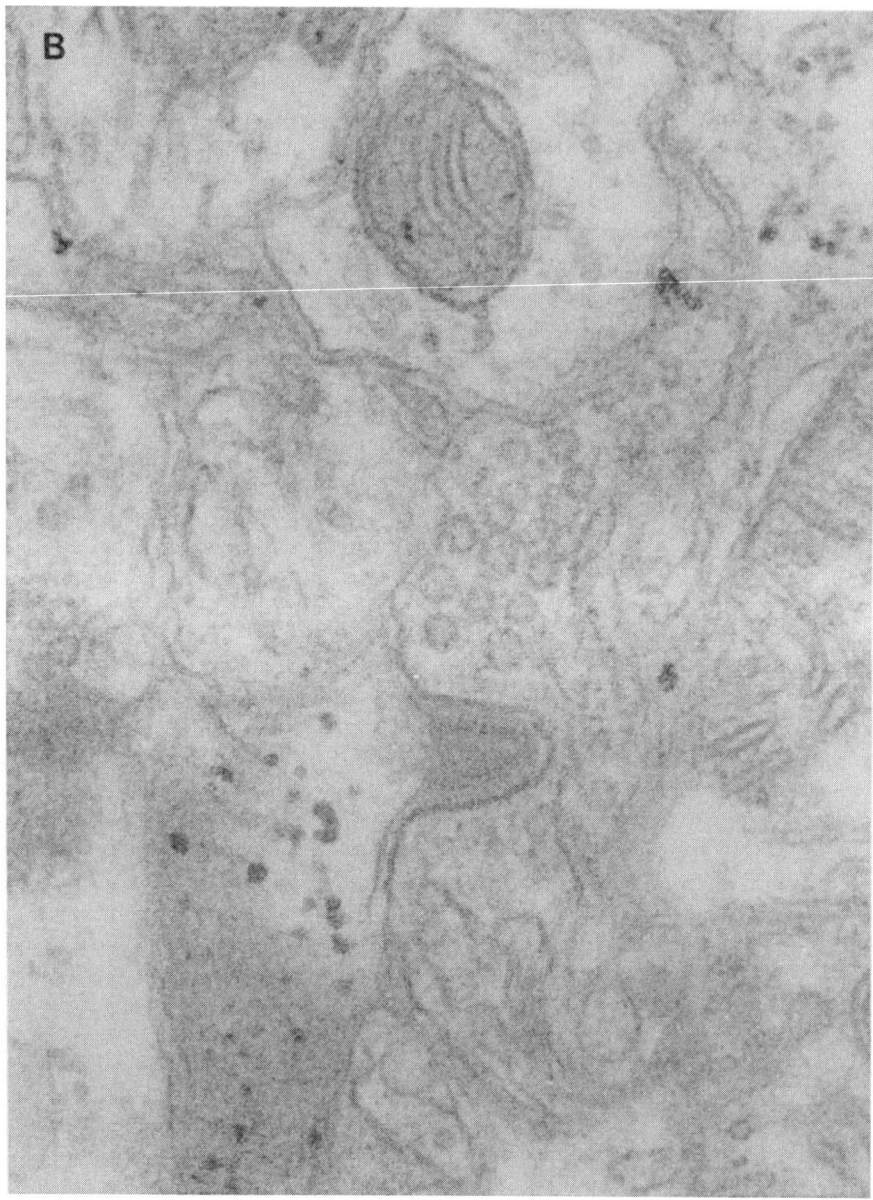

FIGURE 16. (Continued)

cation in neurons and transfer of progeny virus to susceptible cells via budding from plasma membrane, release into tissue spaces, and virus reentry by a pinocytotic pathway (Fig. 19) or by direct cell-to-cell transmission (Iwasaki and Clark, 1975). From the studies cited above, it is clear that most virus particles remain trapped within neurons after budding on internal membranes (particularly in the terminal stages of the infection), and relatively small numbers of virions find their way into extracellular spaces after budding on plasma membranes. Iwasaki and Clark (1975) showed that earlier in infection, a high proportion of virus is released into tissue spaces. Once delivered in the extracellular spaces of the brain, virus can presumably spread by *passive transport* over relatively long distances through intercellular spaces large enough to pass

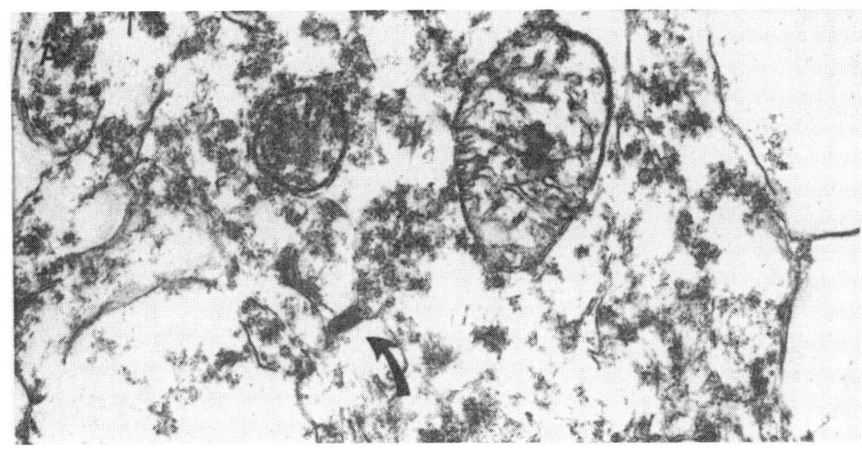

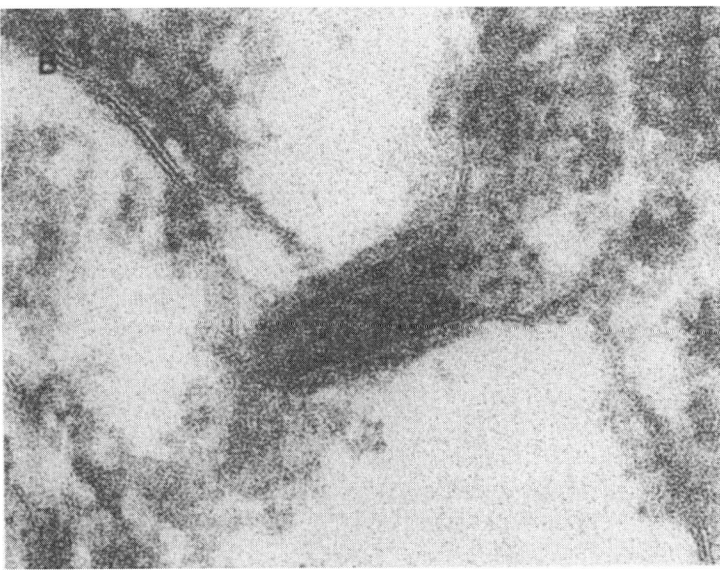

FIGURE 17. Rabies virus in a case of human rabies budding from plasma membrane into intercellular space. (A) The virion (arrow) shows a coat of spikelike projections (presumably surface glycoprotein). ×30,000. (B) Higher magnification of the virion. ×150,000. Supplied by Y. Iwasaki.

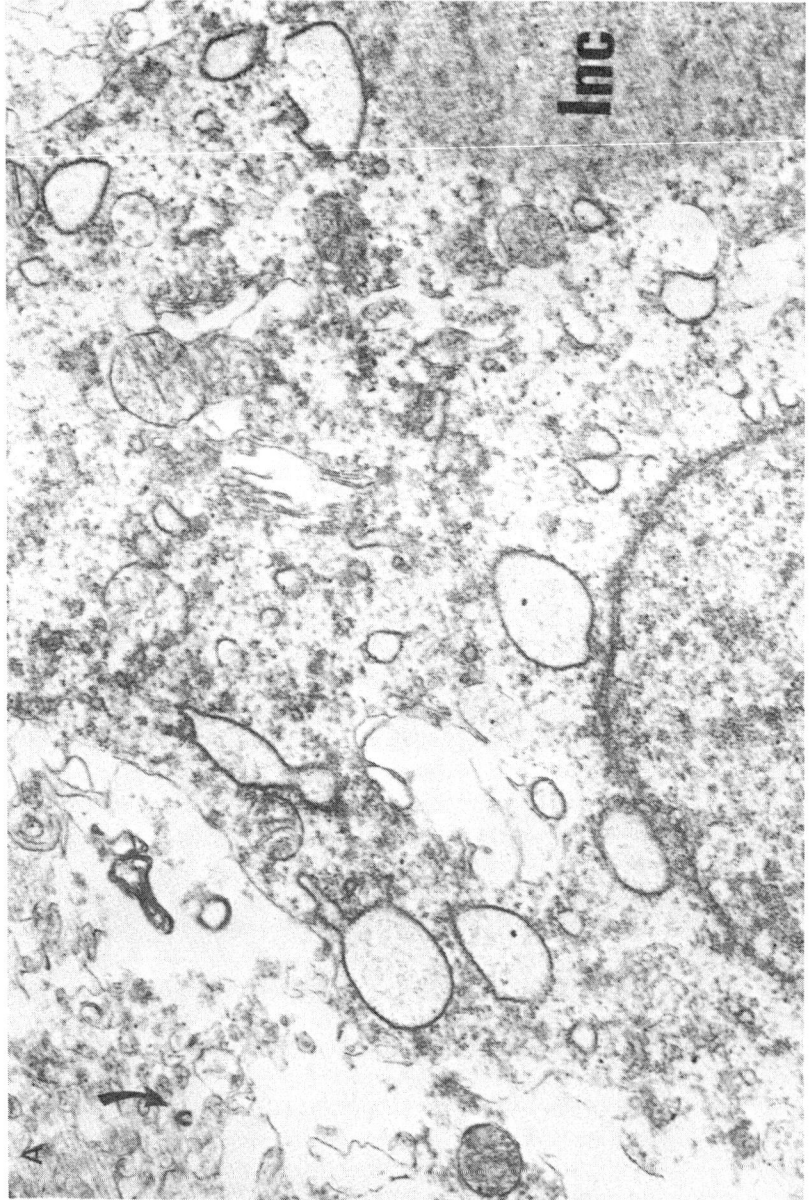

FIGURE 18. (A) An extracellular rabies virion (arrow) in the intercellular space remote from an intracellular inclusion (Inc) within a neuron of the postcentral gyrus of human brain. ×23,000. (B) Higher magnification of the virion. ×100,000. Supplied by Y. Iwasaki.

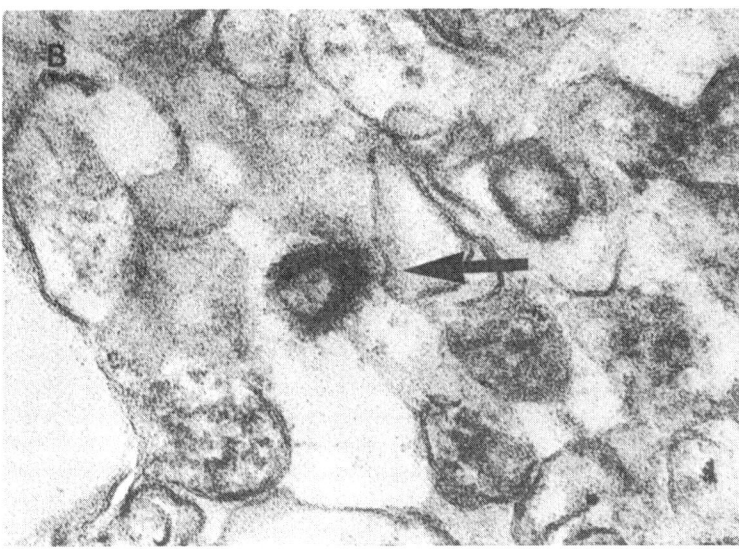

FIGURE 18. (Continued)

virus by interstitial-fluid movement (Blinzinger and Muller, 1971; Cserr and Ostrach, 1974). Virus that otherwise is transmitted directly from one nerve cell to a contiguous one (Iwasaki and Clark, 1975) could also spread and result in the same long-distance movement because of the many long, intertwined neuronal processes that connect in the brain (Murphy, 1977). Still, the most rapid and presumably efficient way to disseminate virus would be to deliver virions by neuronal (passive) transport (just as with peripheral-nerve axons) to distant synaptic endings. There are several reports of finding virus at synapses in the brain (Iwasaki and Clark, 1975; Charlton and Casey, 1979; Iwasaki *et al.*, 1985) that support the theory of transsynaptic propagation and distribution of rabies virions or virion constituents in the brain. Virus budding from postsynaptic plasma membrane and virus entrapment by an invaginated portion of presynaptic plasma membrane of an adjacent axon terminal have also been documented in the dorsal horn of the lumbar spinal cord of skunks (Charlton and Casey, 1979), suggesting that virus can move transneuronally into the presynaptic axon terminal. It is important to note, from the observations of Tsiang (1982) (reviewed by Tsiang, 1985) in this regard, that the impairment of neuron function in the rabies virus-infected rat brain was markedly increased as the symptoms of rabies appeared. The effect of rabies infection on neuronal function was observed by measuring the binding affinity of quinuclidinyl benzylate, an antagonist of muscarinic acetylcholine, to muscarinic acetylcholine receptors. Others studies using a rabies virus-infected hybrid cell line (mouse neuroblastoma × rat glioma) have also suggested that normal cellular receptor functions (at least *in vitro*) can be modified by rabies virus infection (Munzel and Koschel,

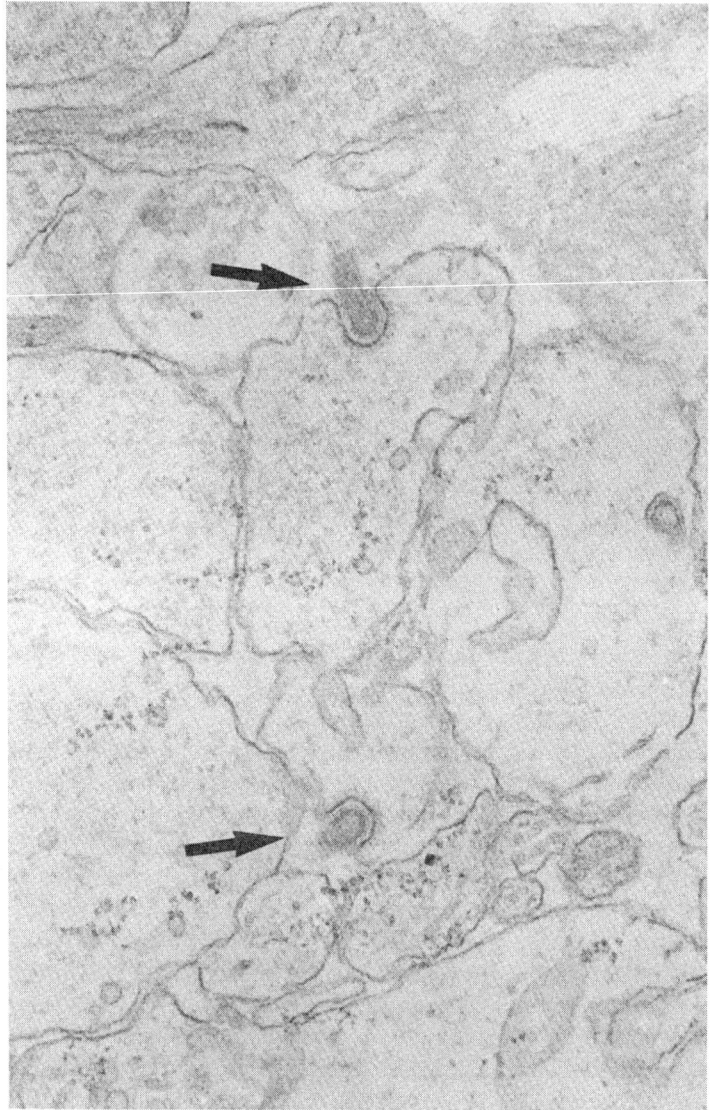

FIGURE 19. Rabies virus entry (arrows) into neuronal cells of hippocampus in CVS virus-infected suckling mouse brain. Note that the plasma membrane is coated with a fringe on the cytoplasmic side at the site of viral entry. ×58,800. Supplied by Y. Iwasaki.

1981; Koschel and Munzel, 1984). It should be realized that the propagation of complete rabies virions may not be absolutely necessary for the spread of infection within the CNS and that viral surface antigens may not be the only viral component that interacts with cellular receptors. The nucleocapsid component of the virus on its own might be sufficient for the transmission of the infection, and this might take place by the passage of the active ribonucleoprotein–transcriptase complex through the synapse (Gosztonyi, 1978; Gosztonyi and Ludwig, 1984).

C. Spread of Rabies Virus from the Central Nervous System: Dissemination to Extraneural and Nonneural Tissues

The virus up to this stage has been specifically neurotropic, with the exception of possible muscle involvement at the inoculation site. Once virus begins to spread from the brain in the final phase of rabies infection, the tropism of infectious virions changes and virus or viral antigens are delivered to a variety of extraneural tissues (Duenas et al., 1973; Howard, 1981). Centrifugal spread of virus generally occurs via the same axoplasmic routes that were used for the centripetal passage of virus to the CNS. The spread of rabies virus in the absence of CNS infection is virtually nil. Among the most heavily infected tissues following propagation of virus in the CNS are the end organs in oral and nasal cavities, and the head and neck (Schindler, 1961, 1965; Kligler and Bernkopf, 1943; Fischman and Schaeffer, 1971; Charlton et al., 1983). The fact that rabies antigen is occasionally detectable in skin biopsy (W. B. Smith et al., 1972; Fekadu and Shaddock, 1984; Bryceson et al., 1975; Blenden, 1981) and in corneal impressions (Baer et al., 1982) is evidence of its nerve-mediated dissemination.

The involvement of salivary glands in the dissemination of virus from the CNS is extremely important because it represents the site of infection required for effective transmission of disease. Virus titers in the salivary glands often exceed those in the brain (Dierks, 1975). In some species, the salivary glands become infected before the signs of CNS infection become apparent (Fekadu et al., 1982; Fekadu and Shaddock, 1984), and at the height of salivary gland infection, large amounts of fluorescent viral antigen may be detected in individual acinar cells, in cell clusters, and occasionally in the entire acinus. Evidence for neuronal centrifugal spread to the salivary glands has been provided in a number of studies (reviewed by Dierks, 1975), the most direct evidence perhaps coming from the studies of Dean et al. (1963). They showed by removing the right lingual nerve and cranial cervical ganglion in mice, foxes, and dogs that infection of salivary glands in these animals was selectively spared. In salivary-gland tissue of hamsters inoculated intramuscularly with vampire bat, arctic fox, or CVS rabies, and the rabies-related (Mokola and Lagos bat) viruses, immunofluorescence observations consistently indicated that mucous acinar cells were involved (Murphy et al., 1973b). "No other cell type within the salivary glands contained antigen, although small nerves within and upon the capsule of glands were commonly infected," according to Murphy et al. (1973b). Detection of viral antigen in the cytoplasm of salivary duct cells and within the duct lumens indicates that virus can be delivered in high titers to the next bite site (Fekadu and Shaddock,

1984). Nearly all virus in the salivary glands is formed on plasma membranes of mucus-producing (mucogenic acinar) cells and is thereby immediately delivered into saliva by normal secretory flows (Dierks *et al.*, 1969; Murphy, 1985). Virus particles amassing in the secretory lumina of salivary gland (Fig. 20) clearly contrasts with the previously discussed sparsity of rabies particles in interstitial spaces in brain. Virus particles may occasionally be found budding from the endoplasmic reticulum into cisternae, but are rarely found directly within the cytoplasm of acinar cells (Dierks *et al.*, 1969). Fixed rabies virus, on the other hand, was not found at all in salivary-gland tissues of hamsters (Murphy *et al.*, 1973b).

Other tissues and organs that become infected late in rabies infection may be responsible for transmitting virus. These include sensory-nerve end organs in the oral and nasal cavities, taste buds in the tongue, and a wide range of other sensory end organs within the mouth and nares (which include "functionally unidentified nerve endings in cornified filiform papillae of the tongue and mucosa and submucosa of the nares, soft palate, and oropharynx") (Murphy *et al.*, 1973b). These sites are well situated to deliver virus directly into oronasal secretions. Since these sites were also found to be infected in bats (Constantine *et al.*, 1972), exposure to aerosol is, then, a plausible mechanism for rabies transmission when a bite is not involved. This mechanism of exposure provides a basis by which spelunkers exploring bat-infested caverns might become infected (Constantine, 1962, 1966; Atanasiu, 1965; Irons *et al.*, 1957; Hronovsky and Benda, 1969; Winkler *et al.*, 1972, 1973) or for human-to-human spread of rabies, particularly among medical and nursing staff who look after rabies patients (Meyer, 1957; Lintjorn, 1982).

Infection sites of less importance to the transmission of virus that are reached by nerves leading centrifugally to nerve endings in contact with susceptible nonneural cells include the adrenal glands, pancreas, kidney, heart muscles, intestine, brown fat, hair follicles, and retina and cornea (Ross and Armentrout, 1962; Dierks *et al.*, 1969; H. J. Johnson, 1965; Cheetham *et al.*, 1970; Cifuentas *et al.*, 1971; Murphy *et al.*, 1973b; Fischman and Schaeffer, 1971; Duenas *et al.*, 1973; Fekadu and Shaddock, 1984). Apart from the unfortunate transfer of rabies by transplanting rabies virus-infected corneas (Houff *et al.*, 1979; *Morbidity and Mortality Weekly Report*, 1980, 1981; Sureau *et al.*, 1981), these sites appear to be of no consequence to the virus growth or cycle of infection.

D. Rabies Virus Entry into Nerves and Cells

1. Pathways of Viral Neural Tropism

The strategies used by rabies virus to gain entry into the body through nerve endings and to spread from cell to cell are perhaps as unique as the mechanisms that allow rabies virus to extend from the entry site to the

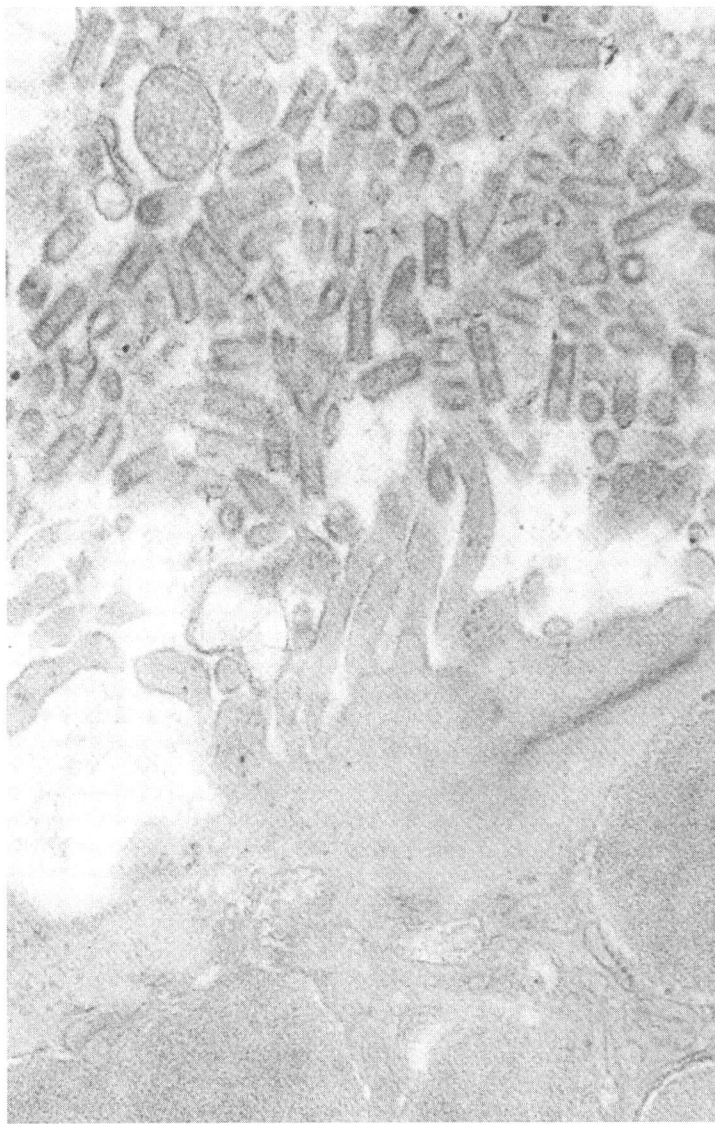

FIGURE 20. Rabies virus amassing in the lumen of fox salivary gland. ×46,500. Supplied by F. A. Murphy and A. K. Harrison.

CNS, to spread to distant target organs, and to exit in a way that favors transmission and perpetuation of the infection. The strict neurotropism of rabies virus in its centripetal and centrifugal dissemination throughout the body immediately suggests that the uptake of virus into nerves occurs, whether through sensory- or motor-nerve endings, by a process involving specific host-cell receptors at nerve endings. Virus transmission through synaptic junctions at nerve terminals may similarly involve receptors interacting with virus, although virus may also passively move through synapses (Gosztonyi, 1978, 1979).

Considering that the terminal portion of the axon at neuromuscular junctions is a possible site of entry for rabies virus into the nervous system, Watson *et al.* (1981), using immunofluorescence techniques, observed the location and uptake of rabies virus at motor-nerve endings. They detected rabies virus antigen 1 hr after intramuscular inoculation of virus into the hindleg at sites comparable in form and distribution to cholinesterase-positive sites, representing motor endplates, in the injected leg. Colocalization of viral proteins and specific anatomical neural structures, cell types, or molecular components of those cells by immunofluorescence and thin-section electron-microscopic techniques has focused attention, in more recent studies, on possible receptor sites in normal tissue for rabies virus attachment and uptake (Lentz *et al.*, 1982; Burrage *et al.*, 1983, 1985). Lentz *et al.* (1982) removed the mouse diaphragm and its attached phrenic nerves and infected several similar preparations with fixed rabies virus. The inoculum virus was then located in the mouse tissue by immunofluorescent staining of viral antigen with anti-rabies virus antibody at different times after the initial exposure to rabies virus. From their observations, they suggested that rabies virus may attach to the acetylcholine receptor at the neuromuscular junctions in the mouse diaphragm. The suggestion was made on the basis of a similarity in the morphological appearance of clustered viral antigen at neuromuscular junctions compared with the distribution of patches visualized on uninfected diaphragm with cholinesterase staining or by a rhodamine-conjugated α-bungarotoxin. When cultured chick myotubes were exposed to rabies virus in a parallel study by Lentz *et al.* (1982), viral antigen was also distributed in patches on the cell surface in a pattern similar to that observed following staining with rhodamine-labeled α-bungarotoxin. By electron microscopy, the infected myotubes further showed association of rabies virus with specialized surface patches previously shown to contain high-density acetylcholine receptors. Finally, treatment of primary cultures of chick myotubes with high concentrations of α-bungarotoxin or D-tubocurarine, both of which specifically compete with acetylcholine for the acetylcholine receptor (Heidmann and Changeux, 1978; Gershoni *et al.*, 1983), reduced the number of myotubes that were infectable by rabies virus. In other studies, the binding of rabies virus to cellular membranes prepared from embryonic-chick myotubes of increasing age closely paralleled the acetylcholine-receptor content of those membranes (Burrage and Lentz, 1981; Lentz *et al.*, 1985). Together, these findings form a hypothesis that the virus's ability to recognize cholinergic binding sites seems to restrict its host-cell range. Moreover, since such terminals are thought to bind acetylcholine at specific sites (Hubbard and Wilson, 1973; Magleby *et al.*, 1981) distinct from the postsynaptic acetylcholine receptor, so they might also be selectively recognized by the virus. On this view, the ability of the rabies virus to mimic acetylcholine-binding properties would have two functions: Virus particles would be concentrated in the nerve-terminal region by binding to the abundant postsynaptic acetylcholine receptors, and this increased concentration would enhance the chances of virus interaction (if it does interact directly) with the rare presynaptic receptors (Wunner, 1982).

The generality of the foregoing observations, however, has been examined in a variety of cell-culture systems including cells that lack high-density acetylcholine receptors, all of which are susceptible to infection by rabies virus (Reagan and Wunner, 1985; Tsiang, 1985). Absence of a total correlation of the presence of acetylcholine receptors with susceptibility to rabies virus infection and failure to demonstrate a dose dependence of toxin-mediated inhibition of rabies virus infection in cultured rat myotubes from 18-day-old embryos or in a differentiated rat cell line that elaborates high-density acetylcholine receptors argue strongly that the susceptibility of various cell types probably does not depend solely on the presence of a single specific receptor (Tsiang, 1985).

There is increasing evidence for the participation of lipids in the early interaction and binding of rabies virus to cell membranes (Perrin et al., 1982; Wunner et al., 1984; Superti et al., 1984). Phospholipids and lipoproteins present in sera have been shown to have inhibitory activity on the infectivity of rabies virus (Halonen et al., 1974; Seganti et al., 1983; Suzuki et al., 1977). Populations of glycolipids found predominantly in nervous tissues (Karlsson, 1982) give special significance to these activities in virus binding. Gangliosides may play a role in the binding of rabies virus to membrane surface by providing sialic acid moieties for attachment, although these gangliosides may also serve a more general role in cell-surface events (Superti et al., 1986). These data indicate that the receptor on the cell surface for the binding of rabies virus is probably not a unique specific molecule; rather, it may be a complex structure comprising several cellular components with specific moieties and structural presentations. The isolation and characterization of this receptor "complex" is not complete, and the relationship between such membrane entities and the known cellular tropism of rabies virus has yet to be determined.

2. Differential Spread of Fully Virulent and Nonvirulent Rabies Viruses in Vitro and in Vivo

The mechanism of rabies virus spread during natural infection was recently reexamined with variant viruses that have a nonvirulent phenotype—i.e., they are unable to kill adult mice by intracerebral virus challenge—and differ from the virulent virus in their ability to infect neurons in vivo and neuroblastoma cells in vitro (Kućera et al., 1985; Dietzschold et al., 1985). The variant viruses were selected on the basis of resistance to neutralizing monoclonal antibody (Flamand et al., 1980;

Coulon *et al.*, 1982; Dietzschold *et al.*, 1983). The glycoprotein comprising the surface spikes of the viral envelope of these viruses is 505 amino acids long (Anilionis *et al.*, 1981; Yelverton *et al.*, 1983), but is modified from the glycoprotein of virulent rabies virus by a single amino acid substitution at position 333; arginine is replaced by isoleucine in the ERA strain and by glutamine or glycine in the CVS strain (Dietzschold *et al.*, 1983; Seif *et al.*, 1984, 1985). An amino acid substitution at position 333 has also been identified in the attenuated (nonvirulent) fixed rabies virus Flury HEP and Kelev strains (Smith and Wunner, unpublished data). Thus, the amino acid arginine at position 333 appears to be a molecular marker of the virulent phenotype and an integral part of the mechanism that determines the extent of virus spread in the CNS; the nonvirulent virus is significantly less neuroinvasive (Kućera *et al.*, 1985; Dietzschold *et al.*, 1985). Penetration of the CNS of the adult rat by the CVS strain of rabies virus inoculated into the anterior chamber of the eye has been compared to the penetration from the same site by two non-virulent derivatives of the CVS strain (Kućera *et al.*, 1985). On examination of the primary sites of entry, i.e., via the intraocular parasympathetic oculomotor fibers, the retinopteral fibers of pretectal origin, and the intraocular fibers at the ophthalmic nerve, it was found that the mutant viruses lost the capacity to invade the first two groups of fibers. The mutant viruses, however, were able to infect the lens, whereas the virulent CVS strain was not. The mutants penetrated the brain, but the infection was slower and involved different cerebral structures compared with that of the virulent strain. These results compare favorably with the differential rates of spread of virulent and nonvirulent viruses throughout the brain seen by Dietzschold *et al.* (1985). After intracerebral inoculation, the distribution of infected neurons in the brain was similar for both types of viruses, but the rates of spread, the number of infected neurons, and the degree of cellular necrosis were much lower in mice infected with the nonvirulent virus. What may be happening to slow down the spread of nonvirulent strains within the brain is suggested by the *in vitro* studies of Dietzschold *et al.* (1985), in which it was shown that direct cell-to-cell transmission of nonvirulent virus in infected mouse neuroblastoma cells is blocked. In other words, the presence of arginine at position 333 in the virulent strain of rabies virus may be essential for this particular mode of virus transmission in the CNS previously described by Iwasaki and Clark (1975), although no evidence has been reported suggesting a direct role for arginine at position 333.

E. Clinical Features of Rabies Virus Infection

1. Acute Infection

The clinical features of rabies are generally indicative of one or the other of two distinct clinical patterns of the disease that rabies assumes.

The clinical patterns or forms of rabies are known as "furious" and "paralytic" or "dumb" rabies, although elements of both forms may often occur together, particularly during the latter stages of the disease. Following a typical exposure from the bite of a rabid animal and a variable and often very prolonged incubation period (2 weeks to 1 year and, in rare instances, over 2 years), the clinical pattern of the infection begins to emerge. K. P. Johnson and Swoveland (1984) have described (as others have before) the development of clinical rabies in humans by dividing the course into three major phases.

The first phase is a prodromal period lasting several days, during which patients experience the first symptoms, which are almost entirely nonspecific. These include general malaise, chills, fever, headache, photophobia, anorexia, nausea, sore throat, cough, and musculoskeletal pain. Patients also complain early of having, along with pain, abnormal sensations around the wound site, such as itching, burning, numbness, or paresthesia (Dupont and Earle, 1965).

The illness then progresses to the second stage, which is an acute neurological phase during which patients exhibit signs of increasing anxiety (agitation) and alternating periods of clarity and episodes of delirium. Behavior disturbances occur regularly and range from mild memory disturbance to severe depression and include periods of hyperactivity, aggressiveness, intolerance to stimuli such as noise, and delirium (Cifuentas *et al.*, 1971; Emmons *et al.*, 1973; Bhatt *et al.*, 1974; Sung *et al.*, 1976; Conomy *et al.*, 1977; DeWet, 1980; Lintjorn, 1982). During this phase, the patient may also experience seizures, hallucinations, and hyperventilation. The classic symptoms of hydrophobia—difficulty in swallowing, and choking—are also often evident (D. A. Warrell, 1976; H. G. Johnson, 1965) during this period. The overpowering fear of water, which is virtually pathognomonic of rabies, is precipitated by attempts to drink, or at times by the sight, sound, or even mention of water. Attempts to swallow are defeated by uncontrollable and equally powerful convulsive spells with an increase of coughing, vomiting, retching, grimacing, and asphyxiation. Violent struggling, bolting, and even attacking attendants are usual forms of uncontrollable behavior. There is a steady deterioration of mental status, unless generalized convulsions suddenly precipitate coma or death. The entire clinical picture of rabies in humans has recently been thoroughly reviewed by Nicholson (1983).

The third and final phase of the disease that progressively replaces the acute neurological phase is one of coma, lasting an average 3–7 days (Nicholson, 1983) and resulting in death. The majority of patients who suffer rabies present this three-phase clinical pattern, which is typically referred to as "furious" rabies.

Not all patients, however, develop the clinical picture of furious rabies. A significant number of cases (close to 20%) present a neurological picture comparable to that of the Landry–Guillain–Barré syndrome, which has paralysis as the principal feature (Nicholson, 1983). The term "paralytic" or "dumb" rabies has been given to this clinical form of rabies infection in humans and animals. It most often develops following vampire bat bites (Hurst and Pawan, 1968; Warrell, 1976) and tends to develop after a minor bite (Love, 1944). After a typical prodromal period, the paresis often begins in the bitten extremity and then spreads either symmetrically or asymmetrically, accompanied by fasciculations that rapidly progress to flaccid paralysis. Patients with paralytic rabies generally survive for longer periods than those with furious rabies, but the course of paralytic rabies can be modified at any stage by the appearance of spasms, hydrophobia, and convulsions (described by Nicholson, 1983). Patients who have died unsuspectingly from paralytic rabies have in at least four instances been the cause of human-to-human transmission of infection and disease when the corneas from the fatal cases of dumb rabies were transplanted to susceptible recipients (Houff *et al.*, 1979; *Morbidity and Mortality Weekly Report*, 1980, 1981; Sureau *et al.*, 1981).

2. Chronic Infection

Chronic or "abortive" infections with rabies virus can be distinguished from the acute rabies infection simply in that they represent rare cases in which recovery from rabies virus infection occurs. They are of interest because they also represent the antithesis of the commonly accepted view that rabies is invariably fatal. Clark and Prabhakar (1985) summarize chronic infections as "those in which prolonged salivary shedding of virus occurs in the absence of encephalitic disease" and include those infections in which the "clinical disease progresses over a very prolonged period to an eventual fatal outcome." Abortive rabies infections do not progress to fatal encephalitis and are accompanied by an immune response. Experimentally induced abortive rabies has been described in three animal-model systems using mice (Bell, 1964; Bell *et al.*, 1966; Lodmell *et al.*, 1969; Wiktor *et al.*, 1972a; Baer *et al.*, 1977; Smith, 1981), chickens (Schneider, 1975) and dogs (Fekadu and Baer, 1980). In the animals that appeared to "abort" rabies, virus-specific antibody was detected in brain and serum, yet infectious virus or virus antigen was absent in the brain. Wiktor *et al.* (1972a) developed a model of abortive rabies by inoculating street rabies virus into the hindlimb either 24 hr before or 2 hr after intracerebral inoculation of the mice with the nonvirulent Flury HEP strain of rabies virus. The mice developed permanent paralysis of the inoculated limb and survived for over 1 year. Both virus-neutralizing antibody and interferon were detected in these animals, but interferon appeared earlier (7–14 hr after Flury HEP virus inoculation) than the neutralizing antibody, which was detected on the second day after intracerebral inoculation.

A rise of rabies-specific antibody is indirect but affirmative evidence of previous rabies infection (Baer and Olson, 1972; Bell, 1975). The presence of specific antibody in cerebrospinal fluid (CSF) is also an indication of a previous rabies virus infection (Arko *et al.*, 1973). Recovery from paralysis to a "normal" state has been reported for a number of animal species and man, and yet with our present state of knowledge of the pathogenesis of rabies (or the immune responses to infection), it has not been possible to draw definitive conclusions regarding the mechanisms of rabies-virus infections that either cause paralysis or allow recovery from paralysis. Many of the individual factors that may contribute to the recovery mechanisms or are conducive to nonfatal infection are described in detail by Bell (1975). Those that deserve mention because they comprise the nonimmune mechanisms of recovery are interferon, the febrile response, local acidity and low oxygen tension caused by the inflammatory reactions, exposure to elevated temperatures, and defective interfering (DI) particles. These have been reviewed by Bell (1975) and by Clark and Prabhakar (1985).

Perhaps the most significant factor or factors of all those that lead to abortive infection is the combined effect or effects of virus strain and route of inoculation; all other factors being governed by these two determinants. Attenuated strains (e.g., Flury HEP and ERA) of rabies virus that are used as "live" vaccines do not cause disease when inoculated peripherally in many species of domestic animals (Abelseth, 1964; Fox et al., 1957). Yet Flury HEP rabies virus, when inoculated intracerebrally into newborn mice (but not adult mice) or into adult rhesus monkeys, and into hamsters and guinea pigs (depending on virus dosage), is invariably lethal. Similarly, ERA rabies virus induces fatal rabies in mice inoculated intracerebrally (Abelseth, 1964) and in several species of rodents inoculated orally (Winkler et al., 1976). As Clark and Prabhakar (1985) aptly point out, "the site of peripheral replication of an attenuated 'modified live' rabies virus administered by the standard intramuscular route has never been identified" and there is no way of telling without knowing the fate of such viruses (rate of replication and spread) what their impact might be on the host's immune response. Albeit, even if they should reach the brain, they may not induce fatal disease (Fischman and Strandberg, 1973; Clark and Prabhakar, 1985) unless the infection is made fatal by induction of B-cell-specific immunosuppression (Miller et al., 1977).

It is hoped that biochemical studies of phenotypically "modified" rabies viruses currently being carried out will reveal in structural detail the full biological and immunological implications of attenuation. As mentioned in Section II.D.2, the first correlation of structure with function has been with regard to the pathogenicity of fixed rabies virus strains for adult mice. The virion glycoprotein of those viruses that lost their ability to cause a fatal disease in adult mice after intracerebral inoculation had a critical amino acid substitution at position 333 of their amino acid sequence. When studied in mice, the mutated virus that lost its lethal capacity was also at a distinct disadvantage compared with the virulent form of the same virus when spreading from the inoculation site. The relative ability of these same viruses to spread from cell to cell in cell culture tended to corroborate the distinction observed in vivo.

Other structural changes involving nucleotide substitutions in the viral RNA and amino acid changes in viral proteins may well play key roles in altering the biological, immunogenic, and antigenic properties of the virus and establishing long-term chronic or "persistent" infections in animals or humans. Long-term persistent infections with rabies virus in cell culture have been studied in considerable molecular detail, but so far there is not a clear understanding of the effects of the evolutionary changes found in the viral RNA and protein (Holland *et al.*, 1976; Rowlands *et al.*, 1980), of the role of DI particles (Holland *et al.*, 1980; Kawai *et al.*, 1975; Wiktor *et al.*, 1977b; Clark *et al.*, 1981; Wunner and Clark, 1980), of the effect of temperature-sensitive mutants (Clark and Ohtani, 1976; Aubert *et al.*, 1980; Clark and Wiktor, 1972), or the effect of changes in synthesis of virus-specific RNA and protein (Villerreal and Holland, 1976; Wild and Bijlenga, 1981; Tuffereau *et al.*, 1985) on altering the course of rabies virus in natural infections. It is to be hoped that new approaches involving sophisticated antibody and molecular probing techniques will allow investigators to search for these specific modifications in the virus-infected animal in order to determine their significance in cell culture.

III. IMMUNITY: MECHANISMS OF RESISTANCE AND SURVIVAL

Unlike most viruses that cause acute infections, rabies virus evades or bypasses the host's immune defenses during much of the early course of infection because of its intrinsic neurotropism and exceptional neuroinvasiveness. Once rabies virus enters the nervous system, it is isolated compartmentally from circulating antiviral antibodies and immunocompetent lymphocytes because of their inability to penetrate the nervous system. Moreover, since rabies virus fails to amplify sufficiently in peripheral inoculation sites and infectious-virus titers as well as viral antigenic mass remain extremely low (whether or not virus replicates at all), the host's responses to virus infection before virus invades the nervous system are either absent or too weak to protect against disease. While this might appear to give the virus a distinct advantage, ironically it is the prolonged incubation period so unique to rabies virus infection that also provides an unusual opportunity to use postexposure treatment, including immunization and immunoglobulin therapy, for the prevention of rabies in humans.

When antiviral responses do occur, usually late during infection or following vaccination, a mixture of specific (humoral and cell-mediated) and nonspecific (induction of interferon) immunological responses are observed. The relative importance of the different components of the immune response and the time after infection or induction by vaccine in which each has its influence on the infectious process have been matters of considerable interest and debate for the past 25 years. Recently, with the development and application of more sophisticated techniques in molecular biology and immunology, it has been possible to take new perspectives in examining the complexity of the apparent interdependence among antibody, macrophages, lymphocytes, and interferon in the host's defense against rabies virus infection. This new research effort is still in its infancy but moving rapidly.

A. Host Reticuloendothelial Response

Macrophages represent a single nonspecific mechanism of defense against rabies virus infection that is stimulated by the trauma and infection of foreign substances from the bite of a rabid animal. By engulfing virus particles, the macrophages limit the spread of virus to cells and nerve endings after entry into the host or after release from infected muscle cells at the bite site (Turner and Ballard, 1976; Lagrange et al., 1978). Some of the scavenging monocyte-macrophages may also be of central importance in initiating the specific immune response by degrading and presenting the processed viral antigens on their surface to the T and B lymphocytes (Unanue, 1984), although the level of activity of these rabies immune lymphocytes throughout the early stages of infection appears to be extremely low. Rabies virus does not grow in cultures of macrophages (Turner and Ballard, 1976).

B. Humoral Immunity

1. Response to Viral Infection

While the slow progression of the early stages of natural (and some experimental) infections would seem to provide opportunity for a total spectrum of host defense mechanisms to be initiated, it often does not. Stimulation of B-lymphocyte immunoglobulin synthesis helped by viral-antigen-dependent T lymphocytes usually does not occur until after clinical symptoms appear. In response to the massive amount of viral antigen that is generated through widespread infection of the CNS and made accessible to the host's reticuloendothelial system, virus-neutralizing antibodies are produced. Serum antibody titers, however, often remain low until the terminal phases of the disease and then reach their highest levels at death (Murphy, 1977; Hattwick and Gregg, 1975). In some cases, antiviral antibodies are found in CSF as well as in serum. The presence of any antibody in a nonimmunized patient or animal is still the most practical and useful basis for antemortem diagnosis of rabies, and antibody determination can be made in 24 hr by an immunofluorescence technique (J. S. Smith et al., 1973). The major class of immunoglobulins that is produced and sustained longest in response to the massive increase of antigen that occurs throughout late infection or in cases of prolonged human rabies appears to be IgG; little, or in some instances no, IgM is detected late in infection (Rubin et al., 1971; Schuller et al., 1979).

In animals experimentally infected with large amounts of rabies virus, neutralizing antibodies of the IgM class (19 S) have appeared in the early stage of infection within 3 or 4 days (Fujisake et al., 1968). Infection with a small amount of rabies virus inoculation, however, did not produce 19 S antibody.

2. Response to Viral Immunization

Resistance to rabies has long been associated with the presence of virus-neutralizing antibody. Vaccine-induced antibody (in animals and humans) or passively administered immunoglobulin (for humans) is regarded as a key factor in the prophylactic protection of animals and humans and postexposure treatment of humans (reviewed by Wiktor, 1980), although the evidence shows that administration of antibody alone after infection seldom significantly reduces mortality. The relationship of virus-neutralizing antibody to protection has been established for a variety of animal species, including dogs, cats, cattle, foxes, monkeys, and rodents (Koprowski and Black, 1952; Fenje, 1960; Otto and Heyke, 1962; Dean et al., 1964; Cabasso et al., 1965; Sikes et al., 1971; Baer and Yager, 1977). Vaccine-induced antibody seems to be more protective against homotypic virus strains in some species than it is in others.

Antiviral antibodies with virus-neutralizing capabilities are appropriately directed at the surface glycoprotein of rabies virus, the only envelope protein capable of eliciting a protective response to challenge virus (Schneider et al., 1973; Wiktor et al., 1973; reviewed by Wunner et al., 1983). Neutralizing antibodies that result from vaccinations interact with the virion glycoprotein at three major antigenic sites, each functionally independent of the other (Lafon et al., 1983, 1984). The antigenic relationships among rabies viruses and the relative binding of virus-neutralizing antibodies depend largely on the antigenic similarities and differences within these major sites. Other antigenic sites that have been recognized on the glycoprotein of some virus strains and not others may also play a role, but the significance of strain-specific antigenic sites has not been fully assessed. Wiktor (1985b) recently reported on the comparative antigenic analysis of more than 170 strains of rabies and rabieslike virus isolated from different animal species or humans in the United States, Canada, Argentina, Western Europe, Zimbabwe, and Nigeria. The reactivities of these viruses and that of the rabies virus strain used for the production of antirabies human diploid-cell vaccine (HDCV) against a panel of monoclonal antibodies directed against the glycoprotein were used to determine the percentages of common antigenic determinants (epitopes for antigen–antibody binding) between rabies and rabieslike viruses. Among the rabies viruses tested, the percentages of homology (shared epitopes) between the glycoproteins of the field isolates and the vaccine strain varied from 44% for some strains of bat origin to 100% for viruses from European foxes and Argentinian dogs. The range of antigenic homology between human rabies virus strains from the United States and the vaccine strain spanned from 64 to 100%. These relative homologies are in marked contrast with the relationships between rabies and rabies-like viruses; Duvenhage virus shares 34% of its antigenic determinants with the rabies vaccine strain, while no common determinants have been detected thus far between the rabies vaccine strain and Mokola and Lagos bat viruses (Wiktor, 1985b). Protection of mice immunized with the vaccine prototype strain and subsequently challenged with the various isolates correlated well with the degree of homology between the virus and vaccine strains. Animals challenged with virus strains that showed a high degree of antigenic homology with the vaccine strain were generally all protected, whereas mice challenged with strains with a low degree of antigenic homology were only partially protected (Wiktor and Koprowski, 1980; Koprowski et al., 1985). This correlation of antigenicity between vaccine and challenge virus strains may not, however, hold up in other animal species. Wiktor (1985b) recently pointed this out after showing that guinea pigs vaccinated with standard rabies vaccine were protected equally well regardless of the differences in homology between the vaccine and rabies challenge virus strains.

The primary antibody responses to modern cell-culture vaccines in humans are rapid and substantial, especially when the vaccines are administered in short schedules of 0, 7, and 21 or 28 days. Usually, antibody responses occur within 1 week and reach high titers in nearly all human vaccinees with antirabies HDCV (Turner et al., 1976, 1982; Anderson et al., 1981; M. J. Warrell et al., 1984). The human responses to the recommended doses of HDCV show normal transition from initial IgM to a more durable IgG response between 7 and 14 days (Sureau et al., 1984). The activity and function of IgM may be of limited value, since the spread of rabies is not through viremia, but might neutralize extracellular virus in contact with blood in bite wounds if induced by postexposure treatment with rabies vaccine (Turner, 1985). Whether "secretory" IgA antibodies also play a role in protecting against virus spread, particularly in instances where virus might be transmitted through undamaged mucous membranes which have been contaminated by aerosols or licking with virus laden saliva has not been established (Nicholson, 1983).

The importance of antibody in controlling the spread of rabies virus infection is further realized when one considers the various antibody-dependent mechanisms by which virus may be cleared. Antibody has the capability of effectively "neutralizing" virus that is present in intercellular spaces or in body fluids. It also may bind to virus expressed on the cell surface, allowing complement- or antibody-dependent cytotoxic T lymphocytes to mediate killing of infected cells (Wiktor et al., 1968; Wiktor and Clark, 1972). Evidence of lytic antibody in sera of patients with rabies has provided good indication that antibody-mediated virus clearance has saved victims from rabies (Hattwick et al., 1972). The role of complement-mediated antibody-dependent clearance of virus from animals is also suggested by studies in which depletion to the C3 component of complement in serum for up to 4 days delayed virus clearance (Miller *et al.*, 1977). Antibody-dependent cellular cytotoxicity involving non-immune human lymphoid cells and human antirabies antibody, and resulting in lysis of rabies virus-infected target cells, has also been identified in studies on human vaccines (Harfast *et al.*, 1977; Pereira *et al.*, 1982).

Antibody to rabies virus may also have immunopathological effects *in vivo*, causing in animals an "early-death" phenomenon (Sikes *et al.*, 1971; Baer and Cleary, 1972; Koprowski, 1974; Tignor and Shope, 1972; Tignor *et al.*, 1976). In cases of "early death," animals with low levels of antirabies antibody resulting from immunization with rabies (or rabies-like) viruses die earlier than nonimmunized controls when challenged with lethal doses of rabies virus. Animals with prechallenge antibody, particularly if inadequately vaccinated (with suboptimal doses of vaccine or with less effective vaccines), therefore, would appear to be at greater risk of dying sooner as a reaction to vaccine than if not vaccinated at all (Blancou *et al.*, 1980). Confirmation that the "early-death" reaction to vaccine is caused by antibody was obtained from a number of studies designed to establish the underlying mechanisms of early death. In one approach, plasma adoptively transferred from immune donor mice into unvaccinated rabies virus-infected mice caused recipients to die of early rabies death (Andral and Blancou, 1981). In another study, transfer of syngeneic secondary immune lymphocytes caused as many as a third of the recipients to die earlier than controls (Prabhakar and Nathanson, 1981). To evaluate the functional participation of immune donor B and T lymphocytes in the adoptive cell transfer, preparations enriched with B or T lymphocytes were subsequently transferred into virus-infected irradiated mice and compared with transfer of mouse antirabies antibody or normal mouse serum. The results indicated that both immune B lymphocytes and rabies-specific antibody, similar to secondary immune cells, can induce early death. Clearly, there is a fine balance between the immune donor B lymphocytes that effectively reduce mortality following adoptive transfer (Prabhakar *et al.*, 1981) and those that sometimes enhance the disease. This dual role of the humoral immune response to rabies virus remains a curious phenomenon, and although it has been cited as a possible cause of early death in humans who have received postexposure treatment with different types of vaccines with or without accompanying immune serum therapy (references cited by Prabhakar and Nathanson, 1981), it remains to be proven whether immunopathology is a significant cause of death in rabies virus infection of humans.

C. Cell-Mediated Immunity

The cell-mediated (T-lymphocyte) immune response is perhaps the most important aspect of the host's response to rabies virus. Any modulation of this response, whether by experimental or natural intervention, can affect susceptibility to infection and the pathological consequences of infection (Murphy, 1977). It was first realized that T lymphocytes were playing a role in protection against rabies virus infection when studies using T-lymphocyte-deficient athymic (nude) mice showed that these mice died after being inoculated intracerebrally with rabies Flury HEP virus, while their normal littermates showed no signs of the disease (Kaplan *et al.*, 1975). In another study, nude mice were unable to respond to virus vaccine either by mounting an antibody response or by developing resistance to subsequent challenge with rabies virus (Turner, 1976). Although these initial results established that the immune response to rabies virus was dependent on T lymphocytes, they did not discern whether T lymphocytes (or their effector subsets) were more important for their effector functions on cell-mediated immunity or for their regulation of antibody production by B lymphocytes. Studies that followed attempted to answer this question. Mice were vaccinated with rabies virus and then immunosuppressed with cyclophosphamide to exert differential effects on T- and B-lymphocyte compartments. Measurement of delayed-type hypersensitivity as a function of T lymphocytes (Lagrange *et al.*, 1978) or the development of cytotoxic T lymphocytes (Turner, 1979; J. Smith, 1981) produced stronger arguments for direct participation of cellular immunological mechanisms in protection against rabies virus infection. Evidence for the presence of rabies virus-induced cytotoxic T lymphocytes developing after immunization with live and inactivated rabies virus previously suggested the possible role of these T lymphocytes in cell-mediated immunity to rabies virus infection (Atanasiu *et al.*, 1977; Wiktor *et al.*, 1977a). The appearance of virus-specific cytotoxic T lymphocytes 4 days after immunization, particularly with inactivated virus, and their peaking in activity by 7 days before disappearing on day 10 or 11 were the best evidence at the time to suggest that cell-mediated immunity involving cytotoxic T lymphocytes was correlated with protection (Wiktor, 1978). More recent evidence of cytotoxic-T-lymphocyte involvement in protection was derived from a comparative study of virulent and nonvirulent viruses to determine whether survival from rabies infection could be correlated with particular cellular immune mechanisms. In this study, normal C57BL/6 mice were intracerebrally inoculated with the virulent strain of Flury HEP virus (derived from passage in suckling mouse brain), with street rabies virus, or with nonvirulent Flury HEP virus from tissue-culture (BHK cell) passage. The mice that received the nonvirulent virus successfully cleared the infection and survived, whereas the mice that received the virulent virus failed to clear the infection and died (Wiktor *et al.*, 1985). The mice that died from the intracerebral infection also failed to generate a rabies-specific cytotoxic-T-lymphocyte response, suggesting that the failure to clear virus was correlated with lack of a specific cellular immune response even though these animals showed strong interferon and antibody responses. A strong cytotoxic-T-lymphocyte response is generally found in mice that survive rabies virus infection (Wiktor *et al.*, 1985).

Just as immunoglobulin molecules can attach to extracellular virus, preventing its infective process in potential host cells and augmenting its phagocytosis by macrophages or causing direct virolysis mediated by complement, cytotoxic T lymphocytes can function to lyse virus-infected cells. Activation of virus-immune T lymphocytes with an unusual but substantial cytotoxicity for H-2-incompatible virus-infected target cells as well as for target cells that share genes mapping at the H-2K or H-2D locus clearly indicates that this response is a distinct arm of the host's immune defense against rabies infection (Wiktor et al., 1977a). Not only are cytotoxic T lymphocytes capable of destroying virus-infected cells by direct cytotoxicity without the aid of antibody, but also their cytotoxic activity is not blocked by rabies virus-specific antibody.

Activated T lymphocytes may mediate the clearance of virus by several different mechanisms. Apart from cytotoxic cells that directly kill cells with the foreign (viral) surface antigen, T-helper cells can elaborate the necessary effector molecules (lymphokines) for the stimulation of both cytotoxic T lymphocytes and B lymphocytes or may recruit macrophages to phagocytose the viral antigen. Other T lymphocytes may suppress T or B-cell function or provide memory for persistence or acceleration of the cell-mediated immune response on subsequent challenges with viral antigen.

The target antigens for antirabies virus-immune T lymphocytes have not been completely defined at this time, nor has the predominant viral gene(s) determining the specificity of activated T lymphocytes been fully established. One report that demonstrates the generation of immunoreactive cytotoxic T lymphocytes during infection with rabies virus indicates that cytotoxic T lymphocytes are broadly cross-reactive against serologically distinct rabies and rabieslike viruses (Reddehase et al., 1984). Determining the minimal molecular requirements for elicitation of virus-specific immune cells has nevertheless been an important goal in recent years in the study of rabies virus immunology. Experiments with the glycoprotein isolated from rabies virus and cyanogen bromide (CNBr) cleavage fragments of the glycoprotein were recently carried out to identify domains of the glycoprotein that retain a capacity to stimulate rabies virus-primed T lymphocytes (Macfarlan et al., 1984). Three of eight peptide fragments, all isolated under reducing (denaturing) conditions, stimulated antigen-specific proliferation of virus-primed T lymphocytes in vitro, indicating that at least three T-lymphocyte determinants were present on the viral glycoprotein. Since these CNBr cleavage fragments represented linear or continuous determinants (by the nature of their isolation), it was feasible to prepare synthetic peptides that contained partial amino acid sequences of the larger cleavage fragments that could be tested in the in vitro system. Interestingly, stimulation of virus-primed T lymphocytes was obtained with two synthetic peptides that matched sequences near the carboxy end of the large CNBr fragment that maps to the amino terminus of the glycoprotein molecule (Macfarlan et al., 1984). While identification of amino acid sequence specificity for virus-immune suppressor or cytotoxic T lymphocytes awaits further experimentation, it is evident that the new molecular technologies combined with rapid advances in our understanding of function of T lymphocytes and their multiple subsets (Doherty, 1985) will help to reveal the nature of rabies virus immunity.

The observation that rabies antigen-specific antibodies can also stimulate proliferation of rabies virus-reactive T lymphocytes *in vitro* has given fresh insight into the potentiation of T lymphocytes by antigens other than the surface glycoprotein of rabies virus (Celis *et al.*, 1985). With the use of antiglycoprotein and antinucleocapsid monoclonal antibodies (derived from mice) as well as human rabies-immune plasmas from vaccinees, it has been possible to establish with T-lymphocyte clones that rabies-reactive T lymphocytes are specific for either rabies virus glycoprotein or nucleocapsid antigen. The immune plasma as well as the antiglycoprotein and antinucleoprotein antibodies significantly increased the antigen-induced proliferative responses of these clones. The results also indicate that the *in vitro* proliferative effect of antigen-specific (effector) antibodies on regulatory T lymphocytes could be important in the development of effective immune responses *in vivo* to rabies virus.

D. Interferon

Numerous studies using a variety of experimental animals have shown that rabies can be prevented by interferon, provided it is induced or administered at or near the time of infection (see reviews by Tsiang, 1974; Sulkin and Allen, 1975; Turner, 1972, 1977, 1983; Baer, 1978). The mechanisms of action of interferon and its relative importance in postexposure protection from rabies, however, remain unclear.

The protective capacity of inactivated rabies virus vaccine produced from cell cultures has long been considered as partly due to interferon induction. Following studies in rabbits in which endogenous interferon induced by the polyribonucleotide complex poly(I)–poly(C) appeared to play a role in the inhibition of rabies virus infection and resistance to challenge with rabies virus (Fenje and Postic, 1970, 1971), evidence was obtained that rabies virus vaccine as well as immunologically unrelated viral vaccines administered shortly before or immediately after challenge with street virus in hamsters were able to protect the animals against death (Wiktor *et al.*, 1972b). The study showed that each of the vaccines stimulated circulating interferon and that live rabies virus vaccine induced more interferon than inactivated vaccine, but that interferon was not induced by the administration of rabies vaccine to rabies-immune animals. Most important, however, all vaccine preparations that induced interferon protected the hamsters against rabies; vaccine preparations that did not induce detectable levels of interferon in serum did not protect animals, even though the antigenic values of the two types of vaccines (those that induce interferon and those that do not) were similar. In an extension of these studies carried out in rhesus monkeys, protection from rabies by a single dose of highly concentrated cell-culture-produced rabies virus vaccine again seemed to be associated, in part, with the ability of the concentrated (vs. less concentrated) vaccine to induce interferon (Wiktor et al., 1976). Interferon induction has also been recorded in humans vaccinated with cell-culture vaccines (Nicholson et al., 1979a,b, 1981); however, as these studies indicate, not all vaccine preparations have been constant inducers of interferon, and this raises a serious question as to whether interferon is as important in preventing rabies by postexposure treatment as might have been originally anticipated (Turner, 1985).

Induction of interferon by vaccine is often rapid, but unfortunately does not last and cannot be restimulated to previous levels by additional doses of vaccine (Mifune et al., 1980). Interferon produced by virus infection also appears to be short-lived and to have little or no protective activity. In recent studies in which mice were inoculated in the footpad with a mouse brain homogenate containing the CVS strain of rabies virus, two episodes of endogenously produced interferon were detected; the first appeared 24–48 hr after inoculation and the second on the 6th or 7th day of infection (Marcovistz et al., 1984, 1986). The early induction of interferon activity in plasma was not necessarily in response to rabies virus, since it was detected in the control mice, which were injected with uninfected mouse brain homogenate. This interferon, however, was capable of modifying the development of rabies, as shown by morbidity changes in infected animals when its activity was blocked by anti-α/β-interferon neutralizing antibody. Rabies virus is capable, however, of inducing interferon at the site of virus inoculation, before its migration to the CNS. This is perhaps the only time when the effects of interferon (endogenous or exogenous) play any significant role in modifying the disease. Interferon produced late in infection seems to be ineffective in inhibiting replication of virus, yet it is not at all clear why this should be so. At this time, the type and effect of local interferon production have not been measured, but only observed as a peak of circulating interferon activity in plasma late in infection. It is also uncertain whether the interferon produced during infection acts directly to clear virus or indirectly through the virus-immune response (Doherty, 1985).

VI. CONCLUDING REMARKS

This review of the pathogenic pathway of rabies viruses and the immune response to rabies virus antigens is of necessity an abridged account of nearly 100 years of intensive study. There have been many reports (too many to cite in this chapter) dealing with the viral pathogenesis of rabies, and the majority of these reports focus on the general mechanisms involved in the spread of virus from the site of virus inoculation to the brain. There are still many puzzling aspects of rabies virus infection, such as the long and varied incubation period, the highly selective mechanism of neuronal-cell invasion, and the physical form (whole virion or nucleocapsid) of rabies virus that successfully invades the peripheral and central nervous systems, to mention only a few. The mechanisms of rabies virus transmission from the bite or wound site, where rabies virus gains access to the peripheral nerves and thence to the central nervous system, as well as its direct cell-to-cell transit and neuronal spread in the brain, have major implications for the lack of effective immune responses to prevent a central nervous system infection. Nevertheless, a clearer understanding that has resulted from past studies of the pathways involved in developing infection in the brain and the temporal aspects as well as the relative importance of the cellular and humoral immune responses to rabies virus has made it possible to introduce effective prophylactic and postexposure treatments where and when they are appropriate.

The challenge of defining more precise mechanisms of rabies virus pathogenesis and of the specific and nonspecific immune responses to infection is now rapidly becoming the task of molecular biologists. Working in close association with virologists, immunologists, and pathologists, the molecular biologist has the task of determining structural correlates of the biological, pathogenic, and immunogenic (antigenic) properties of rabies virus. In general, and at this early stage of the developing interdependence among these various disciplines, the hope of finding answers for the questions that remain is an optimistic one. Comparative analysis of linear sequences at the nucleotide and amino acid levels mark the new beginning. Interpretations of these sequences in their various orders of structure have already begun to bring greater clarity to the mechanisms of antibody and T-lymphocyte function and of virus–cell interactions. Much more needs to be done to precisely define how rabies develops and how it may be controlled.

ACKNOWLEDGMENTS. The contributions of Drs. Frederick A. Murphy, Alyne K. Harrison, and Yuzo Iwasaki are gratefully acknowledged. The author's work is supported by Grants AI-18562 and AI-18883 from the National Institutes of Health.

REFERENCES

Abelseth, M. K., 1964, An attenuated rabies vaccine for domestic animals produced in tissue culture, *Can. Vet. J.* **5**:279.

Abelseth, M. K., 1967, Further studies on the use of ERA rabies vaccine in domestic animals, *Can. Vet. J.* **8**:221.

Afshar, A., 1979, A review of non-bite transmission of rabies virus infection, *Br. Vet. J.* **135**:142.

Almeida, J. D., Howatson, A. F., Pinteric, L., and Fenje, P., 1962, Electron microscopic observations on rabies virus by negative staining, *Virology* **18**:47.

Anderson, L. J., Baer, G. M., Smith, J. S., Winkler, W. G., and Holman, R. C., 1981, Rapid antibody response to human diploid rabies vaccine, *Am. J. Epidemiol.* **113**:270.

Andral, B., and Blancou, J., 1981, Study of the mechanisms of early death occurring after vaccination in mice inoculated with street rabies virus, *Ann. Virol. (Inst. Pasteur)* **132E**:503.

Anilionis, A., Wunner, W. H., and Curtis, P. J., 1981, Structure of the glycoprotein gene in rabies virus, *Nature (London)* **294**:275.

Arko, R. J., Schneider, L. G., and Baer, G. M., 1973, Nonfatal canine rabies, *Am. J. Vet. Res.* **34**:937.

Atanasiu, P., 1965, Transmission de la rage pour la voie respiratoire aux animaux de laboratoire, *C. R. Acad. Sci.* **261**:277.

Atanasiu, P., Lepine, P., Sisman, J., Daugnet, J. C., and Wetton, M., 1963a, Étude morphologique du virus rabiques des rues en culture de tissue, *C. R. Acad. Sci.* **256**:3219.

Atanasiu, P., Orth, G., Sisman, J., and Barreau, C., 1963b, Identification immunologique du virion rabique au cultures cellulaires par les anticorpos spécifiques conjugués a la farritina, *C. R. Acad. Sci.* **257**:2204.

Atanasiu, P., Nosaki-Renard, J., Savy, V., and Eyquem, A., 1977, Évaluation de l'immunité cellulaire apré vaccination chez l'homme, *C. R. Acad. Sci. (D)* **285**:1187.

Aubert, M. F. A., Bussereau, F., and Blancou, J., 1980, Pathogenic, immunogenic and protective powers of ten temperature-sensitive mutants of rabies virus in mice, *Ann. Virol. (Inst. Pasteur)* **131**:217.

Baer, G. M., 1975, Pathogenesis to the central nervous system, in: *The Natural History of Rabies*, Vol. I (G. M. Baer, ed.), pp. 181–196, Academic Press, New York

Baer, G. M., 1978, Advances in postexposure rabies vaccination: A review, *Am. J. Clin. Pathol.* **70**:185.

Baer, G. M., and Cleary, W. F., 1972, A model in mice for the pathogenesis and treatment of rabies, *J. Infect. Dis.* **125**:520.

Baer, G. M., and Olson, H. R., 1972, Recovery of pigs from rabies, *J. Am. Vet. Med. Assoc.* **160**:1127.

Baer, G. M., and Yager, P. A., 1977, A mouse model for postexposure rabies prophylaxis: The comparative efficacy of two vaccines and of antiserum administration, *J. Gen. Virol.* **36**:51.

Baer, G. M., Shanthaveerappa, T. R., and Bourne, G. H., 1965, Studies on pathogenesis of fixed rabies virus in rats, *Bull. W.H.O.* **33**:783.

Baer, G. M., Shantha, T. R., and Bourne, G. H., 1968, Pathogenesis of street rabies virus in rats, *Bull. W.H.O.* **38**:119.

Baer, G. M., Cleary, W. F., Diaz, A. M., and Perl, D. F., 1977, Characteristics of 11 rabies virus isolates in mice: Titers and relative invasiveness of virus, incubation period of infection, and survival of mice with sequelae, *J. Inf. Dis.* **136**:336.

Baer, G. M., Shaddock, J. H., Houff, S. A., Harrison, A. K., and Gardner, J. J., 1982, Human rabies transmitted by corneal transplant, *Arch. Neurol.* **39**:103.

Bell, J. F., 1964, Abortive rabies infection. I. Experimental production in white mice and general discussion, *J. Infect. Dis.* **114**:249.

Bell, J. F., 1975, Latency and abortive rabies, in: *The Natural History of Rabies*, Vol. I (G. M. Baer, ed.), pp. 331–354, Academic Press, New York.

Bell, J. F., Lodmell, D. L., Moore, G. J., and Raymond, G. H., 1966, Brain neutralization of rabies virus to distinguish recovered animals from previously vaccinated animals, *J. Immunol.* **97**:747.

Bernkopf, H., and Kligler, I. J., 1940, Characteristics of a fixed rabies virus cultivated on developing chick embryos, *Proc. Soc. Exp. Biol. Med.* **45**:332.

Bhatt, D. R., Hattwick, M. A. W., Gerdsen, R., Emmons, R. W., and Johnson, H. N., 1974, Human rabies: Diagnosis, complications and management, *Am. J. Dis. Child.* **127**:862.

Bijlenga, G., and Heaney, T., 1978, Post-exposure treatment of mice infected with rabies with two axonal flow inhibitors, colchicine and vinblastin, *J. Gen. Virol.* **39**:381.

Blancou, J., Andral, B., and Andral, L., 1980, A model in mice for the study of the early death phenomenon after vaccination and challenge with rabies virus, *J. Gen. Virol.* **50**:433.

Blenden, D. C., 1981, Rabies in a litter of skunks predicted and diagnosed by skin biopsy, *J. Am. Vet. Med. Assoc.* **179**:789.

Blinzinger, K., and Muller, W., 1971, The intercellular gaps of neuropil as possible pathways for virus spread in viral encephalomyelitides, *Acta Neuropathol.* **17**:37.

Brown, F., and Crick, J., 1979, Natural history of the rhabdoviruses of vertebrates and invertebrates, in: *Rhabdoviruses,* Vol. I (D. H. L. Bishop, ed.), pp. 1–22, CRC Press, Boca Raton, Florida.

Bryceson, A. D. M., Greenwood, B. M., Warrell, D. A., Davidson, N. McD., Pope, H. M., Lawrie, L. H., Barnes, H. J., Bailie, W. E., and Wilcox, G. E., 1975, Demonstration during life of rabies antigen in humans, *J. Infect. Dis.* **131**:71.

Burrage, T. G., and Lentz, T. L., 1981, Ultrastructure characterization of surface specialization containing high-density acetylcholine receptors on embryonic chick myotubes *in vivo* and *in vitro, Dev. Biol.* **85**:267.

Burrage, T. G., Tignor, G. H., and Smith, A. L., 1983, Immunoelectron microscopic localization of rabies virus antigen in central nervous system and peripheral tissue using low-temperature embedding and protein A–gold, *J. Virol. Methods* **7**:337.

Burrage, T. G., Tignor, G. H., and Smith, A. L., 1985, Rabies virus binding at neuromuscular junctions, *Virus Res.* **2**:273.

Cabasso, V. J., Stebbins, M. R., Douglas, B. A., and Sharpless, G. R., 1965, Tissue-culture rabies vaccine (Flury LEP) in dogs, *Am. J. Vet. Res.* **26**:24.

Celis, E., Wiktor, T. J., Dietzschold, B., and Koprowski, H., 1985, Amplification of rabies virus-induced stimulation of human T-cell lines and clones by antigen-specific antibodies, *J. Virol.* **56**:426.

Charlton, K. M., and Casey, G. A., 1979, Experimental rabies in skunks: Immunofluorescence, light and electron microscopic studies, *Lab Invest.* **41**:36.

Charlton, K. M., Casey, G. A., and Campbell, J. B., 1983, Experimental rabies in skunks: Mechanisms of infection of the salivary glands, *Can. J. Comp. Med.* **47**:363.

Cheetham, H. D., Hart, J., Coghill, N. F., and Fox, B., 1970, Rabies with myocarditis: Two cases in England, *Lancet* **1**:921.

Cifuentas, E., Calderon, E., and Bijlenga, G., 1971, Rabies in a child diagnosed by a new intra-vitam method, the cornea test, *J. Trop. Med. Hyg.* **74**:23.

Clark, H F., 1979, Systems for assay and growth of rhabdoviruses, in: *Rhabdoviruses,* Vol. 1 (D. H. L. Bishop, ed.), pp. 23–41, CRC Press, Boca Raton, Florida.

Clark, H F., and Ohtani, S., 1976, Temperature-sensitive mutants of rabies virus in mice: A mutant (*ts2*) revertant mixture selectively pathogenic by the peripheral route of inoculation, *Infect. Immun.* **13**:1418.

Clark, H F., and Prabhakar, B. S., 1985, Rabies, in: *Comparative Pathobiology of Virus Diseases,* Vol. II (R. G. Olsen, S. Krakowka, and J. R. Blakeslee, eds.), pp. 165–214, CRC Press, Boca Raton, Florida.

Clark, H F., and Wiktor, T. J., 1972, Temperature-sensitivity characteristics distinguishing substrains of fixed rabies virus: Lack of correlation with plaque-size markers or virulence for mice, *J. Infect. Dis.* **125**:637.

Clark, H F., Parks, N. F., and Wunner, W. H., 1981, Defective interfering particles of fixed rabies viruses: Lack of correlation with attenuation or autointerference in mice, *J. Gen. Virol.* **52**:245.

Conomy, J. P., Leibovitz, A., McCombs, W., and Stinson, J., 1977, Airborne rabies encephalitis: Demonstration of rabies virus in the human central nervous system, *Neurology* **37**:67.

Constantine, D. G., 1962, Rabies transmission by non-bite route, *U. S. Public Health Rep.* **77**:387.

Constantine, D. G., 1966, Transmission experiments with bat rabies isolates: Responses of certain carnivora to rabies virus isolated from animals infected by nonbite route, *Am. J. Vet. Res.* **27**:13.

Constantine, D. G., Emmons, R. W., and Woodie, J. D., 1972, Rabies virus in nasal mucosa of naturally infected bats, *Science* **175**:1255.

Coulon, P., Rollin, P., Aubert, M., and Flamand, A., 1982, Molecular basis of rabies virus virulence. I. Selection of avirulent mutants of the CVS strain with anti-G monoclonal antibodies, *J. Gen. Virol.* **61**:97.

Cserr, H. F., and Ostrach, L. H., 1974, Bulk flow of interstitital fluid after intercranial injection of blue Dextran 2000, *Exp. Neurol.* **45**:50.

Davies, M. C., Englert, M. E., Sharpless, G. R., and Cabasso, V. J., 1963, The electron microscopy of rabies virus in cultures of chicken embryo tissues, *Virology* **21**:642.

Dawson, J. R., 1939, Infection of chicks and chick embryos with rabies, *Science* **89**:300.

Dean, D. J., Evans, W. M., and McClure, R. C., 1963, Pathogenesis of rabies, *Bull. W.H.O.* **29**:803.

Dean, D. J., Evans, W. M., and Thompson, W. R., 1964, Studies on the low egg passage Flury strain of modified live rabies virus produced in embryonating chicken eggs and tissue culture, *Am. J. Vet. Res.* **25**:756.

DeBrito, T., Aranjo, M.de F., and Tiriba, A., 1973, Ultrastructure of the Negri body in human rabies, *J. Neurol. Sci.* **20**:363.

DeWet, J. S. D. T., 1980, Rabies presenting as an acute psychiatric emergency, *S. Afr. Med. J.* **58**:297.

Dierks, R. E., 1975, Electron microscopy of extraneural rabies infection, in: *The Natural History of Rabies*, Vol. I (G. M. Baer, ed.), pp. 303–318, Academic Press, New York.

Dierks, R. E., Murphy, F. A., and Harrison, A. K., 1969, Extraneural rabies virus infection: Virus development in fox salivary gland, *Am. J. Pathol.* **54**:251.

Dietzschold, B., Wunner, W. H., Wiktor, T. J., Lopes, A. D., Lafon, M., Smith, C., and Koprowski, H., 1983, Characterization of an antigenic determinant of the glycoprotein which correlates with pathogenicity, *Proc. Natl. Acad. Sci. U.S.A.* **80**:70.

Dietzschold, B., Wiktor, T. J., Trojanovski, J. Q., Macfarlan, R. I., Wunner, W. H., Torres-Anjel, M. J., and Koprowski, H., 1985, Differences in cell-to-cell spread of pathogenic and apathogenic rabies virus *in vivo* and *in vitro*, *J. Virol.* **56**:12.

Doherty, P. C., 1985, T cells and viral infections, *Br. Med. Bull.* **41**:7.

Duenas, A., Belsey, M. A., Escobar, J., Medina, P., and Sanmartin, C., 1973, Isolation of rabies virus outside the human central nervous system, *J. Infect. Dis.* **127**:702.

Dupont, J. R., and Earle, K. M., 1965, Human rabies encephalitis: A study of forty-nine fatal cases with a review of the literature, *Neurology* **15**:1023.

Emmons, R. W., Leonard, L. L., DeGenaro, F., Protas, E. S., Bazely, P. L., Giammona, P. L., and Sturkow, K., 1973, A case of human rabies with prolonged survival, *Intervirology* **1**:60.

Fekadu, M., and Baer, G. M., 1980, Recovery from clinical rabies of 2 dogs inoculated with a rabies virus strain from Ethiopia, *Am. J. Vet. Res.* **41**:632.

Fekadu, M., and Shaddock, J. H., 1984, Peripheral distribution of virus in dogs inoculated with two strains of rabies virus, *Am. J. Vet. Res.* **45**:724.

Fekadu, M., Shaddock, J. H., and Baer, G. M., 1982, Excretion of rabies virus in the saliva of dogs, *J. Infect. Dis.* **145**:715.

Fenje, P., 1960, A rabies vaccine from hamster kidney tissue culture: Preparation and evaluation in animals, *Can. J. Microbiol.* **6**:605.

Fenje, P., and Postic, B., 1970, Protection of rabbits against experimental rabies by poly I–Poly C, *Nature (London)* **226**:171.

Fenje, P., and Postic, B., 1971, Prophylaxis of experimental rabies with the polyriboinosinic–polyribocytidylic acid complex, *J. Infect. Dis.* **123**:426.

Fischman, H. R., and Schaeffer, M., 1971, Pathogenesis of experimental rabies as revealed by immunofluorescence, *Ann. N. Y. Acad. Sci.* **177**:78.

Fischman, H. R., and Strandberg, J. D., 1973, Inapparent rabies virus infection of the central nervous system, *J. Am. Vet. Med. Assoc.* **163**:1050.

Flamand, A., Wiktor, T. J., and Koprowski, H., 1980, Use of hybridoma monoclonal antibodies in the detection of antigenic differences between rabies and rabies-related viral proteins. II. The glycoprotein, *J. Gen. Virol.* **48**:105.

Fox, J. P., Koprowski, H., Conwell, D. P., Black, J., and Gelfand, H. M., 1957, Study of antirabies immunization of man: Observations with HEP Flury and other vaccines, with and without hyperimmune serum, in primary and recall immunizations, *Bull. W.H.O.* **17**:869.

Fujisake, Y., Sekiguchi, K., and Hirasawa, K., 1968, 19S Neutralizing antibody produced in guinea pigs by rabies virus, *Nat. Inst. Anim. Health Q. (Tokyo)* **8**:132.

Galtier, V., 1881, Physiologie pathologique—les injections de virus rabique dans le torrent circulatoire ne provoquent pas l'éclosion de la rage et somblent conférer l'immunité: La rage peut-être transmisse par l'infestion de lay matière rabique, *C. R. Acad. Sci.* **93**:284.

Gershoni, J. M., Hawrot, E., and Lentz, T. L., 1983, Binding of α-bungarotoxin to isolated a subunit of the acetylcholine receptor of *Torpedo californica:* Quantitative analysis with protein blots, *Proc. Natl. Acad. Sci. U.S.A.* **80**:4973.

Goldwasser, R. A., and Kissling, R. E., 1958, Fluorescent antibody staining of street and fixed rabies virus antigens, *Proc. Soc. Exp. Biol. Med.* **98**:219.

Goodpasteur, E. W., 1925, A study of rabies with reference to a neural transmission of the virus in rabbits, and the structure and significance of Negri bodies, *Am. J. Pathol.*, **1**:547.

Gosztonyi, G., 1978, Axonal and transsynaptic spread of viral nucleocapsids in fixed rabies encephalitis, *J. Neuropathol. Exp. Neurol.* **37**:618.

Gosztonyi, G., 1979, Possible mechanisms of spread of fixed rabies virus along neural pathways, in: *Mechanisms of Viral Pathogenesis and Virulence* (P. A. Bachmann, ed.), pp. 323–345, Proceedings of the IVth Munich Symposium on Microbiology, Munich.

Gosztonyi, G., and Ludwig, H., 1984, Neurotransmitter receptors and viral neurotropism, *Neuropsychiatr. Clin.* **3**:107.

Habel, K., 1940, Factors influencing the efficiency of phenolized rabies vaccines. 1. Strains of fixed virus, *U. S. Public Health Rep.* **55**:1619.

Halonen, P. E., Toivanen, P., and Nikkari, T., 1974, Non-specific serum inhibitors of activity of hemagglutinins of rabies and vesicular stomatitis viruses, *J. Gen. Vir.* **22**: 309.

Harfast, B., Andersson, T., and Grandien, M., 1977, Enhanced cytotoxicity of human lymphocytes against rabies-infected cells by rabies-specific antibodies, *Scand. J. Immunol.* **6**:1107.

Harrison, A. K., and Murphy, F. A., 1978, Lyssavirus infection of muscle spindles and motor end plates in striated muscle of hamsters, *Arch. Virol.* **57**:167.

Hattwick, M. A. W., and Gregg, M. B., 1975, The disease in man, in: *The Natural History of Rabies,* Vol. II (G. M. Baer, ed.), pp. 281–305, Academic Press, New York.

Hattwick, M. A. W., Weiss, T. T., Stechschulte, C. J., Baer, G. M., and Gregg, M. B., 1972, Recovery from rabies: A case report, *Ann. Intern. Med.* **76**:431.

Heaney, T., Bijlenga, G., and Joubert, L., 1976, Traitment preventif et curatif locale de l'infection a virus rabique fixe (CVS) chez la souris par des alcaloides (colchicine et vinblastin) inhibiteurs du flux axoplasmique: Blocage apparent de la neuroprobasie virale, *Med. Mal. Infect.* **6**:39.

Heidmann, T., and Changeux, J.-P., 1978, Structural and functional properties of the acetylcholine receptor protein in its purified and membrane-bound states, *Annu. Rev. Biochem.* **47**:317.

Herzog, E., 1965, Histologic diagnosis of rabies, *Arch. Pathol.* **39**:279.

Holland, J. J., Villarreal, L. P., Welsh, R. M., Oldstone, M. B. A., Kohne, D., Lazzarini, R., and Scolnick, E., 1976, Long-term persistent vesicular stomatitis virus and rabies virus infection of cells *in vitro, J. Gen. Virol.* **33**:193.

Holland, J. J., Kennedy, S. I. T., Semler, B. L., Jones, C. L., Roux, L., and Grabau, E. A., 1980, Defective interfering RNA viruses and the host-cell response, in: *Comparative Virology,* Vol. 16 (H. Fraenkel-Conrat and R. R. Wagner, eds.), pp. 137–192, Plenum Press, New York.

Houff, S. A., Burton, R. C., Wilson, R. W., Henson, T. E., London, W. T., Baer, G. M., Anderson, L. J., Winkler, W. G., Madden, D. L., and Sever, J. L., 1979, Human to human transmission of rabies virus by corneal transplant, *N. Engl. J. Med.* **300**:603.

Howard, D. R., 1981, Rabies virus tropism in naturally infected skunks (*Mephitis mephitis*), *Am. J. Vet. Res.* **42**:2187.

Hronovsky, V., and Benda, R., 1969, Experimental inhalation infection of laboratory rodents with rabies virus, *Acta Virol.* **13**:193.

Hubbard, J. I., and Wilson, D. F., 1973, Neuromuscular transmission in a mammalian preparation in the absence of blocking drugs and the effect of D-tubocurarine, *J. Physiol.(London)* **228**:307.

Hummeler, K., Koprowski, H., and Wiktor, T. J., 1967, Structure and development of rabies virus in tissue culture, *J. Virol.* **1**:152.

Humphrey, G. L., Kemp, G. E., and Wood, E. G., 1960, A fatal case of rabies in a woman bitten by an insectivorous bat, *U. S. Public Health Rep.* **75**:317.

Hurst, E. W., and Pawan, J. L., 1968, An outbreak of rabies in Trinidad without history of bites, and with the symptoms of acute ascending myelitis, *Caribb. Med. J.* **30**:17.

Innes, J. R. M., and Saunders, L. Z., 1962, *Comparative Neuropathology*, pp. 384–394, Academic Press, New York.

Irons, J. V., Eads, R. B., Grimes, J. E., and Conklin, A., 1957, The public health importance of bats, *Tex. Rep. Biol. Med.* **15**:292.

Iwasaki, Y., Wiktor, T. J., and Koprowski, H., 1973, Early events of rabies virus replication in tissue cultures: An electron microscopic study, *Lab. Invest.* **28**:142.

Iwasaki, Y., Ohtani, S., and Clark, H F., 1975, Maturation of rabies virus by budding from neuronal cell membrane in suckling mouse brain, *J. Virol.* **15**:1020.

Iwasaki, Y., and Clark, H. F., 1975, Cell to cell transmission of virus in the central nervous system. II. Experimental rabies in the mouse, *Lab. Invest.* **33**:391.

Iwasaki, Y., Liu, D., Yamamoto, T., and Konno, H., 1985, On the replication and spread of rabies virus in the human central nervous system, *J. Neuropathol. Exp. Neurol.* **44**: 185.

Jensen, A. B., Rabin, E. R., Bentinck, D. C., and Melnick, J. L., 1969, Rabies virus neuronitis, *J. Virol.* **3**:265.

Johnson, H. G., 1965, Rabies virus, in: *Viral and Rickettsial Infections of Man*, 4th ed. (F. L. Horfall and I. Tamm, eds.), pp. 814–840, J. B. Lippincott, Philadelphia.

Johnson, K. P., and Swoveland, P. T., 1984, Rabies, *Neurol. Clin.* **2**:255.

Johnson, R. T., 1965, Experimental rabies: Studies of cellular vulnerability and pathogenesis using fluorescent antibody staining, *J. Neuropathol. Exp. Neurol.* **24**:662.

Johnson, R. T., 1971, The pathogenesis of experimental rabies, in: *Rabies* (Y. Nogano and F. M. Davenport, eds.), pp. 59–75, University Park Press, Baltimore.

Johnson, R. T., 1982, Rabies, in: *Viral Infections of the Nervous System* (R. T. Johnson, ed.), pp. 159–167, Raven Press, New York.

Jubb, K. V., and Kennedy, P. C., 1963, *Pathology of Domestic Animals*, Vol. 2, pp. 352–357, Academic Press, New York.

Kaplan, C., 1983, The epidemiology of rabies, in: *Rabies: A Growing Threat* (J. R. Pattison, ed.), pp. 1–5, Van Nostrand Reinhold (UK), Wokingham, England.

Kaplan, M. M., Wiktor, T. J., and Koprowski, H., 1975, Pathogenesis of rabies in immunodeficient mice, *J. Immunol.* **114**:1761.

Karlsson, K. A., 1982, Glycosphingolipids and surface membranes, in: *Biological Membranes*, Vol. 4 (D. Chapman, ed.), pp. 1–74, Academic Press, London.

Kawai, A., Matsumoto, S., and Tanabe, K., 1975, Characterization of rabies virus recovered from persistently infected BHK cells, *Virology* **67**:520.

Kligler, I. J., and Bernkopf, H., 1943, The path of dissemination of rabies virus in the body of normal and immunized mice, *Br. J. Exp. Med.* **24**:15.

Koprowski, H., 1974, Immunopathology of rabies virus infection, in: *Symposium Series in Immunobiological Standardization*, Vol. 21, pp. 86–101, S. Karger, Basel.

Koprowski, H., and Black, J., 1952, Studies on chick embryo adapted rabies virus. III. Duration of immunity in vaccinated dogs, *Proc. Soc. Exp. Biol. Med.* **80**:410.

Koprowski, H., and Cox, H. R., 1948, Studies on chick embryo adapted rabies virus, *J. Immunol.* **60**:553.

Koprowski, H., Wiktor, T. J., and Abelseth, M., 1985, Cross-reactivity and cross-protection: Rabies variants and rabies-related viruses, in: *Rabies in the Tropics* (E. Kuwert, C. Merieux, H. Koprowski, and K. Bögel, eds.), pp. 30–39, Springer-Verlag, Berlin.

Koschel, K., and Munzel, P., 1984, Inhibition of opiate receptor-mediated signal transmission by rabies virus in persistently infected NG-108-15 mouse neuroblastoma–rat glioma hybrid cell, *Proc. Natl. Acad. Sci. U.S.A.* **81**:950.

Kristensson, K., and Olsson, Y., 1973, Diffusion pathways and retrograde axonal transport of protein tracers in peripheral nerves, *Prog. Neurobiol. (Oxford)* **1**:85.

Kućera, P. Dolivo, M., Coulon, P., and Flamand, A., 1985, Pathways of the early propagation of virulent and avirulent rabies strains from the eye to the brain, *J. Virol.* **55**:158.

Lafon, M., Wiktor, T. J., and Macfarlan, R. I., 1983, Antigenic sites on the CVS rabies virus glycoprotein: Analysis with monoclonal antibodies, *J. Gen. Virol.* **64**:843.

Lafon, M., Ideler, J., and Wunner, W. H., 1984, Investigation of the antigenic structure of rabies virus glycoprotein by monoclonal antibodies, in: *International Symposium on Monoclonal Antibodies: Standardization of Their Characterization and Use, Developments in Biological Standardization,* Vol. 57 (M. Barme, W. Hennesson, eds.), pp. 219–225, S. Karger, Basel.

Lagrange, P. H., Tsiang, H., Hurtrel, B., and Ravisse, P., 1978, Delayed type hypersensitivity to rabies virus in mice: Assay of active or passive sensitization by footpad test, *Infect. Immun.* **21**:931.

Leech, R. W., 1971, Electron-microscopic study of the inclusion body in human rabies, *Neurology* **21**:41.

Lentz, T. L., Burrage, T. G., Smith, A. L., Crick, J., and Tignor, G. H., 1982, Is the acetylcholine receptor a rabies virus receptor?, *Science* **215**:182.

Lentz, T. L., Chester, J., Benson, R. J. J., Hawrot, E., Tignor, G. H., and Smith, A. L., 1985, Rabies virus binding to cellular membranes measured by enzyme immunoassay, *Muscle Nerve* **8**:336.

Lintjorn, B., 1982, Clinical features of rabies in man, *Trop. Doctor* **12**:9.

Lodmell, D. L., Bell, J. F., Moore, G. J., and Raymond, G. H., 1969, Comparative study of abortive and nonabortive rabies in mice, *J. Infect. Dis.* **119**:569.

Love, S. V., 1944, Paralytic rabies: Review of the literature and report of a case. *J. Pediatr.* **24**:312.

Macfarlan, R. I., Dietzschold, B., Wiktor, T. J., Kiel, M., Houghton, R., Lerner, R. A., Sutcliff, J. G., and Koprowski, H., 1984, T cell responses to cleaved rabies virus glycoprotein and to synthetic peptides, *J. Immunol.* **133**:2748.

Magleby, K. L., Palotta, B. S., and Terrer, D. A., 1981, The effect of (+)-tubocurarine on neuromuscular transmission during repetitive stimulation in the rat, mouse, and frog, *J. Physiol. (London)* **312**:97.

Marcovistz, R., Tsiang, H., and Hovanessian, A. G., 1984, Production and action of interferon in mice infected with rabies virus, *Ann. Virol. (Inst. Pasteur)* **135E**:19.

Marcovistz, R., Galabru, J., Tsiang, H., and Hovanessian, A. G., 1986, Neutralization of interferon produced early during rabies virus infection in mice, *J. Gen. Virol.* **67**:387.

Matsumoto, S., 1962, Electron microscopy of nerve cells infected with street rabies virus, *Virology,* **17**:198.

Matsumoto, S., 1963, Electron microscope studies of rabies virus in mouse brain, *J. Cell Biol.* **19**:565.

Matsumoto, S., 1970, Rabies virus, *Adv. Virus Res.* **16**:257.

Matsumoto, S., 1975, Electron microscopy of central nervous system infection, in: *The Natural History of Rabies,* Vol. I (G. M. Baer, ed.), pp. 217–233, Academic Press, New York.

Matsumoto, S., and Kawai, A., 1969, Comparative studies on development of rabies virus in different host cells, *Virology* **39**:449.

Matsumoto, S., Schneider, L. G., Kawai, A., and Yonezawa, T., 1974, Further studies on the replication of rabies and rabies-like viruses in organized cultures of mammalian neural tissues, *J. Virol.* **14**:981.

Meyer, K. F., 1957, Man contacting rabies from man, *J. Am. Med. Assoc.* **165**:158.

Mifune, K., Shiehijo, A., Makino, Y., Takeuchi, E., Yamada, A., and Sakamoto, K., 1980, A mouse model for the pathogenesis and post exposure prophylaxis of rabies, *Microbiol. Immunol.* **24**:835.

Miller, A., and Nathanson, N., 1977, Rabies: Recent advances in pathogenesis and control, *Ann. Neurol.* **2**:511.

Miller, A., Morese, E. H., Winkelstein, J., and Nathanson, N., 1977, The role of antibody in recovery from experimental rabies, *J. Immunol.* **121**:321.

Miyamoto, K., and Matsumoto, S., 1965, The nature of the Negri body, *J. Cell Biol.* **27**:677.

Miyamoto, K., and Matsumoto, S., 1967, Comparative studies between pathogenesis of street and fixed rabies infection, *J. Exp. Med.* **125**:447.

Morbidity and Mortality Weekly Report, 1977, Rabies in a laboratory worker—New York, **26**:183.

Morbidity and Mortality Weekly Report, 1980, Human to human transmission of rabies via a corneal transplant—France, **29**:25.

Morbidity and Mortality Weekly Report, 1981, Human to human transmission of rabies via a corneal transplant—Thailand, **30**:473.

Morecki, R., and Zimmerman, H. M., 1969, Human rabies encephalitis: Fine structure study of cytoplasmic inclusions, *Arch. Neurol.* **20**:599.

Morrison, T. G., 1980, Rhabdoviral assembly and intracellular processing of viral components, in: *Rhabdoviruses*, Vol. II (D. H. L. Bishop, ed.), pp. 45–114, CRC Press, Boca Raton.

Munzel, P., and Koschel, K., 1981, Rabies virus decreases agonist binding to opiate receptors of mouse neuroblastoma–rat glioma hybrid cells 108.CC-15, *Biochem. Biophys. Res. Commun.* **101**:1241.

Murphy, F. A., 1975, Rabies, morphology and morphogenesis, in: *The Natural History of Rabies*, Vol. I (G. M. Baer, ed.), pp. 33–61, Academic Press, New York.

Murphy, F. A., 1977, Rabies pathogenesis: Brief review, *Arch. Virol.* **54**:279.

Murphy, F. A., 1982, Rabies: Its pathogenesis predicts its ecological entrenchment, in: *Viral Diseases in South-East Asia and the Western Pacific* (J. S. Mackenzie, ed.), pp. 553–562, Academic Press, New York.

Murphy, F. A., 1985, The pathogenesis of rabies virus infection, in: *World's Debt to Pasteur* (S. A. Plotkin and H. Koprowski, eds.), pp. 153–169, Alan Liss, New York.

Murphy, F. A., and Bauer, S. P., 1974, Early street rabies virus infection in striated muscle and later progression to the central nervous system, *Intervirology* **3**:256.

Murphy, F. A., and Bauer, S. P., Harrison, A. K., and Winn, W. C., Jr., 1973a, Comparative pathogenesis of rabies and rabies-like viruses: Viral infection and transit from inoculation site to the central nervous system, *Lab. Invest.* **28**:361.

Murphy, F. A., Harrison, A. K., Washington, W. C., and Bauer, S. P., 1973b, Comparative pathogenesis of rabies and rabies-like viruses: Infection of the central nervous system and centrifugal spread of virus to peripheral tissues, *Lab. Invest.* **29**:1.

Murphy, F. A., Bell, J. F., Bauer, S. P., Gardner, J. J., Moore, G. J., Harrison, A. K., and Coe, J. E., 1980, Experimental chronic rabies in the cat, *Lab. Invest.* **43**:231.

Negri, A., 1903, Contributo allo studio dell'eziologia della rabia (German translation), *Z. Hyg. Infektionskr.* **44**:519.

Nicholson, K., 1983, Human rabies, in: *Rabies: A Growing Threat* (J. R. Pattison, ed.), pp. 6–17, Van Nostrand Reinhold (UK), Wokingham, England.

Nicholson, K. G., Kuwert, E. K., Werner, J., and Harrison, P., 1979a, Interferon response to human diploid cell strain vaccines in man, *Arch. Virol.*, **61**:35.

Nicholson, K. G., Cole, P. J., Turner, G. S., and Harrison, P., 1979b, Immune responses of humans to a human diploid cell strain of rabies virus vaccine: Lymphocyte transformation, production of virus neutralizing antibody and induction of interferon, *J. Infect. Dis.* **140**:176.

Nicholson, K. G., Prestage, H., Cole, P. J., Turner, G. S., and Bauer, S. P., 1981, Multisite intradermal antirabies vaccination: Immune responses in man and protection of rabbits against death from street virus by postexposure administration of human diploid cell strain rabies vaccine, Lancet 2:915.

Otto, G. L., and Heyke, B., 1962, Propagation of rabies virus: Evaluation of a vaccine, Vet. Med. 57:613.

Pasteur, L., Chamberland, S., Roux, E., and Thuillier, 1881, Note sur la rage, C. R. Acad. Sci. 92:1259.

Pasteur, L., Roux, E., Chamberland, C., and Thuillier, L., 1882, Nouveaux faits pour servir a la connaissance de la rage, C. R. Acad. Sci. 95:1187.

Pereira, C. A., Nosaki-Renard, J. N., Schwartz, J., Eyquem, A., and Atanasiu, P., 1982, Cytotoxicity reactions against target cells infected with rabies virus, J. Virol. Methods 5:75.

Perl, D. P., 1975, The pathology of rabies in the central nervous system, in: The Natural History of Rabies, Vol. I (G. M. Baer, ed.), pp. 235–272, Academic Press, New York.

Perrin, P., Portnoi, D., and Sureau, P., 1982, Étude de l'adsoprtion et de la pénétration du virus rabique: Interactions avec les cellules BHK 21 et des membranes artificielles, Ann. Virol. (Inst. Pasteur) 133E:403.

Pinteric, L., Fenje, P., and Almeida, J. D., 1963, The visualization of rabies virus in mouse brain, Virology 20:208.

Prabhakar, B. S., and Nathanson, N., 1981, Acute rabies death mediated by antibody, Nature(London) 290:590.

Prabhakar, B. S., Fischman, H. R., and Nathanson, N., 1981, Recovery from experimental rabies by adoptive transfer of immune cells, J. Gen. Virol. 56:25.

Reagan, K. J., and Wunner, W. H., 1985, Rabies virus interaction with various cell lines is independent of the acetylcholine receptor, Arch. Virol. 84:277.

Reddehase, M. J., Cox, J. H., and Koszinowski, U. H., 1984, Frequency analysis of cytolytic T lymphocyte precursors (CTL-P) generated in vivo during lethal rabies infection of mice. II. Rabies virus genus specificity of CTL-P, Eur. J. Immunol. 14:1039.

Remlinger, M. P., 1903, Isolement du virus rabique par filtration, C. R. Acad. Sci. 55:1433.

Ross, E., and Armentrout, S. A., 1962, Myocarditis associated with rabies: Report of a case, N. Engl. J. Med. 266:1087.

Rowlands, D., Grabau, E. A., Spindler, K., Jones, C., Semler, B., and Holland, J. J., 1980, Virus protein changes and RNA termini alterations evolving during persistent infection, Cell 19:871.

Rubin, R. H., Gough, P., Gerlach, E. H., Dierks, R. E., Gregg, M. B., and Sikes, R., 1971, Immunoglobulin response to rabies vaccine in man, Lancet 2:625.

Schindler, R., 1961, Studies on the pathogenesis of rabies, Bull. W.H.O. 25:119.

Schindler, R., 1965, Pathogenesis of rabies infections, in: Symposium Series in Immunobiological Standardization, Vol. I, pp. 147–152, S. Karger, Basel.

Schneider, L. G., 1975, Spread of virus within the central nervous system, in: The Natural History of Rabies, Vol. 1 (G. M. Baer, ed.), pp. 199–216, Academic Press, New York.

Schneider, L. G., Dietzschold, B., Dierks, R. E., Matthaeus, W., Enzmann, P. J., and Strohmaier, K., 1973, Rabies group-specific ribonucleoprotein antigen and a test system for grouping and typing of rhabdoviruses, J. Virol. 11:748.

Schuller, E., Helary, M., Allinquant, B., Gibert, C., Vachau, F., and Atanasiu, P., 1979, IgM and IgG responses in rabies encephalitis, Ann. Microbiol. (Inst. Pasteur) 130A:365.

Seganti, L., Grassi, M., Mastromarino, P., Pana, A., Superti, F., and Orsi, N., 1983, Activity of human serum lipoproteins on the infectivity of rhabdoviruses, Microbiologica 6:91.

Seif, I., Pepin, M., Blancou, J., Coulon, P., and Flamand, A., 1984, Change in pathogenicity and amino acid substitution in the glycoprotein of several spontaneous and induced mutants of the CVS strain of rabies virus, in: Nonsegmented Negative Strand Viruses (D. H. L. Bishop and R. W. Compans, eds.), pp. 295–300, Elsevier/North-Holland, New York.

Seif, I., Coulon, P., Rollin, P., and Flamand, A., 1985, Rabies virulence: Effect on pathogenicity and sequence characterization of rabies virus mutations affecting the antigenic site III of the glycoprotein, *J. Virol.* **53**:926.

Sikes, R. K., Cleary, W. F., Koprowski, H., Wiktor, T. J., and Kaplan, M. M., 1971, Effective protection of monkeys against death from street virus by post-exposure administration of tissue culture rabies vaccines, *Bull. W.H.O.* **45**:1.

Smith, J., 1981, Mouse model for abortive rabies infection of the central nervous system, *Infect. Immun.* **31**:297.

Smith, J. S., Yager, P. A., and Baer, G. M., 1973, A rapid reproducible test for determining rabies neutralizing antibody, *Bull. W.H.O.* **48**:535.

Smith, W. B., Blenden, D. C., Fuh, T. H., and Hiller, L., 1972, Diagnosis of rabies immunofluorescent staining of frozen sections of skin, *J. Am. Vet. Med. Assoc.* **161**:1495.

Steele, J. H., 1975, History of rabies, in: *The Natural History of Rabies*, Vol. I (G. M. Baer, ed.), pp. 1–29, Academic Press, New York.

Sulkin, S. E., and Allen, R., 1975, Interferon and rabies virus infection, in: *The Natural History of Rabies*, Vol. I (G. M. Baer, ed.), pp. 355–369, Academic Press, New York.

Sung, H. H., Hayano, M., Mastri, A. R., and Okagaki, T., 1976, A case of human rabies and ultrastructure of the Negri body, *J. Neuopathol. Exp. Neurol.* **35**:541.

Superti, F., Seganti, L., Tsiang, H., and Orsi, N., 1984, Role of phospholipids in rhabdovirus attachment to CER cells, *Arch. Virol.* **81**:321.

Superti, F., Hauttaseur, B., Morelee, M.-J., Boldoni, P., Bizzini, B., and Tsiang, H., 1986, Involvement of gangliosides in rabies virus infection, *J. Gen. Virol.* **67**:47.

Sureau, P., Portnoi, D., Rollin, P., Lapresle, C., and Chaouni-Berbich, A., 1981, Prevention de la transmission inter-humaine de la rage aprés greffe de cornée, *C. R. Acad. Sci.* **293**:689.

Sureau, P., Rollin, P. E., and Loucq, C., 1984, Étude de la réponse immunitaire humorale au "vaccin rabique purifié Pasteur," aprés traitement post-exposition, *Ann. Virol. (Inst. Pasteur)* **135E**:277.

Suzuki, M., Kotano, T., and Yamamoto, K., 1977, Interaction of nonspecific inhibitor and rabies virus hemagglutinins, *J. Gen. Virol.* **36**:31.

Szlachta, H. L., and Habel, R. E., 1953, Inclusions resembling Negri bodies in the brains of nonrabid cats, *Cornell Vet.* **43**:207.

Tangchai, P., Yenbutr, D., and Vejjajjva, A., 1970, Central nervous system lesions in human rabies: A study of twenty-four cases, *J. Med. Assoc. Thailand* **53**:472.

Tierkel, E. S., 1959, Rabies, *Adv. Vet. Sci.* **5**:183.

Tignor, G. H., and Shope, R. E., 1972, Vaccination and challenge of mice with viruses of the rabies serogroup, *J. Infect. Dis.* **125**:322.

Tignor, G. H., Mifune, K., and Smith, A. L., 1976, Immunopathology of rabies, in: *Symposium on Advances in Rabies Research*, pp. 13–19, Centers for Disease Control, Atlanta.

Tsiang, H., 1974, Interféron et rage, in: *Symposium Series in Immunobiological Standardization*, Vol. 21, pp. 257–264, S. Karger, Basel.

Tsiang, H., 1978, Evidence for an intraaxonal transport of fixed and street rabies virus, *J. Neuropathol. Exp. Neurol.* **38**:286.

Tsiang, H., 1982, Neuronal function impairment in rabies-infected rat brain, *J. Gen. Virol.* **61**:277.

Tsiang, H., 1985, An *in vitro* study of rabies pathogenesis, *Bull. Inst. Pasteur* **83**:41.

Tsiang, H., Koulakoff, A., Bizzini, B., and Berwald-Nette, Y., 1983, Neurotropism of rabies virus: An *in vitro* study, *J. Neuropathol. Exp. Neurol.* **42**:439.

Tuffereau, C., Lafay, F., and Flamand, A., 1985, Biochemical analysis of rabies virus proteins in three persistently infected BHK-21 cell lines, *J. Gen. Virol.* **66**:159.

Turner, G. S., 1972, Rabies vaccine and interferon, *J. Hyg. (London)* **70**:455.

Turner, G. S., 1976, Thymus dependence of rabies vaccine, *J. Gen. Virol.* **33**:535.

Turner, G. S., 1977, Interferon und Tollwut—eine Ubersicht, *Immunitaet Infekt.* **5:1208.**

Turner, G. S., 1979, Recovery of immuno-responsiveness to rabies vaccine after treatment with cyclophosphamide, *Arch. Virol.* **61**:321.

Turner, G. S., 1983, Current concepts of immunology and pathogenesis of rabies, in: *Rabies: A Growing Threat* (J. R. Pattison, ed.), pp. 27–32, Van Nostrand Reinhold (UK), Wokingham, England.

Turner, G. S., 1985, Immune response after rabies vaccination: Basic aspects, *Ann. Virol.(Inst. Pasteur)* **136E**:453.

Turner, G. S., and Ballard, R., 1976, Interaction of mouse peritoneal macrophages with fixed rabies virus *in vivo* and *in vitro, J. Gen. Virol.* **30**:223.

Turner, G. S., Aoki, F. Y., Nicholson, K. G., Tyrell, D. A. J., and Hill, L. E., 1976, Human diploid cell strain rabies vaccine: Rapid prophylatic immunization of volunteers with small doses, *Lancet* **1**:1379.

Turner, G. S., Nicholson, N. G., Tyrrell, D. A. J., and Akoi, F. Y., 1982, Evaluation of a human diploid cell strain rabies vaccine: Final report of a three year study of pre-exposure immunization, *J. Hyg.* **89**:101.

Unanue, E. R., 1984, Antigen-presenting function of the macrophage, *Annu. Rev. Immunol.* **2**:395.

Villerreal, L. P., and Holland, J. J., 1976, RNA synthesis in BHK21 cells persistently infected with vesicular stomatitis virus and rabies virus, *J. Gen. Virol.* **33**:213.

Wagner, R. R., Heine, J. W., Goldstein, G., and Schnaitman, C. A., 1971, Use of anti-viral–anti-ferritin hybrid antibody for localization of viral antigen in plasma membrane, *J. Virol.* **7**:274.

Warrell, D. A., 1976, The clinical picture of rabies in man, *Trans. R. Soc. Trop. Med. Hyg.* **70**:188.

Warrell, M. J., Suntharasami, P., Nicholson, K. G., and Warrell, D. A., 1984, Multisite intradermal and multisite subcutaneous rabies vaccination: Improved economical regimen, *Lancet* **1**:874.

Watson, H. D., Tignor, G. H., and Smith, A. L., 1981, Entry of rabies virus into the peripheral nerves of mice, *J. Gen. Virol.* **56**:371.

Wiktor, T. J., 1978, Cell-mediated immunity and postexposure protection from rabies by inactivated vaccines of tissue culture origin, *Dev. Biol. Stand.* **40**:255.

Wiktor, T. J., 1980, Virus vaccines and therapeutic approaches, in: *Rhabdoviruses*, Vol. III (D. H. L. Bishop, ed.), pp. 99–112, CRC Press, Boca Raton, Florida.

Wiktor, T. J., 1985a, Historical aspects of rabies treatment, in: *World's Debt to Pasteur* (H. Koprowski and S. A. Plotkin, eds.), pp. 141–151, Alan R. Liss, New York.

Wiktor, T. J., 1985b, Is a special vaccine required against rabies-related viruses and variants of rabies?, in: *Improvements in Rabies Postexposure Treatment* (I. Vodopija, N. G. Nicholson, S. Smerdel, and U. Bijok, eds.), pp. 9–14, "Liber" University Press, Zargeb.

Wiktor, T. J., and Clark, H. F., 1972, Chronic rabies virus infection of cell cultures, *Infect. Immun.* **6**:988.

Wiktor, T. J., and Clark, H. F., 1975, Growth of rabies virus in cell cultures, in: *The Natural History of Rabies*, Vol. I (G. M. Baer, ed.), pp. 155–179, Academic Press, New York.

Wiktor, T. J., and Koprowski, H., 1980, Antigenic variants of rabies virus, *J. Exp. Med.* **152**:99.

Wiktor, T. J., Kuwert, E. K., and Koprowski, H., 1968, Immune lysis of rabies virus-infected cells, *J. Immunol.* **101**:1271.

Wiktor, T. J., Koprowski, H., and Rorke, L. B., 1972a, Localized rabies infection in mice, *Proc. Soc. Exp. Biol. Med.* **140**:759.

Wiktor, T. J., Postic, B., Ho, M., and Koprowski, H., 1972b, Role of interferon induction in the protective activity of rabies vaccine, *J. Infect. Dis.* **126**:408.

Wiktor, T. J., Gÿorgy, E., Schlumberger, H., Sokol, F., and Koprowski, H., 1973, Antigenic properties of rabies virus components, *J. Immunol.* **110**:269.

Wiktor, T. J., Koprowski, H., Mitchell, J. R., and Merigan, T. C., 1976, Role of interferon in prophylaxis of rabies after exposure, *J. Infect. Dis.* **133**:260.

Wiktor, T. J., Doherty, P. C., and Koprowski, H., 1977a, *In vitro* evidence of cell-mediated immunity after exposure of mice to both live and inactivated rabies virus, *Proc. Natl. Acad. Sci. USA* **74**:334.

Wiktor, T. J., Dietzschold, B., Leamnson, R. N., and Koprowski, H., 1977b, Induction and biological properties of defective interfering particles of rabies virus, *J. Virol.* **21**:626.

Wiktor, T. J., Macfarlan, R. I., and Koprowski, H., 1985, Rabies virus pathogenicity, in: *Rabies in the Tropics* (E. Kuwert, C. Mérieux, H. Koprowski, and K. Bögel, eds.), pp. 21–29, Springer-Verlag, Berlin.

Wild, T. F., and Bijlenga, G., 1981, A rabies virus persistent infection in BHK21 cells, *J. Gen. Virol.* **57**:169.

Winkler, W. G., 1968, Airborne rabies virus isolation, *J. Wildlife Dis.* **4**:37.

Winkler, W. G., Baker, E. F., and Hopkins, C. C., 1972, An outbreak of nonbite transmitted rabies in a laboratory animal colony, *Am. J. Epidemiol.* **95**:267.

Winkler, W. G., Fashinell, T. R., Leffingwell, L., Howard, P., and Conomy, J. P., 1973, Airborne rabies transmission in a laboratory worker, *J. Am. Med. Assoc.* **226**:1219.

Winkler, L. G., Shaddock, J. H., and Williams, L. W., 1976, Oral rabies vaccine: Evaluation of its infectivity in three species of rodents, *Am. J. Epidemiol.* **104**:294.

Wunner, W. H., 1982, Is the acetylcholine receptor a rabies virus receptor?, *Trends NeuroSci.* **5**:413.

Wunner, W. H., and Clark, H F., 1980, Regeneration of DI particles of virulent and attenuated rabies virus: Genome characterization and lack of correlation with virulence phenotype, *J. Gen. Virol.* **51**:69.

Wunner, W. H., Dietzschold, B., Curtis, P. J., and Wiktor, T. J., 1983, Rabies subunit vaccines, *J. Gen. Virol.* **64**:1649.

Wunner, W. H., Reagan, K. J., and Koprowski, H., 1984, Characterization of saturable binding sites for rabies virus, *J. Virol.* **50**:691.

Yelverton, E., Norton, S., Obijeski, J. F., and Goeddel, D. V., 1983, Rabies virus glycoprotein analysis: Biosynthesis in *Escherichia coli*, *Science* **219**:614.

CHAPTER 10

Biology, Structure, and Replication of Plant Rhabdoviruses

A. O. JACKSON, R. I. B. FRANCKI,
AND DOUWE ZUIDEMA

I. INTRODUCTION

Of all the taxonomic groups of viruses recognized, only the families Rhabdoviridae and Reoviridae include members that can infect either vertebrates or plants (Matthews, 1982). Furthermore, members of both these groups are transmitted by insects, in which they also multiply. The rhabdoviruses have complex bacilliform or bullet-shaped virions composed of RNA, protein, carbohydrate, and lipid. All these viruses have striking structural similarities and, for this reason, have been classified as a single family by the International Committee on Taxonomy of Viruses (Matthews, 1982). The importance of the rhabdoviruses as disease agents and their potential danger to human, livestock, and wildlife health has been repeatedly documented (Brown and Crick, 1979). However, it is less generally recognized that many serious diseases caused by rhabdoviruses also plague plants, causing substantial crop losses. Thus, the family as a whole presents such a serious threat to the welfare of man, both directly and indirectly, that it is surprising that relatively little is known about their comparative biology.

A. O. JACKSON AND DOUWE ZUIDEMA • Department of Plant Pathology, University of California at Berkeley, Berkeley, California 94720.   R. I. B. FRANCKI • Department of Plant Pathology, Waite Agricultural Research Institute, The University of Adelaide, Glen Osmond, South Australia 5064.

Several reviews describing the general properties of rhabdoviruses that infect plants have appeared over the past 15 years (Francki, 1973, 1984; Francki and Randles, 1975, 1979; Francki et al., 1981, 1985; Hull, 1970, 1976; Jackson et al., 1981; Knudson, 1973; Lastra, 1977; Martelli and Russo, 1977; Russo and Martelli, 1974). In addition, a group description (Peters, 1981) and descriptions of several individual plant rhabdoviruses have appeared in the *CMI/AAB Descriptions of Plant Viruses* (Black, 1970; Campbell and Lin, 1972; Francki and Randles, 1970; Greber, 1982b; Herold, 1972; Jackson and Christie, 1979; Jones et al., 1977; Martelli and Russo, 1973; Peters, 1971; Proeseler, 1983; Shikata, 1972, Sinha and Behki, 1972; Sylvester et al., 1976). These reviews and descriptions provide valuable reference sources for specialists interested in the pathology, epidemiology, and etiology of the diseases caused by the viruses, as well as information on the properties of the viruses themselves. In this chapter, we have emphasized primarily the biology, structure, and replication of a few of the more thoroughly studied plant rhabdoviruses and have selectively compared this information with some of the more extensive data available on the animal rhabdoviruses. Our intent has been to give a general overview of the plant rhabdoviruses in order to illustrate their pathology and biology, and to emphasize the detrimental effect that they can have on plant productivity. We have also highlighted some of the more recent findings on the biochemistry and replication of the plant rhabdoviruses with the hope of stimulating others to study this important group of viruses.

II. BIOLOGY AND ECOLOGY OF PLANT RHABDOVIRUSES

Rhabdoviruses induce a variety of symptoms similar to those caused by infection with viruses belonging to numerous other groups. These symptoms usually have no diagnostic value, but rhabdoviruses can be readily recognized by electron-microscopic observation of sap from diseased tissue or thin sections of infected cells (Fig. 1). Consequently, numerous preliminary reports describe possible rhabdovirus diseases in many different plants. Table I lists some properties of the better-established plant rhabdoviruses. Unfortunately, the communications describing many other rhabdoviruses do not include sufficient details to enable assessment of their relationships. A number of articles merely report the presence of rhabdoviruslike particles in infected plants without providing appropriate information implicating them as the cause of the disease or their relationship to other rhabdoviruses. Therefore, there is a pressing need to conduct more definitive biochemical, physicochemical, and serological studies of plant rhabdoviruses in order to more fully characterize them as specific disease agents and to establish their taxonomic affinities.

A multidisciplinary approach could help expedite investigations of most plant rhabdoviruses, and several areas suitable for collaborative

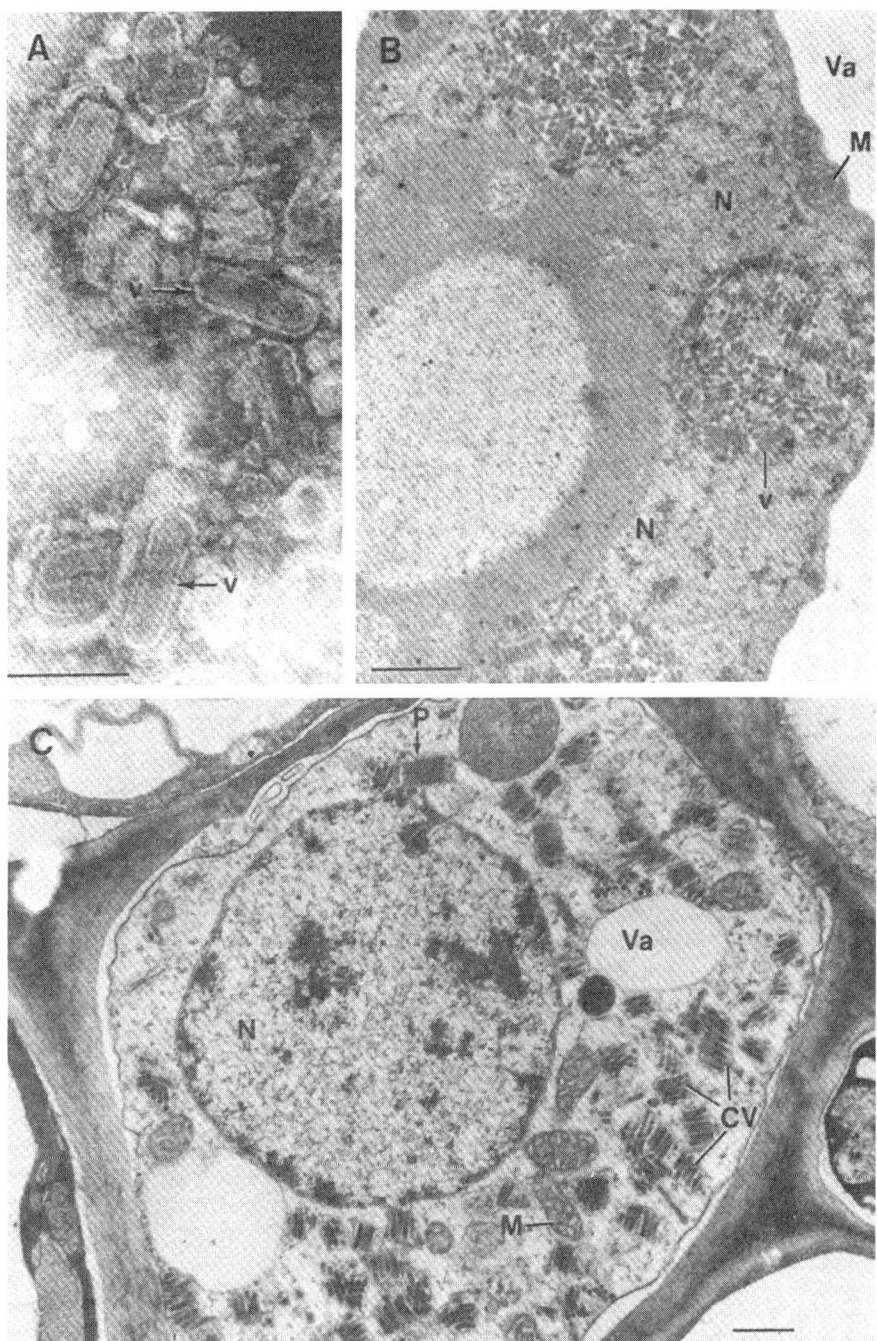

FIGURE 1. (A) Electron micrograph of a leaf dip preparation from a tobacco leaf cell infected with sonchus yellow net virus (SYNV). The preparations were fixed with glutaraldehyde before negative staining with phosphotungstate. Scale bar: 200 nm. (B) Thin section of an SYNV-infected tobacco cell with clumps of virus particles in the nucleus near its periphery. Scale bar: 1 μm. (A, B) Modified from Christie *et al.* (1974). (C) Thin section of a maize cell infected with maize sterile stunt virus, showing virus particles in cytoplasmic vesicles (CV) and in an extrusion of the outer nuclear membrane (P). (N) Nucleus; (M) mitochondrion; (v) virus particle; (Va) vacuole. Scale bar: 500 nm. Modified from Greber (1982a).

TABLE I. Rhabdoviruses of Plants

Virus	Vectors	Sap transmission	Distribution	In vivo site of assembly[a]	Particle dimensions (nm)[b]	References
Aphid-transmitted viruses						
Broccoli necrotic yellows virus (BNYV)	*Brevicoryne brassicae* (L.)	+	Great Britain, Australia	Cyt.	64 × 297	Campbell and Lin (1972)
Carrot latent virus (CaLV)	*Semiaphis heraclei*	−	Japan	Nuc.	70 × 220	Ohki *et al.* (1978)
Lettuce necrotic yellows virus (LNYV)	*Hyperomyzus lactucae* (L.), *Hyperomyzus carduellinus* (Theob)	+	Australia, New Zealand	Cyt.	52 × 360	Francki and Randles (1970)
Lucerne enation virus (LuEV)	*Aphis craccivora* (Koch)	−	France	Nuc.	82–89 × 250	Leclant *et al.* (1973)
Parsley rhabdovirus (PRV)[c]	*Cavariella aegopodii* (Scopoli)	+	Great Britain	Cyt.	87 × 214	Tomlinson and Webb (1974)
Raspberry vein chlorosis virus (RVCV)	*Aphis idaei* (v.d.G.)	−	Europe	Cyt.	65 × 430	Jones *et al.* (1977)
Sonchus yellow net virus (SYNV)	*Aphis coreopsidis* (Thomas)	+	North America	Nuc.	94 × 248[d]	Jackson and Christie (1977, 1979)
Sowthistle yellow vein virus (SYVV)	*Hyperomyzus lactucae* (L.), *Macrosiphum euphorbiae* (Thos.)	−	North America, Europe	Nuc.	80 × 220	Peters (1971) Schultz and Peters (1976), Credi *et al.* (1982)

Virus	Vector		Distribution	Location	Dimensions	Reference
Strawberry crinkle virus (StCV)	*Chaetosiphon fragaefolii* (Cock), *Chaetosiphon jacobi* (H.R.L.)	—	Americas, Europe, South Africa, Australia, New Zealand	Cyt.	69 × 190–380	Sylvester *et al.* (1976)
Leafhopper- or planthopper-transmitted viruses						
Barley yellow striate mosaic virus (BYSMV)	*Laodelphax striatellus* (Fallen)	—	Europe	Cyt. (Vpl)	45 × 330	Conti and Appiano (1973)
Cereal chlorotic mottle virus (CeCMV)	*Nesoclutha pallida* (Evans)	—	Australia	Nuc.	63 × 230	Greber (1979b), Greber and Gowanlock (1979)
Colocasia bobone disease virus (CBDV)	*Torophagus proserpina* (Kirk.)	—	Pacific Islands	Nuc.	65 × 380–335	James *et al.* (1973), Gollifer *et al.* (1977)
Digitaria striate virus (DSV)	*Sogatella kalophon* (Kirkaldy)	—	Australia	Cyt.	55 × 280[d]	Greber (1979a)
Finger millet mosaic virus (FMMV)[e]	*Sogatella longifurcifera* (Esaki and Ishihara), *Peregrinus maidis* (Ashm.)	—	India	Nuc.	80 × 285	Maramorosch *et al.* (1977)
Maize mosaic virus (MMV)	*Peregrinus maidis* (Ashm.)	—	Americas, Caribbean, Hawaii, India, Mauritius, Africa	Nuc.–Cyt.	48 × 240	Herold (1972)

(continued)

TABLE I. (*Continued*)

Virus	Vectors	Sap trans- mission	Distribution	*In vivo* site of assembly[a]	Particle dimensions (nm)[b]	References
Maize sterile stunt virus (MSSV)	*Sogatella longifurcifera* (Esaki and Ishihara), *Peregrinus maidis* (Ashm.), *Sogatella kolophon* (Kirk.)	−	Australia	Cyt.	45 × 255	Greber (1977, 1982a)
Northern cereal mosaic virus (NCMV)	*Laodelphax striatellus* (Fallen), *Ribantodelphax albifascia* (Mats.), *Unkanodes sapporanus* (Mats.), *Meulterianella fairmairei*, (Perris)	−	Japan	Cyt. (Vpl.)	60 × 300– 350	Toriyama (1976b)
Oat striate mosaic virus (OSMV)	*Graminella nigrifrons* (Forbes)	−	North America	Nuc.–Cyt.	100 × 400	Jedlinski (1976)
Potato yellow dwarf virus (PYDV)	*Aceratagallia sanguinolenta* (Provancher), *Aceratagallia lyrata* (Baker), *Aceratagallia obscura* (Oman), *Aceratagallia curvata* (Oman),	+	North America	Nuc.	75 × 380	Black (1970)

Virus	Vector		Geographical distribution	Site of accumulation	Particle size	Reference
Rice transitory yellowing virus (RTYV)	Nephotettix spicalis (Metsch), Nephotettix cincticeps (Uhler), Nephotettix impicticeps (Ish.)	Aggalia constricta (van Duzee), Agallia quadripunctata (Pronvancher), Agalliopsis novella (Say)	Taiwan	Nuc.	94 × 180–210	Shikata (1972), Chiu et al. (1965)
Sorghum stunt mosaic virus (SSMV)	Graminella sonora (Ball)	—	North America	Nuc.	68 × 220	Mayhew and Flock (1981)
Shiraz maize rhabdovirus (SMRV)	Ribautodelphax notabilis (Loguinenko)	—	USSR, Iran	Unknown	70–85 × 150–250	Izadpanah et al. (1983), Milne et al. (1986)
Wheat chlorotic streak virus (WCSV)	Laodelphax striatellus (Fallen)	—	France	Cyt.	55 × 355	Signoret et al. (1978)
Wheat rosette stunt virus (WRSV)	Laodelphax striatellus (Fallen)	—	China	Cyt.	40–54 × 320–400	Tien et al. (1980)
Wheat striate (American) mosaic virus (WSMV)	Endria inimica (Say), Elymana virescens (F.)	—	North America	Cyt.	75 × 250	Sinha and Behki (1972)
Winter wheat (Russian) mosaic virus (WWMV)	Psamotettix striatus (L.), Psamotettix alienus (Dhlb.)	—	Europe	Cyt.	60 × 260	Razvjaskina and Poljakova (1967)

(continued)

TABLE I. (*Continued*)

Virus	Vectors	Sap trans-mission	Distribution	*In vivo* site of assembly[a]	Particle dimensions (nm)[b]	References
Lacebug-transmitted virus						
Beet leafcurl virus [Beta virus 3] (BLCV)	*Piesma quadratum* [Fieb.]	–	Europe	Nuc.	80 × 225–350	Eisbein (1976)
Mite-transmitted virus						
Coffee ringspot virus (CRV)	*Brevipalpus phoenicis* (Geijskes)	+	Brazil	Nuc.	59–76 × 1178–224	Chagas (1980)
Viruses with no known vector						
Cow parsnip mosaic virus (CoPMV)		+	Finland	Nuc.	90 × 265	Polak *et al.* (1977)
Cynara virus (CV)		+	Spain	Cyt.	75 × 260	Pena-Iglesias *et al.* (1972), Russo *et al.* (1975)
Eggplant mottled dwarf virus (EMDV)		+	Italy	Nuc.	66 × 220	Martelli and Russo (1973)
Festuca leaf streak virus (FLSV)		–	Denmark	Cyt.	61 × 330	Lundsgaard and Albrechtsen (1976, 1979)
Gomphrena virus (GV)		+	Brazil	Nuc.	75 × 230–250	Kitajima and Costa (1966)

Virus		Location	Site	Size	Reference
Moroccan wheat rhabdovirus (MWRV)	−	Morocco	Unknown	50–60 × 220–240	Lockhart and Elyamani (1983)
Ivy vein-clearing virus (IVCV)	+	Italy	Nuc.	55 × 325	Castellano and Rana (1981)
Melilotus latent virus (MeLV)	+	North America	Nuc.	80 × 300–350	Kitajima et al. (1969)
Orchid fleck virus (OFV)	+	Japan, Europe, South America	Nuc.	32–35 × 100–140	Dio et al. (1977)
Pelargonium vein-clearing virus (PLVCV)	+	Mediterranean	Nuc.	70 × 250	Di Franco et al. (1979)
Pisum virus (PV)	+	Brazil	Cyt.	45 × 240	Caner et al. (1976)
Pittosporum vein-clearing virus (PVCV)	+	Italy	Nuc.	80 × 245	Rana and Di Franco (1979)
Raphanus virus (RV)	+	Brazil	Cyt.	50–70 × 250–300	Kitajima and Costa (1979)
Sonchus virus (SonV)	+	Argentina	Cyt.	45–57 × 270–350	Vegas et al. (1976)
Tomato vein-yellowing virus (TVYV)	+	Africa	Nuc.	86 × 265	El Maataoui et al. (1985)

[a] (Cyt.) Particles bud from endoplasmic reticulum and accumulate in membrane-bound vesicles; [Cyt. (Vpl)] particles bud from viroplasms located in the cytoplasm and accumulate in vacuolelike sacs; [Nuc.] particles bud from the inner nuclear membrane and accumulate in the perinuclear space.
[b] Measurements were made on particles in thin sections of infected plant cells.
[c] Francki et al. (1981) referred to this virus as parsley latent virus (ParLV).
[d] Measurements were made on particles after negative staining.
[e] Jackson et al. (1981) referred to this virus as Ragi disease rhabdovirus.

efforts are ripe for future studies. For example, investigations of the biology and transmission of many rhabdoviruses can proceed most efficiently by collaboration of entomologists and plant virologists. Most rhabdoviruses that infect plants are transmitted by aphids or leafhoppers, but some are transmitted by other agents such as lacebugs and mites (Table I). The virus–vector interaction is highly specific, and in all cases that have been examined, rhabdoviruses have been shown to replicate in their vectors (for comprehensive reviews, see Francki, 1973; Harris, 1979; Jackson *et al.*, 1981). Viruliferous insects are undoubtedly the principal, if not the sole, source of transmission in nature, but unfortunately, the ecology and epidemiology of only a few plant rhabdoviruses have been studied in any detail.

The dependence in nature on transmission of plant rhabdoviruses by insect vectors, the specificity of the virus–vector relationship, and possible pathological effects of the virus on the insect imply that disease development may be dependent on a delicate balance of several interacting factors. The most obvious of these is a reservoir of virus in weed hosts, in volunteer plants that bridge the season between crops, or in vectors that survive from one crop generation to the next. For rapid disease development, adequate populations of viruliferous vectors must be present sufficiently early in the season to spread the virus into the field from external sources or throughout the field from volunteer plants when crops are most susceptible to infection. With this reliance on the vector, efficient spread of the virus is probably most dependent on environmental conditions conducive to rapid buildup of migratory insect populations, short virus incubation periods in insects after acquisition feeding, and rapid disease development in plants after virus transmission. However, before we can understand how these factors interact to result in disease outbreaks, seemingly minor changes in climate, cropping practices, crop varieties, or vector populations must be analyzed in detail. It is likely that any one of many factors can alter disease development, so it is crucial that each disease be thoroughly investigated with a view toward control.

An example that illustrates how seemingly unrelated ecological events may alter the course of a plant disease has been related in a previous review (Francki, 1984). This example concerns the interdependence of sowthistle (*Sonchus oleraceous*), in which lettuce necrotic yellows virus (LNYV) is latent, *Hyperomyzus lactucae*, its aphid vector, and rabbits. It is generally accepted that the aphids and sowthistle were present in Australia long before 1950, when the first epidemics of LNYV appeared (Randles and Carver, 1971). However, the epidemics must have resulted either from the introduction of the virus into Australia or from the initiation of some ecological imbalance. It is unlikely that the virus was introduced into Australia, because it has not been recorded anywhere outside Australia and New Zealand (Fry *et al.*, 1973). Stubbs, in a personal communication to Matthews (1970), suggested that the introduction of myxomatosis virus into Australia to control rabbits precipitated the events that led to the LNYV epidemics in lettuce. According to this hypothesis, rabbits kept the sowthistle under control until myxomatosis decimated rabbit populations. Then, the weed became more prevalent, resulting in an increase of the reservoir of both LNYV and its vector. This provided conditions suitable for disease epidemics in lettuce. Although it is impossible to determine whether this hypothesis is correct, it has merit because there is a correlation between the time at which the rabbit populations were drastically reduced by myxomatosis and the sudden upsurge of disease in lettuce caused by LNYV (Fenner and Ratcliffe, 1965; Stubbs and Grogan, 1963).

III. PATHOLOGY OF PLANT RHABDOVIRUSES

Rhabdoviruses are known to infect most major crop plants, and many instances of serious disease outbreaks have been reported. Information available on these diseases has been compiled in several previous review articles (Francki, 1973; Jackson et al., 1981; Francki et al., 1981), so we are restricting our discussion on pathology to four examples for which detailed information on disease development has accumulated. These are maize mosaic, rice transitory yellowing, potato yellow dwarf, and lettuce necrotic yellows.

A. Maize Mosaic

Maize mosaic has been known to be a serious disease for more than 50 years, and it was one of the first diseases known to be caused by a rhabdovirus (Herold et al., 1960). The disease was first observed in Hawaii, but has since been reported in Central and South America, Cuba, India, Mauritius, Africa, and recently in the southern part of the United States (Bradfute and Tsai, 1983). Maize mosaic has also been reported in Australia, but recent tests have shown that the Australian maize diseases are caused by other rhabdoviruses (Greber, 1984).

Plants infected with maize mosaic virus (MMV) initially develop stripes between the leaf veins, and the leaves may later turn yellow and become necrotic. If infection occurs early, the plants may be stunted, with shortened internodes and deformed cobs. Little is known about different strains of MMV, but various isolates reportedly cause different symptoms (Lastra, 1977). The virus is not seed or mechanically transmitted, but is spread only by the planthopper *Peregrinus maidis*. MMV undoubtedly multiplies in the vector, because characteristic bacilliform particles accumulate in the salivary glands, nerve ganglia, intestinal cells, and several other tissues of viruliferous planthoppers (Ammar and Nault, 1985). The cytopathology of both plant and insect cells is similar in that MMV particles are located within perinuclear spaces as well as in vacuolelike structures in the cytoplasm and are also associated with the endoplasmic reticulum (Fig. 2).

Major yield losses often occur when susceptible corn is planted in areas where MMV occurs. Several grasses are known to be reservoirs of the virus, and leafhoppers migrating from these grasses appear to transmit the virus efficiently to maize. However, the disease is most severe in areas where corn is grown continuously throughout the year. In some cases, 100% yield losses have been reported (Brewbaker, 1981).

Cultural methods of control that include altered planting and cropping schedules have been reasonably successful in reducing losses caused by MMV in Hawaii (Brewbaker, 1981). However, the most useful control measure has been through incorporation of a single gene for disease tolerance into various maize inbred lines and cultivars. This gene is so effective that even in cases of severe maize mosaic epiphytotics, resistant plants rarely exhibit severe stunting symptoms or yield losses.

B. Rice Transitory Yellowing

Rice transitory yellowing was first observed in Taiwan in 1960 (Chiu and Jean, 1969). It is a serious disease of rice, and especially severe losses may occur in the central and southern regions of Taiwan, where two rice crops are grown in a season. The lower two to three leaves of infected plants first turn yellow, then plants become stunted and tillering is reduced. However, the leaf symptoms gradually disappear, hence the name of the disease.

Rice transitory yellowing virus (RTYV) is transmitted to rice in a persistent manner by three leafhopper species of *Nephotettix* (Chiu *et al.*, 1968). The virus can be detected in the vector by infectivity assays and by electron microscopy (Chen and Shikata, 1972). However, infected plant and vector cells differ considerably in their ultrastructural appearance. In the leafhopper, bacilliform particles are observed in the salivary glands, but the virions accumulate only in cytoplasmic vesicles. Rod-shaped structures similar in appearance to viral nucleocapsids are also present in the nuclei and cytoplasm of intestinal epithelial cells. In contrast, virions are localized primarily in the perinuclear spaces of infected rice cells, and naked particles are not readily detectable.

Damage to rice by RTYV depends to a great extent on the age of the plants at the time of inoculation (R. J. Chiu, personal communication). In field trials, yield reductions are greatest in susceptible varieties infected 15–20 days after germination. Plants infected at subsequent stages have progressively reduced losses, and those infected 45 days after germination usually have little reduction in yield.

Rice transitory yellowing is of considerable importance in Taiwan. The disease has also been reported from the Island of Okinawa (Saito *et al.*, 1978). In China, the disease known as suetou wangai bindou is almost certainly caused by RTYV (S.-X. Chen, C.-W. Huang, and X.-Y. Liang,

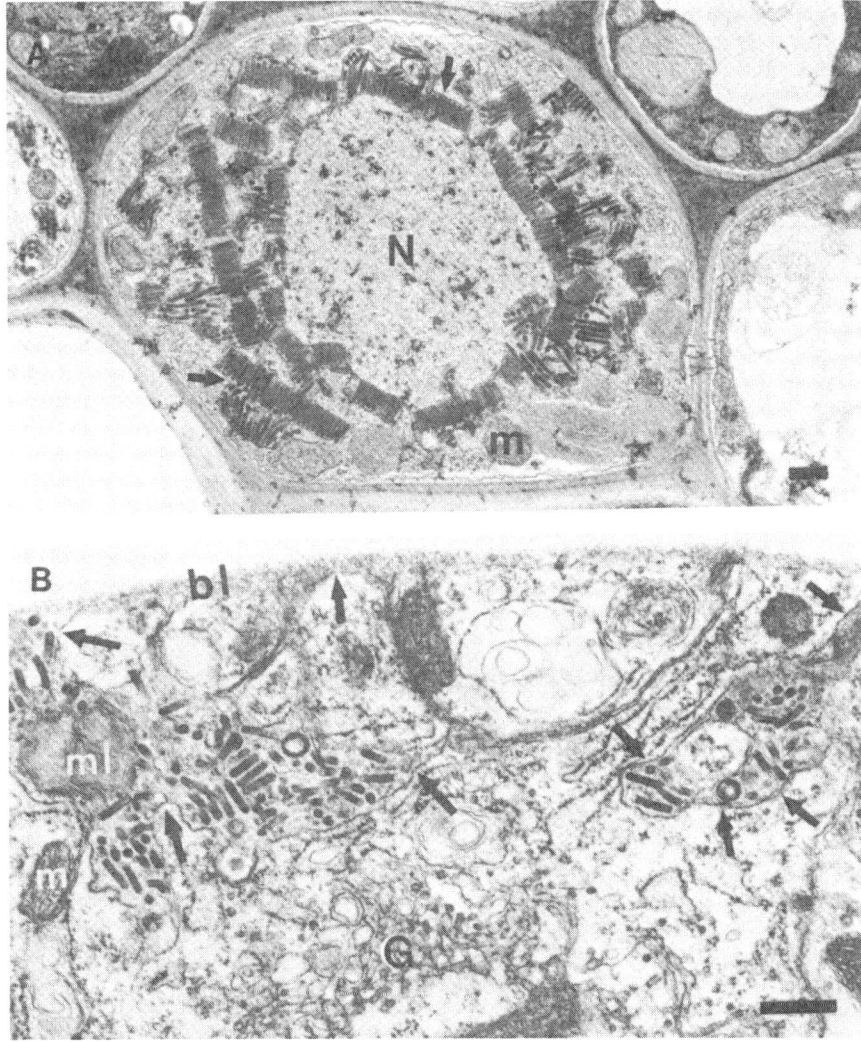

FIGURE 2. Plant and insect vector cells infected with a Hawaiian isolate of maize mosaic virus (MMV). (A) Maize (*Zea mays*) phloem parenchyma cell containing large aggregates of MMV particles (arrows) within an expanded perinuclear space and the cytoplasm. Note the translucent appearance of the nucleus (N). Courtesy of L. L. McDaniel and D. T. Gordon. (B) Small aggregates of MMV particles in the basal periphery of a secretory cell in acinus E of the principal salivary gland cell of the planthopper vector, *Peregrinus maidis*. Particles have accumulated in a space between the basal lamina (bl) and basal infoldings of the plasma membrane (arrows). (G) Golgi; (m) mitochondrion; (ml) multilaminar structures. Courtesy of E. D. Ammar and L. R. Nault. (A, B) Scale bars: 0.5 μm.

personal communication). This disease is transmitted by *N. cincticeps* and appears to cause significant yield losses in some areas.

In Taiwan, rice transitory yellowing has been brought under satisfactory control by a number of different approaches. Protection of plants with the insecticides carbaryl and malathion is an effective and practical control measure. An appropriate schedule of seedbed treatment followed by sprays after transplanting can increase yields nearly 5-fold. Use of resistant cultivars has also contributed to yield increases. Replacement of susceptible lines with resistant cultivars may eventually permit a return to previous cropping practices and may also eliminate the need for costly insecticide applications (R. J. Chiu, personal communication).

C. Lettuce Necrotic Yellows

Lettuce necrotic yellows virus (LNYV) is responsible for a serious disease of lettuce in Australia (Francki, 1973; Randles, 1983). Lettuce plants infected with LNYV become chlorotic and develop a flattened appearance during the acute phase of the disease. They often become necrotic, and many of the plants die. Those plants that survive are stunted and fail to produce a marketable head of lettuce.

LNYV is transmitted in a persistent manner by two aphid species, but *Hyperomyzus lactucae* appears to be the only important vector (Randles and Carver, 1971). LNYV undoubtedly multiplies in *H. lactucae*, because particles of the virus are present in a variety of tissues (O'Loughlin and Chambers, 1967), and the virus can occasionally be transmitted transovarially (Boakye and Randles, 1974).

LNYV is dependent on infection of sowthistle for its survival in nature. Infected sowthistle plants do not show symptoms, and infection also seems not to adversely affect the aphid vector. The aphids breed on sowthistle, and after high population densities develop, migrant winged forms are produced. At this stage, migration occurs, and outbreaks of lettuce necrotic yellows are correlated with flights of viruliferous aphids. During migration, the aphids land on lettuce plants, on which they probe. These probes are sufficient for virus transmission, even though the aphids fail to colonize lettuce and cannot survive unless they find a more suitable host. Major outbreaks of lettuce necrotic yellows occur about a month after extensive aphid migrations (Boakye and Randles, 1974). Since sowthistle is the principal host for both the virus and the aphid, effective control of the disease can be achieved by eradicating sowthistle weeds in and around lettuce fields (Stubbs et al., 1963).

D. Potato Yellow Dwarf

Potato yellow dwarf was discovered in 1917 in the northeastern United States. Infected plants develop a leaf chlorosis that causes the plant to have a yellowish cast. Other symptoms include stem necrosis, stunting, and reduced tuber production. Those tubers that are produced become necrotic and have difficulty sprouting. Potato yellow dwarf virus (PYDV) can be transmitted by several leafhopper species, but the principal vector in nature is the clover leafhopper (*Aceratagallia constricta*). The leafhopper has a wide host range, but breeds primarily on red clover (*Trifolium pratense* L.), which is a symptomless host of PYDV. Several other weed species can provide reservoirs for both the virus and the vector (Black, 1970).

The ability of perennial weeds to serve as hosts for PYDV provides the potential for regular annual outbreaks of the disease. Serious crop losses occurred in the 1930s in dry years, when the vector migrated from native vegetation into potato fields. Since that time, however, the incidence of the disease has decreased, and major losses have not occurred since the 1940s. The absence of epidemics in recent years has probably been due to planting of certified seed and use of tolerant cultivars. The decline in incidence of the disease possibly also resulted from reduced numbers of leafhoppers after the advent of widespread use of insecticides. As a consequence of these practices, the level of PYDV has been reduced to such an extent that the virus is no longer commonly found in nature in the eastern United States (Black, 1979). However, PYDV has recently been isolated from periwinkle in California (Falk *et al.*, 1981), but its presence is primarily of academic interest, since it appears not to cause serious disease.

IV. ULTRASTRUCTURE OF RHABDOVIRUS-INFECTED PLANTS

Although most animal rhabdoviruses replicate and assemble in the cytoplasm (Wagner, 1975), plant rhabdoviruses differ markedly in their morphogenesis and sites of accumulation. The characteristic rhabdovirus structure is easily distinguished from normal cellular components by electron microscopy. This technique has permitted several detailed ultrastructural studies of the cytopathology of infected plants. These studies show that the plant rhabdoviruses can probably be divided into at least three groups depending on the site of nucleocapsid formation and assembly of virions (Francki and Randles, 1980; Francki *et al.*, 1981). One group includes viruses such as sonchus yellow net virus (SYNV), potato yellow dwarf virus, and eggplant mottled dwarf virus (EMDV) that mature in association with the inner nuclear membrane and accumulate in the perinuclear spaces. A second group of viruses, including lettuce necrotic yellows virus (LNYV), broccoli necrotic yellows virus, and maize sterile streak virus, appear to mature in association with the endoplasmic reticulum and accumulate in vesicles of the endoplasmic reticulum. A third group of viruses, represented by barley yellow striate mosaic virus (BYSMV) and northern cereal mosaic virus (NCMV), mature in association with membrane-bound granular structures called "viroplasms." After virus particles bud from membranes associated with the viroplasm, they accumulate in vacuolelike spaces. The latter two groups of viruses may be difficult to distinguish during routine electron microscopy.

Accumulation of plant rhabdoviruses in association with the nucleus is illustrated in Fig. 1B, which shows a tobacco cell infected with SYNV. The most striking abnormality of these cells is the pronounced swelling of the nucleus and the nucleolus, accompanied by the presence of numerous virus particles in the perinuclear spaces. This abnormality is so obvious that it can be visualized by light microscopy (Christie *et al.*, 1974). In the case of SYNV, virions often accumulate in disordered arrays around the nucleus. The swelling of the nucleus combined with the presence of virions in the perinuclear spaces suggests that the nucleus has a crucial role in SYNV assembly. However, we know from studies of putative messenger RNAs (mRNAs) found in SYNV-infected tobacco that the cytoplasm contains a considerable proportion of RNA complementary to the genome of SYNV (Milner and Jackson, 1979, 1983; Milner *et al.*, 1979). Thus, many steps involved in the synthesis, transport, and assembly of polypeptides into virions that accumulate in the perinuclear spaces need to be investigated.

Nascent virions budding from the inner nuclear envelope are not prevalent in tobacco infected with SYNV. However, virus particles that appear to be at various stages of morphogenesis have been observed in plants infected with several other viruses. This is well illustrated in the sections of EMDV-infected plants shown in Fig. 3. Budding of rhabdovirus particles through the inner nuclear envelope has also been carefully documented in rice plant cells infected with rice transitory yellowing virus (Chen and Shikata, 1968, 1971). In both cases, many bullet-shaped particles appear to be at various stages of budding from inside the swollen nuclei. Enveloped viruses are often aligned at the periphery of the nucleus with one end of the viral envelope attached to the inner nuclear envelope (Fig. 3B). Other particles that appear to have completed their envelopment are also present. As infection progresses, the number of mature particles appears to increase and virions accumulate around the nucleus. The interpretation of these observations is that the nucleocapsids are assembled in the nucleoplasm, push outward toward the cytoplasm, and subsequently are pinched off from the inner lamella of the nuclear membrane into the perinuclear space. This general scenario is believed to be the case for most rhabdoviruses that are found in close association with the host nucleus.

Several rhabdoviruses (designated Nuc.–Cyt. in Table I) vary slightly from the pattern of assembly and accumulation described above. Virions of these viruses appear to be assembled in or near the nucleus, but virions may also be found in the cytoplasm close to the cytoplasmic membranes. This type of morphogenesis has been most extensively studied with wheat striate (American) mosaic virus (WSMV). In WSMV-infected wheat leaf

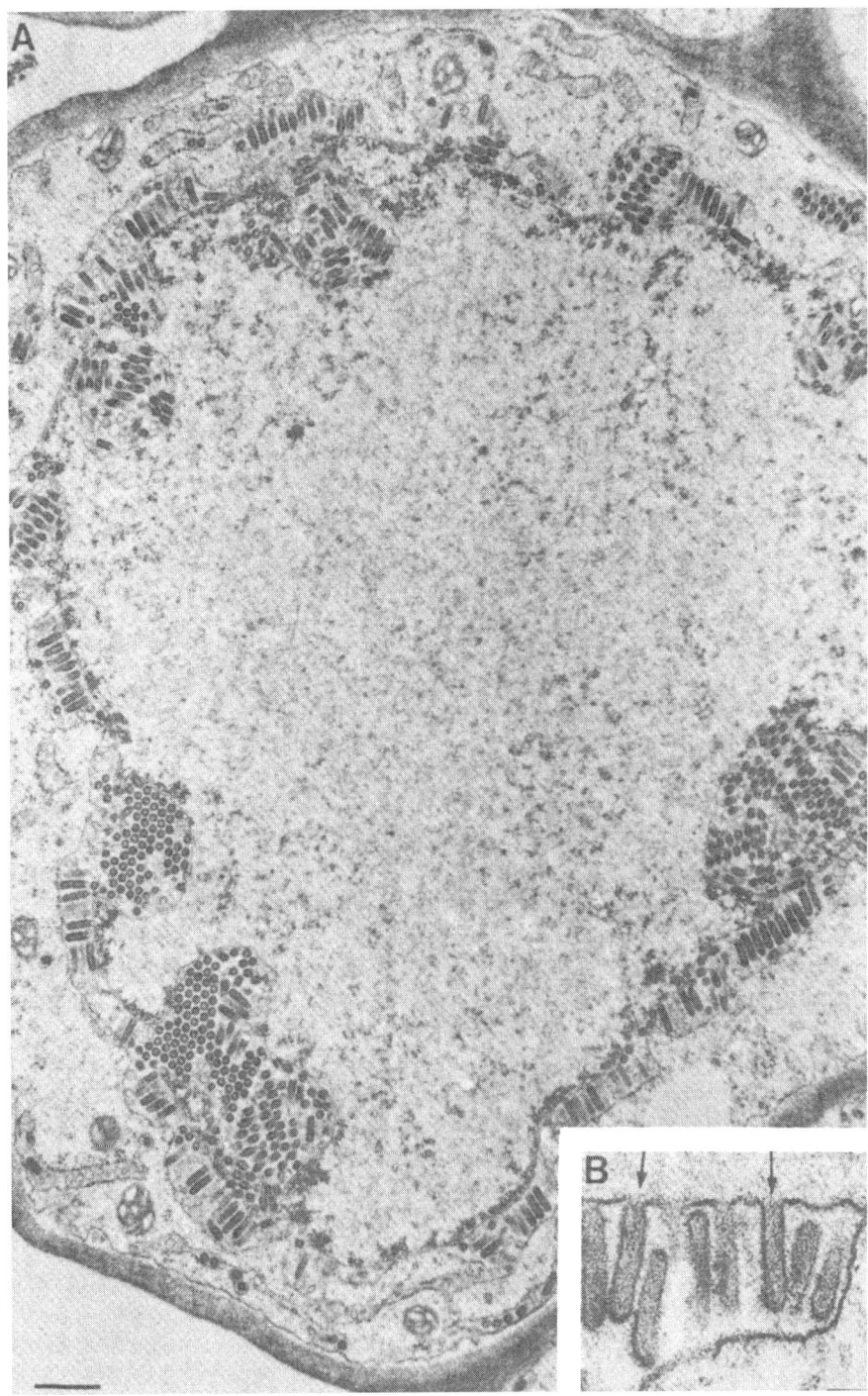

FIGURE 3. (A) Thin section of a nucleus of a *Nicotiana tabacum* leaf cell infected with eggplant mottled dwarf virus (EMDV), showing virus particles accumulating in the perinuclear space. Scale bar: 500 nm. (B) EMDV particles budding (↓) from the inner nuclear membrane to be released into the perinuclear space. Scale bar: 100 nm. Courtesy of G. P. Martelli and M. Russo.

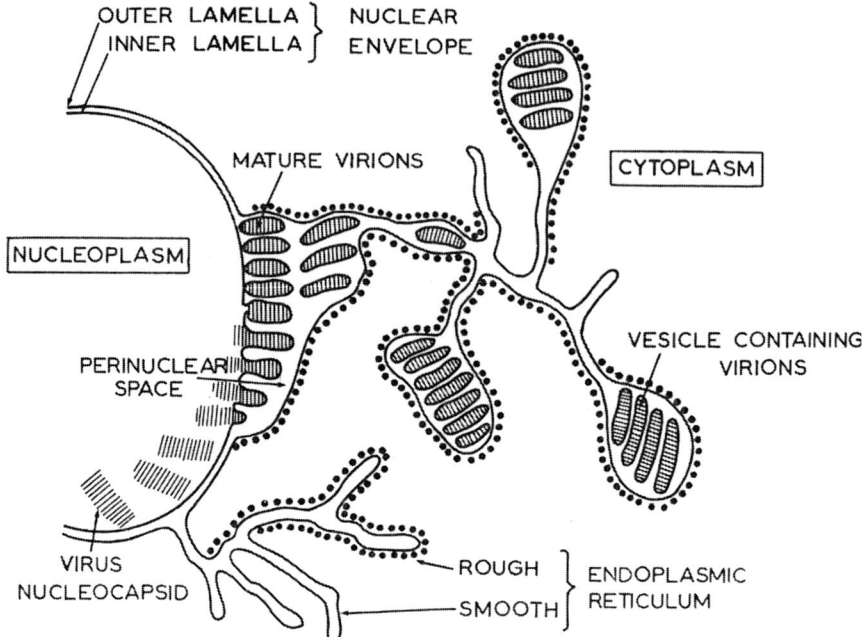

FIGURE 4. Diagrammatic representation of morphogenesis and cellular distribution of a rhabdovirus such as lettuce necrotic yellows virus or maize mosaic virus. Modified from Francki (1973).

cells, virus particles are prevalent in the perinuclear spaces, but virus particles can occasionally be seen in the cytoplasm, sometimes associated with membranes (Lee, 1970). In some instances, virions appear to be budding both from the inner nuclear envelope and from membranes within the cytoplasm. Despite these observations, however, it may be premature to conclude that assembly of WSMV and other viruses found in both the nucleus and the cytoplasm differs appreciably from that of viruses whose particles are predominantly restricted to the perinuclear spaces. In both these instances, it is important to realize that the endoplasmic reticulum is continuous with the outer nuclear membrane. This continuity provides the opportunity for viruses that bud through the inner nuclear envelope to subsequently be transported into cytoplasmic vesicles formed during morphogenesis of the endomembrane system (Fig. 4).

The morphogenesis of the second group of plant rhabdoviruses that assemble at the endoplasmic reticulum has been most thoroughly investigated in *Nicotiana glutinosa* infected with LNYV. The biochemical events thought to occur during the course of infection have been described by Francki and Randles (1979) and thus will not be extensively discussed here. However, the evidence that has been accumulated suggests that the nucleus may be involved in the early stages of infection. In cells that appear to have been recently infected with LNYV, the outer lamellae of the nuclei develop blisters that contain small vesicles and a few virus particles (Wolanski and Chambers, 1971). Thus, during the initial stages of infection, some of the virus particles may acquire their envelopes from the outer nuclear membrane, which is continuous with the endoplasmic reticulum. However, most of the budding at later stages of infection occurs from membranes of the endoplasmic reticulum and cytoplasmic vesicles in which virus particles accumulate (Fig. 5). Masses of threadlike structures that have been called "viroplasms" (Francki, 1973; Francki and Randles, 1979) also appear in the cytoplasm, and these viroplasms are sometimes in close association with virus particles. It must be emphasized, however, that the exact structure or role of these "viroplasms" is obscure. The viroplasms found in LNYV-infected cells appear to be distinct from the granular viroplasms found in cells infected with viruses such as BYSMV. They also lack the extensive proliferation of membranes with associated budding virus particles that surround viroplasms of BYSMV-infected cells.

BYSMV, which undergoes cytoplasmic morphogenesis in association with a membrane-bound viroplasm, differs from LNYV because there appears to be no nuclear phase in the replication cycle (Bassi et al., 1980). BYSMV particles are detected only in the cytoplasm (Conti and Appiano, 1973). The particles are consistently found in association with a viroplasm composed of electron-dense masses of granular material (Fig. 6). Membranes surrounding the viroplasm contain numerous budding particles that presumably originate from the viroplasm (Fig. 6B). Often the infected cells also contain numerous large aggregates of bacilliform particles enclosed in vacuolelike vesicles that may have originated from the endoplasmic reticulum (Conti and Appiano, 1973). Similar structures are present in NCMV-infected cells (Toriyama, 1976b) and festuca leaf streak virus (FLSV)-infected cells (Lundsgaard and Albrechtsen, 1979), so these two viruses probably have the same pattern of morphogenesis as BYSMV. Recently, the morphogenesis of FLSV has been investigated in infected cowpea protoplasts (van Beek et al., 1985c). The ultrastructure of these protoplasts is discussed in Section XII.D.

The ultrastructural differences in the morphogenesis of LNYV and BYSMV are further substantiated by differences in the distribution of [³H]uridine in infected cells. In the case of LNYV-infected cells, radioactivity first accumulates in the nucleus at about the same time that the nuclei become blistered. Later, the radioactivity appears in the cytoplasm, and this appearance is correlated with the appearance of mature virions in the cytoplasm. In the case of BYSMV-infected barley, a different pattern of accumulation is observed (Bassi et al., 1980). In this case, [³H]uridine incorporation is pronounced only in the cytoplasm and even at early stages of infection, whereas RNA synthesis is depressed in the nucleus. Wolanski and Chambers (1971) suggest that synthesis of LNYV RNA occurs in the nucleus, but that nucleoprotein assembly and membrane envelopment occur in the outer nuclear envelope and the endoplasmic reticulum.

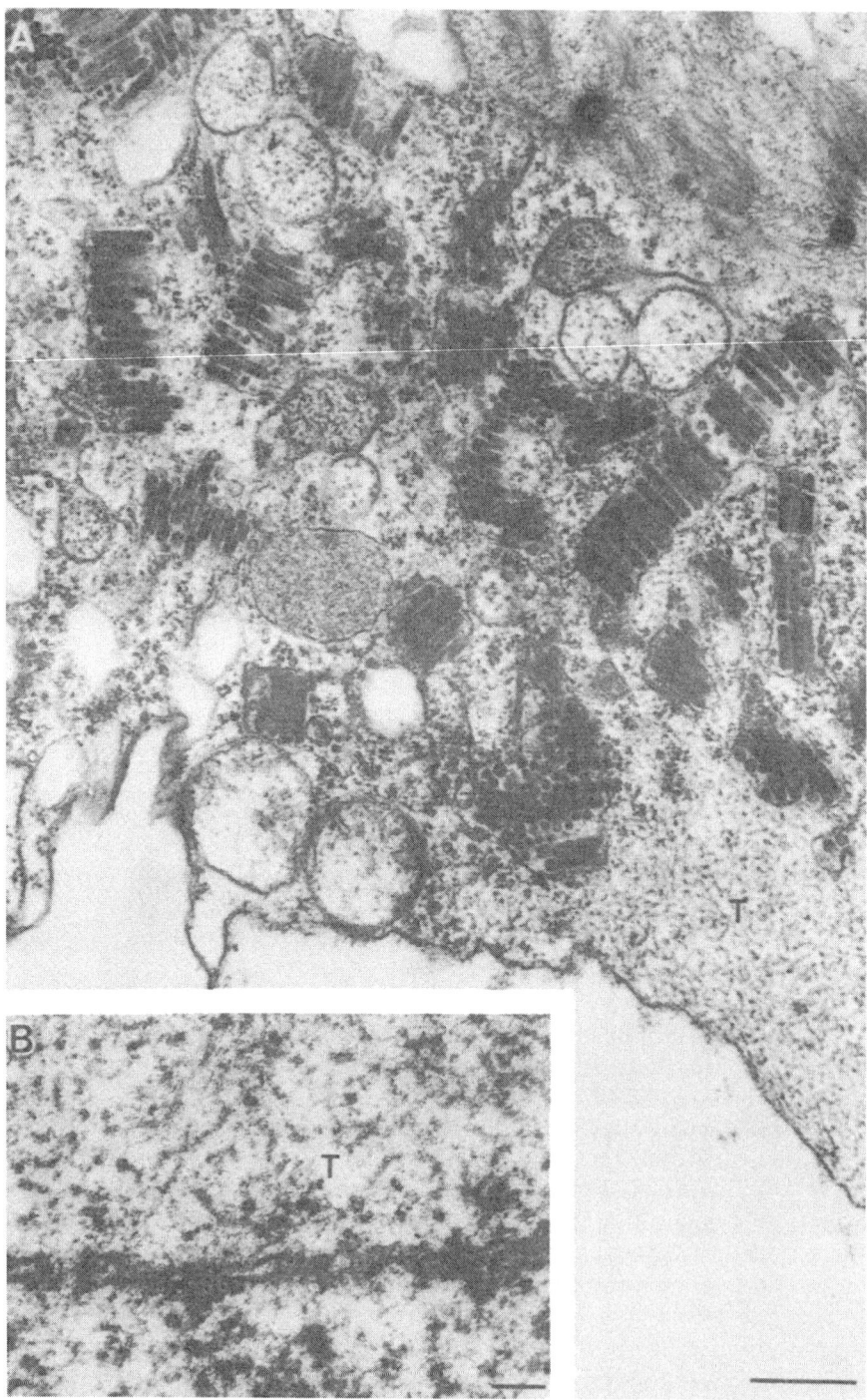

FIGURE 5. (A) Thin section of a *Nicotiana clevelandii* leaf cell infected with lettuce necrotic yellows virus, showing numerous particles in vesicles of the endoplasmatic reticulum and a region with virus-specific threadlike structures (T). Scale bar: 500 nm. (B) Higher magnification of the cytoplasmic threadlike structures (T) adjoining a nucleus. Scale bar: 100 nm. From Francki *et al.* (1985).

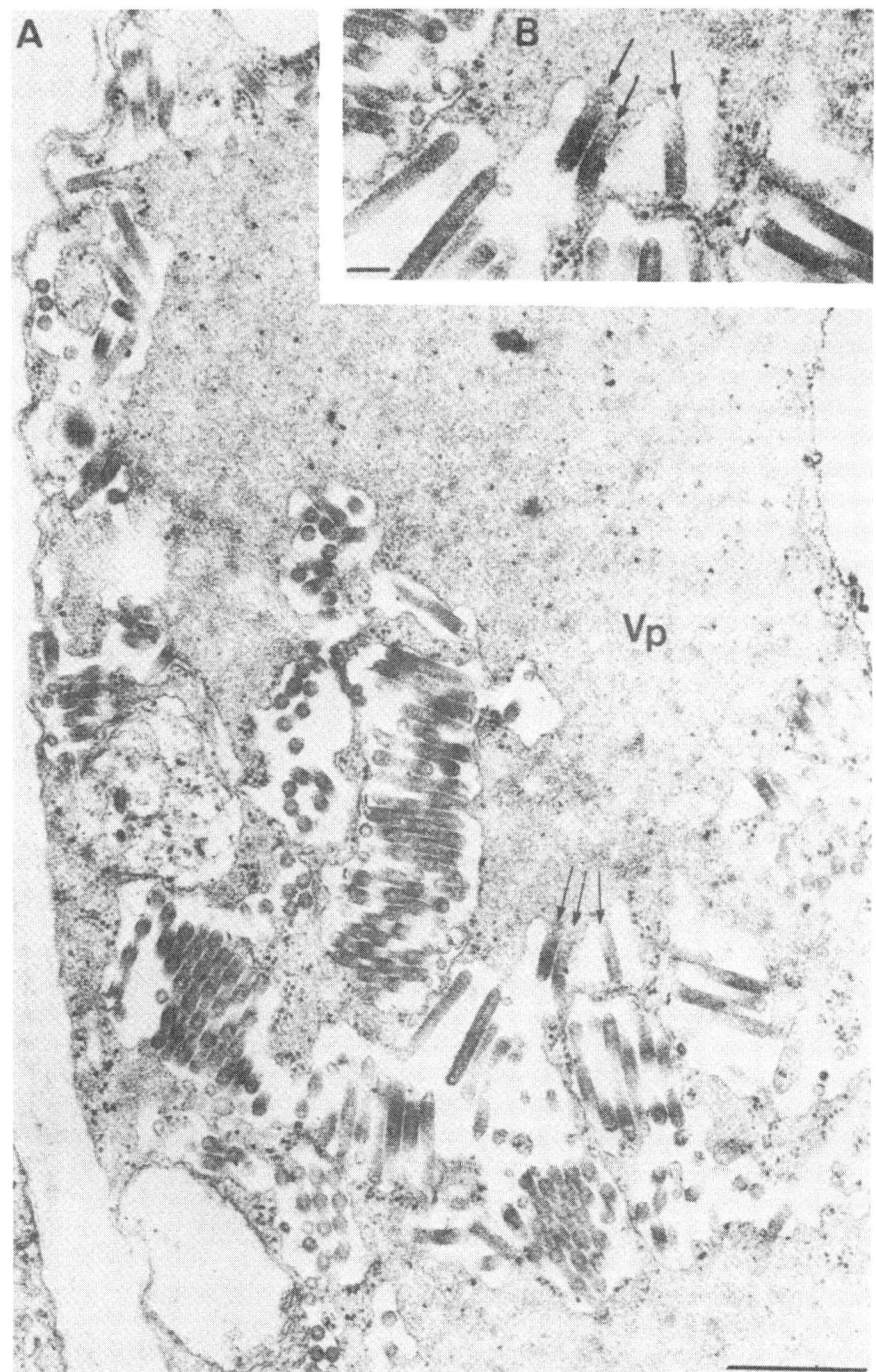

FIGURE 6. (A) Thin section of a barley leaf cell infected with barley yellow striate mosaic virus (BYSMV), showing a granular inclusion [viroplasm (Vp)] and virus particles, some of which appear to be in the process of budding from the viroplasm (↓). Scale bar: 500 nm. (B) Higher magnification of BYSMV particles budding (↓) from the viroplasm. Scale bar: 100 nm. From Francki *et al.* (1985); courtesy of A. Appiano.

It is possible that viral RNA may be synthesized in the nucleus and transported to the cytoplasmic membranes for assembly and envelopment. In contrast, BYSMV RNA synthesis and nucleocapsid assembly may occur in viroplasms, and maturation of virions may occur by budding from membranes associated with the viroplasms. Fortunately, recent advances in cytological techniques that rely on the specificity of monoclonal antibodies and recombinant DNA clones provide us with new tools that should soon enable precise analysis of the subcellular localization of specific components of the virus. Moreover, the ability to infect protoplasts with virus should provide a synchronous system for study of the sequence of events in replication and assembly.

In addition to enlarged nuclei, viroplasms, virus particles, and narrow tubular structures thought to be nucleoprotein, other ultrastructural changes occur in infected cells. Plants infected with viruses that accumulate in the cytoplasm usually have a well-developed endoplasmic reticulum. These membranes may vary in appearance from slight proliferations to massive honeycomblike structures containing large numbers of aggregated virions (Herold et al., 1960; Sylvester et al., 1976; Jones et al., 1977). Other cytopathological abnormalities include changes in the appearance of mitochondria (Vega et al., 1976), as well as swelling and disorganization of chloroplasts (Herold et al., 1960; Lee, 1967; Wolanski, 1969; Martelli et al., 1975) and proliferation of the Golgi apparatus (Toriyama, 1976b).

V. RELATIONSHIPS OF PLANT RHABDOVIRUSES WITH INSECT VECTORS

Table I lists 9 rhabdoviruses known to have aphid vectors, 17 that are transmitted by leafhoppers or planthoppers, 1 by a lacebug, and 1 by a mite. The remaining 15 viruses have no known vectors. The transmission of several of the aphid and leafhopper- and planthopper-borne viruses has been examined in sufficient detail to enable some general conclusions to be made about their insect–virus relationships. The most extensive studies have been with potato yellow dwarf virus (PYDV), sowthistle yellow vein virus (SYVV), lettuce necrotic yellows virus (LNYV), rice transitory yellowing virus (RTYV), and wheat striate (American) mosaic virus (WSMV). Some of these studies have already been discussed in detail (Francki, 1973; Francki and Randles, 1979; Jackson et al., 1981), so only the most significant work will be reviewed here.

A. Vector Specificity

A series of classic studies conducted with PYDV (Black, 1959) revealed two strains of the virus that differed in vector transmissibility. The New York or SYDV strain was transmitted specifically by the leafhopper *Aceratagallia sanguinolenta* and three other related species. However, the New Jersey or CYDV strain could not be transmitted by species of *Aceratagallia*, but was readily transmitted by *Agallia constricta* (Black, 1940). Similar vector specificity has now been documented for a number of other leafhopper- or planthopper-transmitted rhabdoviruses that infect cereals (for reviews, see Shikata, 1979; Jackson *et al.*, 1981). Vector specificity has also been observed with aphid-transmitted rhabdoviruses such as SYVV, LNYV, strawberry crinkle virus (StCV), broccoli necrotic yellows virus (BNYV), and sonchus yellow net virus (SYNV) (Duffus, 1963; Stubbs and Grogan, 1963; Frazier, 1968; Tomlinson *et al.*, 1972; Christie *et al.*, 1974).

Leafhopper transmission of PYDV is genetically determined, and highly efficient, as well as inefficient, races have been selected by breeding (Black, 1943a). Isolates of PYDV that cannot be transmitted by the vector have also been selected by repeated transfer of the virus by mechanical inoculation (Black, 1953). Such mutations may account for the existence of virus strains with no known vector in vegetatively propagated plants (Black, 1969).

B. Multiplication in Vectors

The possibility that a rhabdovirus can multiply in its vector was first investigated by Black and his co-workers. These studies, initiated in the 1940s, first established that PYDV had an incubation period of about 1 week in the leafhopper and was subsequently transmitted by the insect for long periods (Black, 1943b). Later studies showed that the New Jersey strain of PYDV could occasionally be transmitted transovarially (Black, 1970). Recovery of infectious virus from tissue of viruliferous leafhoppers has provided other evidence for multiplication of PYDV in the vector (Sinha, 1968). The pioneering studies of Black and his colleagues have also convincingly demonstrated that PYDV multiplies in cultured cells of its insect vector (Black, 1979).

Direct evidence for virus multiplication in leafhopper vectors has been obtained by serial passage of WSMV (Sinha and Chiykowski, 1969), northern cereal mosaic virus (Yamada and Shikata, 1969), and RTYV (Hsieh *et al.*, 1970) from insect to insect by injection. After a sufficient number of passages through insects maintained on virus-immune plants, the calculated dilution of the initial inoculum injected into the original insect was so great that the only possible conclusion was that the virus had multiplied in the vector. Other studies combining infectivity, serology, and electron microscopy clearly show that leafhopper- and planthopper-transmitted rhabdoviruses multiply in their vectors (see Francki, 1973; Shikata, 1979; Jackson *et al.*, 1981).

Studies with SYVV (Duffus, 1963) also revealed that there is a long latent period before the aphid *Hyperomyzus lactucae* (L.) can transmit the virus. The temperature greatly influenced the length of the latent period, but once the aphids began to transmit the virus, they continued to do so for the remainder of their lives. The observation that plants on which aphids were caged shortly after the latent period took longer to develop symptoms than plants on which the same aphids fed later on suggested that SYVV replicates in the vector (Duffus, 1963). Multiplication of SYVV in *H. lactucae* was subsequently confirmed by electron-microscopic observation of virions in cells of the insect (Richardson and Sylvester, 1968). From a series of dilution experiments in which aphid extracts were serially injected from aphid to aphid, Sylvester and Richardson (1969) calculated that SYVV multiplied in the vector. Sylvester (1969) also observed transovarial transmission of SYVV in the aphid. One study has also suggested that SYVV replicates in cultured aphid explants (Peters and Black, 1970). A number of other observations with aphid-transmitted rhabdoviruses such as LNYV (Boakye and Randles, 1974; O'Loughlin and Chambers, 1967), StCV (Sylvester *et al.*, 1976), and SYNV (Christie *et al.*, 1974) also support the hypothesis that aphid vectors become infected by plant rhabdoviruses and that infection of the insect is important for virus transmission.

The inescapable conclusion of all the studies on the vectors of plant rhabdoviruses is that the virus can multiply in both plants and insects. Other rhabdoviruses that cause serious diseases of vertebrates also multiply in their insect vectors (Brown and Crick, 1979). None of the viruses appears to adversely affect the insects, which suggests a long evolutionary association between them. Thus, rhabdoviruses of both plants and vertebrates may have evolved from common ancestors that infected only insects. This could well account for the many similarities that enable the two groups to be placed in the same family (Matthews, 1982).

C. Cytopathology of Plant and Vector Cells

Rhabdovirus particles have been detected in the cells of several insect vectors of plant rhabdoviruses (Table II). These studies have shown that rhabdoviruses frequently induce similar cytopathic abnormalities in plant and insect cells (Francki *et al.*, 1981; Jackson *et al.*, 1981; Peters, 1981). In the case of infected maize, virions of maize mosaic virus (MMV) bud through the nuclear envelope, and they accumulate in the perinuclear spaces (See Fig. 2) where the outer nuclear envelope proliferates and forms large cisternae with many branches. Virus particles often appear to accumulate in the cytoplasm in such high concentrations that they occupy much of the cell lumen (Lastra, 1977 Martelli *et al.*, 1975; McDaniel *et al.*, 1985). MMV is also found in high concentrations in the intestinal cells and salivary glands of *P. maidis* (Ammar and Nault, 1985; Herold and Munz, 1965). In addition to the perinuclear spaces, virions accumulate in vacuolelike structures and in tubules that appear to have formed from

TABLE II. Records of Rhabdovirus Particles Detected in Cells of Insects

Virus	Vector	Organs in which virus was detected	References
In aphids			
Broccoli necrotic yellows virus (BNYV)	Brevicoryne brassicae (L.)	Most organs except gut	Garrett and O'Loughlin (1977)
Lettuce necrotic yellows virus (LNYV)	Hyperomyzus lactucae (L.)	Alimentary canal, salivary glands, fat body, brain, mycetome, trachea	O'Loughlin and Chambers (1967)
Sowthistle yellow vein virus (SYVV)	Hyperomyzus lactucae (L.)	Salivary glands, fat body, brain, subesophageal ganglion, esophagus, ventriculum, ovary, muscle (not in hindgut)	Sylvester and Richardson (1970)
Strawberry crinkle virus (StCV)	Chaetosiphon jacobi (H.R.L.)	Salivary glands, subesophageal ganglion	Richardson et al. (1972)
Sonchus yellow net virus (SYNV)	Aphis coreopsidis (Thomas)	Not indicated	Christie et al. (1974)
In leafhoppers or planthoppers			
Barley yellow striate mosaic virus (BYSMV)	Laodelphax striatellus (Fallon)	Salivary glands	Conti and Plumb (1977)
Cereal chlorotic mottle virus (CeCMV)	Nesoclutha pallida (Evans)	Salivary glands, brain	Greber and Gowanlock (1979)
Maize mosaic virus (MMV)	Peregrinus maidis (Ashm.)	Salivary glands, gut, fat body, muscle, axon	Harold and Munz (1965), Ammer and Nault (1985)
Maize sterile stunt virus (MSSV)	Peregrinus maidis (Ashm.)	Salivary glands, brain	Greber (1982a)
Potato yellow dwarf virus (PYDV)	Agallia constricta (van Duzee)	Tissue-culture cells	Chiu et al. (1970)
Rice transitory yellowing virus (RTYV)	Nephotettix cincticeps (Uhler)	Salivary glands and gut	Chen and Shikata (1972)
Wheat striate (American) mosaic virus (WSMV)	Endria inimica (Say)	Salivary glands	Bell et al. (1978).

the endoplasmic reticulum. Some virions are found at the plasma membrane (Ammar and Nault, 1985).

Rhabdoviruslike particles of cereal chlorotic mottle virus (CeCMV) are also easily detected in cells of the viruliferous leafhopper *N. pallida* (Greber and Gowanlock, 1979). The association of rhabdoviruslike particles with the nucleus of infected vector cells is generally similar to that of infected plant cells (Fig. 7). In the vector, virions are organized in a palisade arrangement on the inner nuclear envelope of salivary and brain cells. However, the nuclei of CeCMV-infected insect cells appear to differ slightly from those of CeCMV-infected plant cells because virions are embedded in a granular ground mass within the insect cell, but the perinuclear space of infected plant cells has a clear appearance. Leafhopper cells infected with WSMV also have a similar ground mass associated with virions (Bell *et al.*, 1978).

A few rhabdoviruses appear to have different patterns of replication and accumulation in plant and insect cells. These differences are most evident with RTYV, where the virions are primarily confined to the perinuclear spaces in plant cells (Chen and Shikata, 1971), but accumulate mainly in cytoplasmic vacuoles of the leafhopper cells (Chen and Shikata, 1972). The reverse situation occurs with LNYV (O'Loughlin and Chambers, 1967) and possibly with BNYV (Garrett and O'Loughlin, 1977). In these examples, virions are localized in the perinuclear spaces of the aphid vector, whereas in plants they are found predominantly in the cytoplasm. The studies discussed above are all rather limited in scope, but they do suggest that in some instances the cytopathology of virus-infected vector cells may differ from that of infected plants. Even so, the significance of these studies remains uncertain because replication is asynchronous in both the plant and the insect. Therefore, it is not possible to determine whether some of the minor differences in ultrastructure are a consequence of variation in the stage of infection of the examined cells. It is even possible that some of the reported differences may be artifacts resulting from fixation and staining of different tissues. These ultrastructural studies should become more precise in the future when modern techniques of molecular biology are combined with ultrastructural studies.

VI. PURIFICATION OF PLANT RHABDOVIRUSES

The single most important factor limiting studies of plant rhabdoviruses is the difficulty of devising simple and reproducible purification protocols suitable for recovery of adequate amounts of virus of sufficient purity for biochemical analysis. Rhabdoviruses are rather unstable in comparison to most other plant viruses. The presence of a viral envelope precludes the use of organic solvents, heating, freezing, and pH adjustments, which are the common means of removing host materials during virus purification (Francki, 1972). Moreover, the concentration of rhab-

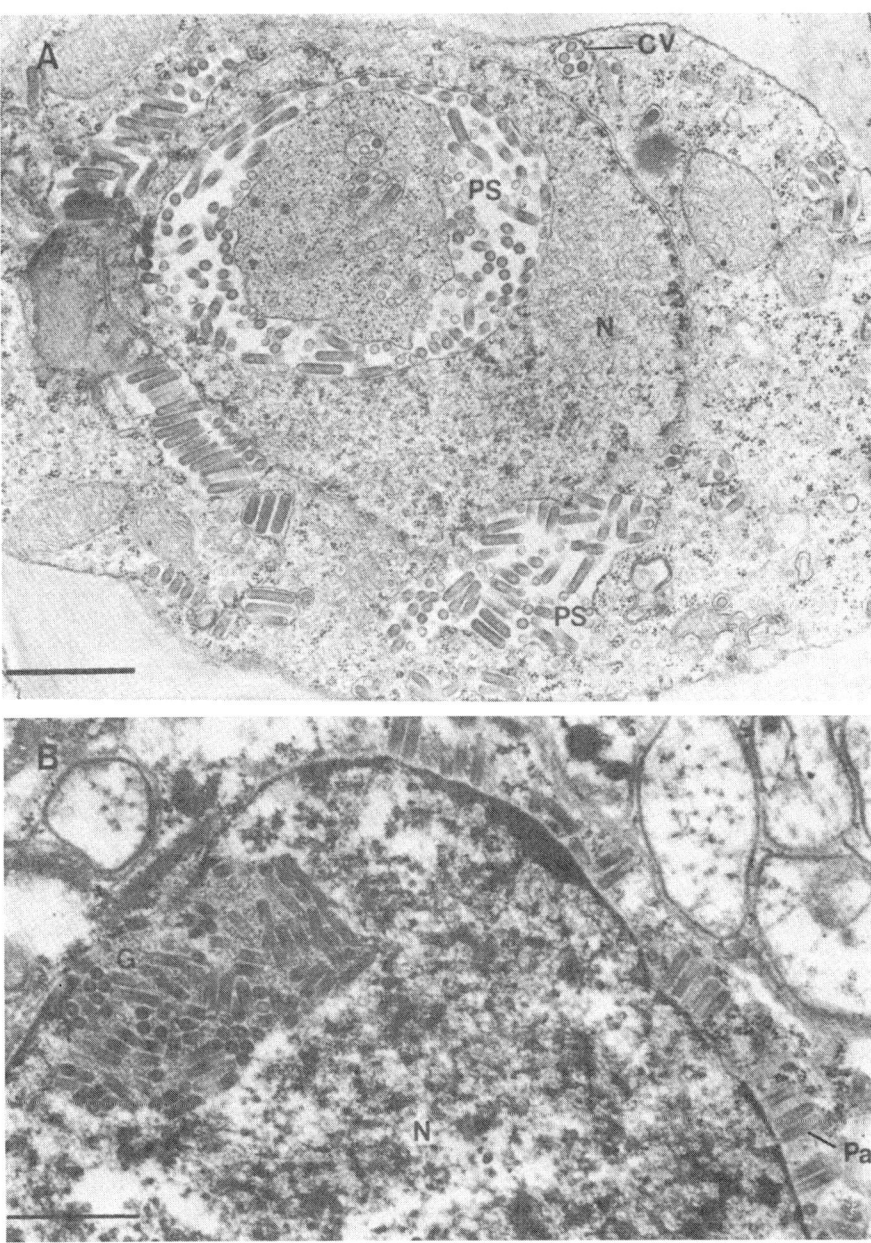

FIGURE 7. (A) Thin section of a plant cell (*Dinebra retroflexa*) infected with cereal chlorotic mottle virus (CeCMV), showing virus particles in both the perinuclear space (PS) and cytoplasmic vesicles (CV). The cytoplasm shown in the center of the section is apparently the result of indentation of the nuclear membrane in this area. Scale bar: 500 nm. (B) Thin section of a leafhopper cell (*Nesoclutha pallida*) infected with CeCMV, showing particles in a palisade (Pa) between the lamellae of the nuclear membrane and some that appear to be in a granular matrix (G). (N) Nucleus. Scale bar: 500 nm. Modified from Greber and Gowanlock (1979).

doviruses in infected plants is usually lower than that of many other plant
viruses. These problems are compounded by lack of efficient and quantitative bioassays for estimating recovery of infectious virus. Over the
past 25 years, however, several important techniques for purifying plant
rhabdoviruses have been developed that may be of general applicability
to many plant rhabdoviruses. Therefore, we are devoting considerable
space to this topic.

A. Assay of Plant Rhabdoviruses

Because of the unique morphology of rhabdoviruses, different steps
in their purification can be followed qualitatively by electron microscopy,
but this method is time-consuming and expensive. Spectrophotometric
determination of rhabdovirus concentration and purity is unreliable because it fails to discriminate between viral and nonviral nucleoprotein
and because the size and shape of the virions cause excessive light-scattering. Moreover, infectivity must ultimately be measured if one is to
assess the effect of different extraction procedures on the biological integrity of the virus. Unfortunately, such assays are difficult to conduct
with many plant rhabdoviruses, especially those that are not sap-transmissible.

Most of the progress in the past has been made with sap-transmissible
viruses such as potato yellow dwarf virus (PYDV), lettuce necrotic yellows
virus (LNYV), and sonchus yellow net virus (SYNV). When inoculated
into indicator plants, these viruses induce a variety of symptoms that
can be used for detection (Table III). Some of the indicator plants respond
to infection by formation of local lesions. These local-lesion hosts are
extremely useful for following the distribution of virus at different purification steps. However, if a virus is mechanically transmissible but has
no local-lesion host, a systemic assay may be used, although such an
assay is much less quantitative and more time-consuming. If the virus
is not sap-transmissible, as is the case with many rhabdoviruses that
infect dicots and all the cereal rhabdoviruses (Jackson et al., 1981), the
vector must be used as an intermediate in the bioassay. For this procedure,
virus extracts are injected into the abdomen of the insect, and the ability
of the insect to transmit the virus to indicator plants is subsequently
measured. Unfortunately, the method is extremely laborious and imprecise.

In a few instances, cultured insect-cell lines are available for assays
of plant rhabdoviruses (Black, 1979). These cells provide a quantitative
and extremely sensitive bioassay. Even though such cultures are difficult
to maintain and have not been established from vectors of most rhabdoviruses, they have considerable potential for bioassays and for studies
of the replication of those viruses that multiply in both plants and insects
(Nuss, 1984; Black, 1979).

TABLE III. Indicator Plants for Detection of Some Sap-Transmissible Rhabdoviruses

Virus	Host plant species[a]	Symptoms	References
Broccoli necrotic yellows virus (BNYV)	Datura stramonium (L.)[b]	Chlorotic to necrotic local lesions, mosaic with necrosis of fine veins	Campbell and Lin (1972)
Eggplant mottled dwarf virus (EMDV)	Nicotiana glutinosa (L.) Solanum melongena (L.)[b]	Mosaic (infection erratic) Chlorotic local lesions, systemic mottle, severe vein clearing, and leaf crinkling	Martelli and Rana (1970), Martelli and Russo (1973)
	Nicotiana glutinosa	Chlorotic local lesions, vein clearing, flecking, and stunting	
	Nicotiana tabacum (L.) (cv. White Burley or Xanthi n.c.)	Chlorotic local lesions, vein clearing, flecking, and stunting	
Lettuce necrotic yellows virus (LNYV)	Nicotiana glutinosa (L.)[b]	Epinasty, mosaic and stunting (some isolates induce chlorotic local lesions, vein chlorosis, and general chlorosis)	Stubbs and Grogan (1963), Crowley (1967), Francki and Randles (1970)
	Datura stramonium (L.)	Faint interveinal chlorosis	
Potato yellow dwarf virus (PYDV)	Nicotiana rustica (L.)[b]	Chlorotic local lesions, mottle, and leaf chlorosis	Black (1940, 1970)
	Nicotiana glutinosa (L.)	Vein clearing and mosaic	
Sonchus virus (SonV)	Nicotiana glutinosa (L.)[b]	Chlorotic lesions, vein clearing, leaf curling, and stunting of plants	Vega et al. (1976), D. Peters (unpublished data)
	Datura stramonium (L.)	Chlorotic spots and mosaic	
	Nicotiana clevelandii (Gray)[b]	Mosaic, chlorotic and necrotic rings, leaf deformation	
	Petunia hybrida (Vilm.)	Chlorotic spots, mosaic, and vein clearing	
Sonchus yellow net virus (SYNV)	Nicotiana edwardsonii (N. clevelandii × N. glutinosa)[b]	Vein clearing and leaf cupping	Christie et al. (1974), Jackson and Christie (1977)
	Chenopodium quinoa (Willd.)	Local lesions	
	Nicotiana glutinosa (L.)	Chlorotic lesions, vein clearing, and leaf cupping	
Tomato vein-yellowing virus (TVYV)	Nicotiana rustica (L.)[b]	Chlorotic local lesions	El Maataoui et al. (1985)

[a] Plants are listed in decreasing order of their susceptibility to the viruses. For further details of symptoms and other host species, consult the references cited.
[b] Most suitable hosts to maintain virus cultures.

B. Effect of Hosts and Environmental Factors on Yield of Virus

Several factors related to the interaction of a virus with its host must be considered in order to develop an efficient purification procedure. If a virus is virulent on several hosts, the host or cultivar that gives the highest yield of virus should be selected. The age of the plants, their light and temperature requirements, and the length of infection are critical factors that can drastically alter the amount of virus recovered. Determination of these variables may be time-consuming, especially with the difficulty of assaying many rhabdoviruses. However, such preliminary studies may save much time in the long run. For instance, PYDV infects a large number of species in several different plant families (Black, 1970), but substantial yields of virus have been obtained only from *Nicotania rustica*. Even in this host, recovery is highly dependent on age and physiology of the plants, the stage of infection, temperature, and light. In fact, for optimum recovery of most plant rhabdoviruses, one must consider interaction of a number of host and environmental variables. For example, adequate amounts of PYDV (H. T. Hsu and Black, 1973b), LNYV (Wolanski and Chambers, 1971), and SYNV (Jackson and Christie, 1977) can be recovered only between 10 and 20 days after infection, since the amount of virus in tissue declines rapidly thereafter. Poor lighting conditions and temperature fluctuations on either side of the optimum can markedly reduce virus yields. The time of the year also has considerable influence on the recovery of all three viruses when plants must be grown in the greenhouse. Generally, those conditions that promote the most luxurious plant growth also result in the highest recovery of virus.

C. Requirements for Extraction of Infected Plants

All rhabdoviruses that have been studied have thermal inactivation points below 60°C and are unstable at room temperature; thus, extraction should be conducted near 0°C. Most rhabdoviruses have pH optima near neutrality *in vitro*, so extraction should be conducted at or slightly above pH 7. The concentration and composition of the extraction medium have a major influence on the infectivity of purified virus. Brakke (1956) found that PYDV was stabilized by high concentrations of sucrose, by plant sap, and by some combinations of amino acids and salts. In general, the substances fall into two classes; (1) those effective at high concentrations, such as sucrose and monovalent salts, and (2) those effective at low concentrations, such as proteins, amino acids, and divalent salts. However, some of the divalent-salt requirements may vary depending on the virus and the host (Francki, 1973). For instance, LNYV is stabilized by Cu^{2+} and Zn^{2+}, but Mg^{2+} and Mn^{2+} have little preservative effect (Atchison *et al.*, 1969). This finding is puzzling because even though both LNYV and SYNV are purified from closely related tobacco species, the former two metals reduce and the latter two enhance infectivity of SYNV (A. O. Jackson, unpublished data). Reducing agents are also necessary to preserve infectivity of some rhabdoviruses. Infectivity of SYNV is not retained *in vitro* if reducing agents are absent (Jackson and Christie, 1977). Reducing agents have also been used in the purification of sowthistle yellow vein virus (SYVV) (Peters and Kitajima, 1970) and PYDV (MacLeod, 1967; Falk and Weathers, 1983). However, several reducing agents enhance the rate of LNYV inactivation (Atchison *et al.*, 1969). El Maataoui *et al.* (1985) recently reported that the method used to purify SYNV (Jackson and Christie, 1977) and several other viruses (Falk and Tsai, 1983; Falk and Weathers, 1983) was unsatisfactory for purification of tomato vein-yellowing virus. Apparently, Tris and phosphate buffers disrupted the virus particles, whereas they were stable in 200 mM sodium citrate. Thus, some agents may influence the stability of rhabdoviruses differently.

D. Clarification and Concentration of Rhabdoviruses from Plant Sap

Efficient separation of rhabdoviruses from host components can be particularly difficult, and the procedures required may vary with different host material. The most troublesome contaminants to remove from such preparations are chloroplast and other membrane fragments. However, these may sometimes be selectively removed by careful filtration through thin pads of Celite before the density-gradient step (Ahmed *et al.*, 1970; Greber and Gowanlock, 1979; Jackson and Christie, 1977; Falk and Tsai, 1983; Peters and Kitajima, 1970). LNYV (Francki and Randles, 1975) and northern cereal mosaic virus (Toriyama, 1972) have also been clarified by shaking extracts with DEAE–cellulose and decolorizing charcoal or bentonite before Celite filtration. Macoloid has been used to clarify SYVV extracts (Ziemiecki and Peters, 1976a). Unfortunately, these clarification procedures are not equally effective with all rhabdoviruses because DEAE–cellulose and bentonite both adsorb SYNV and PYDV (A. O. Jackson, unpublished observations).

Rhabdoviruses are usually concentrated from plant sap by centrifugation, but this procedure may also have some adverse effects. Rhabdoviruses aggregate very readily and are difficult to resuspend when pelleted from crude plant extracts, so it is probably best to avoid ultracentrifugation until as much host material as possible has been removed. The compression forces that result during ultracentrifugation are thought to reduce infectivity of SYNV (Jackson and Christie, 1977) and eggplant mottled dwarf virus (Russo and Martelli, 1973). Fortunately, ultracentrifugation can be avoided by polyethylene glycol (PEG) precipitation of plant rhabdoviruses (H. T. Hsu and Black, 1973c; Jackson and Christie, 1977). Thus, PEG precipitation should be considered in preference to differential centrifugation when developing purification schemes.

Most purification schemes devised for the rhabdoviruses have employed some combination of density-gradient centrifugation. Brakke (1951) actually developed density-gradient centrifugation to aid in purification of PYDV. With this versatile tool, he was able to correlate infectivity with a light-scattering band in the density gradient. Unfortunately, rhabdoviruses are normally too contaminated with host components for refined chemical analysis of preparations if density-gradient centrifugation is performed before intermediate clarification steps have been conducted. Therefore, it is best to utilize density-gradient centrifugation at the later stages of infection after most of the host components have been removed.

Two alternative approaches to density-gradient centrifugation have been applied as final steps in the purification and concentration of plant rhabdoviruses: chromatography on calcium phosphate gels (Lin and Campbell, 1972; McLean and Francki, 1967) and electrophoresis in sucrose density gradients (Ahmed *et al.*, 1970; Brakke, 1955; Peters and Kitajima, 1970; Sinha *et al.*, 1976; Toriyama, 1972). Unfortunately, neither of these methods is really satisfactory for general use with rhabdoviruses. Recovery of rhabdoviruses may vary with different batches of calcium phosphate (Francki, 1973), and adsorption of different viruses by calcium phosphate probably also varies (Luisoni, 1969). The disadvantages of electrophoresis are the time required and the resultant loss in infectivity. Therefore, these procedures must all be explored for each individual virus.

E. Purification of Sonchus Yellow Net Virus

The techniques and concepts discussed above were used to devise a purification scheme for SYNV (Jackson and Christie, 1977). This procedure was initially developed with the aid of the local-lesion host *Chenopodium quinoa* to assay virus recovery after each purification step. Subsequently, more attention was directed to removal of host contaminants to obtain homogeneous preparations suitable for analysis.

To extract SYNV, leaves of systemically infected *Nicotiana edwardsonii* (Christie and Hall, 1979) are harvested 9–13 days after inoculation. The leaves may be extracted immediately or stored for several days at 4°C. The tissue is blended at 4°C in 2 volumes of an extraction medium consisting of 100 mM Tris-HCl (pH 8.4), containing 10 mM Mg acetate, 40 mM Na_2SO_3, and 1 mM $MnCl_2$. The brei is removed by low-speed centrifugation, and the virus is clarified by filtration through a Celite pad. The virus may then be concentrated by centrifuging the filtrate at 50,000g for 30 min. Alternatively, the virus may be precipitated by adjusting the filtrate to 100 mM NaCl and 9% PEG and then stored for 1 hr at 4°C before low-speed centrifugation. The virus is then resuspended in 2 ml of extraction buffer for each 100 g of leaves and purified further by centrifugation into rate zonal and quasi-equilibrium sucrose gradients (Jackson and Christie, 1977). The virus sediments as a major light-scattering band at about 1050 S in the rate zonal gradients and has a density of about 1.18 g · ml⁻¹ in the equilibrium gradients (Fig. 8). The virus recovered from the sucrose gradients is concentrated by centrifugation. The pellet is then resuspended in a small amount of buffer and stored at $-20°C$.

The final preparations should have a milky appearance and should be free of host membranes and ribosomes. Yields of up to 5 mg of virus can sometimes be recovered from 100 g of leaves, but the recovery is very dependent on the age of the plants when inoculated, light and temperature in the greenhouse, and time after inoculation. A particularly critical variable is the thickness of the Celite pads used for filtration. Pads thicker than 7.5 mm seriously diminish yield of virus, but pads less than 2.5 mm thick result in serious contamination with chloroplast fragments.

The most useful indicators of host contaminants have been discoloration of virus pellets caused by chloroplast fragments, the presence of host-membrane fragments as revealed by electron microscopy, and ribosomal RNA contamination of viral RNA preparations. Pellets of reasonably pure SYNV preparations have a light tan color, but even slight contamination with chloroplast membranes produces green discoloration. Homogeneous preparations are also free of other membrane fragments when negatively stained samples are examined by electron microscopy, and sucrose-density gradients of nucleic acid preparations reveal a single sedimenting component of 44 S without contaminating ribosomal RNA (Fig. 9). The latter criterion is particularly sensitive, because ribosomes contaminating viral preparations have about 65% RNA, whereas rhabdovirus particles contain about 2–3% RNA.

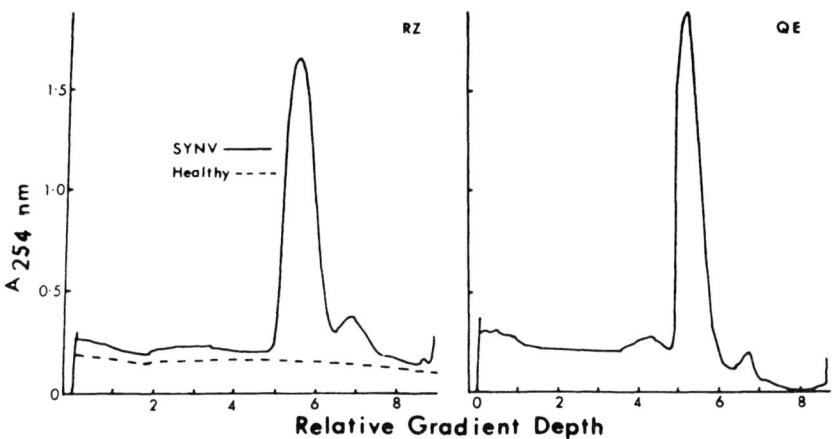

FIGURE 8. Photometric scanning patterns of preparations of sonchus yellow net virus (SYNV) centrifuged in rate zonal (RZ) and quasi-equilibrium (QE) sucrose gradients. Preparations from SYNV-infected tobacco (—) and healthy tobacco (---) were centrifuged in the gradients after clarification by Celite filtration and concentration by high-speed centrifugation. Note that there is only one major light-scattering component separated on the gradients. From Jackson and Christie (1977).

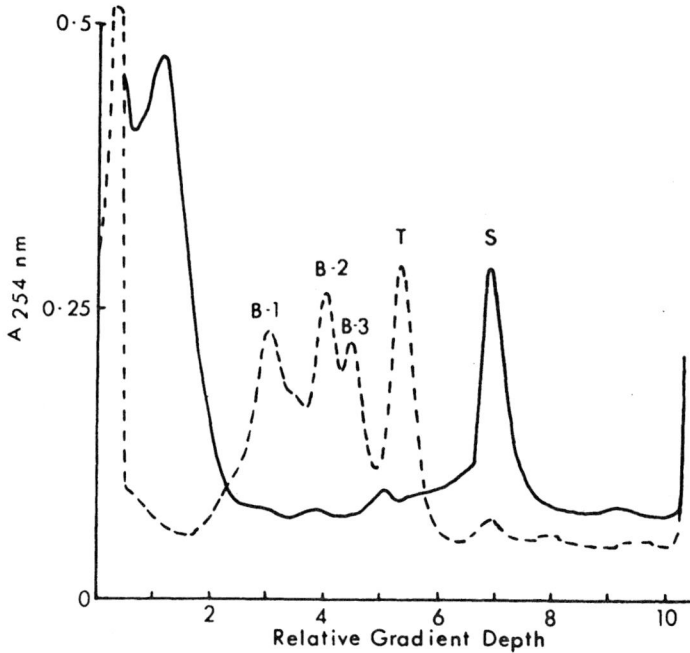

FIGURE 9. Comparison of the sedimentation rates of sonchus yellow net virus RNA [(S) —], brome mosaic virus (BMV) RNAs [(B1–3) ---], and tobacco mosaic virus RNA [(T) ---] in linear-log sucrose gradients. Note that the purity of the preparation can be determined from the amount of ribosomal RNA sedimenting at the position of BMV RNAs 2 and 3. From Jackson and Christie (1977).

VII. SEROLOGY

Generally, plant rhabdoviruses are poor immunogens, and the titer of antibodies raised in rabbits is usually low. In addition, many antiserum preparations also react with host antigens due to impurities in the virus preparations used for immunization. The size of the virion also presents difficulties when gel-diffusion tests are used to determine virus–antibody reactions. Nevertheless, serological procedures have been used successfully for several different purposes (Table IV).

Although gel-diffusion tests require that virus preparations be degraded to enable diffusion into the agar, such tests have provided some useful diagnostic information (Fig. 10). A variety of detergents, enzymes, and lipid solvents have been used for virus disruption. However, the degradation products may cause bizarre results that can complicate interpretation of the precipitin bands. Such effects have been well illustrated with lettuce necrotic yellows virus (LNYV) by McLean *et al.* (1971). In these tests, when virus was disrupted with Tween–ether, sonication, or lipase treatment or some combination thereof, the degradation products reacted with antisera to form several bands when allowed to diffuse into

TABLE IV. Records of Serological Studies with Plant Rhabdoviruses

Virus	Purpose of studies	References
Barley yellow striate mosaic virus (BYSMV)	Virus relationships	Lundsgaard (1984), Greber (1984), Milne et al. (1986)
Broccoli necrotic yellows virus (BNYV)	Basic immunochemical	Lin and Campbell (1972)
	Virus detection	Feldman et al. (1978)
Cereal chlorotic mottle virus (CeCMV)	Virus detection, basic immunochemical	Greber and Gowanlock (1979)
	Virus relationships	Lundsgaard (1984)
Festuca leaf streak virus (FLSV)	Virus relationships	Lundsgaard (1984)
Lettuce necrotic yellows virus (LNYV)	Virus detection	Harrison and Crowley (1965), Chu and Francki (1982)
	Basic immunochemical	McLean et al. (1971)
Maize mosaic virus (MMV)	Virus identification	Lastra (1977)
	Virus relationships	Greber (1984), Milne et al. (1986)
Maize sterile stunt virus (MSSV)[a]	Virus relationships	Greber (1984), Milne et al. (1986)
Moroccan wheat rhabdovirus (MWRV)	Virus detection	Lockhart and Elyamani (1983)
	Virus relationships	Milne et al. (1986)
Northern cereal mosaic virus (NCMV)[b]	Virus identification	Toriyama (1972, 1976b)
	Virus relationships	Lundsgaard (1984), Lundsgaard et al. (1984)
Potato yellow dwarf virus (PYDV)	Virus identification	Wolcyrz and Black (1956, 1957), Falk et al. (1981)
	Virus detection	Chiu et al. (1970)
	Basic immunochemical	Knudson and MacLeod (1972)
Shiraz maize rhabdovirus (SMRV)	Virus identification	Izadpanah et al. (1983)
	Virus relationships	Milne et al. (1986)
Sonchus yellow net virus (SYNV)	Virus relationships	Jackson and Christie (1977)
Sowthistle yellow vein virus (SYVV)	Viral relationships	Hackett et al. (1968)
	Virus detection	Peters and Black (1970)
Tomato vein-yellowing virus (TVYV)	Basic immunochemical	El Maataoui et al. (1985)
Wheat chlorotic streak virus (WCSV)[a]	Virus relationships	Milne et al. (1986)
Wheat rosette stunt virus (WRSV)[b]	Virus relationships	Lundsgaard et al. (1984), Milne et al. (1986)
Wheat striate (American) mosaic virus (WSMV)	Virus detection	Sinha (1968), Thottappilly and Sinha (1973)
	Basic immunochemical	Thottapilly and Sinha (1973), Trefzger-Stevens and Lee (1977)

[a] Milne et al. (1986) report that MWRV and WCSV are strains of BYSMV. MSSV may be considered a strain of BYSMV.
[b] Lundsgaard et al. (1984) and Milne et al. (1986) report a close relationship between NCMV and WRSV.

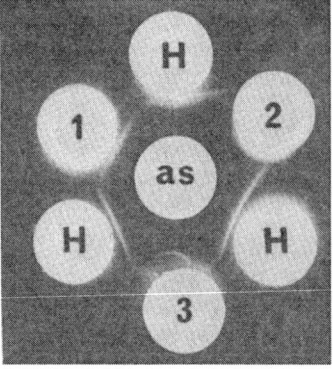

FIGURE 10. Serological reactions of lettuce necrotic yellows virus (LNYV). Antiserum to the S.E. 3 virus strain was placed in the center well (as) and reacted against partially purified preparations of the S.E. 3 strain [homologous antigen (H)] and three different field isolates of LNYV (1–3). From Francki *et al.* (1981); courtesy of J. W. Randles.

the gels for some time before antiserum was added to the wells. These bands probably resulted from dissociation of the virus into different-size structures that migrated into the gels at different rates. Similar results have been reported with several other rhabdoviruses (Greber and Gowanlock, 1979; Lin and Campbell, 1972; Thottappilly and Sinha, 1973). The use of a nonionic detergent such as Triton X-100 or Nonidet P-40 appears to be more satisfactory than the degradation procedures described above (Francki and Randles, 1975). These detergents probably completely solubilize the viral envelope without affecting the protein–RNA interactions (Francki and Randles, 1975). More complete disruption with ionic detergents such as sodium dodecyl sulfate (SDS) usually yields more readily interpretable data, but the titer of the antiserum preparations is usually reduced to less than 1 : 32. For example, even though a single band is produced when sonchus yellow net virus (SYNV) is disrupted with SDS, the best titers ever produced in these tests never exceeded 1 : 16 (A. O. Jackson, unpublished data). Peters and Black (1970) also found that antisera to sowthistle yellow vein virus (SYVV) had a titer of 1 : 5000 in ring precipitin tests, but only 1 : 32 in gel-diffusion tests. Likewise, Thottappilly and Sinha, (1974) observed that the titer of antisera to wheat striate (American) mosaic virus (WSMV) was 1 : 160 in precipitin tests, but was only 1 : 20 in gel-diffusion tests. Moreover, MSMV was detectable at concentrations as low as 15 μg/ml in the precipitin tests, but the detectable level was reduced 75-fold in gel-diffusion tests with detergent. These results obviously preclude use of gel-diffusion assays with SDS for routine diagnosis.

More sensitive methods, such as the enzyme-linked immunosorbent assay (ELISA), avoid many of the problems encountered with other tests and are especially suitable for diagnostic application. This assay has been used to distinguish LNYV- from lettuce mosaic virus-, cucumber mosaic virus-, and tomato spotted wilt virus-infected lettuce and sowthistle plants (Chu and Francki, 1982). LNYV antisera with a titer of 1 : 64 in gel-diffusion tests had a titer of 1 : 78,000 in ELISA tests with extracts from infected *Nicotiana glutinosa* leaves, whereas a 1 : 10 dilution of the serum failed to react with extracts from healthy plants. Moreover, detection of LNYV by ELISA in leaf extracts was at least 8000 times more sensitive than with immunodiffusion tests. The ELISA tests were also 600 times more sensitive than infectivity tests for the detection of LNYV and had an additional advantage over infectivity assays in that the plant extracts could be stored for several days at 40°C without loss of activity. ELISA tests were also used to detect LNYV in individual aphids. In Florida, ELISA also provides a sensitive test for detecting SYNV in plants as well as in aphids (B. Falk, personal communication). ELISA tests thus have great potential for studies of the distribution and epidemiology of rhabdoviruses and should greatly increase the reliability and ease of such studies.

Serology has also been used to assess relationships of plant rhabdoviruses. Immunodiffusion tests in the presence of SDS (Jackson and Christie, 1977) have shown that purified SYNV preparations react with antisera to SYNV, but not with antisera to potato yellow dwarf virus (PYDV), LNYV, broccoli necrotic yellows virus (BNYV), or SYVV. Unpublished data suggesting that SYNV, LNYV, and eggplant mottled dwarf virus (EMDV) are antigenically distinct has also been cited by Francki et al. (1981). Falk et al. (1981) reported that PYDV fails to react with antisera to SYNV, BNYV, and WSMV. Recently, tomato vein-yellowing virus (TVYV) was also clearly distinguished from PYDV by serology (El Maataoui et al., 1985). Thus, the serological data indicate that SYNV, SYVV, PYDV, BNYV, LNYV, EMDV, WSMV, and TVYV are distinct viruses that may be distinguished serologically.

Serological relationships have also been investigated among plant rhabdoviruses that infect cereals. Greber (1984) has reported that maize sterile stunt virus reacted with an antiserum to barley yellow striate mosaic virus (BYSMV) in gel-diffusion tests, but not with an antiserum to cereal chlorotic mottle virus or maize mosaic virus (Greber, 1982a). Northern cereal mosaic virus, which is thought to be closely related or identical to BYSMV (Jackson et al., 1981), also appears to be serologically related to wheat rosette stunt virus (WRSV) from China (Lundsgaard et al., 1984). There also appears to be a weak cross-reaction between WSMV and oat striate mosaic virus in gel-diffusion tests (Milbrath, unpublished data). These studies are all limited in scope, but they illustrate that more extensive investigations will be extremely useful for extending our knowledge of the relationships among plant rhabdoviruses.

VIII. MORPHOLOGY OF VIRIONS

Virions of rhabdoviruses are easily altered *in vitro* by negative-staining procedures. Many structural artifacts have been detected and their possible origins discussed (Francki, 1973; Francki and Randles, 1975). Even when the fragile virions are stabilized with fixatives such as glutaraldehyde prior to negative staining, they may still break or swell. These problems undoubtedly contribute to the variation in size estimates ob-

TABLE V. Physical Properties of the Better-Characterized Plant Rhabdoviruses[a]

Property	Lettuce necrotic yellows virus (LNYV)	Potato yellow dwarf virus (PYDV)	Sonchus yellow net virus (SYNV)	Wheat striate (American) mosaic virus (WSMV)	Broccoli necrotic yellows virus (BNYV)	Maize mosaic virus (MMV)
Virus particles						
Dimensions (nm)						
In sections	360 × 52	380 × 75	—	250 × 75	297 × 64	240 × 48 / 244–253 × 62
In negative stain	227 × 66	—	250 × 94	245 × 75	275 × 75	224 × 68 / 204–245 × 67–80
Sedimentation coefficient (S)	945	880	1045	875	874	774
Buoyant density (g/cm³)	1.20	1.17	1.18	1.22	1.19	—
Nucleocapsids						
Sedimentation coefficient (S)	260	250	250	—	—	—
Infectivity	Yes	Yes	Yes	—	—	—
Ribonucleic acid						
Sedimentation coefficient (S)	43	45	44	32	—	—
Strandedness	Single	Single	Single	Single	—	Single
Infectivity	No	No	No	No	—	—
M_r (× 10⁶)	4.2	4.6	4.4	2.2	—	4.2

[a] References: (LNYV) Francki and Randles (1970, 1973), Chambers et al. (1965); (PYDV) Black (1970), Reeder et al. (1972); (SYNV) Jackson and Christie (1977, 1979); (WSMV) Sinha and Behki (1972), Sinha et al. (1975, 1976); (BNYV) Ohki et al. (1978); (MMV) Herold (1972), Falk and Tsai (1983), McDaniel et al. (1985). (—) Data not available.

tained in different laboratories. Although there appears to be some variation among the different viruses, some common morphological and physical properties of rhabdovirus particles have been repeatedly observed (Table V). The morphology of most of the plant rhabdoviruses examined after careful fixation of the preparations prior to negative staining is bacilliform (Fig. 11). However, omission of the fixation procedure invariably results in deformed structures such as bullet-shaped particles similar to those observed with unfixed animal rhabdoviruses (Orenstein *et al.*, 1976).

A. Rhabdovirus Structure

The structure of the rhabdoviruses is complex, and a number of different models of the physical organization of the virion have been proposed from electron micrographs (Francki *et al.*, 1981; Hull, 1976). Three distinct layers of varying electron density are observed in cross sections of the virion. These layers are thought to be composed of surface projections, a membrane, and a helical ribonucleoprotein (nucleocapsid) surrounding a central canal (Fig. 12). The spikes protruding from the surface of some plant rhabdoviruses such as cereal chlorotic mottle virus (CeCMV) (Greber, 1979b) are arranged in a hexagonal array. This pattern is less pronounced in virions of other plant rhabdoviruses, but they all have a complex array of surface projections protruding 6–10 nm from the lipid membrane that surrounds a nucleoprotein that is about 15–20 nm thick. The nucleocapsid consists of an orderly helix of about 40–50 turns with

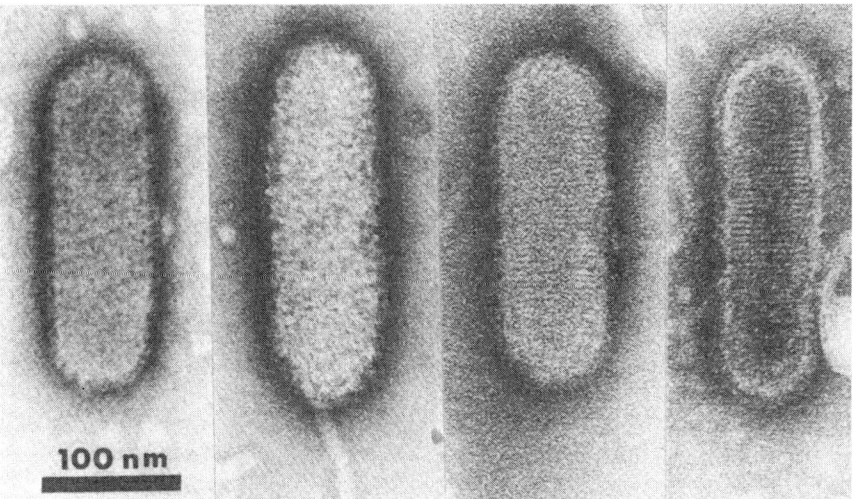

FIGURE 11. Electron micrographs of particles in preparations from tobacco infected with sonchus yellow net virus that was clarified by Celite filtration and concentrated by PEG precipitation. The virions were fixed with 3% glutaraldehyde and negatively stained with 2% uranyl acetate. Individual preparations with varying extents of stain penetration were selected to illustrate the fine structure of the particles. The preparations were examined in a Philips EM 400 electron microscope operating at 60 kV.

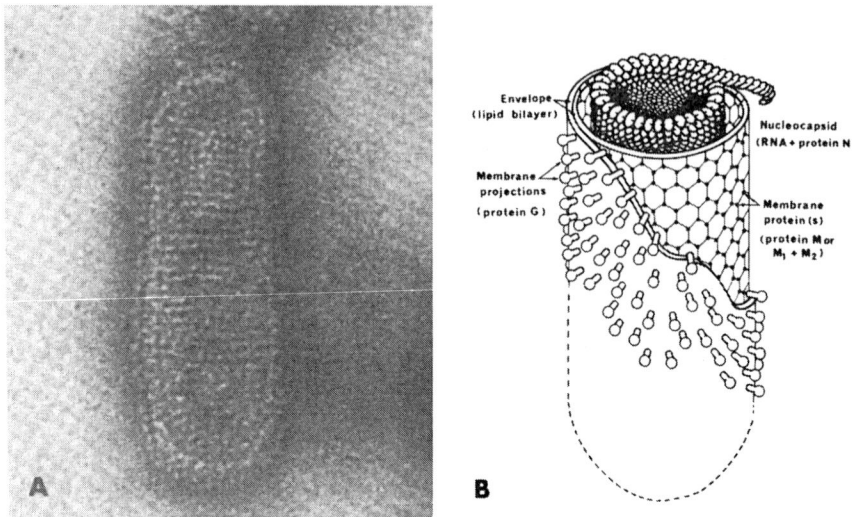

FIGURE 12. (A) Electron micrograph of a negatively stained particle of lettuce necrotic yellows virus showing the radiating spikes on the surface of the virion and the nucleocapsid structure inside the envelope. Courtesy of T. C. Chambers. (B) Model of a rhabdovirus particle cut open to illustrate the internal structure. From Francki and Randles (1980).

a pitch of approximately 4.0–4.5 nm, which produces a tubular structure 35–50 nm in diameter. The precise organization of the nucleocapsid core and the structure of the envelope at the ends of the particles is obscure, but several arrangements are possible (Francki, 1973; Martelli and Russo, 1977; Peters and Schultz, 1975; Hull, 1976). In at least some rhabdoviruses, the most plausible structure is one in which the nucleocapsid is bullet-shaped, with one end rounded and the other end flat. The envelope surrounding the nucleocapsid, when undamaged, is rounded at both ends and thus has a bacilliform exterior. When virions are disrupted with a nonionic detergent such as Triton X-100 or Nonidet P-40, the nucleoprotein is liberated to form a coil (Randles and Francki, 1972; Jackson, 1978; Ziemiecki and Peters, 1976b) with a helical twist (Fig. 13).

Considerable variation has been observed in both the length and the width of plant rhabdovirus particles (see Table I). The widths range from 32–35 nm for orchid fleck virus particles, which appear to lack a membrane (Doi *et al.*, 1977), to 94 nm for sonchus yellow net virus and rice transitory yellowing virus (RTYV). Each rhabdovirus undoubtedly has a characteristic width, as has been demonstrated by Greber (1979a), who was able to show reproducible differences in negatively stained preparations of mixtures containing CeCMV and digitaria striate virus (DSV). Even though these differences in the widths of CeCMV and DSV are sufficiently distinct to be used for diagnosis, less reliability can be placed on individual measurements of other rhabdoviruses (Table I). Variation in preparative procedures for electron microscopy can contribute considerably to the relative degree of flattening, shrinking, or swelling of the

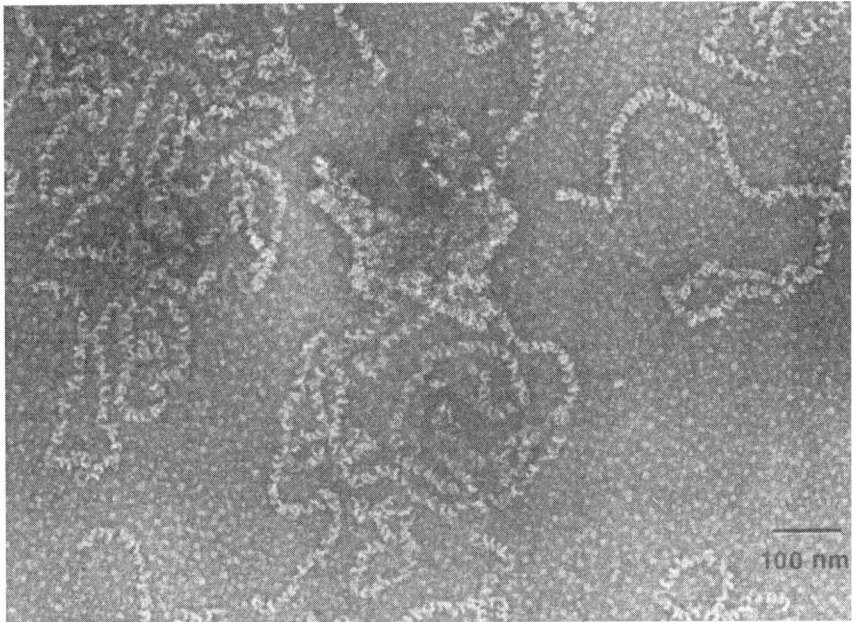

FIGURE 13. Electron micrograph of nucleocapsid cores of sonchus yellow net virus after high-speed centrifugation, fixation, and negative staining. From Jackson (1978).

fragile particles. The particles may also become distorted or may even fragment during fixing and staining. This is particularly evident with partially purified RTYV prepared by Chen and Shikata (1971), where almost all of the particles appear to have been broken. The average length of 129 nm reported for these particles contrasts with the presence of particles ranging from 180 to 210 nm in cells of infected rice. Also, the accuracy of measurements obtained in different laboratories is difficult to evaluate, but the results do indicate that different plant rhabdoviruses vary markedly in length and width (Table I). However, many of these measurements need to be repeated under carefully controlled conditions before we can be confident of their reliability.

B. Defective Interfering Particles

Defective interfering (DI) particles are common in animal rhabdovirus preparations that are maintained by repeated transfer at high multiplicity of infection [for reviews, see Lazzarini *et al.* (1981) and Chapter 8]. These DI particles are dependent for replication on the wild-type helper virus, and they often decrease the titer of the helper virus. The RNAs of most of these DI particles have complementary termini and are derived largely from the 3' or the 5' half of the genome. The mechanisms by which these particles arise is not clear, but a variety of internal deletions have been identified, and there may be more than one pathway leading to their formation. Once generated, the defective genome has a distinct replicative advantage over the complete genome, and thus DI particles accumulate at the expense of the wild-type virus. DI particles are common among several different animal-virus groups, and it is thought that they may have some natural role in attenuation of infection. DI particles are not normally found in plant-virus preparations. This probably is a consequence of the difficulty in obtaining a high multiplicity of infection in plants. Only one case of putative DI particles has been reported in a plant rhabdovirus (Adam *et al.*, 1983). In this instance, potato yellow dwarf virus (PYDV) was maintained under conditions shown to result in most rapid development of symptoms in tobacco. After about 30 successive mechanical transfers at high inoculum concentrations, recovery of virus in purification trials decreased markedly. It was subsequently found that the preparations contained slowly sedimenting entities with a lower density than the normal virus particles. Plants failed to develop lesions typical of infection with normal PYDV when inoculated with the variant form, so the variant appeared not to be infectious. When the variant was mixed with normal virus, the infectivity of the latter was drastically reduced, suggesting that the variant contained DI particles. When plants were inoculated with dilute preparations of the variant, the number of variant particles recovered appeared to be greatly reduced. This result thus appears to be similar to the results obtained when dilute mixtures of wild-type and DI particles of vesicular stomatitis virus (VSV) are inoculated into cells.

Although no detailed structural data on the putative DI particles were presented, the particles appeared to be derived from PYDV, since they contained a normal compliment of PYDV proteins (Adam *et al.*, 1983). These preliminary findings suggest that plant rhabdoviruses have the potential for forming DI particles. The results also imply that the genomes of plant rhabdoviruses may undergo the same rapid evolution observed with VSV and other animal rhabdoviruses (Holland *et al.*, 1982). The details of such evolutionary alterations might provide useful information concerning the infectious process and the ability of plants to recover from infection.

IX. CHEMICAL COMPOSITION OF PLANT RHABDOVIRUSES

Because of difficulties in separating plant rhabdoviruses from host contaminants, the chemical composition is known for only a few members of the group. The best characterized are sonchus yellow net virus (SYNV), lettuce necrotic yellows virus (LNYV), and potato yellow dwarf virus (PYDV). These viruses have been estimated to contain about 70% protein, 20–25% lipid, and 2–3% RNA. The virions also contain carbohydrate, but no detailed studies have been conducted to investigate the carbohydrate composition.

A. Lipid Composition

The lipoprotein envelope of the Rhabdovirus particle is an important structural feature that is necessary for maximum infectivity. For example, removal of the LNYV envelope with Nonidet P-40 resulted in a 20-fold reduction in infection (Francki and Randles, 1975), and purified nucleo-capsids of SYNV have considerably lower infectivity than intact virus (Jackson, 1978). Thus, the lipid envelope appears to be essential for main-taining the integrity of the virion and is probably also important in its maturation. Unfortunately, electron-microscopic studies provide the only available clues concerning the sequence of events that result in *in vivo* assembly of plant rhabdoviruses. However, we presume that, as is the case with vesicular stomatitis virus (VSV) (David, 1977; Morrison and McQuain, 1978), the first stages of plant rhabdovirus assembly are de-pendent on the availability of a host membrane that can be modified by insertion of virus-specific proteins. Therefore, before we can understand the role of the envelope in the structure of the virion or the choice of specific sites for encapsidation, we must obtain a detailed knowledge of the lipid composition of both the host and the virion.

The lipid compositions of four plant rhabdoviruses have been inves-tigated. These are SYNV (Selstam and Jackson, 1983), PYDV (Ahmed *et al.*, 1964), wheat striate (American) mosaic virus (WSMV) (Sinha *et al.*, 1976), and northern cereal mosaic virus (NCMV) (Toriyama, 1976a). The distribution of lipids in PYDV and WSMV appears to be similar to that in SYNV (Table VI). In all three cases, fatty acids and sterols predominate, while the proportion of triglycerides is low. The composition of individual fatty acids in NCMV and WSMV has not been investigated, but palmitic, stearic, oleic, linoleic, and linolenic acids are the most abundant fatty acids in both PYDV and SYNV (Table VII). Phosphatidylcholine is the

TABLE VI. Lipid Composition of Purified Sonchus Yellow Net Virus[a]

Lipid	Lipid (µg/mg protein)	Weight percentage of total lipids	Lipid (nmoles/mg protein)	Mole percentage of total lipids
Total phospholipids	156[b]	62%	186	47%
Free sterols and esterified sterols[d]	68[b]	27%	173	43%
Sterol glycosides and esterified sterol glycosides[e]	10[b]	4%	17	4%
Triglycerides	18[c]	7%	22[b]	6%

[a] From Selstam and Jackson (1983). [b] Mean of four analyses. [c] Mean of two determinations.
[d] Calculated as though the fraction were composed entirely of free sterols.
[e] Calculated as though the fraction were composed entirely of sterol glycosides.

TABLE VII. Fatty Acid Composition of Individual Phospholipids of Sonchus Yellow Net Virus[a]

Lipid	Mole percentage of total fatty acid[b]					Ratio of saturated to unsaturated	Weight percentage of total phospholipids
	16:0	18:0	18:1	18:2	18:3		
Phosphatidylcholine[c]	22%	5%	4%	38%	31%	0:62	53%
Phosphatidylethanolamine	33%	6%	3%	35%	24%	0:63	22%
Phosphatidylinositol	50%	5%	4%	20%	21%	1:22	16%
Phosphatidylglycerol	61%	5%	4%	14%	16%	1:94	9%
Unidentified acyl lipid	60%	11%	10%	15%	11%	1:97	—

[a] Modified from Selstam and Jackson (1983). Values are the means of four determinations.
[b] 16:0 Palmitic acid; 18:1, oleic acid; 18:2, linoleic acid; 18:3, linolenic acid.
[c] The phosphatidylcholine fraction was not separated by two-dimensional chromatography before analysis of fatty acids; therefore, it includes about 10% phosphatidylserine.

major phospholipid constituent of SYNV, but phosphatidylethanolamine, phosphatidylinositol, phosphatidylglycerol, and phosphatidylserine are also present. Sterols are present in all four plant rhabdoviruses, although the proportions of individual sterols have been determined only for NCMV and SYNV, in which cholesterol, campesterol, stigmasterol, and β-sitosterol predominate (Table VIII). The proportions of cholesterol and stigmasterol in SYNV are 2-fold greater than those reported for NCMV. Because of the difficulty in freeing virus preparations from host membranes, it is possible that some of the differences in the proportion of lipids found in different viruses may be partly due to host contaminants. For this reason, several different criteria were used to monitor the preparations of SYNV used for lipid analysis (Selstam and Jackson, 1983). Hence, the authors were confident that the lipids detected in these preparations were primarily virus-specific.

All the lipid studies show that plant rhabdoviruses differ considerably from animal rhabdoviruses in composition and proportion of individual lipids. This probably reflects chemical differences between the lipids of plant and animal membranes. Such differences are most clearly illustrated by the higher levels of unsaturated fatty acids in the plant viruses (Selstam and Jackson, 1983). For example, the ratios of palmitic, stearic, oleic, and

TABLE VIII. Composition (Weight Percentage) of Free Sterols in Isolated Sonchus Yellow Net Virus and Northern Cereal Mosaic Virus[a]

Lipid	SYNV	NCMV
Cholesterol	21%	11%
Campesterol	15%	13%
Stigmasterol	30%	26%
β-Sitosterol	34%	50%

[a] Data from Selstam and Jackson (1983) (SYNV) and Toriyama (1976a) (NCMV). Values are the averages of three determinations.

linoleic acids of SYNV show little similarity to those of VSV. Linolenic acid, which represents 17% of the fatty acids of SYNV, is absent in VSV. In contrast, arachidonic acid, present in VSV (McSharry and Wagner, 1971), has not been detected either in SYNV or in higher plants. The variation in the composition of phospholipids in SYNV and VSV also correlates well with the phospholipid composition of their hosts. This is especially evident with sphingomyelin, which constitutes more than 20% of the phospholipids of VSV (McSharry and Wagner, 1971) and rabies virus (Diringer *et al.*, 1973; Schlesinger *et al.*, 1973), but is absent from SYNV. These latter differences correlate well with the lipid composition at the subcellular site of morphogenesis because VSV and rabies virus both assemble on membranes rich in sphingomyelin, whereas SYNV appears to assemble at plant nuclear membranes that lack sphingomyelin (Philipp *et al.*, 1976). However, the phospholipids of SYNV do show some similarities with those of animal nuclear membranes (Franke, 1974) as well as with those of nuclear membranes of plants (Philipp *et al.*, 1976).

Sterol composition is another major difference between plant- and animal-rhabdovirus lipids. The overall proportion of sterols in the virion is estimated to be slightly higher for SYNV than for either VSV (Patzer *et al.*, 1978) or rabies virus (Diringer *et al.*, 1973). However, both animal viruses contain primarily cholesterol and cholesterol esters (McSharry and Wagner, 1971; Diringer *et al.*, 1973), while SYNV has larger amounts of stigmasterol and β-sitosterol than of cholesterol (Table VIII). In tobacco leaves, the cholesterol content has been reported to be about 5% of the free sterols (Trevathan *et al.*, 1979), whereas the cholesterol content of purified SYNV is about 20% of the total. We do not know whether this higher level in the virion reflects selective assembly within the nuclear envelope at specific sites high in cholesterol or whether infection results in enhanced deposition of cholesterol at the assembly sites.

These differences in composition of the envelopes of plant and animal rhabdoviruses raise questions about the role of particular lipids in maintaining the structural integrity and physical properties of the envelope. For example, evidence from animal viruses shows that the composition and distribution of lipids can greatly influence the rigidity and fluidity of the viral envelope (Compans and Klenk, 1979). Cholesterol is a wedge-shaped molecule that is thought to mediate formation of nonlamellar phases within membranes (Khan *et al.*, 1981; Wieslander *et al.*, 1980). Thus, the distribution of cholesterol may be important for the final step in budding when the host membrane acquires viral polypeptides and envelops the nucleocapsid. Sterols may also influence the subsequent reinfection stage, in which the virus may fuse with the host membrane at specific sites. The increased strength of the viral envelope over the host plasma membrane is also thought to be enhanced by a high ratio of cholesterol to phospholipid combined with strong interactions of the transmembrane glycosylated protein and the membrane matrix protein in the bilayer (Morrison and McQuain, 1978; Petri and Wagner, 1979). The differences in the composition of the envelopes of the plant and animal rhabdoviruses imply that the molecular interaction of the lipid components with the viral polypeptides may markedly influence the architecture and biological properties of the virion. Additional comparative studies of the organization of these components coupled with biophysical studies should reveal more detailed information about the rules that govern the structure and function of these complex viruses.

B. Viral RNA

RNAs of the best-characterized plant rhabdoviruses are single-stranded, and none of those that have been tested is infectious (Francki and Randles, 1980). They are similar in size to the RNAs of the animal rhabdoviruses, since they sediment as a single component with a value of 40–45 S. This corresponds to a molecular weight slightly greater than 4×10^6, or 11,000–13,000 nucleotides (Fig. 14). The sole exception appears to be WSMV, the virions of which have been reported to contain 5% RNA with a sedimentation value of 28 S (Sinha *et al.*, 1976), or a size estimated to be about half that of other rhabdoviruses (see Table V). Since these values deviate so markedly in both the proportion of RNA in WSMV and the size of the RNA, they need reinvestigation.

Very little is known of either the order of the genes on the plant rhabdovirus genome or the primary structure of the RNA. However, analysis of recombinant DNA clones suggests that the organization of the SYNV genome is similar to that of the VSV and rabies virus genomes (Rezaian *et al.*, 1983; Heaton *et al.*, in prep.). The nucleotide sequence of

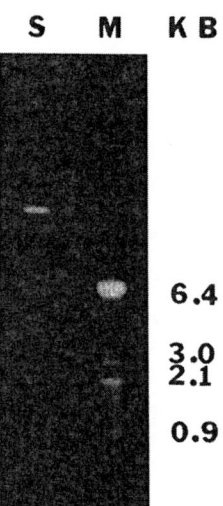

S   M   K B

6.4

3.0
2.1

0.9

FIGURE 14. Electrophoretic patterns of sonchus yellow net virus (SYNV) RNA (S), tobacco mosaic virus RNA plus brome mosaic virus RNA (M) after electrophoresis on an agarose gel. This photograph illustrates the homogeneity and the high molecular weight of SYNV RNA. (KB) Kilobases.

SYNV RNA is also currently being determined in our laboratory. The results of these studies are discussed in Section XII.A.

C. Structural Polypeptides

A proposal for uniform nomenclature of rhabdovirus structural proteins was based on the composition and arrangement of VSV proteins (Wagner et al., 1972). This nomenclature has proved to be useful for both plant and animal rhabdovirus proteins.

The polypeptides of plant rhabdoviruses differ considerably in their size as well as their patterns. This difference is clearly illustrated by the comparison of the relative sizes and intensities of the proteins of SYNV, PYDV, and LNYV separated by electrophoresis in polyacrylamide gels (Fig. 15). A more extensive direct comparison of the sizes of the polypeptides of several plant rhabdoviruses (Table IX) has revealed appreciable differences in their sizes and in their electrophoretic patterns (Dale and Peters, 1981). More recently, the polypeptides of different strains of PYDV have been shown to differ slightly among themselves in electrophoretic mobility (Adam and Hsu, 1984; Falk and Weathers, 1983). Strain differences in polypeptides have also been detected by proteolytic digestion of individual PYDV polypeptides (Adam and Hsu, 1984).

The location of polypeptides in the virions of several plant rhabdoviruses has been investigated by differential detergent solubility, proteolytic digestion, and radioactive labeling of intact and detergent-degraded virions (Schultz and Peters, 1976; Dale and Peters, 1981; Jackson, 1978; Ziemiecki and Peters, 1976a,b; Adam and Hsu, 1984). Although these

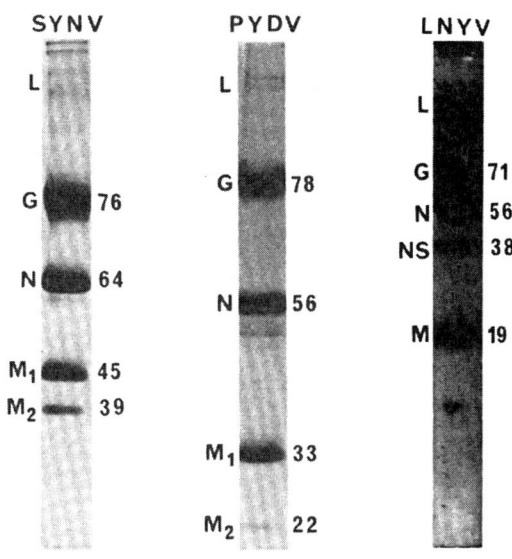

FIGURE 15. Structural proteins of three rhabdoviruses separated by polyacrylamide gel electrophoresis. The letters at left refer to the proteins using the nomenclature introduced by Wagner et al. (1972); the numbers at right are the molecular weights of the proteins ($\times 10^3$) as determined by Jackson and Christie (1977) for sonchus yellow net virus (SYNV), by Knudson and MacLeod (1972) for potato yellow dwarf virus (PYDV), and by Dale and Peters (1981) for lettuce necrotic yellows virus (LNYV).

TABLE IX. Comparison of the Protein Species of Several Plant and Animal Rhabdoviruses

Virus	Viral proteins (mol. wt. $\times 10^3$)[a]							References
	L	G	N	NS	M1	M2	Nonviral	
Plant rhabdoviruses								
Eggplant mottled dwarf virus (EMDV)	–	83	61	–	27	21		Dale and Peters (1981)
Lettuce necrotic yellows virus (LNYV)	170	71	56	38	19	–		Dale and Peters (1981)
Maize mosaic virus (MMV)	–	75	54	–	30	–		Falk and Tsai (1983)
Potato yellow dwarf virus (PYDV)	+	78	56	–	33	22		Knudson (1973)
Sonchus virus (SonV)	170	72	55	38	19	–		Dale and Peters (1981)
Sonchus yellow net virus (SYNV)	(125–150)	77	64	–	45	39		Jackson and Christie (1977)
Sowthistle yellow vein virus (SYVV)	150	83	60	–	44	36		Ziemiecki and Peters (1976a,b)
Wheat rosette stunt virus (WRSV)	140	66	46	40	19	–		Tsuhsun and Qiao-Xi (1984)
Wheat striate (American) mosaic virus (WSMV)	145	92	59	–	25	–		Trefzger-Stevens and Lee (1977)
Animal rhabdoviruses								
Vesicular stomatitis virus (VSV) [Indiana strain]	241[b]	57.4[b]	47.4[b]	25.1[b]	26.1[b]			Gallione et al. (1981), Rose and Gallione (1981), Schubert et al. (1984)
Cocal virus	+	65	45	33	26			Wagner (1975)
Piry virus	160	54	54	–	29			Wagner (1975)
Rabies virus	170	58[c]	58.5		39.5	25		Coslett et al. (1980), Anilionis et al. (1981)
Spring viremia of carp virus (SVCV)	160	85	50	50/40	25.6[b]			Roy (1981), Kiuchi and Roy (1984)
Infectious hematopoietic necrosis virus (IHNV)	150	55	40.5	–	22.5	17	12	Kurath and Leong (1985)

[a] L proteins are high-molecular-weight proteins; some of the L proteins of plant rhabdoviruses may be contaminating proteins of host origin located on the outside of the virions. NS, M_1, and M_1 of the plant rhabdoviruses are listed as designated in the original publications; these assignments have been based primarily on protein patterns revealed by gel electrophoresis, because no data comparing the chemical properties are available. NV indicates nonviral. (−) Protein was not detected. (+) protein was present, but molecular weight was not estimated.

[b] Molecular weight was calculated from nucleic acid sequence data. All other size estimates were obtained by polyacrylamide gel electrophoresis.

[c] The figures 50/40 reflect the apparent molecular weights of two phosphorylated forms of NS.

studies vary, some in approach and others in experimental detail, they all indicate that the structural proteins of the plant rhabdoviruses can be classified according to the nomenclature proposed for the animal rhabdoviruses (Wagner et al., 1972). These designations are described below for those who may be unfamiliar with rhabdovirus terminology.

1. L Protein

A large (L) protein of 241,000 daltons is present in small amounts in the VSV virion (Harmison et al., 1984). An L protein is also associated with the nucleocapsid of rabies virus and other animal rhabdoviruses (Heyward et al., 1979). Reconstitution studies of transcriptase complexes have shown that the L protein is essential for RNA transcriptase activity (Belle-Isle and Emerson, 1982; De and Banerjee, 1984). Complementation analysis of VSV mutants (Hunt et al., 1976) has provided additional evidence that the L protein is involved in transcription (Hunt et al., 1979). Kinetic studies also suggest that the L protein component of the transcriptase complex contains both methyltransferase and transcriptase activities (Morgan and Kingsbury, 1981). These studies therefore encourage our belief that the L protein is an integral structural requirement of all rhabdoviruses.

Preparations of plant rhabdoviruses usually contain two or three high-molecular-weight polypeptides present in small amounts. However, an L protein associated with isolated nucleocapsid preparations has actually been detected only in LNYV (Dale and Peters, 1981). Parallel tests with sowthistle yellow vein virus (SYVV), SYNV, and eggplant mottled dwarf virus failed to reveal the presence of an L protein in nucleocapsid preparations, but showed that high-molecular-weight proteins are present in the soluble fraction after dissociation of the viral envelope with nonionic detergent. These proteins appear to be located on the exterior of the virion because they are readily labeled by iodination of intact virus (Fig. 16), selectively released by treatment with nonionic detergents in the presence of low salt (Fig. 17), and degraded by protease treatment of intact virus particles (Dale and Peters, 1981; Jackson, 1978; Ziemiecki and Peters, 1976a). These external L proteins may be either contaminating host proteins or aggregates of smaller viral structural proteins such as the G protein. Evidence favoring the latter hypothesis is that the proteins are approximately twice the size of the G protein and some of them stain positively for carbohydrate (Dale and Peters, 1981; Ziemiecki and Peters, 1976b). The failure to detect an L protein associated with nucleocapsid preparations of these plant rhabdoviruses may be a consequence of experimental conditions. It is also possible that the L proteins of these viruses have lower binding affinities to the nucleocapsids than do the L proteins of animal rhabdoviruses and LNYV.

Despite the lack of direct evidence for the presence of an L protein associated with the nucleocapsid of plant rhabdoviruses, we anticipate that such a protein will eventually be identified and shown to have a

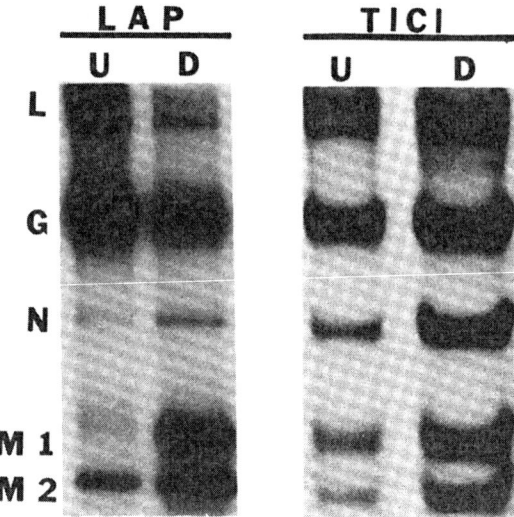

FIGURE 16. Fluorography of sonchus yellow net virus polypeptides after labeling with ^{125}I and separation on polyacrylamide gels. LAP: (U) Undisrupted virus labeled with lactoperoxidase; (D) virus disrupted with Triton X-100 prior to labeling with peroxidase. These data indicate that the L and G proteins are exposed on the surface of the undisrupted virions. The data also indicate that M2 can be labeled slightly when the virions are intact, whereas M1 is almost completely protected from labeling. After dissociation of the virus with Triton X-100, both M-1 and M2 become more highly labeled. N protein appears to be only slighly more accessible to labeling after dissociation, suggesting that the tyrosine residues are protected in the core protein. TlCl: The L and G proteins of undisrupted virions appear to be more accessible to labeling than the N, M1, and M2 polypeptides. After Triton X-100, the N, M1, and M2 proteins become more readily labeled, suggesting that they are located on the interior of the virion. From Jackson (1978).

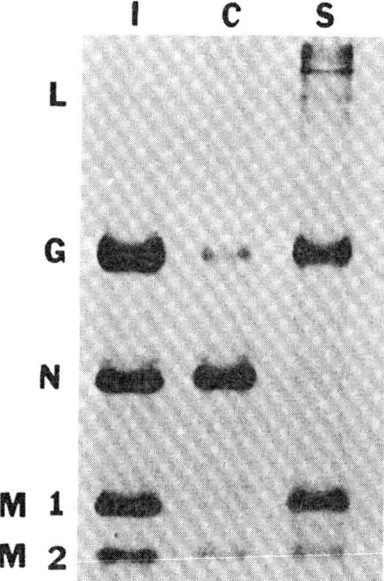

FIGURE 17. Separation of sonchus yellow net virus into soluble and core fractions after dissociation in Triton X-100. Virions were dissociated with 1% Triton X-100 and separated by sucrose density-gradient centrifugation into a nucleocapsid core fraction sedimenting at 250 S and a soluble fraction. The patterns show polypeptides separated by polyacrylamide gel electrophoresis of 1% SDS-treated preparations. (I) Intact virus; (C) 250 S nucleoprotein cores; (S) Triton-X-100-soluble supernatant. Note that the nucleocapsids contain predominantly N protein, and that the L, G, M1, and M2 proteins are released from the virions by Triton X-100 treatment. Modified from Jackson (1978).

functional role in replication. Some indirect evidence favoring this contention has already been obtained with SYNV, where plants infected with this virus contain a 6600-nucleotide virus-specific transcript (Rezaian *et al.*, 1983). This putative mRNA has sufficient coding capacity to encode an L protein. The presence of transcriptase activity in virions of LNYV and broccoli necrotic yellows virus (Toriyama and Peters, 1981) also implies that the proteins of plant and animal rhabdoviruses have similar functional activities.

2. G Protein

The glycosylated (G) protein is the spike protein that protrudes from the envelope of the virion (Wagner, 1975). In animal viruses, G protein functions in attachment of the virion to host receptor sites on the plasma membrane during the early stages of infection (Repik, 1979). Other evidence suggests that G protein also functions in assembly of virions (Wagner *et al.*, 1984). This transmembrane protein contains about 10% carbohydrate, and its structure has been extensively investigated in both rabies virus and VSV (Etchison and Summers, 1979).

A G protein has been identified in all plant rhabdoviruses tested. The G proteins of LNYV (Francki and Randles, 1975), SYVV (Ziemiecki and Peters, 1976b), SYNV (Jackson, 1978), and PYDV (Adam and Hsu, 1984; Falk and Weathers, 1983) have molecular weights varying from 70,000 to 90,000 (Table IX) and stain positively for carbohydrate. The G protein is thought to be part of the viral envelope because it is readily released from the virion (Fig. 17) by mild nonionic detergent treatment (Dale and Peters, 1981; Jackson, 1978; Ziemiecki and Peters, 1976b). G protein appears to be exposed on the surface of virions, since it becomes heavily labeled (Fig. 16) when intact virions are iodinated by various *in vitro* labeling techniques (Jackson, 1978; Ziemiecki and Peters, 1976b). Further evidence for external location of the G proteins is their high susceptibility to protease when intact virions are treated with this enzyme (Ziemiecki and Peters, 1976a).

The G proteins of three strains of PYDV have been shown to be different both in molecular weight and in the polypeptides produced after partial proteolysis (Falk and Weathers, 1983; Adam and Hsu, 1984). However, the G proteins of the SYDV and CYDV strains cross-reacted in ELISA tests, and the antisera raised against the G protein of the SYDV strain neutralized the infectivity of both SYDV and CYDV in tests with insect-cell lines (G. Adam, personal communication). This result provides evidence that the G protein of PYDV functions in binding to host receptor sites on the leafhopper vector cell surface.

3. N Protein

The nucleocapsid (N) protein is a structural protein that is tenaciously associated with the viral RNA and is an integral component of the transcriptase complex in VSV (Hunt *et al.*, 1979). A temperature-sensitive mutation in the N protein of VSV results in failure of the nucleocapsids to assemble and also abolishes synthesis of genome-size RNA when infected cells are incubated at the restrictive temperature (Marks *et al.*, 1985). Replication of VSV genomic RNA in an *in vitro* system programmed by addition of viral mRNA depends on synthesis of the N protein (Patton *et al.*, 1984a). These results thus suggest that the N protein has a direct role in replication of genomic RNA. In fact, several investigators have postulated an important role for the N protein in regulation of the switch from transcription to replication (Blumberg *et al.*, 1981; Perrault *et al.*, 1983). This hypothesis is supported by recent results obtained from both *in vitro* and *in vivo* experiments that strongly suggest that the replication of the genome of VSV is controlled by the availability of N protein (Arnheiter *et al.*, 1985). Therefore, N protein appears to have an important regulatory role as well as a structural function.

All the plant rhabdoviruses that have been studied have an N protein that is tightly complexed with the viral RNA (Francki and Randles, 1975; Ziemiecki and Peters, 1976b; Jackson, 1978; Dale and Peters, 1981). Dissociation of the virus with nonionic detergents releases a core particle that sediments at 200–250 S (Jackson, 1978). The core can be visualized in the electron microscope as a loosely coiled structure (see Fig. 13) that consists of RNA and N protein (Jackson, 1978). With SYNV and LNYV, this particle is less infectious than intact virus, but the nucleocapsids clearly retain some infectivity when assayed on local-lesion hosts (Jackson, 1978; Randles and Francki, 1972). The nucleocapsids of LNYV (Randles and Francki, 1972) and broccoli necrotic yellows virus (BNYV) (Toriyama and Peters, 1981) also have transcriptase activity with *in vitro* requirements similar to those of VSV transcriptase.

4. NS Protein

The nonstructural (NS) protein—which is actually a misnomer [see Chapter 2 (Section II.B)]—is a minor phosphoprotein found in VSV virions. Several different phosphorylated derivatives of NS have been identified in VSV-infected cells (Bell and Prevec, 1985; Clinton and Huang, 1981; C.-H. Hsu *et al.*, 1982). These derivatives differ in transcriptase activity and in their ability to bind to the nucleoprotein complex (Kingsbury *et al.*, 1981; Kingsford and Emerson, 1980). NS protein appears to promote binding of the L protein to the nucleocapsid (Williams and Emerson, 1984) and may also interact with the RNA–N protein complex of VSV to promote RNA unwinding during transcription (De and Banerjee, 1985). Genetic studies have implicated NS mutants with altered electrophoretic mobilities in the replication of the genomic RNA of VSV (Lesnaw *et al.*, 1979). In this regard, NS protein modulates replication of the genome in an *in vitro* system (Patton *et al.*, 1984b). Thus, NS protein has an important role in RNA transcription and probably also functions in the replication of the genome in infected cells.

A phosphorylated protein with the properties of the NS protein has not been unequivocally identified in any plant rhabdovirus. However, several viruses have protein patterns in polyacrylamide gels that resemble the protein profile of VSV, and a protein thought to be an NS protein is associated with purified nucleocapsids of LNYV (Dale and Peters, 1981). Because of this, Dale and Peters (1981) suggested that some other plant rhabdoviruses have an NS protein analogous to that of VSV (Table IX). Viruses that have now been speculatively assigned to this class include LNYV, BNYV, sonchus virus, maize mosaic virus, and tomato vein-yellowing virus (Dale and Peters, 1981; Falk and Tsai, 1983; El Maataoui *et al.*, 1985). Unfortunately, no structural data are available on any of these putative plant rhabdovirus NS proteins. It is not even known whether they are phosphorylated, since the ^{32}P incorporation into virus in infected plants is too low for analysis of phosphoproteins. Because of this lack of information, we believe it is premature to designate NS-protein assignments until additional structural data become available.

5. M Proteins

Two apparently different classes of proteins have been designated as membrane matrix (M) proteins. One class includes the protein of VSV, the prototype of the vesiculovirus group (McSharry, 1979). The M protein of VSV is hydrophobic and is intimately associated with the lipids of the viral envelope (McSharry, 1979). Thus, along with the G protein, the M protein is important in maintaining viral structure. M proteins may also be involved in regulation of assembly (Wagner *et al.*, 1984) and possibly are involved in regulation of replication of the genome (Patton *et al.*, 1984b).

VSV appears to have only one M protein, while rabies virus, belonging to the lyssavirus group, has been assigned two M proteins. This assignment was based on treatments with deoxycholate that released the proteins M1, M2, and G from rabies-virus particles (Sokol *et al.*, 1971). These experiments alone leave considerable doubt concerning the actual location and function of the two "membrane" proteins. Reevaluation of these structural proteins has now provided several different lines of evidence suggesting that the M2 protein of rabies virus and the M protein of VSV have a similar location in the virion and an analogous function. However, the M1 protein of rabies virus appears to be similar to the VSV NS protein (Cox *et al.*, 1981).

Although the rabies virus M2 protein appears to be closely associated with the viral envelope, as is also the case with the VSV M protein, only a small portion of the protein is exposed at the surface of the virion. In contrast, the rabies M1 protein appears to be closely associated with the nucleoprotein complex. The G and M2 proteins are both localized on the surface of cells infected with rabies virus, whereas the NS and M_1 proteins are located within the cells. An additional similarity of the VSV NS and rabies virus M1 proteins is that both are phosphorylated. Moreover, the

NS protein of VSV and the M1 protein of rabies virus are also similar in that each can be resolved into two fractions depending on the extent of phosphorylation (Dietzschold *et al.*, 1979). All these data strongly suggest that the M1 protein of rabies virus is structurally equivalent to the NS protein of VSV. Even though functional information is lacking, it seems likely that the M1 protein of rabies virus may be involved in RNA transcription and replication.

Reassessment of the location and significance of the M proteins of rabies virus necessitates some rethinking of assignments made for M proteins of the plant rhabdoviruses. Basically, we have no structural data available about the M proteins of any plant rhabdovirus that could permit us to distinguish proteins that correspond to the NS and M proteins of VSV. Except possibly for LNYV (Dale and Peters, 1981), detergent-solubility studies have failed to distinguish between M proteins on the basis of their association with the soluble and nucleocapsid fractions (Jackson, 1978; Dale and Peters, 1981). The *in vitro* labeling studies with SYNV (Fig. 16) show that the M1 protein is protected from lactoperoxidase labeling in the intact virion and that M2 is partly accessible to labeling. This result suggests that M1 of SYNV might correspond to M1 of rabies virus. Additional structural information is needed, however, before we can assign NS and M functions to the proteins of any of the plant rhabdoviruses.

X. TRANSCRIPTASE ACTIVITY OF PURIFIED VIRUS

The discovery of an RNA-dependent RNA polymerase in virions of vesicular stomatitis virus (VSV) (Baltimore *et al.*, 1970) greatly accelerated research on the rhabdoviruses because it provided a mechanism to explain how the genome was expressed during replication. The properties of the transcriptase of VSV have been analyzed in detail in a number of studies [for reviews, see Hunt *et al.* (1979) and Chapter 6]. The amount of transcriptase activity, however, is variable among the rhabdoviruses, and it is difficult to detect in rabies virus (Flamand *et al.*, 1978; Kawai, 1977). Plant rhabdoviruses, like the animal viruses, also vary in their virion-associated transcriptase activity. Activity similar to that of VSV is readily detected in preparations of lettuce necrotic yellows virus (LNYV) (Francki and Randles, 1972, 1973; Toriyama and Peters, 1980) and broccoli necrotic yellows virus (BNYV) (Toriyama and Peters, 1981). Transcriptase activity in sonchus virus and sonchus yellow net virus (SYNV) (Peters *et al.*, 1978; Flore and Peters, 1981) has been mentioned in abstracts, but the details of these studies have not yet been published. Transcriptase activity was reportedly detected in SYNV preparations, but only at low levels and only after several hours incubation of virus in transcriptase reactions (Flore and Peters, 1981). However, under assay conditions similar to those used with LNYV (A. O. Jackson, unpublished results), activity was not detectable with either SYNV or potato yellow dwarf virus preparations. It may be possible that enzyme activity of these viruses requires the action of some host factor such as a specific protease or some other activating agent. On the other hand, poor transcriptase activity in these viruses could be due to a number of other factors, including contamination of virus preparations with inhibitory host components or nucleases (Francki and Peters, 1978).

The transcriptase of LNYV and BNYV is activated on disruption of the virus with nonionic detergents. The *in vitro* system requires the presence of the four nucleotide triphosphates and Mg^{2+}. The salt, pH, and temperature optima for enzyme activity are similar to those of VSV. Transcriptase activity of LNYV and broccoli necrotic yellows virus is associated with an infectious nucleocapsid fraction similar to that of VSV (Toriyama and Peters, 1981). When purified preparations of LNYV or BNYV are dissociated in high salt, the viral proteins can be separated by density-gradient centrifugation into supernatant fractions containing glycosylated (G) and matrix (M) proteins and pellet fractions containing nucleocapsid protein complexed to viral RNA (Toriyama and Peters, 1980, 1981). Neither of the fractions from dissociated LNYV contains transcriptase activity. The supernatant fraction of BNYV is also devoid of activity, although the pellet fraction contains about 20% of the original activity. Homologous reconstitution of the fractions derived from LNYV and BNYV restores most of the activity. Interestingly, heterologous mixtures of the two virus fractions are inactive. These experiments provide evidence that LNYV and BNYV are distinct and that the virus-specific polypeptides must interact in a specific manner to form an active transcriptase complex.

The product RNAs generated during *in vitro* transcription by LNYV and BNYV transcriptase are single-stranded and range in size up to about 2000 nucleotides. Although the largest products are of sufficient size to function as RNAs for the most abundant viral proteins (G and M), considerable degradation is evident in the preparations. The transcripts are complementary to the viral genome and are virus-specific, because more than 95% of the newly synthesized RNA anneals to viral RNA. Although the LNYV and BNYV genomic RNAs hybridize to their respective transcripts, they fail to cross-hybridize. These results indicate that both viruses have negative-strand genomes and that they have little sequence homology (Toriyama and Peters, 1981).

XI. TAXONOMY OF PLANT RHABDOVIRUSES

Comparative structural data are not available for analysis of the relationships of plant and animal rhabdoviruses. Nevertheless, several different authors have already separated the plant rhabdoviruses into two subgroups analogous to those of vesicular stomatitis virus (VSV) and rabies virus. These groupings were prompted primarily by the suggestion of Dale and Peters (1981) that division of plant rhabdoviruses could be based on cytopathology, ability to detect a transcriptase *in vitro*, and association of proteins within the virion. According to this proposed nomenclature, viruses in subgroup I, the phytovesiculoviruses, would include those viruses that undergo morphogenesis in the cytoplasm, possess transcriptase activity that is readily detectable *in vitro*, and contain only one matrix (M) protein. The viruses initially placed in this group include lettuce necrotic yellows virus (LNYV), sonchus virus (SonV), and broccoli necrotic yellows virus (BNYV) (Peters, 1981). Subgroup II, the phytolyssaviruses, would include those viruses that accumulate in the perinuclear spaces, have poor transcriptase activity *in vitro*, and possess proteins M1 and M2. Eggplant mottled dwarf virus, potato yellow dwarf virus, sonchus yellow net virus, and sowthistle yellow vein virus have been assigned to subgroup II (Peters, 1981).

Although this proposal provides a useful framework for further experiments, more comprehensive comparisons of the plant rhabdoviruses are essential before such taxonomic divisions can be accepted. Information derived from electron microscopy does suggest that plant rhabdoviruses can be separated into groups corresponding to the site of morphogenesis; however, more detailed examinations of infected cells need to be conducted with most of the plant rhabdoviruses to clarify the details of the budding process. As discussed in Section X, the function of polypeptides within particles is unclear and hence should not be used for classification purposes at present. In fact, some anomalies already exist among the members that have been assigned to these subgroups. For example, both wheat striate (American) mosaic virus (Trefzger-Stevens and Lee, 1977) and maize mosaic virus (Falk and Tsai, 1983) appear to have protein patterns resembling those of VSV, but both viruses appear to differ from BNYV, LNYV, and SonV in their site of morphogenesis. Moreover, the transcriptase activities of members of the groups as a whole have not been adequately tested. Therefore, until more structural and biochemical data become available, it is probably wise to postpone assignment of the plant rhabdoviruses to subgroups.

XII. REPLICATION OF PLANT RHABDOVIRUSES

Four experimental systems currently used have potential for exploring the replication of plant rhabdoviruses. These techniques include systematically infected plants or tissue fragments, insect vectors, cultured vector cells, and plant-cell protoplasts. Each of these systems has distinct advantages for study of various aspects of replication, but none has been fully exploited because of limited resources and because each has some inherent drawbacks.

A. Whole-Plant Studies

Intact plants are useful for many purposes because large amounts of tissue can be conveniently harvested and used for recovery of virus, protein, and nucleic acid. For these reasons, most of the available information on replication of plant rhabdoviruses comes from experiments on systemically infected plants. However, such experiments have distinct disadvantages because infection of individual cells occurs at different times and systemic spread of virus is difficult to control. These problems are much more pronounced with those viruses that can be transmitted only by the insect vector. Moreover, inefficient delivery and uptake of isotopes prevent certain types of experiments, and the pool sizes of precursors create special problems in interpreting pulse–label studies. Even in systems in which the virus replicates to reasonably high concentrations, such as tobacco plants infected with sonchus yellow net virus (SYNV), synthesis of host proteins obscures incorporation of amino acid precursors into viral proteins (A. O. Jackson, unpublished results). In other studies in which attempts have been made to label lettuce necrotic yellows virus (LNYV) with ^{32}P, no incorporation could be detected in purified virus even though a considerable amount of label accumulated in positive-strand RNA viruses such as tobacco mosaic virus or turnip yellow mosaic virus under similar conditions (R. Francki, unpublished results).

Until recently, very little direct information had been accumulated on the virus-specific events involved in infection of plants by rhabdoviruses. Most of the work prior to 1975 involved investigation of the host-specific events that occur during infection and systemic spread of LNYV in *N. glutinosa*. That work has been discussed in detail in earlier reviews (Francki, 1973; Francki and Randles, 1979). At the cellular level, experiments with LNYV revealed that virus accumulated rapidly about 3 days prior to the appearance of visible symptoms. Subsequently, the concentration of virus, as measured by infectivity, declined rapidly (Randles and Coleman, 1970). Similar increases and decreases in infectivity have been observed in tobacco leaves infected with both potato yellow dwarf virus (PYDV) (H. T. Hsu and Black, 1973b) and SYNV (Jackson and Christie, 1977). The reasons for these dramatic changes are not known, but they may be due to activation of some unidentified host defense response or accumulation of materials that can interfere with virus recovery or infectivity assay or both.

LNYV infection has been shown to have a pronounced effect on ribosome metabolism (Randles and Coleman, 1970, 1972). At the stage of most rapid virus synthesis, the concentration of chloroplast (70 S) ribosomes decreased and incorporation of ^{32}P into the chloroplast ribosomal RNA virtually ceased. Shortly after the virus concentration peaked, the 70 S ribosomes disappeared. In contrast, the concentration of cytoplasmic (80 S) ribosomes was not altered appreciably during infection.

Moreover, the incorporation of ^{32}P into the cytoplasmic ribosomal RNA was similar in both healthy and infected leaves. It is not known whether these changes in the ribosomal profiles were a specific consequence of LNYV replication or a reflection of early leaf senescence resulting from the infection.

Changes in the incorporation of [^{3}H]uridine have also been observed in both the nucleus and the cytoplasm of plants sampled at various times after infection with LNYV (Wolanski and Chamber, 1971). Autoradiography of ^{3}H-labeled cells of leaf-tissue fragments that had been treated with actinomycin D revealed accumulation of radioactivity in the nucleus during virus-induced blistering of the nuclear envelope. Subsequently, aggregates of virus particles were detected in the perinuclear spaces, but as infection progressed, virions began to appear in cytoplasmic vesicles near the endoplasmic reticulum. During this time, ^{3}H incorporation decreased in the nucleus and increased in the cytoplasm.

Although these isotope-incorporation experiments are indirect, they suggest that synthesis and encapsidation of LNYV RNA occurs in the nucleus or the perinuclear spaces and that the virions later accumulate in the cytoplasm. More sensitive technology using recombinant DNA probes and monoclonal antibodies is now available to monitor the distribution of viral nucleic acids and proteins in cells. The precision of these techniques combined with the ability to obtain synchronously infected protoplasts should enable direct tests of the sites of virus synthesis and should permit us to follow accumulation of virus-specific products at different stages in replication.

Little is known about the metabolism of host RNA in tobacco infected with SYNV, but information has accumulated on the appearance of virus-specific transcripts in infected leaves. These studies show that infected tobacco leaves contain transcripts that are complementary to more than 95% of the viral RNA (Milner and Jackson, 1979). The complementary transcripts are associated with polyribosomes, are polyadenylated, and are thought to represent mRNAs. The kinetics of hybridization suggest that these transcripts represent between 2 and 5% of the polyadenylated RNAs in infected leaves and that a substantial portion of the RNAs are associated with membrane-bound polyribosomes (Milner *et al.*, 1979; Milner and Jackson, 1983). When the polyadenylated RNAs are separated by electrophoresis in agarose gels, transferred to nitrocellulose, and hybridized with ^{32}P-labeled viral RNA, four distinct fractions ranging from 1200 to 6600 nucleotides (NT) can be detected (Rezaian *et al.*, 1983). These four fractions actually represent five unique RNAs designated SYNV-complementary RNAs (scRNAs) 1–5 (Fig. 18). The 1200-NT band contains two RNAs of similar size, since two unique recombinant DNA clones with combined insert sizes greater than 1200 NT hybridize to the RNAs. The four smaller RNA species are thought to be RNAs corresponding to the four most abundant viral polypeptides because they are of the appropriate size to encode these proteins. The 6600-NT

scRNA   pSYNV

KR   1a   2a   5a   24a

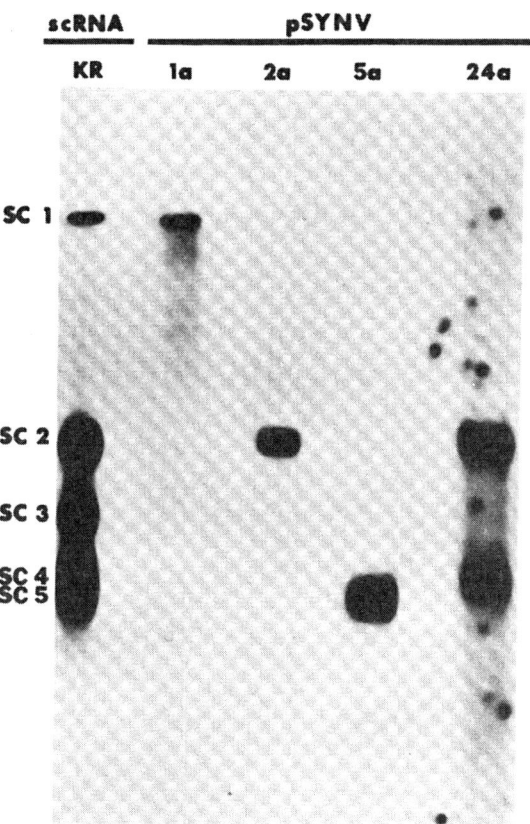

SC 1

SC 2

SC 3

SC 4
SC 5

FIGURE 18. Identification of scRNAs isolated from infected tobacco leaves, electrophoresed on agarose gels, and transferred to nitrocellulose. Lane KR represents scRNAs visualized by hybridization with randomly ^{32}P-labeled sonchus yellow net virus RNA. The succeeding lanes to the right were hybridized to nick-translated ^{32}P-labeled pSYNV clones 1a, 2a, 5a, and 24a. The designations SC 1–5 indicate individual scRNAs. The results show that the individual sc-RNAs have unique sequences. Note that SC 4 and SC 5 hybridize to clones with individual inserts exceeding 1200 NT, and these represent unique coelectrophoresing RNAs. SC 2 and SC 4, visualized by the clone designated pSYNV 24a, are adjacent because both scRNAs are visualized by a single clone. From Rezaian *et al.* (1983).

scRNA has sufficient coding capacity to direct synthesis of a large (L) protein. As discussed in Section IX.C.1, an L protein has not yet been unambiguously identified in purified virions (Jackson, 1978), but the presence of the 6600-NT transcript provides indirect evidence that such a protein does exist and is encoded by SYNV. These tentative coding assignments are illustrated in Fig. 19. Thus, it appears that SYNV is a negative-strand virus and that it has replicative strategies similar to those of animal rhabdoviruses such as vesicular stomatitis virus (VSV).

A genetic map has not yet been completed for SYNV or other plant rhabdoviruses. However, the genes that encode the glycosylated (G) protein and one of the matrix (M) proteins appear to be adjacent to each other. One class of recombinant DNA clones constructed from SYNV RNA hybridizes to both scRNA 2, the putative G mRNA, and scRNA 4, thought to be one of the M mRNAs (Fig. 18). Another clone hybridizes to scRNA 3, thought to be the nucleocapsid (N) protein mRNA, and to scRNA 5, the other putative M mRNA (L. A. Heaton, in prep). The scRNA 3 transcript is located at the 3′ end of the genome because SYNV RNA labeled specifically at the 3′ end hybridizes only with scRNA 3. Other

RNA   **Proteins**

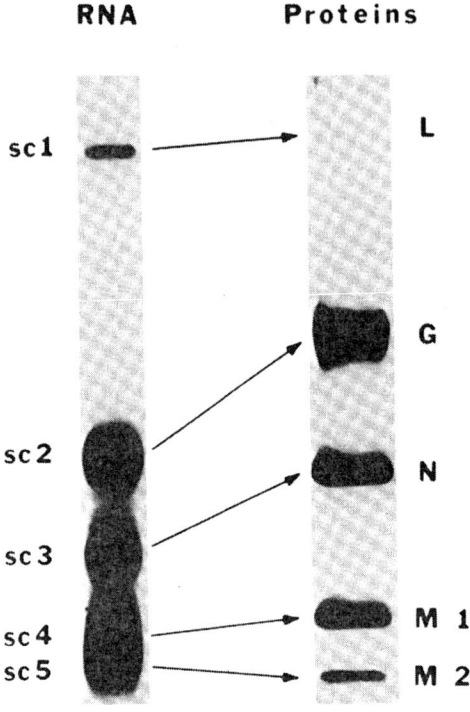

FIGURE 19. Electrophoretic identification of polypeptides and scRNAs. The labels sc 1–5 indicate individual mRNAs that are thought to encode the sonchus yellow net virus polypeptides L, G, N, M1, and M2, respectively. Each of these putative mRNAs is of the appropriate size to encode each individual viral polypeptide.

clones hybridize to scRNAs 4 and 5, so these two RNAs must be encoded by adjacent regions on the genome. These preliminary data indicate that plant and animal rhabdovirus genomes have the same general organization (Ball and Wertz, 1981).

In addition to putative mRNAs, plants infected with rhabdoviruses also contain short transcripts derived from the 3′ end of the genome (Fig. 20). These transcripts appear to correspond to the plus-strand "leader" RNAs transcribed by VSV strains (Giorgi *et al.*, 1983; Kurilla *et al.*, 1982; Kurilla and Keene, 1983), rabies virus (Kurilla *et al.*, 1984), and spring viremia of carp virus (Roy *et al.*, 1984). The leader RNA is thought to have an important role in regulating transcription and replication of the rhabdovirus genome (Colonno and Banerjee, 1976; McGowan *et al.*, 1982; Perrault *et al.*, 1983), so comparison of the leader RNAs of plant and animal rhabdoviruses may provide important clues about common and unique regulatory aspects of the two groups. Therefore, it is of interest to find that the leader RNAs present in plants infected with SYNV differ markedly from those in cells infected with animal rhabdoviruses (Zuidema *et al.*, 1986). The SYNV leader RNA is nearly three times the length of the leader RNAs of the three animal rhabdoviruses currently investigated (Fig. 21). Moreover, the SYNV leader RNA does not contain sequences corresponding to the VSV leader sequences (Wilusz *et al.*, 1985) that are thought to be involved in binding N protein (Blumberg *et al.*,

G T1a T1b OH A⁺ A⁻ H

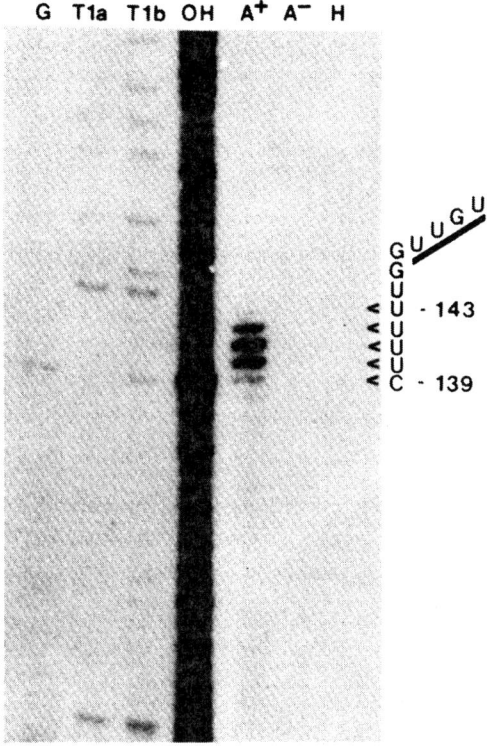

FIGURE 20. Detection and sizing of leader RNA(s) of sonchus yellow net virus (SYNV) by electrophoresis on a polyacrylamide gel. Lanes A⁺, A⁻, and H: RNA from SYNV-infected tobacco leaves, separated into a poly(A)⁺ and a poly(A)⁻ RNA fraction, and RNA from healthy tobacco plants, respectively, were hybridized with 3'-³²P-end-labeled SYNV viral RNA, followed by RNase treatment prior to gel electrophoresis. Only the poly(A)⁺ RNA fraction shows four to five discrete bands. Lane G: 3'-³²P-end-labeled SYNV viral RNA untreated (the majority of the label was at the top of the gel). Lanes T1a and T1b: partially digested 3'-³²P-end-labeled SYNV viral RNA with two different ribonuclease T1 concentrations. Lane OH⁻: a partial alkali degradation of 3'-³²P-end-labeled SYNV viral RNA. From Zuidema *et al.* (1986).

1983) and a host transcription factor (Kurilla *et al.*, 1984; Keene *et al.*, 1984; Kurilla and Keene, 1983). A potential site for initiation of transcription of mRNAs is located on SYNV RNA adjacent to the leader template. This putative initiation site contains the tetranucleotide sequence (UUGU) that complements the consensus sequence (AACA) found at the 5' end of the animal rhabdovirus mRNAs (Fig. 21). These results suggest that the signals that control transcription and replication have diverged markedly during evolution of plant and animal rhabdoviruses, while the tetranucleotide sequence involved in initiation of transcription of individual mRNAs appears to have been conserved (Zuidema *et al.*, 1986).

B. Use of Insect Vectors for Studies of Replication

The replication of plant rhabdoviruses in their insect vectors has not been extensively investigated. Such studies have enormous potential because of the unique opportunity for direct comparison of the responses of plants and animals to infection by the same virus. Most of the available information has been derived by electron microscopy, serology, and infectivity to determine the kinetics of appearance of virus in various organs

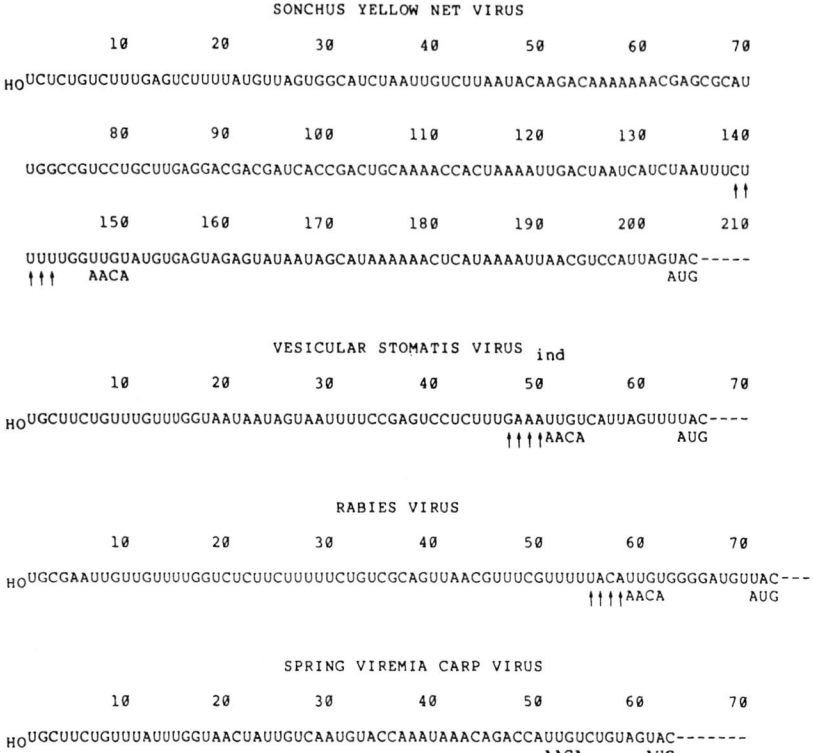

FIGURE 21. Comparison of the genomic templates that encode the plus-strand leader RNAs of plant and animal rhabdoviruses. The leader-RNA "genes" start at the 3' termini of their respective genomes, and their respective ends are indicated by arrows. Note that the sonchus yellow net virus (SYNV) leader RNA is about three times the length of the three animal rhabdovirus leader RNAs and contains little sequence homology with animal rhabdovirus leader RNAs. The sequence UUGU (AACA) is of special significance because it appears to initiate mRNA transcription in all three animal rhabdoviruses. Indirect evidence is obtained that this is also the case of SYNV. It is noted that little is known about a leader RNA or the putative start of the first mRNA of spring viremia of carp virus (SVCV). Data from McGeoch *et al.* (1980) for VSV, Kurilla *et al.* (1984) for rabies virus, and Roy *et al.* (1984) for SVCV.

at different times after acquisition (Francki, 1973; Shikata, 1979). These studies all suggest that similar events occur during infection of these hosts. However, additional studies are needed to determine the molecular interactions that result during replication and the comparative pathological responses of the plant and insect hosts to infection.

C. Insect-Cell Cultures

Cultured insect cells are more attractive than intact insects for virus replication studies because they can be synchronously infected and single-hit kinetics of infection can be obtained (Hsu and Black, 1973a; Black,

1979). Unfortunately, the available cell lines are fastidious and difficult to maintain. In addition, their unusually rich nutrient requirements limit their usefulness for isotope precursor studies. Moreover, a high background of host protein synthesis tends to obscure synthesis of viral polypeptides. These problems have been encountered with leafhopper cell lines infected with a plant reovirus, with wound tumor virus (Nuss, 1984), and with PYDV (Adam, 1982). In the latter case, the five structural proteins thought to be encoded by the genome could be detected after separation of the proteins on polyacrylamide gels, but the high background of host proteins could not be suppressed to acceptable levels with actinomycin D. These difficulties, coupled with the slow replication of the virus, complicated interpretation of the results. Therefore, even though insect-cell cultures hold promise for studies of the molecular biology of replication, some technical obstacles need to be overcome before their potential can be fully realized.

D. Infection of Plant Protoplasts

Development of methods for synchronously infecting protoplasts with plant rhabdoviruses could also enable more precise evaluation of specific events that occur in their replication. Suspensions of protoplasts should enable uniform application of isotopes for pulse–label studies and could be more practical for addition and removal of various inhibitors suitable for probing different aspects of replication. These cell suspensions would also overcome many other disadvantages of whole plants.

The recent infection of cowpea protoplasts with SYNV (van Beek *et al.*, 1985a) and festuca leaf streak virus (FLSV) (van Beek *et al.*, 1985c) with the aid of polyethylene glycol (PEG) now provides a system that should be extremely useful for studies of the replication of plant rhabdoviruses. The use of inhibitors appears to be feasible because infected protoplasts appear to be responsive to at least one inhibitor that blocks virus morphogenesis (van Beek *et al.*, 1985b). With the PEG procedure, more than 90% of the surviving protoplasts become infected even though cowpea is not a systemic host of either virus. The level of infection using the PEG procedure is considerably higher than previously reported for infection of tobacco protoplasts with PYDV (Riesterer and Adam, 1981). The PEG inoculation procedure should have general applicability to most plant rhabdoviruses and would probably be readily adaptable to protoplasts from plants that are systemic hosts. However, even if some unforeseen obstacles prevent the general application of the technique to protoplasts of other species, cowpea protoplasts appear to be useful for answering many questions even though SYNV and FLSV are unable to systemically invade cowpea.

Under the conditions used by van Beek *et al.* (1985a), infectivity of SYNV was detected in extracts of cowpea protoplasts 11–12 hr after inoculation (Fig. 22). The infectivity increased until about 30 hr after in-

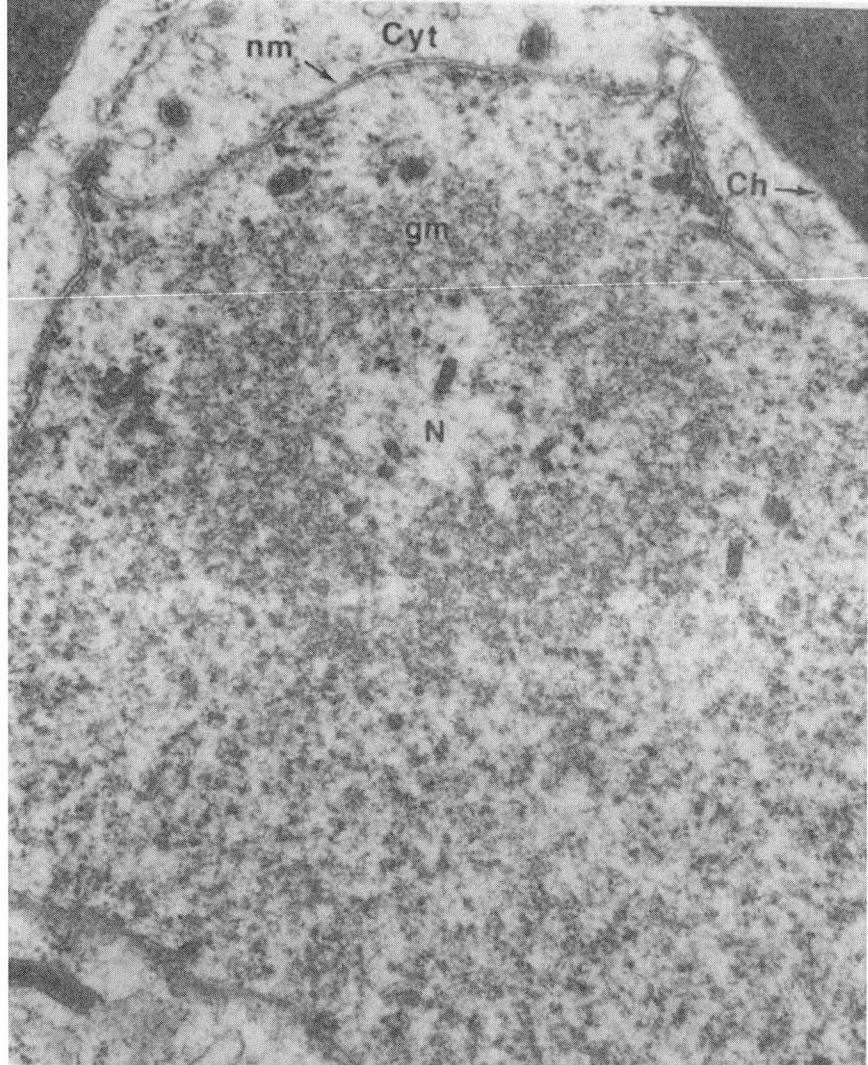

FIGURE 22. A cowpea protoplast, 11 hr after inoculation with sonchus yellow net virus. The nucleus (N) contains a granular matrix (gm) with core formation at the inner and outer edges. (Cyt) Cytoplasm; (Ch) chloroplast; (nm) nuclear membrane. Modified from van Beek *et al.* (1985b).

oculation and subsequently began to decline. Uninoculated protoplasts or protoplasts stained with fluorescent antibody immediately after inoculation gave a negative reaction, whereas 97% of the protoplasts tested 56 hr postinoculation stained positively. Numerous rhabdovirus particles were present in the perinuclear space and in the cytoplasm of infected protoplasts at 67 hr after inoculation (Fig. 23). These particles could be

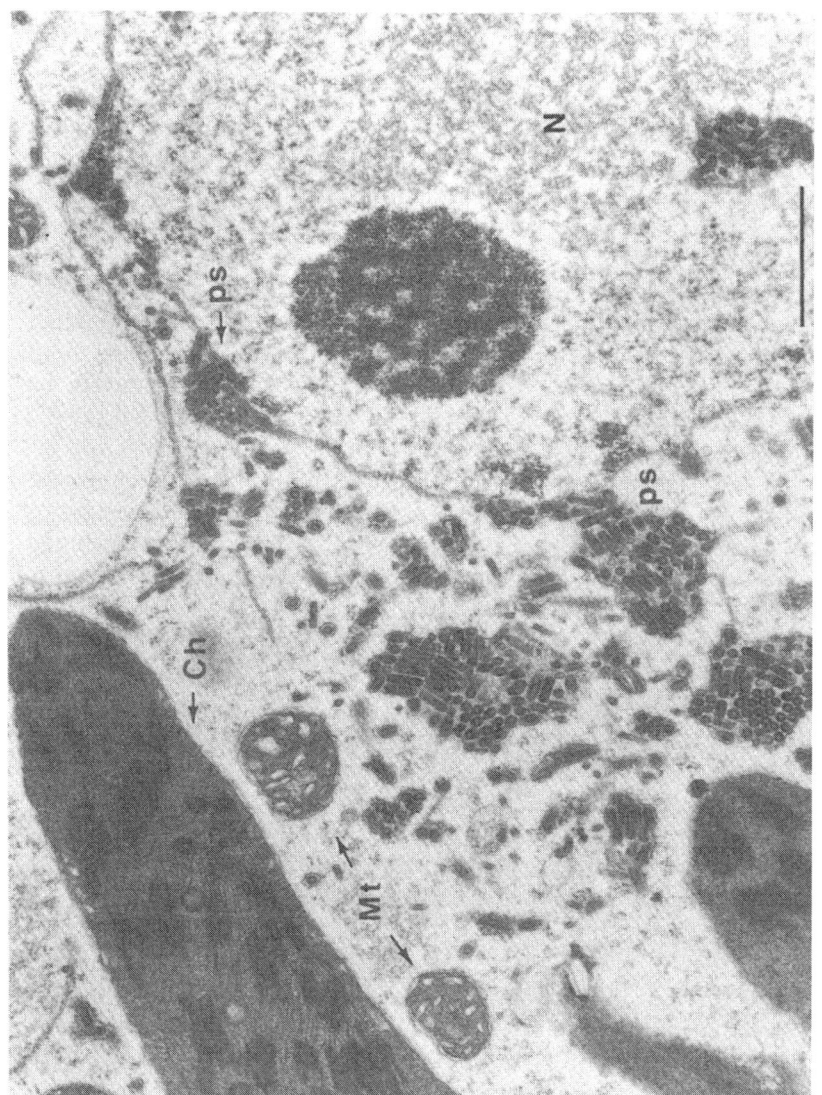

FIGURE 23. Electron micrograph of a cowpea protoplast infected with sonchus yellow net virus, 67 hr after inoculation. Mature particles are present in the perinuclear space (ps) and in the cytoplasm. (N) Nucleus; (Mt) mitochondrion; (Ch) chloroplast. Scale bar: 1 μm. Modified from van Beek *et al.* (1985a).

separated into two different classes. One class averaged 191 nm in length and 41 nm in width and the second class was 201 by 67 nm. The class with the smaller diameter had the appearance of nucleocapsids, whereas particles with the larger diameter were thought to be enveloped virions.

A study of the time–course of SYNV infection (van Beek *et al.*, 1985b) revealed no cytopathological changes for 7 hr after inoculation. However, the cytoplasm of infected cells appeared to have increased numbers of polysomes 9 hr after inoculation. After 10 hr, cores thought to be nucleocapsids had appeared at the edge of a granular matrix in the nucleus and particles that appeared to be in the process of budding were attached to the inner nuclear membrane (Fig. 22). The matrix, the putative nucleocapsids, and budding virions were all observed at all subsequent periods up to 70 hr after inoculation (Fig. 23). The particles began to accumulate in the perinuclear spaces and to enter the lumen of the endoplasmic reticulum by 12 hr after inoculation. At subsequent periods, the virions appeared in the cytoplasm in association with a loose-fitting membrane thought to be part of the rough endoplasmic reticulum. Both enveloped virions and nucleocapsids were present in the cytoplasm by 24 hr after inoculation. However, it could not be ascertained whether the nucleocapsids were destined for morphogenesis at the endoplasmic reticulum or whether they resulted from degradation of the viral envelope. Throughout the course of infection, the number of particles appeared to increase in the cytoplasm, suggesting that virus replication continued throughout the 3-day period, despite an observed drop in extractable virus during the last part of the experiments.

Direct evidence for involvement of the nucleus in SYNV assembly was obtained by incubating infected protoplasts with tunicamycin, a specific inhibitor of glycosylation. In the presence of the inhibitor, a striking array of cores accumulated at the periphery of the nucleus and the granular matrix almost completely filled the nucleus (Fig. 24). No enveloped virus particles were observed in the presence of the inhibitor, and no nucleocapsids were evident in the cytoplasm 24 and 47 hr after inoculation. This result suggests that glycosylation is mandatory for morphogenesis of SYNV and that cores are unable to emerge from the nucleus into the cytoplasm without first having become enveloped at the nuclear membrane.

The assembly of FLSV, which occurs in the cytoplasm of its natural host, *Festuca gigantea*, was also studied in cowpea protoplasts (van Beek *et al.*, 1985c). Replication of FLSV appeared to be slower than that of SYNV, because virus particles did not appear in the protoplasts until 26 hr after inoculation. However, at that time, small viroplasms were observed in the cytoplasm and particles that appeared to be at various stages of morphogenesis were also seen budding from cytoplasmic membranes (van Beek *et al.*, 1985c). The viroplasms were composed of strands of granular material that were often seen in the cytoplasm adjacent to the nucleus. Membranes with budding particles, as well as mature enveloped

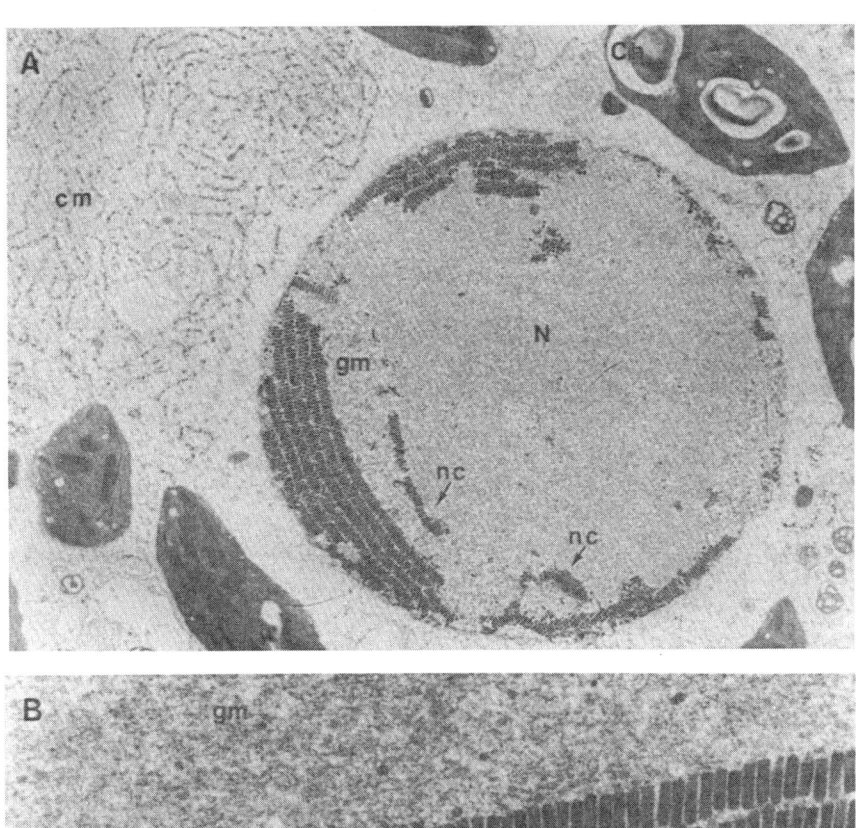

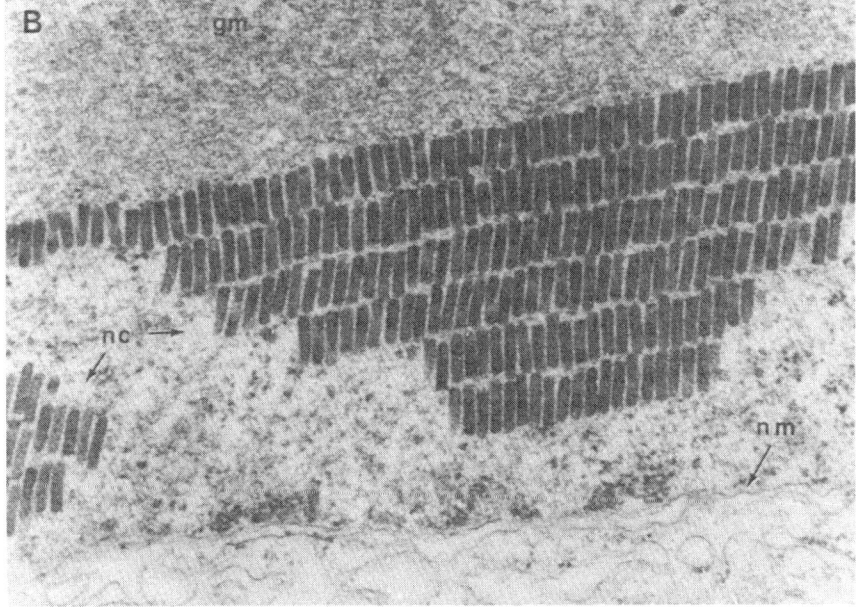

FIGURE 24. (A) Electron micrograph of a tunicamycin-treated cowpea protoplast 47 hr after inoculation with sonchus yellow net virus (SYNV), illustrating cores aggregated close to the nuclear membrane and at the edge of the granular matrix (gm) and deformation of cytoplasmic membranes (cm). (N) Nucleus; (Ch) chloroplast; (nc) nucleocapsids. (B) Nucleus of a tunicamycin-treated cowpea protoplast infected with SYNV, showing accumulation of SYNV cores at the edge of the granular matrix (gm) and blistering of the nuclear membrane (nm). (nc) Nucleocapsids. Modified from van Beek et al. (1985b).

particles within loose-fitting membranes, were present at the edge of the viroplasm. Virus particles surrounded by endoplasmic reticulum membranes were also scattered throughout the cytoplasm. After 34 hr, nucleocapsids that appeared to originate from strands at the edge of the viroplasms were found in the cytoplasm. These cores were frequently seen close to the tonoplast or chloroplast membranes and sometimes were located between the chloroplast and cytoplasmic membranes. At later stages of infection, the viroplasms increased in size and occupied much of the cytoplasm (Fig. 25). In contrast to cells infected with SYNV, particles of FLSV never became abundant in the cytoplasm and no ultrastructural changes were ever observed in the nucleus.

These results clearly show that FLSV, which has a monocotyledonous host, is capable of replicating in protoplasts from dicotyledonous plants. Moreover, some cytopathic alterations such as the appearance of the granular matrix in the nuclei of SYNV-infected protoplasts and the viroplasm induced in the cytoplasm by FLSV seem similar in both hosts. Nevertheless, the site of morphogenesis of FLSV differs markedly from that of SYNV. FLSV also appears to replicate inefficiently in the cowpea protoplasts, because the numbers of virus particles were much lower than those observed in protoplasts infected with SYNV and were also lower than the concentration observed in cells of the native grass host (van Beek *et al.*, 1985c; Lundsgaard and Albrechtsen, 1979). It is possible that synthesis of nucleocapsids and formation of the viroplasma are relatively independent of the host, but that some step in envelopment of the nucleocapsids may be blocked in the cowpea protoplasts.

XIII. FUTURE PROSPECTS

It is evident from the information summarized in this chapter that plant rhabdoviruses provide excellent material for future research. Over the past few years, a number of advances have been made in our understanding of the relationships, physiocochemical properties, and replication of several different members of the group. However, there are still many glaring gaps in our knowledge that need to be filled. Fortunately, several new experimental approaches, including molecular biological techniques and plant protoplast systems, are now available to help answer many important questions. Thus, the time is now ripe for plant rhabdovirus research to change from a predominantly descriptive phase to one that will involve an experimental approach employing modern molecular biological techniques.

One major area that needs urgent attention is the study of the relationships among different plant rhabdoviruses. These studies can now be conducted more efficiently than in the past because purification procedures have been improved. Moreover, a variety of sensitive analytical methods including serology, analysis of structural polypeptides, and nu-

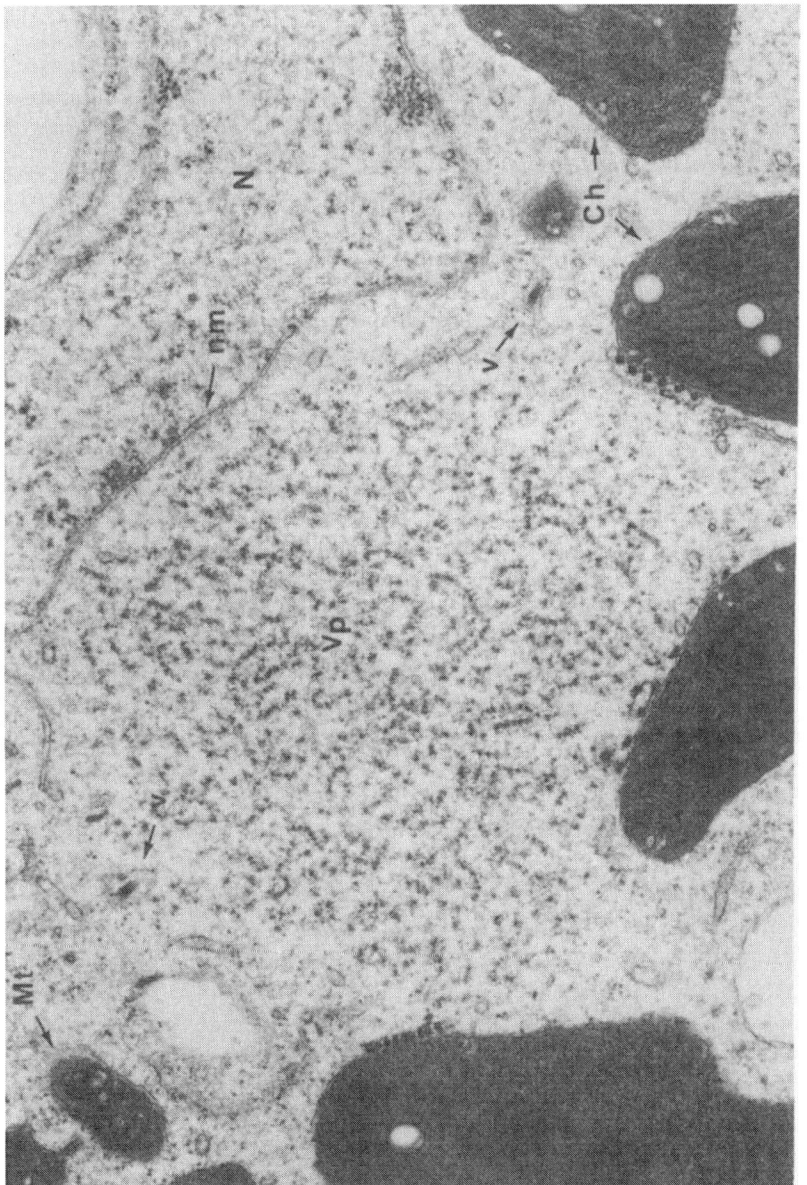

FIGURE 25. A cowpea protoplast infected with festuca leaf streak virus. The cytoplasm contains a viroplasm (Vp) surrounded by cores close to chloroplast membranes and two obliquely sectioned virus particles (v) 48 hr after inoculation. (N) Nucleus; (nm) nuclear membrane; (Ch) chloroplast; (Mt) mitochondrion. modified from van Beek et al. (1985c).

A. O. JACKSON et al.

cleic acid sequencing techniques can be applied to differentiate and group the numerous plant rhabdoviruses described in the literature. Although serological data already provide some information on plant rhabdovirus relationships, these need to be extended. Differentiation of plant rhabdoviruses on the basis of the properties of their polypeptides has not yet progressed beyond simple gel analysis. More sophisticated chemical analyses of viral polypeptides are needed for structural comparisons. Application of recombinant DNA techniques will also increase our understanding of the relationships among the members of the group. The nucleotide and protein sequence information generated by these techniques can be of immense value to comparative virology and taxonomy. The limited information already acquired is revealing intriguing differences between the structure of the genomes of sonchus yellow net virus (SYNV) and rhabdoviruses that infect vertebrates.

Another area that needs urgent attention is that of ecological investigations necessary for control of some of the serious plant diseases caused by rhabdoviruses. Much of the information needed to answer ecological questions should now be more readily obtained with newly available techniques. An excellent example of such a technique is the very sensitive and rapid method of enzyme-linked immunosorbent assay for the routine detection and diagnosis of rhabdoviruses. This technique permits accurate detection of viruses in both plant hosts and their insect vectors. Hence, it should be possible to readily obtain the data necessary to identify virus sources and to predict the course of disease development so that more efficient and timely control measures can be implemented.

The fundamental advances made in understanding the replication and structure of plant rhabdoviruses with model systems such as lettuce necrotic yellows virus and SYNV need to be extended and complemented by studies with other rhabdoviruses if we are to obtain a more comprehensive picture of the uniformity and diversity within the group. Our understanding of the replication of plant rhabdoviruses should accelerate in the near future through more extensive use of molecular biological approaches. Use of nucleic acid probes and monoclonal antibodies can circumvent some of the major obstacles previously encountered with other techniques for investigating virus replication.

The techniques recently used to infect cowpea protoplasts with SYNV and festuca leaf streak virus should be generally applicable to other rhabdoviruses. This approach should facilitate more effective use of inhibitors of viral replication and should simplify application of isotope tracer techniques. Overall, these developments should lead to great improvement in our understanding of the events during infection of plant cells.

Genetic engineering of plants is another application that should have enormous utility for plant rhabdovirus research. Plant transformation systems are already available for studies of gene expression and molecular biology of some dicotyledons (De Block et al., 1984). One obvious outgrowth of this new technology will be the construction of plants that express various viral gene products. Such transformed plants have the long-range potential for a number of new and exciting applications to plant rhabdovirus studies. These studies may include utilization of plants expressing viral proteins for identification and complementation of virus mutants. Transformed plants should be especially useful for exploring the regulatory effect of various viral proteins on different aspects of rhabdovirus replication. Ultimately, these studies should lead to construction of plants resistant to infection by rhabdoviruses. We expect that these and many other unique applications will have greatly extended our knowledge of the plant rhabdoviruses within the next decade of plant molecular biology research.

ACKNOWLEDGMENTS. We thank Doctors G. Adam, D. T. Gordon, and L. R. Nault and their associates, as well as Mr. Louis Heaton at Purdue, for providing us with unpublished data and manuscripts prior to publication; Doctors E. D. Ammar, A. Appiano, S. R. Christie, J. L. Dale, D. H. Gordon, R. S. Greber, T. Hatta, G. P. Martelli, L. L. McDaniel, L. R. Nault, J. W. Randles, M. Russo, and N. A. M. van Beek for illustrations; and Doctor C. E. Bracker, Mrs. Caroline Logan, Mrs. L. Wichman, and Mr. Howard D. Laffoon for assistance with figure preparation. A. O. Jackson wishes to express particular appreciation to Mr. H. D. Laffoon, whose 12 years of competent and dedicated technical support made much of the work on sonchus yellow net virus possible. Research on lettuce necrotic yellows virus by R. I. B. Francki has been generously supported by the Austrailian Research Grants Scheme. Research on sonchus yellow net virus by A. O. Jackson has been supported by a grant from the United States National Science Foundation (PCM-8318007).

REFERENCES

Adam, G., 1982, Plant virus studies in insect vector cell cultures, in: *Vectors in Virus Biology* (M. A. Mayo and K. A. Harrapp, eds.), pp. 37–62, Prathear, London.

Adam, G., and Hsu, H. T., 1984, Comparison of structural proteins from potato yellow dwarf viruses, *J. Gen. Virol.* **65**:991–994.

Adam G., Gaedigk, K., and Mundry, K. W., 1983, Alterations of a plant rhabdovirus during successive mechanical transfers, *Z. Pflanzenkr. Pflanzenschutz* **90**:28–35.

Ahmed, M. E., Black, L. M., Perkins, E. G., Walker, B. L., and Kummerow, F. A., 1964, Lipid in potato yellow dwarf virus, *Biochem. Biophys. Res. Commun.* **17**:103–107.

Ahmed, M. E., Sinha, R. C., and Hochester, R. M., 1970, Purification and some morphological characters of wheat striate mosaic virus, *Virology* **41**:768–771.

Ammar, E.-D., and Nault, L. R., 1985, Assembly and accumulation sites of maize mosaic virus in its planthopper vectors, *Intervirology* **24**:33–41.

Anilionis, A., Wunner, W. H., and Curtis, P., 1981, Structure of the glycoprotein gene in rabies virus, *Nature (London)* **294**:275–278.

Arnheiter, H., Davis, N. L., Wertz, G., Schubert, M., and Lazzarini, R. A., 1985, Role of the nucleocapsid protein in regulating vesicular stomatitis virus RNA synthesis, *Cell* **41**:259–267.

Atchison, B. A., Francki, R. I. B., and Crowley, N. C., 1969, Inactivation of lettuce necrotic yellows by chelating agents, *Virology* 37:396–403.

Ball, L. A., and Wertz, G. W., 1981, VSV RNA synthesis: How can you be positive? *Cell* 26:143–144.

Baltimore, D., Huang, A. S., and Stampfer, M., 1970, Ribonucleic acid synthesis of vesicular stomatitus virus. II. An RNA polymerase in the virion, *Proc. Natl. Acad. Sci. U.S.A.* 66:572–576.

Bassi, M., Barbierie, N., Appiano, A., Conti, M., D'Agostino, G., and Caciagli, P., 1980, Cytochemical and autoradiographic studies on the genome and site(s) of replication of barley yellow striate mosaic virus in barley plants, *J. Submicrosc. Cytol.* 12:201–207.

Bell, J. C., and Prevec, L., 1985, Phosphorylation sites on phosphoprotein NS of vesicular stomatitis virus, *J. Virol.* 54:697–702.

Bell, C. D., Omar, S. A., and Lee, P. E., 1978, Electron microscopic localization of wheat striate mosaic virus in its leafhopper vector, *Endria inimica, Virology* 86:1–9.

Belle-Isle, H. D., and Emerson, S. U., 1982, Use of a hyrid infectivity assay to analyze primary transcription of temperature-sensitive mutants of the New Jersey serotype of vesicular stomatitis virus, *J. Virol.* 43:37–40.

Black, L. M., 1940, Strains of potato yellow dwarf virus, *Am. J. Bot.* 27:386–392.

Black, L. M., 1943a, Genetic variation in the clover leafhopper's ability to transmit potato yellow dwarf virus, *Genetics* 28:200–209.

Black, L. M., 1943b, Some relationships between potato yellow dwarf virus and the clover leafhopper, *Phytopathology* 33:363–371.

Black, L. M., 1953, Loss of vector transmissibility by viruses normally insect transmitted, *Phytopathology (Abstr.)* 43:466.

Black, L. M., 1959, Biological cycles of plant viruses in insect vectors, in: *The Viruses*, Vol. 2 (F. M. Burnet and W. M. Stanley, eds.), pp. 157–185, Academic Press, New York.

Black, L. M., 1969, Insect tissue cultures as tools in plant virus research, *Annu. Rev. Phytopathol.* 7:73–100.

Black, L. M., 1970, Potato yellow dwarf virus, *CMI/AAB Descriptions of Plant Viruses*, No. 35.

Black, L. M., 1979, Vector cell monolayers and plant viruses, *Adv. Virus Res.* 25:192–271.

Blumberg, B. M., Leppert, M., and Kolakofsky, D., 1981, Interaction of VSV leader RNA and nucleocapsid protein may control VSV genome replication, *Cell* 23:837–845.

Blumberg, B. M., Giorgi, C., and Kolakofsky, D., 1983, N protein of vesicular stomatitis virus selectively encapsidates leader RNA *in vitro, Cell* 32:559–567.

Boakye, D. B., and Randles, J. W., 1974, Epidemiology of lettuce necrotic yellows virus in South Australia. III. Virus transmission promoters and vector feeding behaviors on host and non-host plants, *Aust. J. Agric. Res.* 25:791–802.

Bradfute, O. E., and Tsai, J. H., 1983, Identification of maize mosaic virus in Florida, *Plant Dis.* 67:1339–1342.

Brakke, M. K., 1951, Density gradient centrifugation: A new separation technique, *J. Am. Chem. Soc.* 73:1847–1848.

Brakke, M. K., 1955, Zone electrophoresis of dyes, proteins and viruses in density gradient columns of sucrose solutions, *Arch. Biochem. Biophys.* 55:175–190.

Brakke, M. K., 1956, Stability of potato yellow dwarf virus, *Virology* 2:463–476.

Brewbaker, J. L., 1981, Resistance to maize mosaic virus, in: *Virus and Viruslike Diseases of Maize in the United States* (D. T. Gordon, J. K. Knoke, and G. E. Scott, eds.), pp. 145–151, *South. Coop. Ser. Bull. 247.*

Brown, F., and Crick, J., 1979, Natural history of the rhabdoviruses of vertebrates and invertebrates, in: *Rhabdoviruses*, Vol. I (D. H. L. Bishop, ed.), pp. 1–22, CRC Press, Boca Raton, Florida.

Campbell, R. N., and Lin, M. T., 1972, Broccoli necrotic yellows virus, *CMI/AAB Descriptions of Plant Viruses*, No. 85.

Caner, J., July, J. R., and Vincente, M., 1976, Caracteristicas de um rhabdovirus isolads de plantas de ervilha, *Summa Phytopathol.* 2:264–269.

Castellano, M. A., and Rana, G. L., 1981, Transmission and ultrastructure of ivy vein clearing virus infections, *Phytopathol. Mediterr.* **20**:199–205.

Chagas, C. M., 1980, Morphology and intracellular behaviour of coffee ringspot virus (CRV) in tissues of coffee (*Coffee arabica* L.), *Phytopathol. Z.* **99**:301–309.

Chambers, T. C., Crowley, N. C., and Francki, R. I. B., 1965, Localization of lettuce necrotic yellows virus in host leaf tissue, *Virology* **27**:320–328.

Chen, M. J., and Shikata, E., 1968, Electron microscopy of virus-like particles associated with transitory yellowing virus-infected rice plants in Taiwan, *Plant Prot. Bull. (Taiwan)* **10**:19–28.

Chen, M. J., and Shikata, E., 1971, Morphology and intracellular localization of rice transitory yellowing virus, *Virology* **46**:786–796.

Chen, M. J., and Shikata, E., 1972, Electron microscopy and recovery of rice transitory yellowing virus from its leafhopper vector, *Nephotettix cincticeps, Virology* **47**:483–486.

Chiu, R.-J., and Jean, J.-H., 1969, Leafhopper transmission of transitory yellowing of rice, in: *The Virus Diseases of the Rice Plant, Proceedings of International Symposium of the Rice Research Institute, 1967* (R.-J. Chen, ed.), pp. 131–137, Johns Hopkins Press, Baltimore.

Chiu, R.-J., Lo, T.-C., Pi, C.-L., and Chen, M.-H., 1965, Transitory yellowing of rice and its transmission by the leafhopper, *Nephotettix apicalis* (Motsch.), *Bot. Bull. Acad. Sinica* **6**:1–18.

Chiu, R.-J., Jean, J.-H., Chen, M.-H., and Lo, T.-C., 1968, Transmission of transitory yellowing virus of rice by two leafhoppers, *Phytopathology* **58**:740–747.

Chiu, R.-J., Liu, H.-Y., MacLeod, R., and Black, L. M., 1970, Potato yellow dwarf virus in leafhopper cell culture, *Virology* **40**:387–396.

Christie, S. R., and Hall, D. W., 1979, A new hybrid species of *Nicotiana* (Solanaceae), *Baileya* **20**:133–136.

Christie, S. R., Christie, R. G., and Edwardson, J. R., 1974, Transmission of a bacilliform virus of sowthistle and *Bidens pilosa, Phytopathology* **64**:840–845.

Chu, P. W. G., and Francki, R. I. B., 1982, Detection of lettuce necrotic yellows virus by an enzyme-linked immunosorbent assay in plant host and the insect vector, *Ann. Appl. Biol.* **100**:149–156.

Clinton, G. M., and Huang, A. S., 1981, Distribution of phosphoserine, phosphothreonine and phosphotyrosine in proteins of vesicular stomatitis virus, *Virology* **108**:510–514.

Colonno, R. J., and Banerjee, A. K., 1976, A unique RNA species involved in initiation of vesicular stomatitis virus RNA transcription *in vitro, Cell* **8**:197–204.

Compans, R. W., and Klenk, H. D., 1979, Viral membranes, in: *Comprehensive Virology,* Vol. 13 (H. Fraenkel-Conrat and R. R. Wagner, eds.), pp. 293–407, Plenum Press, New York.

Conti, M., and Appiano, A., 1973, Barley yellow striate mosaic virus and associated viroplasm in barley cells, *J. Gen. Virol.* **21**:315–322.

Conti, M., and Plumb, R. T., 1977, Barley yellow striate mosaic virus in the salivary glands of its planthopper vector, *Laodelphax striatellus* Fallen, *J. Gen. Virol.* **34**:107–114.

Coslett, G. D., Holloway, B. P., and Obijeski, J. F., 1980, The structural proteins of rabies virus and evidence for their synthesis from separate monocistronic RNA species, *J. Gen. Virol.* **49**:161–180.

Cox, J. H., Weiland, F., Dietzschold, B., and Schneider, L. G., 1981, Reevaluation of the structural proteins M1 and M2 of rabies virus, in: *The Replication of Negative Strand Viruses* (D. H. L. Bishop and R. W. Compans, eds.), pp. 639–645, Elsevier/North-Holland, New York.

Credi, R., Giunchedi, L., and Bertaccini, A., 1982, Sowthistle yellow vein virus in Italy, *Phytophathol. Mediterr.* **21**:23–26.

Crowley, N. C., 1967, Factors affecting the local lesion response of *Nicotiana glutinosa* to lettuce necrotic yellows virus, *Virology* **31**:107–113.

Dale, J. L., and Peters, D., 1981, Protein composition of the virions of five plant rhabdoviruses, *Intervirology* **16**:86–94.

David, A. E., 1977, Assembly of the vesicular stomatitis virus envelope: Transfer of viral polypeptides from polysomes to cellular membranes, *Virology* **76**:98–108.

De, B. P., and Banerjee, A. K., 1984, Specific interactions of vesicular stomatitis virus L and NS proteins with heterologous genome ribonucleoprotein template lead to mRNA synthesis *in vitro*, *J. Virol.* **51**:628–634.

De, B. P., and Banerjee, A. K., 1985, Requirements and functions of vesicular stomatitis virus L and NS proteins in the transcription process *in vitro*, *Biochem. Biophys. Res. Commun.* **126**:40–49.

De Block, M., Herrera-Estrella, L., Van Montagu, M., Schell, J., and Zambryski, P., 1984, Expression of foreign genes in regenerated plants and in their progeny, *Eur. Mol. Biol. Org. J.* **3**:1681–1689.

Dietzschold, B., Cox, J. H., and Schneider, L. G., 1979, Rabies virus strains: A comparison study by polypeptide analysis of vaccine strains with different pathogenic patterns, *Virology* **98**:63–75.

Di Franco, A., Russo, M., and Martelli, G. P., 1979, Isolation and some properties of pelargonium vein clearing virus, *Phytopathol. Mediterr.* **18**:41–47.

Diringer, H., Kulas, H. P., Schneider, L. G., and Schlumberger, H. D., 1973, The lipid composition of rabies virus, *Z. Naturforsch* **28c**:90–93.

Doi, Y., Chang, M. U., and Yora, K., 1977, Orchid fleck virus, *CMI/AAB Descriptions of Plant Viruses*, No. 183.

Duffus, J. E., 1963, Possible multiplication in the aphid vector of sowthistle yellow vein virus, a virus with an extremely long insect latent period, *Virology* **21**:194–202.

Eisbein, K., 1976, Untersuchungen zum elektronenmikroskopischen Nachweis des Rubenkrausel-Virus (Beta virus 3) in *Beta vulgaris* L. und *Piesma quadratum Fieb.*, *Arch. Phytopathol. Pflanzenschutz* **12**:299–313.

El Maataoui, M., Lockhart, B. E. L., and Lesemann, D.-E. 1985, Biological, serological, and cytopathological properties of tomato vein-yellowing virus, a rhabdovirus occurring in tomato in Morocco, *Phytopathology* **75**:109–115.

Etchison, J. R., and Summers, D. F., 1979, Structure, synthesis, and function of the vesicular stomatitis virus glycoprotein, in: *Rhabdoviruses*, Vol. I (D. H. L. Bishop, ed.), pp. 151–160, CRC Press, Boca Raton, Florida.

Falk, B. W., and Tsai, J. H., 1983, Physicochemical characterization of maize mosaic virus, *Phytopathology* **73**:1536–1539.

Falk, B. W., and Weathers, L. G., 1983, Comparison of potato yellow dwarf virus serotypes, *Phytopathology* **73**:81–85.

Falk, B. W., Weathers, L. G., and Greer, F. C., 1981, Identification of potato yellow dwarf virus occurring naturally in California, *Plant Dis.* **65**:81–83.

Feldman, J. M., Vega, J., and Gracia, O., 1978, Studies of weed plants as sources of virus. VI. Mixed infection of broccoli necrotic yellows and cauliflower mosaic viruses on *Sisymbrium irio* in Argentina, *Phytopathol. Z.* **93**:187–190.

Fenner, F. J., and Ratcliffe, F. N., 1965, *Myxomatosis*, Cambridge University Press, London.

Flamand, A., Delagneau, J. R., and Bussereau, F., 1978, An RNA polymerase activity in purified rabies virions, *J. Gen. Virol.* **40**:233–238.

Flore, P. H., and Peters, D., 1981, *In vitro* transcription of sonchus yellow net virus RNA, *Abstracts of the Fifth International Congress of Virology*, Strasbourg, France, p. 414.

Francki, R. I. B., 1972, Purification of viruses, in: *Principles and Techniques in Plant Virology* (C. I. Kado and H. O. Agrawal, eds.), pp. 295–335, Van Nostrand Reinhold, New York.

Francki, R. I. B., 1973, Plant rhabdoviruses, *Adv. Virus Res.* **18**:257–345.

Francki, R. I. B., 1984, Viral diseases of plants of economic importance and their control, in: *Control of Virus Diseases* (E. Kurstak, ed.), pp. 239–264, Marcel Dekker, New York.

Francki, R. I. B., and Peters, D., 1978, The interference of cytoplasmic membrane-bound material from plant cells with the detection of a plant rhabdovirus transcriptase, *J. Gen. Virol.* **41**:467–478.

Francki, R. I. B., and Randles, J. W., 1970, Lettuce necrotic yellows virus, *CMI/AAB Descriptions of Plant Viruses*, No. 26.

Francki, R. I. B., and Randles, J. W., 1972, RNA-dependent RNA polymerase associated with particles of lettuce necrotic yellows virus, *Virology* **47**:270–275.

Francki, R. I. B., and Randles, J. W., 1973, Some properties of lettuce necrotic yellows virus RNA and its *in vitro* transcription by virion-associated transcriptase, *Virology* **54**:359–368.

Francki, R. I. B., and Randles, J. W., 1975, Composition of the plant rhabdovirus lettuce necrotic yellows virus in relation to its biological properties, in: *Negative Strand Viruses* (B. W. J. Mahy and R. D. Barry, eds.), pp. 223–242, Academic Press, London.

Francki, R. I. B., and Randles, J. W., 1980, Rhabdoviruses infecting plants, in: *Rhabdoviruses*, Vol. III (D. H. L. Bishop, ed.), pp. 135–165, CRC Press, Boca Raton, Florida.

Francki, R. I. B., Kitajima, E. W., and Peters, D., 1981, Rhabdoviruses, in: *Handbook of Plant Virus Infections and Comparative Diagnosis* (E. Kurstak, ed.), pp. 455–489, Elsevier/North-Holland, Amsterdam.

Francki, R. I. B., Milne, R. G., and Hatta, T., 1985, Plant Rhabdoviridae, in: *Atlas of Plant Viruses*, Vol. I, pp. 73–100, CRC Press, Boca Raton, Florida.

Franke, W. W., 1974, Structure, biochemistry and functions of the nuclear envelope, in: *Aspects of Nuclear Structure and Function* (G. H. Bourne, J. F. Danielli, and K. W. Leon, eds.), *International Review of Cytology*, Supplement 4, pp. 71–236, Academic Press, London.

Frazier, N. W., 1968, Transmission of strawberry crinkle virus by the dark strawberry aphid *Chaetosiphon jacobi, Phytopathology* **58**:165–172.

Fry, P. R., Close, R. C., Procter, C. H., and Sunde, R., 1973, Lettuce necrotic yellows virus in New Zealand, *N. Z. J. Agric. Res.* **16**:143–146.

Gallione, C. J., Greene, J. R., Iverson, L. E., and Rose, J. K., 1981, Nucleotide sequences of the mRNAs encoding the vesicular stomatitis virus N and NS proteins, *J. Virol.* **39**:529–535.

Garrett, R. G., and O'Loughlin, G. T., 1977, Broccoli necrotic yellows virus in cauliflower and in the aphid, *Brevicoryne brassicae* L., *Virology* **76**:653–663.

Giorgi, C., Blumberg, B., and Kolakofsky, D., 1983, Sequence determination of the (+) leader RNA regions of the vesicular stomatitis virus Chandipura, Cocal, and Piry serotype genomes, *J. Virol.* **46**:125–130.

Gollifer, D. E., Jackson, G. V. H., Dabek, A. J., Plumb, R. T., and May, Y. Y., 1977, The occurrence and transmission of viruses of edible aroids in the Solomon Islands and the South West Pacific, *Pest Artic. News Summ.* **23**:171–177.

Greber, R. S., 1977, A severe stunting virus disease of maize in Queensland, *Aust. Plant Pathol. Soc. Newslett.* **6**:18.

Greber, R. S., 1979a, Digitaria striate virus—a rhabdovirus of grasses transmitted by *Sogatella kolophon* (Kirk.), *Aust. J. Agric. Res.* **30**:43–51.

Greber, R. S., 1979b, Cereal chlorotic mottle virus—a rhabdovirus of Gramineae in Australia transmitted by *Nesoclutha pallida* (Evans), *Aust. J. Agric. Res.* **30**:433–443.

Greber, R. S., 1982a, Maize sterile stunt—a delphacid transmitted rhabdovirus disease affecting some maize genotypes in Australia, *Aust. J. Agric. Res.* **33**:13–23.

Greber, R. S., 1982b, Cereal chlorotic mottle virus, *CMI/AAB Descriptions of Plant Viruses*, No. 251.

Greber, R. S., 1984, Relationships of rhabdoviruses reported on maize in Australia, *Maize Virus Dis. Newslett.* **1**:46–48.

Greber, R. S., and Gowanlock, D. H., 1979, Cereal chlorotic mottle virus purification, serology and electron microscopy in plant and insect tissues, *Aust. J. Biol. Sci.* **32**:399–408.

Hackett, A. J., Sylvester, E. S., Richardson, J., and Wood, P., 1968, Comparative electron micrographs of sowthistle yellow vein and vesicular stomatitis viruses, *Virology* **36**:693–696.

Harmison, G. G., Meier, E., and Schubert, M., 1984, The polymerase gene of VSV, in: *Nonsegmented Negative Strand Viruses* (D. H. L. Bishop and R. W. Compans, eds.), pp. 35–40, Academic Press, New York.

Harris, K. F., 1979, Leafhoppers and aphids as biological vectors: Vector–virus relationships, in: *Leafhopper Vectors and Plant Disease Agents* (K. Maramorosch and K. F. Harris, eds.), pp. 217–308, Academic Press, New York.

Harrison, B. D., and Crowley, N. C., 1965, Properties and structure of lettuce necrotic yellows virus, *Virology* **26**:297–310.

Herold, F., 1972, Maize mosaic virus, *CMI/AAB Descriptions of Plant Viruses*, No. 94.

Herold, F., and Munz, K., 1965, Electron microscopic demonstration of virus-like particles in *Peregrinus maidis* following acquisition of maize mosaic virus, *Virology* **25**:412–417.

Herold, F., Bergold, G. H., and Weibel, J., 1960, Isolation and electron microscopic demonstration of a virus infecting corn (*Zea mays* L.), *Virology* **12**:335–347.

Heyward, J. T., Holloway, B. P., Cohen, P. and Obijeski, J. F., 1979, Rhabdovirus nucleocapsid, in: *Rhabdoviruses*, Vol. I (D. H. L. Bishop, ed.), pp. 137–149, CRC Press, Boca Raton, Florida.

Holland, J., Spindler, K., Horodyski, F., Grabau, E., Nichol, S., and Van de Pol, S., 1982, Rapid evolution of RNA genomes, *Science* **215**:1577–1585.

Hsieh, S. P. Y., Chiu, R. J., and Chen, C. C., 1970, Transmission of rice transitory yellowing virus by *Nephotettix impicticeps*, *Phytopathology (Abstr.)* **60**:1534.

Hsu, C.-H., Morgan, E. M., and Kingsbury, D. W., 1982, Site-specific phosphorylation regulates the transcriptase activity of vesicular stomatitis virus, *J. Virol.* **43**:104–112.

Hsu, H. T., and Black, L. M., 1973a, Inoculation of vector cell monolayers with potato yellow dwarf virus, *Virology* **52**:187–198.

Hsu, H. T., and Black, L. M., 1973b, Comparative efficiencies of assays of a plant virus by lesions on leaves and on vector cell monolayers, *Virology* **52**:284–286.

Hsu, H. T., and Black, L. M., 1973c, Polyethylene glycol for purification of potato yellow dwarf virus, *Phytopathology* **63**:692–696.

Hull, R., 1970, Large RNA plant-infecting viruses, in: *The Biology of Large RNA Viruses* (R. D. Barry and B. W. J. Mahy, eds.), pp. 153–164, Academic Press, London.

Hull, R., 1976, The structure of tubular viruses, *Adv. Virus Res.* **20**:1–32.

Hunt, D. M., Emerson, S. U., and Wagner, R. R., 1976, RNA-temperature sensitive mutants of vesicular stomatitis virus: L protein thermosensitivity accounts for transcriptase restriction of group I mutants, *J. Virol.* **18**:596–603.

Hunt, D. M., Mellon, M. G., and Emerson, S. U., 1979, Viral transcriptase, in: *Rhabdoviruses*, Vol. I (D. H. L. Bishop, ed.), pp. 169–183, CRC Press, Boca Raton, Florida.

Izadpanah, K., Ahmadi, A. A., Parvin, S., and Jafari, S. A., 1983, Transmission, particle size and additional hosts of the rhabdovirus causing maize mosaic in Shiraz, Iran, *Phytopathol. Z.* **107**:283–288.

Jackson, A. O., 1978, Partial characterization of the structural proteins of sonchus yellow net virus, *Virology* **87**:172–181.

Jackson, A. O., and Christie, S. R., 1977, Purification and some physiochemical properties of sonchus yellow net virus, *Virology* **77**:344–355.

Jackson, A. O., and Christie, S. R., 1979, Sonchus yellow net virus, *CMI/AAB Descriptions of Plant Viruses*, No. 205.

Jackson, A. O., Milbrath, G. M., and Jedlinski, H., 1981, Rhabdovirus diseases of the Gramineae, in: *Virus and Viruslike Diseases of Maize in the United States* (D. T. Gordon, J. K. Knoke, and G. E. Scott, eds.), pp. 51–76, *South. Coop. Ser. Bull.* 247.

James, M., Kenten, R. H., and Woods, R. D., 1973, Virus-like particles associated with two diseases of *Colocasia esculenta* (L.) Schott in the Solomon Islands, *J. Gen. Virol.* **21**:145–153.

Jedlinski, H., 1976, Oat striate, a new virus disease in Illinois spread by the leafhopper, *Graminella nigrifrons* (Forbes), *Proc. Am. Phytopathol. Soc.* **3**:208.

Jones, A. T., Murant, A. F., and Stace-Smith, R., 1977, Raspberry vein chlorosis virus, *CMI/AAB Descriptions of Plant Viruses*, No. 174.

Kawai, A., 1977, Transcriptase activity associated with rabies virion, *J. Virol.* **24**:826–835.

Keene, J. D., Kurilla, M. G., Wilusz, J., and Chambers, J. C., 1984, Interactions between cellular La protein and leader RNAs, in: *Nonsegmented Negative Strand Viruses* (D. H. L. Bishop and R. W. Compans, eds.), pp. 103–108, Academic Press, New York.

Khan, A., Rilfors, L., Wieslander, A., and Lindblom, G., 1981, The effect of cholesterol on the phase structure of glucolipids from *Acholeplasma laidlawii* membranes, *Eur. J. Biochem.* **116**: 215–220.

Kingsbury, D. W., Hsu, C.-H., and Morgan, E. M., 1981, A role for NS-protein phosphory-lation in vesicular stomatitis virus transcription, in: *The Replication of Negative Strand Viruses* (D. H. L. Bishop and R. W. Compans, eds.), pp. 821–827, Elsevier/North-Holland, New York.

Kingsford, L., and Emerson, S. U., 1980, Transcriptional activities of different phosphory-lated species of NS protein purified from vesicular stomatitis virions and cytoplasm of infected cells, *J. Virol.* **33:**1097–1105.

Kitajima, E. W., and Costa, A. S., 1966, Morphology and developmental stages of *Gomphrena* virus, *Virology* **29:**523–539.

Kitajima, E. W., and Costa, A. S., 1979, Rhabdovirus-like particles in tissues of five different plant species, *Fitopatol. Brasil.* **4:**55–62.

Kitajima, E. W., Lauritis, J. A., and Swift, H., 1969, Morphology and intracellular localization of a bacilliform latent virus in sweet clover, *J. Ultrastruct. Res.* **29:**141–150.

Kiuchi, A., and Roy, P., 1984, Comparison of the primary sequence of spring viremia of carp virus M protein with that of vesicular stomatitis virus, *Virology* **134:**238–243.

Knudson, D. L., 1973, Rhabdoviruses, *J. Gen. Virol.* **20:**105–130.

Knudson, D. L. and MacLeod, R., 1972, The proteins of potato yellow dwarf virus, *Virology* **47:**285–289.

Kurath, G., and Leong, J. C., 1985, Characterization of infectious hematopoietic necrosis virus mRNA species reveals a nonvirion rhabdovirus protein, *J. Virol.* **53:**462–468.

Kurilla, M. G., and Keene, J. D., 1983, The leader RNA of vesicular stomatitis virus is bound by a cellular protein reactive with anti-La lupus antibodies, *Cell* **34:**837–845.

Kurilla, M. G., Piwnica-Worms, H., and Keene, J. D., 1982, Rapid and transient localization of the leader RNA of vesicular stomatitis virus in the nuclei of infected cells, *Proc. Natl. Acad. Sci. U.S.A.* **79:**5240–5244.

Kurilla, M. G., Cabradilla, C. D., Holloway, B. P., and Keene, J. D., 1984, Nucleotide sequence and host La protein interactions of rabies virus leader RNA, *J. Virol.* **50:**773–778.

Lastra, R. J., 1977, Maize mosaic and other maize virus and virus-like diseases in Venezuela, in: *Proceedings of the International Maize Virus Disease, Colloquium and Workshop* (1976) (L. E. Williams, D. T. Gordon, and L. R. Nault., eds.), pp. 30–39, Ohio Agricultural Research and Development Center, Wooster.

Lazzarini, R. A., Keene, J. D., and Schubert, M., 1981, The origins of defective interfering particles of the negative-strand RNA viruses, *Cell* **26:**145–154.

Leclant, F., Alliot, B., and Signoret, P. A., 1973, Transmission et epidemiologie de la maladie à enations de la Luzerne (LEV): Premiers resultats, *Ann. Phytopathol.* **5:**441–445.

Lee, P. E., 1967, Morphology of wheat striate mosaic virus and its localization in infected cells, *Virology* **33:**84–94.

Lee, P. E., 1970, Developmental stages of wheat striate mosaic virus, *J. Ultrastruct. Res.* **31:**282–290.

Lesnaw, J. A., Dickson, L. R., and Curry, R. H., 1979, Proposed replicative role of the NS polypeptide of vesicular stomatitis virus: Structural analysis of an electrophoretic var-iant, *J. Virol.* **31:**8–16.

Lin, M. T., and Campbell, R. N., 1972, Characterization of broccoli necrotic yellows virus, *Virology* **48:**30–40.

Lockhart, B. E. L., and Elyamani, M., 1983, Virus and viruslike diseases of Maize in Morocco, in: *Proceedings of the International Maize Virus Disease, Colloquium and Workshop* (1982) (D. T. Gordon, J. K. Knoke, L. R. Nault, and R. M. Ritter, eds.), pp. 127–129, Ohio Agricultural Research and Development Center, Wooster.

Luisoni, E., 1969, Partial purification of barley yellow striate mosaic virus, *Ric. Sci.* **39:** 708–713.

Lundsgaard, T., 1984, Comparison of festuca leaf streak virus antigens with those of three other rhabdoviruses infecting the Gramineae, *Intervirology* **22:**50–55.

Lundsgaard, T., and Albrechtsen, S. E., 1976, Electron microscopy of rhabdovirus-like par-ticles in *Festuca gigantea* with leaf streak mosaic, *Phytopathol. Z.* **87:**12–16.

Lundsgaard, T., and Albrechtsen, S. E., 1979, Ultrastructure of *Festuca gigantea* with rhab-dovirus-like particles, *Phytopathol. Z.* **94:**112–118.

Lundsgaard, T., Tien, P., and Toriyama, S., 1984, The antigens of wheat rosette stunt and northern cereal mosaic viruses are related, *Phytopathol. Z.* **111**:232–235.

MacLeod, R., 1967, The preparation of plant viruses for use as antigens, in: *Methods in Immunology and Immunochemistry*, Vol. 1 (C. A. Williams and M. W. Chase, eds.), pp. 102–115, Academic Press, New York.

Maramorosch, K., Govindu, H. C., and Kondo, F., 1977, Rhabdovirus particles associated with a mosaic disease of naturally infected *Eleusine coracana* (finger millet) in Karnataka State (Mysore), South India, *Plant Dis. Rep.* **61**:1029–1031.

Marks, M. D., Kennedy-Morrow, J., and Lesnaw, J. A., 1985, Assignment of the temperature-sensitive lesion in the replication mutant A1 of vesicular stomatitis virus to the N gene, *J. Virol.* **53**:44–51.

Martelli, G. P., and Rana, G. L., 1970, Transmissione meccanica dell'agente del nanismo maculato della Melanzana (EMDV), *Phytopathol. Mediterr.* **9**:187–191.

Martelli, G. P., and Russo, M., 1973, Eggplant mottled dwarf virus, *CMI/AAB Descriptions of Plant Viruses*, No. 115.

Martelli, G. P., and Russo, M., 1977, Rhabdoviruses of plants, in: *Insect and Plant Viruses: An Atlas* (K. Maramorosch, ed.), pp. 181–214, Academic Press, New York.

Martelli, G. P., Russo, M., and Malaguti, G., 1975, Ultrastructural aspects of maize mosaic virus in the host cells, *Phytopathol. Mediterr.* **14**:140–142.

Matthews, R. E. F., 1970, *Plant Virology*, Academic Press, New York.

Matthews, R. E. F., 1982, Classification and nomenclature of viruses—Fourth Report of the International Committee on Taxonomy of Viruses, *Intervirology* **17**:1–199.

Mayhew, D. E., and Flock, R. A., 1981, Sorghum stunt mosaic, *Plant Dis.* **65**:84–86.

McDaniel, L. L., Ammar, E.-D., and Gordon, D. T., 1985, Assembly, morphology, and accumulation of a Hawaiian isolate of maize mosaic virus in maize, *Phytopathology* **75**:1167–1172.

McGeoch, D. J., Dolan, A., and Pringle, C. R., 1980, Comparisons of nucleotide sequences in the genomes of the New Jersey and Indiana serotypes of vesicular stomatitis virus, *J. Virol.* **33**:69–77.

McGowan, J. J., Emerson, S. U., and Wagner, R. R., 1982, The plus-strand leader RNA of VSV inhibits DNA-dependent transcription of adenovirus and SV40 genes in soluble whole-cell extract, *Cell* **28**:325–333.

McLean, G. D., and Francki, R. I. B., 1967, Purification of lettuce necrotic yellows virus by column chromatography on calcium phosphate gel, *Virology* **31**:585–591.

McLean, G. D., Wolanski, B. S., and Francki, R. I. B., 1971, Serological analysis of lettuce necrotic yellows virus preparations by immunodiffusion, *Virology* **43**:480–487.

McSharry, J. J., 1979, Viral membrane protein structure and function, in: *Rhabdoviruses* Vol. I (D. H. L. Bishop, ed.), pp. 161–168, CRC Press, Boca Raton, Florida.

McSharry, J. J., and Wagner, R. R., 1971, Lipid composition of purified vesicular stomatitis viruses, *J. Virol.* **7**:59–70.

Milne, R. G., Masenga, V., and Conti, M., 1986, Serological relationships between the nucleocapsids of some planthopper-borne rhabdoviruses of cereals, *Intervirology* **25**:83–87.

Milner, J. J., and Jackson, A. O., 1983, Characterization of viral-complementary RNA associated with polyribosomes from tobacco infected with sonchus yellow net virus, *J. Gen. Virol.* **64**:2479–2483.

Milner, J. J., and Jackson, A. O., 1979, Sequence complementarity of sonchus yellow net virus RNA with RNA isolated from the polysomes of infected tobacco, *Virology* **97**:90–99.

Milner, J. J., and Jackson, A. O., 1983, Characterization of viral-complementary RNA associated with polyribosomes from tobacco infected with sonchus yellow net virus, *J. Gen. Virol.* **64**:2479–2483.

Milner, J. J., Hakkaart, M. J. J., and Jackson, A. O., 1979, Subcellular distribution of RNA sequences complementary to sonchus yellow net virus RNA, *Virology* **98**:497–501.

Morgan, E. M., and Kingsbury, D. W., 1981, Association of the transcriptase and RNA methyltransferase activities of vesicular stomatitis virus with the L-protein, in: *The Replication of Negative Strand Viruses* (D. H. L. Bishop and R. W. Compans, eds.), pp. 817–820, Elsevier/North-Holland, New York.

Nuss, D. L., 1984, Molecular biology of wound tumor virus, *Adv. Virus Res.* **29:**57–93.

Ohki, S. T., Doi, Y., and Yora, K., 1978, Carrot latent virus: A new rhabdovirus of carrot, *Ann. Phytopathol. Soc. Jpn.* **44:**202–204.

O'Loughlin, G. T., and Chambers, T. C., 1967, The systemic infection of an aphid by a plant virus, *Virology* **33:**262–271.

Orenstein, J., Johnson, L., Shelton, E., and Lazzarini, R. A., 1976, The shape of vesicular stomatitis virus, *Virology* **71:**291–301.

Patton, J., Davis, N. L., and Wertz, G. W., 1984a, N protein alone satisfies the requirement for protein synthesis during RNA replication of vesicular stomatitis virus, *J. Virol.* **49:**303–309.

Patton, J. T., Davis, N. L., and Wertz, G. W., 1984b, Role of vesicular stomatitis virus proteins in RNA replication, in: *Nonsegmented Negative Strand Viruses* (D. H. L. Bishop and R. W. Compans, eds.), pp. 147–152, Academic Press, New York.

Patzer, E. J., Moore, N. F., Barenholz, Y., Shaw, J. M., and Wagner, R. R., 1978, Lipid organization of the membrane of vesicular stomatitis virus, *J. Biol. Chem.* **253:** 4544–4550.

Pena-Inglesias, A., Rubio-Huertos, M., and Morena San Martin, R., 1972, *Anales del I.N.I.A. (Spain), Serie: Proteccion Vegetal.,* No. 2, pp. 123–137.

Perrault, J., Clinton, G. M., and McClure, M. A., 1983, RNP template of vesicular stomatitis virus regulates transcription and replication functions, *Cell* **35:**175–185.

Peters, D., 1971, Sowthistle yellow vein virus, *CMI/AAB Descriptions of Plant Viruses*, No. 62.

Peters, D., 1981, Plant rhabdovirus group, *CMI/AAB Descriptions of Plant Viruses*, No. 244.

Peters, D., and Black, L. M., 1970, Infection of primary cultures of aphid cells with a plant virus, *Virology* **40:**847–853.

Peters, D., and Kitajima, E. W., 1970, Purification and electron microscopy of sowthistle yellow vein virus, *Virology* **41:**135–150.

Peters, D., and Schultz, M. G., 1975, A model of rhabdovirus morphogenesis, *Proc. K. Ned. Akad. Wet. Ser. C* **78:**172–181.

Peters, D., Toriyama, S., and Terlouw, L., 1978, Composition and polymerase activity of some plant rhabdoviruses, *Abstracts of the Fourth International Congress of Virology*, The Hague, The Netherlands, p. 405.

Petri, W. A., Jr., and Wagner, R. R., 1979, Reconstitution into liposomes of the glycoprotein of vesicular stomatitis virus by detergent dialyses, *J. Biol. Chem.* **254:**4313–4316.

Philipp, E.-I., Franke, W. W., Keenan, T. W., Stadler, J., and Jarsch, E.-D., 1976, Characterization of nuclear membranes and endoplasmic reticulum isolated from plant tissue, *J. Cell Biol.* **68:**11–29.

Polak, Z., Kralik, O., and Limberk, J., 1977, Rhabdovirus-like particles associated with cow-parsnip mosaic, *Acta Phytopathol. Acad. Sci. Hung.* **12:**157–163.

Proeseler, G., 1983, Beet leaf curl virus, *CMI/AAB Descriptions of Plant Viruses*, No. 168.

Rana, G. L., and Di Franco, A., 1979, Mechanical transmission of *Pittosporum* vein clearing virus, *Phytopathol. Mediterr.* **18:**48–56.

Randles, J. W., 1983, Transmission and epidemiology of lettuce necrotic yellows virus, in: *Current Topics in Vector Research*, Vol. 1 (K. F. Harris, ed.), pp. 169–187, Praeger Publishers, New York.

Randles, J. W., and Carver, M., 1971, Epidemiology and lettuce necrotic yellows virus in south Australia. II. Distribution of virus, host plants, and vectors, *Aust. J. Agric. Res.* **22:**231–237.

Randles, J. W., and Coleman, D. F., 1970, Loss of ribosomes in *Nicotiana glutinosa* L. infected with lettuce necrotic yellows virus, *Virology* **41:**459–464.

Randles, J. W., and Coleman, D. F., 1972, Changes in polysomes in *Nicotiana glutinosa* L. leaves infected with lettuce necrotic yellows virus, *Physiol. Plant Pathol.* **2:**247–258.

Randles, J. W., and Francki, R. I. B., 1972, Infectious nucleocapsid of lettuce necrotic yellows virus with RNA-dependent RNA polymerase activity, *Virology* **50:**297–300.

Razvjaskina, G. M., and Poljakova, G. P., 1967, Electron microscopy study of wheat mosaic virus transmitted by the cicada, *Psammottetix striatus* L., *Dokl. Akad. Nauk. SSR* **174:**1435–1436.

Reeder, G. S., Knudson, D. L., and MacLeod, R., 1972, The ribonucleic acid of potato yellow dwarf virus, *Virology* **50:**301–304.

Repik, P., 1979, Adsorption, penetration, uncoating, and *in vivo* mRNA transcription process, in: *Rhabdoviruses*, Vol. II (D. H. L. Bishop, ed.), pp. 1–33, CRC Press, Boca Raton, Florida.

Rezaian, M. A., Heaton, L. A., Pedersen, K., Milner, J. J., and Jackson, A. O., 1983, Size and complexity of polyadenylated RNAs induced in tobacco infected with sonchus yellow net virus, *Virology* **131:**221–229.

Richardson, J., and Sylvester, E., 1968, Further evidence of multiplication of sowthistle yellow vein virus in its aphid vector *Hyperomyzes lactucae, Virology* **35:**347–355.

Richardson, J., Frazier, N. W., and Sylvester, E., 1972, Rhabdovirus-like particles associated with strawberry crinkle virus, *Phytopathology* **62:**491–492.

Riesterer, C., and Adam, G., 1981, Infection of tobacco protoplasts with a plant rhabdovirus, *Abstracts of the Fifth International Congress of Virology*, Strasbourg, France, p. 222.

Rose, J. K., and Gallione, C. J., 1981, Nucleotide sequences of the mRNAs encoding the vesicular stomatitis virus G and M proteins determined from cDNA clones containing the complete coding regions, *J. Virol.* **39:**519–528.

Roy, P., 1981, Nucleotide sequences of the mRNAs encoding the vesicular stomatitis virus G and M proteins determined from cDNA clones containing the complete coding regions, in: *The Replication of Negative Strand Viruses* (D. H. L. Bishop and R. W. Compans, eds.), pp. 623–629, Elsevier/North-Holland, New York.

Roy, P., Gupta, K. C., and Kiuchi, A., 1984, Characterization of spring viremia of carp virus mRNA species and the 3′ sequence of the viral RNA, *Virus* **1:**189–202.

Russo, M., and Martelli, G. P., 1973, A study of the structure of eggplant mottled dwarf virus, *Virology* **52:**39–48.

Russo, M., and Martelli, G. P., 1974, Rhabdovirus parassiti submicroscopici delle piante dalla inconsuèta morfologia, *Ital. Agric.* **111:**86–97.

Russo, M., Martelli, G. P., and Lana, C. L., 1975, A rhabdovirus of cynara in Italy, *Phytopathol. Z.* **83:**223–231.

Saito, Y., Inoue, H., and Satomi, H., 1978, Occurrence of rice transitory yellowing virus in Okinawa, Japan, *Ann. Phytopathol. Soc. Jpn.* **44:**666–669.

Schlesinger, H. R., Wells, H. J., and Hummeler, K., 1973, Comparison of the lipids of intracellular and extracellular rabies viruses, *J. Virol.* **12:**1028–1030.

Schubert, M., Harmison, G. G., and Meyer, E., 1984, Primary structure of the vesicular stomatitis virus polymerase (L) gene: Evidence for a high frequency of mutations, *J. Virol.* **51:**505–514.

Schultz, M. G., and Harrap, K. A., 1976, Structural polypeptides of sowthistle yellow vein virus, *Arch. Virol.* **50:**173–176.

Schultz, M. G., and Peters, D., 1976, Sowthistle yellow vein virus in Europe, *Ann. Phytopathol.* **8:**117–121.

Selstam, E., and Jackson, A. O., 1983, Lipid composition of sonchus yellow net virus, *J. Gen. Virol.* **64:**1607–1613.

Shikata, E., 1972, Rice transitory yellowing virus, *CMI/AB Descriptions of Plant Viruses*, No. 100.

Shikata, E., 1979, Cytopathological changes in leafhopper vectors of plant viruses, in: *Leafhopper Vectors and Plant Disease Agents* (K. Maramorosch and K. F. Harris, eds.), pp. 309–325, Academic Press, New York.

Signoret, P. A., Conti, M., Leclant, F., Alliot, B., and Giannotti, J., 1978, Données nouvelles sur la maladie des striés chlorotiques du ble (wheat chlorotic streak mosaic virus—WCSMV), *Ann. Phytopathol.* **9:**381–385.

Sinha, R. C., 1968, Recent work on leafhopper-transmitted viruses, *Adv. Virus Res.* **13:**181–223.

Sinha, R. C., and Behki, R. M., 1972, American wheat striate mosaic virus, *CMI/AAB Descriptions of Plant Viruses*, No. 99.

Sinha, R. C., and Chiykowski, L. N., 1969, Synthesis, distribution and multiplication sites of wheat striate mosaic virus in a leafhopper vector, *Virology* **38**:679–684.

Sinha, R. C., Sehgal, O. P., and Thottappilly, G., 1975, Effect of temperature on infectivity and some physico-chemical properties of purified wheat striate mosaic virus, *Phytopathol. Z.* **84**:300–306.

Sinha, R. C., Harwalkar, V. R., and Behki, R. M., 1976, Chemical composition and some properties of wheat striate mosaic virus, *Phytopathol. Z.* **87**:314–323.

Sokol, F., Stancek, D., and Koprowski, H., 1971, Structural proteins of rabies virus, *J. Virol.* **7**:241–249.

Stubbs, L. L., and Grogan, R. G., 1963, Necrotic yellows: A newly recognized virus disease of lettuce, *Aust. J. Agric. Res.* **14**:439–459.

Stubbs, L. L., Guy, J. A. D., and Stubbs, K. J., 1963, Control of lettuce necrotic yellows virus disease by the destruction of common sowthistle (*Sonchus oleraceus*), *Aust. J. Exp. Agric. Anim. Husb.* **3**:215–218.

Sylvester, E. S., 1969, Evidence of transovarial passage of the sowthistle yellow vein virus in aphid *Hyperomyzus lactucae*, *Virology* **38**:440–446.

Sylvester, E. S., and Richardson, J., 1969, Additional evidence of multiplication of sowthistle yellow vein virus in an aphid vector—serial passage, *Virology* **37**:26–31.

Sylvester, E. S., and Richardson, J., 1970, Infection of *Hyperomyzus lactucae* by sowthistle vein virus, *Virology* **42**:1023–1042.

Sylvester, E. S., Frazier, N. W., and Richardson, J., 1976, Strawberry crinkle virus, *CMI/AAB Descriptions of Plant Viruses*, No. 163.

Thottappilly, G., and Sinha, R. C., 1973, Serological analysis of wheat striate mosaic virus and its soluble antigen, *Virology* **53**:312–318.

Thottappilly, G., and Sinha, R. C., 1974, Serological analysis of antigens related to wheat striate and mosaic virus, *Endria inimica*, *Acta Virol.* **18**:358–361.

Tien, P., Zhang, Z. Y., Liang, X. X., Shi, C. L., Yang, X. C., Zhao, J. Y., and Zhang, X. H., 1980, Studies on the wheat rosette stunt virus, *Acta Microbiol. Sinica* **20**:289–295.

Tomlinson, J. A., and Webb, M. J. W., 1974, Virus diseases of parsley, *Nat. Veg. Res. Sta. Annu. Rep. 1973*, pp. 100–101.

Tomlinson, J. A., Webb, M. J. W., and Faithfull, E. M., 1972, Studies on broccoli necrotic yellows virus, *Ann. Appl. Biol.* **71**:127–134.

Toriyama, S., 1972, Purification and some properties of northern cereal mosaic virus, *Virus (Tokyo)* **22**:8–18.

Toriyama, S., 1976a, Sterol composition of northern cereal mosaic virus, *Ann. Phytopathol. Soc. Jpn.* **42**:494–496.

Toriyama, S., 1976b, Electron microscopy of developmental stages of northern cereal mosaic virus in wheat plant cells, *Ann. Phytopathol. Soc. Jpn.* **42**:563–577.

Toriyama, S., and Peters, D., 1980, *In vitro* synthesis of RNA by dissociated lettuce necrotic yellows virus particles, *J. Gen. Virol.* **50**:125–134.

Toriyama, S., and Peters, D., 1981, Differentiation between broccoli necrotic yellows virus and lettuce necrotic yellows virus by their transcriptase activities, *J. Gen. Virol.* **56**:59–66.

Trefzger-Stevens, J., and Lee, P. E., 1977, The structural proteins of wheat striate mosaic virus, a plant rhabdovirus, *Virology* **78**:144–149.

Trevathan, L. E., Moore, L. D., and Orcutt, D. M., 1979, Symptom expression and free sterol and fatty acid composition of flue-cured tobacco plants exposed to ozone, *Phytopathology* **69**:582–585.

Tsuhsun, K., and Qiao-xi, Z., 1984, The protein composition of the plant rhabdovirus wheat rosette stunt virus, *Abstracts of the Sixth International Congress of Virology*, Sendai, Japan, p. 327.

Van Beek, N. A. M., Derksen, A. C. G., and Dijkstra, J., 1985a, Polyethylene glycol-mediated infection of cowpea protoplasts with sonchus yellow net virus, *J. Gen. Virol.* **66**: 551–557.

Van Beek, N. A. M., Lohuis, D., Dijkstra, J., and Peters, D., 1985b, Morphogenesis of sonchus yellow net virus in cowpea protoplasts, *J. Ultrastruct. Res.* **90**:294–303.

Van Beek, N. A. M., Lohuis, D., Dijkstra, J., and Peters, D., 1985c, Morphogenesis of festuca leaf streak virus in cowpea protoplasts, *J. Gen. Virol.* **66**:2485–2489.

Vega, J., Gracia, O., Rubio-Huertos, M., and Feldman, J. M., 1976, Transmission of a bacilliform virus of sowthistle: Mitochondria modifications in the infected cells, *Phytopathol. Z.* **85**:7–14.

Wagner, R. R., 1975, Reproduction of rhabdoviruses, in: *Comprehensive Virology*, Vol. 4 (H. Fraenkel-Conrat and R. R. Wagner, eds.), pp. 1–23, Plenum Press, New York.

Wagner, R. R., Prevec, L., Brown, F., Summers, D. F., Sokol, F., and MacLeod, R., 1972, Classification of rhabdovirus proteins: A proposal, *J. Virol.* **10**:1228–1230.

Wagner, R. R., Thomas, J. R., and McGowan, J. J., 1984, Rhabdovirus cytopathology: Effects on cellular macromolecular synthesis, in: *Comprehensive Virology*, Vol. 19 (H. Fraenkel-Conrat and R. R. Wagner, eds.), pp. 223–295, Plenum Press, New York.

Wieslander, A., Christiansson, A., Rilfors, L., and Lindblom, G., 1980, Lipid bilayer stability in membranes: Regulation of lipid composition of *Acholesplasma laidawii* as governed by molecular shape, *Biochemistry* **19**:3650–3655.

Williams, P. M., and Emerson, S. U., 1984, Binding studies of NS1 and NS2 of vesicular stomatitis virus, *Nonsegmented Negative Strand Viruses* (D. H. L. Bishop and R. W. Compans, eds.), pp. 79–85, Academic Press, New York.

Wilusz, J., Youngner, J. S., and Keene, J. D., 1985, Base mutations in the terminal noncoding regions of the genome of vesicular stomatitis virus isolated from persistent infections of L cells, *Virology* **140**:249–256.

Wolanski, B. S., and Chambers, T. C., 1971, The multiplication of lettuce necrotic yellows virus, *Virology* **44**:582–591.

Wolcyrz, S., and Black, L. M., 1956, Serology of potato yellow dwarf virus, *Phytopathology* **46**:32.

Wolcyrz, S., and Black, L. M., 1956, Origins of vectorless strains of potato yellow dwarf virus, *Phytopathology* **47**:38.

Yamada, K., and Shikata, E., 1969, Evidence of multiplication of northern cereal mosaic virus in its insect vectors, *J. Fac. Agric. Hokkaido Univ.* **56**:91–102.

Ziemiecki, A., and Peters, D., 1976a, Selective proteolytic activity associated with purified sowthistle yellow vein virus preparations, *J. Gen. Virol.* **31**:451–454.

Ziemiecki, A., and Peters, D., 1976b, The proteins of sowthistle yellow vein virus: Characterization and location, *J. Gen. Virol.* **32**:369–381.

Zuidema, D., Heaton, L. A., Hanau, R., and Jackson, A. O., 1986, Detection and sequence of plus-strand leader RNA of sonchus yellow net virus, a plant rhabdovirus, *Proc. Natl. Acad. Sci. U.S.A.* **83**: 5019–5023.

CHAPTER 11

The Ecology of Rhabdoviruses That Infect Vertebrates

ROBERT E. SHOPE AND ROBERT B. TESH

I. INTRODUCTION

Rhabdoviruses infect vertebrate and invertebrate animals as well as plants. The life cycle of many rhabdoviruses involves replication in an arthropod and subsequent transmission to either a vertebrate animal or a plant. Thus, arthropods are the unifying life form in the natural history of many of these agents, an observation noted more than 30 years ago (Maramorosch, 1955). Ecological studies are incomplete, but available information suggests that arthropods were in the past essential to the maintenance of rhabdoviruses. It can be argued that rhabdoviruses evolved in arthropods and were originally maintained by vertical transmission, i.e., by passage through the egg. Some, such as Sigma virus, are still maintained exclusively by vertical transmission today (Brun and Plus, 1980). Others, because of the parasitic association of their arthropod host with plants or vertebrates, developed the ability to grow in plants or in vertebrate animals. Consequently, many rhabdoviruses are now maintained in arthropod–vertebrate–arthropod or arthropod–plant–arthropod cycles (Knudson 1973; K. M. Johnson *et al.*, 1969). A few rhabdoviruses, such as rabies and some of the fish viruses, have adapted completely to vertebrates and no longer infect arthropods.

ROBERT E. SHOPE AND ROBERT B. TESH • Yale Arbovirus Research Unit, Department of Epidemiology and Public Health, Yale University School of Medicine, New Haven, Connecticut 06510.

We hasten to emphasize that there is no way to prove the hypothesis of arthropod origin; however, it serves as a useful basis for structuring our thinking throughout the chapter and for considering the ecology of rhabdoviruses. It may also provide insight into some of the genetic and molecular characteristics of the family, including the plasticity (adaptability) of rhabdoviruses, their penchant to grow at widely varying temperatures, and their ability in many instances to persist in vertebrate cells and especially in arthropod host cells without causing apparent damage.

Over 70 rhabdoviruses of vertebrates are recognized. Rabies, vesicular stomatitis, bovine ephemeral fever, and some of the fish rhabdoviruses are known because they cause serious disease in people, domestic animals, or other food sources. Usually, the diseased animal is a dead-end host, and its diseased state is not relevant to the ecology of the virus. Rabies is an exception. Its transmission depends on the furious state of the rabid animal, which in turn is a diabolical natural adaptation of the virus to the host's limbic system, driving the animal to furious biting and successful transmission of rabies virus to a new host (R. T. Johnson, 1970).

There is a high rate of discovery of new rhabdoviruses, and the rate is not diminishing. There are probably hundreds, maybe even thousands, of rhabdoviruses. Their discovery is restricted only by the paucity of scientists searching and our limited technology for their isolation.

This chapter will describe the ecology of the rhabdoviruses of public health, veterinary, or economic importance, with emphasis on vesicular stomatitis viruses (vesiculoviruses), rabies virus and its relatives (lyssaviruses), and bovine ephemeral fever virus and its relatives. Enough is known to suggest concepts of the ecology of these groups of viruses. Much

TABLE I. Currently Recognized Members of the Genus *Vesiculovirus*

Serotype	Known geographic distribution	Source(s) of virus in nature
VSV-New Jersey	North, Central, and South America	Mammals, mosquitoes, midges, blackflies, houseflies
VSV-Indiana	North, Central, and South America	Mammals, mosquitoes, sand flies
VSV-Alagoas	Brazil, Colombia	Mammals, sand flies
Cocal	Argentina, Brazil, Trinidad	Mammals, mosquitoes, mites
Jurona	Brazil	Mosquitoes
Carajas	Brazil	Sand flies
Maraba	Brazil	Sand flies
Piry	Brazil	Mammals
Calchaqui	Argentina	Mosquitoes
Yug Bogdanovac	Yugoslavia	Sand flies
Isfahan	Iran, Soviet Union	Sand flies, ticks
Chandipura	India, Nigeria	Mammals, sand flies
Perinet	Madagascar	Mosquitoes, sand flies
Porton-S	Sarawak	Mosquitoes

less is known of the ecology of other rhabdoviruses of vertebrates. Readers seeking information on the natural history of rhabdoviruses of plants are referred to the review by Francki and Randles (1979).

II. ECOLOGY OF VESICULOVIRUSES

The genus *Vesiculovirus*, or vesicular stomatitis virus (VSV) serogroup, currently consists of 14 distinct virus types (Table I) that show varying degrees of antigenic and biochemical relatedness (Tesh *et al.*, 1983; Travassos *et al.*, 1984; McSharry, 1979; Clewley and Bishop, 1979). Because some of the vesiculoviruses have been commonly used as models in basic virological studies, a great deal is known about their replication and molecular biology. In contrast, relatively little is known about their natural history.

A. Host Range

The vesiculoviruses as a group have an extremely broad host range. They are capable of infecting a wide variety of animal and insect species (Tesh *et al.*, 1970). In addition, they grow in most vertebrate (mammal, bird, reptile, fish) and insect cell lines (Clark, 1979). In general, infection of vertebrate cells with VSV-group viruses results in rapid and massive cell destruction (Marcus and Sekellick, 1980), whereas their growth in insect-cell cultures is characterized by the lack of cytopathic effect (CPE) and by persistent infection (Artsob and Spence, 1974; Mudd *et al.*, 1973; Tesh and Modi, 1983a).

B. Isolation Systems

Vesiculoviruses are relatively easy to isolate. With the exception of Porton-S virus, they grow in most mammalian and avian cell lines, producing rapid CPE in liquid medium and plaques under agar. They are lethal to newborn mice and hamsters by any route of inoculation. During outbreaks of vesicular stomatitis, the causative agent can readily be recovered in cell cultures or in baby rodents inoculated with scrapings of tongue, foot, or teat lesions from affected livestock.

C. Pathogenesis

The pathogenesis of most of the vesiculoviruses is unknown; however, the pathology of VSV-Indiana and VSV-New Jersey infection has been rather extensively studied. The pathogenesis of these two viruses depends on a number of factors such as virus dose, route of infection, and age and species of the vertebrate host. For example, peripheral inoculation of these two agents into newborn mice or hamsters is rapidly lethal, with liver and kidney as key target organs (Murphy et al., 1975; Bruno-Lobo et al., 1968a,b). In contrast, older mice and hamsters inoculated subcutaneously or intramuscularly with the same VSV serotypes generally survive infection and develop protective antibodies. Persistent infection has also been reported in hamsters inoculated intraperitoneally with a mixture of wild-type VSV and defective interfering particles (Fultz et al., 1982). Intracerebral or intranasal inoculation of VSV-Indiana or VSV-New Jersey into laboratory rodents is uniformly lethal and produces an acute necrotizing encephalitis, regardless of the animal's age (Murphy et al., 1975; Bruno-Lobo et al., 1968a,b).

Intralingual inoculation of VSV-Indiana or VSV-New Jersey into susceptible cows, swine, horses, and guinea pigs results in the formation of vesicles and occasionally epithelial slough at the site of injection (Hanson, 1970; Proctor and Sherman, 1975). This procedure has been commonly used by veterinarians to determine the immune status of animals. Intramuscular inoculation of the same animal species with VSV produces inapparent infection and immunity.

Aside from their academic interest as laboratory models, some of the VSV-group viruses (Indiana, New Jersey, Cocal, and Alagoas) are of considerable economic and veterinary importance, because they produce in cattle and swine a disease (vesicular stomatitis) that is clinically indistinguishable from foot-and-mouth disease and vesicular exanthema. Affected animals develop fever, lethargy, decreased appetite, and vesicular lesions of the mouth, teats, and coronary band of the feet (Hanson, 1970). The vesicles rupture easily, leaving raw, reddened erosions surrounded by bits of torn epithelium. Complete healing usually occurs within 7–10 days. Mortality due to vesicular stomatitis is rare; however, lameness, weight loss, decreased milk production, and mastitis are fairly common sequelae in affected animals. During large epizootics of the disease, the economic loss due to these sequelae may be substantial (Alderink, 1984; Ellis and Kendall, 1964).

Five of the VSV-group viruses (Indiana, New Jersey, Alagoas, Piry, and Chandipura) are known to produce disease in humans (Tesh and Johnson, 1975; Karabatsos, 1985; Federer et al., 1967). The disease is characterized by an acute, self-limited, influenzalike illness with fever, myalgia, headache, and malaise of 3–6 days duration. Antibodies to Isfahan and Calchaqui viruses have also been reported in humans (Tesh et al., 1977; Obukhova and Gaidamovich, 1981; Calisher et al., 1987), but it is unknown whether or not these infections resulted in clinical illness. The human disease potential of the other vesiculoviruses is unknown.

VSV-group viruses produce no obvious pathology in infected insects, although CO_2 sensitivity has been reported in mosquitoes and in Drosophila melanogaster experimentally infected with several of these agents (Rosen, 1980; Bussereau, 1971).

D. Reservoir Cycle or Maintenance Mechanism

The reservoir cycles for all the vesiculoviruses are poorly understood. Serological evidence of natural infection with Indiana, New Jersey, Cocal, Alagoas, Calchaqui, Piry, Chandipura, and Isfahan has been demonstrated in a wide variety of wild and domestic mammals, including humans (Tesh et al., 1969, 1977, in press; Bhatt and Rodrigues, 1967; Karabatsos, 1985; Jonkers et al., 1965; Pinheiro et al., 1974; Calisher et al., 1987). However, the exact role that mammals play in the natural cycle of these viruses is uncertain. The available data suggest that mammals may only be indicators of vesiculovirus activity and that they are actually dead-end hosts in the natural cycle of these agents. This conclusion is based on results of experimental studies with Indiana, New Jersey, and Cocal viruses demonstrating that direct animal-to-animal transmission is rare and that the viremia associated with these virus infections is extremely transient and low-level (Jonkers, 1967; Tesh et al., 1970; Tesh and Johnson, 1975). In view of these findings, it seems unlikely that infected mammals could serve as important reservoirs or amplifying hosts of Indiana, New Jersey, or Cocal viruses. Several authors (K. M. Johnson et al., 1969; Hanson, 1968) have hypothesized that some of these agents might actually be plant viruses and that mammals become infected by ingesting infected plant material or that blood-sucking insects might acquire the virus by feeding on infected plants and then subsequently transmit the virus by bite to susceptible mammals.

There is considerable evidence suggesting that biting arthropods play a role in the natural history of some of the vesiculoviruses. The evidence consists of the following observations:

1. Most of the vesiculoviruses have been recovered in nature from blood-sucking arthropods (Karabatsos, 1985; Tesh and Johnson, 1975; Tesh et al., 1977, 1987; Clerc et al., 1983; Dhanda et al., 1970; Gligic et al., 1983; Travassos et al., 1984; Alkhutova et al., 1981; Webb, personal communication). Seven VSV serotypes have been associated with mosquitoes, eight with phlebotomine sand flies, and one each with Culicoides midges, blackflies, ticks, and mites (Table I).

2. Bite transmission of five viruses (Indiana, New Jersey, Cocal, Jurona, and Chandipura) has been demonstrated by allowing experimentally infected mosquitoes to feed on laboratory animals (Rao et al., 1967; Bergold et al., 1968; Jonkers et al., 1964; Tesh, unpublished data). Bite transmission of Indiana, Alagoas, and Chandipura viruses has also been shown with experimentally infected

phlebotomine sand flies (Tesh *et al.*, 1971, 1987; Tesh and Modi, 1983b).

3. Transovarial (vertical) transmission of Indiana, Carajas, Maraba, and Chandipura viruses has been demonstrated in experimentally infected sand flies (Tesh *et al.*, 1972; Tesh and Modi, 1983b; Travassos *et al.*, 1984). The ability of a virus or other parasite to be transovarially transmitted in an insect suggests that the agent and the arthropod have had a long evolutionary relationship (Maramorosch, 1955).

4. Natural infection with three VSV serotypes (Indiana, New Jersey, and Cocal) has occurred in caged sentinel animals exposed in areas of known virus activity (Jonkers *et al.*, 1965; Tesh *et al.*, 1970).

E. Epizootic or Epidemic Cycle

To date, four VSV-group viruses (Indiana, New Jersey, Cocal, and Alagoas) have been associated with epizootics of vesicular disease among bovines, equines, and swine (Tesh and Johnson, 1975). These outbreaks have been largely restricted to the New World. Epizootics of vesicular stomatitis are sporadic in occurrence, having a marked seasonal incidence (Hanson, 1952; Jonkers, 1967). In the United States, major epizootics have occurred in roughly 10-year cycles, although smaller outbreaks have taken place between these occurrences. The disease typically has appeared during summer months, with activity usually ceasing after the onset of freezing weather. In tropical America, epizootics occur with much greater frequency and seem to be associated with the transition from wet to dry season. Vesicular stomatitis epizootics characteristically appear suddenly and rapidly affect a large proportion of a herd, often within 24–48 hr. Spread of the disease is spotty rather than circumferential; often no cases are observed on farms adjacent to the affected farms. During epizootics, human cases of infection with these viruses have also been reported, usually among veterinarians, livestock handlers, and other persons in contact with sick animals (Tesh and Johnson, 1975).

F. Mechanisms of Transmission

The mechanisms of vesiculovirus transmission are poorly understood. The limited available data on this aspect of their ecology have been obtained from experimental studies with the Indiana, New Jersey, and Cocal serotypes (Hanson, 1952; Tesh and Johnson, 1975). Attempts to demonstrate virus transfer from infected to susceptible animals caged together have not been successful. Vesiculoviruses are not excreted in urine, feces, or milk. Direct transmission has been reported in domestic animals, but has usually involved transfer from teat to teat during milking or infection of a cow by her sick, nursing calf. In these cases, contamination of abraded epithelium with virus from vesicular fluid seems most likely.

Most mammals are susceptible to Indiana, New Jersey, and Cocal viruses following parenteral inoculation. Intralingual inoculation of these viruses into cattle, horses, and pigs produces characteristic vesicular mouth lesions followed by antibody formation and immunity. Rubbing of virus onto abraded skin on the snout or teats of susceptible animals produces the same result at the site of entry. In contrast, intramuscular or intravenous inoculation of the same viruses produces infection and subsequent immunity, but without vesicle formation (Tesh and Johnson, 1975). As noted in Section II.D, these viruses can also be transmitted to susceptible animals by the bite of infected insects (Rao *et al.*, 1967; Bergold *et al.*, 1968; Jonkers *et al.*, 1964; Tesh *et al.*, 1971). However, since the aforementioned arthropod-transmission studies were performed using laboratory rodents, it is unknown whether an infected insect feeding on the snout or teats of an infected bovine would transmit the virus and produce the characteristic lesions of vesicular stomatitis.

Human laboratory infections with Indiana, New Jersey, and Piry viruses have also been reported due to aerosol inhalation, splashing virus in the eye, and contamination of a cut or scratch with infected material (Tesh and Johnson, 1975).

In summary, the available data indicate that mammals are susceptible to vesiculovirus infection via a number of different routes: inoculation, entry of virus through abraded or disrupted epithelium, insect bite, and aerosol. The data also suggest that the route of infection determines whether or not an animal develops subclinical infection or frank vesicular disease. However, until the source or sources of virus in nature can be identified, discussions on the mechanism of transmission will remain purely speculative.

G. Mechanisms of Geographic Spread

Since the modes of virus transmission are unknown, the mechanisms of geographic spread are likewise not understood. Retrospective studies of vesicular stomatitis outbreaks among domestic animals in North America suggest that the virus moves in concentric circles away from the site of first appearance (Hanson, 1968; Jonkers, 1967). For example, in 1982–1983, a major epizootic of vesicular stomatitis due to the New Jersey serotype occurred in the western United States. The initial case was diagnosed in Arizona in June 1982 (Buisch, 1983). During the next 11 months, the disease spread northward through bovine and equine herds in New Mexico, Colorado, Wyoming, Utah, Idaho, Montana, Nebraska, and South Dakota.

Another characteristic of vesicular stomatitis epizootics is their spotty distribution. Not all farms within the epidemic zone are affected, and the incidence of clinical disease can vary widely among affected herds. It is unknown whether this is a reflection of the previous immune status of the animals or of the route of infection. It has also been observed that introduction of an infected cow with teat lesions into a milking herd can rapidly produce an epizootic, probably as a result of virus transfer via contaminated milking equipment.

H. Economic Importance

Vesicular stomatitis is a disease of considerable economic importance to livestock producers and dairy farmers. Most important, the disease mimics foot-and-mouth disease and is always an urgent problem of differential diagnosis. An outbreak of vesicular stomatitis in dairy cattle can result in significant financial loss due to reduced milk production, mastitis in affected cows, and sometimes quarantine of the herd (Alderink, 1984; Ellis and Kendall, 1964; Goodger et al., 1985). Weight loss in beef cattle and lameness in horses are other consequences of the disease. An epizootic of vesicular stomatitis can also result in international embargoes on the shipment of animals. For example, during the 1982–1983 vesicular stomatitis outbreak in the western United States, the government of Taiwan imposed a 6-month embargo on the shipment of swine from the United States. Furthermore, several European countries now require that horses from the United States be serologically tested and certified free of VSV antibodies before the animals can enter. All the costs must be absorbed by the farmers and animal producers or by the United States government.

I. Vaccines

In 1928, Olitsky et al. (1928) first reported that cattle inoculated intramuscularly with VSV did not develop visible lesions and subsequently developed immunity to local challenge. Since that time, live unmodified virus, inactivated virus, subunits [glycoprotein (G)] of the virus, and the incorporation of the G-protein gene of VSV into vaccinia virus have all been tried as vaccines, with varying degrees of success (Lauerman, 1967; Holbrook and Geleta, 1957; Mackett et al., 1985). Of these methods, the vaccinia virus recombinant may prove to be the most useful; at present, however, no commercial vesiculovirus vaccine is yet available.

J. Other Control Measures

Until the ecology and mode or modes of transmission of these viruses are better understood, little can be done to prevent or to control the diseases that they cause. Good sanitary practices (disinfecting hands and milking machines) have been recommended to control mechanical spread during vesicular stomatitis outbreaks in milking herds. Isolation of infected animals would also seem prudent, as would the control of biting arthropods during epizootics.

III. ECOLOGY OF LYSSAVIRUSES

The genus *Lyssavirus,* or rabies virus serogroup, consists of six distinct serotypes (Table II). Rabies, Mokola, Lagos bat, and Duvenhage are closely related viruses that infect vertebrate animals in nature. The other two, kotonkan and Obodhiang, are found in insects and are only distantly related to each other and to the four lyssaviruses that infect vertebrates. Rabies is the best-studied lyssavirus, and its natural history is described here with reference, where information is available, to the other members of the genus.

A. Host Range and Geographic Distribution

Rabies virus is capable of infecting virtually all vertebrate species under experimental conditions. Natural infections, however, are much more restricted. Infections of populations are compartmentalized. Rabies occurs in foxes in Europe and in the eastern mountain region of North

TABLE II. Members of the Genus *Lyssavirus*

Serotype	Known geographic distribution	Source(s) of virus in nature
Rabies	Worldwide except Australia, New Zealand, Japan, United Kingdom, Antarctica, parts of Scandinavia, Hawaii, and some other islands	Dogs, cats, wild carnivores, bats, cattle, humans
Lagos bat	Africa: Nigeria, Central African Republic, South Africa	Bats
Mokola	Africa: Nigeria, Cameroons, Zimbabwe	Shrews, humans, cats, dogs
Duvenhage	South Africa, Europe	Humans, bats
Obodhiang	Sudan	*Mansonia* mosquitoes
Kotonkan	Nigeria	*Culicoides* midges

America. The virus infects skunks in California and much of the central plains of North America and raccoons in the southeastern United States and in West Virginia, Virginia, Maryland, and Pennsylvania. Rabies is enzootic in wolves in the Arctic region, meercats in southern Africa, mongooses in Grenada, and vampire bats in South and Central America, where transmission to cattle is a major economic problem. In any given geographic area, the compartmental behavior of rabies may be accompanied by spillover into other species.

Dogs and cats form the prime reservoir for transmitting the virus to human beings. There are large rabies-infected canine populations with concentrations in less-developed areas of the world, including South America, Africa, Asia, and the Philippines.

Bats are a major reservoir of rabies virus in the New World. The bats of Africa, on the other hand, harbor the rabies-related viruses Lagos bat and Duvenhage. Duvenhage has been transmitted by bats to a human (Meredith *et al.*, 1971). Mokola virus has been isolated on numerous occasions in Africa from *Crocidura* shrews, from people, and from cats and a dog (Shope, 1975; Foggin, 1982).

The natural history of the insect-infecting members of the genus remains unknown. Kotonkan virus has been isolated only once from *Culicoides* midges near Ibadan, Nigeria (Kemp *et al.*, 1973a), and Obodhiang virus twice from *Mansonia* mosquitoes near Malakal in central Sudan (Schmidt *et al.*, 1965).

Rabies exists worldwide except for Australia, Antarctica, parts of Scandinavia, and some islands such as New Zealand, the United Kingdom, Japan, and Hawaii, where the virus either has not yet found its way or has been eradicated.

B. Isolation Systems

All the lyssaviruses can be isolated by intracerebral inoculation in baby mice and all except Obodhiang and kotonkan in adult mice. Laboratory strains of rabies have been adapted by passage to BHK-21 cells, and vaccines are produced in brains of baby mice, as well as in duck embryo, human diploid, and dog kidney cells. Until recently, it was thought that cell cultures were not suitable for primary isolation of field rabies virus. Smith *et al.* (1978), however, showed that mouse neuroblastoma and CER cells (a rodent cell line of uncertain ancestry) were highly efficient for isolation of the virus from field materials.

C. Pathogenesis

The unique pathogenesis of rabies is a major determinant governing the maintenance and spread of the virus in nature. Rabies virus is found in the saliva of rabid animals and, sometimes, in the saliva of apparently healthy animals that will later become rabid (Bell, 1975) or, rarely, remain healthy carriers (Fekadu et al., 1981). The virus usually enters the host through wounds or directly through mucous membranes, but in unusual circumstances, it can enter through the olfactory mucosa (Constantine, 1962) or the gut (Fischman and Ward, 1968).

Rabies virus replicates in muscle cells (Murphy et al., 1973) and possibly in other cell types as well, but it must enter nerve cells to cause disease. A rational explanation for the neurotropism of rabies is provided by the experiments of Tignor et al. (1984). They demonstrated attachment of rabies virus at or near the acetylcholine receptor at the neuromuscular junction. The virus is presumed to attack and enter the nerve where it, its nucleocapsid, or its RNA traverses the axon and enters the central nervous system (CNS). Once in the CNS, the virus spreads rapidly either through nerve synapses or by cerebrospinal fluid. Finally, virus spreads centrifugally in nerves to multiple parts of the body including the salivary glands, where large numbers of infective particles are produced and shed in the saliva. Infection of the brain, especially the limbic system, induces the furious state in a vertebrate host (R. T. Johnson, 1970); thus, the infection is ideally suited to maintain the transmission cycle by bite of rabid animals.

Under experimental conditions, the insect-associated lyssaviruses, Obodhiang and kotonkan, replicate in mosquito cells and in mosquitoes. In mosquitoes, they pass to the salivary glands (Buckley, 1973; Aitken et al., 1984); Mokola virus also replicated experimentally in mosquitoes, but required 6 weeks to achieve maximum titers (Aitken et al., 1984). Arthropods are apparently not harmed by the infecting lyssaviruses.

D. Reservoir Cycle or Maintenance Mechanism

Carnivores constitute the major reservoir of rabies virus. Throughout most of the world, dogs and cats are responsible for its maintenance, although significant wildlife reservoirs also exist. The principal reservoirs vary in different parts of the world (see Section III.A). Carnivores such as meercats in South Africa, foxes in Europe, and foxes, coyotes, raccoons, wolves, and skunks in North America are basic reservoir hosts. In South and Central America, the vampire bat, Desmodus rotundus, forms an important reservoir that transmits mainly to cattle. In North America, insect-eating bats, and in Latin America, insect- and fruit-eating bats, also maintain rabies virus. The bat cycles appear to be separate from the carnivore cycle, although this question may be settled only after rabies is controlled in terrestrial animals (Baer, 1975).

Mokola virus is maintained in Crocidura spp. shrews in Africa and sometimes spills over into human, cat, and dog populations. Lagos bat virus is maintained in the African bat populations, and Duvenhage virus is also found in bats of Europe and South Africa, with spillover to human beings.

Duvenhage virus was originally described in South Africa (Meredith *et al.*, 1971); in the past 7 years, however, isolates have also been reported from *Nyctalus noctula*, *Eptesicus serotinus*, and *Rhinolophus ferrumequinum* bats in northern Germany, Denmark, and Poland (World Health Organization, 1986). Although it was postulated that the infected bats in Europe were imported by boat from Africa (Schneider and Meyer, 1981), it seems more likely that Duvenhage virus is enzootic in both South Africa and northern Europe

E. Epizootic or Epidemic Cycle

According to H. N. Johnson (1959), rabies was historically endemic in parts of Europe. Epizootics started in these foci and then periodically moved across Europe in susceptible wolves and foxes. The epizootics followed natural landscape patterns along rivers and valleys. The current epizootic in foxes started at the end of World War II in eastern Europe and has since moved steadily westward at the rate of 30–60 km per year, reaching Germany in 1947, Bulgaria in 1965, and France in 1968.

Within epizootics, circumscribed epidemics also occur when a rabid animal bites several people in a single episode. The best-publicized example is the classic description by Baltazard and Bahmanyar (1955) of a rabid wolf in Iran that bit 29 persons, of whom 4 developed rabies (all 29 persons received vaccine or vaccine plus immune serum). This type of epidemic is highly visible and is limited in time and place.

Epizootics have not been described with Mokola, Duvenhage, or Lagos bat viruses.

F. Mechanisms of Transmission

Rabies virus is classically transmitted by the bite of rabid animals. The virus is present in the saliva and is introduced into the wound. It is believed, but not completely documented, that transmission among wild animals is also by biting, although the animal transmitting may not necessarily be furious. Wild animals bite each other during play and may fight to protect their territory; they also lick mucous membranes while grooming mates, offspring, or siblings and may well transmit during this type of contact.

Transmission also occurs, albeit much less commonly, by aerosol (Constantine, 1962), through surgical procedures such as corneal transplants (Houff *et al.*, 1979), and orally (Soave, 1966).

The transmission of Mokola virus by bite of shrews has been shown in the laboratory (Kemp *et al.*, 1973b), but the mode of natural transmission among shrews and to people is not known.

G. Mechanisms of Geographic Spread

During epizootics, rabies spreads as infected animals range to establish new territory or infect another animal at the periphery of their home range. Rabies in European foxes has spread from east to west at roughly the equivalent to the home range of the fox, from 30 to 60 km per year.

There is evidence that rabies also spreads over long distances by introduction of infected animals into virgin susceptible populations. A striking example was the apparent transport of infected raccoons from the epizootic zone of Georgia and Florida into West Virginia in 1977. The following year, 3 rabid raccoons were found in Virginia, and by 1982, the virus had spread to Maryland and Pennsylvania. In 1982, there were 837 rabid raccoons reported in the four-state area (Witte *et al.*, 1983), and the epizootic continues to enlarge in a radius of 25–50 miles per year. Rabies was found in raccoons captured in Florida and readied to ship to hunting camps in North Carolina in 1979 (Nettles *et al.*, 1979), offering an explanation for the sudden appearance of rabies in West Virginia 2 years earlier. Monoclonal-antibody studies showed that the antigenic composition of the rabies strains from the four-state area was similar to that of strains from Florida and Georgia (Baer, 1985).

H. Economic Importance

The annual loss from bovine paralytic rabies in Latin America in 1967 was estimated at $47,592,000 (Acha, 1967). The cost of rabies in other forms is also significant, not so much from loss of human life as from the medical and public health expense that vaccination and control of dogs and cats entails. A single rabid dog or wild animal may expose a large number of persons, necessitating expenditure of several hundred dollars per person for vaccine and medical services.

I. Vaccines

Vaccine for rabies has been used since 1885, when 9-year-old Joseph Meister was bitten by a dog believed rabid and was inoculated by Louis Pasteur with 13 injections of desiccated rabbit cord. The boy survived. Vaccine currently in use for humans in the United States and Europe is prepared from human diploid-cell culture and administered with rabies immune globulin (Centers for Disease Control, 1984); for dogs, cats, and cattle, the vaccine is prepared from tissue-culture substrates (Centers for Disease Control, 1986). Utilization of these vaccines has no effect, however, on the basic ecology of wildlife rabies.

The challenge is to find a way to intervene in the basic cycle. Live attenuated oral vaccines for wildlife have been used with modest success (Steck et al., 1982), but the wisdom of delivering live rabies virus in any form to wildlife has been questioned. Now, a gene that codes for the protective glycoprotein has been engineered by recombinant DNA technology. This gene was spliced into the genome of vaccinia virus and produced protective antibody in animals (Kieny et al., 1984). It has the promise of being fed safely to both dogs and wildlife as a live replicating rabies immunogen.

There are no vaccines for Mokola and Duvenhage viruses, and the numbers of human cases are too small to warrant use of vaccines. With the spillover of Mokola virus to the dog and cat populations in Zimbabwe and the recent discovery of Duvenhage virus in the bat populations of northern Europe, surveillance should be intensified, and if more cases occur, the need for vaccines should be reassessed.

J. Other Control Measures

Several other methods have been instituted to control rabies. Each is based on depletion of reservoir-animal populations. The most effective method is dog control by leash laws and destruction of stray dogs. Vampire bats have been eliminated by the anticoagulants warfarin and diphenadion. The anticoagulants are applied either to the coat of the bat, which serves to distribute the chemicals to other bats through grooming behavior (Linhart et al., 1972), or directly to the cattle, which circulate sufficient levels in the blood to kill bats feeding on the blood (Thompson et al., 1972). Other rabies-reservoir animals are killed by gassing dens (foxes) or by poison or shooting. Such measures are usually unpopular and only temporarily effective unless drastic numbers of animals are killed.

IV. ECOLOGY OF BOVINE EPHEMERAL FEVER VIRUS AND RELATED VIRUSES

Bovine ephemeral fever is an acute febrile disease of cattle and water buffalo. The illness usually lasts only 3 days and is accompanied by torpor of the animal and lameness. Death may occur suddenly, but bovines that may appear very sick tend to recover abruptly, hence the name "ephemeral" (St. George, 1984). Complete recovery is the rule, although milk production usually remains depressed for the lactating period and paralysis may persist. Animals have metabolic abnormalities, including hypocalcemia, which responds temporarily to intravenous calcium (St. George et al., 1984). The syndrome also responds to the antiinflammatory drug phenylbutazone.

The illness is caused by bovine ephemeral fever virus. There are at least three viruses from Australia [Kimberley, Berrimah, and Adelaide River viruses (St. George, 1984)] and Fukuoka virus from Japan (Kaneko et al., 1986) (Table III) that are related by immunofluorescence tests but distinct by neutralization. These ephemeral-fever-related viruses do not cause disease in livestock, but are troublesome because they result in false-positive seroconversion to bovine ephemeral fever virus (St. George, 1984). In addition, the serologically unrelated kotonkan virus of the *Lyssavirus* genus has proteins of the same mobility as those of bovine ephemeral fever virus (Della-Porta and Brown, 1979) and causes a disease of cattle in Africa similar to bovine ephemeral fever (Kemp et al., 1973a).

A. Host Range and Isolation Systems

Bovine ephemeral fever virus is found in the blood of naturally infected cattle (Burgess, 1971) and buffalo during the acute phase of the illness. Antibody to the virus has also been detected in waterbuck, hartebeest, and wildebeest (Davies et al., 1975). Other animals, including sheep, do not appear to be involved in natural cycles of bovine ephemeral fever virus. Primary isolation of the virus is accomplished by intravenous inoculation of whole blood or buffy coat into cattle. Virus also replicates in baby mouse brain, hamster kidney-cell culture, and *Aedes aegypti* cells, although these systems are less efficient than cattle for primary isolation from field-collected material (St. George, 1984).

B. Pathogenesis

Animals acquire the virus by the bite of an infected arthropod. The site of primary replication in cattle is not known, although the virus is

TABLE III. Bovine Ephemeral Fever and Serologically Related Viruses[a]

Serotype	Known geographic distribution	Source(s) of virus in nature
Bovine ephemeral fever	Africa, southern Asia, Australia	*Culicoides* midges, mosquitoes, bovids
Kimberley	Australia	Mosquitoes, midges, cattle
Berrimah	Australia	Cattle
Adelaide River	Australia	Mosquitoes, midges, cattle
Fukuoka	Japan	Mosquitoes, midges

[a] Oak-Vale, another bovine-ephemeral-fever-related Australian virus isolated by T. St. George from *Culex* spp. mosquitoes, has not yet been published. Mention here is not intended to constitute priority.

found in high titer in the neutrophils (Young and Spradbrow, 1985). The basic lesion appears to be an arteritis of small vessels, accompanied by infiltration of neutrophils, and edema of the lymph nodes, pericardium, lungs, and other tissues. Hemorrhage may also be present (Basson et al., 1970; Mackerras et al., 1930). Young and Spradbrow (1980) treated cattle with antineutrophil serum, which depressed circulating neutrophils. When challenged with live virus, these animals developed viremia, but no illness or serum antibody. The animals were again fully susceptible to disease after their neutrophil populations returned to normal.

C. Reservoir Cycle or Maintenance Mechanism

The reservoirs of bovine ephemeral fever virus and the other members of the serogroup are not known. There are relatively few isolates of members of the group, but of those that exist, strains from mosquitoes and *Culicoides* midges predominate. Thus, these insects must be considered as potential vectors in any hypothesis of a maintenance mechanism.

Several lines of evidence in Kenya favor infection of both wild and domestic bovids by the bite of *Culicoides* midges (Davies et al., 1975): The virus was isolated from *Culicoides*; antibody was found in both cattle and wild ruminants; antibody was detected in animals that had been born since the previous epizootic, indicating participation in an enzootic cycle; and these infections occurred without apparent serious disease. In addition, the distribution of antibodies in cattle of Kenya was widespread and corresponded to the distribution of *Culicoides*. The factor that triggers the change from an enzootic to an epizootic situation is still unknown.

As for vesicular stomatitis, there is evidence that bovine ephemeral fever is enzootic in the tropical regions and periodically spreads to more temperate zones. Bovine ephemeral fever has been studied intensively in Australia and Papua New Guinea by use of 51 sentinel cattle herds distributed in each state and territory (St. George et al., 1977). Outbreaks have characteristically started in the northern part of Australia, then swept southward through Queensland and into New South Wales. Infection is enzootic in cattle in northern Australia. The sentinel herds seroconverted to bovine ephemeral fever virus throughout the year in three foci: the Kimberley region, the far north of the Northern Territory, and the Gulf of Carpentaria region of Queensland. For reasons that are not entirely clear, the disease is mild in these areas. Cases are sporadic at any one time, probably because most of the cattle and buffalo are immune.

D. Epizootic Cycle and Mechanisms of Geographic Spread

Epizootics of bovine ephemeral fever occur in parts of Africa, the Middle East, South Asia, Southeast Asia, and eastern Australia. The disease in the more temperate zones is a summer–fall illness coincident with high population levels of biting insects. Large outbreaks have been described in Japan in 1949–1950 (Inaba, 1973), in Australia in 1936–1937, 1955–1956, 1967–1968, and 1975–1976 (St. George *et al.*, 1977), and in Kenya in 1968 and 1972–1973 (Davies *et al.*, 1975). The major epizootics may be preceded and followed by smaller outbreaks, sometimes annually. Outbreaks in Japan prior to 1968, when specific diagnostic tests for bovine ephemeral fever first came into use, may have been mixtures of Ibaraki disease (caused by an orbivirus) and bovine ephemeral fever (Inaba, 1973).

In Australia, the disease has characteristically started in the enzootic foci of the northern tropical regions and moved rapidly south. Animals in herds in southern Queensland and New South Wales are seronegative until the virus is introduced; then epizootics and seroconversion occur. The virus is believed to spread from the northern enzootic foci in waves when ecological conditions are favorable. The spread is rapid and may possibly be by windborne infected insects (Murray, 1970). In 1967–1968, ephemeral fever covered 3200 km north to south in 4 months.

In the past 5 years, the pattern of origin of epizootics from northern Australian enzootic foci appears to have changed. Outbreaks that cannot be traced to a north–south wave have been reported in Queensland and New South Wales. This phenomenon will be watched closely; it may mean that bovine ephemeral fever virus has been established in new enzootic foci in eastern Australia.

E. Mechanisms of Transmission

Bovine ephemeral fever virus is not transmitted directly from animal to animal. All evidence points to an insect vector, although there is no direct proof. The virus has been isolated from *Culicoides* midges in Kenya, and the distribution of antibody in cattle in Kenya coincides with that of *Culicoides* (Davies and Walker, 1974). By contrast, the virus was isolated from several species of mosquitoes in Australia (Standfast and Muller, 1985). Standfast and Muller point out that the only potential vector with distribution in Australia the same as the virus is *Culex annulirostris*, and this mosquito could be infected in the laboratory by feeding on a virus suspension. It would be a first for an arbovirus to utilize in propagative transmission two such diverse vectors as midges and mosquitoes; only further careful experimentation will give the answer.

F. Economic Importance

Ephemeral fever causes a drop in milk production that accounts for as much as a 10% loss for the animal through the affected milk cycle, causing a significant loss of revenue for the farmer (Theodoridis *et al.*,

1973a). In a broader sense, ephemeral fever is a tremendous burden to livestock-exporting countries such as Australia, which must demonstrate seronegativity for ephemeral fever in cattle sent to New Zealand, the United States, and other nations that are currently free of the infection.

G. Vaccines and Other Control Measures

Vaccines have been developed for ephemeral fever (Theodoridis *et al.*, 1973b; Inaba *et al.*, 1974; Vanselow *et al.*, 1985), but for each, questions have been raised about efficacy under field use. The product used in Japan utilizes a live attenuated vaccine that by itself does not immunize, but primes the host to respond to a second inoculation of inactivated vaccine in adjuvant. The Australian vaccine is a live attenuated strain that is incorporated into Quil A adjuvant just before inoculation. Combination with the adjuvant effectively inactivates the virus so that the final vaccine is essentially a killed product. Efficacy is claimed in the initial field trial of the Quil A vaccine.

Strict control of importation of animals from ephemeral fever enzootic areas is practiced to prevent introduction of bovine ephemeral fever virus to the United States, New Zealand, and other nations currently free of the disease. Other methods applicable during outbreaks, such as vector control to kill the putative insect transmitting ephemeral fever, have not been instituted and would be difficult to evaluate, since the vector is not known.

V. ECOLOGY OF OTHER RHABDOVIRUSES OF VERTEBRATES

A. Rhabdoviruses of Fish

There are at least four different rhabdoviruses that infect fish (Roy, 1979) (Table IV). Viral hemorrhagic septicemia virus (also called Egtved after the locality in Denmark where it was first noted) infects rainbow trout in Europe (Schaperclaus, 1953) and causes either a hemorrhagic illness or anemia with signs involving the nervous system. The infection is contagious, recurs in some infected fisheries annually, and is believed to be maintained from one year to the next in a carrier state in recovered fish. The virus was transmitted experimentally by infected water to rainbow trout that shed virus in the urine and not in the feces. Fish that survived the infection continued to shed virus in the urine for 30 days (Neukirch and Glass, 1984). The disease has its onset in the spring when water temperatures are between 6 and 12°C and disappears or becomes sporadic when temperatures are above 15°C.

Spring viremia of carp virus causes infectious dropsy and swim-bladder inflammation of carp in Europe. Infectious dropsy is prevalent in the

TABLE IV. Rhabdoviruses Isolated from Fish and Lizards

Serotype	Known geographic distribution	Source(s) of virus in nature
Spring viremia of carp	Europe	Carp
Infectious hematopoietic necrosis	United States	Trout, salmon
Pike fry disease	Europe	Pike
Egtved (hemorrhagic septicemia)	Europe	Trout
Marco	Brazil	*Ameiva* lizards
Timbo	Brazil	*Ameiva* lizards
Chaco	Brazil	*Ameiva* and *Kentropyx* lizards
Sena Madureira	Brazil	*Ameiva* lizards
Almpiwar	Australia	*Ablepharus* skink
Charleville	Australia	*Gehyra* lizards, phlebotomine sand flies

spring, while swim-bladder inflammation is a disease of the summer when water temperatures are about 17°C. Both diseases are caused by the same virus serotype (Bachman and Ahne, 1974). Although this virus is also presumed to be water-borne, there is experimental evidence that a hematophagous water arthropod, *Argulus foliaceus*, can transmit the virus by bite (Pfeil-Putzien, 1978).

Pike fry disease virus is found in the Netherlands in two forms, both of which are often fatal. One form, "head disease," is characterized by hydrocephalus and the other form, "red disease of pike," by red and swollen areas of the body and tail.

Infectious hematopoietic necrosis virus occurs in rainbow trout and salmon in the United States, affecting sockeye salmon and Kokanee salmon in Oregon and Chinook salmon in California and Washington State. Fish become lethargic, swim in erratic fashion, and display hemorrhagic signs around the fins and throat. A water temperature of 13°C or lower is associated with epizootics; disease can be prevented by rearing the young fish at 15°C (Wolf, 1966).

Temperature is the common ecological parameter that appears to control the presence or absence of disease in rhabdovirus infections of fish. While one might think that these viruses grow best at relatively cold temperatures, the optimum temperature for virus replication is not always that associated with disease. Roy (1979) has postulated, therefore, that the fish host defense mechanism is slowed at low temperatures, and hence the virus takes over.

B. Rhabdoviruses of Lizards

Marco, Timbo, and Chaco viruses (Table IV) were isolated between 1955 and 1963 from organs of lizards (*Ameiva ameiva* and *Kentropyx*

TABLE V. Rhabdoviruses of Other Vertebrates and Arthropods[a]

Serotype	Known geographic distribution	Source(s) of virus in nature
Kern Canyon	California	Bats
Gossas	Senegal	Bats
Mount Elgon bat	Kenya	Bats
Oita-296	Japan	Bats
Keuraliba	Senegal	Rodents
Klamath	Oregon, Alaska	Rodents
Barur	Somalia, Kenya, India	Rodents, mosquitoes, ticks
Cuiaba	Brazil	Toads
Navarro	Colombia	Birds
Mossuril	Central and southern Africa	Birds, mosquitoes
Bangoran	Central African Republic	Birds, mosquitoes
Hart Park	United States	Birds, mosquitoes
Flanders	United States	Birds, mosquitoes
Tupaia	Germany	Tree shrew
Le Dantec	Senegal	Humans
Mosqueiro	Brazil	Mosquitoes
Kwatta	Surinam	*Culex* mosquitoes
Kamese	Uganda, Central African Republic	*Culex* mosquitoes
Aruac	Trinidad	Mosquitoes
Gray Lodge	California	*Culex* mosquitoes
Joinjakaka	New Guinea	Mosquitoes
Kununurra	Australia	*Aedeomyia* mosquitoes
La Joya	Panama	*Culex* mosquitoes
Nkolbisson	Ivory Coast	Mosquitoes
Xiburema	Brazil	*Sabethes* mosquitoes
Yata	Central African Republic	*Mansonia* mosquitoes
Bahia Grande	Texas	*Aedes* mosquitoes
Muir Springs	Colorado	*Aedes* mosquitoes
Reed Ranch	Texas	*Culex* mosquitoes
Inhangapi	Brazil	Phlebotomine sand flies
Sripur	India	Phlebotomine sand flies
Tibrogargan	Australia	*Culicoides* midges
Sawgrass	Florida	Ticks
New Minto	Alaska	Ticks
Connecticut	Connecticut	Ticks

[a] The viruses listed in this table are registered in the Catalogue of Arthropod-Borne Viruses (Karabatsos, 1985) or otherwise published (Kerschner *et al.*, 1986; Kurz *et al.*, 1986). Oita-296 virus was isolated by A. Oya and is not registered; mention here is not intended to constitute priority.

calcaratus) near Belem, Brazil (Causey *et al.*, 1966). Timbo and Chaco are serologically related and Marco is ungrouped. The three viruses replicate in mosquitoes in the laboratory, but their mode of transmission in nature is not known. Sena Madureira, a fourth rhabdovirus from the lizard *Ameiva*, was more recently isolated in northwestern Brazil and was shown to be related to Timbo and Chaco viruses (Tesh *et al.*, 1983).

Almpiwar virus was isolated from 3 of 75 skunks sampled at the Mitchell River Aboriginal Community in northern Australia (Graf *et al.*, 1967). Antibody to Almpiwar virus is widespread in skinks and other animals of northern Australia, but its natural history is otherwise not well studied. Charleville virus was isolated from phlebotomine sand flies in Charleville and from a *Gehyra australis* lizard at Mitchell River, Queensland, Australia (Karabatsos, 1985).

C. Rhabdoviruses of Other Vertebrates and Arthropods

In addition to members of the genus *Lyssavirus*, 4 rhabdoviruses have been isolated exclusively from bats; 3 from rodents; 1 from a toad; 1 from a tree shrew; 1, Le Dantec virus, from the blood of a febrile child in Dakar, Senegal (Karabatsos, 1985); 5 from birds, and 4 of these also from mosquitoes; 14 from mosquitoes only; 2 from sand flies; 1 from *Culicoides*; and 3 from ticks (Table V). None of these agents except Le Dantec causes disease in nature as far as we know. Each, however, kills baby mice after inoculation intracerebrally or causes cytopathology in vertebrate-cell culture or both. For the most part, their ecology is a mystery and that of rhabdoviruses that do not cause disease in man or domestic animals will probably remain so, because intellectual curiosity is the only likely impetus that could lead to their study. Given that other rhabdoviruses are major pathogens of humans and domestic animals, however, these orphan rhabdoviruses should at least be tested as potential causes of the many diseases for which the etiology remains a mystery.

REFERENCES

Acha, P. N., 1967, Epidemiology of paralytic bovine rabies and bat rabies, *Bull. Off. Int. Epizoot.* **67**:343.

Aitken, T. H. G., Kowalski, R. W., Beaty, B. J., Buckley, S. M., Wright, J. D., Shope, R. E., and Miller, B. R., 1984, Arthropod studies with rabies-related Mokola virus, *Am. J. Trop. Med. Hyg.* **33**:945.

Alderink, F. J., 1984, Vesicular stomatitis epidemic in Colorado: Clinical observations and financial losses reported by dairymen, *Prev. Vet. Med.* **3**:29.

Alkhutova, L. M., Sadykov, V. G., Ponirovsky, E. N., and Listovskaya, E. D., 1981, Isolation of strains identical to Isfahan virus from *Hyalomma asiaticum* ticks in Turkmenistan, *Sb. Tr. Inst. Virus. imeni D. I. Ivanovsky, Akad. Med. Nauk SSSR* 29–32 (in Russian).

Artsob, H., and Spence, L., 1974, Persistent infection of mosquito cell lines with vesicular stomatitis virus, *Acta Virol.* **18**:331.

Bachman P. A., and Ahne, W., 1974, Isolation and characterization of agent causing swim
    bladder inflammation in carp, *Nature (London)* **244**:235.
Baer, G. M., Rabies in non-hematophagous bats, in: *The Natural History of Rabies*, Vol. 2
    (G. M. Baer, ed.), Chapter 5, Academic Press, New York.
Baer, G. M., 1985, Rabies virus, in: *Virology* (B. M. Fields, ed.), Chapter 49, Raven Press,
    New York.
Baltazard, M., and Bahmanyar, M., 1955, Essai pratique du serum antirabique chez les
    mordus par loups enrages, *Bull. W.H.O.* **13**:747.
Basson, P. A., Pienaar, J. G., and van der Westhuizen B., 1970, The pathology of ephemeral
    fever: A study of the experimental disease in cattle, *J. S. Afr. Vet. Med. Assoc.* **40**:385.
Bell, J. F., 1975, Latency and abortive rabies, in: *The Natural History of Rabies*, Vol. 1 (G.
    M. Baer, ed.), Chapter 17, Academic Press, New York.
Bergold, G. H., Suarez, O. M., and Munz, K., 1968, Multiplication in and transmission by
    Aedes aegypti of vesicular stomatitis virus, *J. Invert. Pathol.* **11**:405.
Bhatt, P. N., and Rodrigues, F. M., 1967, Chandipura: A new arbovirus isolated in India
    from patients with febrile illness, *Ind. J. Med. Res.* **55**:1295.
Brun, G., and Plus, N., 1980, The viruses of *Drosophila*, in: *Genetics and Biology of Dro-
    sophila* 2nd ed. (M. Ashburner and E. Novitski, eds.), pp. 625–702, Academic Press,
    New York.
Bruno-Lobo, M., Peralta, P. H., Bruno-Lobo, G., and de Paola, D., 1968a, Pathogenesis of
    vesicular stomatitis virus infection in the infant hamster and mouse, *Ann. Microbiol.*
    15:53.
Bruno-Lobo, M., Peralta, P. H., Bruno-Lobo, G., and de Paola, D., 1968b, Pathogenesis of
    vesicular stomatitis virus in New Jersey infection in the adult hamster and mouse,
    Ann. Microbiol. **15**:69.
Buckley, S. M., 1973, Singh's *Aedes albopictus* cell cultures as helper cells for the adaptation
    of Obodhiang and kotonkan viruses of the rabies serogroup to some vertebrate cell
    cultures, *Applied Microbiol.* **25**:695.
Buisch, W. W., 1983, Fiscal year 1982–83 vesicular stomatitis outbreak, *Proceedings of the
    87th Annual Meeting of the U. S. Animal Health Association*, p. 78.
Burgess, G. W., 1971, Bovine ephemeral fever: A review, *Vet. Bull.* **41**:887.
Bussereau, F., 1971. Etude de symptome de la sensibilité au CO_2 produit par le virus de la
    stomatite vesiculaire chez *Drosophila melanogaster*, *Ann. Inst. Pasteur* **121**:223.
Calisher, C. H., Monath T. P., Sabattini, M. S., Mitchell, C. J., Lazuick, J. S., Tesh, R. B.,
    and Cropp, C. B., 1987, Isolation of a newly recognized vesiculovirus, Calchaqui virus,
    and subtypes of Melao and Maguari viruses from Argentina, with serologic evidence
    for infections of humans and horses, *Am. J. Trop. Med. Hyg.* **36**:114.
Causey, O. R., Shope, R. E., and Bensabath, G., 1966, Marco, Timbo, and Chaco, newly
    recognized arboviruses from lizards of Brazil, *Am. J. Trop. Med. Hyg.* **15**:239.
Centers for Disease Control, 1984, ACIP. Rabies prevention—United States, 1984, *Morbid-
    ity and Mortality Weekly Report* **33**:393.
Centers for Disease Control, 1986, Compendium of animal rabies vaccines, 1986; prepared
    by the National Association of State Public Health Veterinarians, Inc., *Morbidity and
    Mortality Weekly Report* **34**:770.
Clark, H. F., 1979, Systems for assay and growth of rhabdoviruses, in: *Rhabdoviruses*, Vol.
    1 (D. H. L. Bishop, ed.), pp. 23–41, CRC Press, Boca Raton, Florida.
Clerc, Y., Rodhain, F., Digoutte, J. P., Tesh, R., Heme, G., and Coulonges, P., 1983, Le virus
    Perinet du genre *Vesiculovirus* (Rhabdoviridae) isolé de culicides à Madagascar, *Ann.
    Virol.* **134E**:61.
Clewley, J. P., and Bishop, D. H. L., 1979, Rhabdoviral RNA structure, in: *Rhabdoviruses*,
    Vol. I (D. H. L. Bishop, ed.), pp. 119–135, CRC Press, Boca Raton, Florida.
Constantine, D. G., 1962, Rabies transmission by non-bite route, *U. S. Pub. Health Rep.*
    77:287.
Davies, F. G., and Walker, A. R., 1974, The isolation of ephemeral fever virus from cattle
    and *Culicoides* midges in Kenya, *Vet. Rec.* **95**:63.

Davies, F. G., Shaw, T., and Ochieng, P., 1975, Observations on the epidemiology of ephemeral fever in Kenya, *J. Hyg. (Cambridge)* **75**:231.

Della-Porta, A J., and Brown, F., 1979, The physico-chemical characterization of bovine ephemeral fever virus as a member of the family Rhabdoviridae, *J. Gen. Virol.* **44**:99.

Dhanda, V., Rodrigues, F. M., and Ghosh, S. N., 1970, Isolation of Chandipura virus from sand flies in Aurangabad, *Ind. J. Med. Res.* **58**:179.

Ellis, E. M., and Kendall, H. E., 1964, The public health and economic effects of vesicular stomatitis in a herd of dairy cattle, *J. Am. Vet. Med. Assoc.* **144**:377.

Federer, K. E., Burrows, R., and Brooksby, J. B., 1967. Vesicular stomatitis virus: The relationship between some strains of the Indiana serotype, *Res. Vet. Sci.* **8**:103.

Fekadu, M., Shaddock, J. H., and Baer, G. M., 1981, Intermittent excretion of rabies virus in the saliva of a dog two and six months after it had recovered from experimental rabies, *Am. J. Trop. Med. Hyg.* **30**:1113.

Fischman, H. R., and Ward, F. E., III, 1968, Oral transmission of rabies virus in experimental animals, *Am. J. Epidemiol.* **88**:132.

Foggin, C. M., 1982, Atypical rabies virus in cats and a dog in Zimbabwe, *Vet. Rec.* **110**:338.

Francki, R. I. B., and Randles, J. W., 1979, Rhabdoviruses infecting plants, in: *Rhabdoviruses*, Vol. III (D. H. L. Bishop, ed.), pp. 135–165, CRC Press, Boca Raton, Florida.

Fultz, P. N., Shadduck, J. A., Kang, C. Y., and Streilein, J. W., 1982, Vesicular stomatitis virus can establish persistent infections in Syrian hamsters, *J. Gen. Virol.* **63**:493.

Gligic, A., Tesh, R. B., Miscevic, Z., Travassos da Rosa, A., and Zivkovic, V., 1983, Jug Bogdanovac virus—A new member of the vesicular stomatitis virus serogroup (Rhabdoviridae: *Vesiculovirus*) isolated from phlebotomine sandflies in Yugoslavia, *Acta Biol. Iugos.* **20**:97.

Goodger, W. J., Thurmond, M., Nehay, J., Mitchell, J., and Smith, P., 1985, Economic impact of an epizootic of bovine vesicular stomatitis in California, *J. Am. Vet. Med. Assoc.* **186**:370.

Graf, P. A., Whitehead, R. H., and Symons, M. H., 1967, Studies of reptiles and amphibians, *Rep. Queensl. Inst. Med. Res.* **22**:4.

Hanson, R. P., 1952, The natural history of vesicular stomatitis virus, *Bacteriol. Rev.* **16**:179.

Hanson, R. P., 1968, Discussion of the natural history of vesicular stomatitis, *Am. J. Epidemiol.* **87**:264.

Hanson, R. P., 1970, Vesicular stomatitis, in: *Diseases of Swine* (H. W. Dunne, ed.), pp. 292–308, Iowa State University Press, Ames.

Holbrook, A. A., and Geleta, J. N., 1957, Vesicular stomatitis immunization with inactivated vaccines of chicken embryo origin, *Proc. U. S. Livestock Sanitary Assoc.* **61**:308.

Houff, S. A., Burton, R. C., Wilson, R. W., Henson, T. E., London, W. T., Baer, G. M., Anderson, L. J., Winkler, W. G., Madden, D. L., and Sever, J. L., 1979, Human-to-human transmission of rabies virus by corneal transplant, *N. Engl. J. Med.* **300**:603.

Inaba, Y., 1973, Bovine ephemeral fever (three-day sickness) stiff sickness, *Bull. Off. Int. Epizoot.* **79**:627.

Inaba, Y., Kurogi, H., Takahashi, A., Sato, K., Omori, T., Goto, Y., Hanaki, T., Yamamoto, M., Kishi, S., Kodama, K., Harada, K., and Matumoto, M., 1974, Vaccination of cattle against bovine ephemeral fever with live attenuated virus followed by killed virus, *Arch. Gesamte Virusforsch.* **44**:121.

Johnson, H. N., 1959, Rabies, in: *Viral and Rickettsial Infections of Man*, 3rd ed. (T. M. Rivers and F. L. Horsfall, Jr., eds.), Chapter 21, Lippincott, Philadelphia.

Johnson, K. M., Tesh, R. B., and Peralta, P. H., 1969, Epidemiology of vesicular stomatitis virus: Some new data and a hypothesis for transmission of the Indiana serotype, *J. Am. Vet. Med. Assoc.* **155**:2133.

Johnson, R. T., 1970, The pathogenesis of experimental rabies, in: *Rabies* (Y. Nagano and F. M. Davenport, eds.), pp. 59–75, University of Tokyo Press, Tokyo.

Jonkers, A. H., 1967, The epizootiology of the vesicular stomatitis virus: A reappraisal, *Am. J. Epidemiol.* **86**:286.

Jonkers, A. H., Shope, R. E., Aitken, T. H. G., and Spence, L., 1964, Cocal virus: A new agent in Trinidad related to vesicular stomatitis virus, type Indiana, *Am. J. Vet. Res.* **25**:236.

Jonkers, A. H., Spence, L., and Aitken, T. H. G., 1965, Cocal virus epizootiology in Bush Bush forest and the Nariva Swamp, Trinidad, W. I.: Further studies, *Am. J. Vet. Res.* **26**:758.

Kaneko, N., Inaba, Y., Akashi, H., Miura, Y., Shorthose, J., and Kurashige, K., 1986, Isolation of a new bovine ephemeral fever group virus, *Aust. Vet. J.* **63**:29.

Karabatsos, N., 1985, *International Catalogue of Arboviruses Including Certain Other Viruses of Vertebrates*, 3rd ed., American Society of Tropical Medicine and Hygiene, San Antonio, Texas.

Kemp, G. E., Lee, V. H., Moore, D. L., Shope, R. E., Causey, O. R., and Murphy, F. A., 1973a, Kotonkan, a new rhabdovirus related to Mokola of the rabies serogroup, *Am. J. Epidemiol.* **98**:43.

Kemp, G. E., Moore, D. L., Isoun, T. T., and Fabiyi, A., 1973b, Mokola virus: Experimental infection and transmission studies with the shrew, a natural host, *Arch. Gesamte Virusforsche.* **43**:242.

Kerschner, J. H., Calisher, C. H., Vorndam, A. V., and Francy, D. B., 1986, Identification and characterization of Bahia Grande, Reed Ranch and Muir Springs viruses, related members of the family Rhabdoviridae with widespread distribution in the United States, *J. Gen. Virol.* **67**:1081.

Kieny, M. P., Lathe, R., Drillien, R., Spehner, D., Skory, S., Schmitt, D., Wiktor, T., Koprowski, H., and Lecocq, J. P., 1984, Expression of rabies virus glycoprotein from a recombinant vaccinia virus, *Nature (London)* **312**:163.

Knudson, D. L., 1973, Rhabdoviruses, *J. Gen. Virol.* **20**:105.

Kurz, W., Gelderblom, H., Flugel, R. M., and Darai, G., 1986, Isolation and characterization of a tupaia rhabdovirus, *Intervirology* **25**:88.

Lauerman, L. H., 1967, Vesicular stomatitis in temperate and tropical America, Ph.D. thesis, University of Wisconsin, Madison.

Linhart, S. B., Flores Crespo, R., and Mitchell, G. C., 1972, Control of vampire bats by topical application of an anticoagulant, chlorophacinone, *Bull. Pan Am. Health Org.* **6**:31.

Mackerras, I. M., Mackerras, M. J., and Burnet, F. M., 1940, Experimental studies of ephemeral fever in Australian cattle, *Bulletin of the Council of Scientific Industrial Research*, No. 136, Melbourne, Australia.

Mackett, M., Yilma, T., Rose, J. K., and Moss, B., 1985, Vaccinia virus recombinants: Expression of VSV genes and protective immunization of mice and cattle, *Science* **227**:433.

Maramorosch, K., 1955, Multiplication of plant viruses in insect vectors, *Adv. Virus Res.* **3**:221.

Marcus, P. I., and Sekellick, M. J., 1980, Cell killing by vesicular stomatitis virus: The prototype rhabdovirus, in: *Rhabdoviruses*, Vol. III (D. H. L. Bishop, ed.), pp. 13–49, CRC Press, Boca Raton, Florida.

McSharry, J. J., 1979, The lipid envelope and chemical composition of rhabdoviruses, in: *Rhabdoviruses*, Vol. I (D. H. L. Bishop, ed.), pp. 107–117, CRC Press, Boca Raton, Florida.

Meredith, C. D., Rossouw, A. P., and Van Praag Koch, H., 1971, An unusual case of human rabies thought to be of Chiropteran origin, *S. Afr. Med. J.* **45**:767.

Mudd, J. A., Leavitt, R. W., Kingsbury, D. T., and Holland, J. J., 1973, Natural selection of mutants of vesicular stomatitis virus by cultured cells of *Drosophila melanogaster*, *J. Gen. Virol.* **20**:341.

Murphy, F. A., Bauer, S. P., Harrison, A. K., and Winn, W. C., Jr., 1973, Comparative pathogenesis of rabies-like viruses, *Lab. Invest.* **28**:361.

Murphy, F. A., Harrison, A. K., and Bauer, S. P., 1975, Experimental vesicular stomatitis virus infection: Ultrastructural pathology, *Exp. Mol. Pathol.* **23**:426.

Murray, M. D., 1970, The spread of ephemeral fever of cattle during the 1967–68 epizootic in Australia, *Aust. Vet. J.* **46**:77.

Nettles, V. F., Shaddock J. H., Sikes, R. K., and Reyes, C., 1979, Rabies in translocated raccoons, *Am. J. Public Health* **69**:601.

Neukirch, M., and Glass, B., 1984, Some aspects of virus shedding by rainbow trout (*Salmo gairdneri* Rich.) after waterborne infection with viral haemorrhagic septicaemia (VHS) virus, *Zentralbl. Bakteriol. Parasitenkd. Infektionskr. Hyg. Abt. 1: Orig. Reihe A* **257**:433.

Obukhova, V. R., and Gaidamovich, S. Ya., 1981, Human infections with Isfahan virus, *Sb. Tr. Inst. Virus. imeni D. I. Ivanovsky, Akad. Med. Nauk SSR* 97–101 (in Russian).

Olitsky, P. K., Traum, J., and Schoening, H. W., 1928, Report of Foot-and-Mouth Disease Commission of the U. S. Department of Agriculture, *U. S. Dept. Agric. Tech. Bull.* **76**:1–168.

Pfeil-Putzien, C., 1978, Experimentelle Ubertragung der Frühjahrsvirämie (spring viremia) der Karpfen durch Karpfenläuse (*Argulus foliaceus*), *Abh. Vet Med. B* **25**:319.

Pinheiro, F. P., Bensabath, G., Andrade, A. H. P., Lins, Z. C., Fraika, H., Tang, A. T., Lainson, R., Shaw, J. J., and Azevedo, M. C., 1974, Infectious diseases along Brazil's Trans-Amazon highway: Surveillance and research, *Pan. Am. Health Org. Bull.* **8**:111.

Proctor, S. J., and Sherman, K. C., 1975, Ultrastructural changes in bovine lingual epithelium infected with vesicular stomatitis virus, *Vet. Pathol.* **12**:362.

Rao, T. R., Singh, K. R. P., Dhanda, V. and Bhatt, P. N., 1967, Experimental transmission of Chandipura virus by mosquitoes, *Ind. J. Med. Res.* **55**:1306.

Rosen, L., 1980, Carbon dioxide sensitivity in mosquitoes infected with sigma, vesicular stomatitis, and other rhabdoviruses, *Science* **207**:989.

Roy, P., 1979, Fish rhabdoviruses, in: *Rhabdoviruses*, Vol. III (D. H. L. Bishop, ed.), pp. 193–208, CRC Press, Boca Raton, Florida.

Schaperclaus, W., 1953, *Fischkrankheiten*, 3rd ed., Akademie Verlag, Berlin.

Schmidt, J. R., Williams, M. C., Lule, M., Mivule, A., and Mujomba, E., 1965, Viruses isolated from mosquitoes collected in the southern Sudan and western Ethiopia, *East Afr. Virus Res. Inst. Rep.* **15**:24.

Schneider, L. G. and Meyer, S., 1981, Antigenic determinants of rabies virus as demonstrated by monoclonal antibody, in: *The Replication of Negative Strand Viruses* (D. H. L. Bishop and R. W. Compans, eds.), pp. 947–953, Elsevier/North-Holland, New York.

Shope, R. E., 1975, Rabies virus antigenic relationships, in: *The Natural History of Rabies*, Vol. 1 (G. M. Baer, ed.), Chapter 8, Academic Press, New York.

Smith, A. L., Tignor, G. H., Emmons, R. W., and Woodie, J. D., 1978, Isolation of field rabies virus strains in CER and murine neuroblastoma cell cultures, *Intervirology* **9**:359.

Soave, O. A., 1966, Transmission of rabies to mice by ingestion of infected tissue, *Am. J. Vet. Res.* **27**:44.

Standfast, H. A., and Muller, M. J., 1985, Vectors of bovine ephemeral fever, in: *Veterinary Viral Diseases* (A. J. Della-Porta, ed.), pp. 394–397, Academic Press, Sydney.

Steck, F., Wandeler, A., Bichsel, P., Capt, S., Hafliger, U., and Schneider, L., 1982, Oral immunization of foxes against rabies, *Comp. Immunol. Microbiol. Infect. Dis.* **5**:165.

St. George, T. D., 1984, Bovine ephemeral fever, in: *Foreign Animal Diseases* (G. S. Trevino and J. L. Hyde, eds.), pp. 110–119, U. S. Animal Health Association, Richmond, Virginia.

St. George, T. D., Standfast, H. A., Christie, D. G., Knott, S. G., and Morgan, I. R., 1977, The epizootiology of bovine ephemeral fever in Australia and Papua-New Guinea, *Aust. Vet. J.* **53**:17.

St. George, T. D., Cybinski, D. H., Murphy, G. M., and Dimmock, C. K., 1984, Serological and biochemical factors in bovine ephemeral fever, *Aust. J. Biol. Sci* **37**:341.

Tesh, R. B. and Johnson, K. M., 1975, Vesicular stomatitis, in: *Diseases Transmitted from Animals to Man*, 6th ed. (W. T. Hubbert, W. F. McCullock, and R. P. Schnurrenberger, eds.), pp. 897–910, Springfield, Illinois.

Tesh, R. B., and Modi, G. B., 1983a, Development of a continuous cell line from the sand fly *Lutzomyia longipalpis* (Diptera: Psychodidae), and its susceptibility to infection with arboviruses, *J. Med. Entomol.* **20**:199.

Tesh, R. B., and Modi, G. B., 1983b, Growth and transovarial transmission of Chandipura virus (Rhabdoviridae: *Vesiculovirus*) in *Phlebotomus papatasi*, *Am. J. Trop. Med. Hyg.* **32**:621.

Tesh, R. B., Peralta, P. H., and Johnson, K. M., 1969, Ecologic studies of vesicular stomatitis virus. 1. Prevalence of infection among animals and humans living in an area of endemic VSV activity, *Am. J. Epidemiol.* **90**:255.

Tesh, R. B., Peralta, P. H., and Johnson, K. M., 1970, Ecologic studies of vesicular stomatitis virus. 2. Results of experimental infection in Panamanian wild animals, *Am. J. Epidemiol.* **91**:216.

Tesh, R. B., Chaniotis, B. N., and Johnson, K. M., 1971, Vesicular stomatitis virus, Indiana serotype: Multiplication in and transmission by experimentally infected phlebotomine sandflies (*Lutzomyia trapidoi*), *Am. J. Epidemiol.* **93**:491.

Tesh, R. B., Chaniotis, B. N., and Johnson, K. M., 1972, Vesicular stomatitis virus (Indiana serotype): Transovarial transmission by phlebotomine sandflies, *Science* **175**:1477.

Tesh, R., Saidi, S., Javadian E., Loh, P., and Nadim, A., 1977, Isfahan virus, a new *Vesiculovirus* infecting humans, gerbils, and sand flies in Iran, *Am. J. Trop. Med. Hyg.* **26**:299.

Tesh, R. B., Travassos da Rosa, A. P. A., and Travassos da Rosa, J. S., 1983, Antigenic relationship among rhabdoviruses infecting terrestrial vertebrates, *J. Gen. Virol.* **64**:169.

Tesh, R. B., Boshell, J., Modi, G., Morales, A., Young, G., Corredor, A., Ferro, C., and Rodriguez, C., 1987, Natural infection of humans, animals and phlebotomine sand flies with the Alagoas serotype of vesicular stomatitis virus in Colombia, *Am. J. Trop. Med. Hyg.* (in press).

Theodoridis, A., Boshoff, S. E. T., and Botha M. J., 1973a, Studies on the development of a vaccine against bovine ephemeral fever, *Onderstepoort J. Vet. Res.* **40**:77.

Theodoridis, A., Giesecke, W. H., and Du Toit, I. J., 1973b, Effect of ephemeral fever on milk production and reproduction of dairy cattle, *Onderstepoort J. Vet. Res.* **40**:83.

Thompson, R. D., Mitchell, G. C., and Burns, R. J., 1972, Vampire bat control by systemic treatment of livestock with an anticoagulant, *Science* **177**:806.

Tignor, G. H., Smith, A. L., and Shope, R. E., 1984, Utilization of host proteins as virus receptors, in: *Concepts in Viral Pathogenesis* (A. L. Notkins and M. B. A. Oldstone, eds.), pp. 109–116, Springer-Verlag, New York.

Travassos, A. P. A., Tesh, R. B., Travassos, J. F., Herve, J. P., and Main, A. J., 1984, Carajas and Maraba viruses, two new vesiculoviruses isolated from phlebotomine sand flies in Brazil, *Am J. Trop. Med. Hyg.* **33**:999.

Vanselow, B. A., Abetz, I., and Trenfield, K., 1985, A bovine ephemeral fever vaccine incorporating adjuvant Quil A: A comparative study using adjuvants Quil a, aluminium hydroxide gel and dextran sulfate, *Vet. Rec.* **117**:37.

Witte, E. J., Hays, C. W., Miller, G. B., Jr., Haddy, L. E., Levy, M. E., and Israel, E., 1983 Update: Raccoon rabies—mid-Atlantic states, *Morbidity and Mortality Weekly Report*, **32**:97.

Wolf, K., 1966, The fish viruses, *Adv. Virus Res.* **12**:35.

World Health Organization, 1986, Rabies, *Weekly Epidemiol. Rec.* **15**:109.

Young, P. L., and Spradbrow, P. B., 1980, The role of neutrophils in bovine ephemeral fever virus infection of cattle, *J. Infect. Dis.* **142**:50.

Young, P. L., and Spradbrow, P. B., 1985, Transmission of virus from serosal fluids and demonstration of antigen in neutrophils and mesothelial cells of cattle infected with bovine ephemeral fever virus, *Vet. Microbiol.* **10**:199.

Index

535